# SOUTHERN AFRICAN MOTHS & THEIR CATERPILLARS

Hermann Staude
Mike Picker
Charles Griffiths

Published in 2023 by
Pelagic Publishing
20–22 Wenlock Road
London N1 7GU, UK

www.pelagicpublishing.com



https://doi.org/10.53061/KUDW7054

British Library Cataloguing in Publication Data
A catalogue record for this book is available from the British Library

ISBN 978-1-78427-347-7 Pbk
ISBN 978-1-78427-348-4 ePub
ISBN 978-1-78427-349-1 PDF

Printed and bound in China by 1010 Printing International Ltd

Cover images: Front: (top) *Epiphora mythimnia; Plectranthus barbatus* (Alexxlga-stock.adobe.com); (bottom) *Pericyma mendax.*
Back: (top) *Zutulba namaqua* (Ian Sharp); (centre left) *Amata cerbera* (Steve Woodhall); (centre right) *Asota speciosa.* Page 3: (top) *Zamarada adiposata* (Ian Sharp); (bottom) *Heniocha dyops* (Ian Sharp).

# CONTENTS

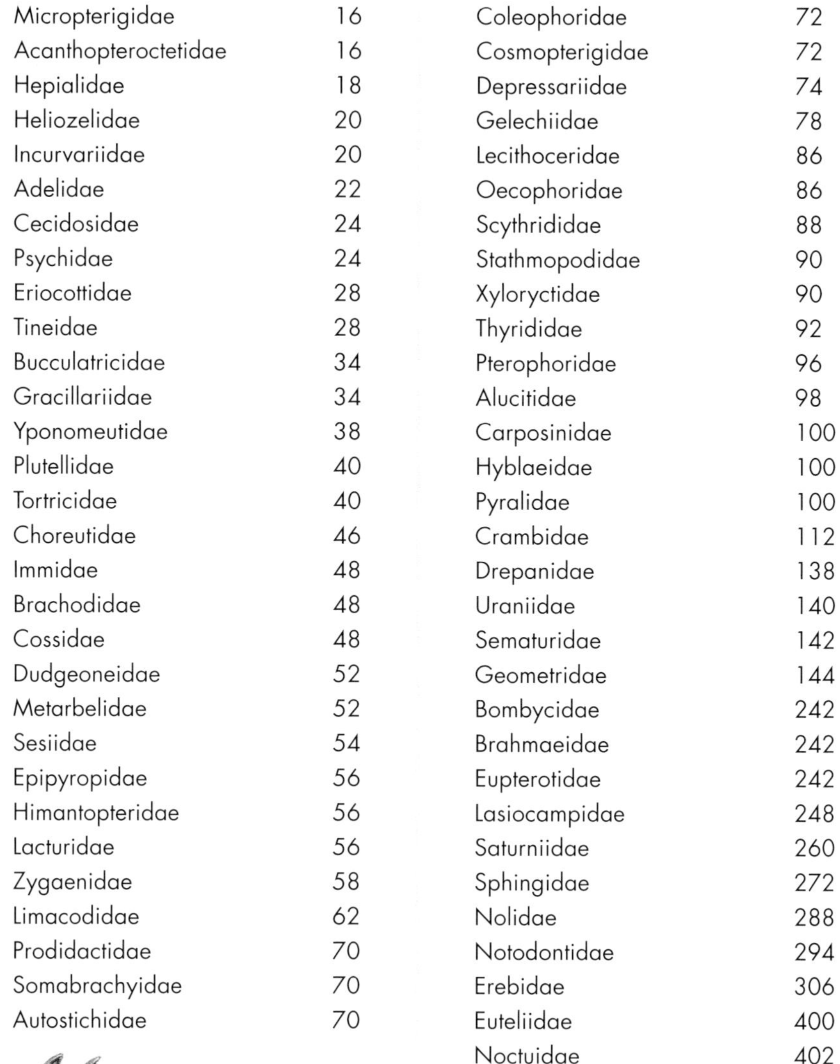

# ACKNOWLEDGEMENTS

The production of this book over the last decade necessitated many hours of field work, collection of images and data and then many hours of processing voucher specimens, identifying specimens, building databases, basic research and finally writing of the actual book. During this process an untold number of people encouraged us and helped us in many ways whenever called upon. We apologise profusely to those we may have unintentionally left out of these acknowledgements.

First and foremost, we thank our families for allowing us the time to embark on and complete this project and the many photographers who provided images, as acknowledged on the images in the book.

We have taken every care to ensure the accuracy of information provided in this book, but take full responsibility for any errors that may have slipped through. We are grateful to the following colleagues who greatly improved the accuracy of information provided by helping with identifications and by reviewing sections of this book: Marc Epstein, Alexey Prozorov, Pasi Sihvonen, Josef de Freina, Rolf Oberprieler, Rudolphe Rougeri, David Agassiz, Graziano Bassi, Koen Maes, Oleksiy Bidzilya, Hermann Hacker, John Brown, Todd Gilligan, Axel Hausmann, Carlos Lopez Vaamonde, Axel Hoffman, Wolfram Mey, Ingo Lehman, Roman Yakovlev, Leif Aarvik, Alexander Schintlmeister, Michal Rindos, the late Douglas Kroon, Lukasz Przybylowicz, David Lees, and the late John Joannou.

Property owners and managers too numerous to mention all over the region provided permission to work on their properties and have been helpful in many ways. In particular, there are places that we visited several times and we would like to thank Willem Prinsloo and Alister Tuckett of Mogale's Gate Biodiversity Centre, Hekpoort; Brenda Cauldwell and Lynn Katsoulis of Happy Acres Environmental Education Centre, Magaliesburg; Nelis Moll of the farms Langverwacht, Die Kom and Chaos, Vryheid and Mkuze; Frits Hunlun of the farm Zandrivier, Ladismith; Errol Sanders of the farm Nooitgedacht, Hekpoort and Louis Gunther of Entumeni Guest Farm for their hospitality.

We thank the following for their companionship on field trips, for providing valuable specimens and for their insights during interesting discussions about moths: Louisa, Heiku, Pierre and Rozanna Staude, Louél and Heinz Schnülzer, Wolfgang Staude, Steve Collins, the late John Joannou, the late Neville Duke, the late Douglas Kroon, the late Peter and Sue Webb, Pasi Sihvonen, Adrian Armstrong, Jonathan Ball, Reinier Terblanche, Richard Stephen, Alf Curle, Steve Woodhall, Colin Congdon, Peter and Allison Ward, Julio Balona, David Agassiz, Todd Gillingham, Axel Hausmann, Manfred Sommerer, Axel Hofmann, Harald Sulak, Dave Edge, Andries de Vries, Marion Maclean, Silvia Mecenero, Andy Mayer, Andre and Bennie Coetzer, Augustine Morkel, Anton, Johann and Braam Bosman and the late Ian Bosman, Kevin Cockburn, Danie Human, Stefan Meyer, Andrew Morton, Andre Grobler, JP Greeff, Altha Liebenberg, Frans Pretorius, Quartus Grobler, Johan Heyns, Wayne Forrester, Franscois Swart, Alan Gardiner, Cameron Scott, Kate Braun, Graham Henning, Hossein Rajaei, Peter Hawkes, Mervyn Mansell, Clarke Scholtz, Hennie de Klerk, Chris Willis, Jeremy Dobson, Mark Williams, Jurgen Lenz, Justin and Yolande Bode, James Lawrence, Pierre le Roux, Martin Lunderstead, Susan Malan, Nelis Moll, Owen Garvie, Nina Parry, Peter Sharland, Jan Praet, Ernest Pringle, Ricardo Riddles, Hanna and Wolfachim Roland, Rodolphe Rougerie, Jonathan Colville, Raimund Schutte, Ian and Allison Sharp, Dylan Smith, Bill Steele, Magriet Brink, Lucia Phillips, Suncana Bradley, Magda Botha, Etienne Terblanche, Les Underhill and Wynand Uys.

For giving access to collections under their care and providing specimen information, our thanks go to: Peter Radebe and the late Martin Krüger, Ditsong Museums of South Africa, Pretoria; Simon van Noort, Iziko South African Museum, Cape Town; Malcolm Scoble and Martin Honey, Natural History Museum, London; Axel Hausmann, ZSM, Munich.

The following nature conservation authorities are thanked for permission to conduct research on properties under their jurisdiction: CapeNature, Ezemvelo KZN Wildlife, iSimangaliso Wetland Park Authority, South African National Parks, Northern Cape Nature Conservation, Eastern Cape Parks and Tourism Agency and Limpopo Provincial Agency. In particular, the following conservators are thanked for their help and assistance in the field: Adrian Armstrong, Ezemvelo KZN Wildlife; Sharon Louw, Ezemvelo KZN Wildlife; Mark and Amida Johns, CapeNature, and Henk Nieuwoudt, CapeNature.

Rio Button is thanked for drawing the maps used in this book.

Finally, thanks to our editor Colette Alves and designer Dominic Robson, who put an enormous effort into the production of this book. Pippa Parker, Publisher at Struik Nature, is thanked for her enthusiasm and support for this book.

# PREFACE

Although southern Africa has several excellent regional guides to specific groups of Lepidoptera, not a single modern, comprehensive popular guide exists covering the Lepidoptera as a whole. There are a few recent publications covering specific groups, such as the Papilionoidea (butterfies) (Woodhall, 2005 and Mecenero *et al.*, 2013), Thyretini (Przybylowicz, 2009), Notodontidae (Schintlmeister and Witt, 2015), Boletobiinae (Hacker, 2019), Rivulinae, Hypeninae, Hermiinae and Hypenodinae (Hacker, 2021) and the ongoing *Esperiana* book series (Hacker *et al.*, Vols 14–21, 2008–2019), but these are mostly of a technical and specialised nature and unsuitable for popular reference. Recent popular coverage of African moths is to be found within broader regional insect field guides, but of course those are only able to cover a small fraction of the diverse moth fauna. The absence of up-to-date regional moth guides is especially surprising given that several substantive books on southern Africa moths were published in the last century. The most notable of these were the compehensive eight-volume *The Moths of South Africa* (Janse, 1930–1964), *Hawk Moths of Central and Southern Africa* (Pinhey, 1960), *Emperor Moths of Central and Southern Africa* (Pinhey, 1972), *The Emperor Moths of Namibia* (Oberprieler,1993), and, most significantly, the more comprehensive reference *Moths of Southern Africa* (Pinhey, 1975). These are all long out of print, and are severely dated in terms of content, presentation style, photography, taxonomy and systematics. Like older butterfly and moth guides worldwide, they also mostly depict pinned museum specimens, which often bear little resemblance to the live animals observed by modern enthusiasts and ecologists, who encounter, photograph and identify living moths in the wild. Another important shortcoming of earlier books is that they included only brief mention and few illustrations of the caterpillar stages and their host plants. Due largely to recent efforts by the Caterpillar Rearing Group of the Lepidopterists' Society of Africa, the association between the caterpillar and adult life history stages of a considerable number of regional moth species has now been established. There are also now far more numerous and accurate distributional and plant host records available from both museum collections and citizen science initiatives, allowing greatly improved (point) distributional maps to be generated.

This book, therefore, aims to provide a fairly comprehensive account of the moth fauna of southern Africa, with the focus on the biology of live rather than pinned moths, and their caterpillars.

*Oaracta auricincta*

# INTRODUCTION

With around one million species, insects are by far the most diverse group of organisms on the planet. Within this class, the order Lepidoptera (comprising moths and butterflies) represents one of the major radiations, with approximately 160,000 described species.

Lepidopterans are ecologically important as pollinators, herbivores, and as food sources for other animals. They are thought to have diversified along with, and possibly as a response to, the radiation of the angiosperms (flowering plants).

Adult Lepidoptera typically have two pairs of large, membranous wings covered by small scales. Their closest living relatives are Caddisflies (Trichoptera), with which they share wing venation patterns and the ability of the larvae to spin silk.

The oldest described lepidopteran fossil dates back about 190 million years, and most of the major lineages present today were already established 100 million years ago.

**Phylogeny of Lepidoptera**

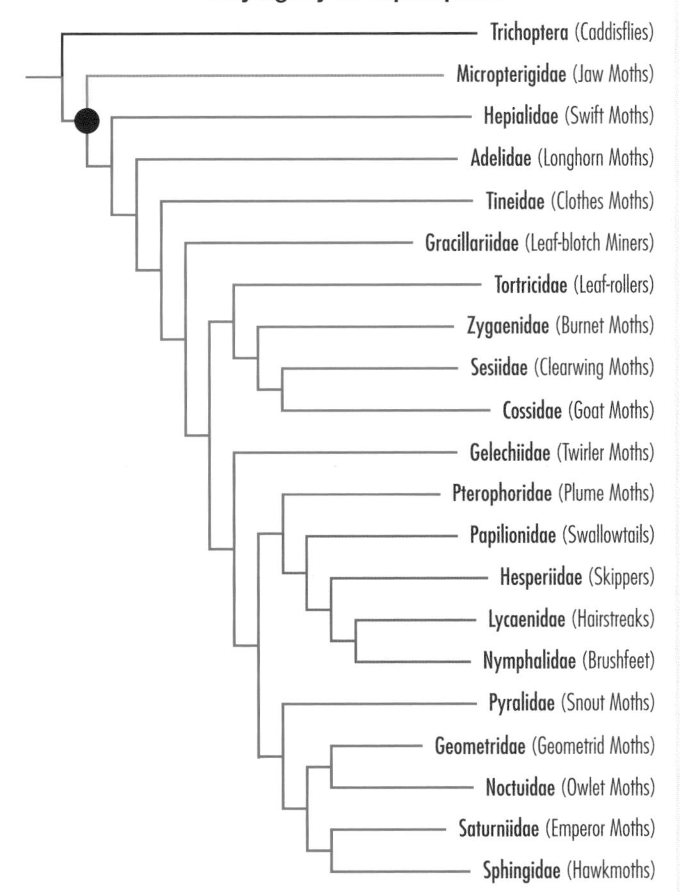


The evolutionary development of Lepidoptera (after Douglas Boyes www.douglasboyes.co.uk).

However, the first moth is thought to have evolved around 300 million years ago. The most ancestral of all Lepidoptera are the ancient Jaw Moths (Micropterigidae, see second lineage in phylogeny), which are the only lepidopterans that have jaws instead of a coiled, nectar-feeding proboscis, and use these to feed on pollen. Their larvae feed on ferns, liverworts and mosses – ancient plants that occur near water, thus representing a transitional stage between the earlier aquatic caddisfly larvae (see first lineage in phylogeny) and the truly terrestrial caterpillars of Lepidoptera. The Jaw Moths can be seen in the phylogeny to be the sister group to all other Lepidoptera (see red dot in phylogeny).

It is thought that early moth caterpillars were internal feeders within plant tissue, and that external feeding on plant foliage only evolved later. Most ancient moth families comprise small moths, due to the size constraints of living within plant tissue (especially leaves). As the caterpillars began to feed externally on leaves, their body size increased, so that we now see moths with wingspans ranging from a few millimetres to more than 250mm.

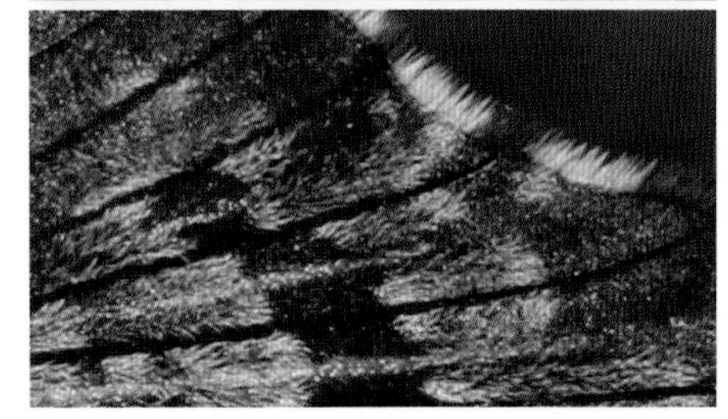

Lepidoptera have wing scales (top), while their closest relatives, Caddisflies, have hairs (above).

By far the largest and most diverse group of extant Lepidoptera is the Ditrysia, with an estimated 150,000 species globally. This group contains many moth families, including those called butterflies. The adult females all share a unique feature: they have separate reproductive openings for mating and egg laying. Butterflies (Papilionoidea, with 19,000 species) are a group that lie deeply nested within the later lineages of the moth-Lepidoptera (see blue lineage in phylogeny). Butterflies, although morphologically fairly distinctive, are simply day-flying moths that evolved around 100 million years ago. The switch from a largely nocturnal existence to a day-time lifestyle has happened in many unrelated lineages and for a variety of reasons.

## Diversity of moths

Lepidoptera are found on all continents except Antarctica, and occur across a wide range of terrestrial habitat types, although they are almost completely absent from freshwater habitats. The majority are specialists, feeding on a specific group of hosts, a single host species, or even just a part of a single host species. While some are predatory, parasitic, or breed in dead organic materials (such as animal horn, wool, bat guano, honeycombs, or stored food products), the vast majority are associated with plants, especially angiosperms. Many species also feed on algae, lichens and mosses.

The lepidopteran fauna of southern Africa comprises well over 11,000 described species, of which fewer than 800 are butterflies. However, only a small proportion of the tiny and more obscure moths have in fact been described scientifically. The regional moth fauna currently encompasses about 90 of the 126 families found globally. Species richness differs greatly among the families, with more than half of the described regional moth species found in just five very species-rich families: Erebidae (1,587 described species in southern Africa), Geometridae (1,563), Noctuidae (983), Pyralidae (631) and Gelichiidae (598). The majority of the remaining families typically have fewer than 30 species each.

## Wing features

Various wing features are useful in distinguishing different moth species. These include wing shape, colour pattern and distribution and length of the wing scales. There is a general groundplan of transverse lines across the wing, from the base to the apex. These lines often demarcate broader bands, depending on the colouration of the wing. All lines may be present in some species, while in others no lines are evident. The wings also have small spots (stigmata) that may differ between species.

Specialised patches of scales called androconia may also occur on the wings, bodies or legs of male moths, and have evolved to facilitate the distribution of pheromones used to attract mates. In nocturnal moths, dense hairs and scales serve an additional function of trapping body heat during the cooler night temperatures, when the thorax and its musculature need to be considerably warmer than ambient temperature. Heat is generated by thoracic muscle contraction (seen as 'shivering' in large moths such as in *Eutelia adulatrix* p.400), and the dense layer of hairs on the thorax and abdomen help retain this elevated body temperature during flight. Scales may also absorb bat echolocation signals, allowing the moths to escape detection by bat radar systems.

## Life cycle

Moths undergo complete metamorphosis, with eggs hatching into caterpillars, which are quite unlike the adults, and growing through a series of moults into the winged adult form. Although the adult phase is by far the best known and the one used to define and describe species, it is in fact much shorter in duration than that of the caterpillar. This stage of the life cycle is generally reserved for reproduction and dispersal, and many adult moths have non-functional mouthparts, living only long enough to find a mate and lay eggs.

### Mating

Moths rely on fairly complex methods to attract and find a mate. Diurnal moths may attract mates using visual stimuli, but most moths are nocturnal and use pheromones to attract mates. These may be secreted by the male to attract females, sometimes using eversible scent glands or specialised scent scales. However, more often it is the female who produces the pheromones and the males who detect them, using sensitive receptors on their specially modified, comb-like antennae, which are densely covered in exceptionally sensitive chemoreceptors.

These species-specific pheromones released by a female moth are volatile compounds that drift downwind until they are detected by the male of that species. Pheromones can elicit responses in male moths over a distance of several kilometres. If the male receives the correct chemical signal, even at concentrations of only a few molecules, he is stimulated to take flight and follow the odour trail upwind until he encounters the female. In some species, males form a lek that females visit when they want to mate. Lek sites may be an emergent tree above the forest canopy

Male *Diaphania indica* everting specialised scent scales (androconia), which release sex pheromones.

The comb-like antennae of a male emperor moth, which are used to detect pheromones.

(Geometridae: *Callioratis abraxas*) or a rock face (Geometridae: *Callioratis mayeri*) or the top of a ridge (Autostichidae: *Procometis ochricilia*). In some cases, males and females are active at different times of day, such as in Diurnal Phiala (Eupterotidae: *Phiala flavipennis*), where males fly early in the morning low over the grassland searching for calling females, and females fly late in the afternoon depositing their eggs on grass stems.

### Egg laying and the egg phase

Female moths normally attach their eggs, either singly or in groups, directly onto the larval host plant, using an adhesive. The host plant is detected by odour, typically at night. However, some species, particularly those in which the caterpillars feed on plant roots, or are capable of feeding on a wide variety of plants, simply drop their eggs over the vegetation while in flight.

The number of eggs produced per female varies between species, ranging from a few to thousands. Eggs are diverse in form: many are round or oval, but they may be elongated, ribbed, pitted or otherwise decorated. The ribbing is an adaptation to prevent collapse of the egg as it loses moisture.

The period of egg development normally ranges from a few days to a few weeks, but in temperate regions, eggs laid towards the onset of winter may remain dormant and only hatch the next spring.

### The caterpillar (larval) phase

Caterpillars hatch by using their mandibles to bite a hole in the eggshell. They often consume the shell immediately after hatching, since it is rich in proteins and thus provides a nutritious first meal. Caterpillars are extremely diverse in shape and form, but most have cylindrical bodies consisting of a hard, tough (sclerotised) head capsule, followed by three thoracic, limb-bearing segments and 10 abdominal segments. The head capsule bears two groups of simple eyes and ventral mouthparts, dominated by a pair of strong, biting mandibles. The lower jaw or labium is sometimes equipped with a spinneret, which is used to spin silk. With their simple eyes (stemmata), caterpillars can detect different light intensities but are unable to form images.

Abdominal segments 3–6 typically each bear a pair of short, thick prolegs or false legs, each ending in a ring of terminal hooks (crochets), while the last segment has a similar pair of prolegs, called claspers. As seen in the labelled image of *Odontestra bulgeri* (Noctuidae), there are typically eight pairs of walking appendages, three pairs of thoracic legs and five sets of prolegs, but this number varies between groups. For example, the slender, elongate caterpillars of the family Geometridae (loopers or inch-worms, p.144) retain only two widely spaced pairs of abdominal appendages, while many of the diverse superfamily Noctuoidea have lost either the first or first two pairs of prolegs and are known as semi-loopers.

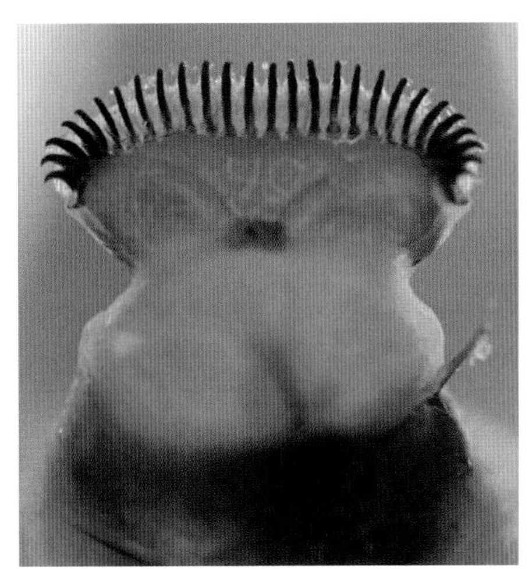

Caterpillars have thick prolegs ending in a ring of crochets, which may be complete or partial.

**Anatomical features of a typical caterpillar**

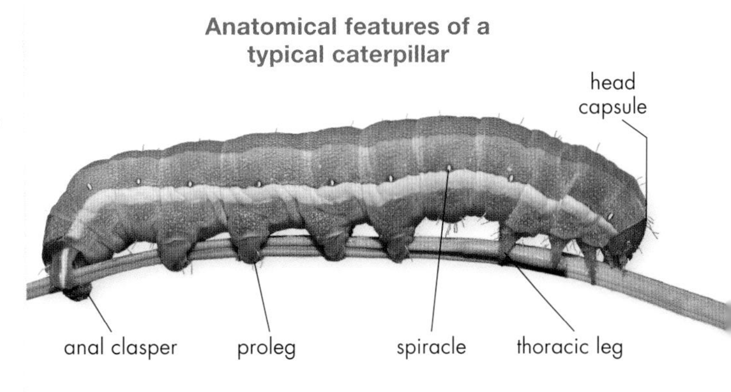


Caterpillars have chewing mouthparts and feed on a wider range of materials than do adult moths. Some species are highly specialised to consume only a single food species, or only a part of it (monophagous). Others will exclusively utilise hosts from a closely related group of plants (oligophagous), and yet others feed on a wide range of unrelated plants (polyphagous) in order to obtain all their nutritional requirements.

The caterpillar phase of the life cycle typically lasts from a few weeks to a few months, but can vary considerably within a species. Where food resources are very scarce, caterpillars may take more than a year to develop fully. Caterpillars that bore into wood, which is a particularly poor food source, often take up to eight years to reach maturity. Univoltine species complete one generation per year, whereas multivoltine species complete numerous generations annually.

### The adult phase

In contrast to their caterpillars, most moths are fairly short-lived and are adapted to feed on a liquid diet, typically nectar, but some imbibe tree sap, the juices from rotting fruit, bird droppings, or animal urine or dung. Moths can also be seen drinking water from wet sand, or along the margins of pools or streams where salts accumulate. Occasionally the proboscis has been modified with sharp barbs for feeding on fruit or even mammalian lachrymal fluids or blood.

### Adult emergence and seasonality

Most moths, especially those in regions with very seasonal climates, show strong and very predictable temporal patterns of adult emergence (phenology). In southern Africa, moth emergence is mainly synchronised with rainfall. In winter-rainfall areas, peak emergences are March–April and then again August–September, with generally higher emergences

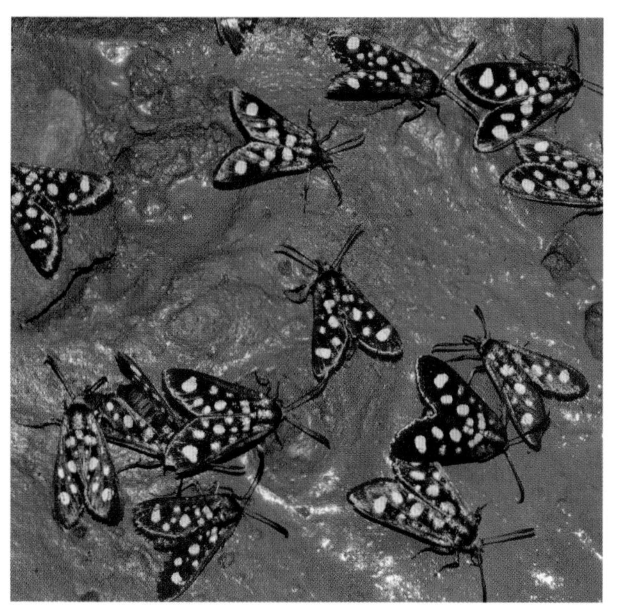

Many Lepidoptera, such as the Gold-spotted Burnet (*Arniocera auriguttata*), imbibe dissolved salts from mud.

during winter than summer. In summer-rainfall areas, peak emergences are September–November and February–March, with generally higher emergences during summer than winter. In areas where rainfall patterns are erratic, peak emergences occur a week or so after good rains. In subtropical regions, where rainfall occurs throughout the year, emergences are spread throughout the year. At high elevations, peak emergences occur December–January. These are general trends, and adult moths can be found throughout the year in all but the coldest and driest of habitats. As adult emergence is also a clue to identification, this information is provided in the species entries in this book. Flight periods can vary from site to site and from year to year depending upon climatic conditions. Many species time the hatching of their eggs to match the brief period in which their host trees have nutritious buds, on which they feed. These species 'outbreaks' may cause considerable damage to their host trees.

## Defence adaptations

Both caterpillars and adult moths exhibit a wide range of defence mechanisms that help protect them from predators and parasitoids.

### Caterpillar defence

The most common defence mechanism used by caterpillars is concealment, and many are convincingly disguised as leaves or twigs, or as unpalatable objects, such as bird droppings. Many hide away and remain motionless during daylight hours, when most predators are active, and only emerge at night to feed. Some also live underground or within wood, horn or other protective materials, where they are physically protected from potential predators. Caterpillars living on vegetation, where birds are constantly searching for food, have some of the most perfect forms of camouflage known, and often mimic the leaf structure of their host plant in fine detail.

Caterpillars also avoid predator detection by flicking faecal pellets some distance away from their feeding sites. However, chemicals given off by chewed leaves attract parasitic wasps, which are probably the major cause of mortality in lepidopterans. A variety of physical defences are also used by caterpillars to deter parasitic wasps, the most common being dense coverings of hairs or setae

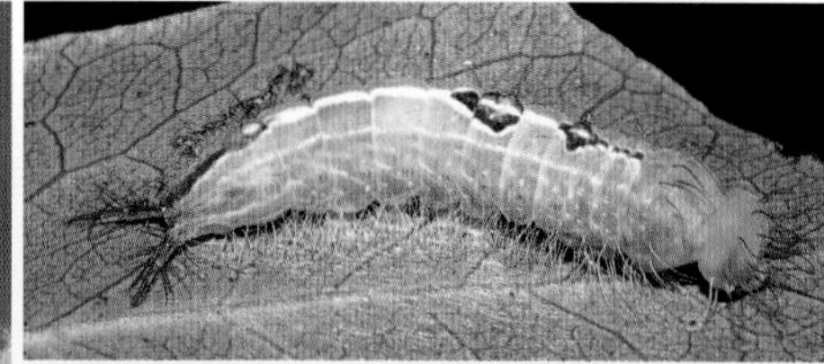

As a defence strategy, caterpillars may resemble leaves (*Acrasia crinita*, left; *Pseudobunaea tyrrhena*, bottom right) or bird droppings (*Janthinisca joannoui*, top right).

The bright aposematic colouration of *Holocerina smilax* caterpillars warns predators of its toxicity.

(stouter hairs). These may simply serve as a physical barrier, making it hard for wasps to penetrate the hairs to reach the caterpillar's skin with their ovipositors. However, the hairs often also contain irritating chemicals to ward off predators.

Caterpillars may also construct protective shelters, their form often being characteristic of different families. For example, Leafrollers (Tortricidae p.40) often form shelters from a rolled leaf, whereas Bagworms (Psychidae p.24) use thin sticks, thorns or plant debris to form characteristic tubular cases, and Clothes Moths (Tineidae p.28) often make characteristic flattened silken cases, in which the larvae are concealed. Another similar but communal form of defence is to spin large silken webs that cover entire tree branches and within which whole broods and groups of caterpillars feed and grow together.

Some caterpillars use bright (aposematic) colours as signals, warning of their toxicity. A large number of moth caterpillars harvest and store (sequester) defensive chemicals from plants. These caterpillars usually have bright warning colouration – yellow, red and black being particularly common.

## Moth defence

Adult moths are often exquisitely camouflaged, mimicking leaves, bark, broken twigs, or bird droppings. Many species are toxic or distasteful and have aposematic colouration. Mimicry of toxic or stinging insects is also common, such as in the bee- and wasp-mimicking Clearwing Moths (Sesiidae p.54).

Night-flying moths have evolved behavioural patterns such as evasive flight patterns or sound emission in response to bat predation. Different groups of moths have evolved ears (tympanal organs) on various parts of their bodies. These are used to detect bat sonar, allowing the moth to take evasive flight manoeuvres on detecting a bat. Some moths, including Tiger Moths (Arctiinae, p.364), Ermine Moths

Some moths rely on camouflage to avoid detection (*Cleora oligodranes*, left), while others use aposematic colouration to deter predators (*Diaphone lampra*, right).

(Yponomeutidae, p.38) and others, may in fact emit their own calls within the same frequency range as those of the bats, thus jamming the bats' echolocation mechanism, or possibly advertising their own toxic nature (most Tiger Moths are toxic and in this sense the call they produce may be an auditory form of aposematism).

## Regional distribution patterns

Some moth species have very wide natural distributions, occurring all over southern Africa, and indeed extending into the rest of Africa and even further afield. Others have been recently introduced to the region from distant lands, either accidentally along with other imported products, or deliberately as biocontrol agents. Most moth species, however, have relatively restricted ranges that are often associated with those of particular host plants and, hence, vegetation types. Thus the distribution of major vegetation types can be used as a good proxy to predict the types and numbers of moth species likely to be found in any particular area. Some moths, mostly migratory species, have surprisingly wide distributions, extending across a range of vegetation types, biomes and countries, or even continents.

The dominant biome and vegetation type in the northern parts of southern Africa is savanna (or bushveld), a mixed system where the trees are widely spaced, such that they do not form a closed canopy and thus support a ground cover of grass. The species associated with this biome often have very wide distributions, and are generally subtropical forms whose presence in the region reflects a southern extension of a range that is primarily Afrotropical. The fauna of this biome is both diverse and rich in species.

Southwards, the savanna merges into grassland, which covers much of the Free State, Mpumalanga, Gauteng, Lesotho and the Eastern Cape. This biome lies largely at higher altitudes and can be subject to cold temperatures in winter. The region has a distinctive moth fauna, less diverse than that of savanna, and includes many grass mimics and endemics.

The few areas where the climate is suitable for natural forest occur mostly along the eastern mountains and southern coastal regions and are associated with relatively high rainfall, spread throughout the year. Montane forests are rich in plant species and thus support a diverse fauna of moths and other insects. This is especially true of the more tropical forest patches in the north, which are richer than those in the south, although the latter have many endemics.

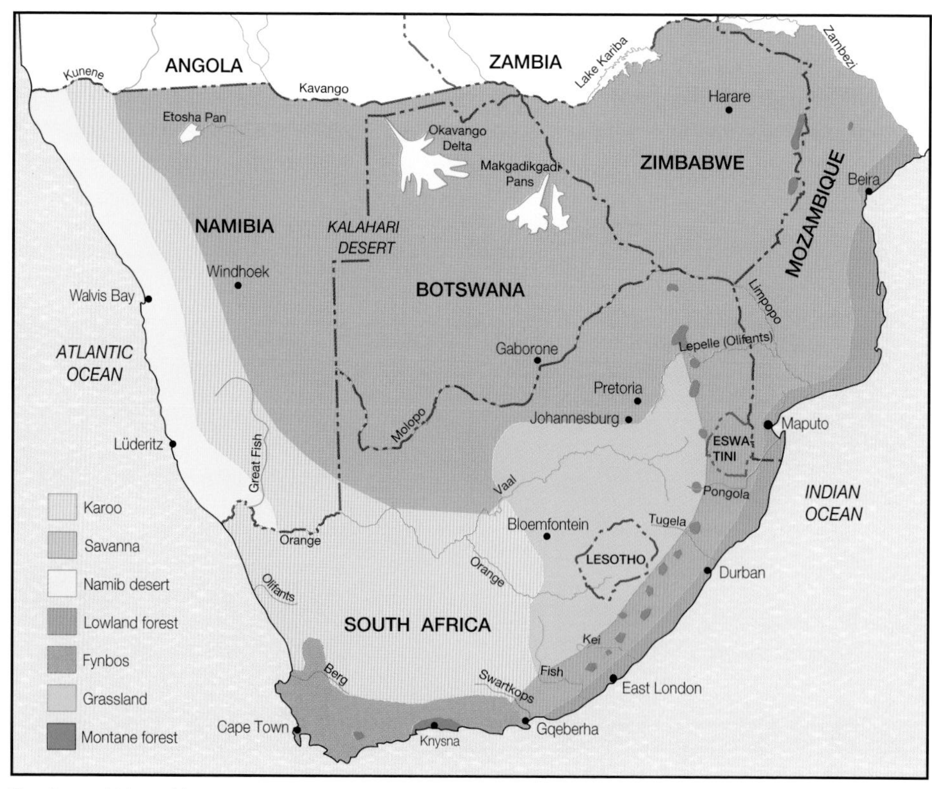

Southern African biomes

The more arid karoo biomes encompass the hotter inland Nama karoo, which experiences mostly summer rainfall, and the cooler, drier, winter-rainfall succulent karoo, which runs along the Namaqualand coast up into Namibia. These are areas of low rainfall and low vegetation with unique moth assemblages, characterised by high levels of endemism and adaptive radiation.

Northwards into Namibia, the succulent karoo merges into true desert, which has very little rainfall and hardly any vegetation. This biome supports a very sparse moth fauna, mostly in the few vegetated areas.

The fynbos biome of the Western Cape, Cederberg and the southern Cape forms the only major global biome that is largely confined to a single country and supports a unique assemblage of both plants and insects of considerable global evolutionary interest. This biome is characterised by hot, dry summers and cold, wet winters, and has a moderate rainfall of over 500mm per annum. The moth fauna is rich in endemic species and local radiations. Remnants of fynbos exist in isolated pockets along the Great Escarpment of Africa and many typical fynbos moth lineages can be found in these habitats north to Angola in the west and Ethiopia in the east.

## Ecological importance

Many caterpillar populations can fluctuate, reaching high enough densities to defoliate trees. Mopane caterpillars, for example, may have significant impact as defoliators of Mopane trees in Limpopo, helping to keep these trees, which tend to form dense monocultures, in check. In the

Defoliation of White Milkwoods (*Sideroxylon inerme*) in Noordhoek, Western Cape, by *Laelia* caterpillars.

## GETTING INVOLVED IN CITIZEN SCIENCE AND CONSERVATION OF MOTHS

Observing and photographing moths is a rewarding activity in its own right, but many enthusiasts like to use the observations they make to contribute towards scientific knowledge and conservation efforts. There are several ways this can be done. Perhaps the simplest way is to join a citizen science initiative, such as **LepiMAP**, run by the Biodiversity and Development Institute at the University of Cape Town (**https://vmus.adu.org.za/**) or **iNaturalist** (**www.inaturalist.org**), run by the South African National Biodiversity Institute (SANBI). The aim of these projects is to determine the distribution patterns of and conservation priorities for wildlife, including moths, on the African continent. Contributors register on the website and then load their photographs, which are then identified by others. Since a large group of citizens can gather far more distribution records than the few researchers who focus on Lepidoptera in the region, these sites have already significantly contributed to our understanding of the distribution patterns and relative abundances of moth species across the region. The best way to get involved is to join **The Lepidopterists' Society of Africa** (LepSoc Africa ) and share your interest in the group with other enthusiasts. There are LepSoc branches in the main urban centres in Gauteng, KwaZulu-Natal and the Western Cape, and these host a variety of activities, including field trips, branch meetings and talks by guest speakers.

One of the specialist projects of LepSoc Africa (**https://lepsocafrica.org/**) is the **Caterpillar Rearing Group** (CRG), which aims to document the life history patterns of regional species and, in particular, to establish the associations between caterpillars and the moths they turn into, their hosts, and their parasitoids – those insects that feed in or on other insects without initially killing them. The majority of images, associations and host plant records in this book have been drawn from the CRG. Details on how to collect and rear caterpillars can be found on the CRG page.

Western Cape, coastal White Milkwood (*Sideroxylon inerme*) forests can be stripped bare of leaves by caterpillars of the *Laelia* Gypsy moths. Another example is Michii's Prominent (*Rhenea michii*), which from time to time defoliates Cape Ash (*Ekebergia capensis*).

These explosions in caterpillar populations mostly occur when periods of drought or extreme cold have had an adverse effect on the natural parasitoid that usually keeps populations of these herbivorous larvae in check. Such ecological imbalances are usually temporary, and balance is quickly restored, mostly with no long-term effects on the ecosystem. However, in cases where humans interfere and use pesticides to 'control' the herbivores, the imbalance can remain for extended periods. This is because parasitoids are also affected by pesticides and hence regular spraying maintains a perpetual cycle of imbalance, with long-term adverse effects on the ecosystem.

## Conservation status

There has been a concerted attempt by LepSoc Africa, supported by SANBI, to assess the conservation status of southern African Lepidoptera species. This assessment, known as the Southern African Lepidoptera Conservation Assessment or SALCA project, revealed that three of the region's butterfly species are already extinct, and a further 77 species or subspecies are threatened with extinction. Due to data deficiency, almost all

## About this book

This guide is designed for use in southern Africa, defined here as the region incorporating South Africa, Lesotho, Eswatini, Namibia, Botswana, Zimbabwe and southern Mozambique. However, many of the species featured have distribution ranges that extend into equatorial Africa and beyond. Over 10,000 moth species have been described from southern Africa, and many more remain undescribed, making it impossible to include all of them in a book of this size. This guide now concentrates on providing information on each of the regional families and, within these, featuring those species most likely to be encountered by readers. These include species that are most abundant, widespread, economically or ecologically important, conspicuous, large or unusual.

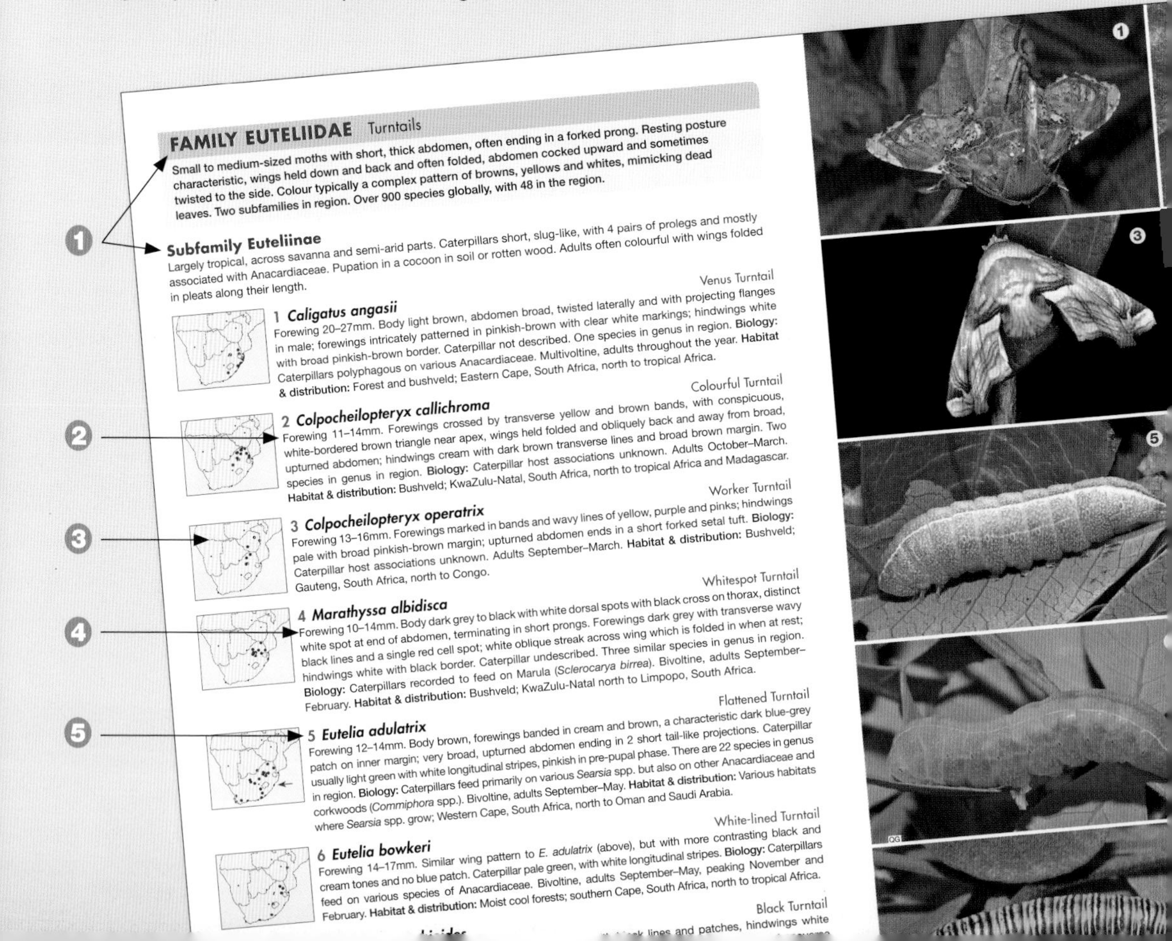

**FAMILY EUTELIIDAE** Turntails

Small to medium-sized moths with short, thick abdomen, often ending in a forked prong. Resting posture characteristic, wings held down and back and often folded, abdomen cocked upward and sometimes twisted to the side. Colour typically a complex pattern of browns, yellows and whites, mimicking dead leaves. Two subfamilies in region. Over 900 species globally, with 48 in the region.

**Subfamily Euteliinae**

Largely tropical, across savanna and semi-arid parts. Caterpillars short, slug-like, with 4 pairs of prolegs and mostly associated with Anacardiaceae. Pupation in a cocoon in soil or rotten wood. Adults often colourful with wings folded in pleats along their length.

Venus Turntail
**1 *Caligatus angasii***
Forewing 20–27mm. Body light brown, abdomen broad, twisted laterally and with projecting flanges in male; forewings intricately patterned in pinkish-brown with clear white markings; hindwings white with broad pinkish-brown border. Caterpillar not described. One species in genus in region. **Biology:** Caterpillars polyphagous on various Anacardiaceae. Multivoltine, adults throughout the year. **Habitat & distribution:** Forest and bushveld; Eastern Cape, South Africa, north to tropical Africa.

Colourful Turntail
**2 *Colpocheilopteryx callichroma***
Forewing 11–14mm. Forewings crossed by transverse yellow and brown bands, with conspicuous, white-bordered brown triangle near apex, wings held folded and obliquely back and away from broad, upturned abdomen; hindwings cream with dark brown transverse lines and broad brown margin. Two species in genus in region. **Biology:** Caterpillar host associations unknown. Adults October–March. **Habitat & distribution:** Bushveld; KwaZulu-Natal, South Africa, north to tropical Africa and Madagascar.

Worker Turntail
**3 *Colpocheilopteryx operatrix***
Forewing 13–16mm. Forewings marked in bands and wavy lines of yellow, purple and pinks; hindwings pale with broad pinkish-brown margin; upturned abdomen ends in a short forked setal tuft. **Biology:** Caterpillar host associations unknown. Adults September–March. **Habitat & distribution:** Bushveld; Gauteng, South Africa, north to Congo.

Whitespot Turntail
**4 *Marathyssa albidisca***
Forewing 10–14mm. Body dark grey to black with white dorsal spots with black cross on thorax, distinct white spot at end of abdomen, terminating in short prongs. Forewings dark grey with transverse wavy black lines and a single red cell spot; white oblique streak across wing which is folded in when at rest; hindwings white with black border. Caterpillar undescribed. Three similar species in genus in region. **Biology:** Caterpillars recorded to feed on Marula (*Sclerocarya birrea*). Bivoltine, adults September–February. **Habitat & distribution:** Bushveld; KwaZulu-Natal north to Limpopo, South Africa.

Flattened Turntail
**5 *Eutelia adulatrix***
Forewing 12–14mm. Body brown, forewings banded in cream and brown, a characteristic dark blue-grey patch on inner margin; very broad, upturned abdomen ending in 2 short tail-like projections. Caterpillar usually light green with white longitudinal stripes, pinkish in pre-pupal phase. There are 22 species in genus in region. **Biology:** Caterpillars feed primarily on various *Searsia* spp. but also on other Anacardiaceae and corkwoods (*Commiphora* spp.). Bivoltine, adults September–May. **Habitat & distribution:** Various habitats where *Searsia* spp. grow; Western Cape, South Africa, north to Oman and Saudi Arabia.

White-lined Turntail
**6 *Eutelia bowkeri***
Forewing 14–17mm. Similar wing pattern to *E. adulatrix* (above), but with more contrasting black and cream tones and no blue patch. Caterpillar pale green, with white longitudinal stripes. **Biology:** Caterpillars feed on various species of Anacardiaceae. Bivoltine, adults September–May, peaking November and February. **Habitat & distribution:** Moist cool forests; southern Cape, South Africa, north to tropical Africa.

Black Turntail

moth species could not be evaluated in this assessment. However, progress was made in the consolidation and geo-referencing of occurrence data, culminating in some 260,000 records. A further 20,000 records were subsequently added during the recording and evaluation of occurrence data for species included in this book.

Currently, only one regional moth species is considered Critically Endangered, according to IUCN criteria: the day-flying cycad moth Millar's Tiger (*Callioratis millari*, p.184), which is restricted to small grassland patches in and around a single nature reserve in KwaZulu-Natal, South Africa.

*Callioratis millari*, a regional species that is Critically Endangered.

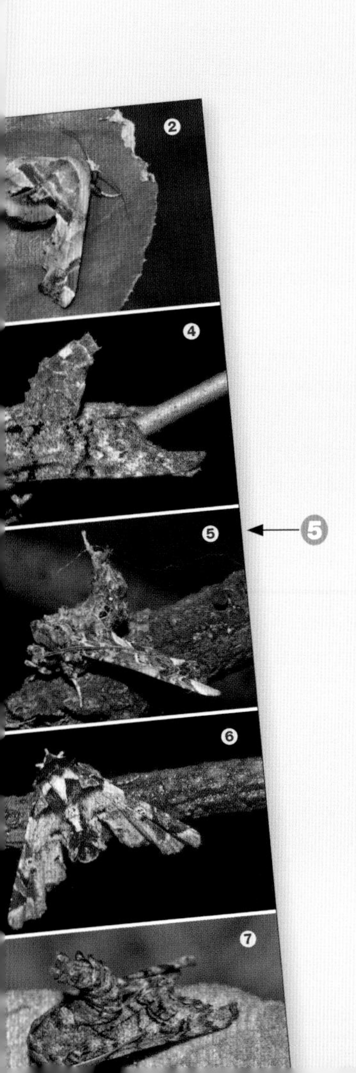


## 1 Family and subfamily

These describe features that identify the family or subfamily and the biology of its members.

## 2 Species or genus

Describes key identification features readily visible without detailed examination. A group may have numerous similar-looking species, not all of which are included in the book. Caution is therefore required when making a positive identification.

Based on available data, the preferred habitat and distribution of the species is given. Flight periods and host associations are provided, where known. Not all known host plants are listed, but the degree of known host specialisation is given. For example, if a moth species feeds on a number of host species in one genus, or other higher taxon, then the genus or higher taxon is provided.

## 3 Distribution maps

These show accurate point localities where the species has been recorded. The distribution of many species is poorly known, so identification should not be dismissed if it falls outside of the range indicated; the specimen may be a new record for the area. Maps are not provided for alien species, as many are ubiquitous and have shifting distributions.

It is notable that climate change has resulted in numerous recent range extensions by moths, which, being highly mobile and having short life cycles, can rapidly occupy and establish in suitable new areas as conditions change.

## 4 Size

The sizes given are forewing lengths. A certain amount of variation around these lengths is to be expected. Sometimes the given length is that of the specimen photographed, but a size range is usually given where the adult size is known to be variable.

## 5 Numbering system

The species or group entries on each page are numbered sequentially to match the photographs on the opposite page. Additional images of larval or other life-history stages are also provided.

## FAMILY MICROPTERIGIDAE Jaw Moths

**Most primitive extant family of moths, thought to have evolved over 100 million years ago, concurrent with early diversification of flowering plants. The only adult moths still equipped with chewing mouthparts; the proboscis being absent. Hindwings similar in size and venation to forewings, wings held tent-like over body, both fringed posteriorly with long setae. Adults usually diurnal; feed on pollen or fern spores. Most known caterpillars of the family feed on non-angiosperm plants, such as liverworts or fungi. In Africa, the caterpillar, which is squat with short dorsal pegs is only known from a single, undescribed South African species. Approximately 160 species globally; many more collected but still undescribed. In the region, 9 described *Agrionympha* species in South Africa, and 1 undescribed species, and an undescribed genus from the southern Cape. The South African species are most closely related to the Madagascan species.**

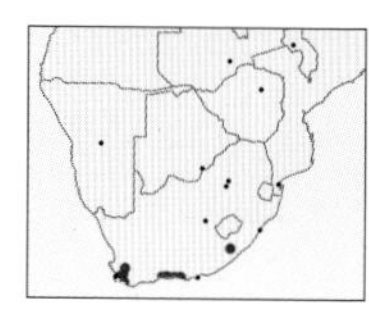

### 1 *Agrionympha capensis* — Silver-L Jaw Moth

Forewing 3mm. Head capped with fuzz of elongate cream scales, eyes very large; forewings elongate and pointed, bright metallic purplish-black with broad silvery white bands, the first of which is distinctively 'L'-shaped. **Biology:** Caterpillars feed on liverworts. Adults November–March. Moisture sensitive and generally active only around dawn before evaporation of overnight dew. Only females attracted to light traps. **Habitat & distribution:** Damp seepage areas in fynbos, among maidenhair ferns on stream banks or around forest margins; most widespread and common species, from Stellenbosch to Mthatha, South Africa.

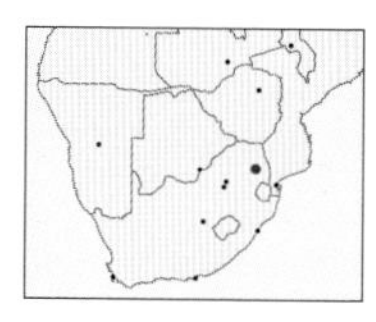

### 2 *Agrionympha kroonella* — Mariepskop Jaw Moth

Forewing 3mm. Iridescent purple wings marked with 1 longitudinal basal, and 2 transverse distal white flashes; unique in the genus is the median white band present only as a spot. **Biology:** Caterpillar foodplant likely liverworts. **Habitat & distribution:** Open and sunny, moist banks between areas of tall forest and fynbos, at high altitude.

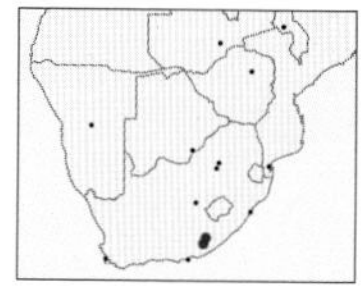

### 3 *Agrionympha fuscoapicella* — Hogsback Jaw Moth

Forewing 3–3.5mm. Has similar white markings to *A. kroonella*, but tips of wings grading to brown. **Biology:** Caterpillar foodplant likely liverworts. **Habitat & distribution:** Collected in February from herbaceous vegetation in sunny places in dense forest; currently known only from Hogsback, South Africa.

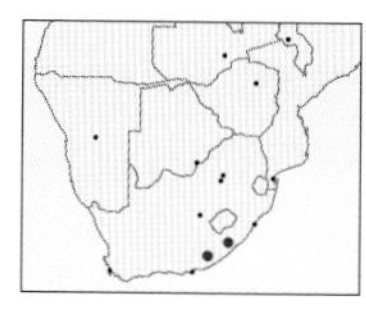

### 4 *Agrionympha sagitella* — Silver-arrow Jaw Moth

Forewing 3–4mm. Base of forewing gives rise to arrow-shaped silvery marking, followed by 2 jagged silvery bands. Caterpillar unknown. **Biology:** Caterpillar foodplants likely liverworts. **Habitat & distribution:** Tall, open forest, where it was swept from damp, sunny banks overgrown with ferns, herbs and liverworts; known from high-altitude sites at Hogsback and Ngadu forests (Mthatha), South Africa.

## FAMILY ACANTHOPTEROCTETIDAE Archaic Sun Moths

**Very small family of fairly small (forewing 2–8mm) and basal diurnal moths with vestigial mandibles. Wings very slender, margin densely fringed by long setae. Most species iridescent. Caterpillars are blotch leaf miners. Only about 7 species globally, most North American, only 1 in the region.**

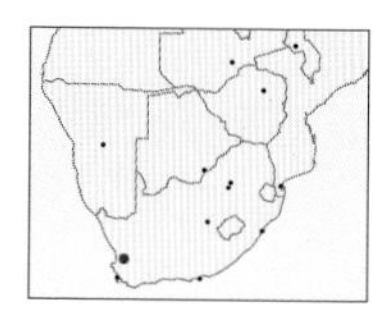

### 5 *Acanthopteroctetes nepticuloides* — African Archaic Sun Moth

Forewing length 2mm. Thorax brown; forewings narrow, iridescent silvery brown with darker scales, especially distally; hindwings pale grey, margins fringed with very long setae. **Biology:** Unknown. Caterpillars possibly feed on plants of the genus *Phylica* (Rhamnaceae). **Habitat & distribution:** Only African species in family; Cederberg, Western Cape, South Africa.

1
2
GG
3

WM
4
5

# FAMILY HEPIALIDAE Swift Moths, Ghost Moths

**Primitive group of medium to very large moths with very short antennae and reduced mouthparts. Wings long and narrow, similarly sized, usually steeply inclined against body when at rest. Adults sexually dimorphic; females larger than males. Some with rapid and erratic flight, hence common name. Females of most species lay eggs in flight, scattering them over suitable vegetation. Grub-like caterpillars form silk-lined tunnels underground, mostly feeding on roots or the fungi associated with roots, development sometimes taking several years. Some caterpillars exploited as food by indigenous peoples. Hepialidae can be found in most relatively moist terrestrial habitats and are common even on the highest mountains over 3,000m, but are rare or absent in arid areas. They form an important part of the food chain, also in newly transformed habitats. Over 600 described species globally, 81 in the region.**

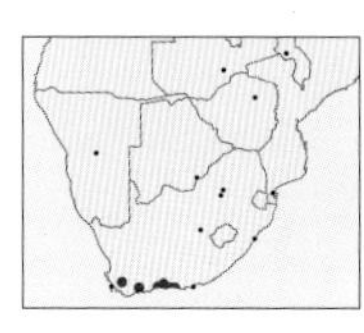

### 1 *Leto venus* Silver-spotted Ghost Moth

Forewing 54–70mm. Distinctive salmon pink forewings patterned with broken silver bars; hindwings plain orange. Caterpillar white and grub-like. One species in genus in region. Thought to be a Gondwana relict with its closest relatives living in Southwest Australia. **Biology:** Caterpillars, unusual for family, burrow into keurboom trees (*Virgilia* spp.) expelling frass from holes in the trunk; also into Honeybush Tea (*Cyclopia subternata*). Larval development may take several years. Adults, February–April. **Habitat & distribution:** Fynbos and forest; Western Cape and Eastern Cape, South Africa.

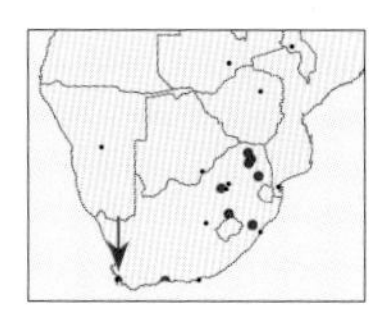

### 2 *Antihepialus antarcticus* Antarctic Swift

Forewing ♂ 23–26mm, ♀ 33–36mm. Thorax covered in shaggy brown setae; forewings elongate, pinkish-brown with darker brown patches and marginal dots, rising into broad hump posteriorly when at rest in male, rounded in female, broken yellow to orange longitudinal band diagnostic. Three species in genus in region, which are the closest African relatives to *Leto venus* (above). **Biology:** Caterpillar host associations unknown. Adults December–May. **Habitat & distribution:** Moist forest and riverine vegetation; Western Cape, South Africa, north to Mozambique and Zimbabwe.

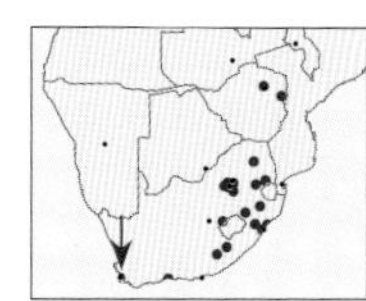

### 3 *Eudalaca ammon* Brown Lawn Swift

Forewing ♂ 12–15mm, ♀ 20–22mm. Male dark brown thorax densely setose, forewings brown with distinctively broad, meandering median white-edged brown band discontinuous to base, dense fringe of setae along distal margin; hindwings uniform dark brown; female uniform grey-brown with markings much reduced or absent. Caterpillar with reddish-brown anterior end and whitish body. Thirty-one similar species in genus in region that differ in the shape and size of the forewing medial band. **Biology:** Caterpillars feed in tunnels on the roots of various grasses. Adults August–April, peaking in late summer. **Habitat & distribution:** Abundant in moist habitats, including transformed habitats, where grass grows, apparently absent in arid areas; Western Cape, South Africa, north to tropical Africa.

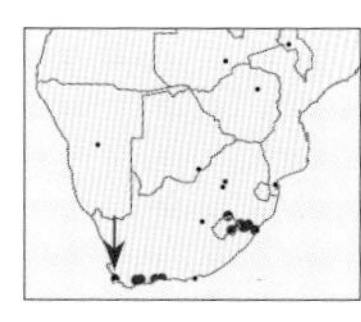

### 4 *Eudalaca exul* Crooked Swift

Forewing ♂13–17mm, ♀ 22–25mm. Male similarly patterned to *E. ammon* (above), but median band silvery white, continuous to base and with bulge toward apex and white terminal spots, band thinner in Western Cape populations; hindwings light brown with dark brown setae; female uniform grey-brown with markings much reduced or absent. **Biology:** Caterpillar host associations unknown. Adults January–March. **Habitat & distribution:** Two separated areas: moist fynbos, Western Cape, South Africa, and mountain grassland, Lesotho and KwaZulu-Natal, South Africa.

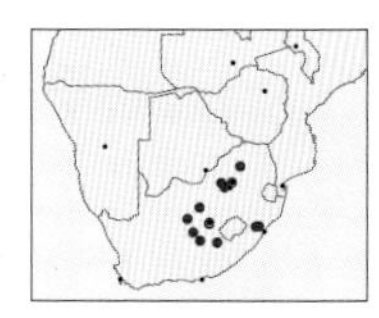

### 5 *Eudalaca ibex* White Waved Swift

Forewing ♂12–14mm, ♀ 22mm. Thorax and head with long clump of greyish-white setae; forewings variably grey-brown with thin, irregular, variably toothed central white band, long white cilia, distinctive white line tapering from base to apex along dark costa; hindwings uniform light grey brown. Similar *E. bacotii* has the median white band more regular and not toothed. **Biology:** Caterpillar host associations unknown. Adults January–April. **Habitat & distribution:** More arid bushveld and grassland than *E. exul*; Eastern Cape north to Limpopo, South Africa.

1
1
IT
2♂
2♀
3
3♂
4
5

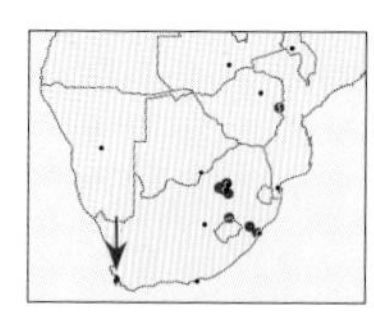

**1 *Gorgopis libania*** Golden Swift

Forewing ♂15–17mm, ♀ 23–27mm. Male thorax densely clothed in very long golden-brown setae, becoming dark toward head; forewings dark brown, densely covered in golden scales, especially in cell area, cilia golden; hindwings uniform golden-brown; female dark brown without golden scaling. Caterpillar elongate yellow grub. Thirty species in genus in region; males and females in genus have well-developed bipectinate antennae. **Biology:** Caterpillars feed on roots of grasses. Adults March–April, when they can be abundant at times. **Habitat & distribution:** Moist grasslands, including transformed habitats; Western Cape, South Africa, north to tropical Africa.

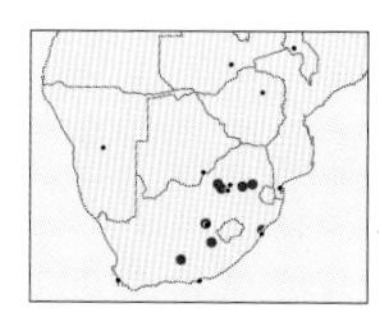

**2 *Gorgopis fuscalis*** Brown-veined Swift

Forewing ♂16–17mm, ♀ 23mm. Similar to *G. libania* (above), but golden scales only between dark brown veins; thorax densely clothed in very long golden-brown setae, becoming dark toward head; hindwings light brown, darker brown on veins; female similar but lighter and without golden scales. **Biology:** Caterpillar host associations unknown. Adults October and March–April. **Habitat & distribution:** Grasslands; Eastern Cape north to Mpumalanga, South Africa.

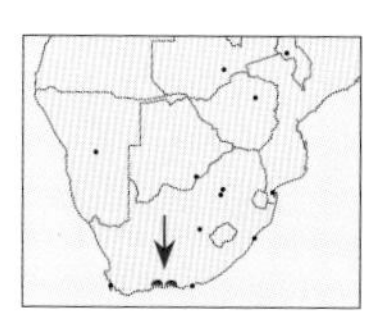

**3 *Metahepialus plurimaculata*** Blotched Swift

Forewing 20mm. Thorax densely covered in dark brown setae; forewings with complex pattern of small patches of different shades of brown, with overlay of shiny purplish scales that fades in dried specimens; hindwings plain brown with some markings on apex. Two species in genus in region. **Biology:** Caterpillar host associations unknown. Adults January–March. **Habitat & distribution:** Temperate forest; southern Cape, South Africa.

## FAMILY HELIOZELIDAE Shield-bearer Moths

**Widely distributed family of minute to small metallic, day-flying moths with very smooth head scaling; labial palps project forward. Caterpillars make meandering mines in leaves of trees and bushes. Vacated mines distinctive, as caterpillars cut oval sections from the leaf at wide end of mine, within which they usually pupate on the ground. About 123 species globally, 5 in the region.**

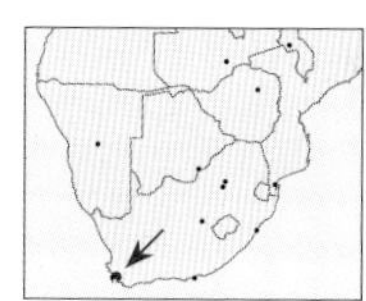

**4 *Holocacista varii*** Pelargonium Leafminer

Forewing 2–3mm. Head and thorax brown, tuft of light brown scales on head; forewings chocolate brown with broad, complete white basal bar, followed by incomplete white bars on trailing and leading edges; hindwings grey, narrow, with broad fringe of marginal hairs. Caterpillar short, flattened, yellowish grub with dark head capsule and constricted abdominal segments. One of several very similar species. *H. capensis* caterpillars mine grapevines (*Vitis vinifera*) and Baboon Grape (*Rhoicissus digitata*); mine begins as a zigzag, later enlarging into wide gallery, pupae attached directly to grapes or trunks of vines. Both species endemic to South Africa. **Biology:** Caterpillars of *H. varii* feed within leaves of the Hooded-leaf Pelargonium (*Pelargonium cucullatum*), where larvae formed meandering mines in leaves in summer, later forming pupae (depicted) protected within a case made of 2 oval sections of the leaf joined with silk, and attached to the plant. Adults September–May. **Habitat & distribution:** Exposed mountain slopes covered in low-growing fynbos; southwestern Cape endemic.

## FAMILY INCURVARIIDAE Leafcutter Moths

**Small to medium-sized moths with narrow wings held steeply roof-like over body when at rest. Adults diurnal. Early instar caterpillars leaf miners, but later stage caterpillars form cases cut from sections of leaf, sometimes continuing to feed from these cases. Some 51 species globally, only 1 in the region.**

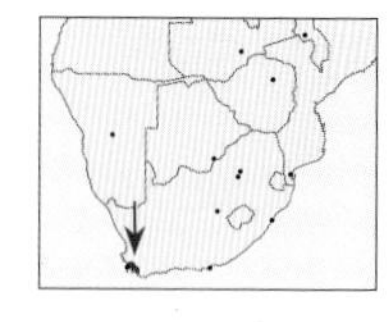

**5 *Protaephagus capensis*** Protea Blotch Leafminer

Forewing 3–5mm. Delicate body, forewings largely metallic blue-black, hindwings brown; tuft of yellow scales on head. Caterpillar flesh-coloured, short and flat, with constricted abdominal segments and conspicuous mandibles. **Biology:** Caterpillars serious pest of proteas, feeding in winter in groups, within leaf tissue of various *Protea* and *Leucadendron* spp. Early tunnels are narrow, later broadening and forming coalesced, grey necrotic areas. The flattened pupal cases are formed in late winter from 2 oval pieces cut from leaf and bound together with silk. These fall to the ground and remain in the leaf litter until the moths emerge in early summer. **Habitat & distribution:** Endemic to mountainous areas of the Western Cape, South Africa.

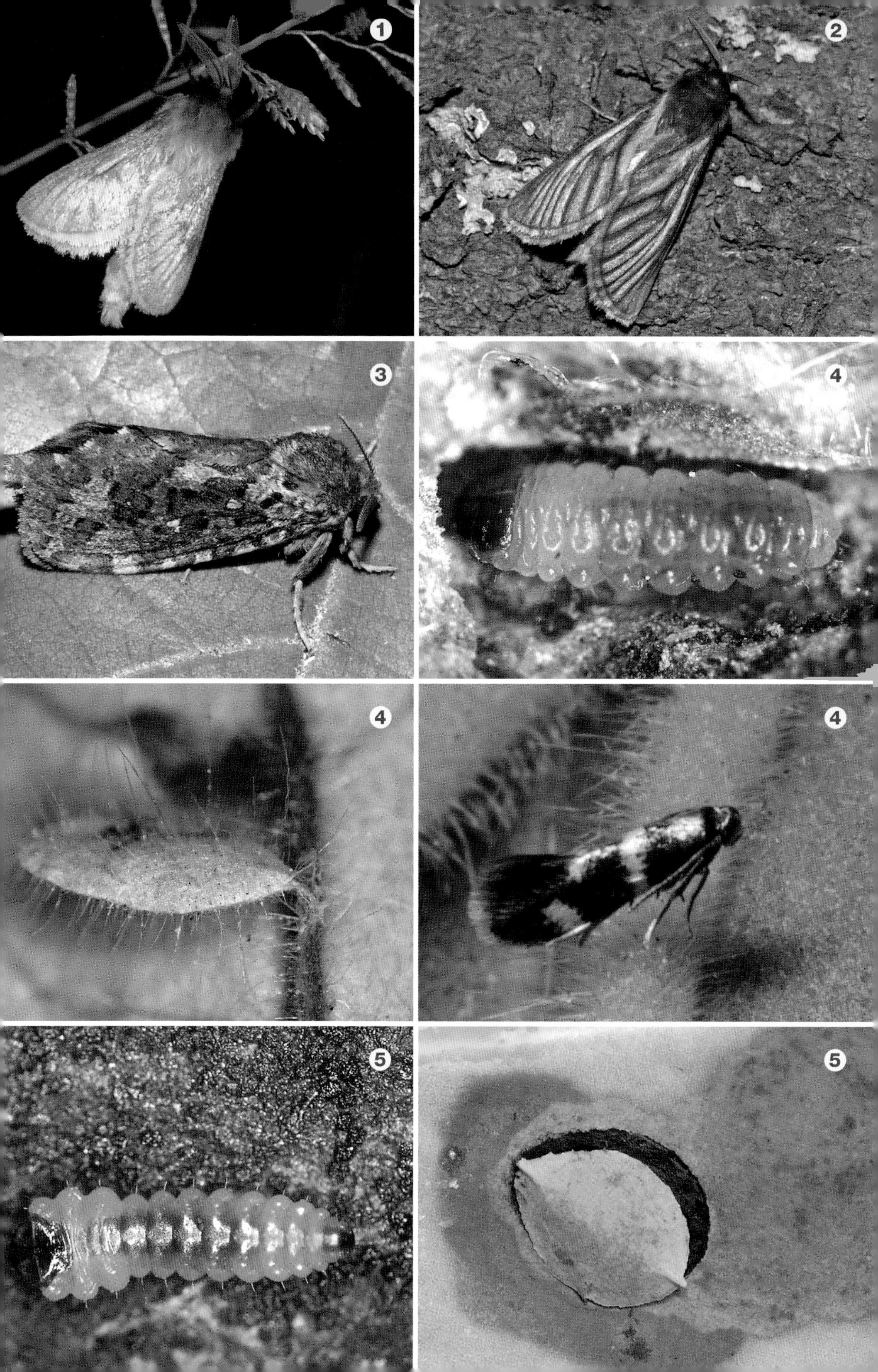

1
2
3
4
4
4
5
5

## FAMILY ADELIDAE Longhorn Moths

**Small to very small moths, easily recognised by their extremely long antennae, which are up to 4 times body length, especially in males. Eyes exceptionally large. Narrow forewing typically off-white and mottled, sometimes banded or metallic, held roof-like against sides of body. Brightly coloured species usually diurnal, drab ones nocturnal. Most caterpillars all leaf miners, some making small bags later. Over 300 species described globally, with 73 in the region, of which 66 are in genus *Ceromitia*.**

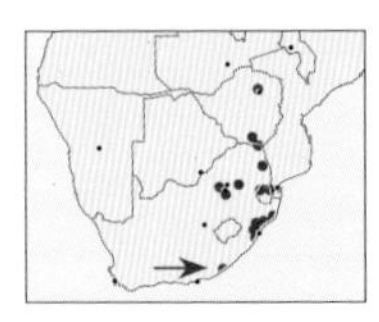

### 1 *Ceromitia turpisella* — Dusted Longhorn

Forewing 7–8mm. Body and wings white; forewings with 2 or 3 oblique brown to black bands, often broken into dots; antennae white, several times length of body. **Biology:** Caterpillars make a flat oval bag using clay particles, feed on Black Wattle (*Acacia mearnsii*). Univoltine, adults active December–February. **Habitat & distribution:** Various vegetation types; in warmer, wet eastern parts of South Africa, Mozambique and Zimbabwe.

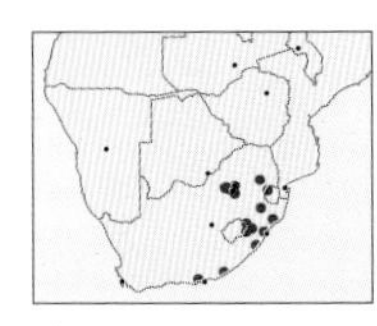

### 2 *Ceromitia wahlbergi* — Wahlberg's Longhorn

Forewing 8–10mm. Forewings white with brown crossbands, which are wider and more distinct than in *C. turpisella* (above), the middle one distinctly branched to form a 'Y'. **Biology:** Host associations unknown. Adults active December–March, can be abundant at times. **Habitat & distribution:** In grassland and wooded habitats, including transformed habitats; northeastern South Africa, Mozambique and Zimbabwe.

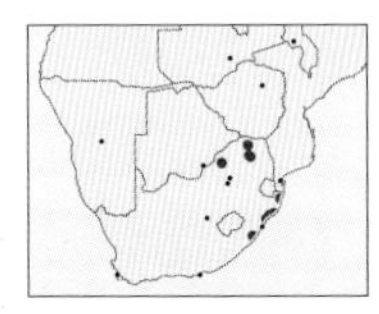

### 3 *Ceromitia trigoniferella* — Wedged Longhorn

Forewing 8–10mm. Body white; forewings white with variable but distinct large black triangular patch medially, wings speckled with black posteriorly. **Biology:** Unknown. **Habitat & distribution:** Afromontane forests and coastal bush; subtropical South Africa, Zimbabwe and Kenya.

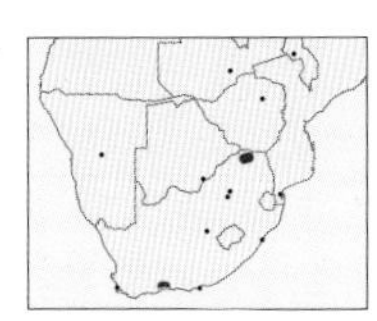

### 4 *Ceromitia punctulata* — Spotted Longhorn

Forewing 8–10mm. Ground colour cream, rather than white; forewings crossed by 2 complete oblique brown bands and with 2 other brown patches, spotted with fine brown dots. **Biology:** Unknown. **Habitat & distribution:** Known from 2 widely separated locations, South Africa.

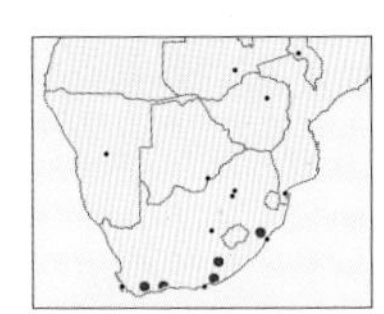

### 5 *Adela natalensis* — Barred Longhorn

Forewing 7–9mm. Narrow, pale cream-yellow band inwardly oblique across the forewings and yellow triangular macula diagnostic. Top of head covered with light orange hairs. Antennae white, ringed with chestnut brown. Six species in genus in region. **Biology:** Unknown. **Habitat & distribution:** Cool, moist, well-wooded habitats; apparently restricted to South Africa.

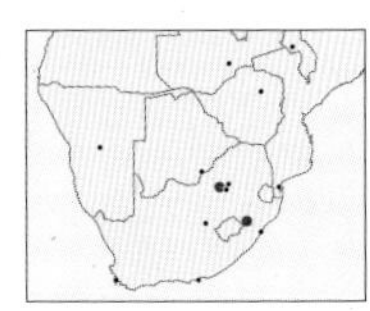

### 6 *Adela cuneella* — Hippie Longhorn

Forewing 4mm. Head and body extremely setose with hairs standing in all directions; basal part of forewing finely striated with black and white scales, diagnostic oblique, black-lined white bar down from costa, black and white circular mark on apex, cilia alternating black and white; hindwings grey with long cilia. **Biology:** Adults diurnal, swarming around flowering *Buddleja* trees in early spring. Closely related species lay their eggs on flowers, later caterpillars drop to the ground, feeding on detritus and building a case from it. Adults August–September. **Habitat & distribution:** Grassland; KwaZulu-Natal and Gauteng, South Africa.

1
2
3 ♀
4
5
6
6

## FAMILY CECIDOSIDAE Gall Moths

**Little-known family of basal moths restricted mainly to South Africa, but also present in New Zealand and South America. Best known for their piercing ovipositors, used to deposit eggs in plant tissue, inducing galls. Approximately 18 species globally, 11 in the region. All regional species belong to genus *Scyrotis*, which form galls on leaves of *Searsia* (Anacardiaceae) species.**

### 1 *Scyrotis athleta* — Jumping-ball Moth

Forewing 6–7mm. Body brown; forewings light grey or brown with darker speckling; hindwings light brown with fringed margins. Caterpillar white and grub-like. **Biology:** Females lay eggs in leaves of Glossy Currant (*Searsia lucida*) and galls form around feeding caterpillar. When mature, in about November, external layer of gall bursts open and oval ball falls to the ground. Balls remain mobile for at least 6 weeks, propelled by movements of caterpillar inside. Moths emerge around March, excising neat hole in pupal case. **Habitat & distribution:** Coastal strandveld; in the Western Cape, South Africa, wherever host plant occurs.

## FAMILY PSYCHIDAE Bagworm Moths

**Caterpillars more conspicuous than the short-lived adults; constructing portable, silk-lined cases from leaves, twigs, thorns or other debris, these being of characteristic shape and form in each species. Adult males have comb-like antennae, filiform in females, lack mouthparts and are often stout and dull, sometimes with transparent wings. In some species females are wingless and never leave their larval cases, attracting males to them using pheromones. Males mate by thrusting their abdomen into the lower opening of the female case; the female later lays her eggs within her case and dies. A few species are said to be parthenogenetic. Caterpillars mostly slow to mature, in many species taking more than a year; can remain dormant in bag during adverse conditions and resume feeding when conditions improve and food is again available. About 1,000 species have been described globally, with some 146 in the region.**

### Subfamily Oiketicinae

Females wingless, with legs reduced or absent and head lacking ocelli. Females remain in the bag for life, where mating takes place.

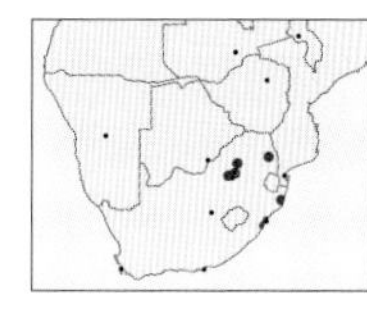

### 2 *Eumeta cervina* — Common Bagworm

Forewing ♂ 15–20mm. Male stocky with large, pectinate antennae and grey thorax with 2 black longitudinal stripes; wings short, grey and partially translucent, marked with black veins; female wingless. Two species in genus in region, very similar *E. hardenbergi* can be separated reliably only by genitalia comparison. **Biology:** Caterpillars build large cylindrical cases of thin twigs or thorns arranged in parallel and bound with strong silk. Caterpillars polyphagous on trees and shrubs. Caterpillars eaten by humans in some parts of the range. Adults seldom encountered, but throughout the year. **Habitat & distribution:** Variety of habitats; KwaZulu-Natal, South Africa, north to tropical Africa.

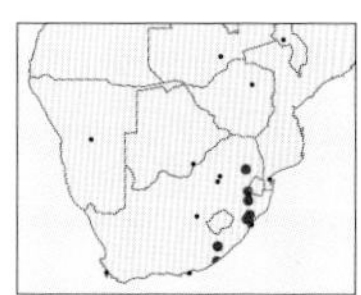

### 3 *Kotochalia junodi* — Wattle Bagworm

Forewing ♂ 9–12mm. Black body diagnostic, very setose, wings broad and almost completely transparent with black marginal veins; female wingless and limbless, grub-like, yellow with black eyes. One species in genus in region. Several very similar species in region best separated by their larval shelters. **Biology:** Caterpillars construct diagnostic pear-shaped bags coated with leaves and twigs. Males do not feed and are short lived, fly rapidly at night in search of females. Caterpillars polyphagous on trees, primarily on thorn trees (*Vachellia* spp.) and Australian wattles (*Acacia* spp.). Adults throughout the year. **Habitat & distribution:** Moist grassland and forest; Eastern Cape, South Africa, north to tropical Africa.

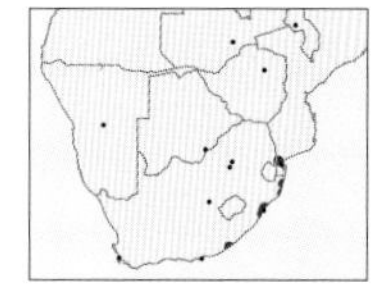

### 4 *Diaphanopsyche rogenhoferi* — Calbas Psyche

Forewing ♂ 6–7mm. Male delicate with slender black body and transparent white wings, antennae dark and strongly pectinate; forewing with large black apical tip, absence of black cell spots on forewing diagnostic. Caterpillars form irregular conical sacs with extended tube, which look like calabash gourds. One species in genus in region, similar *Monda delicatissima* has distinct black spots on forewing. **Biology:** Caterpillars polyphagous, feeding on tree flowers and leaves. Adults throughout the year. **Habitat & distribution:** Coastal forest and bush; Eastern Cape, South Africa, north to Mozambique.

1
2
2
3
3
MG
MA
3
4

## Subfamily Psychinae

Females of most species wingless, lacking ocelli but have fully developed legs. Female emerges from pupa, leaving pupal skin in bag and mating on outside of bag.

### 1 *Gymnelema vinctus* — Cross Stick Bagworm

Forewing ♂ 17–18mm, ♀ 27–30mm. Male has robust comb-like antennae, filiform in female, strongly setose head; forewings cream with variable speckling of brown scales and mottled black crossbands that form a conspicuous network; hindwings pale grey-brown. Caterpillar fashions a bag from short lengths of stick arranged transversely. Nine species in genus in region, similar *G. leucopasta* is smaller and has network much reduced. **Biology:** Caterpillars polyphagous on trees. Adults throughout the year. **Habitat & distribution:** Forest and riverine vegetation; KwaZulu-Natal, South Africa, north to Mozambique and Zimbabwe.

### 2 *Gymnelema stygialis* — Charcoal Bagworm

Forewing ♀ 14mm. Filiform antennae in female, strongly setose head; body and forewings charcoal black with a fine sprinkling of silvery white scales; hindwings dark grey. Caterpillar fashions a bag from short lengths of stick arranged transversely. **Biology:** Caterpillars live on tree trunks feeding on mosses (Bryophyta). **Habitat & distribution:** Afromontane forest; KwaZulu-Natal north to Mpumalanga, South Africa.

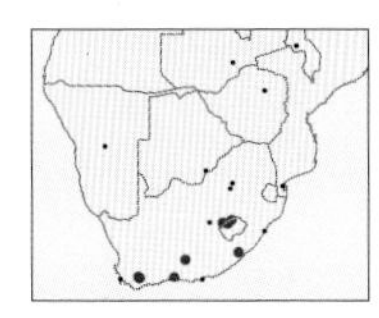

### 3 *Criocharacta amphiactis* — White-streak Bagworm

Forewing 14–16mm. Elongate with wings held roof-like against sides of body; forewings light grey, marked with numerous thin black, and thicker white lines, distinctive black spots beyond cell; hindwings grey, white around veins. Larval case made of neatly arranged, parallel sticks. One described species in genus in region. **Biology:** Caterpillars feed on flowers of various herbaceous plants. Adults October–March. **Habitat & distribution:** Fynbos, shrubland and grassland rich in herbaceous flowers; Western Cape, South Africa, to Lesotho.

## Subfamily Typhoniinae

Females have fully developed legs, and may be winged, wingless or have vestigial wings. When the female ecloses from the pupa the pupal skin does not remain attached to the opening of the bag, and mating occurs some distance from the bag.

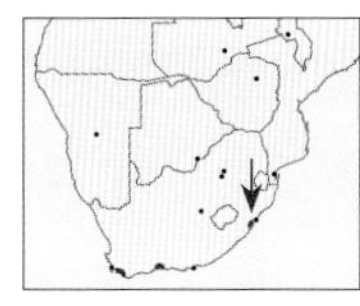

### 4 *Typhonia picea* — Bronze Typhonia

Forewing 5–6mm. Thorax black; forewings fairly uniform bronze-brown with a few black stipples and fringe of bronze setae; held roof-like against body. Caterpillars construct conical case from fine plant debris. Thirty-four diverse species in genus in region, probably polyphyletic. **Biology:** Caterpillars found scraping substrate off leaves of conebushes (*Leucadendron* spp.) also on wattles (*Acacia* spp.). Males diurnal, occur in aggregations, possibly near female. **Habitat & distribution:** Moist fynbos and forest; Western Cape and KwaZulu-Natal, South Africa, probably more widespread.

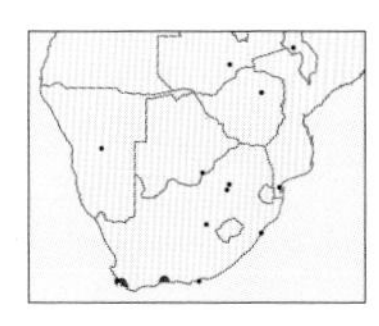

### 5 *Typhonia petrodes* — Woolly Typhonia

Forewing 7–8mm. White with black stippling on forewings forming dark transverse bands of varying intensity, 4 distinct black marks on costa. **Biology:** Caterpillars construct smooth, tapering 'woolly' cases from plant hairs, adding to the case as larva grows, which can change colour when they switch hosts. They feed on stems and leaf surface of various hairy fynbos plants, possibly digesting the long leaf hairs. **Habitat & distribution:** Moist fynbos; Western Cape, South Africa.

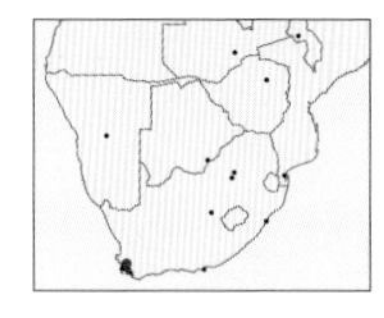

### 6 *Typhonia circophora* — Sooty Typhonia

Forewing ♂ 12mm, ♀ 15mm. Antennae of male finely feathered, filiform in female; head black, body and forewings dark charcoal grey, forewings with sparse sprinkling of lighter scales, 2 irregular black bands in male, absent in female; hindwings lighter with black veins. Caterpillar with black head capsule and with approximately square cases constructed from fine twigs and wood debris. Cases sometimes have collar of fine splayed material, followed by tube of parallel twigs. **Biology:** Caterpillars feed on *Manuela cheiranthus* and Common Button Bush (*Berzelia lanuginosa*). **Habitat & distribution:** Fynbos; Western Cape, South Africa.

1
1 ♂
SB
2
2 ♀
3
3
MM
4
4
5
5
MB
MB
6
6

## Subfamily Placodominae

Females lack ocelli, but have fully developed legs and may be winged or wingless. Mating occurs either on the bag or some distance from it.

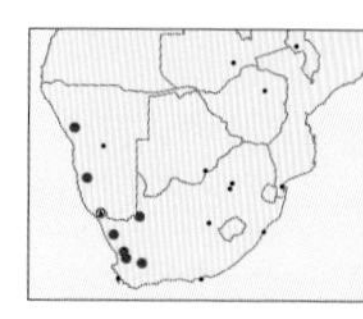

**1 *Australoplacodoma bicolorata*** Banded Bagworm

Forewing 12–14mm. Head and frons white; antennae pectinate in male, filiform in female; forewings banded brown and white; hindwings dark brown. Caterpillars and cases unknown. One species in genus in region. **Biology:** Caterpillar host associations unknown. **Habitat & distribution:** Succulent karoo and desert; Western Cape, South Africa, north to Namibia.

# FAMILY ERIOCOTTIDAE Old-world Spiny-winged Moths

**Adults small to medium-sized (wingspan 5–50mm), with rough head scaling and forward-projecting labial palps; antennae sometimes pectinate (feathery) in males; wings elongate, often tufted, usually dull grey or brown, sometimes banded or spotted. Caterpillars thought to tunnel in soil and feed on roots or detritus. About 212 species globally, 34 in the region.**

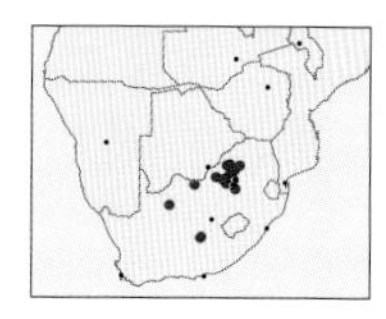

**2 *Compsoctena aedifica*** Pale Baglet

Forewing 10–14mm. Head with prominent tuft of erect yellow scales, thorax with crest of erect scales; forewings off-white variably scattered with brown speckles, sometimes grouped into broken crossbands, distally edged with long chequered setae; hindwings broad, off-white, with long marginal setae. Thirty-one similar regional species in genus. **Biology:** Caterpillar host associations unknown. Adults October–February. **Habitat & distribution:** Dry grassland and bushveld; Eastern Cape north to Limpopo, South Africa.

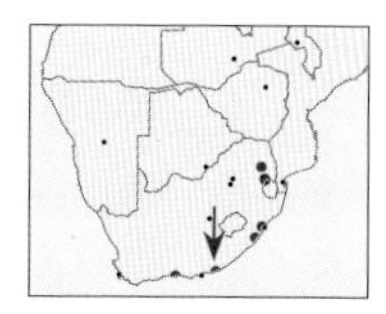

**3 *Compsoctena primella*** Dark Baglet

Forewing 7–8mm. Head with tuft of erect orange setae; forewings dark grey with purplish sheen fading in dried specimens, obscure dark brown patches, legs, forewing costa and cilia chequered orange and black; hindwings dark brown to black. **Biology:** Caterpillar host associations unknown. Adults August–December, males diurnal, congregating around bushes, also attracted to light at night. **Habitat & distribution:** Moist forests and coastal bush; Eastern Cape north to Mpumalanga, South Africa.

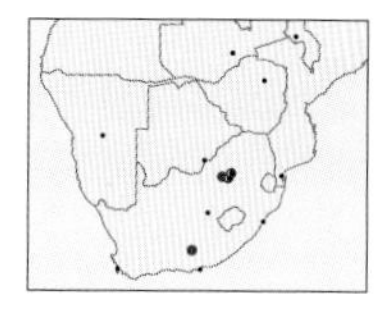

**4 *Cathalistis secularis*** Dusk Baglet

Forewing 10–13mm. Long, prominent, upright grey palps, antennae held at right angle to head, thorax white; forewings creamy white dusted with brown scales, less so in cell area; hindwings hyaline grey. Caterpillar unknown. Three species in genus in region. **Biology:** Caterpillar host associations unknown. Adults start flying at dusk, but are also attracted to light, primarily in winter, April and June–July. **Habitat & distribution:** Grassland; Eastern Cape and Gauteng, South Africa, probably more widespread.

# FAMILY TINEIDAE Fungus Moths, Clothes Moths

**A very diverse family of moths placed in 12 subfamilies, with many species not yet assigned to a subfamily. Small to medium-sized moths with wings held roof-like against the body, most yellow or white with shiny scales, sometimes speckled with brown or orange, proboscis reduced or absent, head usually with fuzzy coating of hairs, labial palps bristly. Wings fairly long and narrow, with fringes of setae. Caterpillars unusual in that they feed on a large variety of organic food sources, few on live plants, instead consuming fungi, lichens, faeces or detritus (often of animal origin), sometimes forming specialised symbiotic relationships with their hosts. Many of their caterpillars are case bearing and some species can digest keratin, the protein in hair. Several preferentially feed on wool carpets and clothes in human dwellings, and have become widely distributed as a result. One unusual genus feeds exclusively on horn, others in the abandoned nests of birds or mammals, feeding on their faeces. More than 3,000 species described globally, with over 275 in the region.**

1
2
2
2
3
4

## Subfamily Tineinae

Proboscis of adult reduced or absent, and the forewing has a clear or partially clear spot. Caterpillars often produce dumbbell-shaped portable cases and often feed in bird and mammal nests on animal tissue, faeces, silk, hair and horn.

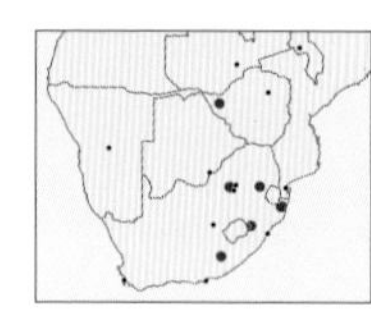

### 1 *Ceratophaga vastellus* — Horn Moth

Forewing 11–14mm. Head covered in prominent fuzz of darkish yellow scales; forewings uniformly golden yellow, upturned at tip and with fringe of terminal setae; hindwings grey with yellow fringes. Caterpillar squat and cream-coloured with brown head capsule and tip of abdomen. Seven similar-looking species in genus in region. Adults October–May. **Biology:** Caterpillars feed on old, detached horns of antelope, cattle or rhino, the larval tunnels emerging in characteristic branched masses covered by faecal pellets. **Habitat & distribution:** Bushveld and grassland; Eastern Cape, South Africa, north to tropical Africa.

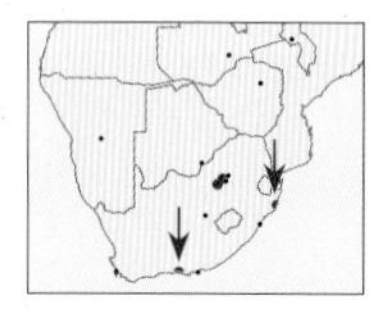

### 2 *Monopis truncata* — Truncate Bat Moth

Forewing 5–6mm. Head with prominent crest of orange scales; forewings dark brown, white along inner margin, prominent large translucent patch on cell, fringe of long, light brown and white setae along distal and posterior margins. Caterpillar pale and translucent with brown head capsule, in case covered in bat guano. Twelve species in genus in region. **Biology:** Caterpillars and adults live in caves to at least 80m below surface, feeding on bat guano. Probably multivoltine, adults recorded February–March. **Habitat & distribution:** Sea caves and inland deep caves where bats roost; Eastern Cape north to Gauteng, South Africa.

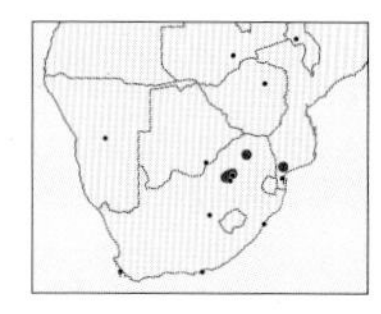

### 3 *Monopis lamprostola* — White-V Moth

Forewing 7–8mm. Head white with crest of orange scales; thorax and forewings dark brown with purplish sheen, prominent large white 'V' mark on costa; hindwings uniform yellow. A similar forest species, *M. megalodelta*, has the hindwings grey; *M. immaculata* has double 'V' mark and thorax white. **Biology:** Caterpillar host associations unknown. Adults December–March. **Habitat & distribution:** Bushveld; Gauteng, South Africa, to Mozambique.

Ubiquitous in homes across region

### 4 *Tinea pellionella* — Case-bearing Clothes Moth

Forewing 5mm. Small brownish moth; forewings marked by black, usually forming 3 spots. Caterpillar pale with dark head and thoracic segments. **Biology:** Caterpillars occupy a flattened case open at both ends, made from fibres and hairs. Feed on a wide range of materials including wool, fur, detritus, cobwebs, wallpaper and stored food products. **Habitat & distribution:** Common in homes and buildings, where the caterpillar in its case is often seen climbing up walls; originally from Europe, but now distributed worldwide.

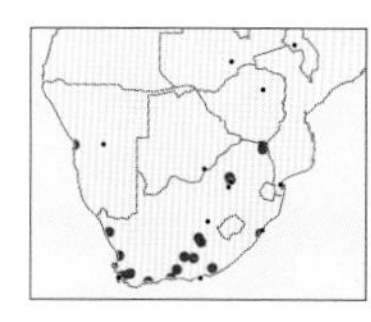

### 5 *Trichophaga cuspidata* — Tapestry Moth

Forewing 7–9mm. Colouration distinctive; head white and tufted; forewings divided into black basal half and creamy white distal half (also some darker speckling toward upturned setose tip); hindwings grey with fringe of long setae. Two species in genus in region, very similar *T. tapetzella* possibly also present in region, introduced from Europe. **Biology:** Caterpillar host associations unknown, *T. tapetzella* feeds on a variety of fibrous organic matter including fur, feathers and stored grain. Adults October–May. **Habitat & distribution:** Generally more arid habitats; Western Cape, South Africa, north to Tanzania.

## Subfamily Perissomasticinae

The antennae are almost as long as the forewing, the first antennal segment with comb-like structure. The proboscis is reduced.

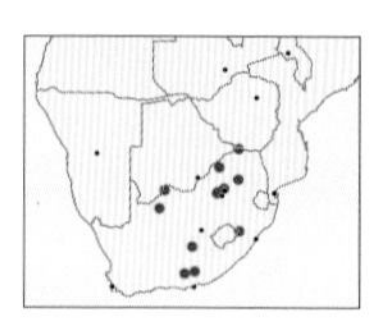

### 6 *Edosa leucastis* — Frosted Tineid

Forewing 8–12mm. Head covered in tuft of orange scales, eyes large and black; forewings white; hindwings grey with cream setae. Thirteen species in genus in region, some very similar, can only be reliably separated by genitalia comparison. **Biology:** Caterpillar host associations unknown. **Habitat & distribution:** Bushveld and grassland; Eastern Cape, South Africa, north to tropical Africa.

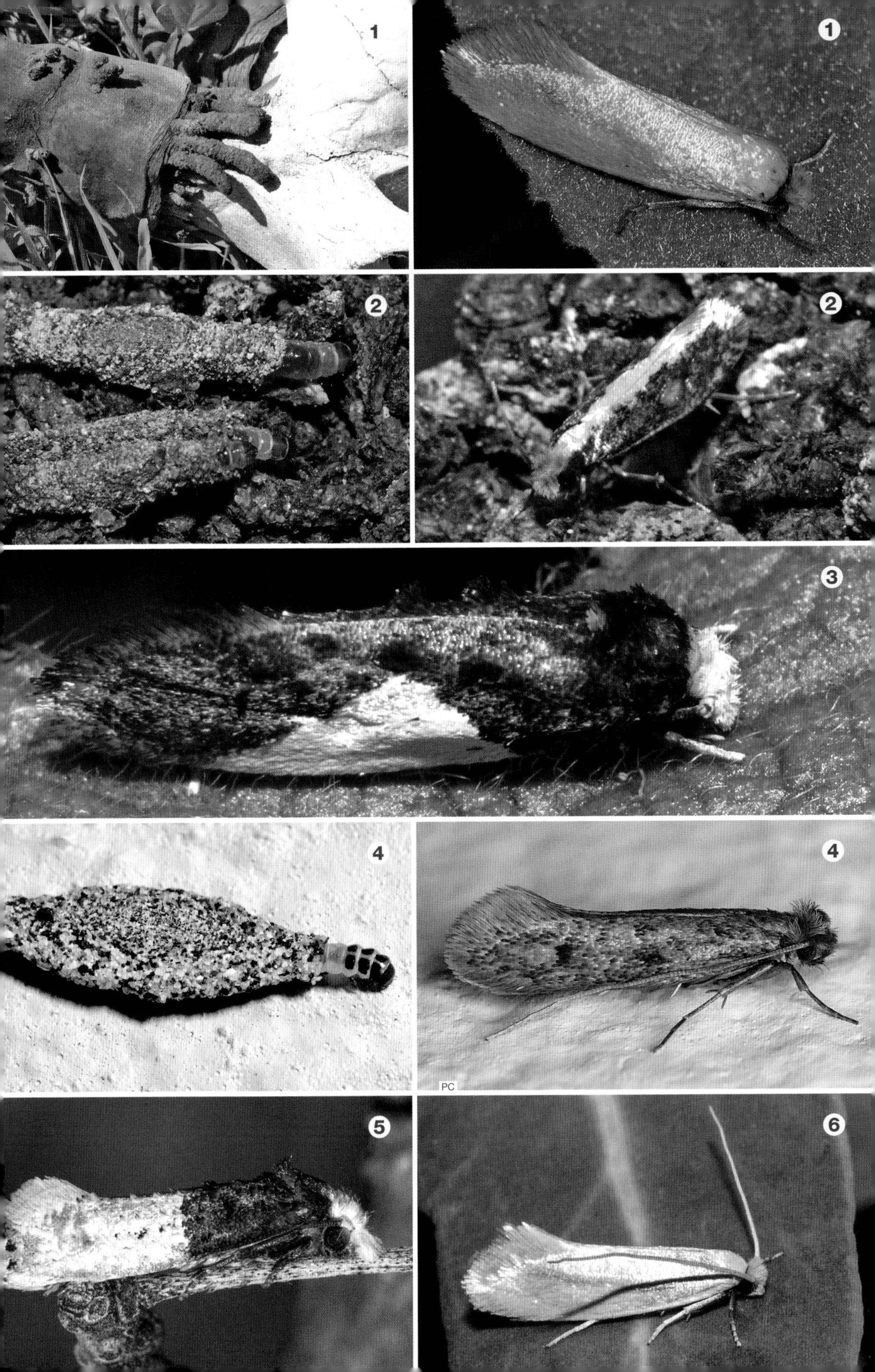
1
1
2
2
3
4
4
PC
5
6

## Subfamily Hieroxestinae

When viewed side-on, the head can be seen to taper toward its top. The bases of the antennae are connected to one another by a compact ridge of forward-projecting scales.

Ubiquitous in banana plantations around region

### 1 *Opogona sacchari* — Banana Moth

Forewing 9–12mm. Forewing yellow-brown with darker brown spots and speckles, males with dark spot at tip of wing; hindwings paler with fringe of yellowish setae. Caterpillar dirty white with black dorsal thoracic and abdominal plates and reddish-brown head capsule. **Biology:** Caterpillars usually feed on decaying plant matter, but can also be a pest of banana fruit and pseudostems, pineapples, bamboo, maize, sugar cane, and a variety of other ornamental plants. **Habitat & distribution:** Ubiquitous where host plants occur; native range spans sub-Saharan Africa, but also now widely introduced into Asia, Europe, North and South America.

## Subfamily Myrmecozelinae

The antennae are approximately three-quarters the length of the forewing. The mouthparts are reduced, with very short maxillary palps and proboscis.

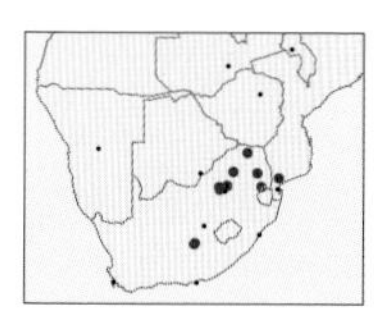

### 2 *Ateliotum crymodes* — Painted Tineid

Forewing 7–8mm. Head and frons extremely hairy, snout-like; body and forewings pearly white; forewings decorated with variable large orange patches and small black spots; hindwings hyaline grey. Two species in genus in region. **Biology:** Caterpillar host associations unknown. Adults December–March. **Habitat & distribution:** Grassy karoo and bushveld; Northern Cape to Limpopo, South Africa, and Mozambique.

## Subfamily Hapsiferinae

The head of the adults have 2 patches of scales extending behind the eyes upward to the top of the head. The antennae are smooth and the first segment is comb-like.

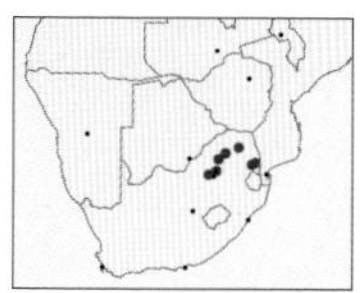

### 3 *Hapsifera meliceris* — Afternoon Hapsid

Forewing 9–11mm. Head and frons orange, body and forewing colour variable, off-white to orange, sprinkled with rows of black elongate spots, head, frons and forewings hairy; hindwings pitch black with orange fringe. Seven species in genus in region. **Biology:** Caterpillar host associations unknown. Adults fly in full sunshine in the afternoon, conspicuously flashing the orange and black wings when in flight, November–April. **Habitat & distribution:** Grassland and bushveld; Gauteng to Limpopo, South Africa.

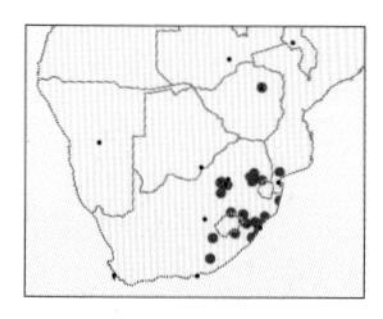

### 4 *Hapsifera glebata* — Evening Hapsid

Forewing 9–11mm. Head, body, frons and forewings orange; irregular dots of brown and raised black scales over forewings; hindwings light grey. **Biology:** Caterpillars found making weak silk tunnels in and feeding on rhino dung. Adults nocturnal, attracted to light, September–April. **Habitat & distribution:** Variety of habitats; Eastern Cape, South Africa, north to tropical Africa.

## *Pseudurgis* group

Species of *Pseudurgis* are in need of a subfamily placement within Tineidae.

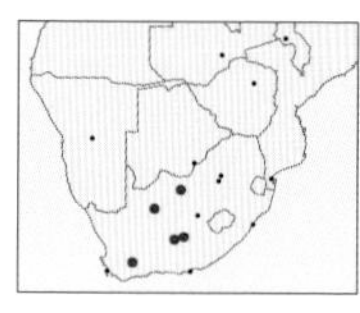

### 5 *Pseudurgis protracta* — Fan Durgis

Forewing 15–16mm. Body and forewings grey with black marks, forewings with thin black lines along wing from base fanning out at apex in a diagnostic pattern, black-lined white stigma in cell and several along wing margin; hindwings grey to white. Seventeen species in genus in region, pectinate antennae in males distinguish genus from other similar species. **Biology:** Caterpillar host associations unknown. Adults September. **Habitat & distribution:** Nama karoo and arid bushveld; Western Cape to North West, South Africa.

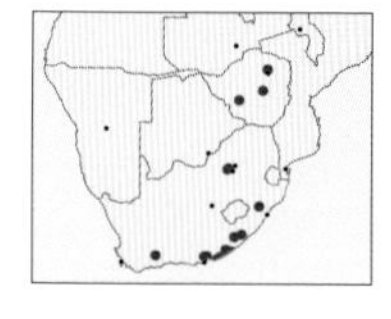

### 6 *Pseudurgis leucosema* — Spotted Durgis

Forewing 8–9mm. Body and forewings white; forewings with several distinctive marks consisting of black and brown scales, median bar variable; hindwings grey. Caterpillar light coffee brown with well-defined dorsal line, shiny pronotal shield. **Biology:** Caterpillars live in a tough tubular construction of silk and frass, incorporating leaves of *Asparagus* spp. Bivoltine, adults October–March. **Habitat & distribution:** Bushveld and grassland; Western Cape, South Africa, north to tropical Africa.

1
2
3
4
5
6

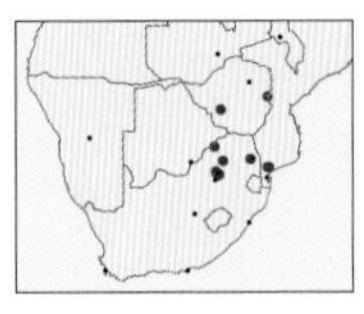

**1 *Pseudurgis polychorda*** Banded Durgis

Forewing 7–8mm. Body and forewings white, several distinctive black and brown bands across body and forewings, sprinkling of grey scales; hindwings uniform grey. Caterpillar light coffee brown with several dorsal and lateral lines along body, shiny pronotal shield. **Biology:** Caterpillars live in a loose silken nest among leaves of *Asparagus suaveolens*. Bivoltine, adults September–February. **Habitat & distribution:** Bushveld; Gauteng, South Africa, north to Mozambique and Zimbabwe.

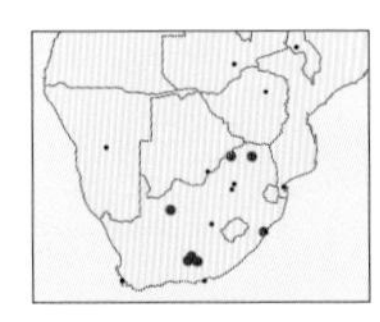

**2 *Pseudurgis undulata*** Mustard Durgis

Forewing 8–9mm. Body and forewings grey, covered in yellow scales that fade in dried specimens to uniform light grey; hindwings dark brownish-grey. **Biology:** Caterpillar host associations unknown. Bivoltine, adults September–February. **Habitat & distribution:** Bushveld; Eastern Cape north to Limpopo, South Africa.

## FAMILY BUCCULATRICIDAE Ribbed Cocoon-maker Moths

**Very small, easily overlooked moths found throughout the world; most species within the very diverse genus *Bucculatrix*. Most distinctive feature is the pupal case, which bears distinctive longitudinal ridges. About 297 species globally, 33 in the region.**

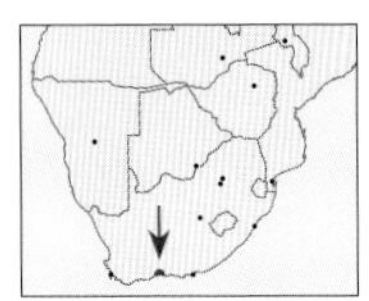

**3 *Bucculatrix* species** Ribbed Cocoon-makers

Forewing length 3–4mm. Wings narrow and wrapped around body when at rest, conspicuous tuft of setae on head; wing colour of regional species varies from dark brown, as in the illustrated species, through various shades to ivory white, some with darker markings on forewings; hindwings sharply pointed with long setae, usually grey. Thirty-two species in genus in region. Caterpillar translucent whitish-grey, with 2 broken lines running along length of body. **Biology:** Early instar caterpillars are leaf miners, leaving distinctive brown blotches on host plants; later instar caterpillars usually excavate the surface of the leaf. The illustrated species feeds on *Hippia frutescens* (Asteraceae). **Habitat & distribution:** Illustrated species southern Cape forests; *Bucculatrix* species occur in a large variety of habitats throughout the region.

## FAMILY GRACILLARIIDAE Leaf-blotch Miner Moths

**Important and widespread family of minute to small (forewing length 2–10mm) moths, best recognised by the characteristic 'tripod' resting posture of most adults (notably in dominant subfamily Gracillariinae), with front of body steeply raised on long legs and wing tips touching substratum. However, members of small subfamily Phyllocnistinae rest with body parallel to the surface, and subfamily Lithocolletinae often rest with head lowered. Caterpillars produce blotch-mines on leaves of trees and shrubs, infected leaves often being folded and secured with silk. Over 1,860 described species, with 175 in region, many more certainly undescribed.**

### Subfamily Gracillariinae

At rest, body held upright, at around 40° to the substrate. Either the fore- and midtibia and last part of midfemur thickened and covered in projecting, pigmented scales, or hindtibia with conspicuous projecting bristles.

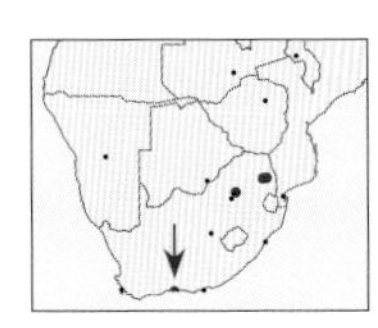

**4 *Dialectica trigonidota*** Trigon Blotchminer

Forewing 3mm. At rest, anterior of body raised up; antennae and legs long, banded brown and white; forewings patterned with broad alternating copper and white bands; hindwings narrow with long posterior setae. Caterpillar smooth, pale translucent brown. **Biology:** Caterpillars create blotch mines in leaves of Silver Cluster-leaf (*Terminalia sericea*), also Bushy Honeycup (*Melhania acuminata*), Large-fruit Bushwillow (*Combretum zeyheri*) and Water Pear (*Syzygium guineense*). Adults November–March. **Habitat & distribution:** Forest and bushveld; Western Cape north to Limpopo, South Africa.

Ubiquitous on Australian Myrtle in W & E Cape

**5 *Aristaea thalassias*** Myrtle Blotchminer

Forewing 3mm. Forewings light brown crossed by series of faint 'V'-shaped white bands; hindwings strongly setose. Caterpillar undescribed. **Biology:** Native to Australia and deliberately introduced to South Africa in 1996 to control Australian Myrtle (*Leptospermum laevigatum*). Eggs laid on young leaves, caterpillars initially making long, serpentine mines, later becoming large blotches and causing premature dropping of leaves. Life cycle rapid, with up to 5 or 6 generations per year. Although common, has not been effective in controlling myrtle, as it only causes minor damage to trees. **Habitat & distribution:** Coastal regions, wherever Australian Myrtle is invasive; South Africa.

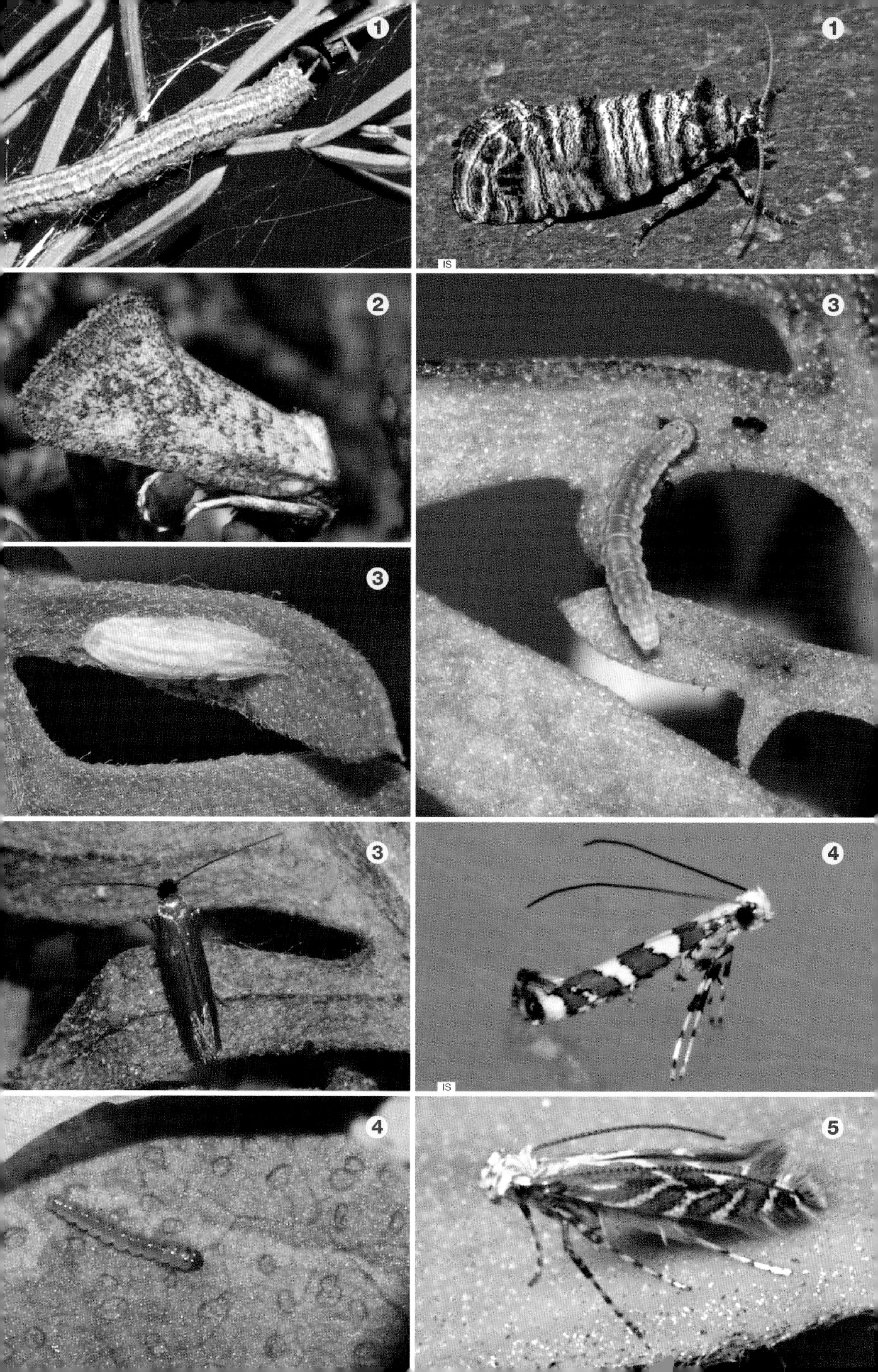
1
1
IS
2
3
3
3
4
IS
4
5

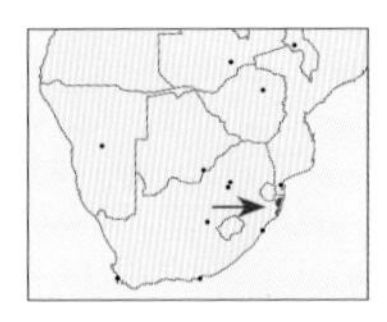

**1 *Caloptilia octopunctata*** Eight-spotted Blotchminer

Forewing 3mm. Upright posture; striking black and white banded legs bearing distinctive tufts of black scales posteriorly; forewings black with 4 white blotches on each and strong posterior fan of setae posteriorly. Caterpillar greenish with orange head. Thirty species of this very diverse genus in region. **Biology:** Caterpillars create blotch mines in leaves of Duiker Berry (*Sclerocroton integerrimus*) and Jumping-seed Tree (*Sapium ellipticum*). **Habitat & distribution:** Coastal forest; KwaZulu-Natal, South Africa, north to tropical Africa and Australia.

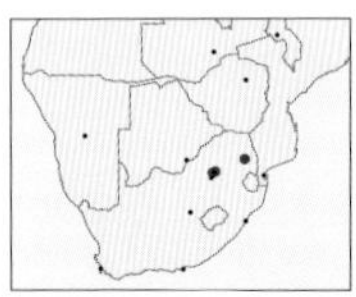

**2 *Caloptilia celtina*** Stinkwood Blotchminer

Forewing 3mm. Nut brown forewings with an elongate undulating yellow band along costa, fringe of long setae along posterior margin become brown distally; hindwings narrow and pointed, with fringe of long, pale setae. Caterpillar light green. **Biology:** Caterpillars feed on White Stinkwood (*Celtis africana*). **Habitat & distribution:** Bushveld; Gauteng to Limpopo, South Africa.

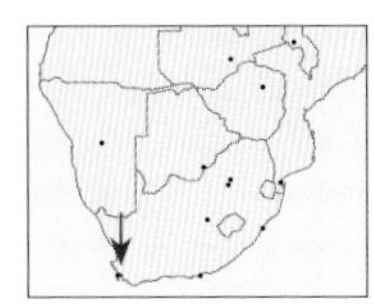

**3 *Caloptilia prosticta*** Pea Blotchminer

Forewing 4mm. Body light brown, legs dark brown basally, white distally; forewings with central brown spot becoming purple-brown distally and ending in dark brown patch. Caterpillar unknown. **Biology:** Caterpillars make meandering burrows on leaves of various pea species (Fabaceae). **Habitat & distribution:** Various agricultural and undisturbed habitats; recorded from Western Cape, South Africa, Madagascar, Seychelles and Nigeria, also Sri Lanka.

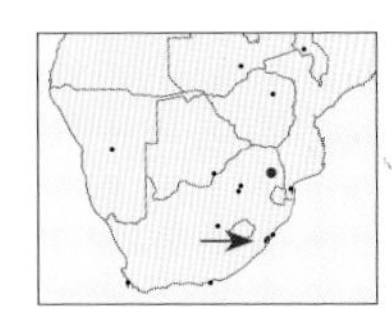

**4 *Macarostola noellineae*** Fiery Blotchminer

Forewing 5mm. Forelegs and most of forewing fiery red, head yellow; forewings with distinctive yellow marks along costa and inner margin, white and black tip on apex, cilia yellow; hindwings white. Caterpillar translucent green. Two species in genus in region. **Biology:** Caterpillars feed in blotch mine made on underside of Water Berry leaves (*Syzygium cordatum*). Adults December. **Habitat & distribution:** Lowland areas where Water Berry grows; KwaZulu-Natal north to Limpopo, South Africa.

## Subfamily Lithocolletinae

Antennal base lacking eyecap (enlargement partially covering eye), hindtibia lacking bristles, flat-lying scales on frons of head, first antennal segment comb-like.

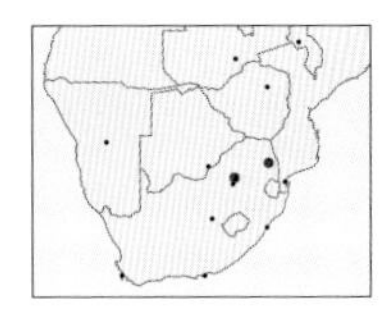

**5 *Phyllonorycter brachylaenae*** Malva Blotchminer

Forewing 3mm. Head white, body light brown; forewings brown, attractively patterned with several transverse white bands each bordered by fine black line; hindwings white, fringed with long setae. Caterpillar smooth and pinkish-yellow. Sixteen species in genus in region. **Biology:** Caterpillars mine the leaves of various Malvaceae, including *Brachylaena* spp. **Habitat & distribution:** Bushveld; Gauteng north to Limpopo, South Africa.

## Subfamily Acrocercopinae

Long (segmental) membrane between abdominal segment 8 and the male genitalia. Final instar caterpillar entirely red, with 2 hairs on the side of each segment.

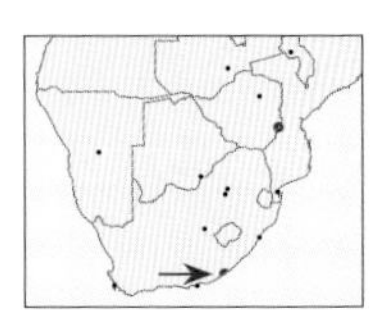

**6 *Acrocercops chrysophylli*** Milkwood Blotchminer

Forewing 3mm. Large, bright red eyes and long antennae reaching end of wings; long legs banded brown and white; forewings narrow, light shiny brown, paler and setose posteriorly, distally with white and dark crossbars and triangles; hindwings with long fringe of setae. Sixteen species in genus in region. **Biology:** Caterpillars create wide blotch mines in leaves of Common Red Milkwood (*Mimusops zeyheri*) and Fluted Milkwood (*Chrysophyllum gorungosanum*). **Habitat & distribution:** Recorded from South Africa and Zimbabwe only.

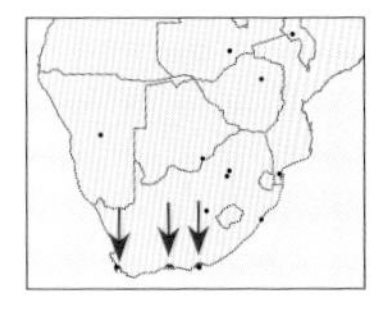

**7 *Amblyptila cynanchi*** Monkey Rope Blotchminer

Forewing 5mm. Eyes bright red in live moth; forewings white with broad light and dark nut brown bands, prominent black dot near tip of wing, followed by thin marginal brown line; hindwings narrow, light brown, fringed by long setae. Caterpillar pale orange. **Biology:** Caterpillars mine leaves of Monkey Rope (*Cynanchum obtusifolium*) and *C. ellipticum*. **Habitat & distribution:** Coastal woodland and forest; Western Cape to Eastern Cape, South Africa.

1
1
SM
2
IS
2
3
3
4
4
IS
5
5
IS
6
QG
6
7
7

## Subfamily Phyllocnistinae

Amongst the smallest gracillariids; rest with body and wings parallel to substrate. Adults usually white; caterpillars translucent, lacking appendages apart from a pair of conspicuous projections near the end side of the body.

Scattered across region where citrus occurs

### 1 *Phyllocnistis citrella* — Citrus Leafminer

Forewing length 3mm. Rests parallel to substratum; silvery white; dark stripes on forewings, which are strongly setose distally and have distinctive black dot at tip. Caterpillar minute, translucent greenish-yellow. **Biology:** Caterpillars make meandering mines in leaves, pupating in pupal cell near leaf margin. Originally Asian, but widely introduced to North and Central America, Australia and Africa. Adults live only a few days and can go through 10 or more generations a year. Associated with all species and varieties of citrus, but also occur on other hosts. **Habitat & distribution:** Eastern Cape and Limpopo, South Africa, with scattered records across sub-Saharan Africa.

# FAMILY YPONOMEUTIDAE Ermine Moths

**Adults small to medium-sized (forewing length 4–16mm), wings elongate, folded around body, hindwings often fringed with long setae. Colour usually white or grey, often with black dots on wings. Caterpillars usually live gregariously in webs, under which they consume leaves. *Yponomeuta* species feed on Celastraceae, primarily on *Gymnosporia* spp. About 395 species globally, 27 in region (see also superficially similar-looking *Ethmia* spp. in family Depressariidae p. 74).**

## Subfamily Yponomeutinae

The crochets (hooks) on the abdominal prolegs of the caterpillars are arranged in multiple rows.

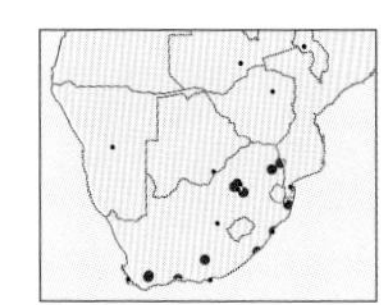

### 2 *Yponomeuta fumigatus* — Speckled Ermine

Forewing 11–13mm. Head and thorax white, diagnostic, abdomen black; forewings silvery grey with scattered small black spots on thorax and forewings; hindwings dark grey, sometimes almost black, fringed by setae of the same colour, a pair of scent brushes near base. Caterpillar mottled orange and blue with transverse rows of black spots, sometimes with dark transverse bands, and sparse long setae; head capsule black. One of several similar grey species in the region, each with different number, size and arrangement of black dots of forewings. Fifteen species in genus in region. **Biology:** Caterpillars feed communally under webs on spikethorns (*Gymnosporia* spp.). Multivoltine, adults throughout the year, peaking in summer months. **Habitat & distribution:** Variety of habitats where spikethorns grow; Western Cape, South Africa, north to tropical Africa, also Madagascar and Comoros.

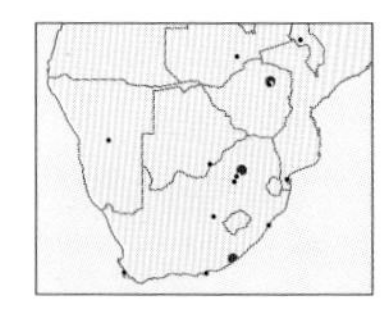

### 3 *Yponomeuta staudei* — Lesser Speckled Ermine

Forewing 10–11mm. Similar to *Y. fumigatus* (above); head pale grey, forewings ash grey, a pair of sub-basal black spots and 4 irregular rows of small spots on forewings; hindwings grey. Caterpillar grey with orange dorsoventral lines, black dorsal spots, head capsule brown and black. **Biology:** Caterpillars feed singly in web spun among leaves of Spikethorn (*Gymnosporia buxifolia*). Adults December–April. **Habitat & distribution:** Bushveld; Eastern Cape, South Africa, north to Zimbabwe and Malawi.

### 4 *Yponomeuta subplumbella* — Narrow Speckled Ermine

Forewing 8–9mm. Smaller and narrower wings than other grey species, with very small black spots arranged in indistinct longitudinal rows; hindwings pale grey, densely fringed with long pale setae. Caterpillar with yellow-orange and black bands, head capsule black. **Biology:** Caterpillars feed on spikethorns (*Gymnosporia* spp.). Adults often rest with posterior part raised. Multivoltine, adults throughout the year. **Habitat & distribution:** Variety of habitats where spikethorns grow; Western Cape, South Africa, north to tropical Africa, also Madagascar.

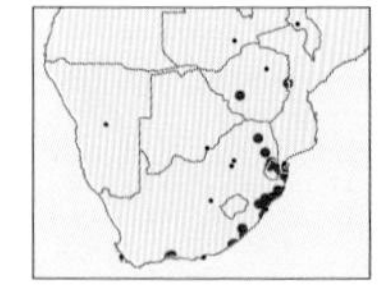

### 5 *Yponomeuta strigillatus* — Streak Ermine

Forewing 11–15mm. Body and forewings white with irregular black spots, row of elongate spots in a direction across wings diagnostic; hindwings smoky grey, becoming darker toward margins, fringed with long dark setae. Caterpillar dark brown around crest of segment, lighter brown between, shiny brown head and capsule. **Biology:** Caterpillars feed communally under webs on leaves and fruit of spikethorns (*Gymnosporia* spp.). Multivoltine, adults throughout the year, peaking in summer months. **Habitat & distribution:** Moist forest and bush; southern Cape , South Africa,north to tropical Africa, also Madagascar and Yemen.

1
1
2
JL
2
2
IS
3
3
4
4
5
5

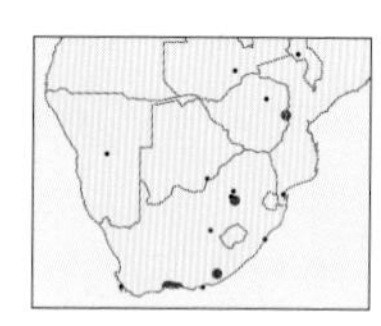

**1 *Yponomeuta africanus*** African Ermine

Forewing 10–11mm. Similar to *Y. strigillatus* (p.38), but smaller and has dark blotches amongst the black dots on the forewings, which are round in outline, but form a band in some individuals; hindwings pale brown with a few tiny black dots toward edge, males have long androconial setae on inner margin. Caterpillar unknown. **Biology:** Caterpillar host associations unknown. Adults September–May. **Habitat & distribution:** Moist forest and kloofs; southern Cape, South Africa, north to Zimbabwe and eastern Africa to South Sudan.

## FAMILY PLUTELLIDAE Diamondback Moths

**Adults small to medium-sized (wingspan 7–55mm), with pronounced antennae held forward; wings elongate, with long fringes of marginal setae, giving them an upturned appearance when at rest. Caterpillars feed mostly on Brassicaceae (cabbage family). About 386 species globally, 8 in the region.**

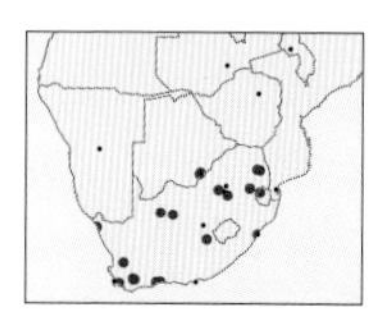

**2 *Plutella xylostella*** Diamondback Moth

Forewing 6mm. Dark brown with distinctive, white-edged buff stripe running along back, upturned ends to wings. Caterpillar green, tapering at both ends. **Biology:** Caterpillars feed on various Brassicaceae. Multivoltine, adults throughout the year. **Habitat & distribution:** Originate from Europe, but now spread globally; only in areas where introduced Brassicaceae grow; reported from many African countries.

## FAMILY TORTRICIDAE Leaf-roller Moths

**Very diverse family of minute to small (forewing length <15mm) moths with broad forewings squared off distally. Labial palps normally project forward and wings are held folded back against the body or form a broad roof to the sides. Colour typically mottled in browns, greys, whites or yellows in patterns that are typical for the family but are found across genera, making it impossible to identify many species reliably without detailed examination, even to genus level, especially in the tribe Archipini. Caterpillars usually polyphagous, cryptic and may form tubes within rolled or folded leaves, bore into fruits, stems, seeds, or roots, or are leaf miners or gall-formers. Many species adapt well to habitats transformed by humans and are often dominant in such areas, others are restricted to pristine untouched environments. Over 11,000 species have been described globally, with about 300 in the region and many more awaiting formal description.**

### Subfamily Chlidanotinae

The apical region of the wing is bent upward at rest, and the male genitalia have deep invaginations, which may house hair pencils. The caterpillars of some species are borers.

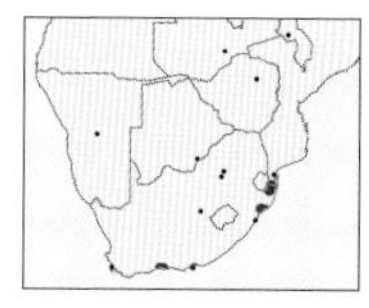

**3 *Trymalitis scalifera*** Colour-tip Gallmoth

Forewing 8mm. Frons, head and forewings pearly white; wings elongate, rounded with falcate tip which is folded in when at rest; broken band of grey scales along inner margin of forewing extending in a postmedian band to apex, row of distinctive metallic blue spots along termen with colourful red and yellow 'tip'; hindwings uniform white. One species in genus in region. **Biology:** Caterpillar host associations unknown. Adults throughout the year. **Habitat & distribution:** Moist forest and coastal bush; southern Cape north to tropical Africa, also Madagascar.

### Subfamily Olethreutinae

Largest of the subfamilies (defined by asymmetrical male genitalia). The caterpillars are either leaf rollers, seed or stem borers.

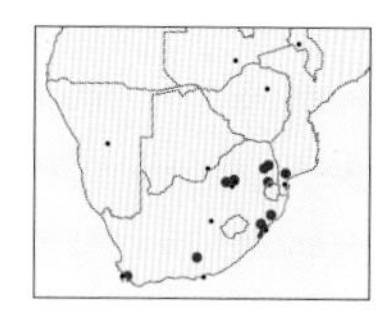

**4 *Crocidosema plebejana*** Cotton Tipworm

Forewing 6–8mm. Brown with black mottling; forewings cream with 2 distinct black patches, the distal one triangular in shape. Caterpillar plump and pinkish with red head capsule and black dorsal plates on thoracic segment 1. Three species in genus in region. **Biology:** Caterpillars polyphagous, primarily on mallows (Malvaceae), including invasive *Sida* weeds, which it helps to control. Adults December–May. **Habitat & distribution:** Variety of relatively moist habitats where Malvaceae grow; Western Cape north to tropical Africa and most warmer regions of the world.

1
2
3
2
IS
4
4

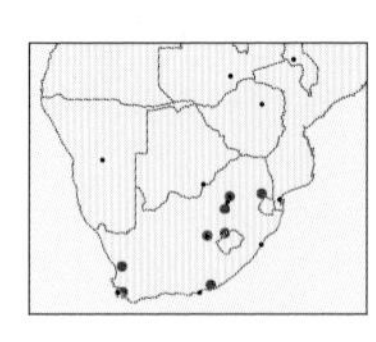

### 1 *Cydia pomonella* — Apple Codling Moth

Forewing 8–10mm. Body and forewings brown with grey and white speckling; a dark brown spot encircled by a gold ring near squared-off tip of forewings; hindwings brownish-white with cream margin. Caterpillar pink. Eighteen species in genus in region. **Biology:** Caterpillars feed on apples, pears, quince and other fruit. Adults September–March. **Habitat & distribution:** Inadvertently introduced to South Africa in 1890s; found in agricultural areas and gardens.

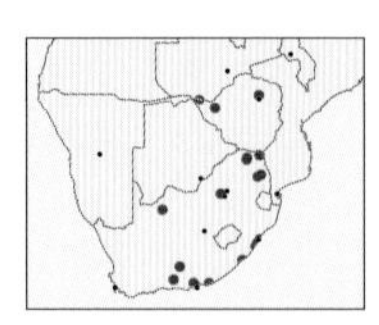

### 2 *Cryptophlebia peltastica* — Pod Moth

Forewing 8–13mm. Mottled reddish-brown with dark brown patches near tip of forewing, dark triangular patch on inner margin diagnostic; hindwings uniform pale brown. Caterpillar reddish-yellow with dark head capsule. Six species in genus in region. **Biology:** Caterpillars feed on seeds, inside seed pods, primarily on various Fabaceae. Multivoltine, adults throughout the year. **Habitat & distribution:** Various habitats; Eastern Cape, South Africa, north across Africa, Indian Ocean islands and on the Pacific island of Guam, where presumed to be accidentally introduced.

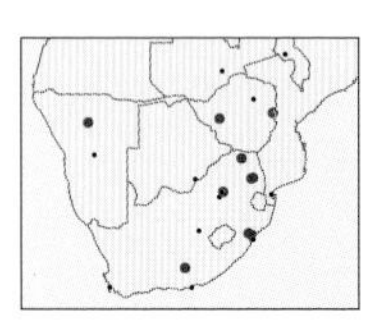

### 3 *Eccopsis incultana* — Flame Tortrix

Forewing 6–7mm. Triangular when at rest with inner margin of wings falcate; body light brown; forewings cream with several flame-like brown streaks emanating from median area; hindwings dark brown. Caterpillar light green with brown head. Nine species in genus in region. **Biology:** Caterpillars polyphagous, primarily on Fabaceae, feeding on leaves and young stems. Adults August and January–May. **Habitat & distribution:** Forest and bushveld; KwaZulu-Natal, South Africa, north to tropical Africa, also Madagascar, La Réunion, Mauritius and Seychelles.

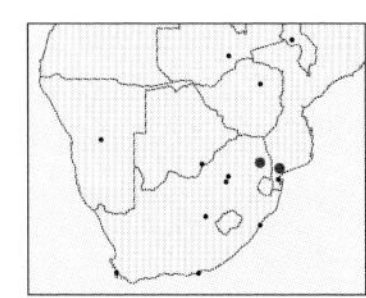

### 4 *Geita micrograpta* — Grey Delta-wing

Forewing 5mm. Strongly triangular with pointed tips to forewings and distinct scallop part way along outer margin of forewing, hind margin concave; plain grey with small black spots midway along forewing and patch near each corner, shiny sheen on wings of live individuals; hindwings uniform light brown. Caterpillar translucent yellow-green with brown head capsule. One species in genus in region. **Biology:** Caterpillars feed on milkplums (*Englerophytum* spp.). Adults March–June. **Habitat & distribution:** Rocky lowveld hillsides where host plants grow; Limpopo, South Africa, and Mozambique.

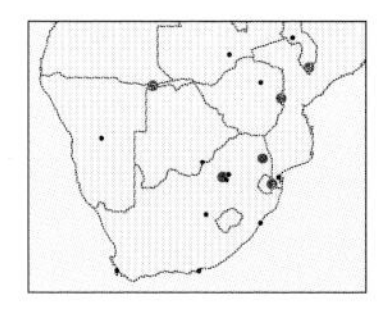

### 5 *Metendothenia balanacma* — Pentagon Leafroller

Forewing 5mm. At rest, forms arrow-head shape, various shades of brown, a pale pentagonal patch with dark brown centre in midline. Caterpillar slender and pale green with dark head capsule and black dorsal plates on thoracic segment 1. One species in genus in region. **Biology:** Caterpillars are polyphagous leaf rollers. Adults February–April. **Habitat & distribution:** Bushveld and forest; Gauteng, South Africa, north to tropical Africa.

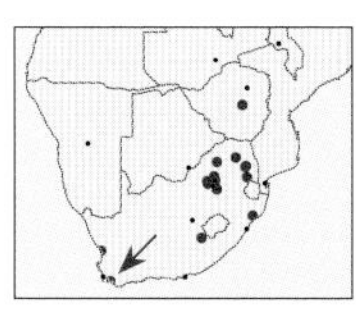

### 6 *Neaspasia orthacta* — Choc-dip Leafroller

Forewing 6–7mm. Basal half of forewings uniform dark chocolate brown, posterior half cream, variably shaded with brown almost obliterating the cream in some specimens, diagnostic darker patch toward termen; hindwings smoky grey. Caterpillar dark green. Two species in genus in region. **Biology:** Caterpillars on various Ebenaceae, feeding between leaves held together with silk. Probably multivoltine, adults September–April. **Habitat & distribution:** Variety of habitats; Western Cape, South Africa, north to tropical Africa, also Australia.

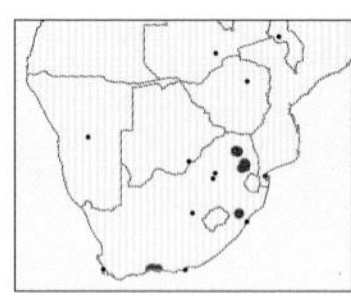

### 7 *Gypsonoma paradelta* — Pearly Leafroller

Forewing 6–8mm. Head capsule white, thorax black, abdomen brown; forewings silvery white with pinkish sheen in some individuals, fading to grey in old specimens, 2 diagnostic black triangular patches on inner margin, a line of smaller black spots along costa, cilia black; hindwings smoky black. Caterpillar translucent with black head capsule. Four species in genus in region. **Biology:** Caterpillars feed on Wild Mulberry (*Trimeria grandifolia*), folding leaf in half. Adults December–June. **Habitat & distribution:** Cool, moist Afromontane forests; southern Cape, South Africa, north to Ethiopia.

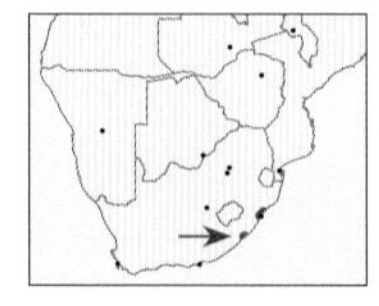

### 8 *Plutographa encharacta* — Choc-ripple Leafroller

Forewing 6mm. Body and forewings cream, densely patterned with broken longitudinal light and dark brown lines, hind margin of forewing with light brown fringe; hindwings plain cream with white fringe. Caterpillar light green with brown head capsule. One regional species in genus. **Biology:** Caterpillars feed on Small Cluster Pear (*Monanthotaxis caffra*). Adults November–January. **Habitat & distribution:** Coastal forests; Eastern Cape, South Africa, north to Mozambique.

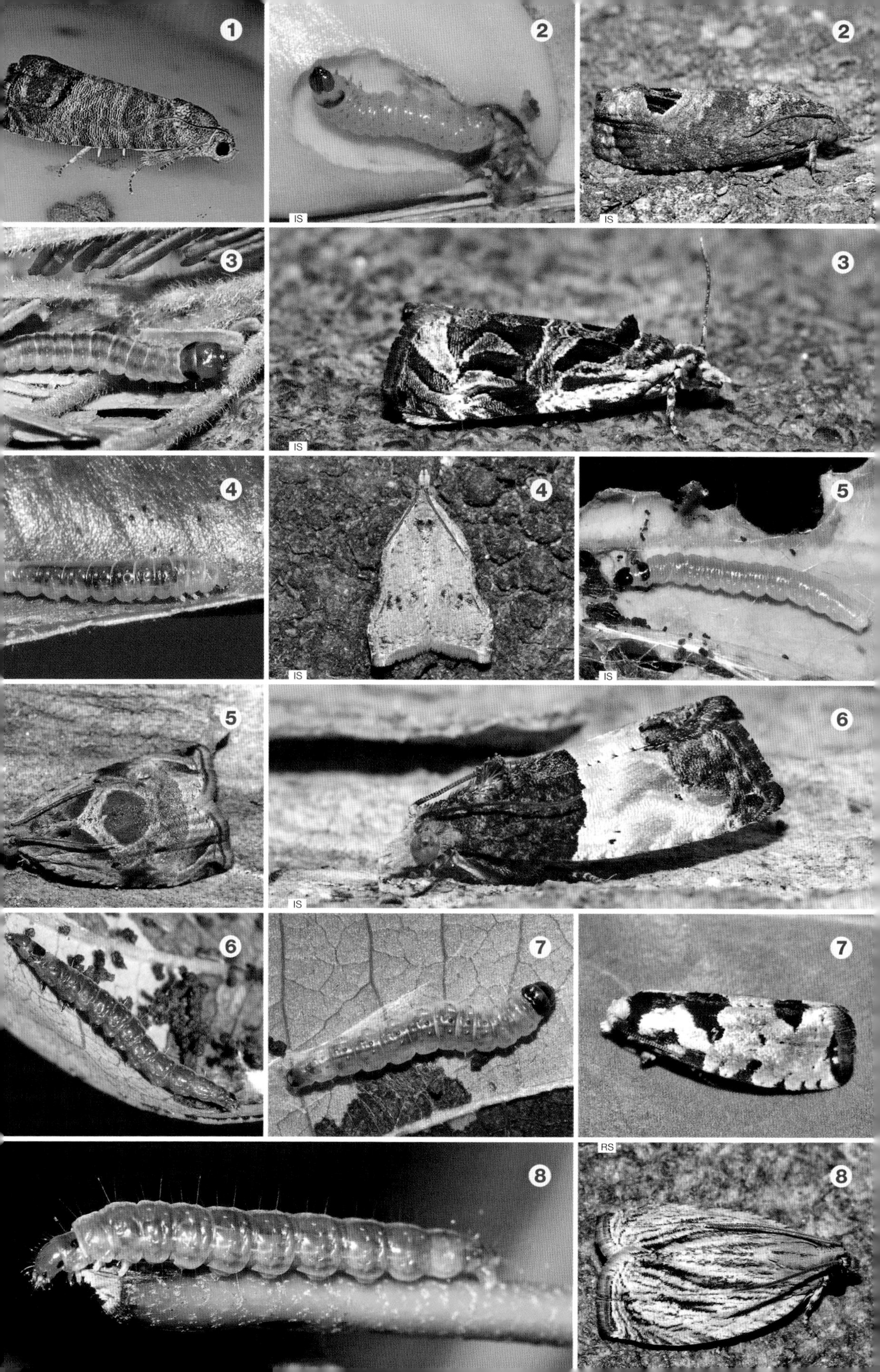
1
2
IS
2
IS
3
3
IS
4
4
IS
5
IS
5
6
IS
6
7
7
RS
8
8

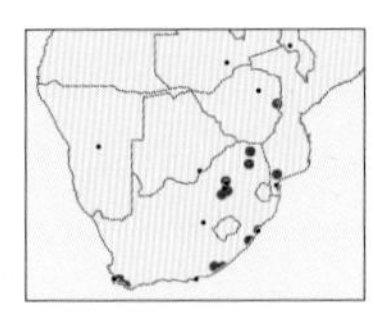

### 1 ***Thaumatotibia leucotreta*** False Codling Moth

Forewing 6–10mm. Female larger than male; mottled grey and brown with distinct dark patch midway along edge of forewings and diagnostic clear white spot in cell. Caterpillar smooth, plump and pink with brown plates on thoracic segment 1. **Biology:** Caterpillars polyphagous, feeding inside many types of fruits and seeds, often inadvertently transported by humans. Multivoltine, adults throughout the year. **Habitat & distribution:** Well-adapted to human habitation but also present in wild areas; natural distribution throughout most of sub-Saharan Africa; has been introduced to Europe and the USA.

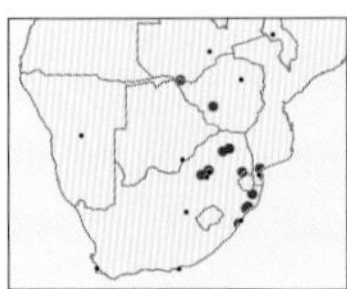

### 2 ***Brachioxena lutrocopa*** Blackspot Leafroller

Forewing 6mm. Folds forewing in when at rest; frons, head and thorax peach; when viewed while moth is at rest, forewings seen to be intricately mottled with peach, brown and silver scales. However, the hidden parts of the forewing are white with 2 distinct black spots, in postmedian area, leading margin of wing marked black and white; hindwings uniform light grey. Caterpillar translucent yellow with sparse setae, brown plates on thoracic segment 1. Two species in genus in region. **Biology:** Caterpillars recorded folding several leaves together and feeding outside of shelter on leaves of *Indigofera setiflora* (Fabaceae). Multivoltine, adults throughout the year. **Habitat & distribution:** Grassland, forest and bushveld; KwaZulu-Natal, South Africa, north to Zimbabwe.

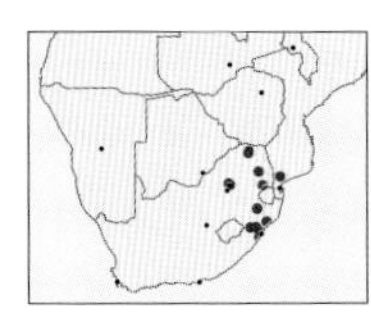

### 3 ***Malolotia galactitis*** Two-tone Leafroller

Forewing 6–7mm. Frons, head and thorax white; folds forewings inward when at rest, area along costa dark brown, area along inner margin white, variably spotted with grey, brown and black scales, terminal area reddish with white stripes at apex; hindwings uniform light grey with white cilia. Four species in genus in region. **Biology:** Caterpillar host associations unknown. Adults November–March. **Habitat & distribution:** Grassland, forest and bushveld; KwaZulu-Natal north to Limpopo, South Africa, and Mozambique.

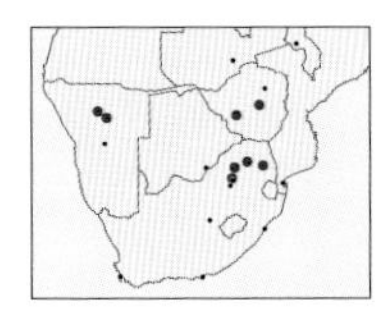

### 4 ***Ancylis halisparta*** White-tip Leafroller

Forewing 6mm. Narrow winged; frons, head and base of forewing bluish-green in live specimens, thorax above rust red, forewings grey with longitudinal darker grey lines, twirly half-round marks on inner margin making a circle when at rest, creamy white terminal cilia make distinctive 'white tip'; hindwings uniform pale brown. Perfectly camouflaged on tree trunks when at rest. Five species in genus in region. **Biology:** Caterpillars roll leaves, feeding on Buffalo Thorn (*Ziziphus mucronata*). Adults October–April. **Habitat & distribution:** Bushveld; Gauteng, South Africa, north to Namibia.

## Subfamily Tortricinae

The male genitalia are the least specialised among the family. Each antennal segment bears 2 rows of scales. The caterpillars typically tie and roll leaves where they live in silken tubes, and feed on a range of plant tissue, with early instars sometimes mining leaves.

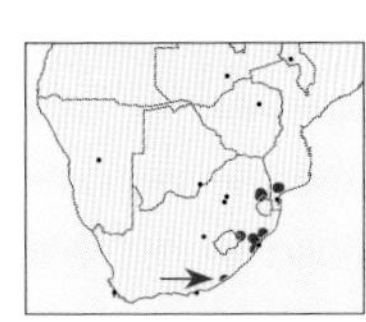

### 5 ***Choristoneura occidentalis*** Citrus Leafroller

Forewing 9–13mm. Broadly bullet-shaped at rest, distal corners of forewings distinctively flared outward; forewings very broad and rectangular, those of female with strongly sinuate hind margin, yellowish-brown with dark reddish-brown central patch and distal spot; hindwings plain orange. Caterpillar translucent green, with dark head capsule and black plate on thoracic segment 1. Three species in genus in region. **Biology:** Caterpillars polyphagous, feeding on leaves. Multivoltine, adults throughout the year. **Habitat & distribution:** Relatively moist areas, including transformed habitats; Eastern Cape, South Africa, north to tropical Africa.

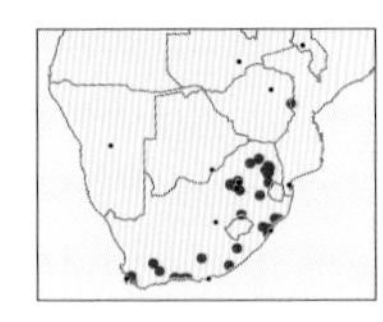

### 6 ***Lozotaenia capensana*** Common Leafroller

Forewing 7–12mm. Extremely variable in markings; forewings broad and rectangular, falcate in females, ground colour light to medium brown with brown postmedian band mostly darker near costa, partially obscured by dark-light brown shading in some and sometimes absent in others. Caterpillar light translucent yellow-green with variably coloured head. This species can only be reliably separated from the very similar *L. dorsiplagana* (p.46) through detailed morphological examination. Thirteen similar species in genus in region, also many similar species in other genera. **Biology:** Caterpillars polyphagous leaf rollers. Adults throughout the year. **Habitat & distribution:** A common species adapted to many different habitats; Western Cape, South Africa, to Kenya.

1
2
2
3
4
4
IS
5
SB
5
6
6
6

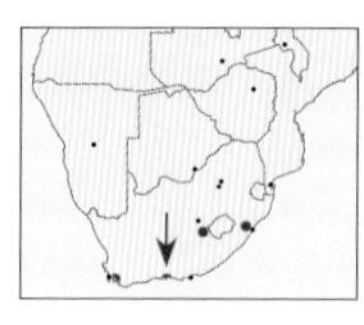

### 1 *Lozotaenia dorsiplagana* — Paleback Leafroller

Forewing 7–8mm. Forewings broad and rectangular, ground colour light to medium brown with central pale oval, and marked with brown scribblings and lighter in colour, postmedian band absent or obscure. Caterpillar light translucent yellow-green. This species can only be reliably separated from the very similar *L. capensana* (p.44) through detailed morphological examination of male genitalia. **Biology:** Caterpillars polyphagous leaf rollers. Adults September–April. **Habitat & distribution:** Moist habitats; Western Cape to KwaZulu-Natal, South Africa.

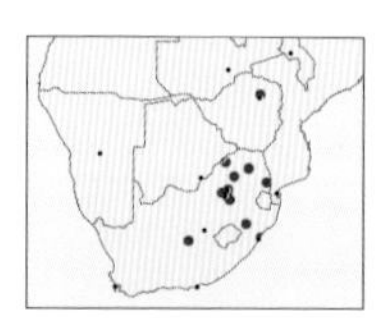

### 2 *Droceta cedrota* — Brownband Leafroller

Forewing 5–6mm. Snout and head setose; forewings narrow; head, thorax and forewing ground colour orange-yellow; forewings with distinct dark brown basal mark, terminal mark and median band; hindwings uniform cream to grey. One species in genus in region. **Biology:** Caterpillar host associations unknown. Adults September–April. **Habitat & distribution:** Variety of habitats; Western Cape north to Zimbabwe.

## FAMILY CHOREUTIDAE Metalmark Moths

**Small, often day-flying moths also attracted to light, with metallic scales. Adults rest in upright posture, wings flat and slightly spread, and exhibit jerky, pivoting movements. Some have eyespots on wings, which further complements the mimicry of jumping spiders. Caterpillars slender with elongate prolegs; mostly feed on leaves from within a silken webbing. About 440 species globally, 12 in the region.**

### Subfamily Choreutinae

Approximately 300 species found in the Holarctic and tropics. In most, the roughly square forewing has a pointed tip.

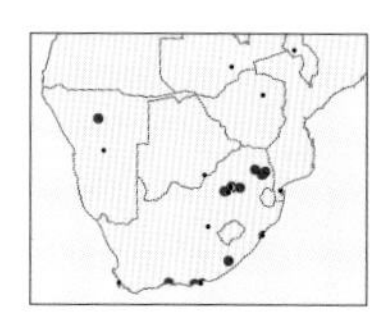

### 3 *Choreutis aegyptiaca* — Egyptian Metalmark

Forewing 5–6mm. Broad wings held flat to the sides, mottled grey with darker brown spots, sometimes forming distinct bands, distal margin red in some specimens, cilia uniform red. Caterpillar slender, green to brown with transverse rows of black dots. Seven species in genus in region. **Biology:** Caterpillars feed on fig trees (*Ficus* spp.). Multivoltine, adults August–April. **Habitat & distribution:** Habitats where fig trees grow; apparently disjunct distribution, populations in South Africa and Namibia, but also Uganda, Egypt, Arabian Peninsula and La Réunion.

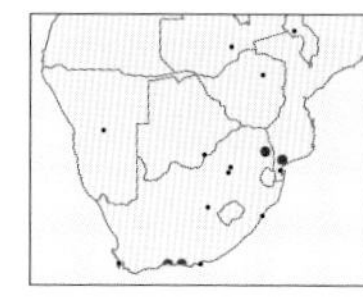

### 4 *Choreutis dryodora* — Key Metalmark

Forewing 3–4mm. Conspicuous coppery spot with white half ring at base of thorax; broad wings held flat to the sides, reddish-brown sprinkled with grey scales and darker brown spots, keyboard shape across wings, cilia alternately light and dark brown. Caterpillar slender, plain green. **Biology:** Caterpillars feed on fig trees (*Ficus* spp.). Adults October–December and March. **Habitat & distribution:** Habitats where fig trees grow; Western Cape to Mozambique.

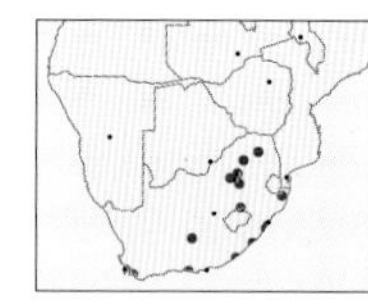

### 5 *Tebenna micalis* — Large-eye Metalmark

Forewing 2–3mm. Tent-like, unlike other species in family; forewings brown, banded with fine white scales and darker distal band, 3 large, darker eye-like patches across forewing. One species in genus in region. **Biology:** Caterpillars are leaf miners, host associations unknown in region. Multivoltine, adults throughout the year. **Habitat & distribution:** Weedy fields and range of open vegetation types; subsp. *dialecta* Western Cape north to Limpopo, South Africa; other subspecies very widely distributed, including Europe, Asia, Australia, New Zealand; scattered records across Africa.

### Subfamily Brenthiinae

Approximately 70 species, most occurring in the tropics of the Oriental, Australian and Nearctic regions. The forewing has a rounded outline.

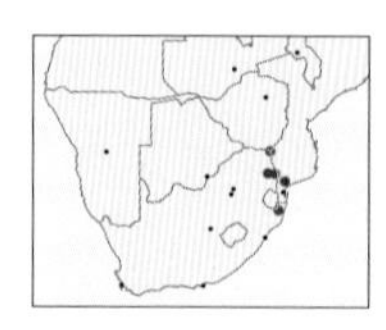

### 6 *Brenthia pleiadopa* — Peacock Metalmark

Forewing 3mm. Forewings black, covered in bright white scales, row of 7 small eyespots on forewing margin; underside with terminal orange band. Caterpillar translucent white with green dorsolateral lines. Three species in genus in region. **Biology:** Caterpillars feed on sycamore fig trees (*Ficus sycomorus*). Adults hold forewings in an unusual upright position and move about in a jerky fashion – likely mimicking jumping spiders, with the eyespots on the wings resembling the spider's eyes. Probably multivoltine, adults January, May and August. **Habitat & distribution:** Lowland bushveld; KwaZulu-Natal north to Limpopo, South Africa, and Mozambique.

1
2
3
3
4
4
5
6
6
6
AS
AS

## FAMILY IMMIDAE Imma Moths

**Minor family of small to medium-sized, often brightly coloured, diurnal or crepuscular moths with elongate forewings. About 250 mostly tropical species globally, about half of them in the genus *Imma*, of which 3 are found in the region.**

### 1 *Imma arsisceles* — Two-tone Imma

Forewing 10mm. Body setose and light brown; forewings marked with basal half pale brown, distal half dark brown, distal section bounded by black lines, short fringes brown setae along distal margin; hindwings pale basally, becoming dark grey distally. **Biology:** Caterpillar host associations unknown. Adults January–April. **Habitat & distribution:** Moist lowland forest and coastal bush; Eastern Cape, South Africa, to Mozambique.

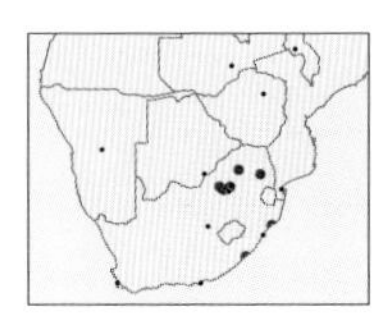

### 2 *Moca tormentata* — Dark Imma

Forewing 8–9mm. Body setose and light brown to black; forewing ground colour light green (fading to straw in dried specimens) variably, but densely, suffused with black scales; hindwings pale grey. Caterpillar variable, green to red with short setae and 'wart-like' green setal bases. **Biology:** Caterpillars feed openly on leaves of the Common Red Milkwood (*Mimusops zeyheri*), pupate in dense silken cocoon. Adults October–March. **Habitat & distribution:** Mountain bushveld and forests where mikwoods grow, and coastal bush; Eastern Cape to Limpopo, South Africa, also recorded from DRC.

## FAMILY BRACHODIDAE Little Bear Moths

**Small to medium-sized moths; adults diurnal and with elongate forewings held curved over the body and often bearing metallic markings. Hindwings often much enlarged and triangular. Caterpillars usually feed on grasses or among seeds under a web. About 135 species globally, 11 in the region.**

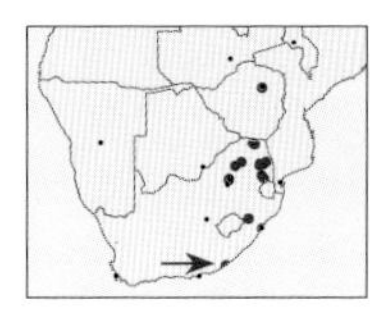

### 3 *Phycodes punctata* — Leaden Grey

Forewing 10–11mm. Body and forewings finely dusted with bright white scales; thorax dark brown, abdomen with black and pale transverse bands; forewings dark mottled grey with several black spots; hindwings dark brown with pale bases and white marginal cilia. One species in genus in region. **Biology:** Caterpillars feed on fig trees (*Ficus* spp.). Bivoltine, adults October–November and January–February. **Habitat & distribution:** Forest and bushveld where fig trees grow; Eastern Cape, South Africa, north to Zimbabwe.

## FAMILY COSSIDAE Goat Moths, Carpenter Moths

**Adults medium-sized to large (forewing 12–68mm, females often much larger than males), with broad body and long, narrow forewings. Colouration usually shades of grey, brown, black or white, many being twig, bark or leaf mimics. Proboscis vestigial or absent, antennae usually short and pectinate, at least along basal portion. Most are nocturnal. Caterpillars resemble stout, cream-coloured grubs with hard brown dorsal plate over prothorax; bore into live wood of trees or shrubs and hence seldom seen, often killing host tree. Development may take several years. About 700 described species globally, with 95 in the region.**

### Subfamily Cossinae

Antennae variable in form (in some bipectinate along entire length), medial tibial spurs present, hindtibia and first tarsal segment swollen. Eurasia, Africa and Americas, most species in Asia.

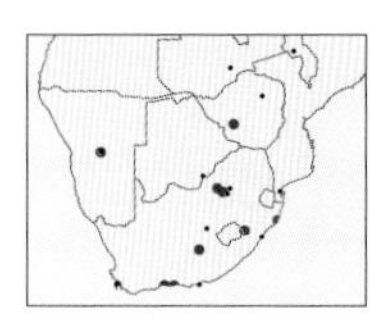

### 4 *Brachylia terebroides* — Charcoal Goat

Forewing 15mm. Body and forewings mottled dark charcoal grey, sometimes with brown patches, patterned with broken and intersecting black lines; hindwings plain light grey. Caterpillar unknown. Twelve species in genus in region with varying patterns on forewings. **Biology:** Caterpillars feed in the trunks of Sweet Thorn (*Vachellia karroo*). Adults September–February. **Habitat & distribution:** Variety moist and dry habitats where host trees grow; Western Cape, South Africa, north to Kenya.

1
2
2
3
3
4

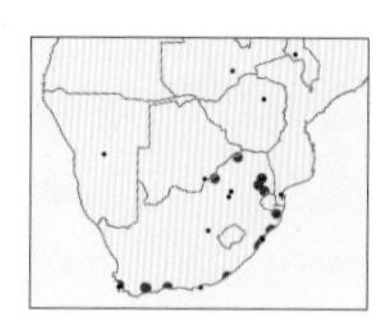

**1 *Coryphodema tristis*** Sad Goat

Forewing 17–25mm. Body and forewings pale brown with white and grey mottling, with broken, wavy black lines over wing veins; hindwings translucent brown. Caterpillar pinkish with broken brown markings. Two species in genus in region. **Biology:** Caterpillars polyphagous, bore into trunks of trees. Adults throughout the year. **Habitat & distribution:** Moist wooded areas; Western Cape, South Africa, north to Botswana.

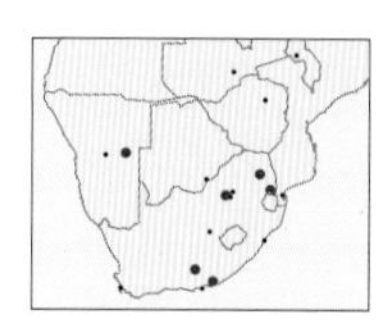

**2 *Macrocossus toluminus*** Giant Goat

Forewing 55mm. Thorax black dorsally; forewings mottled white and brown with a broken darker crossbar; hindwings mottled brown, black patch near base. Caterpillar undescribed. One species in genus in region. **Biology:** Caterpillars bore into *Entandrophragma* spp. (Mahogany family) and Sweet Thorn (*Vachellia karroo*). Adults October–January. **Habitat & distribution:** Variety of moist and dry habitats where host trees grow; Western Cape, South Africa, north to tropical Africa.

## Subfamily Zeuzerinae

Forewings typically spotted ('Leopard Moths'). Antennae of males bipectinate only in basal half basally, remainder thread-like. Occurs globally, but greatest diversity in Africa and Asia.

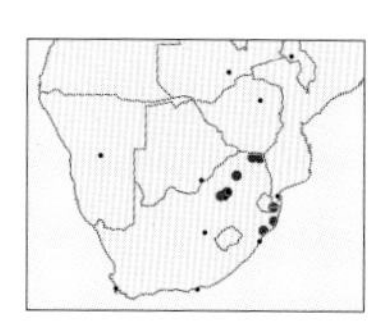

**3 *Aethalopteryx squameus*** Scaly Goat

Forewing 18–32mm. Thorax brown, pale dorsally; forewings white heavily speckled with brown or black scales and with network of fine brown or black lines; large elongate brown or black patches on costa; hindwings cream. Caterpillar undescribed. Six species in genus in region. **Biology:** Caterpillar host plants unknown. Adults November–March. **Habitat & distribution:** Bushveld and forest; KwaZulu-Natal, South Africa, north to tropical Africa.

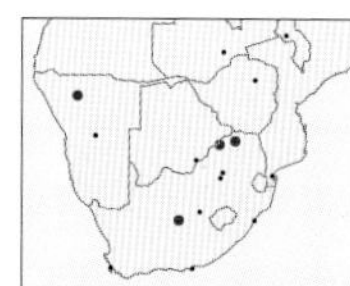

**4 *Aethalopteryx pindarus*** Arrow Goat

Forewing 18–22mm. Thorax greyish-black; forewings white, shaded with brown and with black streaks along inner margin, diagnostic arrow-like black streak along middle of wing; hindwings white with brown borders. **Biology:** Caterpillar host plants unknown. Adults November–March. **Habitat & distribution:** Arid bushveld; Northern Cape, South Africa, to Namibia and Kenya.

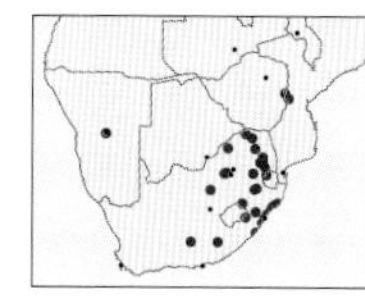

**5 *Azygophleps inclusa*** Inclusive Goat

Forewing ♂ 23–30mm, ♀ 30–40mm. Thorax white with black speckling; forewings mainly cream marked with numerous small broken black and variable orange shadings; hindwings plain cream. Caterpillar undescribed. Sixteen species in genus in region. **Biology:** Caterpillars feed in stems of *Indigofera* spp. (Fabaceae). Adults November–April. **Habitat & distribution:** Grassland, bushveld and forest; Eastern Cape, South Africa, north to tropical Africa.

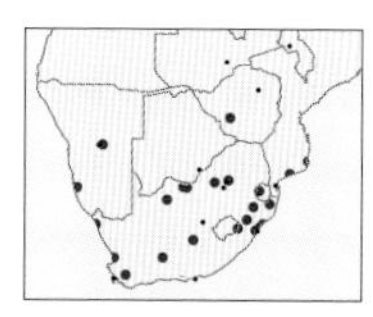

**6 *Azygophleps asylas*** White Striped Goat

Forewing 10–20mm. Similar to *A. inclusa* (above), but somewhat smaller with a pure white body and black stripes only on forewings, arranged to leave a clear white central band along wing. Caterpillar undescribed. **Biology:** Caterpillar host associations unknown. Adults September–April. **Habitat & distribution:** Variety of habitats; Western Cape, South Africa, north to tropical Africa.

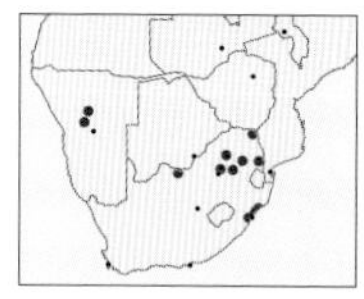

**7 *Azygophleps leopardina*** Leopard Goat

Forewing 19–32mm. White, with thorax and forewings patterned with large shiny black spots; hindwings with much paler 'shadow' spots. Caterpillar undescribed. **Biology:** Caterpillar host associations unknown. Adults November–May. **Habitat & distribution:** Bushveld and woodland; KwaZulu-Natal, South Africa, north to tropical Africa.

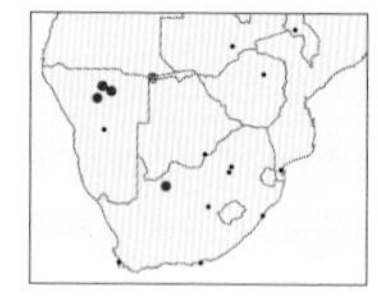

**8 *Azygophleps liturata*** Chain Goat

Forewing 20–27mm. White, with thorax patterned with large shiny black spots, a diagnostic chain-like black pattern along length of forewings. Hindwings plain white with some grey terminal spots. Caterpillar undescribed. **Biology:** Caterpillar host associations unknown. Adults November-February. **Habitat & distribution:** In arid bushveld; Northern Cape, South Africa, to Namibia.

1
1
2
3
4
5
5
5
6
7
7
8

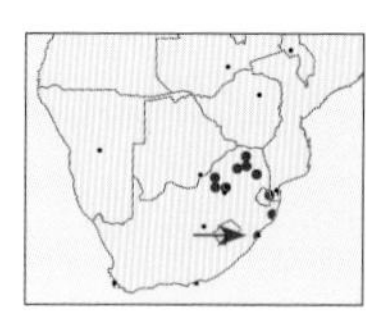

### 1 *Eulophonotus myrmeleon* — Dejected Goat

Forewing ♂ 13–14mm, ♀ 22–24mm. Sexually dimorphic; male with pilose orange thorax, abdomen banded in black and scarlet (scarlet fades in dried specimens), and ending in conspicuous tufts of black setae; wings almost completely translucent with black veins and yellow tinge on costa, some markings along inner edge of wings in certain specimens; female much larger, body and all wings black, forewings with sparse sprinkling of white markings. Two species in genus in region. **Biology:** Caterpillars polyphagous, bore into wood of trees. Adults October–March. **Habitat & distribution:** Forest and bushveld; Eastern Cape, South Africa, north to tropical Africa.

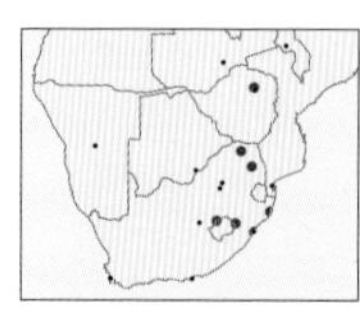

### 2 *Phragmataecia irrorata* — Brindled Goat

Forewing 13–25mm. Off-white to light brown, becoming grey toward outer border of wing with darker speckling; hindwings white to grey. Five species in genus in region. **Biology:** Caterpillar host associations unknown. Adults November–December. **Habitat & distribution:** Woodland and bushveld; KwaZulu-Natal, South Africa, north to Malawi and Zambia.

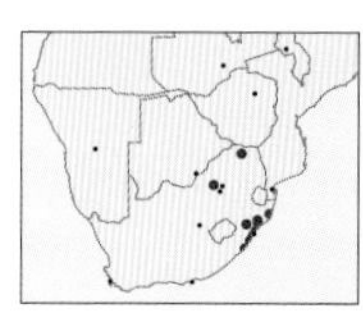

### 3 *Strigocossus capensis* — Pink Goat

Forewing 24–49mm. Thorax grey laterally, brown dorsal patch outlined with very dark band, abdomen uniformly light brown; forewings cream, covered with dense network of fine dark brown lines, a clearer pinkish patch near base; hindwings cream with network of pale brown lines confined to lateral areas. Caterpillar undescribed. Two species in genus in region. **Biology:** Caterpillars polyphagous, bore into stems of various plants. Adults throughout the year. **Habitat & distribution:** Forest and bush; Eastern Cape, South Africa, north to tropical Africa.

## FAMILY DUDGEONEIDAE Dudgeon Carpenterworm Moths

**Minor family containing just one genus of medium-sized moths, sparsely distributed across the Old World. Proboscis absent, caterpillars mostly unknown, but one Australian species is a stem borer on *Canthium* (Rubiaceae). Only 6 species globally, one in the region.**

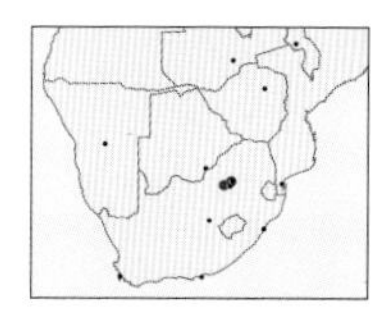

### 4 *Dudgeonea leucosticta* — Spotted Dudgeon

Forewing 16mm. Body and wings rusty red or chocolate brown; forewings with many cream marks of uneven shape and some white round spots, grouped at base and tip of wing; hindwings uniform rufous-brown. Caterpillars unknown. **Biology:** Host plant unknown. Possibly nocturnal, attracted to lights. **Habitat & distribution:** In region, known only from few specimens; Gauteng and Mpumlanga, South Africa.

## FAMILY METARBELIDAE Tropical Carpenterworm Moths

**Small to medium-sized moths with small head bearing short bipectinate antennae; squat body, short legs and elongate, rounded wings that are usually patterned in shades of white, grey or brown. Long abdomen usually protrudes beyond body, with tuft of posterior setae. Caterpillars bore into tree bark or the trunks of trees. About 103 described species globally, with 61 in the region.**

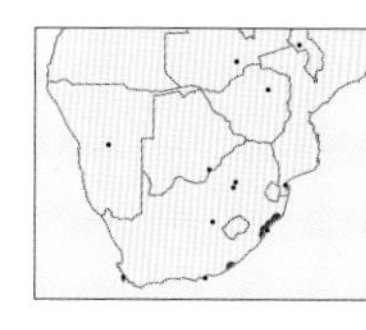

### 5 *Lebedodes rufithorax* — Redcoat Goat

Forewing 13–15mm. Distinctive in appearance and posture; thorax with dense dorsal ridge and crest of long red-brown setae; midlegs strongly setose and held out at right angles to body at rest; forewings creamy brown with network of fine brown lines; abdomen strongly upturned. Two species in genus in region. Probably univoltine, adults October–February. **Biology:** Caterpillars feed on Crossberry (*Grewia occidentalis*). **Habitat & distribution:** Coastal forest and bush; Eastern Cape to KwaZulu-Natal, South Africa.

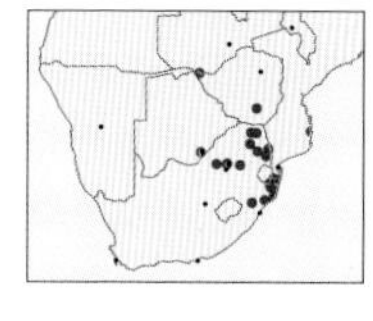

### 6 *Salagena tessellata* — Tessellate Goat

Forewing ♂ 10–11mm, ♀ 15–16mm. Highly distinctive; head and body orange, wings white, boldly patterned with small orange and black bars; legs, thorax and tip of abdomen bearing prominent long tufts of white setae; hindwings white. Caterpillar undescribed. Seven species in genus in region. **Biology:** Caterpillars feed on Common Pear (*Pyrus communis*). In Kenya, a species of *Salagena* damages mangrove trees (*Sonneratia*). Adults September–March. **Habitat & distribution:** Bushveld and subtropical forest; KwaZulu-Natal north to Kenya.

1♂
1♀
2
2
3
4
JH
5
6♂

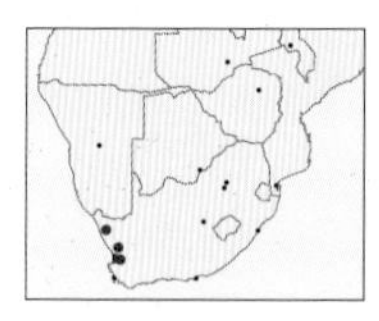

### 1 *Arbelodes haberlandorum* — Brown Kid

Forewing 11mm. Male antennae large, black, dusted with white, strongly pectinate; female antennae less pectinate; thorax very setose, mottled white and brown; forewing ground colour white, dusted with brown and black, 2 distinct orange spots, distinct black spots along costa. *Arbelodes* males have long antennae (more than half forewing length) that bear long narrow branches and most species are brown with a group of small geometric dark or light patches in centre of forewing. Nineteen species in genus in region. **Biology:** Caterpillar host associations unknown. Probably univoltine, adults January–March. **Habitat & distribution:** Endemic to the Cederberg; Western Cape, South Africa.

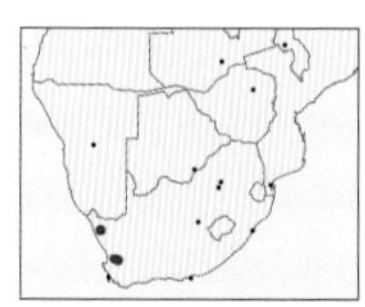

### 2 *Arbelodes franziskae* — Franziska's Kid

Forewing 9mm. Antennae large and strongly pectinate; thorax very setose, mottled grey and brown with black 'collar'; forewings mottled pinkish-brown with a central 3-lobed white marking; hindwings grey-brown. **Biology:** Caterpillar host associations unknown. Probably univoltine, adults March–April. **Habitat & distribution:** Recorded from dry mountain fynbos; Cederberg and Kamieskroon klipkoppe, South Africa.

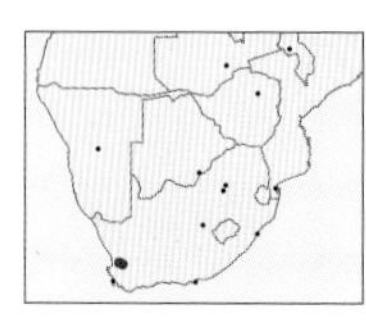

### 3 *Arbelodes shimonii* — Shimoni's Kid

Forewing 12mm. Larger than *A. franziskae* (above), with bigger forewing markings and row of white marks along margin. **Biology:** Caterpillar host associations unknown. Probably univoltine, adults March–April. **Habitat & distribution:** Dry mountain fynbos; Cederberg, South Africa.

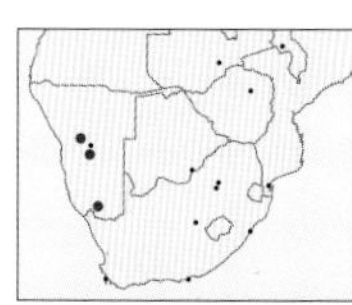

### 4 *Metarbela naumanni* — Golden Kid

Forewing 12mm. Golden-yellow forewings and body (which is strongly setose); forewings crossed by an oblique curved white bar; hindwings pale yellow with a very thin darker line close to margin. Twelve species in genus in region. **Biology:** Caterpillar host associations unknown. Probably univoltine, adults March–May. **Habitat & distribution:** Arid bushveld; Namibia south to South Africa.

## FAMILY SESIIDAE Clearwing Moths

**Distinctive family easily identified by their transparent wings, which often have black, brown, yellow or orange borders, and by their convincing mimicry of bees or wasps. Legs mostly long and thin, abdomen frequently elongate, often with yellow, red or white bands and frequently ending in fan of long hairs. Adults diurnal and fast flying, frequently visiting flowers. Caterpillars typically bore into wood and succulent plants or develop in flower heads; sometimes pests of fruit trees, timber or crop plants. Caterpillar development slow, sometimes taking years. About 1,370 species globally, with 97 known from the region. Although diverse, seldom common, and many species known only from few specimens, or still not formally named. Adult males routinely attracted using pheromone lures.**

### Subfamily Sesiinae

Proboscis reduced, components of male genitalia thickened.

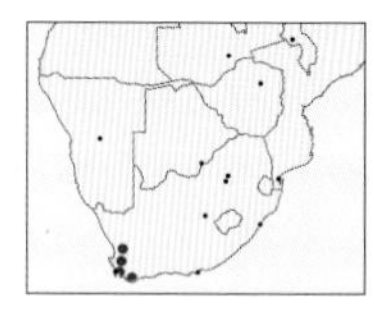

### 5 *Synanthedon platyuriformis* — Dark Clearwing

Forewing 10mm. Body brown with white bands; wings clear with characteristic orange inner margin and distal veins. Twelve species in genus in region. **Biology:** Caterpillars bore into flower heads of Proteaceae, pupating inside seed heads. Adults often seen on white, ball-like flowers of *Berzelia*. Adults August–March. **Habitat & distribution:** Fynbos vegetation; Western Cape, South Africa.

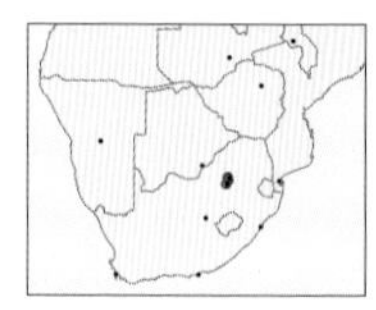

### 6 *Cabomina leucopleura* — White-sided Clearwing

Forewing 9mm. Predominantly brown with characteristic white patches on sides of elongate abdomen; forewings largely opaque brown with white distal patch, sometimes absent; legs fringed with white setae; abdomen tipped by fan of long brown setae. Caterpillar unknown. **Biology:** Caterpillar host associations unknown. Adults March–April. **Habitat & distribution:** Grassland; Gauteng, South Africa.

1
2
3
4
5
PW
6

## FAMILY EPIPYROPIDAE Planthopper Parasite Moths

**Minor family of minute to small (forewing 2–17mm) moths with robust body, conspicuous bipectinate antennae in males, and broadly rounded wings. Most are black or dark brown. Unique among the Lepidoptera in that the slug-like larvae are ectoparasites, typically of Lantern Bugs (Fulgoridae). Some 32 described species, at least another 30 known but awaiting naming, 4 in the region.**

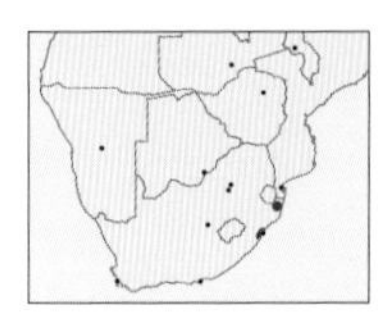

### 1 *Epipyrops fulvipunctata* Coccid Parasite Moth

Forewing ♂ 8mm, ♀ 10mm. Body black, short and rotund; wings short and rounded, dark brown with black markings and scattering of silvery scales along leading margin. Caterpillar not described, probably grub-like and covered in white, fluffy coating, as for other species in the family. **Biology:** Caterpillars are external parasites of the Lantern Bug (*Rhinortha guttata*, Fulgoridae). Adults do not feed, collected in July. **Habitat & distribution:** Coastal bush and forest; endemic to KwaZulu-Natal, South Africa.

## FAMILY HIMANTOPTERIDAE Long-tailed Burnets and False Tussocks

**Adults small to medium-sized (forewing 8–21mm); antennae bipectinate in males, filiform in females. Members of subfamily Himantopterinae easily recognised, as hindwings extended into extremely long tails; subfamily Anomoeotinae have normal rounded wings and superficially resemble Tussock Moths (see Erebidae: Lymantriinae p.384). Mostly diurnal and brightly coloured. Caterpillars slug-like with concealed head; skeletonise leaves, and sometimes communal, massing to move down trunk of host plant to pupate in soil. Some 96 species globally, 15 in the region.**

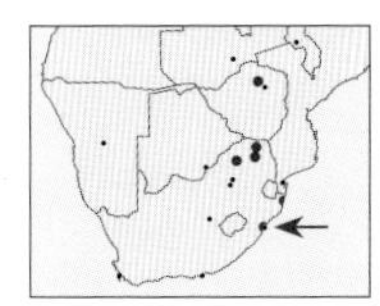

### 2 *Semioptila torta* Sickle Streamer Tail

Forewing 11–13mm. Wings bright orange basally, becoming black distally, each with a large orange dot; hindwings greatly extended into tail with sickle-shaped tip. Eight described regional species in genus. **Biology:** Caterpillar host associations unknown. Adults diurnal, slow flying, but also attracted to light. **Habitat & distribution:** Tropical woodland; KwaZulu-Natal, South Africa, north to tropical Africa.

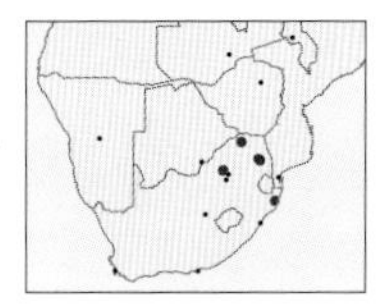

### 3 *Anomoeotes nigrivenosus* Black-veined False Tussock

Forewing ♂ 11–12mm, ♀ 17–18mm. Body and wings uniform orange; thinly scaled, orange scales on wings denser near base becoming slightly transparent distally; veins variably dark brown, especially on forewings. Caterpillar robust and setose, patterned in orange and black. Two species in genus in region, similar *A. levis* is smaller and has wings with basal scales variably brownish on forewings and yellow to orange on hindwings becoming completely transparent toward apex. **Biology:** Caterpillars polyphagous on trees. Adults diurnal, bivoltine, October–March. **Habitat & distribution:** Bushveld; KwaZulu-Natal, South Africa, north to tropical Africa.

## FAMILY LACTURIDAE Tropical Burnet Moths

**Adults small to medium-sized (wingspan 11–65mm), nocturnal or crepuscular, some diurnal, mostly tropical. Wings elongate, usually brightly coloured, often banded and spotted in orange, red and yellow. Caterpillars slug-like with concealed head, often colourful, mostly feed on Anacardiaceae. Closely related to Zygaenidae. About 250 species globally, 10 in the region.**

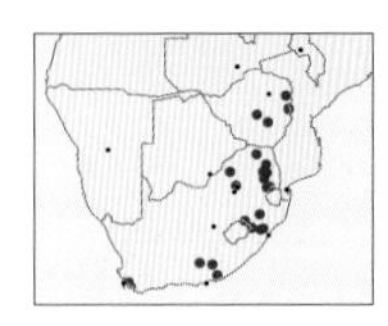

### 4 *Coptoproctis languida* Rosy Flounced

Forewing 7–11mm. Forewing extremely variable in colour, from pure white or yellow with no markings, or pink with yellow band along inner margin to dark purple with yellow band; hindwings white to rose pink with pink cilia posteriorly. Caterpillar slug-like, green with distinctive dorsal lines and dark red marks. One regional species in genus. **Biology:** Caterpillars feed on Real Wild Currant (*Searsia tomentosa*) and other *Searsia* spp. Bivoltine, adults October–May. **Habitat & distribution:** Variety of habitats where *Searsia* spp. grow; Western Cape, South Africa, north to Tanzania.

1
2
3
3
4
MB
4

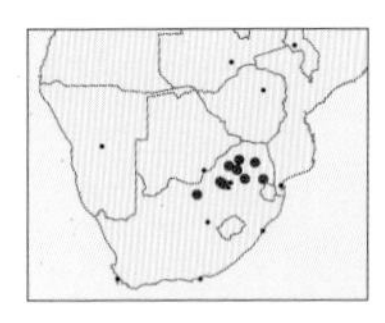

### 1 *Gymnogramma pyrozancla* — Rosy Waves

Forewing 10–12mm. Orange legs and cream forewings with distinctive wavy orange line of varying intensity extending along inner margin of wing, a few orange dashes present in some individuals; hindwings deep yellow. Caterpillar rotund and slug-like, pale green with longitudinal white stripes. Five species in genus in region. **Biology:** Caterpillars feed on Common Resin Tree (*Ozoroa paniculosa*). Bivoltine, adults September–March. **Habitat & distribution:** Bushveld and grassland where host plant occurs; North West to Limpopo, South Africa.

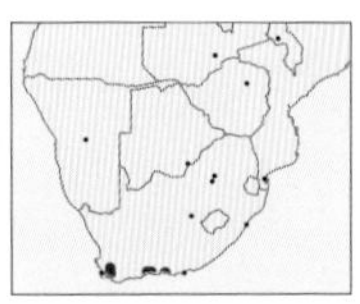

### 2 *Utenikwa rhodoneura* — Rosy Streak

Forewing 8–10mm. Palpi and legs rose pink, head, thorax and abdomen creamy yellow; forewings variably striated with pink, almost completely pink in some individuals; hindwings white with variable pinkish sheen. Caterpillar green with yellow dorsal and ventral lines, prominent 'horns' on prothorax segments. One species in genus in region. **Biology:** Caterpillars feed on *Berzelia* spp. (Bruniaceae). Adults February–March. **Habitat & distribution:** Fynbos; Western Cape and Eastern Cape, South Africa.

# FAMILY ZYGAENIDAE Burnets

**Adults diurnal, with narrow, brightly coloured fore- and hindwings (dark metallic blue, brown or red) bearing red or yellow bands or spots. Antennae thickened toward tip in Zygaeninae (except in *Orna* spp.) and pectinate in Procridinae. Caterpillars slug-like, short and plump with concealed head, feed in exposed positions on plants containing toxic alkaloids. Zygaeninae in the region feed on various Celastraceae. Pupation in tough cocoon attached to host plant. About 1,000 species globally, 37 in the region. Three subfamilies in region, Procridinae (14 species), Zygaeninae (20 species) and Phaudinae (3 species).**

## Subfamily Procridinae

The caterpillars have an unusual gland with a mouth-like opening between the spiracles on abdominal segments 2 and 7, and another simpler gland at the base of the first thoracic leg.

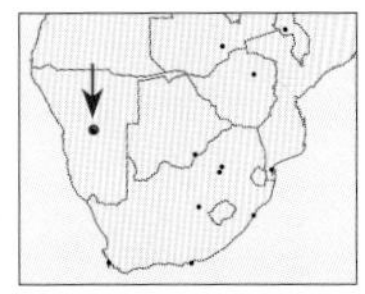

### 3 *Saliunca homochroa* — Ruby Burnet

Forewing 11–15mm. Body and forewings uniform metallic blue-green, thorax with red patch on each side; hindwings transparent basally. Caterpillar red laterally, dorsally black, with paired white patches on each segment. Four similar dark metallic species in genus in region. **Biology:** Caterpillars feed on Grape Ivy (*Cissus*). Adults recorded in December. **Habitat & distribution:** Bushveld; Namibia, Mozambique and Zimbabwe to Ethiopia.

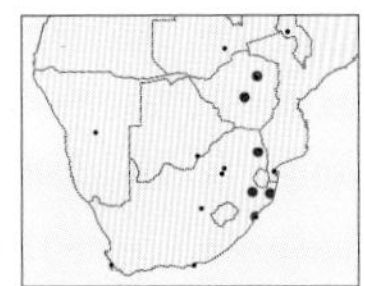

### 4 *Tascia finalis* — Orange-tip Burnet

Forewing 12–17mm. Body greenish-black with variable metallic blue spots, tip of abdomen orange to red in female, 2 or more metallic blue bands across; wings greenish-black with transparent windows, single large one on hindwing and 2 smaller ones on forewing. Caterpillar red with alternating black and white dorsal spots. Three species in genus in region. **Biology:** Caterpillars feed on various wild grape vines (Vitaceae). Probably bivoltine, adults September–December. **Habitat & distribution:** Grassland and bushveld; KwaZulu-Natal, South Africa, north to tropical Africa.

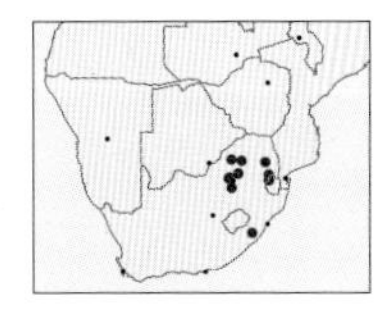

### 5 *Janseola titaea* — Black Burnet

Forewing 6–7mm. Head, thorax and abdomen variably orange, black and red banded; antennae, cilia and all wings charcoal black with no markings. Caterpillar with alternately coloured longitudinal bands and short bristles. One species in genus in region. **Biology:** Caterpillars feed on herbaceous Rubiaceae. Adults diurnal, flying in full sunshine, probably bivoltine, adults September–December. **Habitat & distribution:** Grassland and bushveld; KwaZulu-Natal north to Limpopo, South Africa.

## Subfamily Zygaeninae

Caterpillars have cavities in their skin in which cyanogenic glycosides are stored. These toxins are released as defensive droplets when the caterpillar is disturbed.

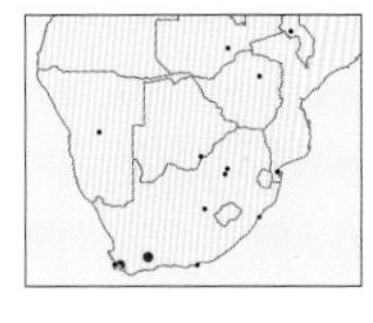

### 6 *Neurosymploca affinis* — Jordan's Burnet

Forewing 15mm. Forewings dark grey with 5 diagnostic black- and white-edged, red spots; hindwings with broad black-brown border; inner part of wing red with clear base. There are 10 species in genus. **Biology:** Caterpillars feed on Candlewood (*Pterocelastrus tricuspidatus*). Adults November–January. **Habitat & distribution:** Coastal forest where host plant is plentiful; southern Cape, South Africa.

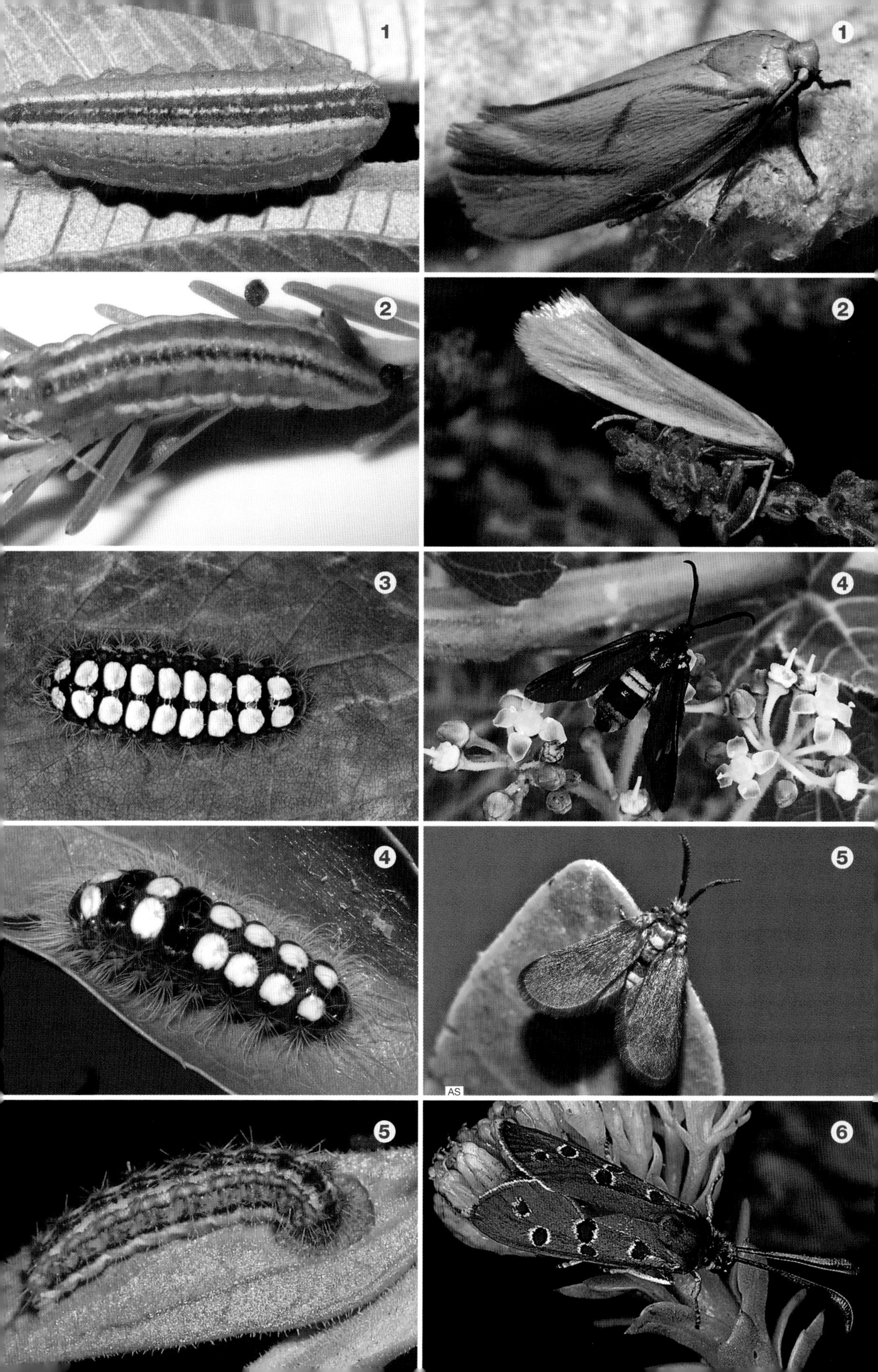
1
1
2
2
3
4
4
5
AS
5
6

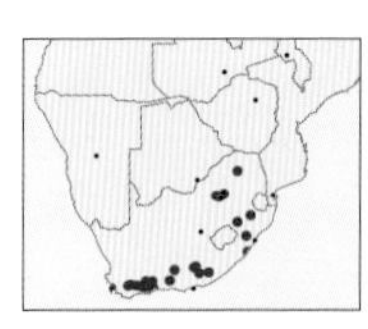

### 1 *Neurosymploca concinna* — Pleasing Burnet

Forewing 9–15mm. Easily distinguishable from other *Neurosymploca* spp. by white- or cream-edged (not black) mostly yellow forewing spots. Variable locally with several subspecies described; hindwings red, yellow in some specimens, with thin black border, basal part of hindwing transparent in some populations. Caterpillar broad and light green with indistinct pale lateral line, conspicuous orange dorsal spot in many specimens, some with white and pink dorsal line. **Biology:** Caterpillars feed on spiny Celastraceae (*Gymnosporia* and *Putterlickia* spp.). Bivoltine, adults August–May. **Habitat & distribution:** Variety of habitats where host plants grow; Western Cape north to Limpopo, South Africa.

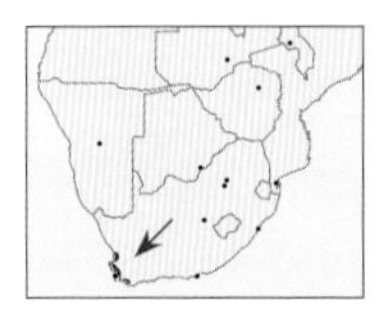

### 2 *Neurosymploca geertsemai* — Henk's Burnet

Forewing 9–10mm. Brown to brownish-grey with orange tip to abdomen; forewings broader than in other *Neurosymploca* spp. with a white longitudinal stripe between red dots in fresh specimens, and 2 distal red dots; hindwings red with thin black border. Caterpillar light green with a black-ringed white dorsal bar on thorax. **Biology:** Caterpillars feed on non-spiny Celastraceae (*Maytenus*, *Cassine* and *Pterocelastrus* spp.). Bivoltine, adults October–April. **Habitat & distribution:** Fynbos and strandveld vegetation; Western Cape, South Africa.

### 3 *Neurosymploca kushaica* — Kusha's Burnet

Forewing 9–13mm. Thorax dark grey-brown, abdomen black with 2 segments red interrupted by black dorsal spots; forewings grey with black-ringed red spots; hindwings scarlet with narrow black border. Caterpillar plump, mostly uniform pale green. **Biology:** Caterpillars feed on non-spiny Celastraceae (*Maytenus*, *Cassine* and *Pterocelastrus* spp.). **Habitat & distribution:** Various habitats where host plants grow; Western Cape to Eastern Cape, South Africa.

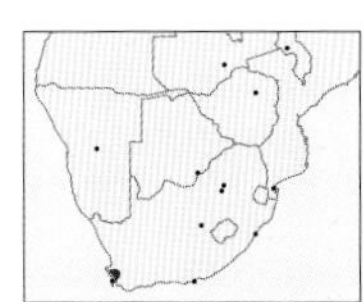

### 4 *Neurosymploca hottentota* — Dark Burnet

Forewing 9mm. Differs from other *Neurosymploca* spp. in lacking the large red spots on forewings; body black except for red neck and wide orange band toward tip of abdomen; forewings brownish-grey with a dusting of white scales in fresh specimens, few (typically 3) small black oval spots and a tiny distal orange or yellow spot; hindwings bright red with narrow brown margin. Caterpillars plump and plain pale green. **Biology:** Caterpillars feed on Spikethorn (*Gymnosporia buxifolia*). Adults in October. **Habitat & distribution:** Seems to be restricted to threatened Swartland renosterveld habitats; Western Cape, South Africa.

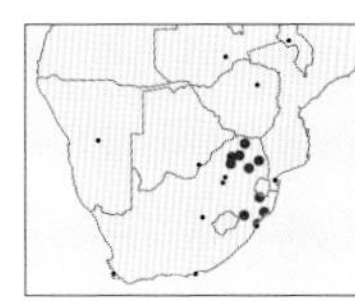

### 5 *Zutulba namaqua* — Brilliant Burnet

Forewing 15–18mm. Thorax black with pale yellow dots; forewings black with 5 or 6 off-white dots; abdomen and hindwings bright red and black (yellow and black in form *zelleri*). Caterpillar slug-like, pale green with white-ringed patches. Two species in genus in region. **Biology:** Caterpillars feed on Bushveld Saffron (*Elaeodendron transvaalense*). Bivoltine, adults October–April. **Habitat & distribution:** Bushveld; KwaZulu-Natal, South Africa, north to tropical Africa.

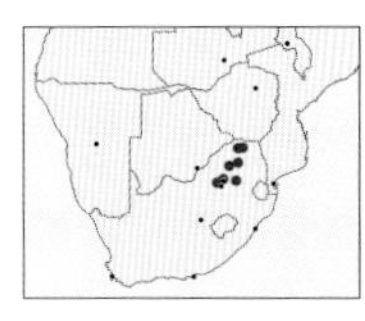

### 6 *Praezygaena agria* — Orange-spotted Burnet

Forewing 12–13mm. Thorax black, orange dorsally, abdomen red; forewings blue-black with large, round, cream-ringed markings; hindwings bright red with black border. Caterpillar light green with yellow markings, densely spinose. Six species in genus in region. **Biology:** Caterpillars feed on Spikethorn (*Gymnosporia buxifolia*). Bivoltine, adults November–March. **Habitat & distribution:** Middle elevation bushveld; Gauteng to Limpopo, South Africa.

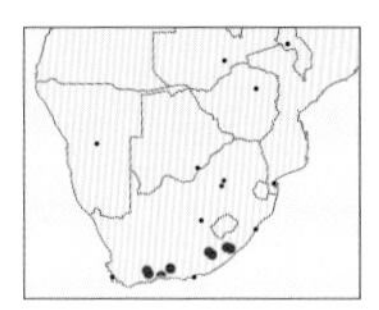

### 7 *Orna nebulosa* — Clouded Burnet

Forewing 10–14mm. Antennae in male bipectinate, in female slender, inflated at tip; thorax grey, abdomen bright green (in live adults); wings translucent black, whitish in females, greyish-brown in males. Caterpillar uniform green without markings. Two species in genus in region. **Biology:** Caterpillars feed on spikethorns (*Gymnosporia* spp.). **Habitat & distribution:** Moist and dry habitats where Spikethorns occur; Western Cape and Eastern Cape, South Africa.

1
1
SJ
1
2
2
PW
3
3
4
4
AH
5
IS
5
6
AS
6
AS
7
AH
7 ♀
7 ♂

## FAMILY LIMACODIDAE Slug Moths

**Short, squat moths with setose body and legs, small head with reduced mouthparts, and short, rounded wings; most show cryptic colouration in browns, yellows or greens. Caterpillars very characteristic: short, plump, often very flattened, with concealed head and reduced thoracic legs; prolegs absent and replaced by suckers, such that caterpillars glide along like slugs. Caterpillars may be smooth and green, or have bright warning colouration and bear conspicuous tubercles armed with urticating hairs and spines, some of which can cause brief but intense pain when brushed against. Cocoon unusual – hard, elliptical and egg-like with a neat circular 'lid' through which adults emerge. The adults of most species are strictly nocturnal and do not move when disturbed by day. Exceptions are the aposematic, diurnal *Caffricola* and *Arctozygaena*. About 1,000 species described globally, with 125 in the region.**

### Subfamily Limacodinae

The body of the caterpillars is covered in a few large tubercles or projections bearing stout, urticating spines, as well as clusters of urticating bristles, and the upper lip (labrum) bears numerous short spines.

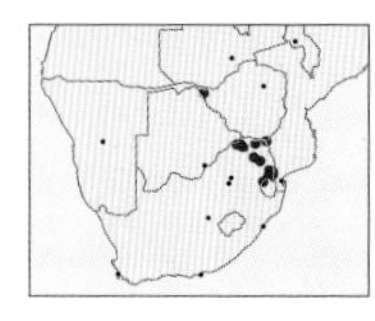

**1 *Afraltha chionostola*** Learner Slug

Forewing 12–15mm. Body and wings white; forewings crossed by a wide, bent, dark green to black band, resembling an inverted 'L'; hindwings white. Caterpillar unknown. **Biology:** Caterpillar host associations unknown. Probably bivoltine, adults November–March. **Habitat & distribution:** Lowland bushveld; Mpumalanga, South Africa, north to tropical Africa.

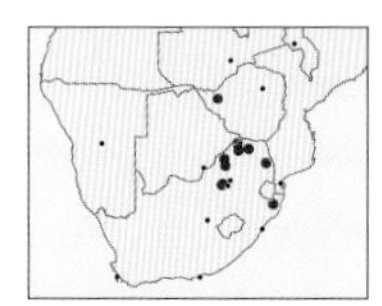

**2 *Afrobirthama flaccidia*** Flaccid Slug

Forewing 10–14mm. Antennae of male bipectinate over basal half only; thorax reddish-brown, abdomen grey-brown; forewings strikingly divided into a dark reddish-brown basal half and light-brown distal half, dark brown, zigzag anteromedially, lightly dusted with grey scales distally; hindwings light brown with darker margin. Three species in genus in region. **Biology:** Caterpillars reported on Coffee plants (*Coffea arabica*). Adults, November–April. **Habitat & distribution:** Bushveld; Zululand, South Africa, north to Tanzania.

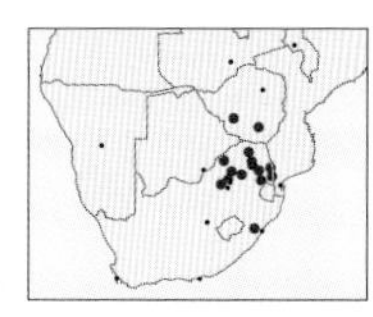

**3 *Afrobirthama reducta*** Reduced Slug

Forewing 13–16mm. Very similar to *A. flaccidia*, but the male antennae are bipectinate throughout their length and the darker proximal section to forewing is far less pronounced, yet still separated from distal section by distinct crenulated, dark antemedial line, heavily dusted with grey scales all over forewing. Similar *A. hobohmi* from Namibia has antemedial line straight, with slight kink. **Biology:** Caterpillar host associations unknown. Adults December-March. **Habitat & distribution:** Bushveld Gauteng, South Africa, north to Tanzania.

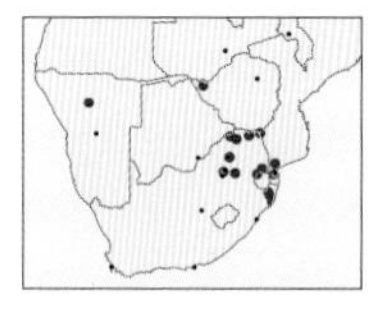

**4 *Scotinochroa inconsequens*** Dusted Slug

Forewing 12–15mm. Body and forewings chocolate brown, median and terminal area of forewings variably cream; hindwings light brown; wings, body and legs lightly dusted with white scales. Three species in genus in region. **Biology:** Caterpillar host associations unknown. Probably multivoltine, adults throughout the year. **Habitat & distribution:** Bushveld; KwaZulu-Natal, South Africa, north to Zambia.

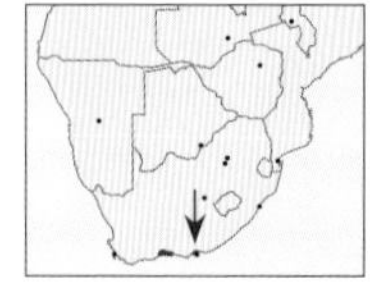

**5 *Omocenoides breijeri*** Breijer's Slug

Forewing ♂ 14mm, ♀ 17mm. Ground colour fawn, thorax furry; forewings traversed by obvious dark diagonal brown band; hindwings uniform pale brown; both wings strongly fringed with setae; male uniform fawn colour apart from dark band; female with dark brown thorax, abdomen paler brown, forewings mixed shades of brown. Caterpillar unknown. Two species in genus in region. **Biology:** Caterpillar host associations unknown. **Habitat & distribution:** Fynbos and forest; southern Cape, South Africa.

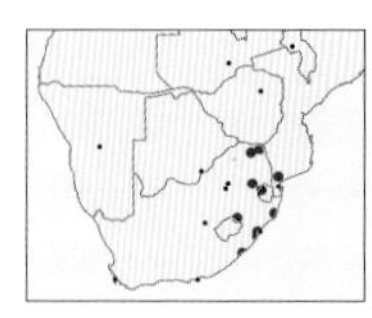

**6 *Omocenoides isophanes*** Sprayed Slug

Forewing 12mm. Thorax brown, broad abdomen banded pale and dark brown; forewings silvery brown, darker proximal portion separated from paler distal portion by dark diagonal line, marginal line straight; hindwings plain brown. Caterpillar green with white lateral stripe and pink, spiny tubercles. Several superficially similar species. **Biology:** Polyphagous, caterpillars recorded on various angiosperm families. Probably bivoltine, adults November–April. **Habitat & distribution:** Forest and coastal bush; Eastern Cape to Limpopo, South Africa.

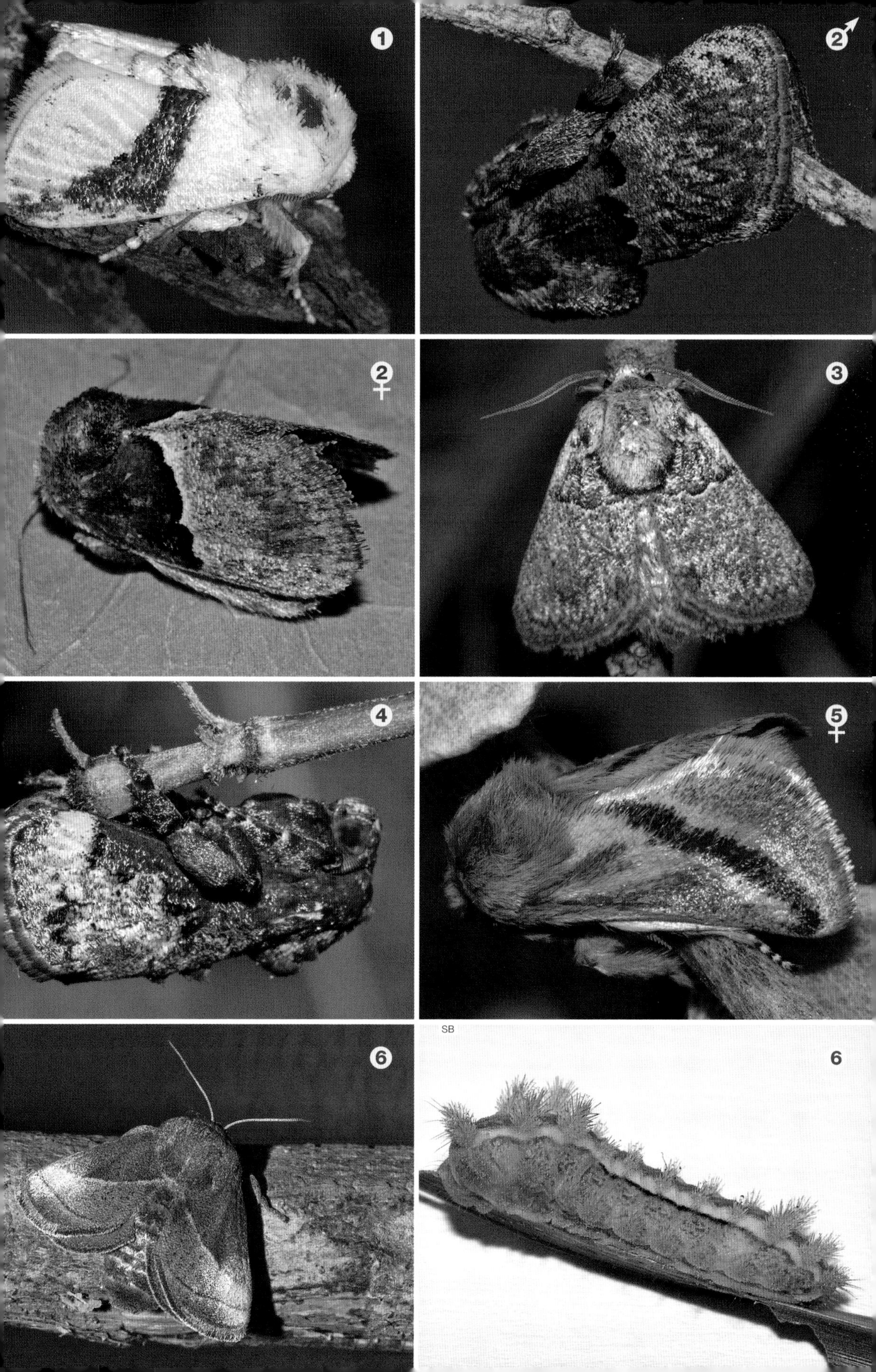
1
2♂
2♀
3
4
5♀
6
SB
6

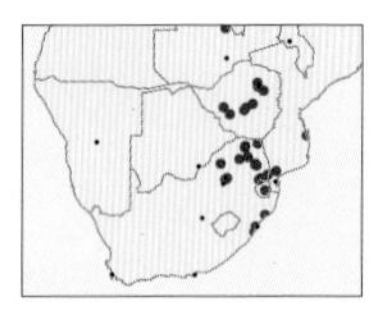

### 1 *Neomocena convergens* — Convergent Slug

Forewing 10–13mm. Plump and fairly plain brown, similar to *Omocenoides isophanes* (p.62) but marginal line distinctly curved; 2 oblique curved lines cross forewings, separating darker proximal section from distal paler one; hindwings uniform pale brown. Caterpillar strongly armoured with spinose projections, reddish-brown to grey dorsal saddle with variable green or white lateral spots or patches. Shape of dorsal saddle diagnostic. Two species in genus in region; similar *N. brunneocrossa* has forewing terminal area dark chocolate brown. **Biology:** Polyphagous, caterpillars recorded on various angiosperm families. Probably bivoltine, adults October–April. **Habitat & distribution:** Forest and thick bush; KwaZulu-Natal, South Africa, north to tropical Africa.

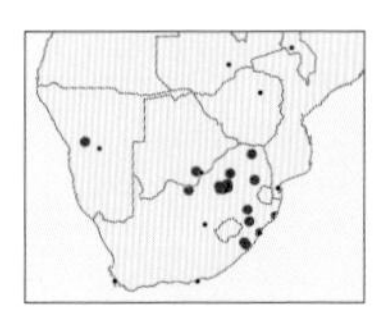

### 2 *Caffricola vicina* — Scarlet Puffball Slug

Forewing 14mm. Brightly coloured, with black head, thorax and cilia, and orange posterior abdomen; basal part of forewings scarlet, distal part black with 4 similarly shaped large white spots; hindwings scarlet. Caterpillar a smooth green slug. Three species in genus in region; *C. cloeckneria* has cilia yellow and basal white spot elongate; *C. kenyensis* has scarlet colour replaced by yellow. **Biology:** Caterpillars feed on Spikethorn (*Gymnosporia buxifolia*). Adults diurnal, flying slowly low amongst vegetation in full sunshine. Probably bivoltine, adults October–March. **Habitat & distribution:** Bushveld; Eastern Cape, South Africa, to Namibia.

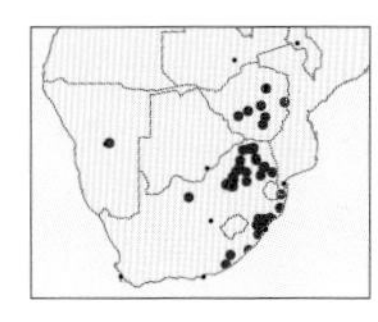

### 3 *Coenobasis amoena* — Rayed Slug

Forewing 9–15mm. Thorax emerald green with yellow tufts of setae, abdomen pale orange with dorsal tufts of setae; forewings emerald green with cream bars and tiny black marginal dots; hindwings straw with black dots along margin. Caterpillar green with white bars and blue dots and large barbed projections. Cocoon a hard 'egg' cemented to branches. Seven species in genus in region, differing in the shape and extent of the cream markings on forewings. **Biology:** Caterpillars feed on thorn trees (*Vachellia* spp.) and wattles (*Acacia* spp.) and should not be handled, as the spines are highly irritant. Probably bivoltine, adults September–April. **Habitat & distribution:** Thorn bushveld and wattle-infested areas; Eastern Cape, South Africa, north to Ethiopia.

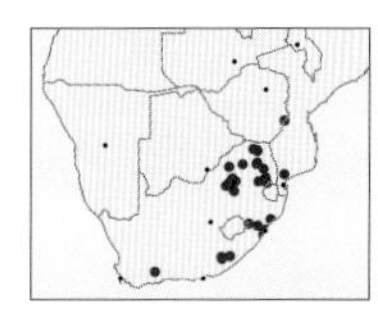

### 4 *Crothaema decorata* — Decorated Crothaema

Forewing 12–15mm. Thorax brown with a characteristic strong upright resting posture with dorsal crest of setae; abdomen pale pink; forewings patterned in patches of pink or light and dark brown, some outlined with white lines, posterior margin strongly fringed with setae; hindwings light pink. Caterpillar a pale green slug with small black dots, perfectly camouflaged on its host's leaves. Four similar species in genus in region. **Biology:** Caterpillars feed primarily on Spikethorn (*Gymnosporia buxifolia*), also on other Celastraceae. Probably bivoltine, adults September–March. **Habitat & distribution:** Bushveld and woodland; Western Cape, South Africa, north to Zimbabwe.

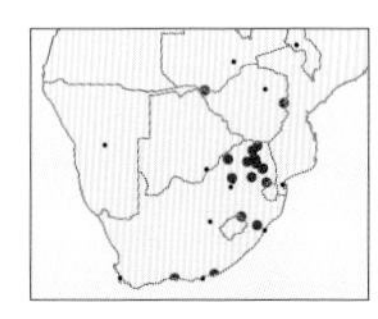

### 5 *Exanthica trigonella* — Pencil Slug

Forewing 8–9mm. Similar resting posture to *C. decorata* (above); slender winged, forewing basal area dark olive green, rest of wing lighter olive green with distinctive triangular dark patch on apex, white line across, colour fading to greyish-brown in dried specimens; hindwings white to grey. Two species in genus in region. **Biology:** Caterpillar host associations unknown. Probably bivoltine, adults September–March. **Habitat & distribution:** Forest and bushveld; Western Cape north to Zimbabwe.

### 6 *Gavara velutina* — Velvet Slug

Forewing 9mm. Body and forewings reddish to brown, thorax with upright dorsal tuft of setae; forewings crossed by faint broken bands of darker colour; hindwings plain straw colour. One species in genus in region. **Biology:** Caterpillar host associations unknown. Adults September–May. **Habitat & distribution:** Forest and thick bush; Eastern Cape north to Limpopo, South Africa.

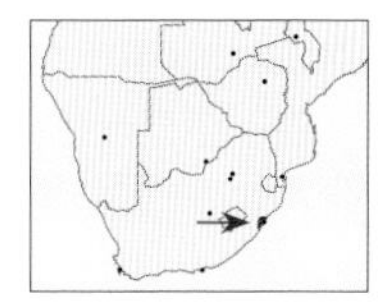

### 7 *Halseyia biumbrata* — Broken Tiptail

Forewing 8mm. Body and forewings coppery brown basally, sharply divided by dark median line, dark brown broken postmedian band diagnostic, terminal area dusted with dark brown scales. Caterpillar white mottled with black, heavily armed with spines. Eight similar species in genus in region. **Biology:** Caterpillars reared on *Achyranthes aspera* (Amaranthaceae). Probably bivoltine, adults October–February. **Habitat & distribution:** Scarp forest and coastal bush; KwaZulu-Natal, South Africa.

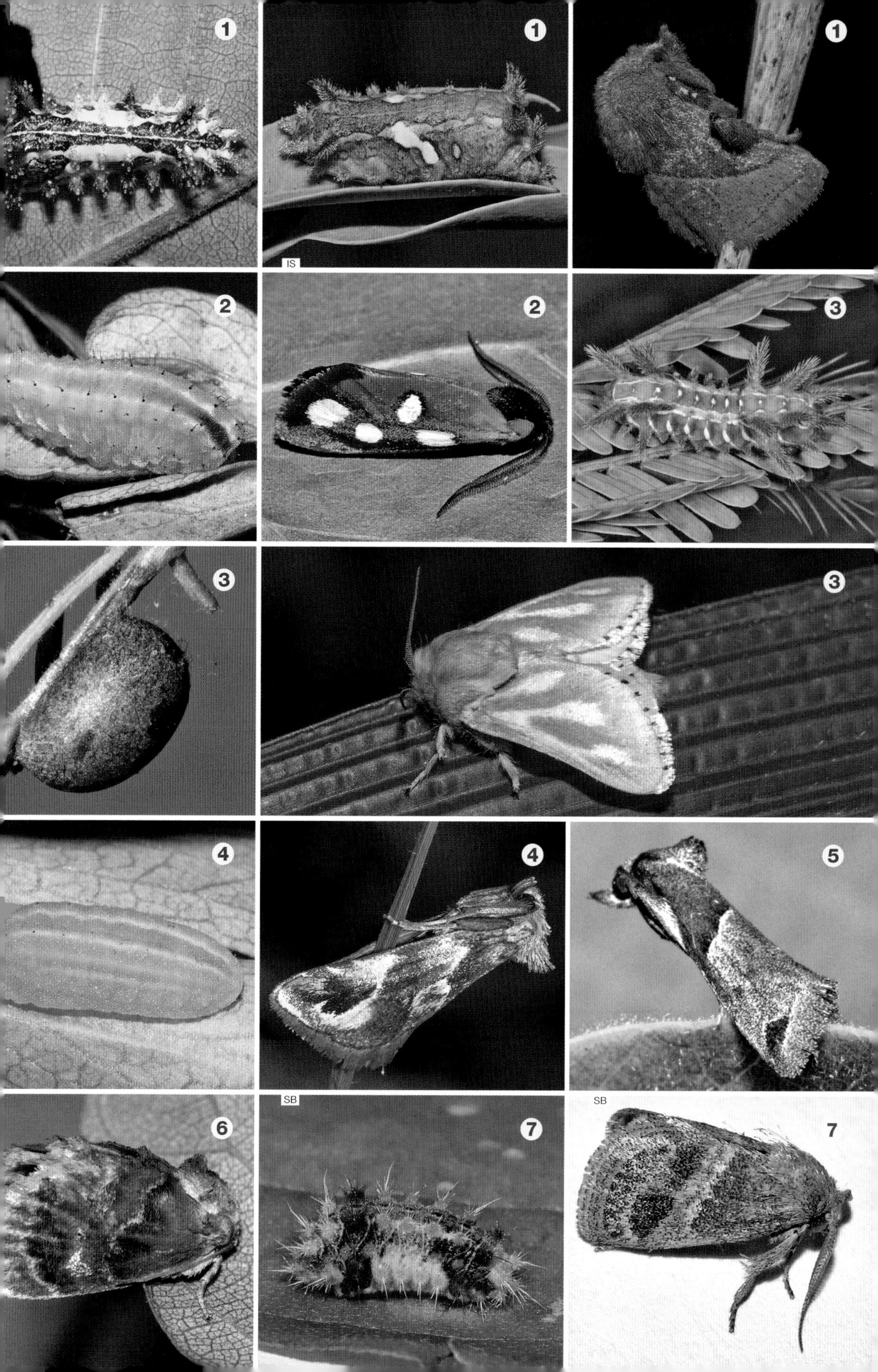

1
1
1
IS
2
2
3
3
3
4
4
5
6
SB
7
SB
7

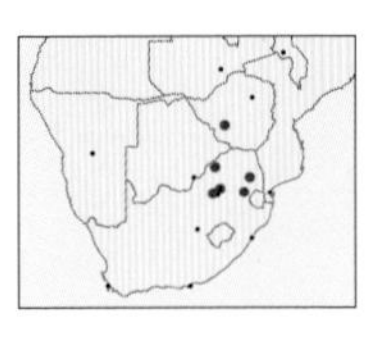

**1 *Halseyia intacta*** Intact Tiptail

Forewing 8–9mm. Body rich brown, abdomen held strongly curled upward when at rest; forewings rich brown basally, sharply divided by dark line from lighter distal half, which is crossed by faint brown bands and diagnostic thin marginal lines. Caterpillar unknown. **Biology:** Caterpillar host associations unknown. Probably bivoltine, adults September–November and February–March. **Habitat & distribution:** Grassland and bushveld; Gauteng, South Africa, north to Zimbabwe.

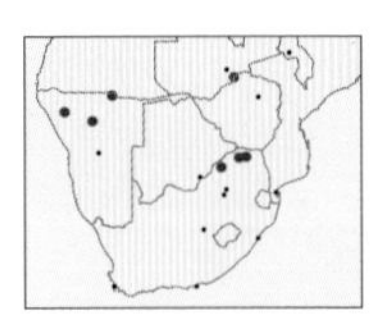

**2 *Isozinara pallidifascia*** White-sided Tiptail

Forewing 7–11mm. Perches in unusual 'U'-shaped posture, with anterior parts strongly elevated and abdomen raised; light to dark brown with large white panel on distal end of forewings; hindwings cream; both fore- and hindwings with dense marginal fringe of pale setae. One species in genus. **Biology:** Caterpillar host associations unknown. Adults November–March. **Habitat & distribution:** Bushveld; Limpopo, South Africa, north to Zimbabwe and Namibia, probably further north.

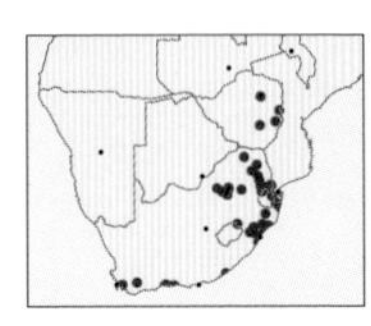

**3 *Latoia latistriga*** Broad-banded Latoia

Forewing 12–19mm. Green thorax and orange abdomen; forewings green with brown triangle at base and broad brown band distally; hindwings yellow. Caterpillar green with blue longitudinal stripes or dots and short spinose tubercles. *L. johannes* very similar but brown triangle more extensive and much larger (forewing 24–28mm). Twelve similar species in genus in region. **Biology:** Caterpillars polyphagous on a wide range of trees and shrubs. Spines give a painful but short-lived sting when touched. Multivoltine, adults throughout the year. **Habitat & distribution:** Moist wooded habitats, including suburban gardens; Western Cape, South Africa, north to Kenya.

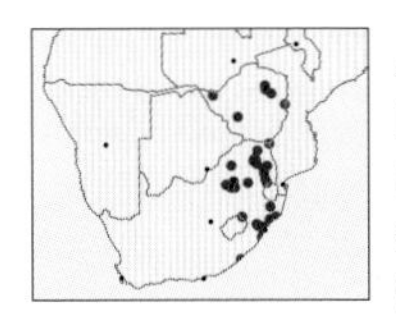

**4 *Latoia vivida*** Vivid Slug

Forewing 12–16mm. Similar to *L. latistriga* (above) but has a larger, more rectangular brown patch at base of forewings and narrower marginal brown band. Caterpillar yellow with green bands and large spinose tubercles. **Biology:** Caterpillars polyphagous on a wide range of trees and shrubs. Spines give a painful but short-lived sting when touched. Multivoltine, adults throughout the year. **Habitat & distribution:** Moist wooded habitats, common in gardens; Eastern Cape, South Africa, north to tropical Africa.

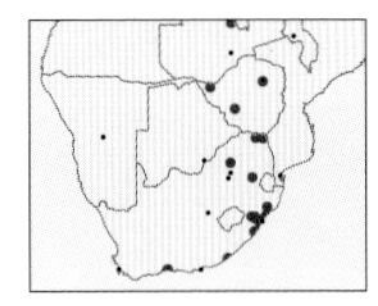

**5 *Latoia albicosta*** White-edged Latoia

Forewing 14–21mm. Thorax setose, reddish-brown, abdomen light orange; forewings divided by oblique line, basal section dark brown, distal section lighter, leading edge of wing distinctively white; hindwings light orange with brown veins. Caterpillar green with bristly tubercles. **Biology:** Caterpillars found on white stinkwoods (*Celtis* spp.), probably polyphagous. Multivoltine, adults September–June. **Habitat & distribution:** Forest and bushveld; southern Cape, South Africa, north to tropical Africa.

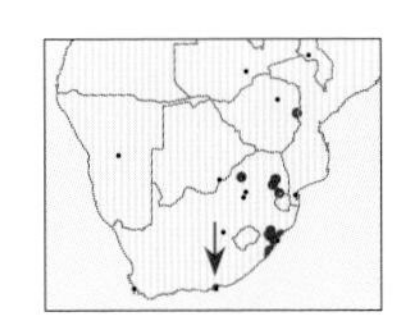

**6 *Stroter intermissa*** Square Stroter

Forewing 24–28mm. Similar to *L. albicosta* (above) but larger, thorax and forewing patch green; forewings dark brown at base, lighter brown distally, with large green rectangular patch on inner margin; hindwings yellow. Caterpillar not described. Three species in genus in region; *S. dukei* has triangular green patch on forewing; *S. capillatus* has green line extending across forewings. **Biology:** Caterpillars feed on privets (*Ligustrum* spp.). Multivoltine, adults throughout the year. **Habitat & distribution:** Forests; Eastern Cape, South Africa, north to tropical Africa.

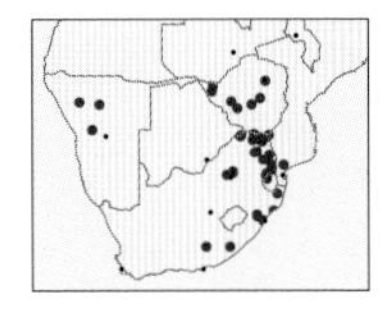

**7 *Micraphe lateritia*** Rosy Slug

Forewing 10–14mm. Body and forewings distinctive bright rosy pink, but sometimes pink and with faint reticulated pattern of brown or grey lines on forewings; hindwings pink. **Biology:** Caterpillars polyphagous on various tree species. Multivoltine, adults throughout the year. **Habitat & distribution:** Bushveld; Eastern Cape, South Africa, north to tropical Africa.

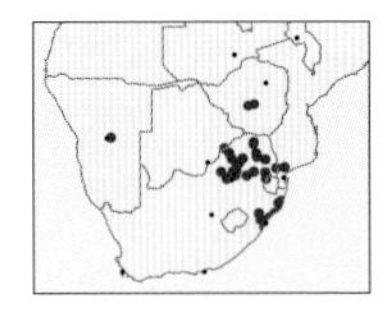

**8 *Parapluda invitabilis*** Netted Slug

Forewing 9–13mm. Thorax white, abdomen orange; forewings attractively subdivided with network of thick brown lines and figure '8' pattern in male, absent in female. Caterpillar green with yellow markings and robust projecting spines. Two species in genus in region; *P. neglecta* similar with only one brown line across forewing, angled at inner margin. **Biology:** Caterpillars feed on Sweet Thorn (*Vachellia karroo*) and Hook Thorn (*Senegalia caffra*). Multivoltine, adults September–May. **Habitat & distribution:** Thorn bushveld; KwaZulu-Natal, South Africa, north to tropical Africa.

1
2
3
IS
3♀
3♂
4
4
5♂
6
7
7
8
8♂

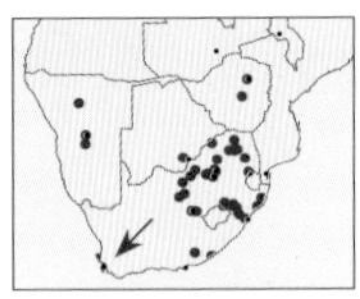

### 1 *Taeda aetitis* — Eagle Slug

Forewing 12–15mm. Thorax densely covered with green hairs, abdomen orange; forewings paler green with a cluster of brown-ringed white dots near base, second cluster at base of postmedian line and brown-ringed white cell spot; hindwings brown with yellow base and margins or all yellow in form *pallida*. Caterpillar green, slug-like, 2 yellow lateral lines, short spines. Four similar species in genus in region; *T. connexa* has forewing cell spot entirely brown; *T. prasinaria* has larger and more extensive brown-ringed dots; *T. aureliae* has no brown-ringed dots, orange cell spot. **Biology:** Caterpillars polyphagous on various tree species. Multivoltine, adults September–May. **Habitat & distribution:** Variety of habitats; Western Cape, South Africa, north to Tanzania.

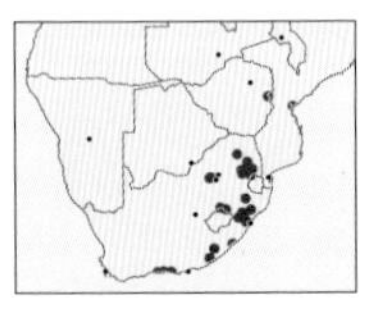

### 2 *Pantoctenia gemmans* — Gem Slug

Forewing 10–13mm. Thorax and forewings bright, light green, abdomen orange; forewing basal area and inner margin brown, often with cream spots, cell spot brown, postmedian line spotted; hindwings yellow. One species in genus in region. **Biology:** Caterpillars feed on *Searsia* and *Myrica* spp. Possibly univoltine, adults November–April, primarily late summer. **Habitat & distribution:** Forest and thick bush; southern Cape, South Africa, north to Zimbabwe.

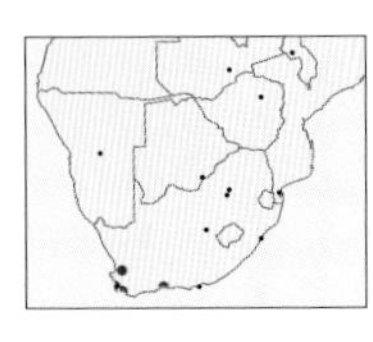

### 3 *Latoiola pusilla* — Marble-dot Slug

Forewing 9mm. Thorax with long dense setae; body and forewings uniform rusty brown, large triangular marbled white patch on forewings; hindwings dark brown with straw-coloured cilia. Caterpillar unknown. One species in genus in region. **Biology:** Caterpillar host associations unknown. Probably univoltine, adults March–April. **Habitat & distribution:** Moist fynbos habitats; Western Cape, South Africa.

## Subfamily Chrysopolominae

The green or pink caterpillars are smooth (i.e. lacking tubercles or projections), usually green or purple and are covered in long, white-tipped hairs. The top of the body has a covering of short spines and the upper lip (labrum) lack spines.

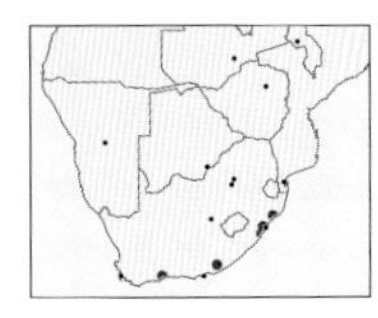

### 4 *Chrysopoloma rudis* — Ruddy Slug

Forewing 18mm. Wings broad, flattened and rounded-oval in outline; body and forewings light reddish-brown, densely peppered with black dots, a conspicuous single white black-ringed cell spot. Caterpillar unknown. Ten species in genus in region. **Biology:** Caterpillar host associations unknown. Probably multivoltine, adults throughout the year. **Habitat & distribution:** Forests along east coast; southern Cape to Zululand, South Africa, also into tropical Africa.

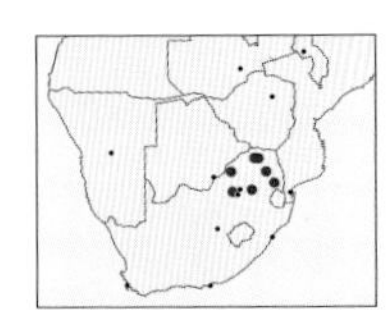

### 5 *Chrysopoloma isabellina* — Isabell's Slug

Forewing 20–23mm. Body orange-red, antennae black; forewings peach to orange-red, large brown-ringed white cell spot on forewing; hindwings pale peach. Caterpillar unknown. Similar *C. pallens* is uniform peach on all wings and body; *C. restricta* is pale peach, without cell spot and with straight postmedian forewing line. **Biology:** Caterpillar host associations unknown. Probably bivoltine, adults November–February. **Habitat & distribution:** Bushveld; Gauteng, South Africa, north to tropical Africa.

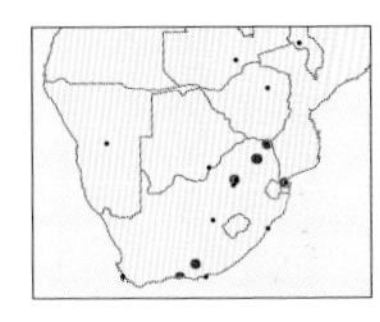

### 6 *Chrysopoloma varia* — Variable Slug

Forewing 18–23mm. Body orange, antennae light brown; forewings orange to light brown variably sprinkled with small black dots, black-ringed white cell spot on forewing may be present; hindwings orange. Caterpillar unknown. **Biology:** Caterpillar host associations unknown. Probably multivoltine, adults throughout the year. **Habitat & distribution:** Bushveld and forest; southern Cape north to Limpopo, South Africa.

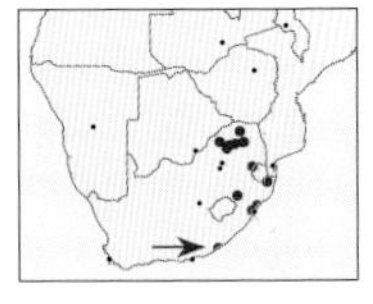

### 7 *Ectropa ancilis* — Angular Slug

Forewing 10–13mm. Squat appearance; wings golden-brown with black flecks; forewings short and broad with faint crosslines, strongly scalloped margins fringed with long golden setae. Caterpillar unknown. One species in genus in region. **Biology:** Adults have a unique resting posture (with rear wings held up), and a strong buzzing flight. Caterpillar host associations unknown. Adults September–May. **Habitat & distribution:** Bushveld and woodland; Eastern Cape to Limpopo, South Africa.

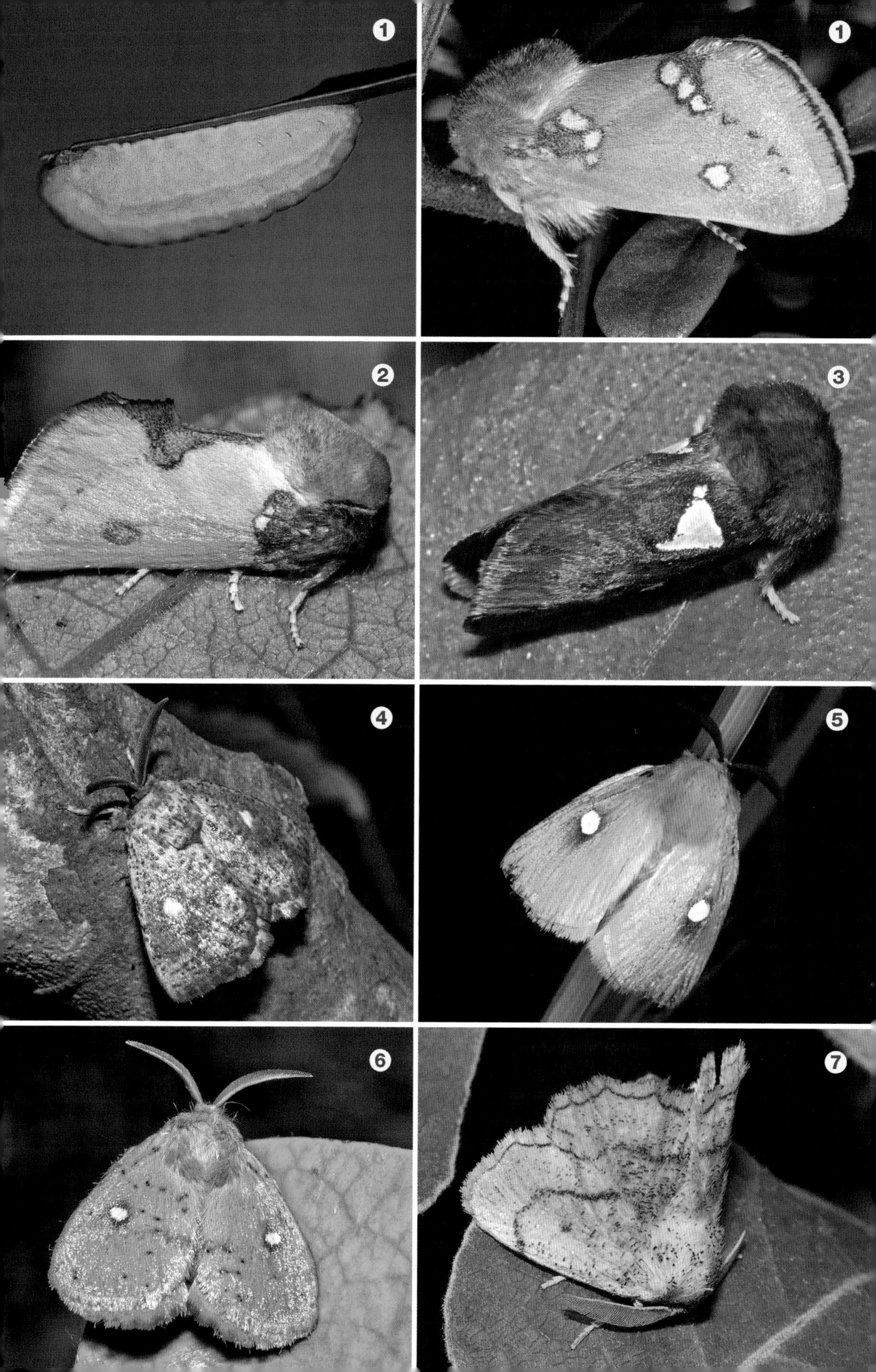

## FAMILY PRODIDACTIDAE

**Enigmatic family containing a single genus and species. Previously assigned to 3 different families before being allocated own family in 2003. Adults distinguished by reduced labial palps and unusual form of male hindleg, which bears an elongate membranous lobe on hind coxa and swollen, pocket-like tibia bearing large hair-pencil.**

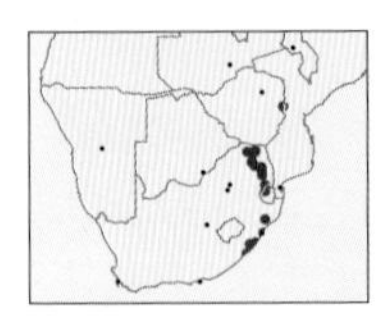

### 1 *Prodidactis mystica* — Mystic Moth

Forewing 18–24mm. Legs red; forewings long and square-ended, white to light brownish-yellow with broken rows of red dots, underside bright red; hindwings orange-yellow. Caterpillar with red-brown head capsule and dark red anterior thoracic segments, rest of body light green with black longitudinal bands, some enclosing green spots. **Biology:** Caterpillars feed on *Nuxia* spp., forming shelters in rolled leaf, pupate in tough cocoon formed between leaves. Adults October–March. **Habitat & distribution:** Moist forests; Eastern Cape, South Africa, north to Kenya.

## FAMILY SOMABRACHYIDAE African Flannel Moths

**Adults lack labial and maxillary palps, and have a hairy and robust body, with broadly rounded, largely transparent wings marked in grey or brown. Caterpillars slug-like with concealed head. Only 2 genera, with highly disjunct distribution. The single species of *Somabrachys* from North Africa and the Mediterranean region has wingless females, while *Psycharium* and *Parapsycharium* from South Africa have winged females. Around 35 largely Afrotropical species, 6 in the region.**

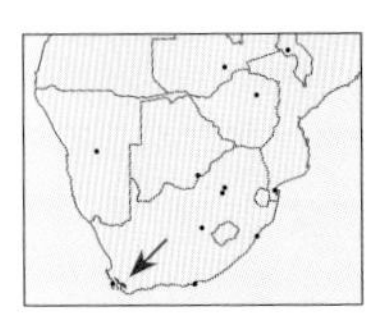

### 2 *Psycharium pellucens* — Table Mountain Flannel Moth

Forewing 9–11mm. Antennae of male bipectinate, those of female filamentous; adults with squat, setose red-brown body crossed by brown transverse bands; wings almost completely transparent, except for yellow-brown veins. Caterpillar brightly coloured in blue, black, yellow and white with red spots laterally and tufts of urticating hairs. **Biology:** Eggs laid in clusters, covered in urticating hairs shed from female's abdomen. Caterpillars present year-round, but most abundant in winter; feed on species of Restionaceae, Cyperaceae, Ericaceae and Bruniaceae, also on *Pinus radiata*; spin cocoons on the ground. Adults short-lived. **Habitat & distribution:** Mostly confined to fynbos vegetation and pine plantations; Western Cape, South Africa.

## FAMILY AUTOSTICHIDAE

**A family of small to relatively large Gelechioidea. Most are nocturnal, but some are active late afternoon before dark. Caterpillar host associations are unknown in the region; elsewhere some feed on leaf litter. More than 150 species globally, with 29 species in the region.**

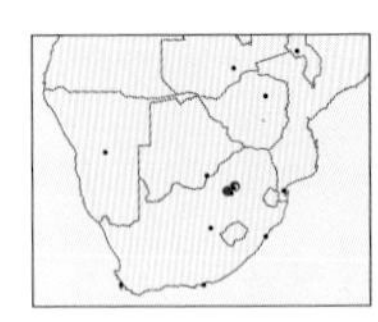

### 3 *Procometis ochricilia* — Spear Fan

Forewing 13–14mm. Slender with labial palps upturned over face, both wings sharply pointed at apex without markings; male forewings pastel cream becoming more rosy toward inner margin, grey shading near costa in some specimens; hindwings dark brown becoming paler toward inner margin, completely surrounded by long, brown fan-like setae, androconial fold on inner margin; female similar, but hindwing cilia shorter and yellow, lacking androconial fold. Six similar species in genus in region. **Biology:** Caterpillar host associations unknown. Males form leks on hilltops late afternoon till dusk, displaying their fan-like cilia (bottom left image shows displaying male releasing pheromones). Adults December. **Habitat & distribution:** Grassland; Gauteng, South Africa.

1
1
2
2
3♂
3♂
3

## FAMILY COLEOPHORIDAE Casebearer Moths

**Diverse and widely distributed family of small to very small, slender moths with very narrow hindwings fringed with long setae. Most are plain grey or brown, and rest with body slightly inclined and antennae projecting forward. Caterpillars initially leaf miners, later carry protective cases, which are often distinctive; cases discarded and rebuilt as caterpillars moult and grow, unlike bagworms, which build similar cases, but which extend as they grow. Females have fully developed wings. Over 1,000 species globally, over 50 in the region, dominated by the genus *Coleophora*.**

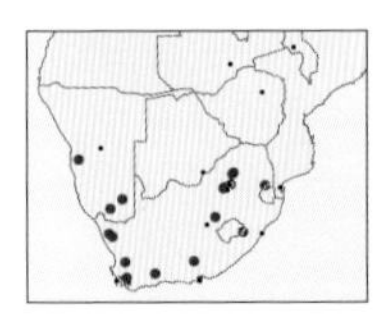

### 1 *Coleophora* species — Casebearer Moths

Forewing 5–7mm. Large genus with 52 species described in region; individual species can only be reliably separated by detailed examination of genitalia. Adults have a distinctive resting posture with long filiform antennae held upright in line with body and wings at an angle of around 30 degrees from substrate. **Biology:** Caterpillars are case bearers feeding on many plant families. Adults can be found throughout the year. **Habitat & distribution:** Variety of moist and dry habitats throughout the world.

## FAMILY COSMOPTERIGIDAE Cosmet Moths

**Very diverse family of small moths with narrow wings held tightly around body. Forewings often with brightly coloured crossbars, long setae distally; hindwings very narrow and pointed, with dense fringes of long setae. Caterpillars leaf miners, or feed internally on seeds and stems of host plants. About 1,500 species described globally, 86 in the region.**

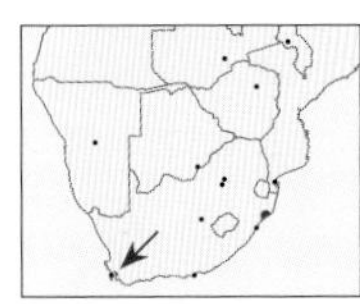

### 2 *Cosmopterix* species — Cosmet Moths

Forewing 3–6mm. Elongate labial palps bent upward to extend well above head; forewings elongate, narrow, pointed; many species in this very diverse and cosmopolitan genus of usually black species with orange or yellow markings midway along forewings. About 40 similar species known from South Africa. Caterpillar undescribed. **Biology:** Caterpillars are blotch leaf miners, host plants include Purple Nutsedge (*Cyperus rotundus*), Molasses Grass (*Melinis minutiflora*) and *Scirpus* spp. Adults diurnal and found on upper surfaces of leaves, where they show 'twirling' behaviour. **Habitat & distribution:** Wide range of habitats, both agricultural and undisturbed. Genus extremely widespread, from North and South America, Australia, New Zealand, East Asia, Saudi Arabia and from scattered records across sub-Saharan Africa, from South Africa north to Cameroon and Uganda, also Madagascar, Mauritius, La Réunion and Seychelles.

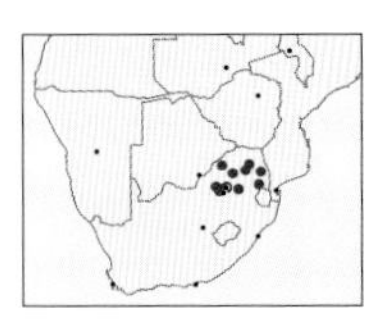

### 3 *Limnaecia neurogramma* — Spear Cosmet

Forewing 9–12mm. Elongate labial palps bent upward to extend well above head; forewings elongate, narrow, pointed; head, thorax and forewings black with many elongate thin white lines, one thicker curved line from base to wing tip diagnostic; hindwings light brown. Caterpillar dark brown to black. Twelve species in genus in region. **Biology:** Caterpillars fold leaf of Monkey's Tail (*Xerophyta retinervis*) with silk, feeding inside. Bivoltine, adults December–March. **Habitat & distribution:** In seepage areas on hills where host plants grow; Gauteng to Limpopo, South Africa.

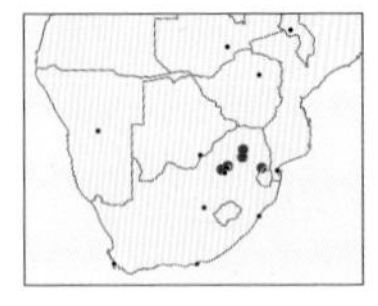

### 4 *Limnaecia ichnographa* — Grizzled Cosmet

Forewing 7–8mm. Smaller than *L. neurogramma* (above), and forewings with short white lines across that vary in brightness; hindwings charcoal grey. Caterpillar unknown. **Biology:** Caterpillar host associations unknown. Bivoltine, adults December–March. **Habitat & distribution:** Bushveld; Gauteng to Limpopo, South Africa.

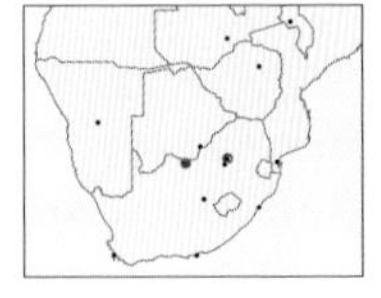

### 5 *Alloclita paraphracta* — Crossline Cosmet

Forewing 6mm. Forewings dark brown with white median band and white postmedian markings, 2 dark spots in cell; hindwing charcoal grey. Caterpillar unknown. **Biology:** Caterpillar host associations unknown. Bivoltine, adults December–March. **Habitat & distribution:** Bushveld; Gauteng to North West, South Africa.

1
2
JB
3
3
4
4
5

## FAMILY DEPRESSARIIDAE Grass Miner Moths

**Small to medium-sized moths (forewing length 3–15mm), of which the taxonomy is not fully resolved. There are well over 2,000 species globally, with 93 described in the region, many other local representatives certainly remain unnamed.**

### Subfamily Ethmiinae

Previously recognised as the family Ethmiidae, these moths have narrow wings held folded backward tightly against the body, and can be easily confused with Ermines (Yponomeutidae, p.38). They are mostly whitish or grey, often with darker dots, bands or blotches; best-known species are in the very diverse genus *Ethmia*. Their caterpillars, which feed in the open on Boraginaceae worldwide, are distinctively marked.

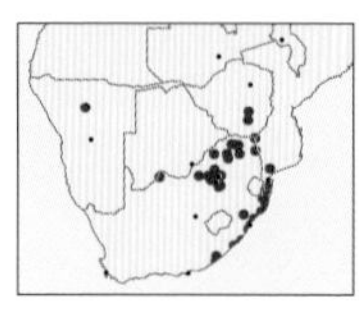

#### 1 *Ethmia concineutis* — Dotted Ethmia

Forewing 11–14mm. Legs banded black and white, body and forewings white, with scattered small black dots; hindwings grey or white with grey front margin; abdomen with black dorsal patches and orange tip. Caterpillar largely black, with orange markings and fairly dense white setae. Twenty-one species in genus in region. **Biology:** Caterpillars feed on Puzzle Bush (*Ehretia rigida*). Bivoltine, adults October–April. **Habitat & distribution:** Forest and bush where host plant grows; Eastern Cape, South Africa, north to Zimbabwe.

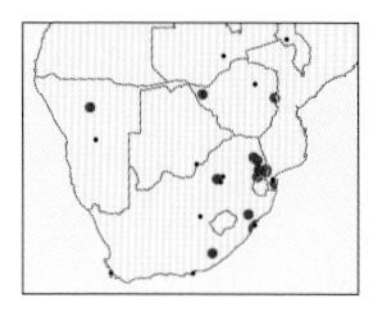

#### 2 *Ethmia rhomboidella* — Grey Ethmia

Forewing 12–14mm. Head and abdomen yellow or orange, thorax and forewings uniform grey with sparse scattering of small black dots, cilia grey; hindwings grey with pale yellow base. Caterpillar black, with either several narrow or many wide transverse cream bands, sparsely setose. **Biology:** Caterpillars feed on Puzzle Bush (*Ehretia rigida*). Bivoltine, adults September–May, peaking November and February. **Habitat & distribution:** Forest and bush where host plant grows; Eastern Cape, South Africa, north to tropical Africa.

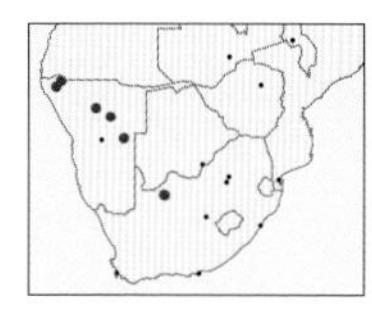

#### 3 *Ethmia kunenica* — Pastel Ethmia

Forewing 10–11mm. Similar to *E. rhomboidella*, but thorax and forewings pastel shades of grey to yellow, cilia yellow; row of black spots on thorax; hindwings pale yellow. **Biology:** Caterpillar host associations unknown. Adults November–April. **Habitat & distribution:** Arid areas; Northern Cape, South Africa, to Namibia.

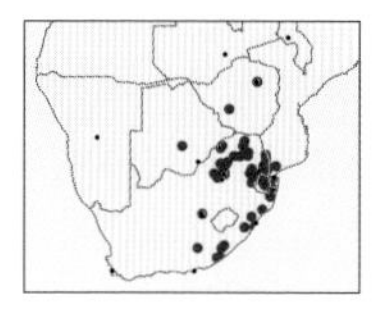

#### 4 *Ethmia livida* — Plain Ethmia

Forewing 10–12mm. Similar to *E. kunenica*, but thorax and forewings plain ivory grey and black spots much smaller, cilia yellow; hindwings light grey. **Biology:** Caterpillar host associations unknown. Adults September–March. **Habitat & distribution:** Bushveld; Eastern Cape, South Africa, north to Zimbabwe.

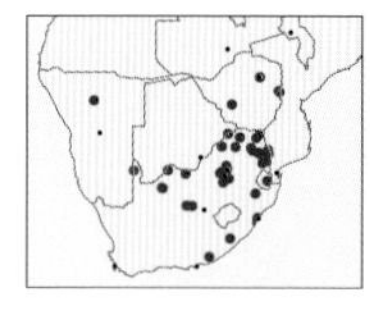

#### 5 *Ethmia oculigera* — Eyed Ethmia

Forewing 8–10mm. Head, thorax and forewings darkish grey with larger black dots than *E. rhomboidella* (above), each dot surrounded by an indistinct white ring; hindwings orange with black tip. Caterpillar black, with longitudinal orange dorsal stripes and lateral patches, sparsely setose. **Biology:** Caterpillars feed on various Boraginaceae. Bivoltine, adults September–April, peaking November and February. **Habitat & distribution:** Bushveld and forest; Eastern Cape, South Africa, north to tropical Africa.

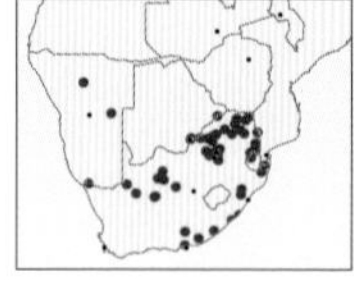

#### 6 *Ethmia circumdatellus* — Ringed Ethmia

Forewing 10–14mm. Similar to *E. oculigera*, but has a yellow head, and paler grey forewings with wider, more distinct white or yellow rings around larger dots; a clear line of smaller regularly spaced dots along terminal margin; hindwings pale yellow, variably shaded with grey. Caterpillar white with black bands and markings, conspicuous orange lateral warts. **Biology:** Caterpillars feed on Puzzle Bush (*Ehretia rigida*). Adults throughout the year, peaking in summer months. **Habitat & distribution:** Bushveld; Eastern Cape, South Africa, north to tropical Africa.

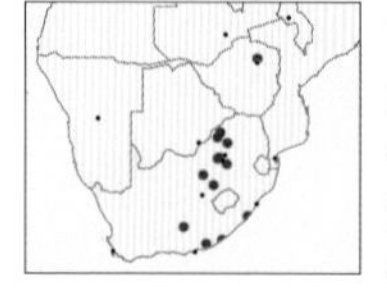

#### 7 *Ethmia sabiella* — Calculator Ethmia

Forewing 9–14mm. Thorax and forewings white to cream with large black dots, those midway along wings merged into larger, sometimes 'V'-shaped black patches and conspicuous '+' mark. Caterpillar yellow with black crossbars and orange head. **Biology:** Caterpillars feed singly on Puzzle Bush (*Ehritia rigida*). Bivoltine, adults October–April. **Habitat & distribution:** Bushveld and forest; Eastern Cape, South Africa, north to tropical Africa.

1
1
2
2
3
IS
4
5
5
6
6
6
7
7

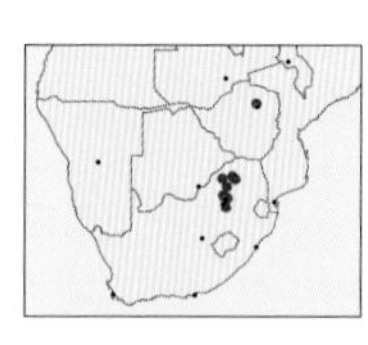

**1 *Ethmia anikini*** Toothed Ethmia

Forewing 10–11mm. Similar to *E. sabiella* (p.74), but thorax and forewings greyish-white with large black 'toothed' patch along costa; hindwings pale yellow to grey at termen. **Biology:** Caterpillar host associations unknown. Adults October–February. **Habitat & distribution:** Grassland and bushveld at higher elevations; Free State, South Africa, north to Zimbabwe.

## Subfamily Stenomatinae

Largely Neotropical; hindwings broader than forewings. Caterpillars usually feed within leaves tied together with silk, but may also mine leaves or bore into stems.

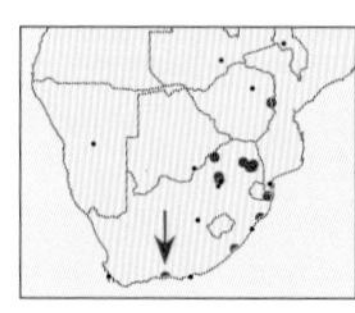

**2 *Orygocera thysanarcha*** Red-horned Warrior

Forewing 7–8mm. Forewings held roof-like against body; labial palps enlarged and strongly upturned into a pronounced 'horn'; body and forewings dark red and in striking contrast to dull green legs, wings crossed by diagonal yellow bars. Caterpillar white with thin, longitudinal brown lines. **Biology:** Caterpillars feed on guarri trees (*Euclea* spp.). Bivoltine, adults December–March and June–July. **Habitat & distribution:** Forest and bushveld; Western Cape, South Africa, to Zimbabwe.

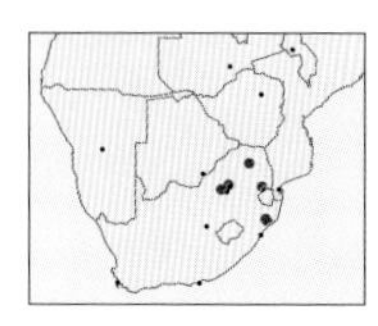

**3 *Stenoma stolida*** Rough Warrior

Forewing 12mm. Compressed species, forewings overlapping, held flat against body; labial palps short, upturned; body and forewings dark grey with raised white hairs, several black lines from postmedian to edge of wing, partially or completely absent in some individuals; hindwings light to dark grey. Caterpillar cream with several reddish-brown lateral lines, dark dorsal line, pronotum black. Five species in genus in region. **Biology:** Caterpillars spin several leaves together, excavating the inside of leaf, on Blue Guarri trees (*Euclea crispa*). Probably univoltine, adults December–January. **Habitat & distribution:** Forest and bushveld; KwaZulu-Natal north to Limpopo, South Africa.

## Subfamily Oditinae

Pupae have spines in a small patch on last abdominal segment.

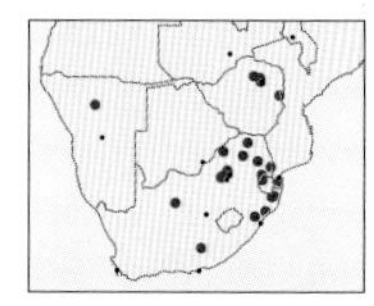

**4 *Odites natalensis*** Cream Dip

Forewing 8mm. Distinctive colour pattern, with front half of body and forewings cream, posterior part of forewings brown, becoming lighter distally, the two sections demarcated by wavy black line; hindwings pale grey to cream with long marginal setae. Twenty-six species in genus in region. **Biology:** Caterpillar host associations unknown. **Habitat & distribution:** Bushveld; Northern Cape north to Congo and Tanzania.

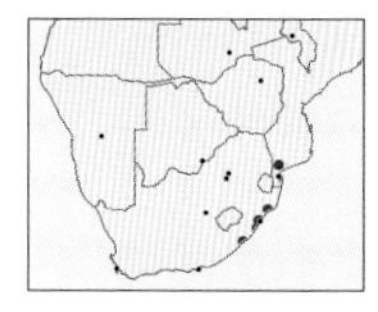

**5 *Odites sucinea*** Brown-edged Dip

Forewing 10–11mm. Head, body and forewings creamy yellow, with variable brown lines across forewings, absent in some individuals, costa edged with brown, diagnostic; hindwings cream, unmarked. Caterpillar yellow with brown head capsule. Several similar yellow species in genus. **Biology:** Caterpillars reared on *Xylotheca kraussiana* (Achariaceae). Multivoltine, adults throughout the year. **Habitat & distribution:** Lowland moist forest; Eastern Cape, South Africa, north to Mozambique.

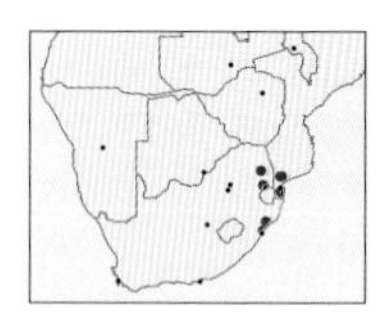

**6 *Odites incolumis*** White Dip

Forewing 8–9mm. Head, body and forewings white, with variable sprinkling of brown scales across forewings, absent in some individuals, 4 distinct black spots and row of small black terminal spots diagnostic; hindwings white, unmarked. Caterpillars pinkish-yellow with black head capsule. **Biology:** Caterpillars reared on *Carissa edulis* (Apocynaceae). Multivoltine, adults throughout the year. **Habitat & distribution:** Moist forest; KwaZulu-Natal, South Africa, north to Mozambique.

## Subfamily Depressariinae

Wings held flat over body, giving a broad, bullet-shaped profile, but some hold wings roof-like. They are poorly known in the region. Caterpillars are mostly leaf miners or leaf rollers, feeding on grasses or reeds, but also other plants.

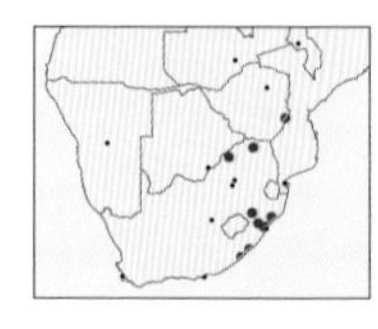

**7 *Agonopterix trimenella*** Black Dot Bullet

Forewing 10–12mm. Head yellow, body and forewings cream, irrorated with brown scales in wavy patterns from base, becoming heavier toward wing margin, conspicuous large black spot on each forewing diagnostic for this species and very similar *Martyrhilda melanarcha*, which has the base of forewings black; hindwings light grey, unmarked. Ten species in genus in region. **Biology:** Caterpillar host associations unknown. Adults throughout the year. **Habitat & distribution:** Forest and riverine vegetation; Eastern Cape, South Africa, north to Zimbabwe.

1
2
IS
2
2
3
3
4
5
SB
5
6
IS
6
7

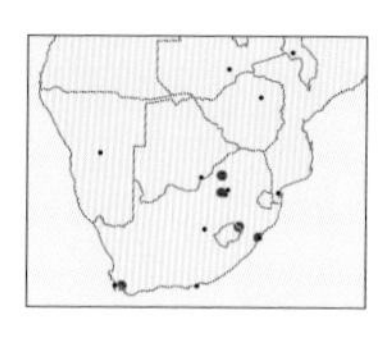

**1** ***Orophia roseoflavida*** Rose Bullet

Forewing 14–16mm. Palps pink, head, anterior part of thorax and costal edge of forewings light to deep pink, extending around termen to inner margin in some individuals, rest of forewings yellow; hindwings uniform light brown. Fourteen species in genus in region. **Biology:** Caterpillars feed on *Protea* spp. Probably univoltine, adults April. **Habitat & distribution:** Fynbos and grassland where proteas grow; Western Cape north to Limpopo, South Africa.

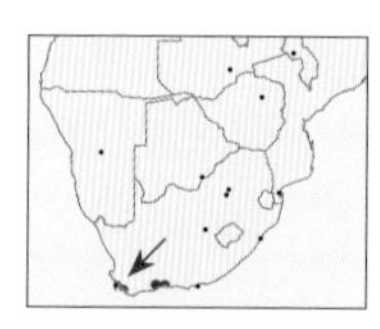

**2** ***Orophia ammopleura*** Camo Bullet

Forewing 12–14mm. Palps dark pink, rest of body and forewings mottled light brown and pinkish-brown in a camouflage pattern; hindwings straw-coloured. Caterpillar light brown dorsally, grey laterally, with all setal bases black. **Biology:** Caterpillars feed on various Proteaceae. Probably univoltine, adults April. **Habitat & distribution:** Fynbos; Western Cape, South Africa.

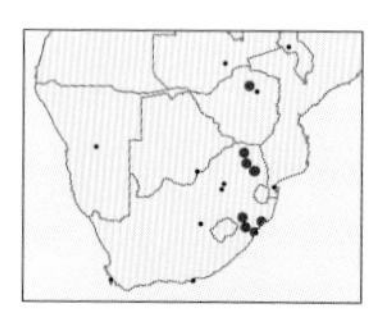

**3** ***Orophia straminella*** Ivory Bullet

Forewing 7–9mm. Head, body and forewings plain ivory yellow with a black cell spot; smaller black dots and grey shading present in some specimens. **Biology:** Caterpillar host associations unknown. **Habitat & distribution:** Moist forests; KwaZulu-Natal, South Africa, north to Zimbabwe.

## FAMILY GELECHIIDAE Twirler Moths

**Large family of tiny to small moths with narrow, fringed wings, the hindwings markedly concave below apex, coming to sickle-shaped, curved point. Hindlegs often with elongate, hair-like scales. Adults mostly sombre in colour, although sometimes brightly coloured, metallic or iridescent. Adults may twirl in circles when disturbed, hence common name. Larval feeding habits vary widely and include leaf miners, leaf-folders, gall-makers and seed feeders. Some species are agricultural pests, others have been used as biological control agents for invasive plants. Subfamily divisions controversial, currently Dichomeridinae, Physoptilinae, Gelechiinae and Pexicopiinae recognised. Some 4,500 described species globally, with about 600 in the region.**

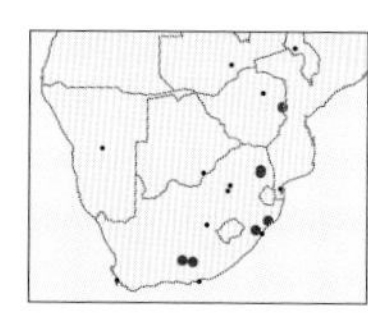

**4** ***Anarsia amalleuta*** Grey Weevil Moth

Forewing 6–8mm. At rest, resembles a weevil, wings curled around body and greatly enlarged labial palps extended forward; forewings elongate-oval with fringes of setae distally, grey with longitudinal brown markings; hindwings lighter with darker veins and completely fringed with elongate setae. Caterpillar light green with black head capsule and black rings around thoracic segments. Twenty-two species in genus in region. **Biology:** Caterpillars feed singly in silken nest on leaves of thorn trees (*Vachellia* spp.). Multivoltine, adults November–March. **Habitat & distribution:** Bushveld; Eastern Cape, South Africa, north to Kenya.

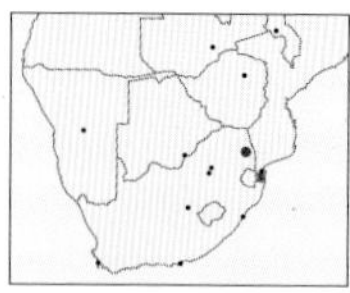

**5** ***Encolapta stasimodes*** False-head Twirler

Forewing 7mm. At rest, looks like it has a head on both ends, wings curled around body and wing scales raised at apex to imitate a false head; forewings white and black with 2 distinct black marks. Caterpillars olive green with brown head capsule. One species in genus in region. **Biology:** Caterpillars feed on Bushveld Mahogany (*Trichilia emetica*). Probably multivoltine, adults recorded in May. **Habitat & distribution:** Lowveld bushveld; Limpopo, South Africa, and Mozambique, also Uganda.

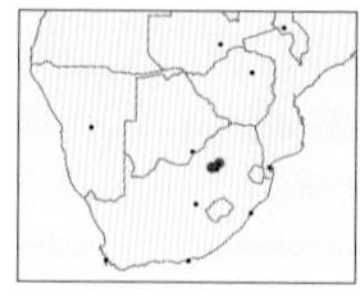

**6** ***Hypatima melanecta*** Black-lined Twirler

Forewing 7–8mm. Elongate; at rest, wings curled around body, enlarged palps held extended forward and antennae extended backward; forewings grey, becoming black at tips, marked with numerous longitudinal fine black lines; hindwings light grey. Caterpillar translucent light green with black head capsule. Seventeen species in genus in region. **Biology:** Caterpillars feed on Dwarf Currant (*Searsia magalismontanum*); they fold leaf with 2 exits, from which they move either out to feed or to escape predators by dangling from silk. **Habitat & distribution:** Rocky ridges where host plants grow; Gauteng and North West, South Africa.

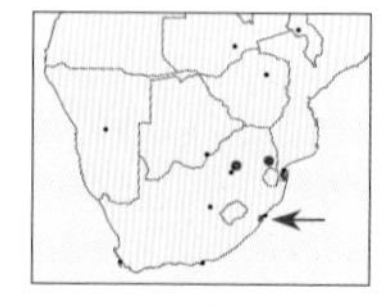

**7** ***Hypatima loxosaris*** Grey Twirler

Forewing 6–9mm. Short; at rest, wings curled around body; forewings grey, marked with indistinct short black streaks; hindwings light grey. Caterpillar dark brown with black head capsule. **Biology:** Caterpillars feed on Sicklebush (*Dichrostachys cinerea*). Adults November–March. **Habitat & distribution:** Bushveld; KwaZulu-Natal north to Mpumalanga, South Africa, and Mozambique.

1
2
MB
2
3
4
IS
4
5
IS
5
5
6
IS
AS
6
AS
7
7

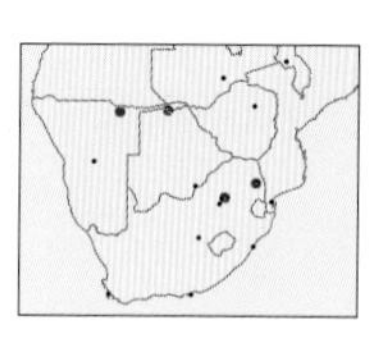

## 1 *Mesophleps kruegeri* — Martin's Twirler

Forewing 6–8mm. Head smooth, light brown; forewings dark brown striated variably with white scales; hindwings dark grey. Caterpillar olive green with black head capsule. One species in genus in region. **Biology:** Caterpillars feed on Cluster Leaf (*Terminalia sericea*). Adults November–December and March. **Habitat & distribution:** Bushveld; Gauteng, South Africa, north to Namibia.

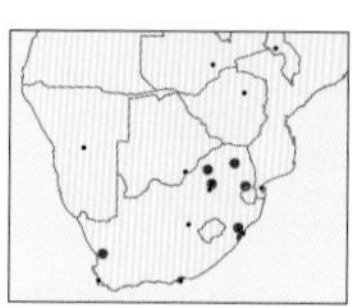

## 2 *Sitotroga cerealella* — Angoumois Grain Moth

Forewing 5–7mm. Dull grey-brown; forewings marked by some small patches of black scales; hindwings greyish-brown, distinctly tapering distally and with greatly produced apex and fringe of long, pale setae. Caterpillar yellowish-white, with darker head capsule that can be retracted into thorax. Three species in genus in region. **Biology:** Primarily found in grains such as wheat, oats, barley, millet, rye and maize, but also found in various stored spices. Caterpillars complete their development within a single cereal grain kernel. May complete 5–12 generations per year, depending on temperature. **Habitat & distribution:** Probably originated in Australia, but accidentally dispersed with stored foods and now widespread across Americas, Europe, Asia and Africa.

Zimbabwe & Mozambique, on cotton crops

## 3 *Pectinophora gossypiella* — Pink Bollworm

Forewing 7–9mm. Head and thorax brown; forewings pale brown with black dots and bands, setose distally; hindwings grey with pointed tips and long fringe of pale setae. Caterpillar pinkish, with hard, brown head capsule. One species in genus in region. **Biology:** Well-known pest of cottons (*Gossypium*); also feeds on Okra (*Abelmoschus esculentus*), Deccan Hemp (*Hibiscus cannabinus*), lucerne, etc. Caterpillars feed internally within cotton boll. Pupation occurs in the ground. **Habitat & distribution:** Widely introduced across Africa in a range of agricultural and undisturbed habitats; on cotton crops in Zimbabwe and Mozambique, also in North and South America, Europe, Asia and Oceania.

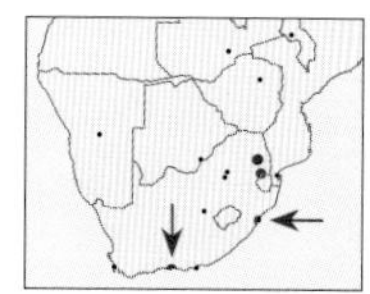

## 4 *Furcaphora caelata* — Chequered Twirler

Forewing 9–10mm. Body white with black spots; forewings narrow and marked with patchwork of black shapes of long, white setae on distal margin; hindwings cream, tapering to point distally, a darker patch anterodistally, margins fringed with long grey to white setae. Caterpillar slender, pale green, with black head capsule. One species in genus in region. **Biology:** Caterpillars feed on Red Currant (*Searsia chirindensis*). Adults recorded March–July, probably throughout the year. **Habitat & distribution:** Moist forests where host tree grows; southern Cape north to Limpopo, South Africa.

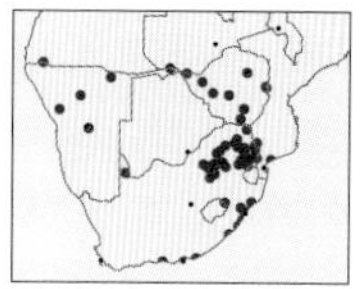

## 5 *Stegasta sattleri* — Twinspot Twirler

Forewing 5–6mm. Head and thorax dark; forewings largely black with 2 distinct white spots on costa and brown patch near inner margin; hindwings light grey with fringe of long grey setae. Caterpillar light green, with dark head capsule. One species in genus in region. **Biology:** Caterpillars feed on Fishbone Dwarf Cassia (*Chamaecrista mimosoides*), probably also other Fabaceae. Multivoltine, adults throughout the year, peaking in late summer. **Habitat & distribution:** Variety of habitats from forest to desert; southern Cape, South Africa, north to Ethiopia, also Madagascar.

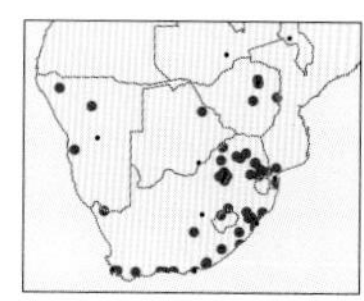

## 6 *Phthorimaea operculella* — Potato Tuber Moth

Forewing 5–7mm. Body light brown; forewings held steeply roof-like against body, grey-brown with darker spots and long marginal setae, forewings of female with 'X'-shaped dorsal marking; hindwings cream with long fringe of setae. Caterpillar white or yellow, with dark head capsule. **Biology:** Eggs laid on leaves or exposed tubers; caterpillars normally leaf miners, but also feed on stems and exposed tubers; pupate in the soil. Caterpillars feed on Solanaceae, including tobacco, eggplant, tomatoes, bell peppers and Common Thorn Apple (*Datura stramonium*), but best known on potato tubers. **Habitat & distribution:** Originates from South America and accidentally introduced to region, now widespread across the globe wherever host plants occur, including warmer parts of North America, Europe, Asia, Oceania and Africa.

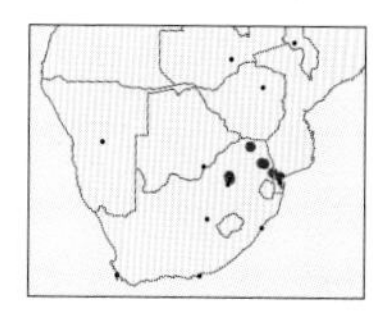

## 7 *Tricerophora commaculata* — Sable Twirler

Forewing 7–8mm. Body white; forewings white, heavily marked with longitudinal black patches; hindwings grey, becoming darker posteriorly, with fringe of grey setae. Caterpillar yellow with black head capsule or white with 2 red longitudinal stripes. Three species in genus in region. **Biology:** Caterpillars feed on various trees in the family Anacardiaceae. Multivoltine, adults throughout the year. **Habitat & distribution:** Bushveld; Gauteng, South Africa, northeast to Zimbabwe.

1
1
AS
2
2
SY
3
4
IS
4
5
IS
5
6
IU
IS
7
7

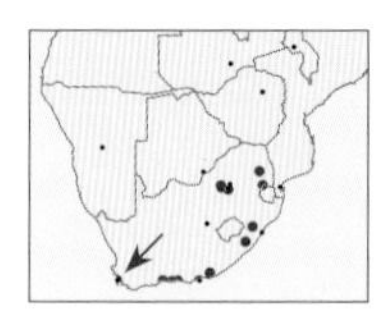

**1 *Syngelechia psimythota*** Eyed Twirler

Forewing 7–9mm. Body white; forewings white, broad black band before middle of wing, 2 triangular patches on costa, and conspicuous 'eyespot' at wing tip; hindwings grey. Caterpillar translucent pinkish with black head capsule. One species in genus in region. **Biology:** Caterpillars feed on Taaibos (*Searsia pyroides*), make silken shelter by spinning several leaves together, feeding outside of shelter. Bivoltine, adults throughout the year, primarily in rainy season. **Habitat & distribution:** Relatively cool moist habitats where host plant grows; Western Cape to Limpopo, South Africa.

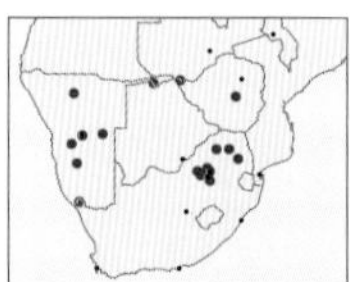

**2 *Armatophallus exoenota*** Brown Twirler

Forewing 6–10mm. Body and forewings light brown to dark brown, forewings mottled in various shades of brown in some specimens, colour uniform in others; head is lighter than rest of body, diagnostic; hindwings grey. Caterpillar light green and yellow banded with brown head, turns pink before pupation. One species in genus in region. **Biology:** Caterpillars feed on Hook Thorn (*Senegalia caffra*), make silken shelter by spinning several leaflets together, feeding outside of shelter. Bivoltine, adults October–March. **Habitat & distribution:** Bushveld; Northern Cape, South Africa, north to tropical Africa.

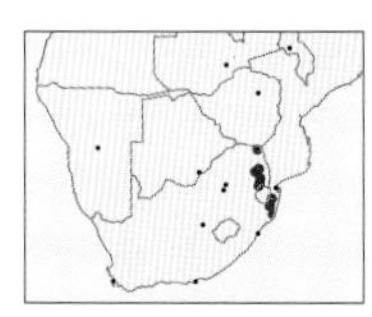

**3 *Neotelphusa craterota*** Black-banded Twirler

Forewing 6–7mm. Body and forewings creamy white, forewings mottled in various shades of brown and black, broad pitch black band across wings diagnostic; hindwing translucent olive grey. Caterpillar light green with orange head capsule. Seventeen species in genus in region. **Biology:** Caterpillars feed on Marula (*Sclerocarya birrea*). Bivoltine, adults November–March. **Habitat & distribution:** Lowland bushveld; KwaZulu-Natal north to Limpopo, South Africa.

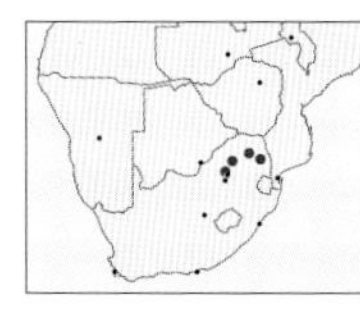

**4 *Neotelphusa tapinota*** Warty Twirler

Forewing 6–7mm. Body and forewings white, forewings with black shading along costa and inner margin, several patches of raised black scales, some ringed with orange, making the adult look 'warty'; hindwings grey. Caterpillar translucent brown. **Biology:** Caterpillars feed on White Seringa (*Kirkia acuminata*). Adults January–February. **Habitat & distribution:** Bushveld; North West and Limpopo provinces, South Africa.

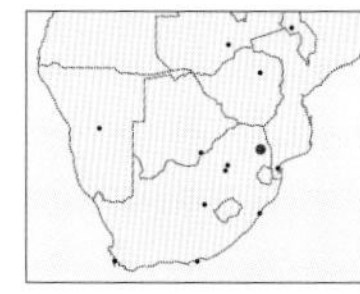

**5 *Teleiopsis sharporum*** Sharp's Twirler

Forewing 7–8mm. Body and forewings greyish-black with distinct ochreous brown markings and light brown irroration; hindwings translucent grey with brown spots. Caterpillar olive grey with black spots, orange head. **Biology:** Caterpillars feed on resin trees (*Ozoroa* spp.). Probably bivoltine, adults June–July and November. **Habitat & distribution:** Bushveld; Limpopo, South Africa.

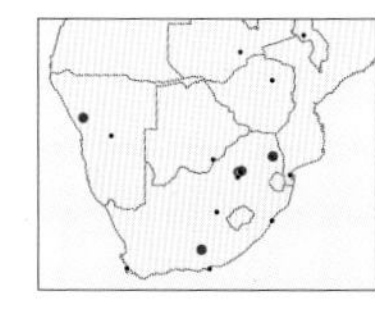

**6 *Scrobipalpa ergasima*** Streaked Twirler

Forewing 5–6mm. Slender, body and forewings greyish-white with several distinct ochreous brown longitudinal streaks on forewings; hindwings grey. Caterpillar olive grey with black head. Twenty-four species in genus in region. **Biology:** Caterpillars feed on various Solanaceae, including cultivated species. Multivoltine, adults throughout the year. **Habitat & distribution:** Various habitats where Solanaceae grow; Eastern Cape, South Africa, north to southern Palaearctic and Asia.

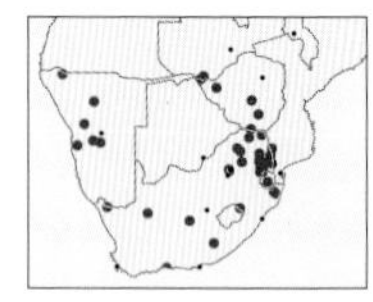

**7 *Deltophora typica*** Diamond Spot Twirler

Forewing 5mm. Body cream, abdomen tip darker brown; forewings cream with light brown and darker mottling, especially along leading margin, and large darker patches merging into diamond shapes along centre of back when at rest; hindwings cream, strongly incised and pointed distally, edged with long, pale setae. Caterpillar with brown head capsule, strongly inflated green thorax and greenish-brown abdomen. Two species in genus in region. **Biology:** Caterpillars feed on *Securinega virosa* (Phyllanthaceae), probably other plants too. Multivoltine, adults throughout the year. **Habitat & distribution:** Various habitats; Western Cape, South Africa, north to tropical Africa.

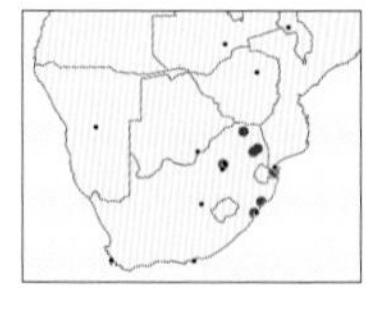

**8 *Aristotelia comis*** Aristotle Twirler

Forewing 6–8mm. Body and abdomen straw-coloured; forewings dark brown with bands of bright white scales, black spots along costa; hindwings uniformly covered with smoke grey scales. Caterpillar green with crenulated longitudinal white lines, distinctive white and black anal claspers, head pink. Seven species in genus in region. **Biology:** Caterpillars feed on *Phyllanthus nummulariifolius* (Phyllanthaceae). Bivoltine, adults November–April. **Habitat & distribution:** Forest and bushveld; KwaZulu-Natal north to Limpopo, South Africa, and Mozambique.

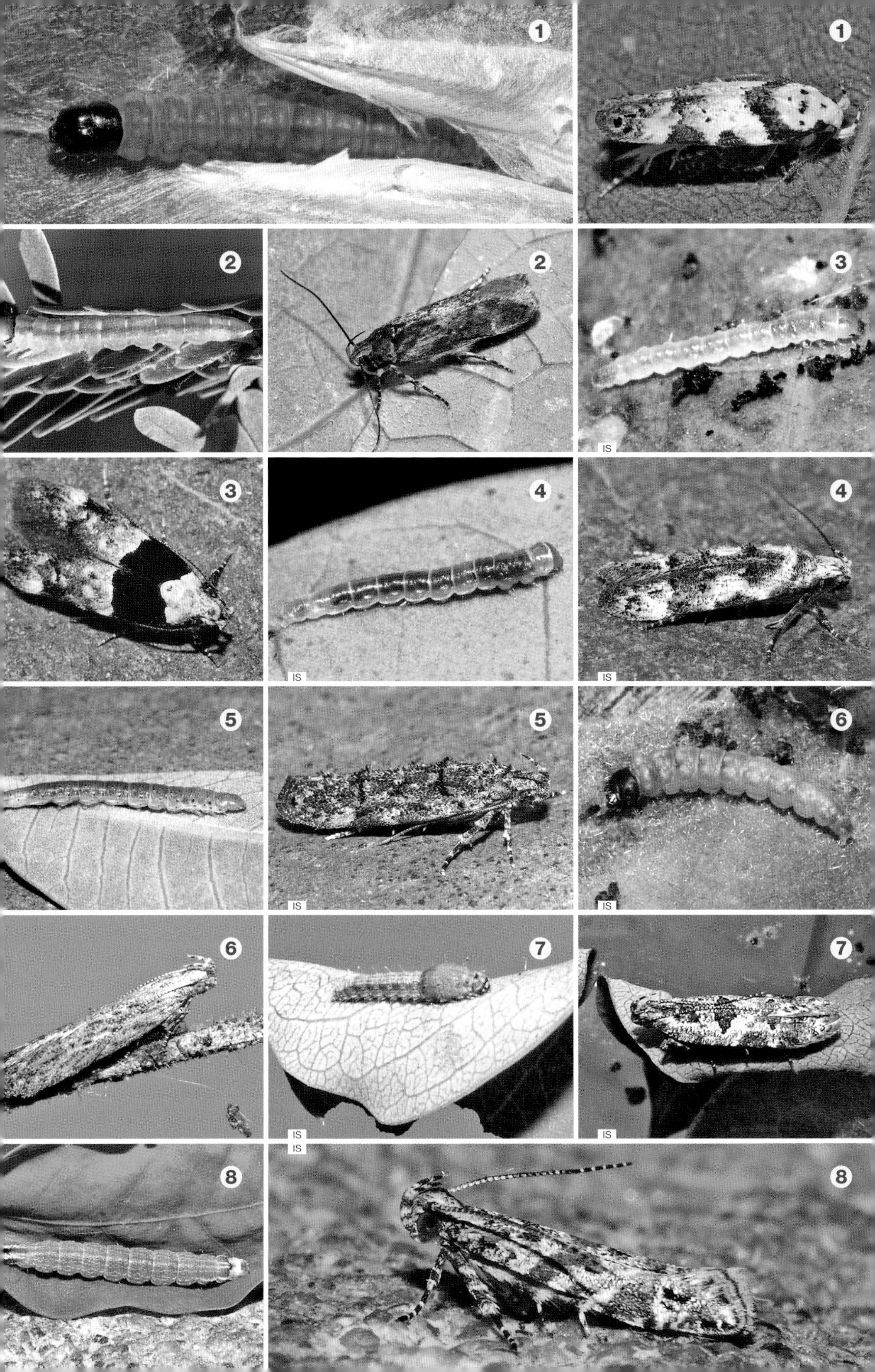

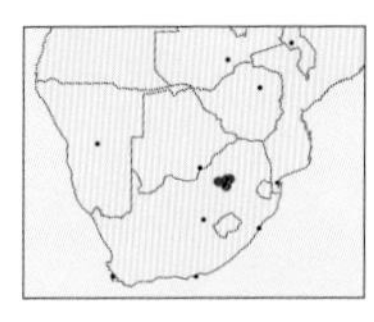

### 1 ***Dichomeris torrefacta*** Two-tone Orange Dichomeris

Forewing 6–8mm. Narrowly triangular at rest, body pale orange; forewings plain yellowish-orange with darker speckles, slightly darker terminal area, orange cilia; hindwings grey. Caterpillar green, becoming red anteriorly, head capsule brown. *D. xanthophylla* similar, but with terminal area dark brown. More than 70 diverse species in genus in region, which have distinctive long snout with protruding palps. **Biology:** Caterpillars feed on Transvaal Milkplum (*Englerophytum magalismontanum*). Multivoltine, adults September–March. **Habitat & distribution:** Rocky ridges where host plant grows; Gauteng, South Africa.

### 2 ***Dichomeris marmorata*** V-Dichomeris

Forewing 7–10mm. Bullet-shaped when at rest, with strongly upturned palps; forewings white, mottled with various shades of brown, conspicuous black bar in basal area that forms a 'V' when at rest diagnostic. Caterpillars pale green with small black dot around setal base, and black head capsule. **Biology:** Caterpillars feed on Limpopo Fever-berry (*Croton madandensis*). Multivoltine, adults throughout the year. **Habitat & distribution:** Bushveld; Gauteng, South Africa, north to Zimbabwe.

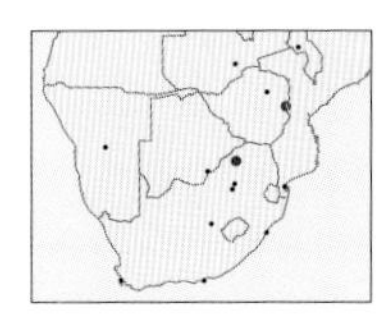

### 3 ***Dichomeris coenulenta*** Four-patch Dichomeris

Forewing 10mm. Bullet-shaped when at rest with wings held horizontal; palps strongly upturned, forewings light brown with black speckling and 4 transverse rows of black patches, distal margin concave; hindwings grey. Caterpillar green or yellow with black markings and orange head capsule. **Biology:** Caterpillars feed on Round-leaved Bloodwood (*Pterocarpus rotundifolia*). Probably multivoltine, adults December–April. **Habitat & distribution:** Bushveld; Limpopo, South Africa, and Zimbabwe.

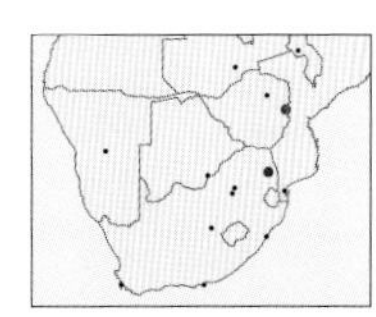

### 4 ***Dichomeris eustacta*** Mustard Dichomeris

Forewing 6mm. Arrow-shaped when at rest with wings held at an angle; forewings mustard yellow to orange, variably shaded black area along inner margin and termen with black patches; hindwings grey. Caterpillar translucent orange with faint brown lateral lines and black head capsule. **Biology:** Caterpillars feed on *Philenoptera violacea* (Fabaceae). Probably multivoltine, adults December–February. **Habitat & distribution:** Bushveld; Limpopo, South Africa, and Zimbabwe.

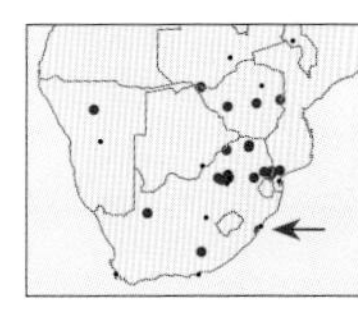

### 5 ***Dichomeris metrodes*** Cream Dichomeris

Forewing 6–7mm. Arrow-shaped when at rest with wings held at an angle; body, wings and legs cream to cinnamon-buff, sparse black markings on forewings with short streak at the base of costa; hindwings grey. Caterpillar olive with dark brown lateral lines and orange head capsule. **Biology:** Caterpillars feed on *Tephrosia rhodesica* (Fabaceae). Multivoltine, adults throughout the year. **Habitat & distribution:** Bushveld; Eastern Cape, South Africa, north to Zimbabwe, also from India.

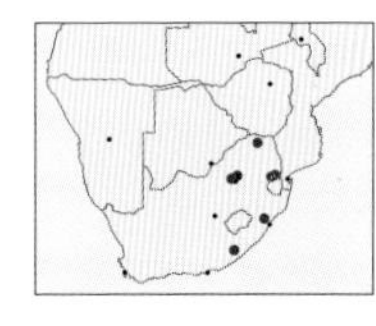

### 6 ***Dichomeris acuminatus*** Orange Dichomeris

Forewing 5–6mm. Arrow-shaped when at rest with wings held at an angle; body, wings and legs orange, sparse black markings on forewings with short streak at termen; hindwings grey. Caterpillar cream with olive longitudinal lines and segmental black dots. **Biology:** Caterpillars feed on *Dolichos angustifolius* (Fabaceae) in leaf fold. Bivoltine, adults January–July. **Habitat & distribution:** Grassland; South Africa, also recorded from Namibia, Kenya, Madagascar and Italy.

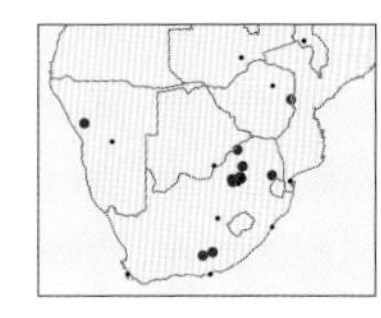

### 7 ***Polyhymno eurydoxa*** Pearly Twirler

Forewing 7–9mm. Slender, with wings wrapped around body; pearly white, orange frons and grey patch on thorax; forewings with dark band along costa terminating in an intricate pattern of black and white, iridescent scales along termen, black streak from middle of wing to apex; hindwings glossy mouse grey. Caterpillar lemon green with brown head capsule. Twenty-five similar species in genus in region, which differ in forewing terminal markings and longitudinal bands. **Biology:** Caterpillars feed on Eland's Wattle (*Elephantorrhiza elephantina*) (Fabaceae). Mutivoltine, adults diurnal, flying in sunshine, September–February. **Habitat & distribution:** Grassland and bushveld where host plants grow; Eastern Cape, South Africa, north to Namibia and Zimbabwe.

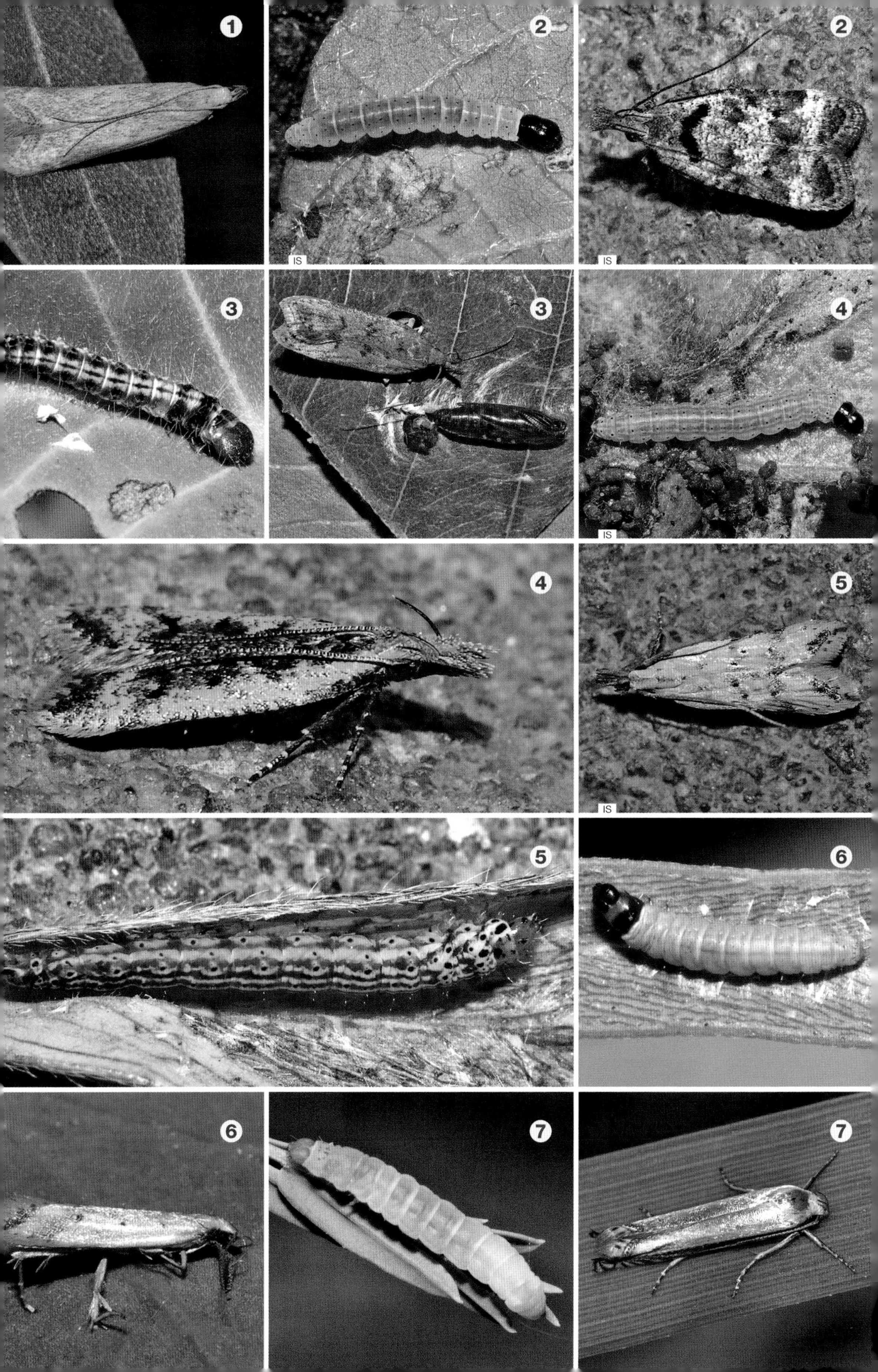
1
2
IS
2
IS
3
3
4
IS
4
5
IS
5
6
6
7
7

## FAMILY LECITHOCERIDAE Long-horned Moths

**Very diverse, but poorly known, family of small, mostly diurnal moths that occur globally. Antennae and labial palps long, antennae usually held forward and together when at rest. Wings long, dull to brightly coloured. Larvae mostly leaf litter feeders. Over 1,200 species globally, 55 in the region.**

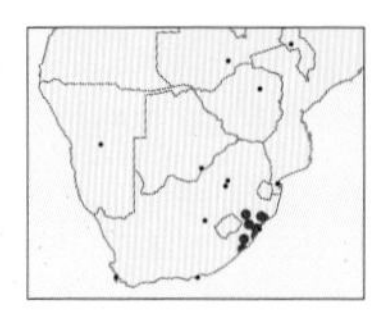

### 1 *Idiopteryx obliquella* — Cream Whirl

Forewing 7–9mm. Body and wings cream, thorax with black longitudinal stripes; forewings with complicated whirl-like brown markings; hindlegs extremely setose and often held extended out to side. **Biology:** Larval host associations unknown. Adults December–January and April. **Habitat & distribution:** Moist subtropical and Afromontane forests; Eastern Cape and KwaZulu-Natal, South Africa.

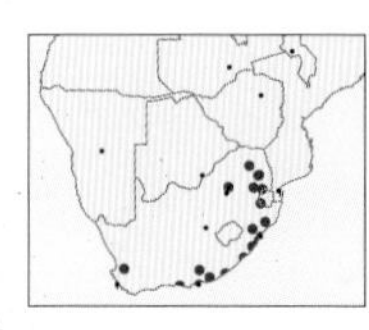

### 2 *Encolpotis xanthoria* — Peach Whirl

Forewing 8mm. Body and wings peach-coloured; forewings slender and falcate, brown mark in cell, brownish toward wing margin with long, bright yellow to orange cilia; hindwings uniform white; legs white. **Biology:** Caterpillars recorded on Sweet Thorn (*Vachellia karroo*) in region, on fruit trees in Kenya. Multivoltine, adults throughout the year. **Habitat & distribution:** Moist habitats; Western Cape, South Africa, north to Kenya.

## FAMILY OECOPHORIDAE Concealer Moths

**Diverse family of small moths of variable shape, some holding the wings roof-like against sides of the body, others flattened; broad forewings end in fringe of long setae. Conspicuous curved labial palps project forward or upward, antennae usually held backward alongside body. Colouration usually drab grey and brown, but some brightly coloured. Caterpillars of most species live concealed in webs or rolled in leaves of host plants, many consume dead plant material or fungi and some are domestic pests, or have been used as biocontrol agents. Over 3,000 described species globally, of which about 90 recorded in the region.**

### Subfamily Oecophorinae

Wings usually held flat (horizontally) over body; adults have slow, waddling gait. Caterpillars have varied feeding habits, often feeding amongst leaf litter in portable cases. Widespread, many species in Australia.

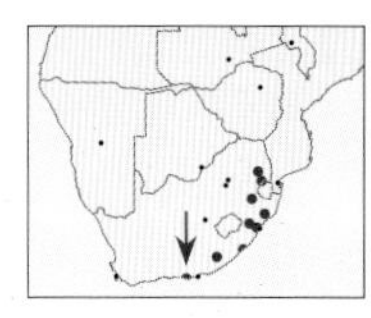

### 3 *Promalactis pedicata* — Orange-marbled Concealer

Forewing 6–7mm. Palps strongly upturned; forewings attractively patterned in marbled patches of orange and brown, these separated by thin white lines, white patch on costa variable but diagnostic; hindwings charcoal grey. **Biology:** Caterpillars feed on Rust-fungus galls (*Ravenelia macowaniana*) growing on Sweet Thorn (*Vachellia karroo*). **Habitat & distribution:** Moist forest; Eastern Cape north to Mpumalanga, South Africa.

Ubiquitous in homes across region

### 4 *Endrosis sarcitrella* — White-shouldered House Moth

Forewing 5–8mm. Head and prothorax white; forewings cream, mottled with black, fringe of cream setae distally; hindwings white with fringe of long pale setae. Caterpillars grub-like, white with brown head. **Biology:** Caterpillars feed on a wide range of products, such as grains, seeds, potatoes, dried fruits, insect specimens, wool, guano. **Habitat & distribution:** Associated with human dwellings; distributed worldwide along with dried foods.

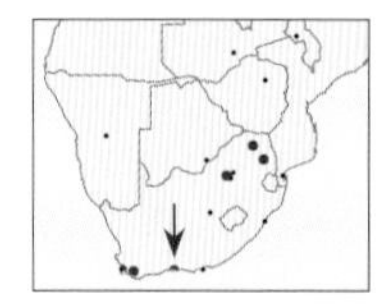

### 5 *Diocosma tricycla* — Three-ring Concealer

Forewing 9mm. Palps pink, spaced apart and strongly upturned; head, thorax and forewings butter yellow, ringed with pink, 3 distinct rings on forewings, partially obscured in some specimens by the variable deep red shading; hindwings uniform silvery white. Caterpillars deeply indented, grey with reddish shading. Seven species in genus in region. **Biology:** Caterpillars feed between 2 leaves spun together, excavating the surface, on the Grey Tree Pincushion *Leucospermum concarpodendron*, also on other Proteaceae. Adults September–April. **Habitat & distribution:** Fynbos and grassland habitats where Proteaceae grow; Western Cape north to Limpopo, South Africa.

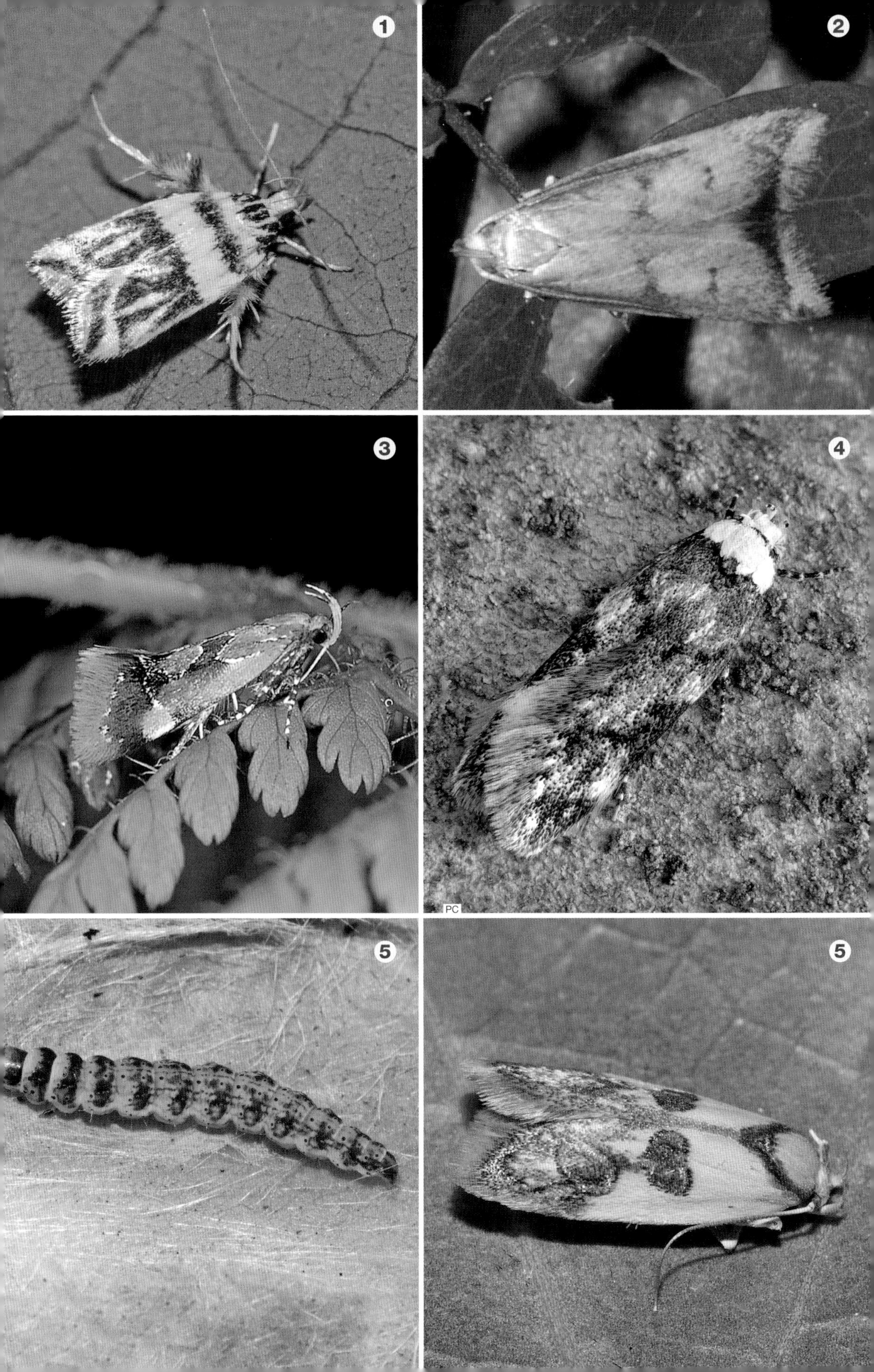
1
2
3
4
PC
5
5

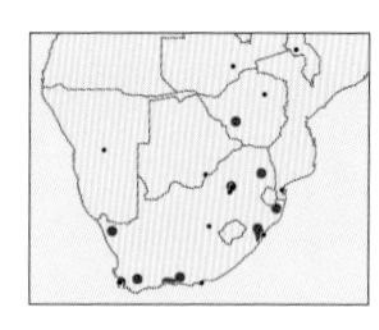

### 1 *Isocrita protophanes* — Crisscross Concealer

Forewing 5–7mm. Head, thorax and forewings cream, grey terminally, criss-crossed by many red lines, darker cross across both forewings; hindwings uniform cream. Caterpillars deeply indented, green. Six species in genus in region. **Biology:** Caterpillars feed on *Searsia* spp. Multivoltine, adults throughout the year. **Habitat & distribution:** Various habitats where *Searsia* spp. grow; Western Cape, South Africa, north to Zimbabwe.

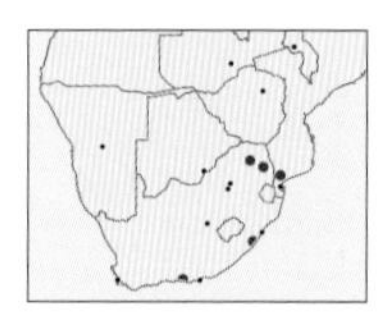

### 2 *Isocrita phlyctidopa* — Eyed Concealer

Forewing 5–7mm; head, thorax and forewings cream, criss-crossed by many red lines, head and basal part of forewings grey in some specimens, 1 or 2 dark shiny spots on forewings; hindwings uniform white. Caterpillars deeply indented, red with conspicuous white and black marks. **Biology:** Caterpillars feed on *Mystroxylon aethiopicum* (Celastraceae). Adults October–March. **Habitat & distribution:** Forest and thick bush; Western Cape north to Limpopo, South Africa.

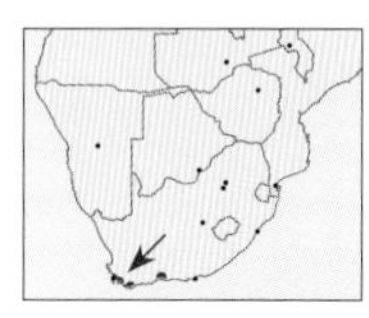

### 3 *Philobota erastis* — Redspot Concealer

Forewing 10mm. Head, thorax and forewings grey, forewings with longitudinal white streaks and basal, median and terminal red spots diagnostic, basal and median spots connected with orange band in some individuals; hindwings grey. Caterpillars grey. Five species in genus in region. **Biology:** Caterpillars feed inside cones of *Leucadendron xanthoconus*, making shelter from dried bracts. Adults December–January. **Habitat & distribution:** Moist fynbos; Western Cape, South Africa.

## FAMILY SCYTHRIDIDAE Flower Moths

**Smallish, usually dull, mostly nocturnal (but sometimes diurnal) moths with elongate wings bearing long setal fringes. Many are black, but members of the most frequently noticed genus, *Eretmocera*, often have bright spots and red or yellow abdomen and hindwings. Caterpillars are skeletonisers of leaves, buds or flowers and feed under silken webbing. These moths are primarily associated with arid and semi-arid areas. There are about 680 species globally, with over 175 recorded in the region.**

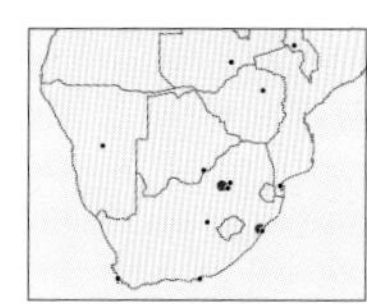

### 4 *Eretmocera laetissima* — Three-spotted Joyful Moth

Forewing 6–7mm. Thorax black with yellow shoulders, abdomen bright red with black spots at tip; antennae bearing prominent setal tufts distally; forewings black, marked with 1 proximal and 2 distal large, round yellow spots; hindwings narrow and strongly fringed with long setae, bright red basally, becoming brown at tip; hindlegs strongly spined. Thirteen species in genus in region. **Biology:** This and other *Eretmocera* spp. are often diurnal, but are also attracted to light at night. Adults perch with hindlegs strongly splayed. Caterpillars feed on *Alternanthera achyrantoides* and *Clerodendrum* spp. **Habitat & distribution:** Widespread across sub-Saharan Africa north to Ethiopia, also Madagascar and Comoros.

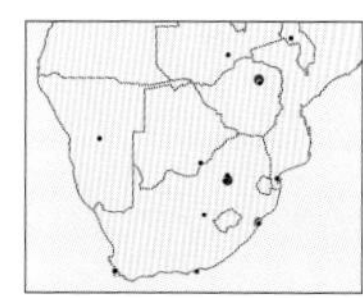

### 5 *Eretmocera fuscipennis* — Banded Joyful Moth

Forewing 6–7mm. Similar in appearance and posture to *E. laetissima* (above), but very variable; abdomen red or yellow with black base and tip; forewing markings comprise an oblique yellow proximal band and 2 distal yellow patches, or uniform brown in some specimens; hindwings brown with yellow or red fringe. **Biology:** Caterpillars feed on *Clerodendrum* spp. Adults throughout the year. **Habitat & distribution:** Open woodland; scattered records from South Africa north to Gambia and Tanzania, also Madagascar.

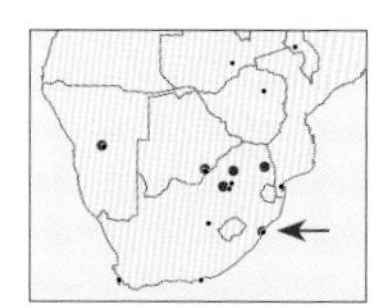

### 6 *Eretmocera scatospila* — Brown-winged Joyful Moth

Forewing 6–7mm. Red abdomen with black base and black distal ring terminating in orange tuft of setae; forewings dark brown with yellow central line and small yellow markings near apex; hindwings brown with yellow anterior and red posterior fringes of long setae. Caterpillar mottled grey-brown; white on thorax and head capsule. **Biology:** Caterpillars polyphagous on various herbaceous plants, often Amaranthaceae. In silken retreat between leaves of host plant. Adults November–May. **Habitat & distribution:** Coastal bush and woodlands; KwaZulu-Natal, South Africa, north to tropical Africa.

1
1
2
2
2
IS
3
3
4
5
6
6

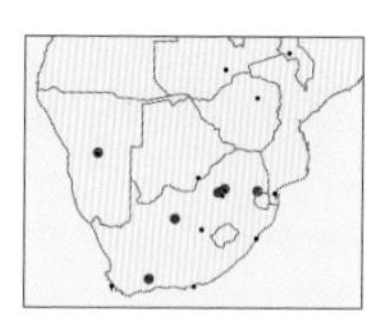

### 1 *Scythris cretiflua* — Marble Scythrid

Forewing 12–14mm. Head smooth, slender with narrow wings held roof-like against body; labial palps upturned; forewings marble white; hindwings dull white, slender with fringe of long setae. This is the most diverse genus in the family. Over 100 recorded species in the region, most are hard to identify to species without dissection to examine genitalia. **Biology:** Host associations unknown. Adults October–December and March. **Habitat & distribution:** Dry areas; Western Cape, South Africa, to Namibia.

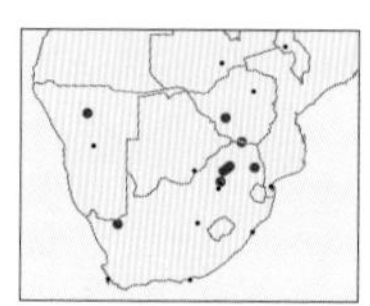

### 2 *Scythris justifica* — Two-spot Scythrid

Forewing 7mm. Head and palpi smooth, orange below and grey above; antennae grey, slender with narrow wings held roof-like against body; forewings light to dark grey with 2 conspicuous black dots; hindwings dark grey, slender with long setae. Caterpillar pinkish-grey with several crenulated white lines. **Biology:** Caterpillars feed in leaf fold of *Pupalia lappacea* (Amaranthaceae). Adults recorded in December. **Habitat & distribution:** Bushveld; Northern Cape, South Africa, north to Namibia and Zimbabwe.

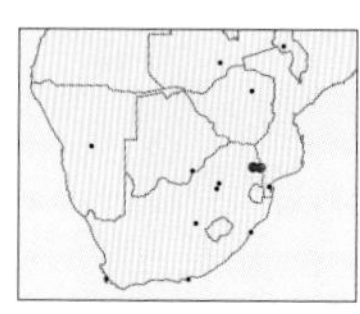

### 3 *Scythris satarensis* — Brown Scythrid

Forewing 9mm. Head smooth, head, palpi and thorax greyish-brown, neck tuft whitish; forewings pale greyish-brown with a whitish streak from base to middle of wing, faint ivory line on costa; hindwings greyish-brown. Body of caterpillar deeply indented between segments, orange with dark spots, whitish lateral lines on thoracic segments, 2 dark spots on head capsule. **Biology:** Caterpillars live in loose silken retreat between leaflets of Flaky-barked Thorn (*Vachellia exuvialis*). Adults recorded March and April. **Habitat & distribution:** Lowland bushveld; Limpopo, South Africa.

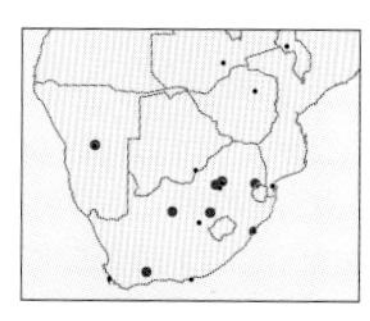

### 4 *Scythris* species — Brown Scythrids

Forewing 9mm. Head and palpi orange; forewings uniform light brown with no markings; hindwings darker brown with no markings. One of several similar species in region that need genitalia dissection for positive identification. **Biology:** Caterpillars of one species in web between leaves of *Faidherbia albida* (Fabaceae). Adults recorded April. **Habitat & distribution:** Uniform brown species occurs in arid areas of southern Africa.

## FAMILY STATHMOPODIDAE Featherfoot Moths

**Small moths, most with characteristic resting posture in which spiny hindlegs are held outstretched and at right angles to body. Wings usually narrow and strongly fringed with setae, often banded yellow or brown, but some brightly coloured. Caterpillars bore into seeds, fruits or buds and sporangia of ferns; a few are scavengers or predators of scale insects. Over 300 species globally, with 23 in the region.**

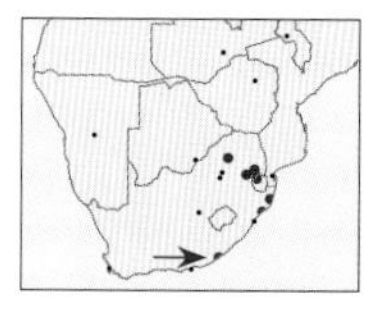

### 5 *Stathmopoda crassella* — Crossed Featherfoot

Forewing 5mm. Orange with dark crossbar at back of thorax and curved brown bars of varying intensity across forewings; hind margin of forewings and hindwings very narrow and strongly fringed with setae; middle pair of legs long, strongly spined and held out extended to sides. Seventeen species in genus in region, several in the *S. crassella* species complex very similar and may represent more than one species. **Biology:** Host associations unknown. Adults September–November. **Habitat & distribution:** Forest and bushveld; Eastern Cape to Limpopo, South Africa.

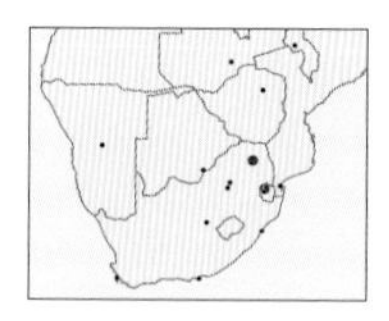

### 6 *Stathmopoda auriferella* — Finger Featherfoot

Forewing 4mm. Body and wings yellow; forewings with wide central brown crossband and brown tips. **Biology:** Caterpillars polyphagous on trees. **Habitat & distribution:** Lowland forest and bush; KwaZulu-Natal, South Africa, north to tropical Africa, also Seychelles and widely distributed across Asia.

## FAMILY XYLORYCTIDAE Timber Moths

**Very diverse family of small to medium-sized moths with erect labial palps, and hindwings broader than forewings. Found in South America, Africa, Asia, with most species in Australia. Larvae often burrow into wood, some having unusual habit of securing leaves to the entrance of burrow, eating them as they dry out. Others feed on bark or lichens. About 1,255 species globally, 16 in the region.**

1
2
IS
2
3
IS
3
4
4
IS
5
6

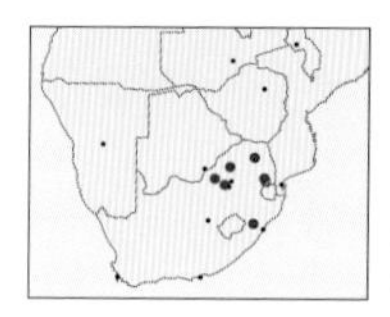

### 1 *Eupetochira xystopala* Grassland Moth

Forewing 13mm. Elongate and cylindrical at rest, frons conspicuous and porrect with long setae, thin pointed palps protruding at an angle diagnostic; forewings white with single golden-brown to almost black longitudinal stripe. Can be confused with several very similar-looking unrelated species in Crambinae (*Caffrocrambus* spp.), which do not have the protruding palps from frons, and Gelechiidae (*Holaxyra* spp.), which are much smaller and have the palps spaced apart and visible from base. Two species in genus in region. **Biology:** Caterpillar host associations unknown. Adults usually found resting on grass stems together with similar-looking Crambinae. **Habitat & distribution:** Grassland; KwaZulu-Natal north to Limpopo, South Africa.

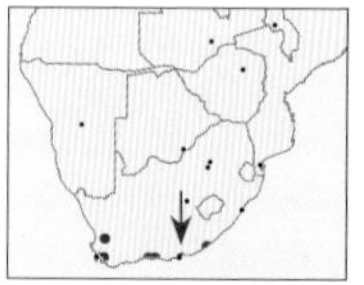

### 2 *Glycynympha roseocostella* Hourglass Moth

Forewing 8mm. Palpi upturned over head; body and wings creamy white, variably irrorated with brown and grey scales, dark shading on costa and oblique line to base of forewing diagnostic; hindwings plain white. One species in genus in region. **Biology:** Caterpillar host associations unknown. Adults throughout the year. **Habitat & distribution:** Fynbos and coastal bush; Western Cape and Eastern Cape, South Africa.

## FAMILY THYRIDIDAE Picture-winged Leaf Moths

**Related to butterflies; primarily tropical family of small to large moths (forewing length 5–45mm), most with stout body and relatively short wings that sometimes have irregular margins and clear windows. Adults nocturnal, some diurnal or crepuscular, most cryptic, resembling dead leaves, but some exhibiting bright warning colouration. At rest, body raised and wings held outspread. Caterpillars usually concealed within rolled leaves or burrow into plant stems, some induce galls. About 795 species globally, with 146 in the region.**

## Subfamily Charideinae

A small group restricted to the Afrotropics. They are active by day, have bright metallic colouration and narrow wings, and superficially resemble some Zygaeninae (p.58).

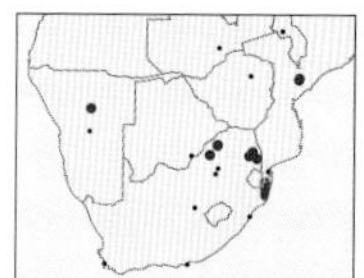

### 3 *Arniocera erythropyga* Fire Grid Burnet

Forewing 13–16mm. Head red, thorax dark blue, abdomen red with black dorsal patch on each segment; forewings dark metallic blue, crossed by 3 wide red stripes; hindwings banded red and black with broad red base, the red replaced by yellow in some specimens. Caterpillar yellow, short white setae, setal bases black, large dorsal spot on thorax and black head. Eight species in genus in region; similar *A. zambesina* differs in the hindwings being uniform purplish-blue; *A. cyanoxantha* has the postmedian and terminal red stripes broken. **Biology:** Caterpillars feed in leaf fold of Marula trees (*Sclerocarya birrea*). Probably univoltine, adults September–January. **Habitat & distribution:** Bushveld; KwaZulu-Natal, South Africa, north to tropical Africa.

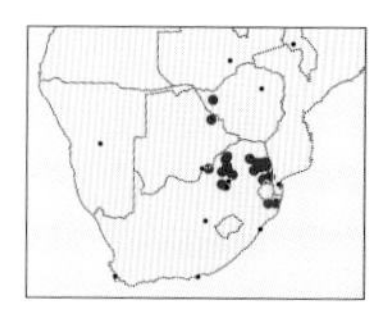

### 4 *Arniocera auriguttata* Gold-spotted Burnet

Forewing 12–16mm. Red head and leg segments, metallic blue body and forewings, and variable number of large round golden to white spots on forewings; hindwings dark blue. **Biology:** Caterpillar host associations unknown. Adults September–December, peaking November, often abundant, gathering in large numbers on mud and flowers. **Habitat & distribution:** Bushveld; KwaZulu-Natal north to tropical Africa.

## Subfamily Siculodinae

Possess a structure in a vein at the base of the wing that may be used for hearing.

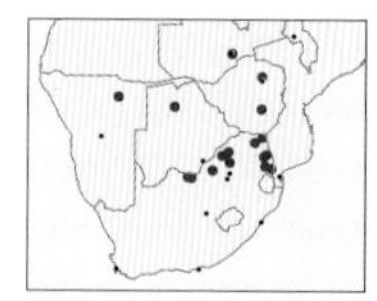

### 5 *Cecidothyris pexa* Clouded Lattice

Forewing 11–14mm. Wings held outspread at rest; body and wings pale apricot; forewings with white, slightly concave leading edge; basal part of hindwings with lattice-work patch of brown lines, which is strongly prevalent on underside of both wings. Caterpillar white, grub-like with pink head capsule. Two species in genus in region, form *guttulata* has underside pattern repeated on upperside. **Biology:** Caterpillars form galls on stems of cluster-leaf trees (*Terminalia* spp.). Probably univoltine, adults December–May. **Habitat & distribution:** Sandy bushveld where host plants grow; North West, north to tropical Africa.

1
2
2
3
3
4
AHa
4
5
IS
5
5

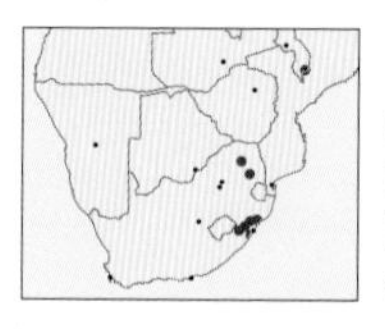

### 1 *Hypolamprus gangaba* — Gangaba Lattice

Forewing 8–9mm. Wings held outstretched; body and wings silvery white, covered in fine lattice of broken brown lines, a more distinct line just before pointed wing tip, white dots on costa. Two species in genus in region. **Biology:** Caterpillar host associations unknown. Adults September–April. **Habitat & distribution:** Moist forest; KwaZulu-Natal, South Africa, north to tropical Africa.

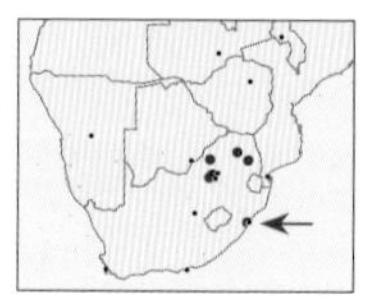

### 2 *Kalenga ansorgei* — Ansorg's Lattice

Forewing 7–9mm. Wings held outstretched; body and wings pale reddish-brown covered in dense reticulated pattern of fine brown lines, a bolder line midway along each wing; darker narrow triangle near apex of forewing, cell spot present in some specimens. Caterpillar pale green with reddish head and black dorsal bar across first thoracic segment. Three species in genus in region. **Biology:** Caterpillars feed in rolled leaves of Forest Bushwillow (*Combretum kraussii*). Bivoltine, adults September–January. **Habitat & distribution:** Forest and riverine bush; KwaZulu-Natal, South Africa, north to tropical Africa.

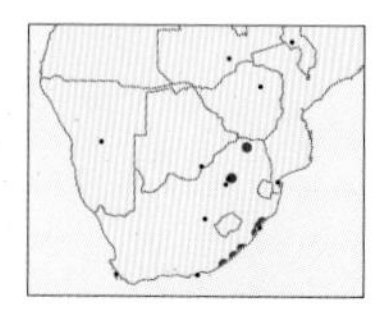

### 3 *Kuja squamigera* — Blotched Lattice

Forewing 11–12mm. Body and wings light brown with brown blotches and fine lines, an uneven dark brown band crosses all wings; forewings held at right angles to body, leading edge slightly hooked at tip. One regional species in genus. **Biology:** Caterpillar host associations unknown. Adults throughout the year. **Habitat & distribution:** Coastal forest and riverine bush; Eastern Cape, South Africa, north to Zambia.

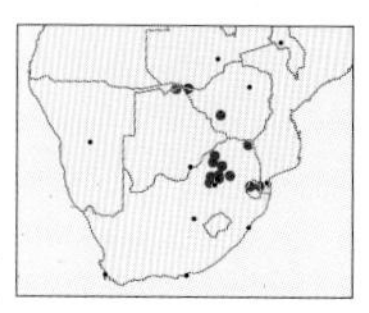

### 4 *Rhodoneura flavicilia* — Pink-edged Lattice

Forewing 8–10mm. Yellowish-brown with lattice of fine broken orange lines over wings, which have pink margins. Caterpillar plump and green with brown head capsule and black dorsal plate on first thoracic segment. Six species in genus in region. **Biology:** Caterpillars feed in leafroll on Red Bushwillow (*Combretum apiculatum*). Multivoltine, adults throughout the year. **Habitat & distribution:** Bushveld; Eswatini, South Africa, north to Tanzania and Zambia.

## Subfamily Thyridinae

This subfamily has a pantropical distribution.

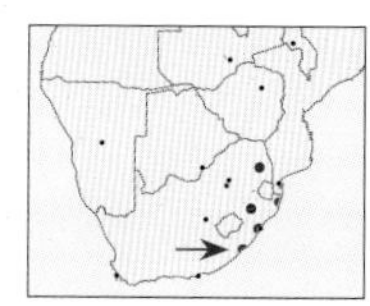

### 5 *Dysodia crassa* — Stout Dysodia

Forewing 8–9mm. Body unusually stout; wings falcate, covered with lattice of fine markings, light brown basally, crossed by thick dark brown band across both wings, distal margin sinuous, round hyaline cell spot on hindwings. Similar *D. subsignata* has toothed hyaline cell spot on hindwings; *D. intermedia* has thinner band across wings. Caterpillar yellow with transverse rows of black spots and bright red head capsule and prothorax. Fifteen species in genus in region. **Biology:** Caterpillars feed on Wild Yam (*Dioscorea cotinifolia*). Probably bivoltine, adults September–February. **Habitat & distribution:** Moist forests; Eastern Cape, South Africa, north to Zambia.

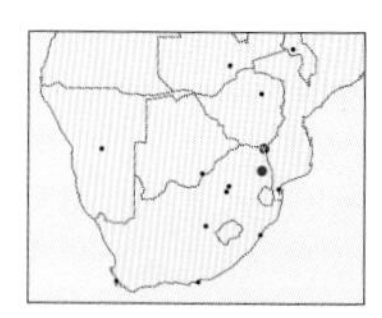

### 6 *Dysodia fenestratella* — Windowed Dysodia

Forewing 8–9mm. Forewings often held partially rolled; both wings falcate, mottled in shades of dark brown with a diagnostic large transparent triangular window near base of hindwings. Caterpillar yellow-green with sparse covering of fine setae, red 'collar' between black head capsule and black prothorax. **Biology:** Caterpillars feed on Shepherd's Tree (*Boscia albitrunca*) and other Capparaceae. Probably bivoltine, adults November–May. **Habitat & distribution:** Bushveld; Limpopo, South Africa, north to Kenya.

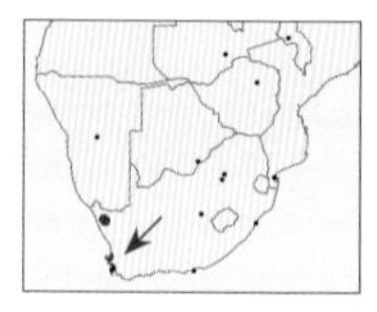

### 7 *Dysodia antennata* — Cape Dysodia

Forewing 9–10mm. Forewings not held partially rolled; body and wings a deep coppery brown, wings rounded, lattice markings regular with a small round hyaline cell spot on all wings, 2 dark spots on abdomen. Similar *D. binoculata* has distinct black postmedian line across all wings and larger hyaline patch. **Biology:** Caterpillar host associations unknown. Possibly univoltine, adults August–September. **Habitat & distribution:** West coast seashore vegetation and succulent karoo; Western Cape and Northern Cape, South Africa.

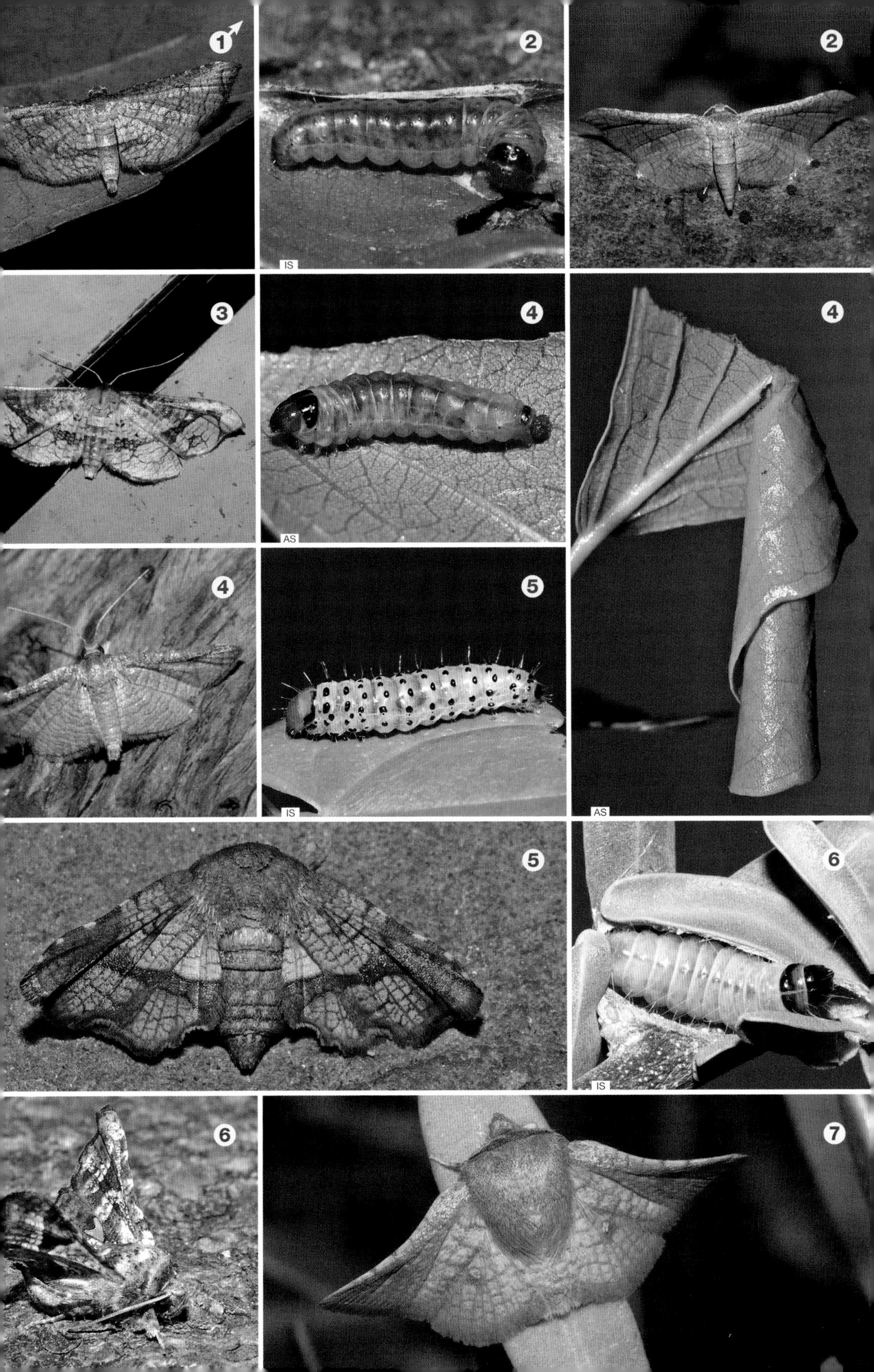
1♂
2
IS
2
3
4
AS
4
4
5
IS
AS
5
6
IS
6
7

## Subfamily Striglininae

All members have a pantropical distribution.

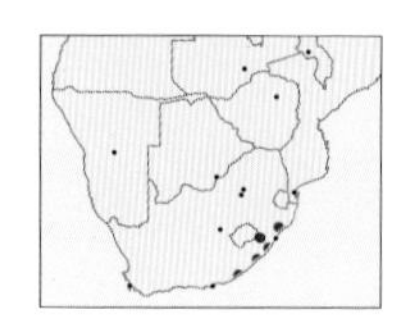

### 1 *Banisia myrsusalis* — Drab Lattice

Forewing 10mm. Wings held outstretched, straight leading edge ending in sharp tip; body reddish-brown with darker brown crossbands; forewings variable reddish-brown to cream, densely marked with fine wavy brown lines, cluster of small transparent windows centrally in some specimens. Caterpillar not described. Four species in genus in region. **Biology:** Caterpillars feed on Sapodilla (*Manilkara zapota*) and Crocodile Bark (*Terminalia tomentosa*) in region, Indian Butter Tree (*Madhuca latifolia*) elsewhere. **Habitat & distribution:** Moist forests; Eastern Cape, South Africa, north to tropical Africa as well as Americas, Australia and South Asia.

# FAMILY PTEROPHORIDAE Plume Moths

**Easily recognisable family of small, delicate moths with elongate, prominently spined legs and characteristic unusual slender wings, which are mostly divided into separate 'plumes' and held at right angles to the body, or projecting forward, so resting moth resembles either a 'T' or a 'Y'. Most are nocturnal and dull-coloured, mottled in brown or grey. Caterpillars are elongate and setose, feeding on foliage or boring into stems, mostly of daisies (Asteraceae). More than 1,000 species described globally, with 125 described in the region, many remain undescribed.**

## Subfamily Agdistinae

Wings not divided into filaments, with markings on the forewings limited to a short row of dots along the tip of the leading (costal) margin. Only 1 genus (*Agdistis*).

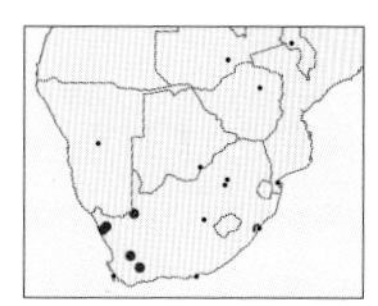

### 2 *Agdistis* species — Plumeless Y-Plumes

Forewing 10–15mm. Unusual in that the long, narrow wings are not divided into plumes, but rolled around one another at rest and held forward, giving the moth a 'Y'-shaped outline; generally brown or grey with darker spots on wings. Caterpillar typically green with silver-white speckling and spiky uniform green shield over head and thorax. **Biology:** Habits poorly known, caterpillars of some species feed on Hondebossie (*Exomis microphylla*). **Habitat & distribution:** Widespread across a wide range of vegetation types. Fifty similar species known from the region, many ranging into Central and East Africa or beyond.

## Subfamily Pterophorinae

Both fore- and hindwings cleft. The third lobe of the hindwing has 2 wing veins. The markings on the costal region of the forewings are more pronounced than in Agdistinae.

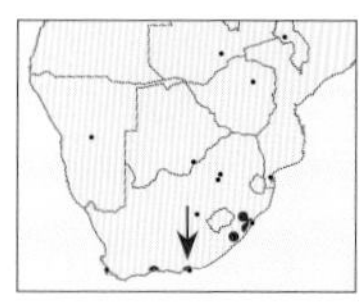

### 3 *Vietteilus vigens* — Mustard T-Plume Moth

Forewing 8–11mm. At rest, wings held at right angles to body, creating 'T'-shaped outline; mustard to brown, sometimes with 2 distinctive black spots on costa and small white apical markings; long, delicate legs bearing long spines; forewings divided only one-third from termen into 2 plumes; hindwings divided into 3 plumes. Caterpillar green, hairy, with 2 conspicuous white dorsal lines. One species in genus in region. **Biology:** Caterpillars feed on various Rosaceae, including Wild Bramble (*Rubus rigidus*) and *Cliffortia odorata*. **Habitat & distribution:** Humid coastal areas in region; widespead across sub-Saharan Africa and La Réunion; also widespread across Asia, Pacific Ocean islands and Australia, recorded from New Zealand.

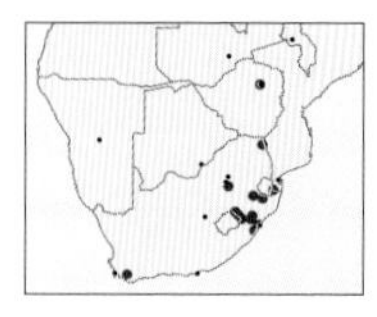

### 4 *Stenodacma wahlbergi* — Orange T-Plume Moth

Forewing 6–7mm. At rest, wings held extended at right angles to body; body and wings orange-brown with white longitudinal lines down length of abdomen; legs longer than body with long, thin projecting spines; forewings divided about midpoint into 2 plumes; hindwings divided closer to base into 3 plumes. Caterpillar grey with pale longitudinal lines and rosettes of white spines. **Biology:** Caterpillars recorded on Sweet Potato (*Ipomoea batatas*), Creeping Grape (*Vitis flexuosa*), *Gymnanthemum appendiculata* and Creeping Woodsorrel (*Oxalis corniculata*). **Habitat & distribution:** Common in rank vegetation in weedy fields and long grass under bushveld trees; widespread across sub-Saharan Africa, Madagascar, Indian Ocean islands, Asia and Middle East.

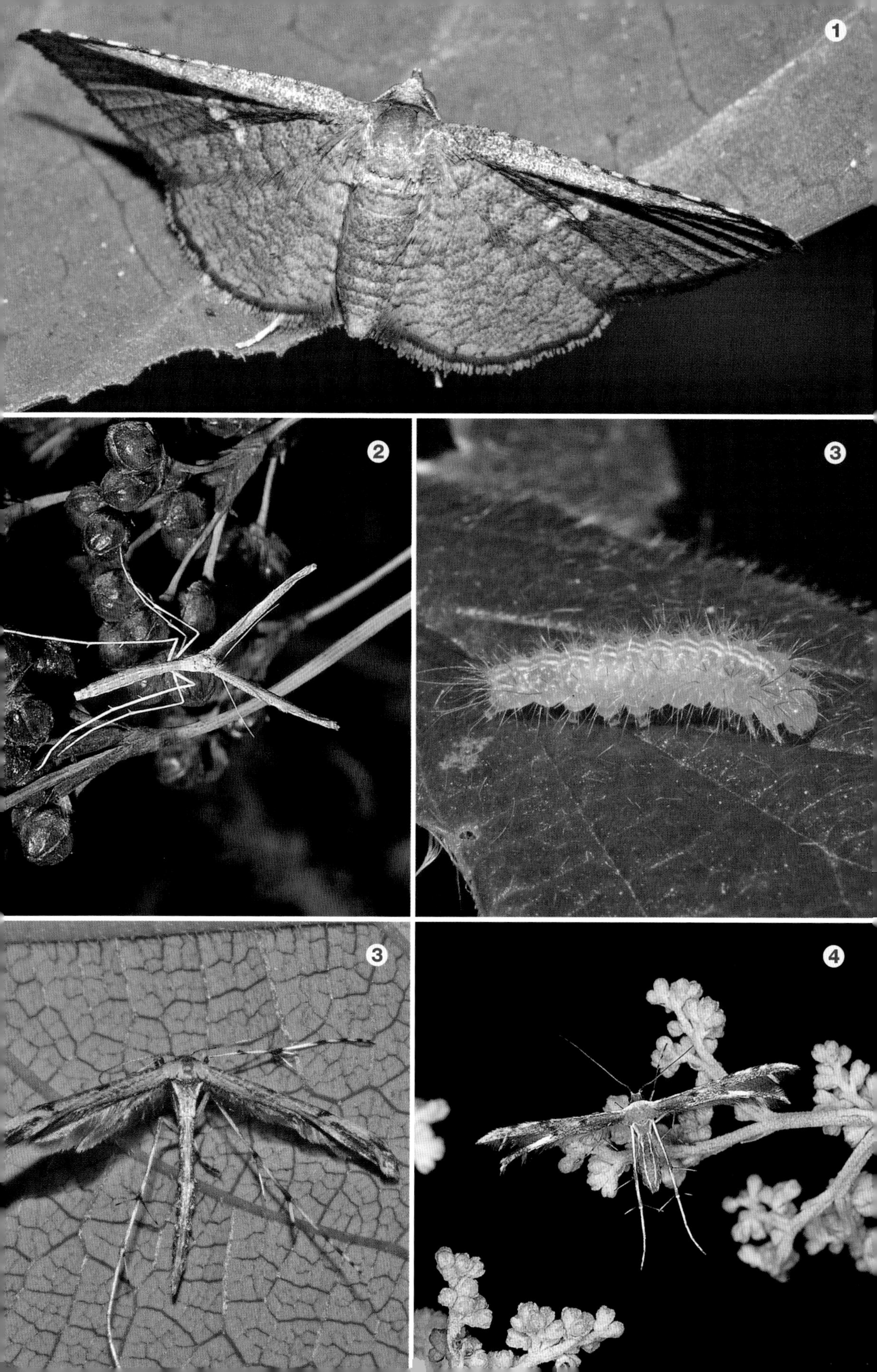
1
2
3
3
4

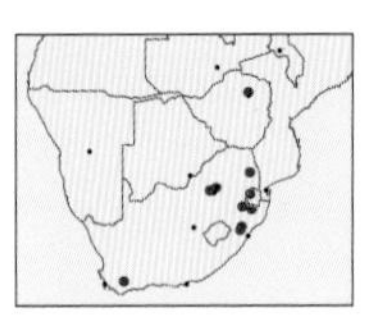

### 1 *Megalorhipida leucodactylus* — Chevron T-Plume Moth

Forewing 8–9mm. At rest, wings held extended at right angles to body; body and wings orange-brown with chevron-like white and brown lines on first 3 abdominal segments; legs longer than body with long, thin projecting spines; forewings divided about midpoint into 2 plumes; hindwings divided closer to base into 3 plumes with black central areas. Caterpillar cream with pale longitudinal lines and long brown spines. One species in genus in region. **Biology:** Caterpillars polyphagous on many plant families. **Habitat & distribution:** Dense grass in open woodland and bushveld; widespread across sub-Saharan Africa, Madagascar and Indian Ocean islands.

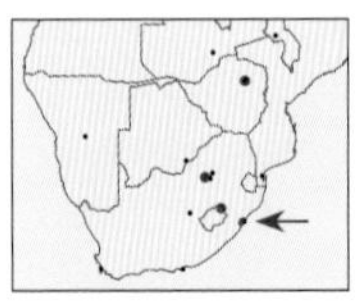

### 2 *Hellinsia madecasseus* — Grey T-Plume Moth

Forewing 8–9mm. At rest, wings held extended at right angles to body; body and wings grey, variably spotted with black dots and spots, dorsal row of black dots on abdomen; legs longer than body with long, thin projecting spines; forewings divided one-third from apex into 2 plumes; hindwings divided closer to base into 3 plumes with black central areas, spotted with cream. Caterpillar green with pale longitudinal lines and dense spines. Twenty species in genus in region. **Biology:** Caterpillars feed on the leaves of various daisies (Asteraceae). **Habitat & distribution:** Relatively moist areas; KwaZulu-Natal, South Africa, north across sub-Saharan Africa, Madagascar and Indian Ocean islands.

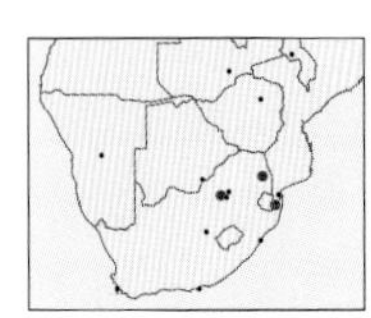

### 3 *Exelastis phlyctaenias* — Cream T-Plume Moth

Forewing 8–9mm. Similar to *H. madecasseus* (above), but ground colour cream and lacks row of black dots on abdomen; sprinkling of black dots and spots very variable, some specimens appear almost entirely black. Caterpillar green with dense short spines and sparse longer setae. Four species in genus in region. **Biology:** Caterpillars feed on the leaves of various Fabaceae. **Habitat & distribution:** Relatively moist areas; KwaZulu-Natal, South Africa, north across sub-Saharan Africa, Madagascar and Indian Ocean islands.

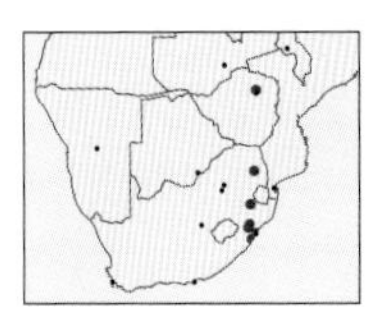

### 4 *Titanoptilus stenodactylus* — Black T-Plume Moth

Forewing 8mm. At rest, wings held extended at right angles to body; thorax and wings black with brown mottling in some specimens; abdomen basally cream becoming variably darker toward apex; legs white and black, longer than body with long thin projecting spines. Caterpillar pale green with dark setal bases, setae white, short, 2 rows of pink dorsal spots. Two species in genus in region. **Biology:** Caterpillars in silken nests feeding on the leaves of Hairy Grape Bush (*Cyphostemma woodii*, Vitaceae). **Habitat & distribution:** Relatively moist areas; KwaZulu-Natal, South Africa, north to tropical Africa, Madagascar and Indian Ocean islands.

## FAMILY ALUCITIDAE Many-plumed Moths

**Family easily recognised by unusual form of wings, the membrane being reduced such that each wing is split into about 6 separate, feather-like plumes, which are generally held open like a delicate miniature fan. Hairy caterpillars bore into buds, flowers or shoots and may cause galls. About 120 species globally, 28 in the region.**

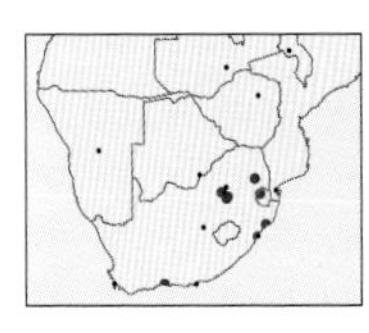

### 5 *Alucita spicifera* — Corn-ear Plumed Moth

Forewing 6–8mm. Dark stripe across thorax, and wing plumes attractively banded in alternating brown and white. Caterpillar grub-like, lives in galls or unripe seed pods. Twenty species in genus in region. **Biology:** Adults diurnal, feeding on nectar. Host plants not identified. Common on annual plants in weedy fields. **Habitat & distribution:** Most habitats, including transformed habitats; South Africa, Tanzania, Malawi and Cameroon.

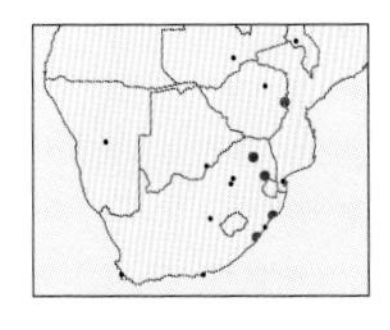

### 6 *Alucita tesserata* — Chequered Many-plumed Moth

Forewing 7–9mm. Body and wings finely banded in yellow and white stripes and dotted with rows of black spots. **Biology:** Host associations unknown. **Habitat & distribution:** South Africa, Malawi and Zimbabwe.

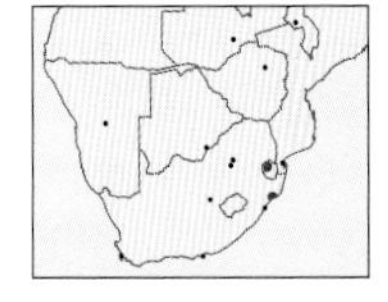

### 7 *Alucita phanerarcha* — Arched Plume Moth

Forewing 15mm. Wings often held separated when at rest; forewings banded brown, black then white distally; hindwings black basally and white distally; male with 2 orange androconial patches on hindwing. **Biology:** Host associations unknown, likely mimics bird droppings. **Habitat & distribution:** Subtropical forests; South Africa, Eswatini, Malawi and DRC.

## FAMILY CARPOSINIDAE Fruitworm Moths

**Small to medium-sized moths characterised by prominent elongate and upcurved labial palps with elongate third segment. Most fairly dull, camouflaged greenish or greyish. Caterpillars typically bore into fruits, seeds or trees, but some are leaf miners. About 280 species, mostly from Australia and the South Pacific, 13 in the region.**

Scattered in W & E Cape where host plant occurs

### 1 *Carposina autologa* — Hakea Seed Moth

Forewing 5mm. Forewings white speckled with brown, with black front margin; hindwings white. **Biology:** Adults lay eggs on Silky Hakea (*Hakea sericea*) fruits, and caterpillars excavate a hole where 2 halves of fruit meet and feed on seeds within; after about 135 days they emerge to pupate in the soil. Introduced from Australia in 1970 as biocontrol agent for *Hakea sericea* and *H. nodosa*; can destroy up to 80% of seeds at some sites. **Habitat & distribution:** Mountain fynbos; Western and Eastern Cape provinces, up to Makhanda, South Africa.

## FAMILY HYBLAEIDAE Sharp Snouts

**Minor family of small moths distinguished by exceptionally small head and sharply pointed labial palps, which are long and project horizontally forward like a beak; thorax short and wings rather rectangular. Only about 20 species globally, with 3 similar *Hyblaea* species in the region.**

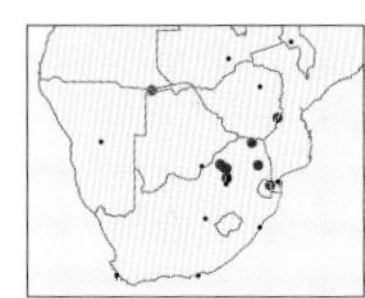

### 2 *Hyblaea puera* — Mottled Sharp Snout

Forewing 9–13mm. Variable; thorax grey to brick red, abdomen banded yellow and brown; forewings variably mottled grey, cream and brown, sometimes speckled; hindwings brown with distinctive yellow or orange or orange-yellow central patch, relatively narrow compared to similar species. Caterpillar with black or brown head capsule, longitudinal black and red bands and narrow white stripes. **Biology:** Caterpillars polyphagous on young leaves of trees and shrubs; cutting flap from edge of leaf, folding it over and fastening it with silk, feeding from within retreat. Adults October–February. **Habitat & distribution:** Bushveld and forest; Eswatini north to tropical Africa, Indian Ocean islands; also widespread in Asia.

## FAMILY PYRALIDAE Snout Moths

**Diverse family of small to large moths with large, straight or curved, forward-projecting labial palps, which form the 'snout'. Some hold wings flat at rest, but others appear more cylindrical, with the wings held tightly against the abdomen. Antennae filiform and held facing backward against the body. Caterpillars mostly concealed feeders, often forming silken tubes and tunnels and consuming a diverse range of foodstuffs, including living leaves, stems and fruits, but also seeds and grains and wax in beehives. More than 6,000 species described worldwide, with 630 in the region, many still undescribed. Five subfamilies in region, dominated by the Phycitinae with 403 species, most of which are difficult to identify, followed by Pyralinae (175 species), Galleriinae (25 species), Epipaschiinae (25 species) and Endotrichinae (4 species).**

### Subfamily Endotrichinae

Small to medium-sized pyralids, with a well-developed proboscis and very reduced maxillary palps. Considered by some as better placed as a tribe within the subfamily Pyralinae.

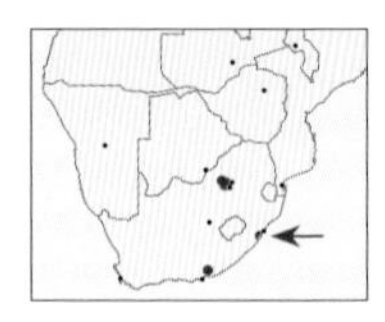

### 3 *Endotricha consobrinalis* — Whiteline Triangle

Forewing 8–9mm. At rest, appears triangular in outline with abdomen curled upward; body light brown with dark brown crossbar on abdomen; forewings brown at base, becoming darker toward tip, basal and subterminal line black and white. Three species in genus in region. **Biology:** Caterpillar host associations unknown. Adults October–April. **Habitat & distribution:** Riverine bush; Eastern Cape north to Gauteng, South Africa.

1
1
TG
2
2
IS
2
3

## Subfamily Epipaschiinae

Largely tropical in distribution. Bear a superficial resemblance to Noctuidae. Scale tufts present on surface of forewings and on membrane separating abdomen and thorax. Males have scale-free window on forewings. Some species have prominent labial palps that extend backward over the head.

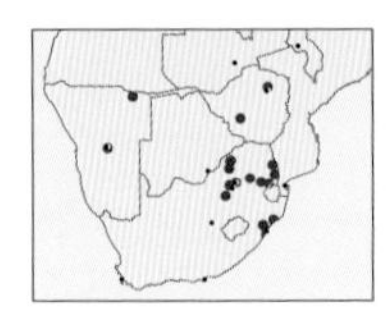

### 1 *Epilepia melanobrunnea* — Two-tone Brown Pyrale

Forewing 7–10mm. Broadly triangular with abdomen upturned at rest; thorax and basal half of forewings dark chocolate brown, distal half paler, varying from medium to very pale brown, with darker distal crossbands and dark spot in corner; hindwings pinkish-white with incomplete subterminal line. Caterpillar undescribed. Five species in genus in region. **Biology:** Caterpillar host associations unknown. Adults November–March. **Habitat & distribution:** Forest and bushveld; KwaZulu-Natal, South Africa, north to Zimbabwe and Namibia.

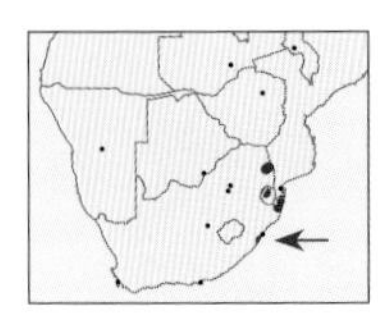

### 2 *Termioptycha fasciculata* — Soot-speckled Pyrale

Forewing 9–13mm. Body and wings variable, heavily marked with bands of black and brown, especially around head and with bands across thorax and abdomen, jagged postmedian line diagnostic, female much darker; forewings held flat against body, triangular in outline. Caterpillar variable, green or grey, sometimes with pink head capsule, with or without longitudinal white or black bands. One species in genus in region. **Biology:** Caterpillars feed on Duiker Berry (*Sclerocroton integerrimus*), Velvet Corkwood (*Commiphora mollis*) and Firethorn Corkwood (*C. pyracanthoides*). Bivoltine, adults October–December and March–June. **Habitat & distribution:** Lowland forest and bushveld; KwaZulu-Natal north to Limpopo, South Africa.

## Subfamily Galleriinae

Adults lack external simple eyes (ocelli). Caterpillars feed on dried plant material, beeswax, sedges and sugar cane, and some predate larvae of wood-boring beetles.

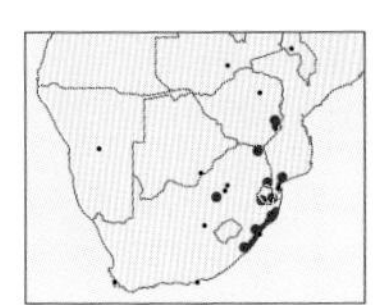

### 3 *Eldana saccharina* — African Sugarcane Stalkborer

Forewing 17mm. Forewings light brown with 2 small dark spots centrally; hindwings whitish with brown longitudinal veins. Caterpillar light brown with small dark spots and dark head capsule. **Biology:** Caterpillars feed mostly on sugar cane, but also maize, sorghum, rice and other grasses and sedges. Eggs laid in batches of 50–100 on dried leaves or behind leaf sheath and hatch in 5 or 6 days. Young caterpillars feed on green leaf material, later burrowing into stems and eating out tunnels, within which they pupate. Holes in tunnelled stems often revealed by frass hanging from them. A major pest of sugar cane. **Habitat & distribution:** Wetlands where host sedges and grasses occur, also cultivated sugar cane fields; widespread across sub-Saharan Africa.

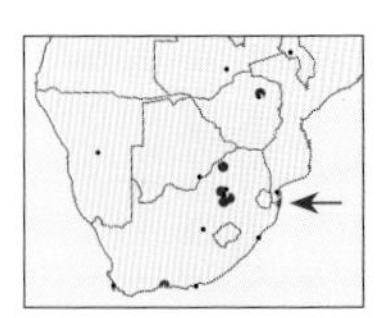

### 4 *Galleria mellonella* — Greater Wax Moth

Forewing 8–17mm. Female larger and heavier; thorax brown, darker patches with raised setal tuft dorsally, abdomen lighter grey-brown; forewings mottled brown, with greasy appearance; hindwings cream. Caterpillar white with brown head capsule and thoracic sclerites. **Biology:** Adults enter beehives and lay their eggs on the comb. Caterpillars feed on the wax, covering the combs with their silken galleries. Males generate ultrasonic sounds to attract females for mating. **Habitat & distribution:** Introduced from Eurasia and now widespread in transformed and natural habitats across Africa, Europe, Australia and North America.

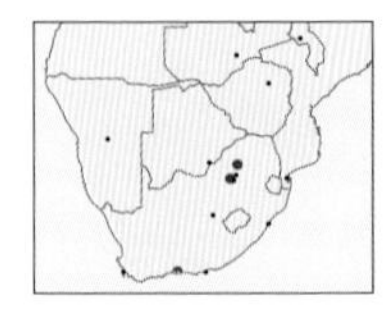

### 5 *Achroia grisella* — Lesser Wax Moth

Forewing 11–13mm. Uniform silver-grey with a distinctive yellow head and dark band across thorax. **Biology:** Males emit sex pheromones and ultrasonic sounds to attract females, who select males largely on the basis of these sounds. Caterpillars feed mostly on deserted honey-bee combs (doing less damage than *Galleria mellonella* above), rarely on dried fruit and vegetables, cork and even sugar. **Habitat & distribution:** Introduced from Europe to Africa, Australia and North America.

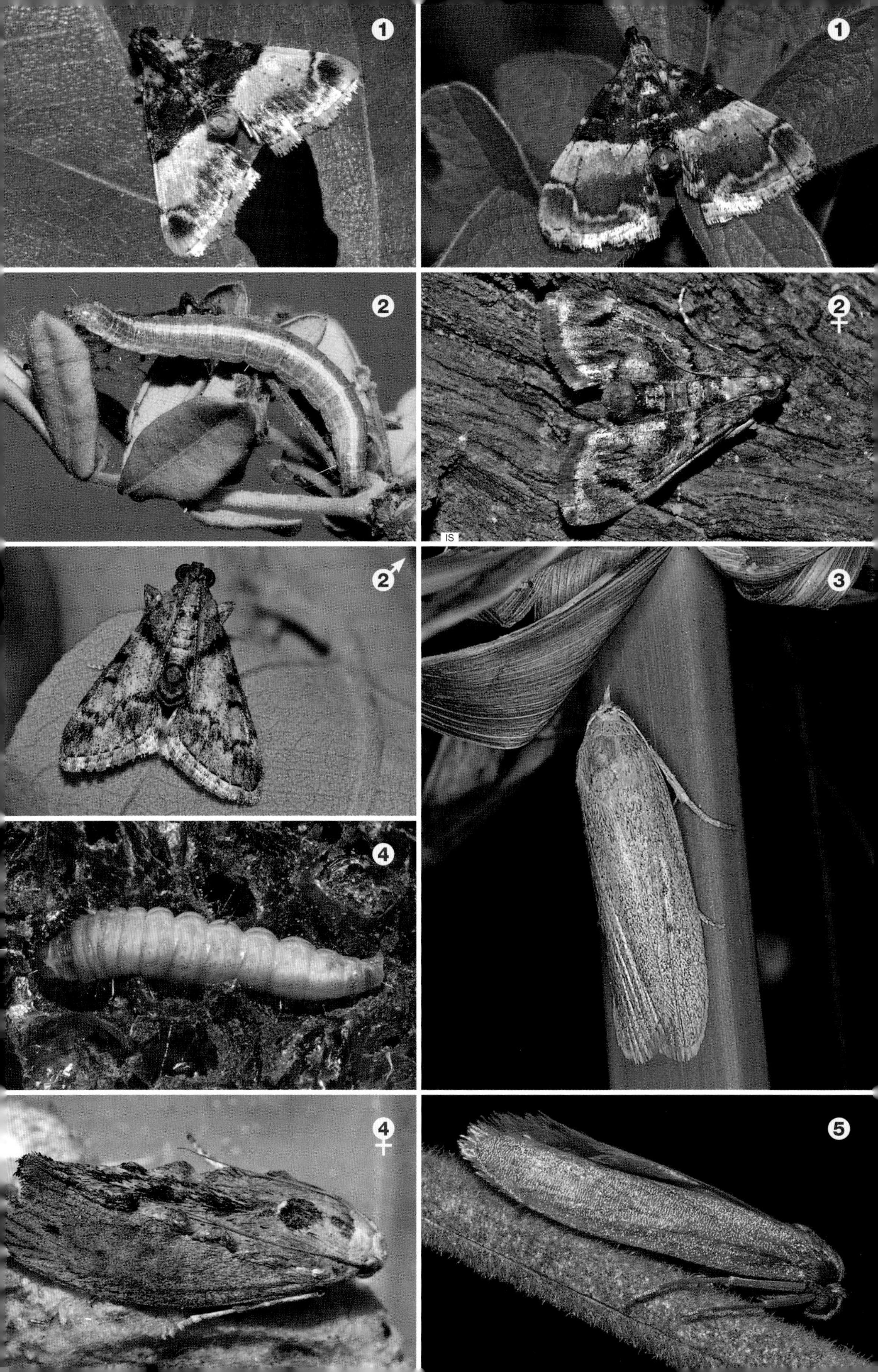

1
1
2
2 ♀
IS
2 ♂
3
4
4 ♀
5

## Subfamily Phycitinae

The antennae are thread-like in females and comb-like in males. Caterpillars have varied feeding habits, typically feeding within webbing and tied leaves, or boring into plants. Some feed on scale insects, or caterpillars of Lymantriidae and Notodontidae moths.

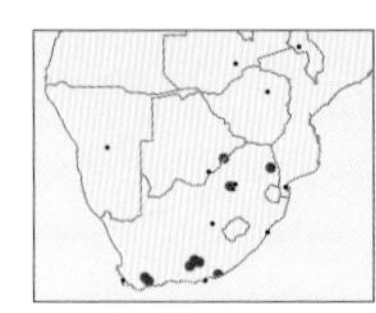

**1 *Cactoblastis cactorum*** Prickly Pear Moth

Forewing ♂ 12–16mm, ♀ 14–20mm. Elongate and silver-grey with strongly projecting palps and poorly defined darker crossbands; ends of forewings with black zigzag line and row of small black dots. Caterpillar bright orange, marked with black dots. **Biology:** Native to South America, deliberately introduced to control species of invasive prickly pear (*Opuntia* spp.); caterpillars feed internally on cactus pads. **Habitat & distribution:** Areas where prickly pear occur; Western Cape, South Africa, to Botswana, Kenya and Tanzania.

Ubiquitous in homes across region

**2 *Ephestia kuehniella*** Mediterranean Flour Moth

Forewing 8–10mm. Pale grey; parallel white and dark zigzag bands across centre of forewings may be obvious, indistinct or absent; hindwings white with thin, brown marginal band. Caterpillar pinkish, with dark head capsule and thoracic sclerites. **Biology:** Primarily infests flour, but also various stored cereal grains, where caterpillars feed within silk webbing. **Habitat & distribution:** Cosmopolitan, in households and food-processing plants; despite common name, thought to originate in India, later introduced to Europe and elsewhere.

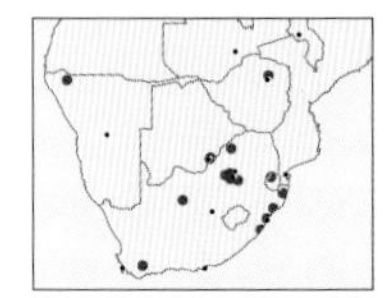

**3 *Etiella zinckenella*** Pulse Pod Borer

Forewing 9–11mm. Distinctive cylindrical posture, wings held closely furled around body and long labial palps angled strongly above and beyond head; forewings grey-brown with white fore-margin and orange transverse bar, bearing raised knobs in male; hindwings grey. Caterpillar fat and pale green. **Biology:** Caterpillars feed on a variety of legumes (Fabaceae). Multivoltine, adults throughout the year. **Habitat & distribution:** Variety of habitats where legumes grow; Western Cape, South Africa, north across Africa, Europe and Asia, with introduced populations in North America and Australia.

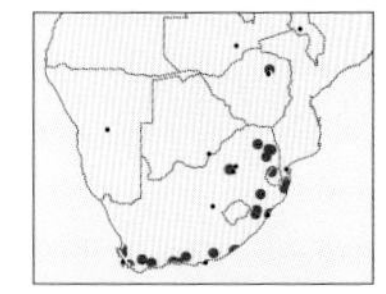

**4 *Acrobasis viridella*** Lichen Knothorn

Forewing 8–10mm. Wings folded against body; thorax black to brown; forewings very variable, pale black to brown or green with dark brown patches or dots, basally and in distal half, a broad white band halfway along and paler area distally; hindwings plain cream with grey border. Caterpillar, early instars yellow with black head capsules, body with fine black and white longitudinal lines and black dots on each segment, final instar uniform green. One described species in genus in region. **Biology:** Caterpillars polyphagous, often on Spikethorns (*Gymnosporia* spp.), gregarious in early instars. Multivoltine, adults throughout the year. **Habitat & distribution:** Moist coastal bush, forest and riverine bush; Western Cape, South Africa, north to Ethiopia, and Madagascar.

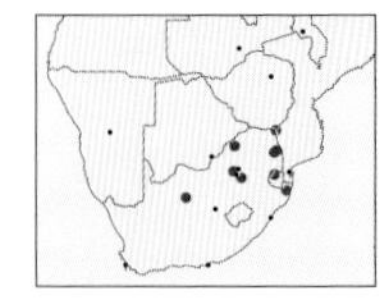

**5 *Hypargyria metalliferella*** Silvered Knothorn

Forewing 6–10mm. Forewings narrow and held curled against body, antennae folded along back; forewings narrow, silvery infused with red, a marked red crossband about one-third from base, underside and hindwings in male coated with metallic silver; hindwings hyaline sometimes with grey margins. Caterpillar pale green or brown with brown thoracic sclerites. Two species in genus in region. **Biology:** Caterpillars feed on Red Spikethorn (*Gymnosporia senegalensis*), also European Olive (*Olea europaea*), Bushman's Tea (*Catha edulis*) and Black Oil Plant (*Celastrus paniculatus*). Bivoltine, adults November–May. **Habitat & distribution:** Bushveld; Northern Cape, South Africa, north to tropical Africa, India, Sri Lanka and Australia.

Ubiquitous in homes across region

**6 *Plodia interpunctella*** Indian Meal Moth

Forewing 6–8mm. Body pale reddish-brown; forewings basally cream with brown distal section separated off by a black crossband. Caterpillar white with yellowish-brown head capsule. **Biology:** Accidentally introduced around the world, where it is a pest in households, food-processing plants and other areas where dry foodstuffs are stored. Caterpillars feed on a wide range of stored products including flour, maize meal, cereals, dried fruit, nuts and beans, binding food together with webbing. Multivoltine in warmer parts, adults throughout the year. **Habitat & distribution:** Human habitation; cosmopolitan.

1
1
IS
2
3
4
4
IS
4
5
IS
5
6

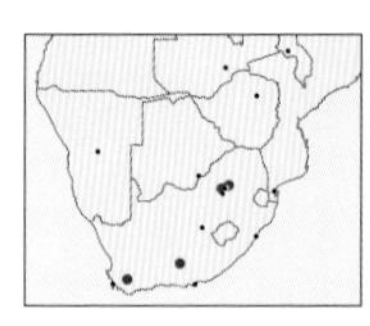

### 1 *Trachypteryx magella* — Magella Thorn

Forewing 9–10mm. Wings held roof-like against body; thick, elevated labial palps, body red-brown with raised dorsal tufts of setae on head and thorax; forewings red-brown, patterned with white diagonal dash and sweeping dark brown lines, distal edge of wing with white marginal line; hindwings white. Seven species in genus in region. **Biology:** Caterpillars feed on Sweet Thorn (*Vachellia karroo*), constructing a smooth case that closely resembles the long thorns of host plant. Probably univoltine, adults September. **Habitat & distribution:** Dry areas where host plant grows; Western Cape to Gauteng, South Africa.

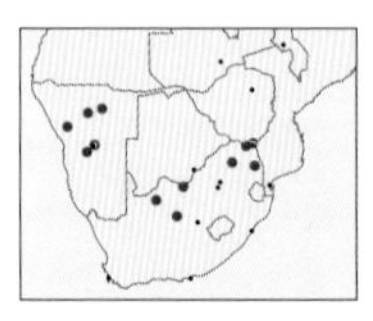

### 2 *Ceutholopha isidis* — Grey Thorn

Forewing 6–7mm. Slender with wings held roof-like against body, labial palps upturned over head; forewings mottled with grey, black and rusty red scales, rusty colour fades quickly in dried specimens; hindwings translucent with distinct long black androconial hairs along inner margin in male. One species in genus in region, but several very similar species in subfamily that are difficult to separate. **Biology:** Caterpillars feed primarily on thorn trees (*Vachellia* and *Senegalia* spp.) also on other Fabaceae. Multivoltine, adults October-March. **Habitat & distribution:** Arid thorn bushveld; Northern Cape, South Africa, north to Egypt.

## Subfamily Pyralinae

The caterpillars are usually associated with decaying plant material. The forewings of males have a subcostal retinaculum (a hook-like structure).

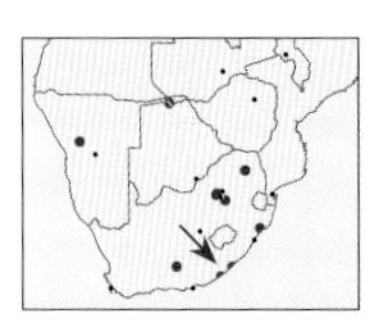

### 3 *Bostra albilineata* — White Line Bostra

Forewing 8–12mm. At rest, legs extended and abdomen held steeply upturned; forewings variably graded pale to dark brown, crossed by 2 sinuous bright white lines, a marginal black dot between the lines and grey dots along border of wing; hindwings cream with brown marginal line. Caterpillar undescribed. Twenty-seven rather diverse species in genus in region. **Biology:** Caterpillars reported to feed on Monterey Cypress (*Cupressus macrocarpa*). Adults September–April. **Habitat & distribution:** Various habitats; Eastern Cape, South Africa, north to Namibia, Mozambique and Zimbabwe.

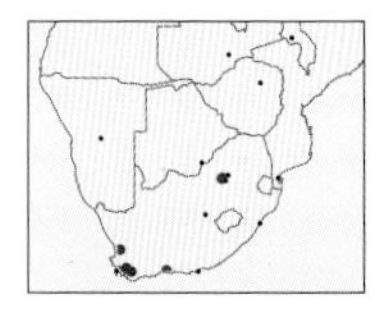

### 4 *Bostra conspicualis* — Conspicuous Bostra

Forewing 13–18mm. Narrowly triangular in outline; thorax purplish-brown; forewings purplish-brown with broad lighter central crossband bearing a small cell spot; hindwings orange-yellow with violet-grey border. **Biology:** Caterpillars feed on Proteaceae. Bivoltine, adults October–March. **Habitat & distribution:** Fynbos and grassland where Proteaceae grow; Western Cape north to Gauteng, South Africa.

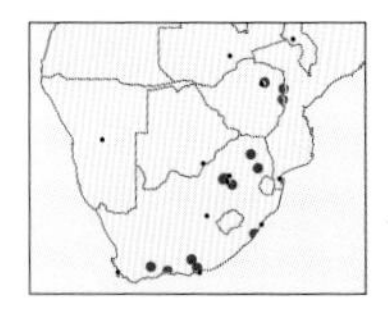

### 5 *Bostra dipectinialis* — Ruddy Bostra

Forewing 13–17mm. Broadly triangular in outline; fairly uniform dark red-brown to grey-brown, forewings crossed by 2 thin wavy crenulated black lines, mixed with white in some specimens, black cell spot, white spots on costa; hindwings grey to cream. Caterpillar unknown. **Biology:** Caterpillar host associations unknown. Probably bivoltine, adults September–March. **Habitat & distribution:** Variety of habitats; Western Cape, South Africa, north to Zimbabwe.

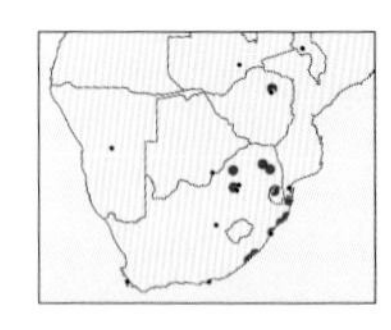

### 6 *Bostra xanthorhodalis* — Yellow-barred Bostra

Forewing 6–8mm. Forewings and hindwings crimson, crossed by broad, wavy, yellow band edged by dark lines. Caterpillar undescribed. **Biology:** Caterpillar host associations unknown. Adults throughout the year. **Habitat & distribution:** Moist forest and bush; Eastern Cape, South Africa, north to Zimbabwe.

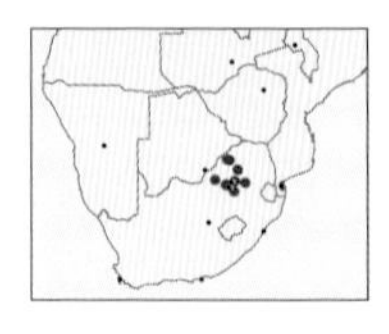

### 7 *Episindris albimaculalis* — Wedge Pyrale

Forewing 12–14mm. Wings held fairly flat against body; orange head and thorax and dark grey-brown forewings, each with a large central white spot and white fringe; hindwings orange. Caterpillar multicoloured, ground colour orange, with red patches and dense pattern of black and white spots, sparsely decorated with long white setae. One species in genus in region. **Biology:** Caterpillars feed on Poison Leaf (*Dichapetalum cymosum*). Bivoltine, adults November–March. **Habitat & distribution:** Bushveld; Gauteng, South Africa, north to tropical Africa.

1
2
2
IS
3
4
5
6
7
7

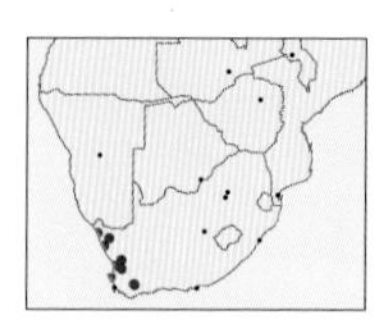

**1 *Hypotia pallidicarnea*** Orange Wave Hypotia

Forewing 10–13mm. Head mustard with forward-facing palps; wings held roof-like against body, tip of abdomen not upturned; forewings mustard with broad wavy red or pink and brown bands outlined by white lines, terminal area silvery white in some specimens. Caterpillar undescribed. Nineteen species in genus in region, all well adapted to arid conditions. **Biology:** Caterpillar host associations unknown. Probably bivoltine, adults September–November and March–April. **Habitat & distribution:** Succulent karoo and desert; Western Cape, South Africa, north to Namibia.

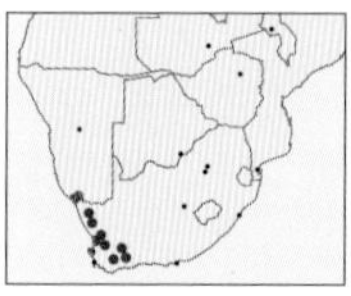

**2 *Hypotia achatina*** Hourglass Hypotia

Forewing 9–11mm. Wings held roof-like against body, tip of abdomen upturned; head, body and wings brown, heavily constricted median band gives an hourglass effect with smoothly curved postmedian line diagnostic, variably mottled brown white and black. Caterpillar undescribed. **Biology:** Caterpillar host associations unknown. Probably bivoltine, adults August–November and March–April. **Habitat & distribution:** Succulent karoo and desert; Western Cape and Northern Cape, South Africa.

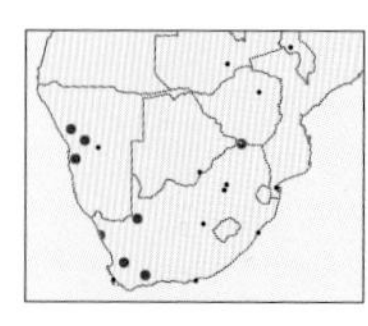

**3 *Hypotia dinteri*** Toothed Hypotia

Forewing 12–13mm. Wings held flat against substrate, tip of abdomen upturned; head, body and wings white, heavily irrorated with grey and brown scales, postmedian line sharply toothed toward costa diagnostic; hindwings brown with cream border. **Biology:** Caterpillar host associations unknown. Probably bivoltine, adults September–October and March–April. **Habitat & distribution:** Succulent karoo, arid bushveld and desert; Western Cape to Namibia and Limpopo, South Africa, also DRC.

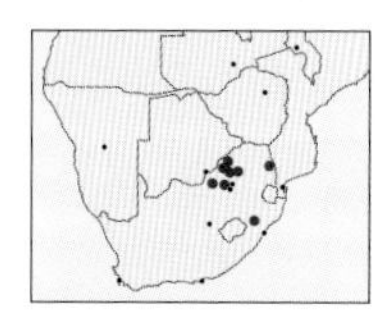

**4 *Hypsopygia sanguinalis*** Sanguine Pyrale

Forewing 7–11mm. Body and wings blood red; forewings with 2 yellow dots on fore-margin; hindwings red and crossed by 2 pale yellow lines; both wings with yellow cilia. Caterpillar undescribed. One of several red species in this genus, but distinguished by yellow spots on forewing costa. Two species in genus in region. **Biology:** Caterpillar host associations unknown. Probably bivoltine, adults October–May, peaking October and March. **Habitat & distribution:** Bushveld; KwaZulu-Natal north to Limpopo, South Africa.

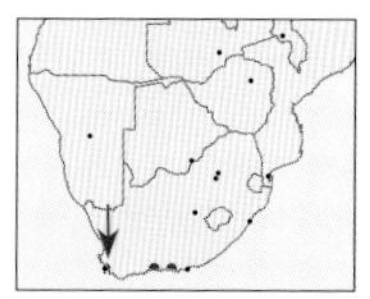

**5 *Lantzodes obovalis*** Brown Oval Pyrale

Forewing 8–9mm. Rests with wings set apart and abdomen curved upward; thorax light brown, abdomen dark brown; forewings mustard brown, distal section comprising a large, uniform dark brown oval patch that extends to costa, separated by curved pale line; hindwing dark brown. Caterpillar undescribed. One species in genus in region, very similar but larger than *Sphalerosticha oblunata* (below). **Biology:** Caterpillar host associations unknown. Adults November–March. **Habitat & distribution:** Moist forest, apparently divergent distribution; Western Cape, South Africa, and Ghana.

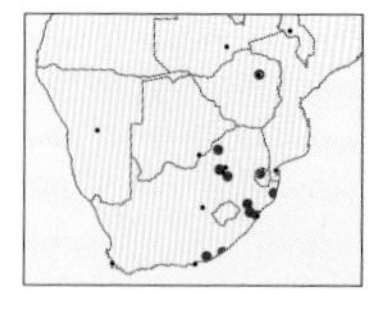

**6 *Sphalerosticha oblunata*** Spectacle Pyrale

Forewing 6–7mm. Rests with wings together and abdomen not curved upward; dark brown patch on forewings not extending to costa; hindwings white. One species in genus in region. **Biology:** Caterpillar host associations unknown. Adults August–March. **Habitat & distribution:** Variety of habitats; Eastern Cape, South Africa, north to Zimbabwe.

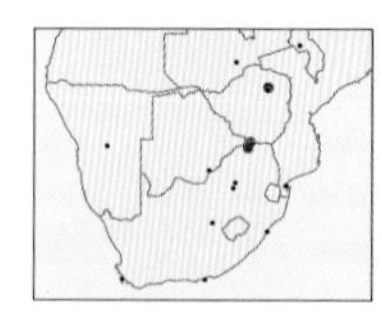

**7 *Aglossa ferrealis*** Rusty Pyrale

Forewing 9mm. Head, thorax and forewings rusty brown with 2 distinct white lines across forewing; hindwings rusty red-brown with indistinct white line across. Thirteen species in genus in region. **Biology:** Caterpillar host associations unknown. Adults in March. **Habitat & distribution:** Bushveld; Limpopo, South Africa, and Zimbabwe.

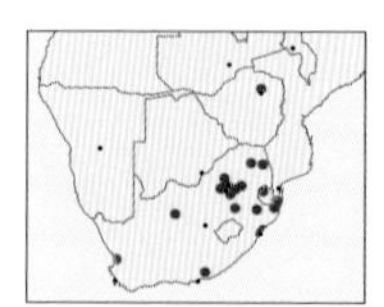

**8 *Aglossa basalis*** Basal Pyrale

Forewing 6–10mm. Wings held fairly flat against body; pale yellow; base of forewings dark brown, distal portion creamy yellow, with a dark spot, faint wavy lines and row of marginal dots; hindwings off-white. **Biology:** Caterpillar host associations unknown. Probably multivoltine, adults throughout the year. **Habitat & distribution:** Variety of habitats; Western Cape, South Africa, north to tropical Africa, also Madagascar.

1
2
3
4
5
6
7
8

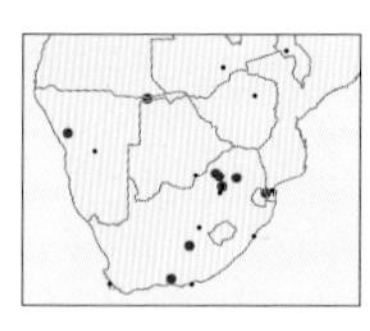

**1 *Delopterus basalis*** Curvy Pyrale

Forewing 9mm. Narrow winged, abdomen extremely curved upward at rest; body and wings dirty grey, dark brown basal and terminal areas, lines white, sometimes indistinct; hindwings greyish-brown. One described species in genus in region, probably a complex of species. **Biology:** Caterpillar host associations unknown. Adults October–February. **Habitat & distribution:** Arid areas, Eastern Cape, South Africa, north to Namibia and Zimbabwe.

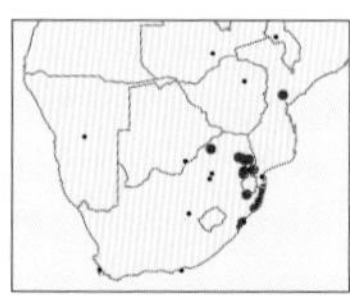

**2 *Mittonia hampsoni*** Hampson's Salad

Forewing 18–27mm. Large, unusual in shape; wings with ragged, strongly concave hind margin; bright iridescent emerald green with black thorax and broad black bands across distal sections of wings. One species in genus in region. **Biology:** Caterpillar host associations unknown. Adults October–February. **Habitat & distribution:** Coastal bush, lowveld forest and bushveld; KwaZulu-Natal, South Africa, north to tropical Africa.

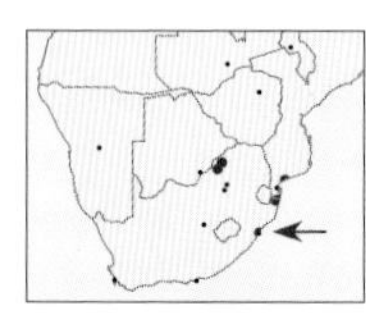

**3 *Pyralis fumipennis*** Flag Pyrale

Forewing 7mm. Small; head orange, body brown; forewings with brown basal and terminal area, median area orange, lines white, hindwings brown with 2 thin white lines. Eleven species in genus in region. **Biology:** Caterpillar host associations unknown. Adults March–October. **Habitat & distribution:** Coastal bush and lowland bushveld; KwaZulu-Natal, South Africa, north to tropical Africa and India.

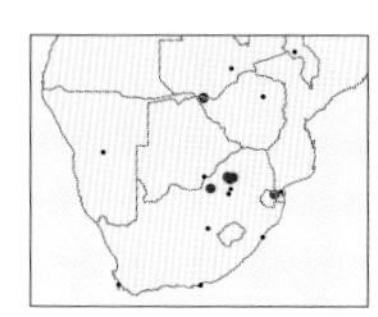

**4 *Stemmatophora chloralis*** Blackbridge Pyrale

Forewing 9–10mm. White with small black bar across thorax; forewings silvery gray with fringe of long white setae, a small black dot at base and conspicuous bridge-shaped black bar centrally; hindwings dark grey basally, silvery white toward termen. Nine species in genus in region. **Biology:** Caterpillar host associations unknown. Adults September–October. **Habitat & distribution:** Bushveld; Eswatini north to Zimbabwe.

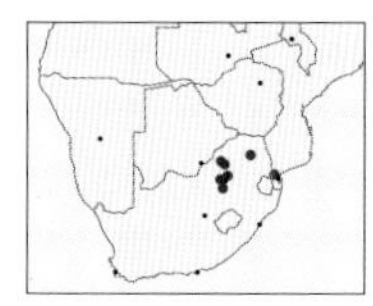

**5 *Stemmatophora perrubralis*** Straight-line Pyrale

Forewing 12–13mm. Body and forewings brick red, anteromedial and postmedian lines white edged black, straight postmedian line with unmarked costa diagnostic; hindwings light brown with faint postmedian line. Similar *Herculia roseotincta* has costa speckled. **Biology:** Caterpillar host associations unknown. Adults October–November. **Habitat & distribution:** Bushveld and riverine forest; Free State north to Limpopo, South Africa, and Mozambique.

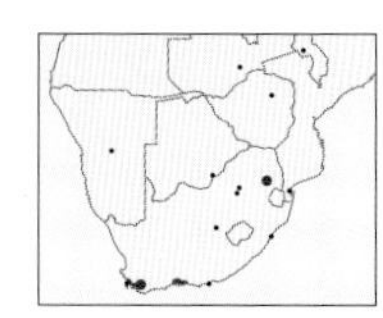

**6 *Triphassa stalachtis*** Spangled Pyrale

Forewing 12–15mm. Head and thorax red-brown; forewings patterned with red-brown bands and lines of black-edged white dots, and row of white dashes along distal margin; hindwings coppery brown, arrangement of dots differ between sexes. Caterpillar brown with darker crossbands. Three species in genus in region. **Biology:** Caterpillars form communal silken nests, polyphagous on fynbos plants. Multivoltine, adults throughout the year. **Habitat & distribution:** Moist fynbos; Western Cape with an isolated population in fynbos remnant vegetation, Mpumalanga, South Africa.

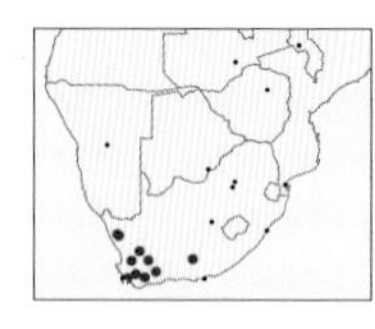

**7 *Triphassa argentea*** Mimic Pyrale

Forewing 12–14mm. Remarkable mimic of the widespread geometrid *Argyrophora trofonia* (p.182); head and thorax brown with white lines across, abdomen cream; forewings brown patterned with rows of silver blocks; hindwings cream with terminal brown shading. **Biology:** Caterpillar host associations unknown. Adults November and January. **Habitat & distribution:** Renosterveld; Western Cape, Eastern Cape and Northern Cape, South Africa.

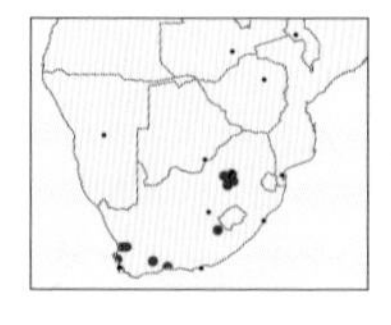

**8 *Zitha laminalis*** Laminated Pyrale

Forewing 10–14mm. Palps strongly projecting; thorax brown, abdomen buff; forewings held somewhat flattened against body, brown at base, cream to pinkish distally, crossed by reddish-brown zigzag lines and bands, single brown cell spot; hindwings cream. Caterpillar grey with white intersegmental markings. Six species in genus in region. **Biology:** Caterpillars feed on aloes, constructing silk nest on leaf surface, from which they burrow into leaf to feed. Adults August–March. **Habitat & distribution:** Fynbos and grassland; Western Cape, South Africa, north to Zimbabwe.

1
2
BW
3
4
5
6
MB
6 ♀
7
MM
8
8

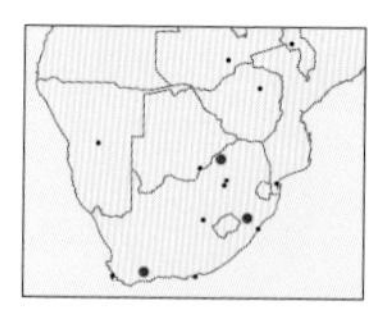

### 1 *Synaphe styphlotricha* — Hieroglyph Pyrale

Forewing 10–11mm. Palps short, abdomen curved over body and head when at rest; forewings cream with hieroglyphic-like pattern of green and red (red turning brown in old specimens); hindwings cream with reddish border. Caterpillar undescribed. Two species in genus in region. **Biology:** Caterpillar host associations unknown. Adults February–March. **Habitat & distribution:** Dry habitats; Western Cape north to Limpopo, South Africa.

## FAMILY CRAMBIDAE Crambid Snout Moths and Pearl Moths

**Variable family of mostly delicate and long-legged moths often with long, forward-directed palps. Members of the subfamily Crambinae (Grass Moths) tend to be cryptic and slender, with the wings closely folded around the abdomen. They often perch on grass stems and, although diverse (140 species), are hard to identify, so only relatively few representatives are featured here. Members of other subfamilies, of which the Spilomelinae are by far the most speciose (over 200 species), have broader, often brightly patterned wings either held to form a triangular outline, or held outstretched. Caterpillars include stem borers, root feeders, leaf miners and leaf tiers, as well as species that feed on aquatic plants. Over 11,600 species have been described worldwide, with at least 526 in the region.**

### Subfamily Acentropinae (= Nymphulinae)

Colourful pyralids that often have a row of metallic dots along the margin of the hindwings. Many are pale and have complex banding patterns on the wings. The caterpillars are mostly aquatic and many have gills. They live in cases and feed on aquatic vegetation. Those inhabiting fast-flowing streams construct silken webs and feed on algae.

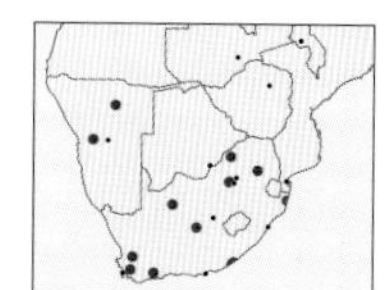

### 2 *Parapoynx diminutalis* — Hydrilla Leafcutter Moth

Forewing ♂ 6–7mm, ♀ 8–12mm. Wings held widely outstretched; body white with dark-edged, light brown crossbars; forewings elongate, white with broad brown bands; hindwings white with brown kinked crossbars. Caterpillar yellowish, with external gills. Five species in genus in region; similarly widespread *P. fluctuosalis* has lines dark brown and hindwing crossbars straight. **Biology:** Unusual in having aquatic larval stage. Caterpillar host plants include Hydrilla (*Hydrilla verticillata*) and waterlilies (*Nymphaea* spp.). Eggs are laid in groups just below the water surface on submerged plants, later larval instars use plant tissue to weave a silken case. Adults October–April. **Habitat & distribution:** Habitats with standing water; widely distributed across Africa, as well as Madagascar and Indian Ocean islands, also Asia and Australia; accidentally introduced to southern USA.

### Subfamily Glaphyriinae

Adults have external ocelli, and the caterpillars have varied diets, some feeding on plants, with other preying on bagworm (Psychidae) caterpillars or even feeding on larvae of paper wasps (Vespidae).

### 3 *Crocidolomia pavonana* — Cabbage Moth

Forewing 11–13mm. Broadly triangular outline at rest; forewings variegated cream and brown with 2 adjacent white dots in cell; hindwings translucent white. Caterpillar with red head and black band on segment one; body with rows of black dots and longitudinal pale and dark bands. One species in genus in region. **Biology:** Eggs laid in groups on underside of leaves, young caterpillars feed gregariously under a silken web. Caterpillars feed on Capparaceae and Brassicaceae, and are an important pest of cabbage. Multivoltine, adults throughout the year. **Habitat & distribution:** Moist habitats; southern Cape, South Africa, north to tropical Africa, also southern Asia and Australia.

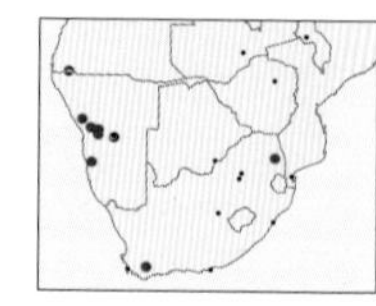

### 4 *Hyperlais conspersalis* — Concorde

Forewing 7–8mm. Palps prominently forward projecting, wings held flat with strongly triangular profile; forewings mottled grey, median lines broken into scattered black spots, 3 white dots on costa, row of subterminal black dots; hindwings plain white, grey at inner margin. Caterpillar flattened, light grey, darker grey laterally, sparse setae with yellow base and black mottled head. Four species in genus in region. **Biology:** Caterpillars feed on Bush Cherry (*Maerua juncea*). Multivoltine, adults September–March. **Habitat & distribution:** Arid bushveld and desert; Namibia, Western Cape and Limpopo, probably more widespread.

1
2
3
3
3
4
4

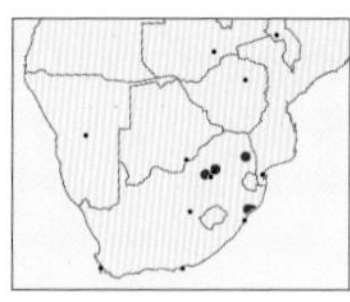

**1 *Trichophysetis krooni*** Marble Tent Pearl

Forewing 6–7mm. Palps prominently forward projecting, wings held tent-like with forewing costa on substrate; wings and body marble white with sprinkling of orange-brown scales all over, double black basal, postmedian and terminal lines across both wings. Caterpillar yellowish-green with black head. Three similar species in genus in region. **Biology:** Caterpillars feed inside the fruit of River Jasmine (*Jasminum fluminense*). Multivoltine, adults throughout the year. **Habitat & distribution:** Bushveld and forest; KwaZulu-Natal to Limpopo, South Africa,probably more widespread.

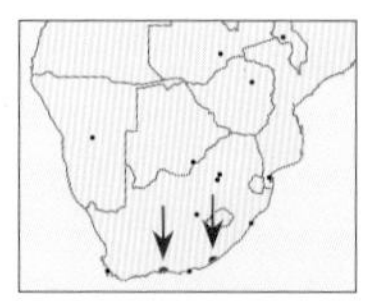

**2 *Goniophysetis lactealis*** Bronze Tent Pearl

Forewing 11–12mm. Wings shiny white shaded with bronze, lines crenulated, silvery cell spot and dark terminal patch. One species in genus in region. **Biology:** Caterpillar host associations unknown. Adults October–December. **Habitat & distribution:** Coastal bush and forest; Eastern Cape, South Africa, north to tropical Africa.

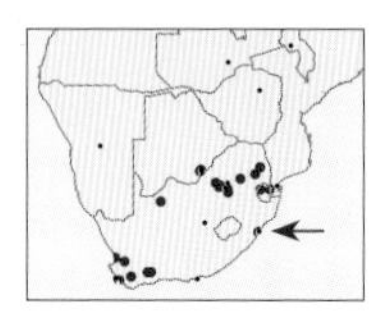

**3 *Hellula undalis*** Cabbage Webworm

Forewing 7–9mm. Forewings whitish-brown with darker central markings and dark kidney-shaped spot, crossed by sinuous white lines; hindwings translucent white. Caterpillar pale yellow-green with black head and faint longitudinal brown bands. **Biology:** Caterpillars feed on range of crucifers including broccoli, cabbage, mustard, radish and turnip, also Spindle-pod (*Cleome monophylla*), Rocket (*Eruca sativa*) and Wild Radish (*Raphanus raphanistrum*). Eggs laid singly, caterpillars feed under a web coated in droppings, pupation occurring within this silken shelter. Multivoltine, adults throughout the year. **Habitat & distribution:** Wide range of vegetation types, including agricultural land; very widespread across Africa and Indian Ocean islands, as well as in Europe and Asia.

## Subfamily Odontiinae

Small to medium-sized, resembling Pyraustinae. Compound eyes often reduced, and the caterpillars feed as web-spinners, leaf miners or within seeds.

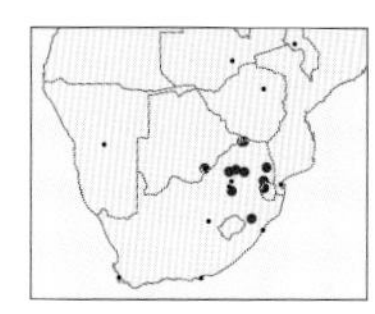

**4 *Autocharis sinualis*** Sinuous Pearl

Forewing 7–9mm. Body white dorsally, grey-brown ventrally, with very prominent forward-projecting palps; forewings with black costal region, otherwise divided by sinuous dark line into white basal portion and brown distal portion; hindwings white with black margin. Caterpillar with light brown longitudinal stripes and several dark brown dots on each segment. Four species in genus in region. **Biology:** Caterpillars feed on Large-fruited Bushwillow (*Combretum zeyheri*). Bivoltine, adults August–March. **Habitat & distribution:** *Combretum* bushveld; KwaZulu-Natal to Limpopo, South Africa, and Botswana.

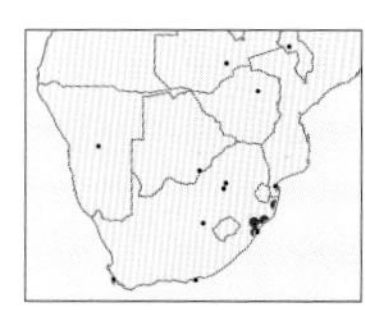

**5 *Autocharis fessalis*** Cream Pearl

Forewing 8–10mm. Body and wings cream bordered with reddish-brown, wide reddish-brown distal section separated by diagnostic darker crenulated brown line; forewings with small but distinctive black cell spot, more black spots present in some specimens; hindwings similar to forewings. Can be confused with *Pseudonoorda rubricostalis* (below). **Biology:** Caterpillar host associations unknown. Adults October–April. **Habitat & distribution:** Moist forests; KwaZulu-Natal, South Africa, north into tropical Africa and India.

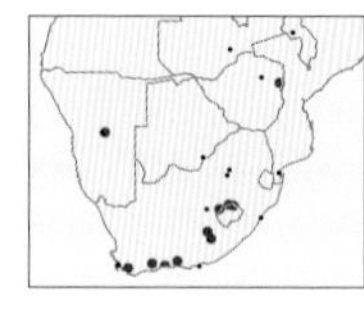

**6 *Blepharucha zaide*** Zigzag Pearl

Forewing 12–15mm. Forewings with zigzag pattern of white, light and dark brown bands, with chequered brown and white fringe; hindwings variable, light to dark brown with darker bands. Isolated populations are regionally variable. Caterpillar cream, grub-like with small black spots on each body segment. One described species in genus in region. **Biology:** Caterpillar host associations unknown. Adults September–May. **Habitat & distribution:** Fynbos and Afroalpine heathland; Western Cape, South Africa, to Lesotho, as well as highlands in Zimbabwe and Namibia.

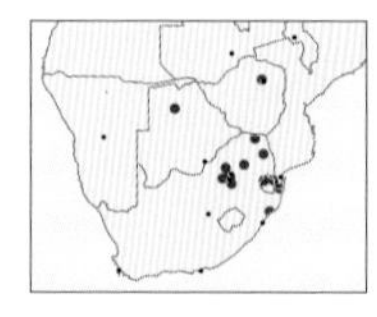

**7 *Pseudonoorda rubricostalis*** Red-edged Pearl

Forewing 6–8mm. Forewings pale yellow variably sprinkled with light brown scales with narrow red border and wide brown distal section separated by darker brown variably crenulated line; hindwings white with grey border. Easily confused with *Autocharis fessalis* (above). Two species in genus in region. **Biology:** Caterpillar host associations unknown. Adults September–March. **Habitat & distribution:** Grassland and bushveld; KwaZulu-Natal, South Africa, north to tropical Africa.

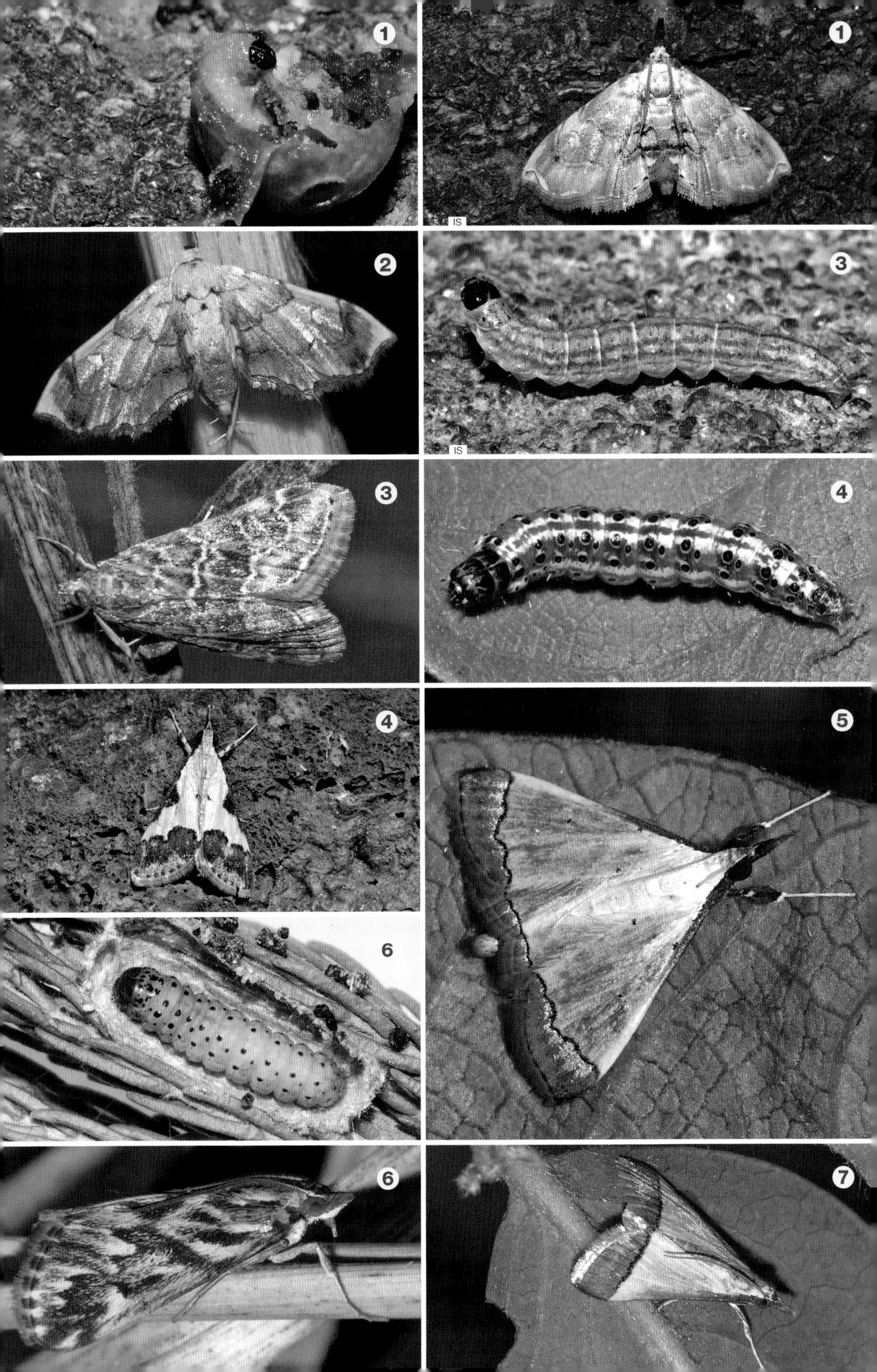
1
1
IS
2
3
IS
3
4
4
5
6
6
7

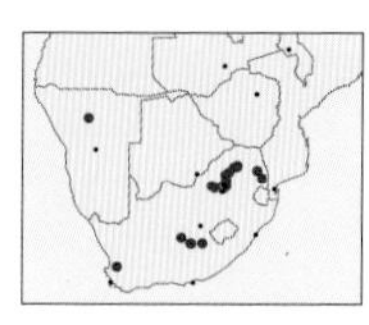

### 1 *Emprepes florilegaria* — Bloodstripe Pearl

Forewing 11–12mm. Forewings bright white to cream with blood-red margin (sometimes red-brown along hind margin), and crossed by diagonal red stripe ending on apex. Caterpillar yellow-green with brown plate on prothorax. Two species in genus in region; similar *E. maesi* is smaller with the diagonal red stripe ending on costa well before apex. **Biology:** Caterpillar feeds on Red Curryflower (*Lasiosiphon rubescens*). Bivoltine, adults September–April. **Habitat & distribution:** Grassland, bushveld and arid habitats; Western Cape, South Africa, north to Namibia.

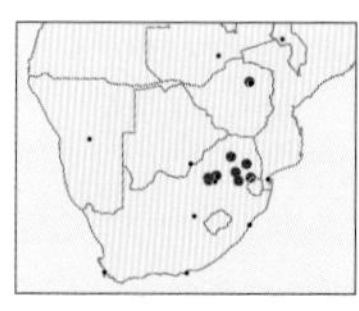

### 2 *Epascestria euprepialis* — Blackline Pearl

Forewing 8–10mm. Similar to *Emprepes florilegaria* (above), but forewing lines and borders orange-red, postmedian line black, black cell spot and a row of black terminal spots; hindwings charcoal black with white cilia. Six species in genus in region. **Biology:** Caterpillar host associations unknown. Adults October–December. **Habitat & distribution:** Grassland and bushveld, Gauteng, South Africa, north to Zimbabwe.

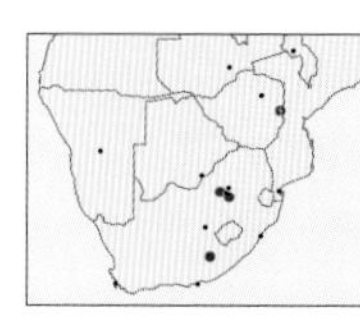

### 3 *Cynaeda distictalis* — Two-spot Pearl

Forewing 11–12mm. Body and wings bright orange, which fades in dried specimens; forewings with 2 distinct black dots, faint curved postmedian line in males, otherwise no markings. Caterpillar reddish-orange with several rows of black dots on each abdominal segment. Six species in genus in region. **Biology:** Caterpillars single, found feeding in silken leaf fold on Christmas Bush (*Pavetta gardeniifolia*). Probably bivoltine, adults November–March. **Habitat & distribution:** Highveld grassland; Eastern Cape, South Africa, north to Zimbabwe.

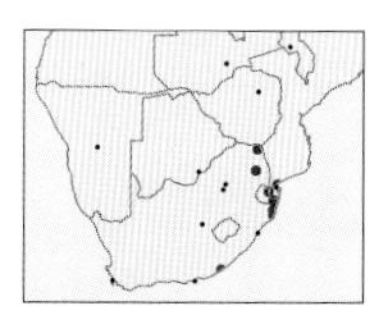

### 4 *Viettessa margaritalis* — Ivory Margined Pearl

Forewing 9–12mm. Thorax white, abdomen brown; forewings ivory white, strikingly patterned with orange to brown wavy stripe along front margin and extending onto prothorax, a broad orange-brown distal panel outlined by dark brown or black line; hindwings white with brown distal patch. One species in genus in region. **Biology:** Caterpillar host associations not recorded in region; recorded elsewhere on Small-leaved Dragon Tree (*Dracaena mannii*), Natal Guarri (*Euclea natalensis*), Snakebite Root (*Synaptolepis alternifolia*) and African Dreamroot (*S. kirkii*). Multivoltine, adults throughout the year. **Habitat & distribution:** Moist forest and coastal bush; Eastern Cape, South Africa, north to tropical Africa.

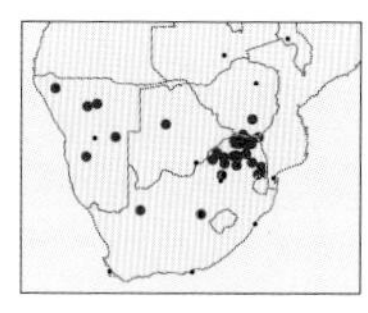

### 5 *Tegostoma bipartalis* — Two-tone Pearl

Forewing 5–6mm. Distinctive species; body, legs and forewings to postmedian white, remainder of wing dark grey sprinkled with reddish scales; hindwings grey. Seven species in genus in region. **Biology:** Caterpillar host associations unknown. Adults throughout the year. **Habitat & distribution:** Bushveld; Free State, South Africa, north to Yemen.

## Subfamily Pyraustinae

A large, mostly tropical subfamily with a well-developed proboscis. The eggs are scale-like, and the caterpillars are either borers, leaf rollers or leaf tiers, and commonly feed on crop plants.

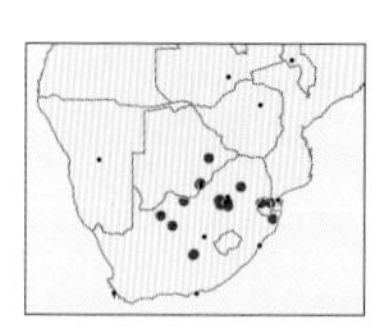

### 6 *Achyra coelatalis* — Graded Brown Pearl

Forewing 6–9mm. Wings held roof-like against body; body light brown; forewings mottled shades of pale to darker brown, with at least 1 (up to 3) dark central spots and lighter distal band bordered by dark wavy lines, longish fringe of marginal setae; hindwings cream. Caterpillar undescribed. Four species in genus in region. **Biology:** Caterpillars reported feeding on sorghum, Asian Rice (*Orzya sativa*), Pearl Millet (*Pennisetum americanum*) and maize. Probably bivoltine, adults October–April. **Habitat & distribution:** Primarily in dry grassland and bushveld; Northern Cape, South Africa, north to tropical Africa; widespread across Africa and Indian Ocean islands, also Asia and Australia.

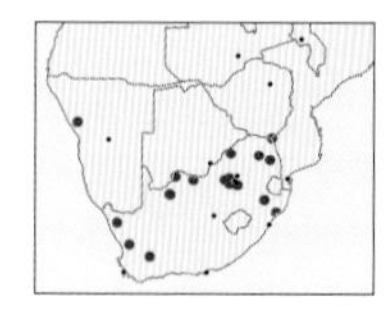

### 7 *Achyra nudalis* — Orange Naked Pearl

Forewing 11–13mm. Profile narrowly triangular; body and wings uniform light orange-brown; forewings marked with up to 6 small darker brown spots and variable orange shading, which fades in dead specimens. Caterpillars variably mottled green and brown, black head. **Biology:** Caterpillars feed in a web on various herbaceous plants (primarily Amaranthaceae and Talinaceae). Adults diurnal, also attracted to lights, bivoltine, August–April, peaking November and February. **Habitat & distribution:** Grassland and bushveld; Northern Cape, South Africa, north to tropical Africa.

1
1
2
3
3
4
5
6
7
7

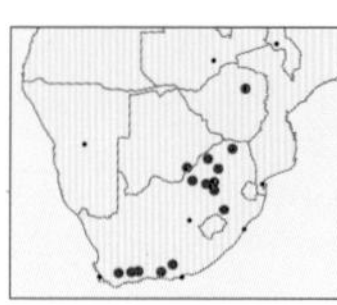

**1 *Euclasta warreni*** Warren's White-line Pearl

Forewing 13–18mm. Palps very large, setose and forward projecting; body and forewings slender and grey-brown; forewings with conspicuous white longitudinal stripe with dark brown margins, small black cell spot; hindwings iridescent white with dark brown apex. Four similar species in genus in region; *E. varii* is lighter brown and has large black cell spot. **Biology:** Caterpillar host associations not recorded in region, recorded elsewhere on Bushman's Poison (*Acokanthera oppositifolia*). Adults throughout the year. **Habitat & distribution:** Karoo, grassland and bushveld; Western Cape, South Africa, north to Mali and Ethiopia.

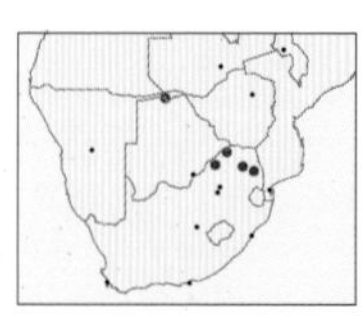

**2 *Hyalobathra filalis*** Chequered Orange

Forewing 9–10mm. Body and wings orange-yellow, crossed by several wavy brown lines, one of which encircles an elongate white stigma, margins of wings with prominent white line and fringe formed of alternating blocks of white and mauve setae. Caterpillar pale green with orange head and conspicuous whitish anal claspers. Three species in genus in region. **Biology:** Caterpillar feeds on Potato Bush (*Phyllanthus reticulatus*). Multivoltine, adults November–June. **Habitat & distribution:** Bushveld; Limpopo, South Africa, north to tropical Africa.

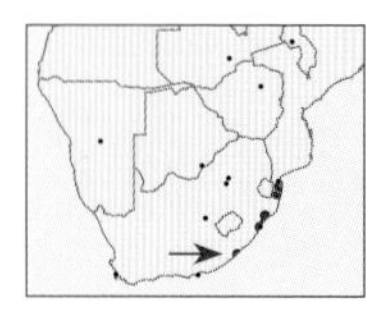

**3 *Lirabotys prolausalis*** Scribbled Pearl

Forewing 11–12mm. Profile triangular; body light brown; palps extended strongly forward; forewings light brown, crossed by 3 thin meandering lines and darkening distally, margin with row of dark purplish dots, postmedian line acutely curved outward; hindwings paler with no markings. Caterpillar pale transparent green, with protruding anal claspers. One species in genus in region. **Biology:** Caterpillars feed on Buckweed (*Isoglossa woodii*) and Tinderwood (*Clerodendrum glabrum*). Probably bivoltine, adults April–May and November–January. **Habitat & distribution:** Coastal bush and forest; Eastern Cape, South Africa, to Mozambique.

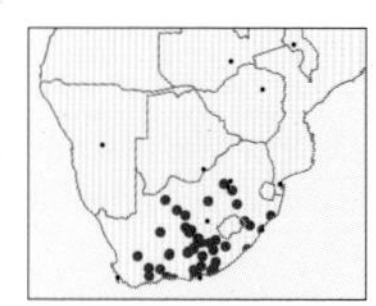

**4 *Loxostege frustalis*** Karoo Moth

Forewing 9–12mm. Wings held roof-like against body; forewings mottled brown and white with zigzag line distally and fringe of long setae with white bases and black tips. Caterpillar green to blackish. Four species in genus in region. **Biology:** Numbers thought to have increased in recent years due to vegetation changes brought about by human activities. Caterpillars ('Karoo caterpillars') feed in communal webs primarily on various Asteraceae species and reputedly compete for food with sheep and cattle. Bivoltine, adults August–May, peaking December and March. **Habitat & distribution:** Primarily in karroid vegetation, where it can be abundant but extending into fynbos and grassland; Western Cape north to Gauteng, South Africa.

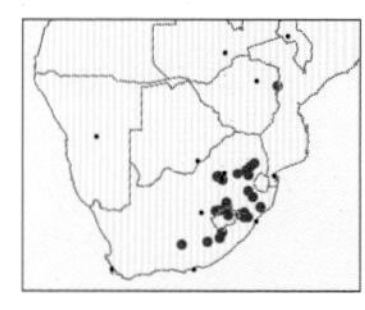

**5 *Loxostege venustalis*** Orange Porphyry

Forewing 8–11mm. Wings held shallowly roof-like against body; thorax orange; forewings black and boldly marked with broad cream to yellow stripes and transverse distal white band; hindwings orange with thin black margin. **Biology:** Caterpillar host associations unknown. Adults diurnal and nocturnal, bivoltine, September–March. **Habitat & distribution:** On Great Escarpment and adjacent highveld grasslands; Eastern Cape to Mpumalanga, South Africa, also highlands of Zimbabwe, Zambia, Mozambique and Malawi.

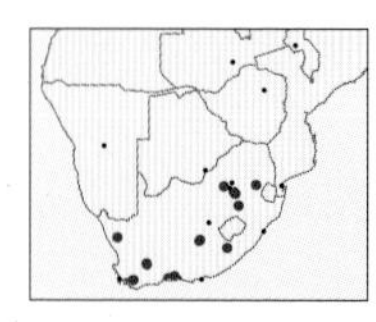

**6 *Loxostege plumbatalis*** Branched Pearl

Forewing 11–13mm. Wings held roof-like against body; thorax and forewings cream, variably suffused with reddish-brown scales, lines branched; hindwings cream to light brown with brown margin. **Biology:** Caterpillar host associations unknown. Adults October–April. **Habitat & distribution:** Fynbos, karoo and grassland; Western Cape to Mpumalanga, South Africa.

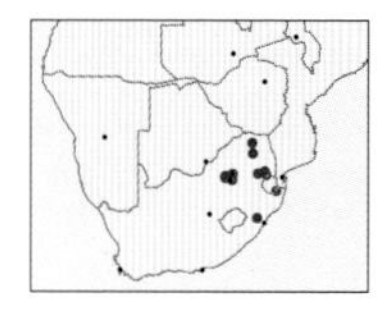

**7 *Daulia auriplumbea*** Silver-spotted Pearl

Forewing 8–10mm. Wings held roof-like against body; thorax and forewings golden orange, interspersed with charcoal grey spots and bands, which are dusted with glittering silver scales; hindwings charcoal grey with orange cilia. Two described species in genus in region. **Biology:** Caterpillar host associations unknown. Adults October–February, active late afternoon but also attracted to light at night. **Habitat & distribution:** Grassland; KwaZulu-Natal north to Limpopo, South Africa.

1
2
IS
2
3
SB
3
4
5
6
7

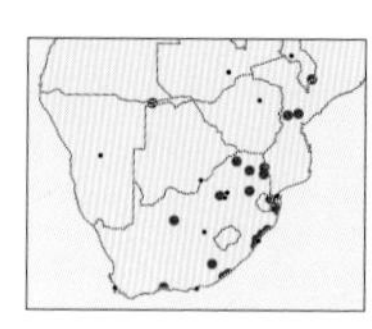

### 1 *Maruca vitrata* — Lace Pearl

Forewing 10–12mm. Wings held extended broadly to sides; forewings brown with transparent, dark-bordered spots and patches; hindwings largely transparent with wavy-edged brown patch distally. Caterpillar pale and transparent with 2 rows of small dark dorsal plates. One species in genus in region. **Biology:** Caterpillar host plants include a wide variety of legumes such as pigeon pea, cowpea, mung bean, green bean and soya bean. Eggs are laid singly, young usually feed on flowers, later instars on young pods. Multivoltine, adults throughout the year. **Habitat & distribution:** Most vegetation types as well as agricultural land; very widely distributed across Africa, Asia, North and South America, Australia and the Pacific Islands; a significant pest of cultivated crops.

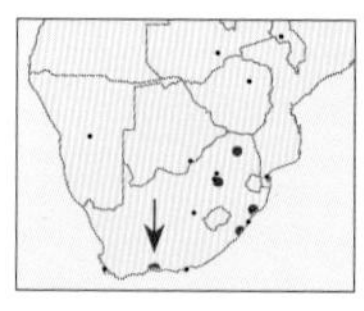

### 2 *Paschiodes mesoleucalis* — Mottled Pearl

Forewing 10–13mm. Forewings patterned with patches of white and brown, crossed by wavy black lines and fringed with brown setae; hindwings white to grey. Caterpillar green and yellow with intricate white and black dorsal markings, orange head. One species in genus in region. **Biology:** Caterpillars feed on Forest Elder (*Nuxia floribunda*). Bivoltine, adults September–May. **Habitat & distribution:** Forests; southern Cape, South Africa, up eastern side of Africa to Eritrea.

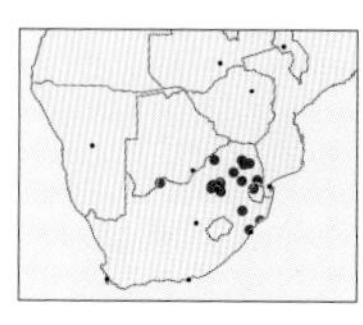

### 3 *Pyrausta phoenicealis* — Perilla Leaf Pearl

Forewing 6–8mm. Triangular in outline; body and wings light to dark red, marked with large paler yellowish to orange patches. Caterpillar green with small black dorsal spots. Twenty-two species in genus in region. **Biology:** Caterpillars feed primarily on herbaceous Lamiaceae, including cultivated herbs. Multivoltine, adults September–May, peaking February–March. **Habitat & distribution:** In various habitats including transformed ones, KwaZulu-Natal, South Africa, north across Africa, including Madagascar and Indian Ocean islands, also America, Australia and Asia.

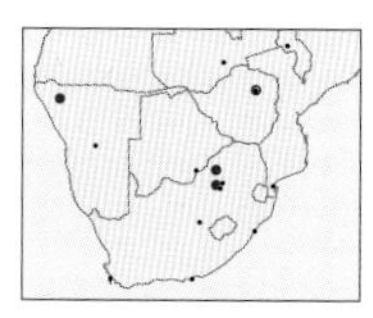

### 4 *Pyrausta tetraplagialis* — Diurnal Pearl

Forewing 6–8mm. Triangular in outline; body and wings black, both wings marked with large conspicuous orange patches. **Biology:** Caterpillar host associations unknown. Diurnal, adults fly actively in full sunshine, fast and low over vegetation, apparently not attracted to light, November–December and April. **Habitat & distribution:** Little known, probably overlooked, records from Gauteng and Limpopo, South Africa, Zimbabwe and Namibia, also DRC.

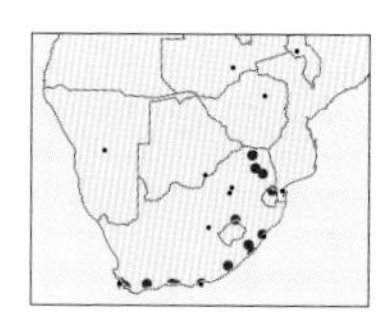

### 5 *Pyrausta procillusalis* — Yellow Pearl

Forewing 10–12mm. Triangular in outline; body and wings deep yellow, both wings with brown margin, large black cell spot on forewing, curved dotted postmedian line on both wings, variable scattered black dots in some individuals. **Biology:** Caterpillar host associations unknown. Adults nocturnal but readily disturbed by day, September–December and March–April. **Habitat & distribution:** In moist fynbos and grasslands with fynbos plants, Western Cape north to Limpopo, South Africa.

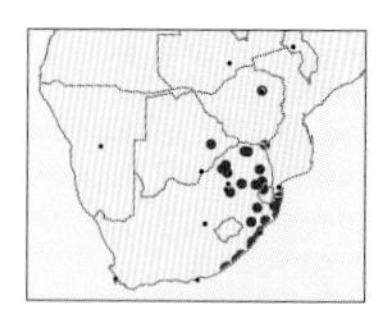

### 6 *Stenochora lancinalis* — Lanced Pearl

Forewing 10–13mm. Triangular in outline; thorax yellow, abdomen orange, long and narrow; forewings pink to orange, strikingly marked with large yellow dots and yellow patches; hindwings pale yellow with brown margin. Two species in genus in region. **Biology:** Caterpillars feed on Tinderwood (*Clerodendrum glabrum*). Adults September–May. In frost-free forest and bushveld, Eastern Cape, South Africa, north to central Africa, including Madagascar and Indian Ocean islands.

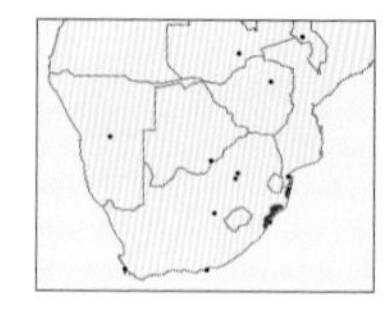

### 7 *Ulopeza conigeralis* — Orange Cone Pyrale

Forewing 10–12mm. Triangular in outline; head and prothorax orange, rest of body brown; wing colour sexually dimorphic; male forewings brown with orange triangular marking, female forewings orange basally and brown distally. Caterpillar initially white, later instars red. Four species in genus in region. **Biology:** Caterpillar host associations not recorded in region, elsewhere on Drumstick Tree (*Moringa oleifera*). Adults throughout the year. **Habitat & distribution:** Lowland forest and coastal bush; KwaZulu-Natal, South Africa, north to tropical Africa.

1
2
2
3
IS
3
3
4
5
6
7♂

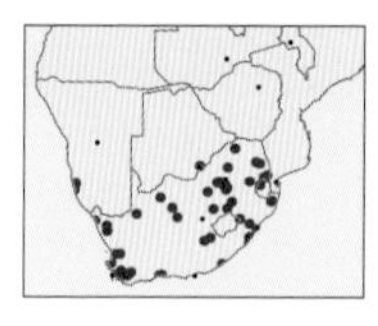

**1 *Uresiphita gilvata*** Yellow Under Pearl

Forewing 13–16mm. Normally narrowly triangular in outline, palps strongly extended forward; thorax brown, abdomen yellow; forewing colour highly variable, ranging from light to very dark brown, sometimes reddish, marked to varying degrees by dark spots and lines; hindwings distinctive, yellow to orange with dark brown margin. Caterpillar green and yellow with black dots or stripes. **Biology:** Caterpillars polyphagous, primarily on Fabaceae, feeding gregariously in webs. Multivoltine, adults throughout the year. **Habitat & distribution:** All habitats; distributed across southern and eastern Africa, also Indian Ocean islands, Europe, Asia, New Zealand and Hawaii.

## Subfamily Crambinae Grass Moths

The wings are folded around the body at rest, and are commonly marked with longitudinal lines. Common in grasslands, where the larvae feed internally on grasses and other monocotyledonous plants. Many species damage pastures and grain crops.

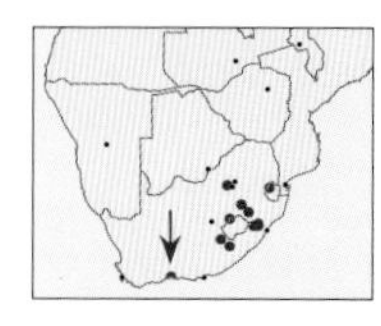

**2 *Ancylolomia prepiella*** Angled Grass Moth

Forewing 14–16mm. Wings held steeply pitched against body; forewings yellowish-brown with 3 longitudinal white streaks diagnostic, their squared off ends marked with light transverse band and dark hyphen-like streaks. Caterpillar not known. Fourteen similar species in genus in region. **Biology:** Caterpillars feed on grasses. Adults November–March, diurnal and nocturnal. **Habitat & distribution:** Moist montane grassland and fynbos; Western Cape north to Gauteng, South Africa, Eswatini and Lesotho.

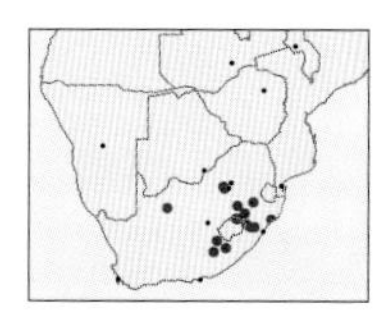

**3 *Aurotalis nigrisquamalis*** Silvered Veneer

Forewing 13–15mm. Ground colour white; forewings with 6 black streaks emanating from base, brown mark in cell, orange postmedian and terminal lines diagnostic for genus; hindwings white to light grey. Six species in genus in region; can only be reliably separated through examination of genitalia. **Biology:** Caterpillars feed on grasses. Adults October–March, diurnal and nocturnal. **Habitat & distribution:** Grassland; Eastern Cape, South Africa, north to Zimbabwe.

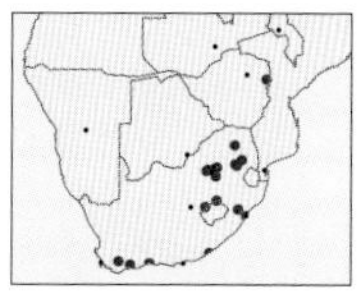

**4 *Crambus sparsellus*** Dotted White Veneer

Forewing 9–13mm. Slender with strongly projecting palps; body and forewings white with diagnostic dark spots on inner margin, indistinct wavy lines; hindwings creamy white. Caterpillar not described. Eighteen diverse species in genus in region. **Biology:** Caterpillars feed on grasses, sheltering within silk webbing. Adults often rest on grass stalks, which they closely mimic. Probably multivoltine, adults throughout the year. **Habitat & distribution:** Variety of habitats, also in gardens; Western Cape, South Africa, north to Zimbabwe.

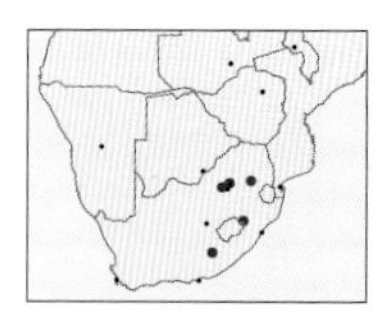

**5 *Crambus leucoschalis*** Arrow Snout

Forewing 10–11mm. One of several very similar species with falcate forewings and arrow-like lines; forewings dark brown, lighter on inner margin, with bright white band along middle of wing from base to apex ending in an arrow formed by the shape of the wing; hindwings light grey to white. Caterpillar not described. **Biology:** Females active during the day, males at night; adults often rest on grass stalks. Probably bivoltine, adults November and March–April. **Habitat & distribution:** Highveld grasslands; Eastern Cape north to Mpumalanga, South Africa.

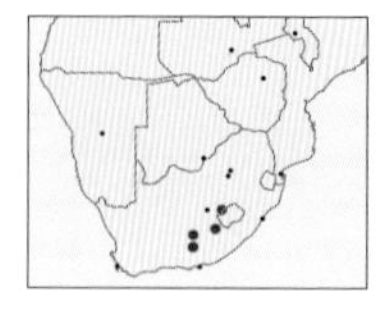

**6 *Caffrocrambus dichotomelus*** Fluffed Snout

Forewing 11–13mm. Slender with strongly projecting palps covered in dense setae giving it a 'fluffy' appearance; forewings white, area along costa and halfway from base along inner margin golden, central black streak; hindwings creamy white. Nine species in genus in region, some very similar, which can only be reliably separated by examining genitalia. **Biology:** Caterpillar host associations unknown. Adults often rest on grass stalks. Adults November–April. **Habitat & distribution:** Dry grassland habitats; Western Cape and Eastern Cape, South Africa, and Lesotho.

**7 *Caffrocrambus albistrigatus*** White-streaked Snout

Forewing 16–17mm. Differs from the *C. dichotomelus* group (above), by being larger with forewings brown with white streaks fanning out toward termen; hindwings hyaline grey. **Biology:** Caterpillar host associations unknown. Adults December–January. **Habitat & distribution:** High-altitude grassland; Lesotho and KwaZulu-Natal, South Africa.

1
1
IS
2
3
4
5
6
7

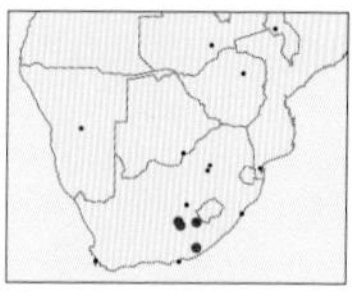

**1 *Prionapteryx molybdella*** Jagged Grass Moth

Forewing 10–12mm. Slender with strongly projecting palps; forewings with various straw, brown and grey markings, jagged lines diagnostic. Twelve similar species in genus in region, in arid areas. **Biology:** Caterpillar host associations unknown. Probably univoltine, adults December–January, can be abundant in habitat. **Habitat & distribution:** Grassy karoo; Northern Cape and Eastern Cape, South Africa.

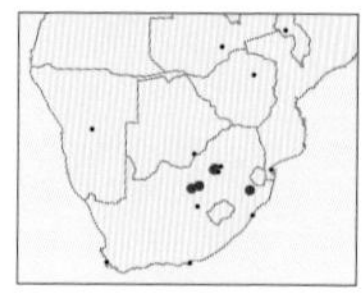

**2 *Haimbachia unipunctalis*** Wedge Grass Moth

Forewing 8–10mm. Wedge-like with long, hairy, projecting palps; forewings light brown with white veins, black cell spot, black spots on termen, brown and white cilia; hindwings white, no markings. Five species in genus in region. **Biology:** Caterpillar host associations unknown. Probably univoltine, adults December–March. **Habitat & distribution:** Highveld grassland; Free State, north to Gauteng, South Africa.

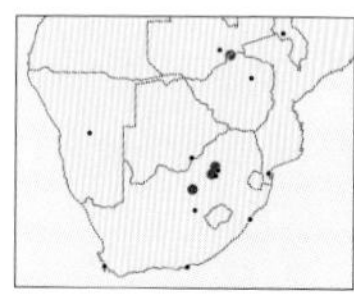

**3 *Euchromius nigrobasalis*** Black-eyed Grass Moth

Forewing 7–8mm. Forewing ground colour white, large diagnostic black basal patch, rest of wing finely striated with black scales, golden postmedian and terminal patches with wavy black lines, row of shiny black 'eye-like' cilia; hindwings grey, no markings. Six similar species in genus in region. **Biology:** Row of eye-like cilia of moth, typical for genus, possibly mimics jumping spider eyes. Caterpillar not known. Adults October–December. **Habitat & distribution:** Grassland and bushveld; South Africa, Botswana and Zimbabwe.

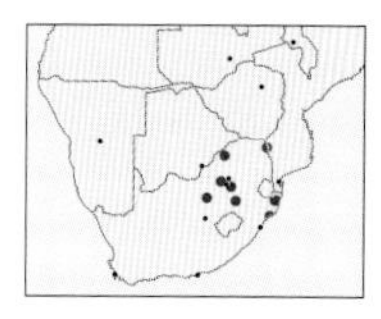

**4 *Chilo partellus*** Chilo Borer

Forewing 8–14mm. Body and wings light brown; forewings variably mottled with greyish scales, variable black spots all over wing, postmedian line when present dotted, row of terminal spots; hindwings greyish in male, white in female, no markings. Seven species in genus in region. **Biology:** Caterpillars bore into the seeds and shoots of various grasses, including sorghum and maize. Adults throughout the year. **Habitat & distribution:** Mainly in cultivated farmland but also grassland and bushveld; species range expanding, displacing similar local species in East Africa, KwaZulu-Natal north to tropical Africa and eastwards up to Indonesia.

## Subfamily Musotiminae Angle Pearls

The adults have reduced acoustic bullae (hearing organs) and the caterpillars feed on ferns.

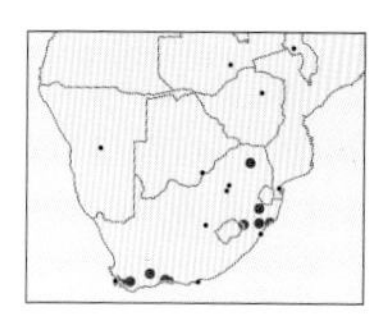

**5 *Panotima angularis*** Acute Angle Pearl

Forewing 9–13mm. Wings angular, like other species in this little-known subfamily; body and forewings chocolate brown with white lines acutely angled outward and white cell mark; hindwings cream with brown border, dark markings at anal angle. Two species in genus in region. **Biology:** Caterpillar host associations unknown. Adults throughout the year. **Habitat & distribution:** Moist fynbos and forests; Western Cape, South Africa, north to tropical Africa and Madagascar.

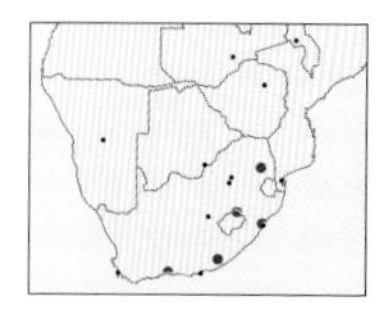

**6 *Ambia chalcichroalis*** Kite Angle Pearl

Forewing 7–9mm. Wings angular, resting posture unique with stretched out wings; body and wings mottled orange and brown with numerous variously shaped silvery white marks on all wings. Two species in genus in region. **Biology:** Caterpillar host associations unknown. Adults October–February. **Habitat & distribution:** Moist, cool places near streams; southern Cape north to Mpumalanga, South Africa.

## Subfamily Spilomelinae Spilomelid Moths

This is the most species-rich subfamily within Crambidae. They lack a retinacular hook (bristle used to couple the fore- and hindwings). The wings are fairly broad and often patterned, and the legs are long. The caterpillars are leaf rollers or web-spinners.

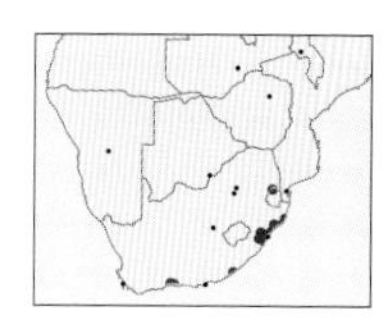

**7 *Aethaloessa floridalis*** Florid Pearl

Forewing 8–9mm. Wings outspread at rest; strikingly coloured with orange body and orange wings marked with uneven iridescent purple-brown borders and crossbars, yellow patches; distinctive black spot on abdomen. One species in genus in region. **Biology:** Caterpillar host associations unknown. Adults fly actively by day under forest canopy, also attracted to light. Probably multivoltine, adults throughout the year. **Habitat & distribution:** Coastal bush and moist forests; southern Cape, South Africa, north to tropical Africa and Asia.

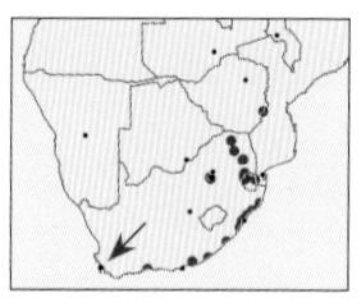

### 1 *Agathodes musivalis* — Painted Pearl

Forewing 14–18mm. Body brown, abdomen crossed by narrow white bands, resting wing posture unusual with oval space left between tips of elongate, hook-tipped forewings; forewings brown with distinctive purple inner margins; hindwings cream, brown anteriorly. Caterpillar green to red with white-edged black dorsal stripe. Four similar species in genus in region. **Biology:** Caterpillars feed on coral trees (*Erythrina* spp.). Multivoltine, adults throughout the year. **Habitat & distribution:** Forest, bushveld, grassland and gardens where *Erythrina* spp. grow; Western Cape, South Africa, north to tropical Africa, including Madagascar and Indian Ocean islands.

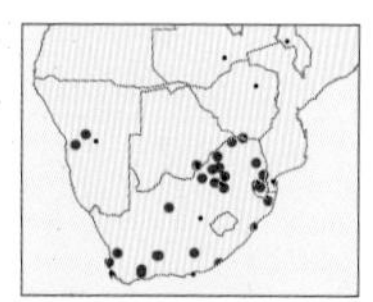

### 2 *Antigastra catalaunalis* — Red-veined Pearl

Forewing 7–11mm. Palps extended strongly forward; triangular in outline with falcate tips to forewings; sexually dimorphic; male forewings orange-brown interspersed with elongate cream patches giving 'red-veined' appearance; female straw brown with reduced cream patches and short distinct brown lines; hindwings creamy white, brown at apex. Caterpillar green with small black dots. Two species in genus in region. **Biology:** Caterpillars feed in shelters spun between leaves on various *Sesame* herbs (family Pedaliaceae). Multivoltine, adults throughout the year. **Habitat & distribution:** Diverse habitats from desert to tropical forest; Western Cape, South Africa, north across Africa, as well as in Europe and Asia.

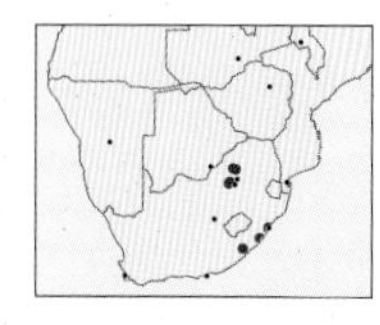

### 3 *Archernis flavidalis* — Brown-spotted Pearl

Forewing 11mm. Triangular in outline; body and wings orange-brown; forewings narrow with many large, dark brown marks that can be faded to almost invisible in some specimens; hindwings plain cream. One species in genus in region. **Biology:** Caterpillar host associations unknown. Adults October–February. **Habitat & distribution:** Grassland and bushveld; Eastern Cape to Limpopo, South Africa, also reported from Kenya.

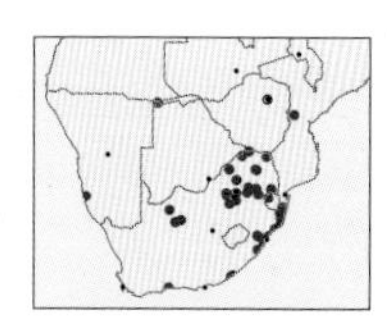

### 4 *Bocchoris inspersalis* — Dotted Sable

Forewing 7–9mm. At rest, wings held outspread; body and wings black with large white dots and bands, wings fringed with chequered black and white setae. Three species in genus in region. **Biology:** Host associations unknown. Multivoltine, adults throughout the year. **Habitat & distribution:** All vegetated habitats; across sub-Saharan Africa, including Indian Ocean islands, also widespread across South and East Asia.

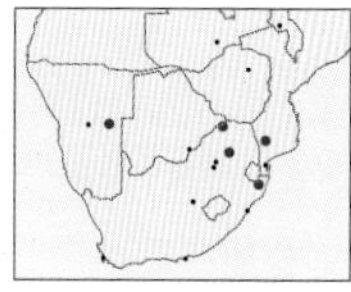

### 5 *Bocchoris nuclealis* — Nuclear Sable

Forewing 6–7mm. At rest, wings held outspread; wings light brown, patterned with darker brown circles and bands and with darker brown broad marginal bands lined with fine brown lines of variable density between specimens, pale setal fringes. **Biology:** Caterpillar host associations unknown. **Habitat & distribution:** Lowland bushveld; South Africa, Namibia and Mozambique.

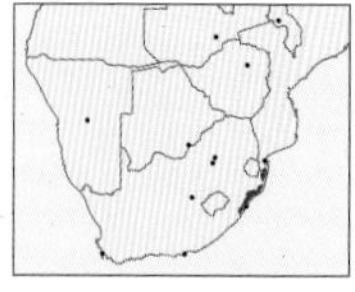

### 6 *Botyodes asialis* — Giant Yellow Pearl

Forewing 22–28mm. At rest, wings held outspread; body yellow and brown; fore- and hindwings dark yellow with broad marginal band, which can be absent on one or both wings in some specimens, and scattered dots and bands, ringed cell spot on all wings. One species in genus in region. **Biology:** Caterpillar host associations not recorded in region, polyphagous on forest trees elsewhere. Adults recorded January–June. **Habitat & distribution:** Moist forest and coastal bush; KwaZulu-Natal, South Africa, north to tropical Africa, including Madagascar and La Réunion, also Asia and Australia.

### 7 *Cadarena pudoraria* — Purple-edged Pearl

Forewing 17–19mm. Wings held outspread at rest; forewings white with broad bright pink to purple wavy margin; hindwings white with 2 pink to purple marginal patches. One species in genus in region. **Biology:** Caterpillar host associations not recorded for region, elsewhere recorded on Arrowleaf Sida (*Sida rhombifolia*), as well as *Gossypium* and *Adenia* spp. Adults January–August, peaking in March. **Habitat & distribution:** Moist forest and coastal bush; KwaZulu-Natal, South Africa, north to tropical Africa, including Madagascar, La Réunion and India.

### 8 *Chalcidoptera thermographa* — Marbled Pearl

Forewing 10–11mm. Triangular in outline; forewings white to cream with network of light yellow and brown markings forming cell-like pattern, leading edge and hind margin darker brown; hindwings white with brown distal margin. One species in genus in region. **Biology:** Caterpillar host associations unknown. Perhaps bivoltine, adults October and April–June. **Habitat & distribution:** Moist forest and coastal bush; KwaZulu-Natal, South Africa, north to tropical Africa.

1
1
2
2♂
3
IS
4
6
5
7
8

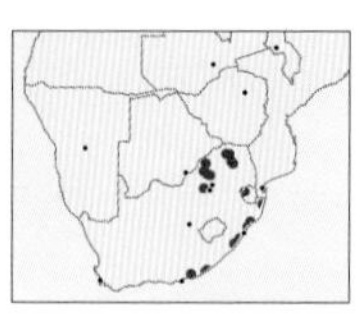

### 1 *Cirrhochrista grabczewskyi* — Marbled Concorde

Forewing 12 –15mm. Strongly triangular in outline with elongate forward-projecting palps, body white with orange dorsal stripe; forewings white, patterned with jagged orange or brown border and crossed by meandering orange or brown lines; hindwings white. Two species in genus in region. **Biology:** Caterpillar host associations unknown. Adults September–May. **Habitat & distribution:** Bushveld and forest; Eastern Cape, South Africa, north to tropical Africa.

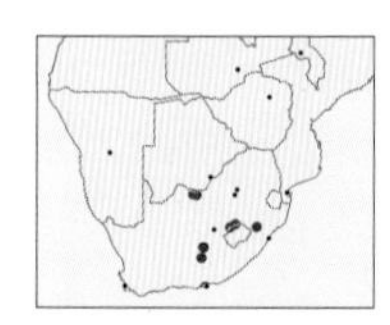

### 2 *Cirrhochrista argentiplaga* — Silver Concorde

Forewing 14–16mm. Wings held tight over body with elongate forward-projecting palps; head and palpi rusty red; body and forewings marble white with antemedial and postmedian lines distinctive, basal and subterminal area variably shaded with rusty red; hindwings white with grey shading and well-developed postmedian line. **Biology:** Caterpillar host associations unknown. Adults September–May. **Habitat & distribution:** Grassy karoo and arid grasslands; Eastern Cape north to KwaZulu-Natal, South Africa.

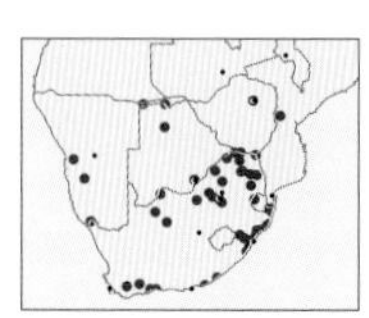

### 3 *Diaphania indica* — Cucumber Moth

Forewing 11–13mm. Outline broadly triangular at rest; thorax and tip of abdomen brown, with white area between; distinctive tuft of light brown hair-like scales project from tip of abdomen, especially of females; wings white with broad brown borders. Caterpillar pale green, often with pair of paler longitudinal stripes. One species in genus in region. **Biology:** Caterpillars polyphagous on soft creepers and herbaceous plants. Adult females rest with abdominal tufts fully spread to disperse pheromones and attract males. Multivoltine, adults throughout the year, peaking in autumn. **Habitat & distribution:** Wide range of vegetation types and agricultural land. Widespread across Africa and Indian Ocean islands, as well as Asia and Australia.

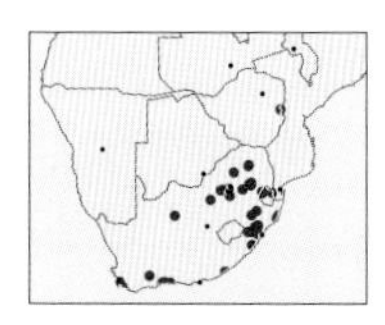

### 4 *Diasemia monostigma* — Black Wedge Pyrale

Forewing 8–11mm. Narrowly triangular in outline with strongly forward-projecting palps; forewings light brown with a distinctive black wedge-shaped marking and darker distal band; hindwings grey-brown with long setal fringe. Three similar species in genus in region; *D. disjectalis* with broadly triangular wedge-shaped mark. **Biology:** Caterpillar host associations unknown. Probably multivoltine, adults throughout the year. **Habitat & distribution:** Variety of habitats, particularly abundant in transformed habitats; Western Cape, South Africa, north to Kenya, also Madagascar and La Réunion.

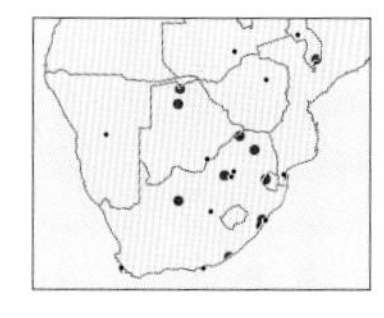

### 5 *Filodes costivitralis* — Window Pearl

Forewing 15–18mm. Very long legs; silvery grey (rarely dark grey) forewings with black-dotted orange basal patch; hindwings grey with clear streak in male. Caterpillar silvery to orange with small black spots and orange head. **Biology:** Caterpillars feed on *Thunbergia* spp. Adults observed drinking tear secretions from the eyes of cattle. Bivoltine, adults November–June. **Habitat & distribution:** Forest and bushveld; Eastern Cape, South Africa, north to tropical Africa, including Madagascar and Indian Ocean islands.

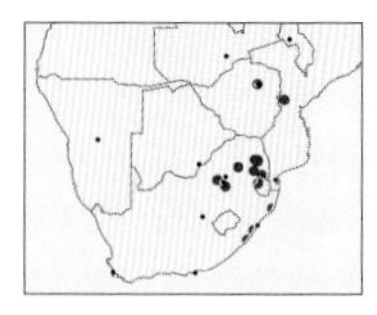

### 6 *Glyphodes bicoloralis* — Bi-coloured Pearl

Forewing 9–12mm. Body dark brown; forewings dark brown with large white spots; hindwings white with broad brown band distally; both wings fringed with long white setae. Caterpillar brown-green. Ten similar species in genus in region. **Biology:** Caterpillars polyphagous on trees. Multivoltine, adults throughout the year. **Habitat & distribution:** Bushveld and forest; Eastern Cape, South Africa, north to tropical Africa, also Asia and Australia.

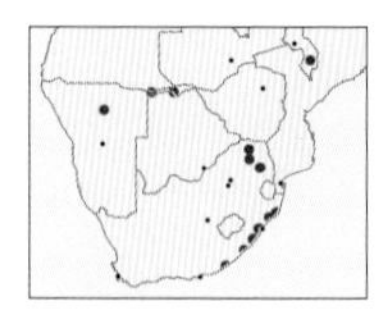

### 7 *Glyphodes bitriangulalis* — Bi-triangle Pearl

Forewing 9–14mm. Similar to *G. bicoloralis* (above), but reddish-brown with larger white forewing patches, the distal one being distinctly triangular, the basal one forming a band right across the wing. **Biology:** Caterpillars feed on fig trees (*Ficus* spp.). Multivoltine, adults August–May. **Habitat & distribution:** Forests and lowland bushveld where fig trees grow; Eastern Cape, South Africa, north to tropical Africa.

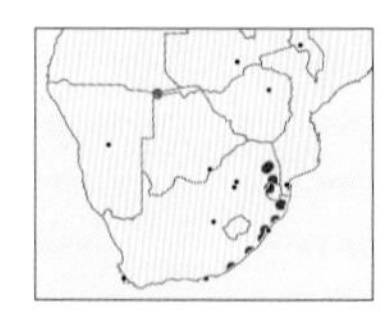

### 8 *Glyphodes stolalis* — Banded Pearl

Forewing 15–16mm. Resembles *G. bicoloralis* (above), but has 4 dark-brown-edged white bands crossing forewings. Caterpillar translucent green-brown. **Biology:** Caterpillars feed on fig trees (*Ficus* spp.). Multivoltine, adults August–May. **Habitat & distribution:** Forests and lowland bushveld where fig trees grow; Eastern Cape, South Africa, north to tropical Africa, western Indian Ocean islands, also Asia and Australia.

1
2
3
IS
3
3
IS
4
5
5
6
6
IS
7
7
8
IS
IS
8
8

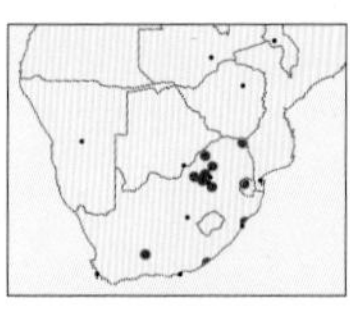

### 1 *Glyphodes negatalis* — Denied Glyphodes

Forewing 12–14mm. Body grey-brown striped; wings translucent, irregularly mottled and banded brown and white, fringed with grey setae. Caterpillar pale translucent brown. **Biology:** Caterpillars feed in leaf fold on fresh leaves of Red-leaved Fig (*Ficus ingens*). Bivoltine, adults September–April. **Habitat & distribution:** Various habitats where figs grow, more common in drier areas; Western Cape, South Africa, north to tropical Africa.

### 2 *Glyphodella flavibrunnea* — Golden Lace-wing

Forewing 8–9mm. At rest, wings held outspread; golden-brown wings marked with large, brown-ringed transparent windows. One species in genus in region. **Biology:** Caterpillar host associations unknown. Probably multivoltine, adults throughout the year. **Habitat & distribution:** Forest and riverine vegetation; Western Cape, South Africa, north to tropical Africa and Madagascar.

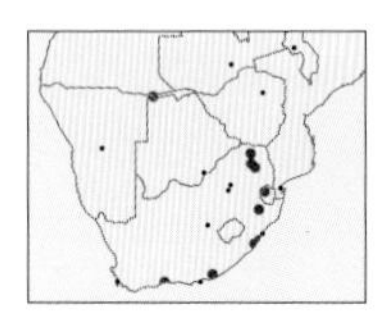

### 3 *Ghesquierellana hirtusalis* — Blotched Pearl

Forewing 14–22mm. Broadly triangular in outline; forewings yellow-brown with numerous, sometimes coalescing dark brown to purplish spots, becoming sparser distally; hindwings paler, with fewer spots. One species in genus in region. **Biology:** Caterpillars feed on figs (*Ficus* spp.). Adults October–May. **Habitat & distribution:** Forest and coastal bush; southern Cape, South Africa, north to tropical Africa, Madagascar and Indian Ocean islands.

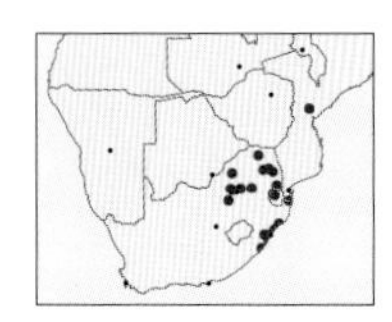

### 4 *Haritalodes polycymalis* — Twinspot Pearl

Forewing 10–15mm. Pale yellow with a complex pattern of darker sinuous brown lines across both wings, creating a chequered pattern, often with darker band near wing tips. Caterpillar green initially with dark brown head, some specimens becoming pink in later instars. One species in genus in region. **Biology:** Caterpillars polyphagous primarily on Malvaceae but also other plants; roll host leaves into tubes, in which they live and pupate. Multivoltine, adults throughout the year. **Habitat & distribution:** Variety of moist habitats, seemingly absent from arid areas; Eastern Cape, South Africa, north to tropical Africa.

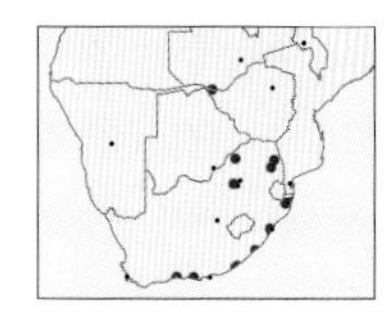

### 5 *Herpetogramma phaeopteralis* — Dusky Pearl

Forewing 9–12mm. Body and wings various shades of greyish-brown crossed by indistinct wavy lines, postmedian line kink with jagged edges diagnostic, black cell spot. Caterpillar green with darker blotches and orange or black head. Three species in genus in region. **Biology:** Caterpillars polyphagous on low-growing plants in shade, including wide-blade grasses, living in silk in folded leaf. Multivoltine, adults throughout the year. **Habitat & distribution:** Shady moist places, including gardens; southern Cape, South Africa, north to tropical Africa, Indian Ocean islands as well as in Asia and the Americas.

### 6 *Herpetogramma morysalis* — Faint Pearl

Forewing 11–12mm. Body and wings shiny white; wings with variably broken golden-brown lines and golden-brown margin. Caterpillar green with distinctive black lateral spots. **Biology:** Caterpillars recorded on *Halleria lucida* and *Tecomaria capensis*. Multivoltine, adults throughout the year. **Habitat & distribution:** Coastal habitats; Western Cape, South Africa, to Mozambique.

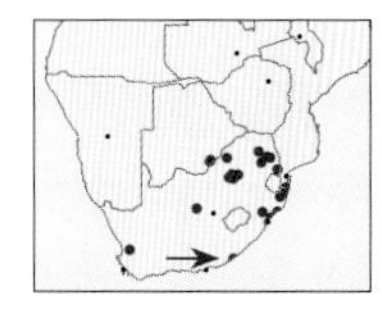

### 7 *Hodebertia testalis* — Incolourous Pearl

Forewing 12–16mm. Body and wings white to cream, wings crossed by 2 very faint wavy lines; front margin of forewings orange, orange band extending across thorax. Caterpillar pale green to yellow. One species in genus in region. **Biology:** Caterpillars feed on various Apocynaceae. **Habitat & distribution:** Various habitats; widespread across Africa, also Yemen and Saudi Arabia, Seychelles, St Helena, Ascension and Tristan da Cunha.

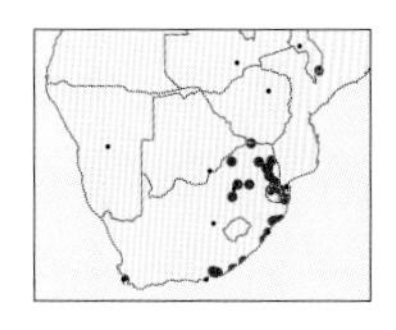

### 8 *Hydriris ornatalis* — Ornate Pearl

Forewing 8–10mm. Wings held widely extended laterally; forewings pale brown, crossed by wavy brown line, area distal to line (and sometimes basal area of wings) darker, row of small dots along border, indistinct kidney-shaped patch centrally; hindwings hyaline basally, light brown distally with marginal dots. Caterpillar transparent pale green with short terminal branched projections. One species in genus in region. **Biology:** Caterpillars polyphagous. Multivoltine, adults throughout the year. **Habitat & distribution:** Frost-free habitats; Western Cape, South Africa, north to tropical Africa, Indian Ocean islands, also Asia and Australia.

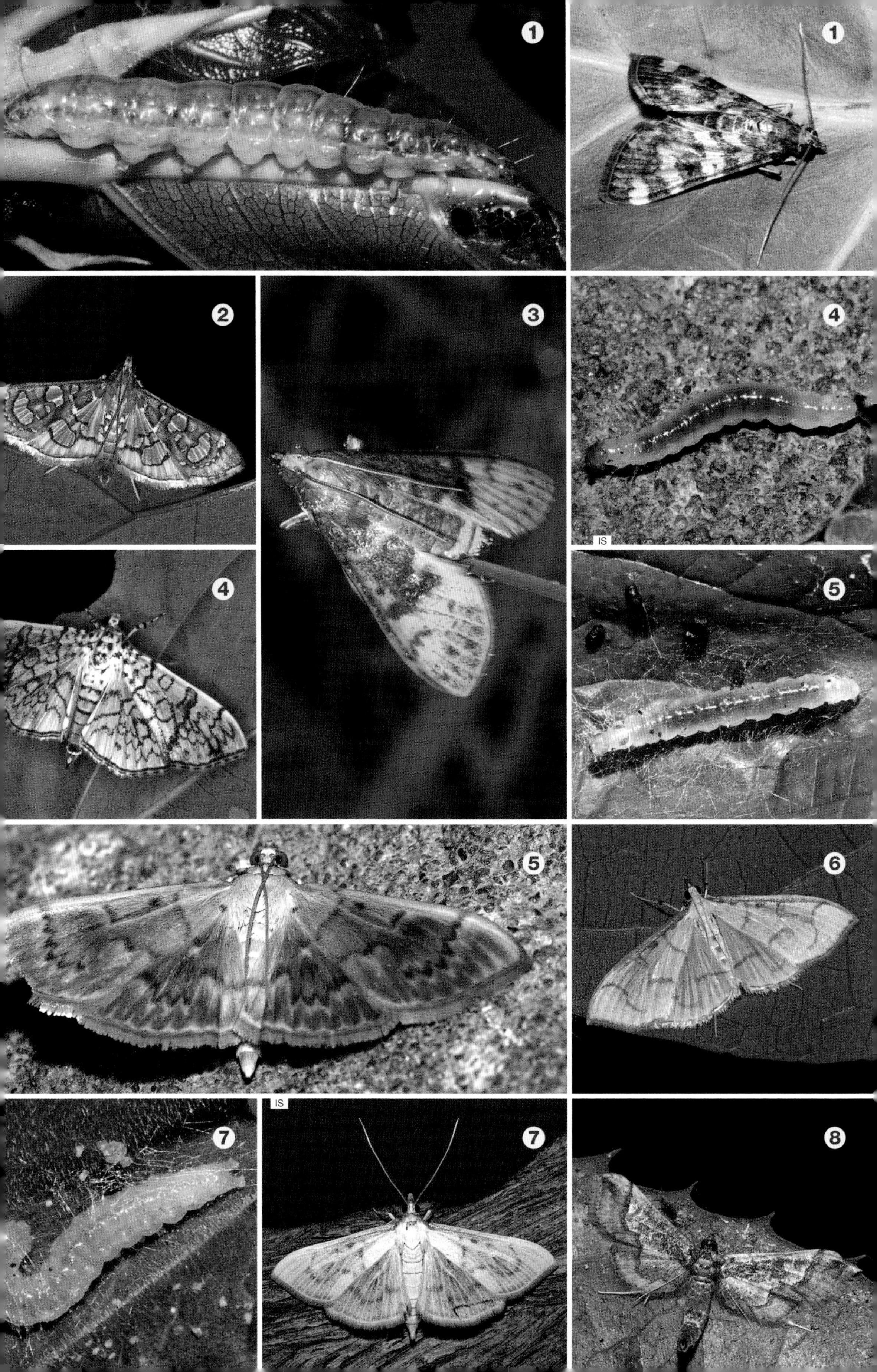

1
1
2
3
4
IS
4
5
5
6
IS
7
7
8

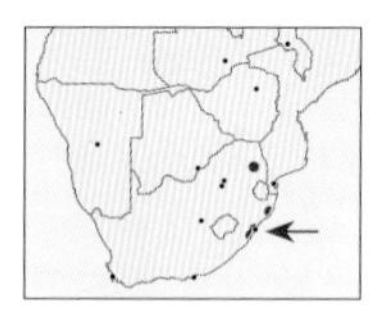

**1** ***Lygropia tetraspilalis*** Fourspot Orange

Forewing 8mm. At rest, wings held outspread; body yellow-orange; wings yellow and crossed by about 3 broad, broken orange bands; forewings with 3 black dots along fore-margin and on cell, wings fringed by long pale setae. Caterpillar pale translucent green with orange head. Six species in genus in region. **Biology:** Caterpillars feed on Crossberry (*Grewia occidentalis*). Adults recorded December–February. **Habitat & distribution:** Coastal bush and bushveld; KwaZulu-Natal north to Limpopo, South Africa.

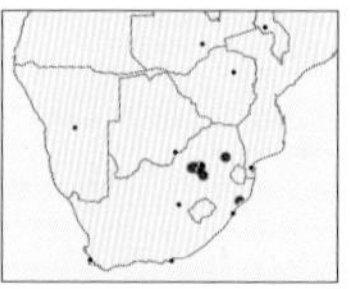

**2** ***Metasia eremialis*** Silver Pearl

Forewing 7–9mm. Outline triangular with labial palps strongly projecting forward and antennae folded back above body; body silvery, crossed by brown bands; forewings silvery, marked with diffuse brown bands and circles, row of longitudinal bars toward distal margin, which is edged by long, banded setae. Six species in genus in region. **Biology:** Caterpillar host associations unknown. Adults November–March. **Habitat & distribution:** Grassland; KwaZulu-Natal to Mpumalanga, South Africa.

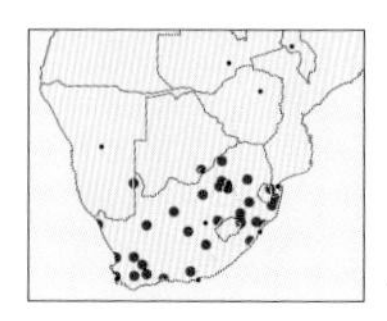

**3** ***Nomophila noctuella*** Rush Veneer

Forewing 11–16mm. Body plain brown; forewings elongate, light brown with indistinct darker-edged, kidney-shaped patches and sometimes zigzag lines; hindwings grey-brown with slightly darker margin. Caterpillar grey-green and spotted. Four species in genus in region; similar *N. africana* has the dark markings pronounced, black. **Biology:** Caterpillars form webs on grasses. Multivoltine, adults throughout the year, sometimes abundant, easily disturbed by day. **Habitat & distribution:** All habitats with grass; across Africa and Indian Ocean islands, Europe, Asia and North America.

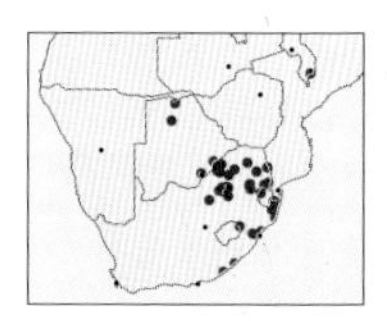

**4** ***Nausinoe geometralis*** Measured Pearl

Forewing 7–11mm. At rest, wings held outspread; body brown with pale spots dorsally; wings orange-brown with complex pattern of large, dark-ringed, pearly spots and streaks. Caterpillar pale green with darker head. Five species in genus in region. **Biology:** Caterpillars feed on Pink Jasmine (*Jasminum polyanthum*) and Tinderwood (*Clerodendrum glabrum*). Multivoltine, adults throughout the year. **Habitat & distribution:** Various habitats, seemingly absent from arid areas; Eastern Cape, South Africa, north to tropical Africa, Madagascar, Mauritius and La Réunion.

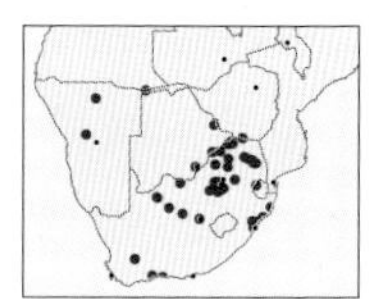

**5** ***Notarcha quaternalis*** Fourth Pearl

Forewing 6–10mm. Body and wings hyaline cream, crossed by about 5 pale orange wavy bands of raised scales (more and narrower than in *Lygropia tetraspilalis*, above, which it otherwise resembles); forewing has 3 black dots along costa and a cell spot; loses cream scales in flight, giving older specimens a hyaline appearance. Caterpillar light green with orange head. Five species in genus in region. **Biology:** Caterpillars polyphagous primarily on Malvaceae, helping to control alien weeds such as Arrowleaf Sida (*Sida rhombifolia*), and Flannel Weed (*S. cordifolia*). Multivoltine, adults throughout the year. **Habitat & distribution:** Various habitats, common in transformed habitats; Africa and Indian Ocean islands.

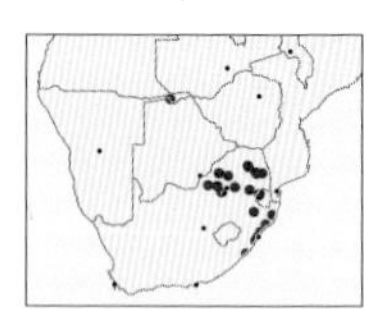

**6** ***Obtusipalpis pardalis*** Snow Leopard

Forewing 9–14mm. Body and forewings yellow-brown, marked by evenly spaced large white spots; hindwings white; a line of small black spots along margin of both wings. One species in genus in region. **Biology:** Caterpillars feed on wild figs (*Ficus* spp.). Probably bivoltine, adults August–April. **Habitat & distribution:** Forest and bushveld where figs grow; Eastern Cape, South Africa, north to tropical Africa.

**7** ***Palpita vitrealis*** White Pearl

Forewing 11–15mm. Body and wings white, wings translucent; forewings with brown or orange line along costa and marked with 2 minute black dots. Caterpillar bright green with orange head. Eleven species in genus in region. **Biology:** Caterpillars feed in leaf fold on various Oleaceae, including cultivated olives. Multivoltine, adults throughout the year. **Habitat & distribution:** Variety of habitats where Oleaceae are found; across Africa, Indian Ocean islands, Europe, the Americas and Australia.

**8** ***Palpita elealis*** Bordered Pearl

Forewing 13–17mm. Similar to *Diaphania indica* (p.128) but larger, wings broader, black margin around thorax and both wings and 2 black dots on each forewing. Caterpillar green with distinctive black warts. **Biology:** Caterpillars feed on African Teak (*Breonadia salicina*), Poison Devil's-pepper (*Rauvolfia vomitoria*) and *Tabernanthe* spp. Bivoltine, adults October–May. **Habitat & distribution:** Moist forests; Eastern Cape, South Africa, north to tropical Africa.

1
1
2
3
4
IS
4
5
5
6
7
7
8
8
IS

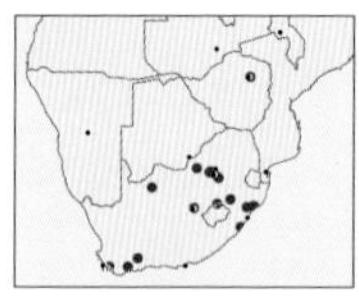

## 1 *Pagyda salvalis* Sage Pearl

Forewing 9–11mm. Broadly triangular in outline; wings pale with orange tint, crossed by several oblique orange or brown bands. Caterpillar grey-green with indistinct longitudinal white stripes, band of eye like spots on each segment. **Biology:** Caterpillars polyphagous on figworts (Scrophulariaceae), partially skeletonising leaf. Bivoltine, adults September–May, peaking October and March. Two species in genus in region. **Habitat & distribution:** Variety of habitats where figworts grow; Western Cape, South Africa, north to Zimbabwe, described from Sri Lanka, also in Australia and widespread across Asia.

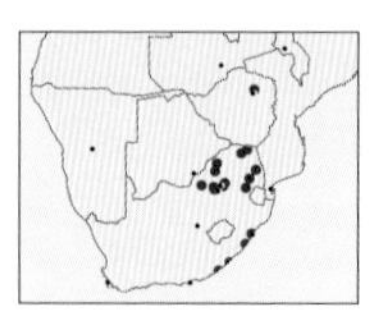

## 2 *Pardomima callixantha* Netted Pearl

Forewing 9–12mm. At rest, wings held outspread; body yellow with brown stripes; wings orange-yellow, crossed by network of wavy, sometimes merging brown lines, terminal area alternating yellow and brown diagnostic. Caterpillar pale green with orange head. Four similar species in genus in region, others have terminal area brown. **Biology:** Caterpillars feed inside folded leaf on bride's bush (*Pavetta* spp.). Multivoltine, adults throughout the year. **Habitat & distribution:** Forest and bushveld where *Pavetta* spp. grow; Eastern Cape, South Africa, north to tropical Africa.

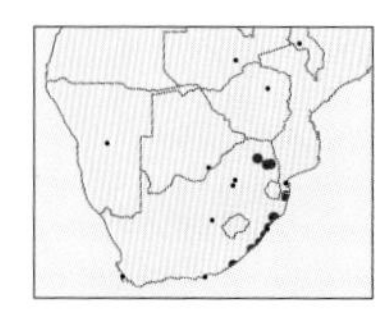

## 3 *Parotis baldersalis* Verdant Pearl

Forewing 13–16mm. Body and wings pale green; forewing costa orange, both wings with small black cell spot and green cilia diagnostic. Caterpillar greenish-yellow. Eight similar species in genus in region; *P. prasinalis* is smaller, blue-green with orange cell spots. **Biology:** Caterpillars mine leaves of Quinine Tree (*Rauvolfia caffra*). Probably multivoltine, adults throughout the year. **Habitat & distribution:** Moist forest and coastal bush; Eastern Cape, South Africa, north to tropical Africa.

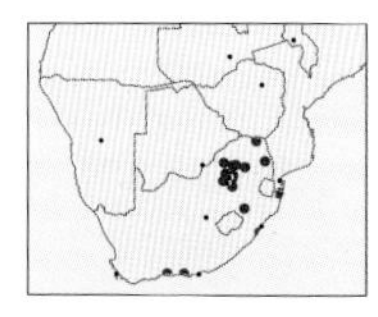

## 4 *Parotis prasinophila* Monkey Pearl

Forewing 11–16mm. Specimens from drier areas smaller; similar to *P. baldersalis* (above); wings with dark brown cilia and some terminal black spots. Caterpillar translucent green with 2 dorsal rows of prominent black spots. **Biology:** Caterpillars feed on monkey apple (*Strychnos* spp.), spinning 2 leaves together. Bivoltine, adults September–March. **Habitat & distribution:** Forest, bushveld and grassland where *Strychnos* spp. grow; southern Cape north to tropical Africa, also Madagascar and Seychelles.

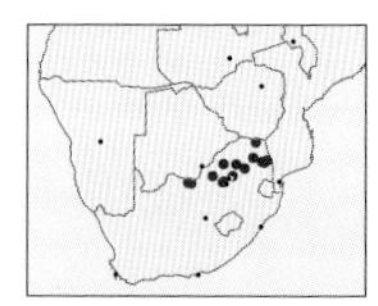

## 5 *Pleuroptya balteata* Orange Leaf Folder

Forewing 12–14mm. Body and wings orange; wings crossed by indistinct wavy dark lines, shape of postmedian line diagnostic. Caterpillar light green with dark head. Three species in genus in region. **Biology:** Caterpillars feed in leaf fold on various Malvaceae in region, recorded on many other plants elsewhere. Bivoltine, adults November–March. **Habitat & distribution:** Bushveld, North West province, South Africa, to tropical Africa, southern Europe, Asia and Australia.

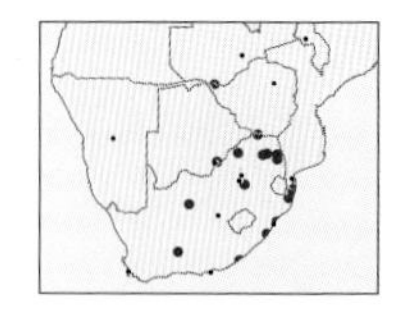

## 6 *Psara atritermina* Grey-waved Pearl

Forewing 8–10mm. Body and wings greyish-cream to light brown, crossed by 2 wavy dark lines and with a few black dots. Caterpillar light green with dark patches on first thoracic segments. Five species in genus in region. **Biology:** Caterpillars reported feeding on Devil's Horsewhip (*Achyranthes aspera*) and *Isoglossa* spp. Bivoltine, adults August–November and March–May. **Habitat & distribution:** Variety of habitats; Western Cape, South Africa, north to tropical Africa.

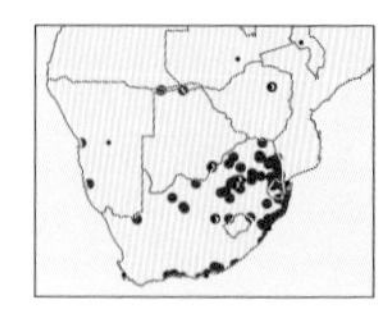

## 7 *Spoladea recurvalis* Beet Webworm

Forewing 11–12mm. Body dark brown with white markings on head and across abdomen; wings dark brown; forewings with 1 complete and 1 partial white band; hindwings with 1 complete band. Caterpillar pale green with pale longitudinal stripes. One species in genus in region. **Biology:** Caterpillars polyphagous on herbaceous plants, rolling the leaves with silk, initially feeding on them from below. Multivoltine in warmer parts, adults throughout the year. **Habitat & distribution:** Many habitats, probably expanding its range following human habitation where it is common; widely distributed across Africa, Asia, North and South America, Oceania and rarely Europe.

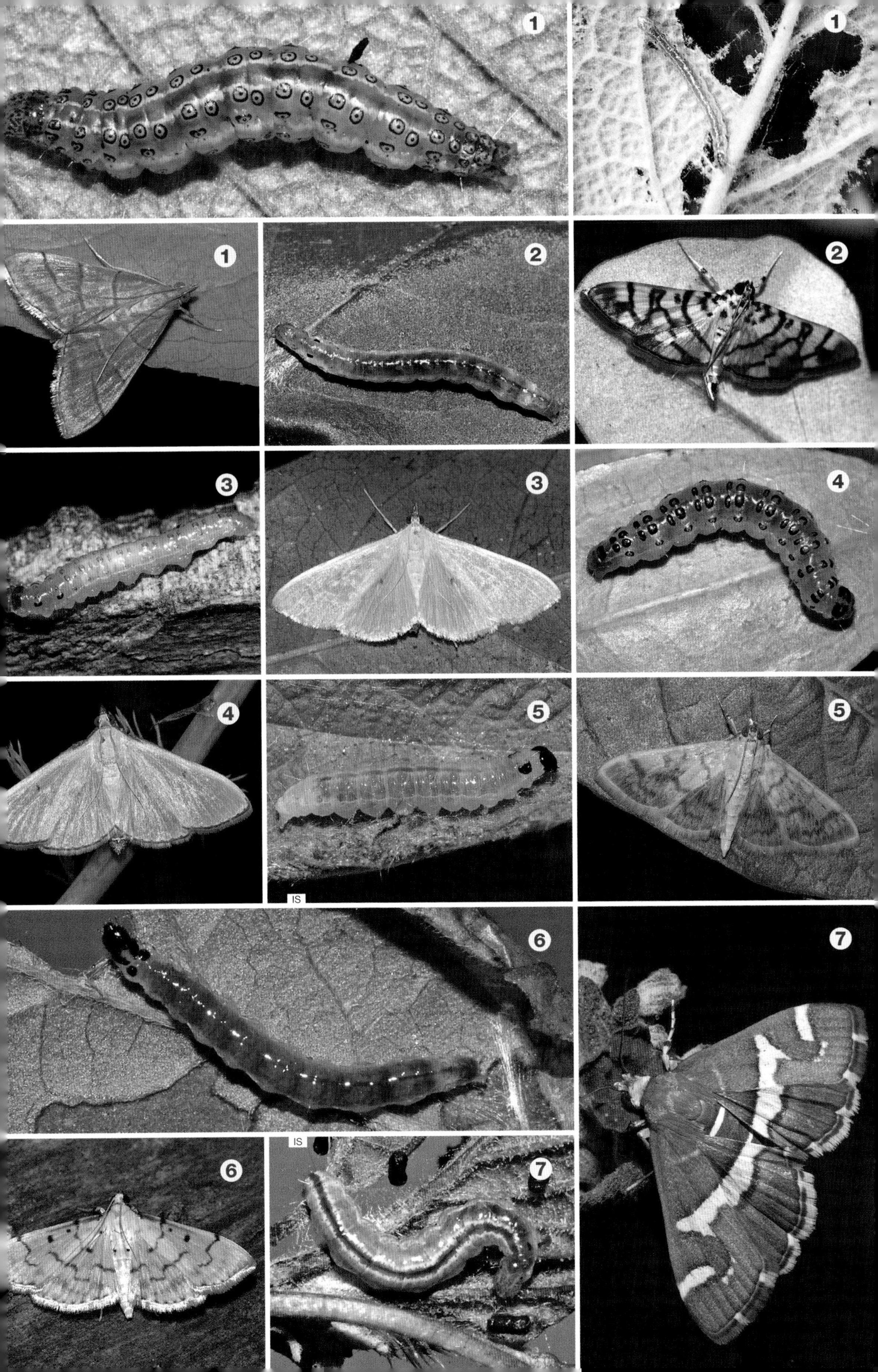

1
1
1
2
2
3
3
4
4
5
5
IS
6
7
IS
6
7

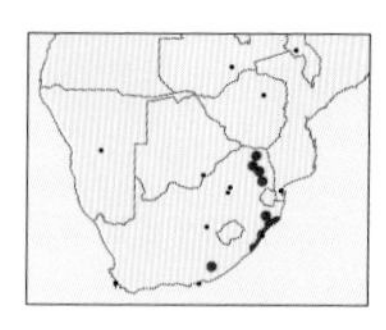

**1 *Stemorrhages sericea*** Large Emerald Pearl

Forewing 23–26mm. Body and forewings pale hyaline bluish-green with a brown edge that continues across sides of thorax; male can evert conspicous tuft of brown setae from end of abdomen. Caterpillar pale green with distinctive rows of black 'warts'. One species in genus in region. **Biology:** Caterpillars feed on broadleaf trees (Apocynaceae and Rubiaceae), folding the leaf into shelter. Multivoltine, adults throughout the year. **Habitat & distribution:** Moist forests; Eastern Cape, South Africa, north to tropical Africa, including Madagascar and Indian Ocean islands.

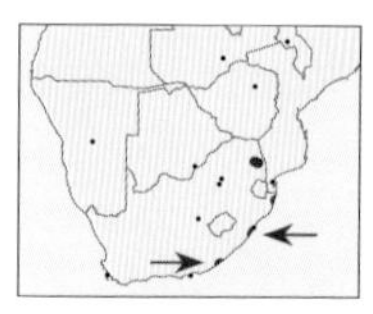

**2 *Syllepte straminea*** Mustard Pearl

Forewing 13–16mm. Could be confused with *Pleuroptya balteata* (p.134) but lines and cell spot dark, also with orange *Psara* spp., which are much smaller; pale orange, wings crossed by 2 or 3 wavy dark brown lines, dark cell spot, fringed with darker setae. Caterpillar green with orange head. Fourteen species in genus in region. **Biology:** Caterpillars feed on Pink Wild Pear (*Dombeya burgessiae*). Bivoltine, adults October–March. **Habitat & distribution:** Thick bush and forest; Eastern Cape to Limpopo, South Africa.

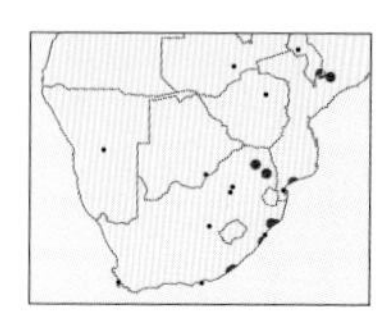

**3 *Syllepte purpurascens*** Twin-tooth Purple Pearl

Forewing 13–15mm. Several similar species with purple sheen that differ in the shape and number of white marks on wings; this species with double tooth mark on forewings and few other marks, lines indistinct. Caterpillar greenish with orange head. **Biology:** Caterpillars feed on Cape Grape (*Rhoicissus tomentosa*). Multivoltine, adults throughout the year. **Habitat & distribution:** Coastal bush and moist forest; Eastern Cape, South Africa, north to tropical Africa.

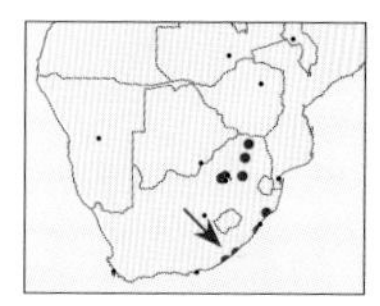

**4 *Syllepte nasonalis*** Tri-tooth Purple Pearl

Forewing 11–12mm. Smaller than other *Syllepte* spp. with purple sheen; markings hyaline with 3 teeth on forewings; hindwings rounded, hyaline with broad brown band. **Biology:** Caterpillar host associations unknown. Adults September–April. **Habitat & distribution:** Forest and bushveld; Eastern Cape north to Limpopo, South Africa.

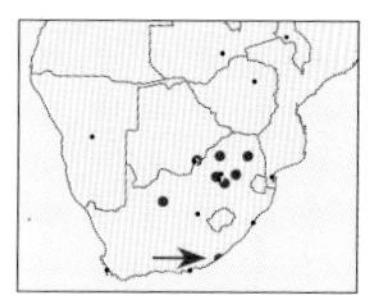

**5 *Syllepte vagans*** Leaden Pearl

Forewing 8–15mm. Triangular in outline with largely hyaline wings; forewings with dark brown to purplish base extending into an elongate blue to purple longitudinal band, hind margin edged with double line. **Biology:** Caterpillar host associations unknown. Moth adopts unique resting posture possibly mimicking a spider with wing lines forming the spider's legs. Adults October–March. **Habitat & distribution:** Coastal bush and bushveld; Eastern Cape, South Africa, north to tropical Africa.

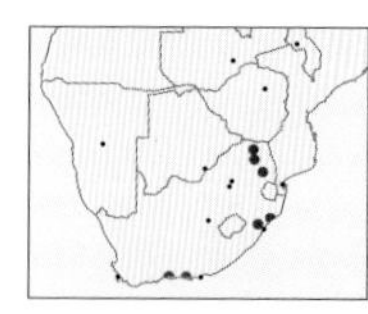

**6 *Syllepte mesoleucalis*** Toothed Pearl

Forewing 11–13mm. Similar to *S. vagans* (above), but with normal resting posture, postmedian line kinked, a series of distinctive teeth beyond postmedian on all wings. **Biology:** Caterpillar host associations unknown. Adults October–March. **Habitat & distribution:** Moist forests; southern Cape north to Limpopo, South Africa.

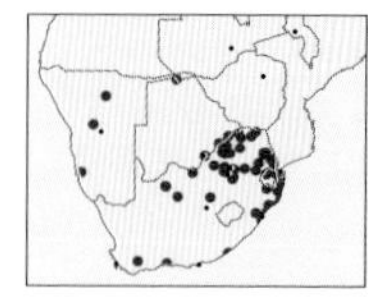

**7 *Synclera traducalis*** Variegated Pearl

Forewing 8–12mm. Unmistakable, at rest, wings held outspread; body and wings white, covered with intricate network of lace-like brown markings; large hyaline patches on all wings. Caterpillar green. One species in genus in region. **Biology:** Caterpillars feed on various Rhamnaceae. Multivoltine in warmer parts, adults throughout the year, peaking January–May. **Habitat & distribution:** Variety of habitats where Rhamnaceae grow; Western Cape, South Africa, north through Africa, Middle East, Europe and India.

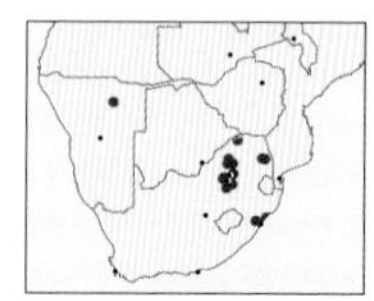

**8 *Terastia margaritis*** Pearl-spotted Porphyry

Forewing 14–18mm. Profile broadly triangular; forewings shaded from pale to dark brown with iridescent pearl spots and wavy lines; hindwings clear white with wide brown border. Caterpillar white dorsally with light and dark brown lateral bands, head capsule dark. Two species in genus in region. **Biology:** Caterpillars feed in webs in leaf fold of coral trees (*Erythrina* spp.). Bivoltine, adults September–March, peaking November and February. **Habitat & distribution:** Forest, bushveld and grassland where *Erythrina* spp. grow; KwaZulu-Natal, South Africa, north to tropical Africa.

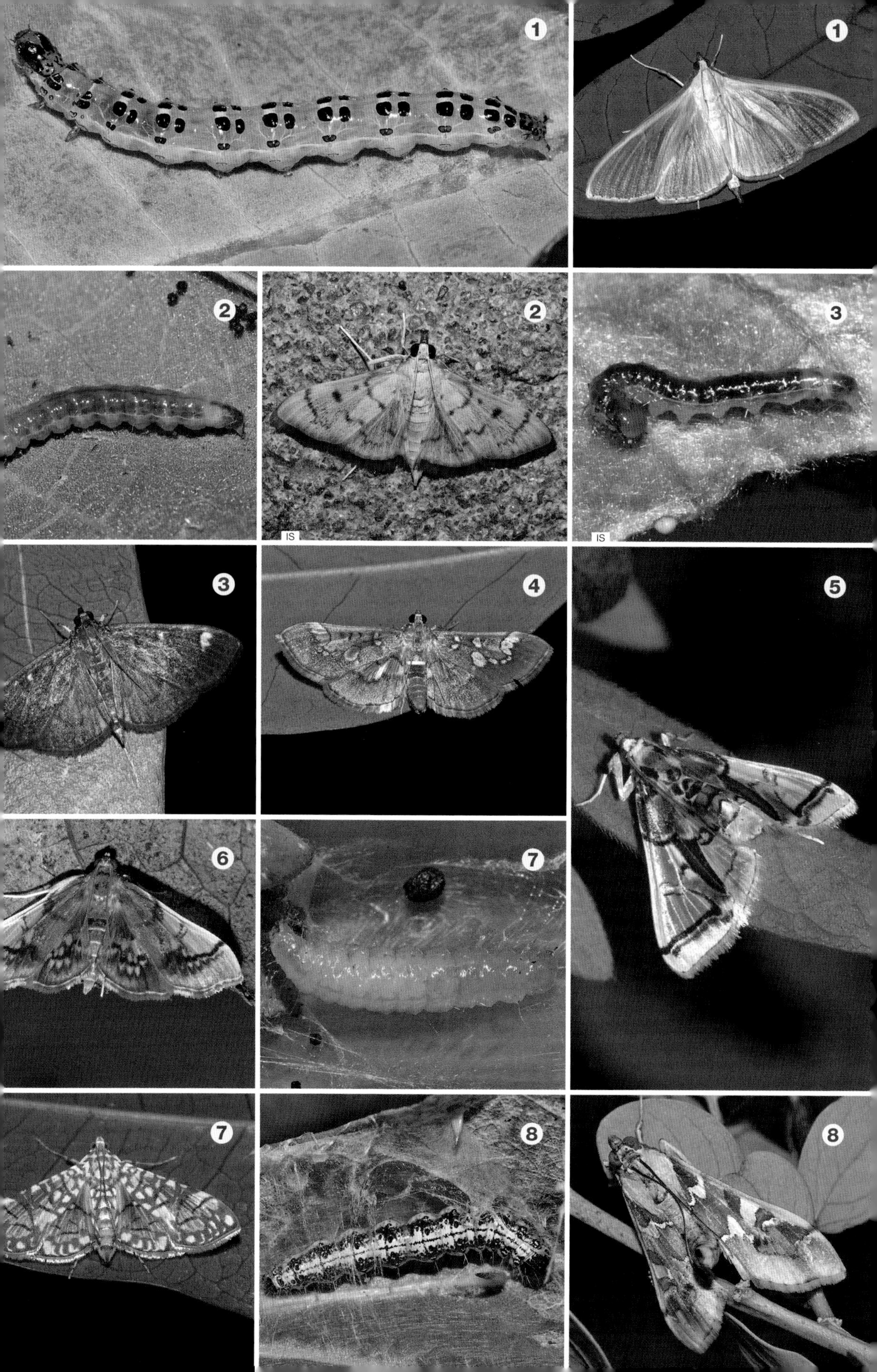
1
1
2
2
3
IS
IS
3
4
5
6
7
7
8
8

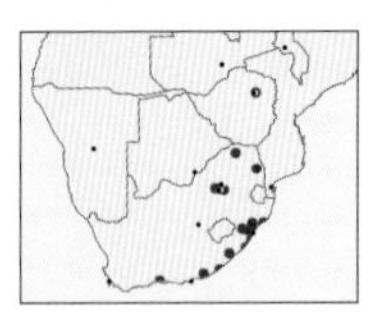

### 1 *Terastia africana* — African Porphyry

Forewing 15–23mm. Wings brown and narrow, partially transparent with irregular brown markings and scalloped margins; abdomen bears unusual side projections and is often held raised. **Biology:** Caterpillars feed in webs or tunnelling in shoots of coral trees (*Erythrina* spp.). Multivoltine if conditions allow, adults throughout the year. **Habitat & distribution:** Forest, bushveld, grassland and gardens where *Erythrina* spp. grow; southern Cape, South Africa, north to tropical Africa.

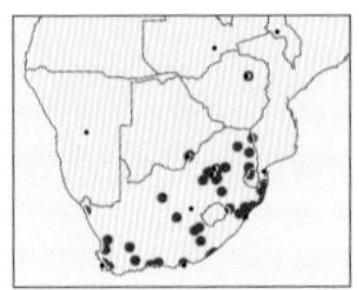

### 2 *Udea ferrugalis* — Rusty Dot Pearl

Forewing 7–10mm. Profile triangular; legs white; forewings orange-brown with darker lines and spots that fade in older specimens, postmedian line dotted and evenly curved; hindwings whitish-grey; fore- and hindwings with minute dots and long marginal setae. Caterpillar pale yellow-green with pale orange head capsule. Five species in genus in region; similar *U. infuscalis* has postmedian line sharply pointed. **Biology:** Caterpillars polyphagous on herbaceous plants in silken leaf fold. Multivoltine if conditions allow, adults throughout the year. **Habitat & distribution:** All habitats with herbaceous vegetation, abundant near human habitation; widely distributed in Africa, Europe and Asia.

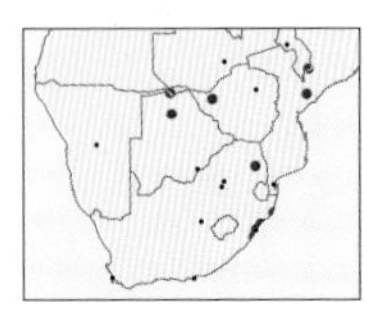

### 3 *Zebronia phenice* — Zebra Pyrale

Forewing 10–15mm. At rest, wings held outspread; body and wings hyaline white, wings marked with dark brown to black border and 3 broad lines. Caterpillar pale green with longitudinal white stripes. One species in genus in region. **Biology:** Caterpillars polyphagous on broadleaf forest trees. Adults diurnal and nocturnal, multivoltine throughout the year. **Habitat & distribution:** Lowland forest and bush; KwaZulu-Natal coast, South Africa, north to tropical Africa, also Madagascar and Indian Ocean islands.

## FAMILY DREPANIDAE Hook-tips

**Small to medium-sized, mostly brown moths with distinctive broad forewings with hook-shaped apex. Proboscis vestigial or absent. Caterpillars very distinctive and sometimes of bizarre appearance, lacking claspers and tapering to a point at rear – at rest, both head and tail are usually raised. Caterpillars feed on leaves of trees and shrubs, and pupate between leaves spun together with silk. About 660 described species globally, 9 in the region.**

### Subfamily Drepaninae

Most of the 800 tropical or subtropical species have hooked forewing tips, and bipectinate antennae. Wings usually held flat (horizontally) at rest. Caterpillar with long hairs extending from rear, pupa often covered in bluish wax.

### 4 *Negera natalensis* — Natal Hook-tip

Forewing 19–24mm. Wings brown with numerous fine, mottled lines, resembling veins on a dead leaf, shape distinctive, with slightly falcate tips to forewings; hindwings strongly squared-off posteriorly and held in an upward-bowed posture when at rest. Caterpillar very unusual, with thoracic segments swollen, and a dorsal hook on the third, abdomen tapering with long, curved terminal process. Two species in genus in region. **Biology:** Caterpillars feed on various forest Rubiaceae. Multivoltine, adults throughout the year. **Habitat & distribution:** Moist forest and coastal bush; Eastern Cape, South Africa, north to tropical Africa.

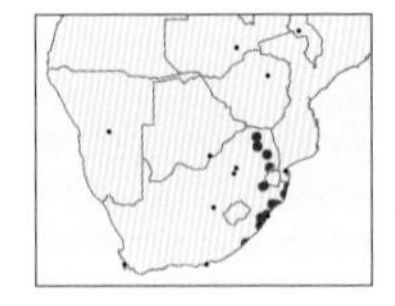

### 5 *Gonoreta opacifinis* — Scalloped Hook-tip

Forewing 15–17mm. Body and wings variable in colour, olive green to light brown with irregular darker mottling and distinct thin brown (or two-tone brown and white) lines, postmedian line ending at apex; forewings with sharp hook, posterior margin strongly scalloped distally. One species in genus in region. **Biology:** Caterpillar host associations unknown. Adults throughout the year. **Habitat & distribution:** Moist forest; Eastern Cape, South Africa, north to tropical Africa.

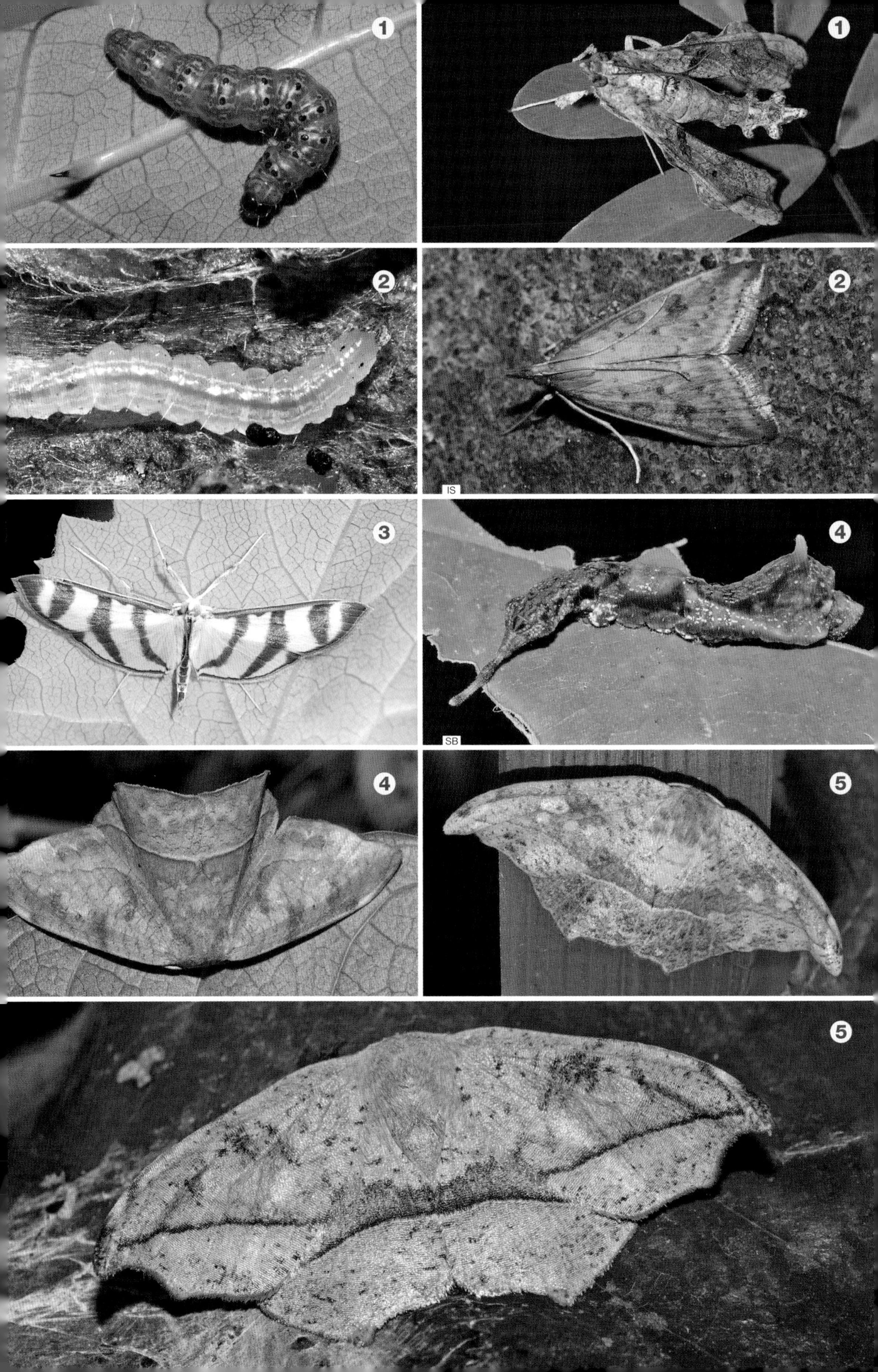
1
1
2
2
IS
3
4
SB
4
5
5

## Subfamily Thyatirinae

The 200-odd species occur mainly in Palaearctic and Oriental regions, 4 species in Africa. Strongly built, resembling drab noctuids. Forewings often spotted and rarely hooked at their tips, wings held roof-like at rest. Antennae usually thread-like. Caterpillars often feed in rolled-up leaves.

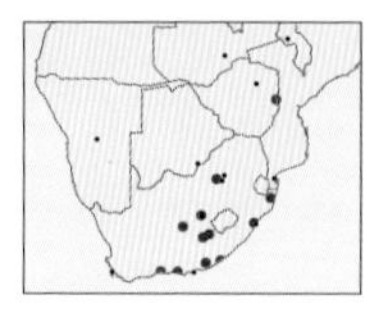

### 1 *Aethiopsestis austrina* — Hooded Hook-tip

Forewing 17–23mm. Thoracic setae upturned, giving a hooded appearance when at rest; forewings greyish-brown, crossed by many curved lines, median area variable, forming a white, black or black and white band in some individuals; 2 sharp black streaks over forewings variably present; hindwings pale dirty brown. Two species in genus in region. **Biology:** Caterpillar host associations unknown. Probably bivoltine, adults August–September and March–April. **Habitat & distribution:** Moist forests and dry wooded kloofs; southern Cape, South Africa, north to Zimbabwe.

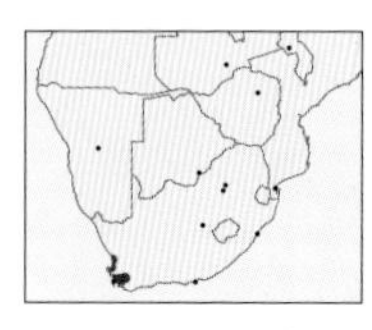

### 2 *Marplena designina* — Double-hooded Hook-tip

Forewing 17–23mm. Thoracic setae separated and upturned, giving the species a double-hooded appearance when at rest; forewings greyish-white, antemedial and postmedian lines double with 2 circular rings in cell; hindwings dark grey, pale at base. One species in genus in region. **Biology:** Caterpillar host associations unknown. Adults August–October. **Habitat & distribution:** Renosterveld and fynbos; Western Cape, South Africa.

# FAMILY URANIIDAE Swallowtail Moths

**Mostly tropical family of small to large moths (forewing length 5–80mm); hindwings usually bear tails. The family includes diurnal species, which are amongst the most brightly coloured Lepidoptera in the world (one species of the subfamily Uraniinae, *Chrysiridia croesus*, occurs in Tanzania and Kenya), as well as nocturnal species. Epipleminae are smaller and mostly nocturnal but some are also active during the day while Microniinae are nocturnal; wings often held in unusual, opened posture. About 700 species globally, with 20 in the region.**

## Subfamily Microniinae

Old World species, usually small, wings white or grey with fine striations, hindwings with short tails and associated black dots. Antennae thread-like, sometimes flattened. Fly weakly at dusk or at night, when disturbed settling on undersides of leaves.

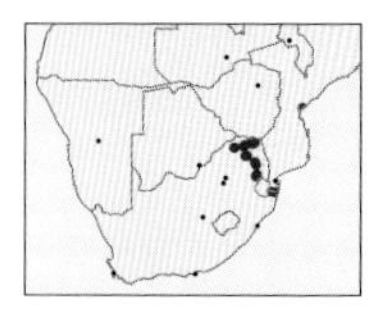

### 3 *Acropteris illiturata* — Plain Swallowtail

Forewing 15–20mm. Body slender and wings white, wings extended laterally, held flat against substratum; forewings with pointed apex; hindwings with short tail, both wings with numerous faint grey scribblings. One species in genus in region. **Biology:** Caterpillar host associations unknown. Elsewhere *Acropteris* caterpillars feed on milkweeds (Apocynaceae). Adults October–April. **Habitat & distribution:** Lowland forest and thick bush; KwaZulu-Natal, South Africa, north to tropical Africa.

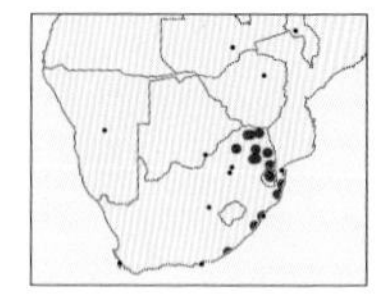

### 4 *Urapteroides recurvata* — Recurved Swallowtail

Forewing 21–26mm. Smoothly convex front margin of forewings bearing tiny brown spots, wings white, variably spotted with fine black spots in some individuals, neatly divided by 3 fine transverse brown lines, bordered by thin brown line; hindwings with short tail and 2 black false eyespots. Two species in genus in region. **Biology:** Caterpillar host associations unknown. Adults September–May, peaking October–November. **Habitat & distribution:** Frost-free wooded areas; Eastern Cape, South Africa, north to East Africa.

1
2
2
3
4

## Subfamily Epipleminae

Pantropical group, although more diverse at altitude. Small, hindwings tailed, fasciae (thin striated lines) parallel with posterior tail. Antennae usually thickened and plate-like, occasionally uni- or bipectinate.

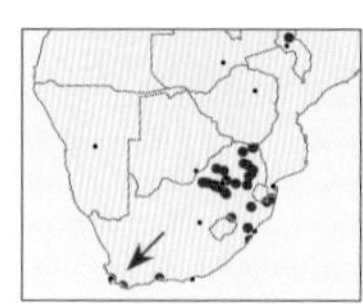

### 1 *Epiplema inconspicua* — Gap-wing Swallowtail

Forewing 8–12mm. Characteristic resting posture; forewings narrowest at base and broader toward apex, held forward at rest, although may also be tightly rolled at rest; widely separated from hindwings, which have 2 small tails on hind margin; postmedian lines well developed over both wings, diagnostically sharply toothed in middle of hindwing. Caterpillar green to yellow, hair bases black and head capsule orange. Seven similar species in genus in region; *E. pulveralis* has hindwing tooth rounded; *E. tristis* has forewing postmedian line straight toward costa. **Biology:** Caterpillars feed on rock elders (*Afrocanthium* spp.). Adults usually found resting on leaves. Multivoltine, adults August–May. **Habitat & distribution:** Forest, grassland and bushveld where rock elders grow; Western Cape, South Africa, north to Malawi.

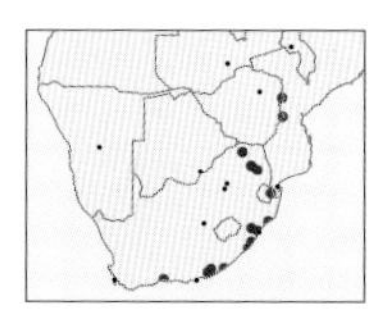

### 2 *Epiplema reducta* — Straight-line Swallowtail

Forewing 12–14mm. Differs from other *Epiplema* spp. by being larger and lighter brown to whitish brown; hindwing postmedian line being straight from inner margin to tooth is diagnostic. Caterpillar black in final instar, orange in earlier instars. **Biology:** Caterpillars feed gregariously on *Kraussia floribunda* (Rubiaceae). Multivoltine, adults throughout the year. **Habitat & distribution:** Forests and coastal bush; Western Cape north to Zimbabwe.

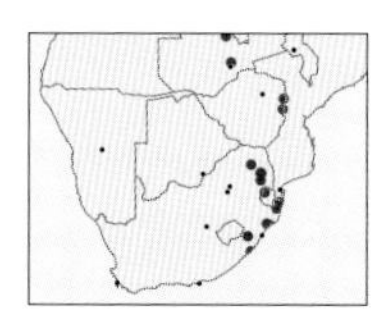

### 3 *Leucoplema dohertyi* — Straight-spots Swallowtail

Forewing 6–8mm. Similar resting posture to *Epiplema* spp. (above) with forewings spread laterally and widely separated from hindwings; creamy white with distinct triangular brown markings on both wings, straight row of black terminal spots on forewing diagnostic. Caterpillar green with black setae. Three species in genus in region. **Biology:** Caterpillars feed on various Rubiaceae, including *Pavetta* and *Coffea*. Multivoltine, adults throughout the year. **Habitat & distribution:** Moist forests; Eastern Cape, South Africa, north to tropical Africa.

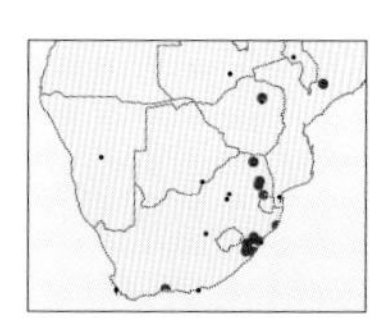

### 4 *Leucoplema triumbrata* — Angled-spots Swallowtail

Forewing 9–11mm. Similar to *L. dohertyi* (above), but larger with markings much reduced; sharply angled row of black terminal spots on forewing diagnostic. **Biology:** Caterpillar host associations unknown. Multivoltine, adults throughout the year. **Habitat & distribution:** Moist forests; Western Cape, South Africa, north to tropical Africa.

## FAMILY SEMATURIDAE Sematurid Moths

**Minor family of mostly large day- or night-flying moths, usually with tailed hindwings. Antennae thickened toward clubbed or hooked ends and tympanal organ (for hearing) always absent. About 41 species globally, most in the Americas. The single African genus and species belongs to its own subfamily (Apoprogoninae).**

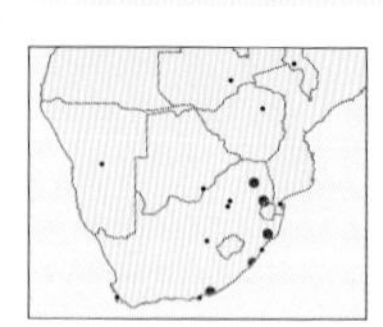

### 5 *Apoprogones hesperistis* — Skipper Sematurid

Forewing 16–22mm. Resembles a skipper butterfly; antennae markedly thickened distally; body brown; wings mottled greyish-brown crossed by darker wavy brown lines, paler patch at base of hindwings, which lack tails (unusually, for members of this family). **Biology:** Host associations unknown. Adults fly fast, low to the ground early in the morning, not attracted to light. **Habitat & distribution:** Moist forests; Eastern Cape to Limpopo, South Africa.

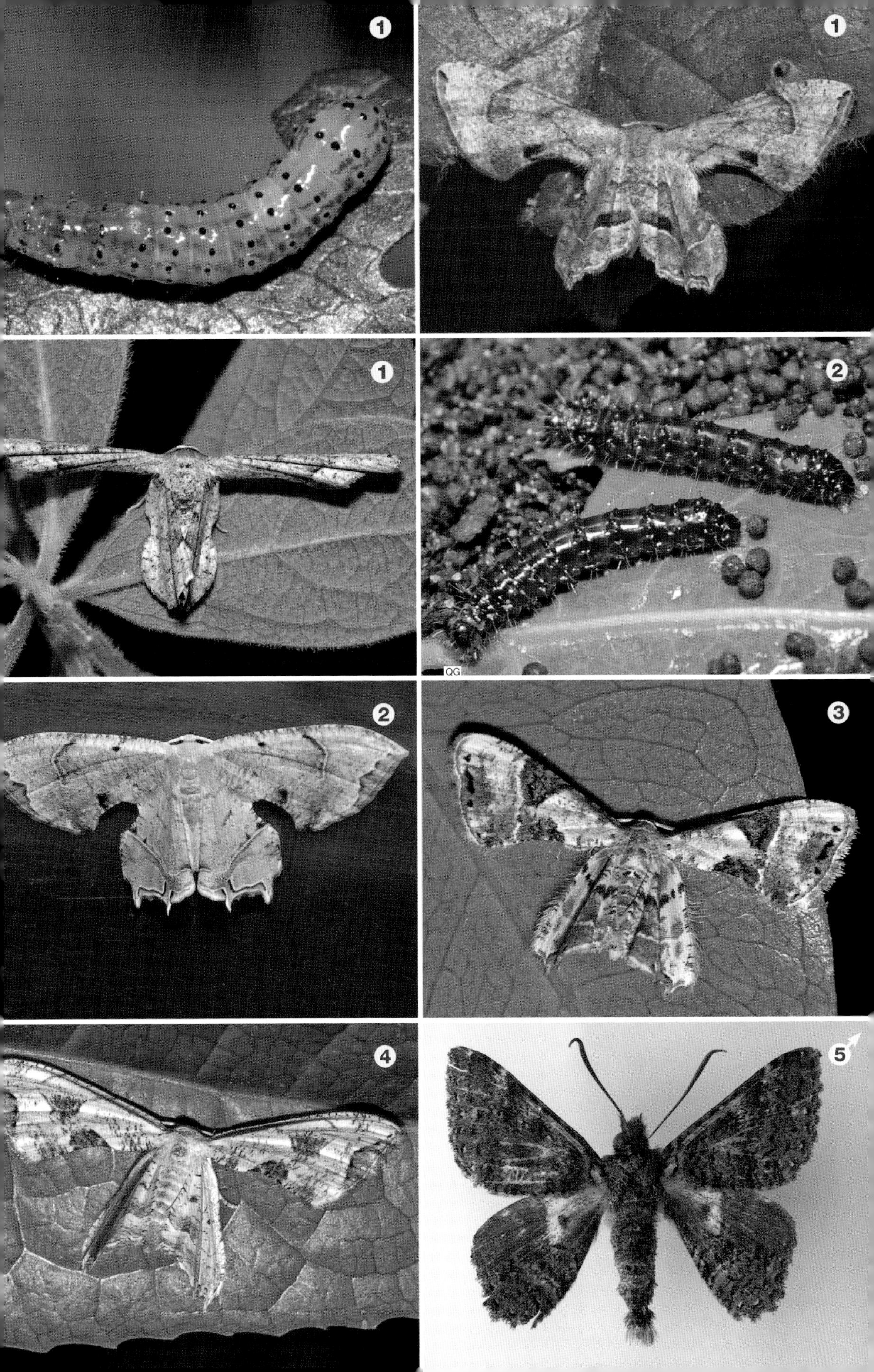
1
1
1
2
QG
2
3
4
5

# FAMILY GEOMETRIDAE Geometrid Moths

**Very large and diverse family, second only to Erebidae in species diversity. Characterised primarily by a pair of tympanal organs situated at the base of their abdomen and the unusual form of the caterpillars, which are usually long and slender, with almost all having only 2 pairs of developed abdominal prolegs. These 'loopers' or 'inchworms' have a deliberate locomotion, bringing the hind end forward, causing the body to form an upright loop, then extending the front, hence the family name Geometridae or 'earth measurers'. Adults show a wide diversity of body form, typically being thin bodied with very broad wings often held flat and held outspread against the substratum. However, some have thicker bodies and have a triangular outline at rest, or have narrow wings held pitched roof-like against the body, or held aloft, as in butterflies. A few species have short-winged females. Adults are mostly nocturnal, cryptic in colouration, blending with tree bark, lichens or vegetation, but a few are diurnal and brightly coloured. Caterpillars usually feed in the open, with very few making larval shelters. They are extremely cryptic, mimicking twigs, often perfectly resembling their host plant; those in the subfamily Geometrinae have the head capsule incised to imitate new shoots. Over 23,000 global species, in 8 subfamilies. More than 1,500 known from the region in 5 subfamilies.**

## Subfamily Sterrhinae Waves

Over 200 regional species. Most are nocturnal, with some exceptions. Wave-like patterns on the wings of most adults. Caterpillars usually long and slender, lacking appendages, with rounded heads and often a 'question mark-like' posture. Many are well-adapted to arid conditions and can be abundant in all habitats, but less so at high altitudes. The cryptic caterpillars feed mostly on low-growing herbaceous plants, a few are host specialists. Most members of the tribe Sterrhini feed on wilted or dried plant matter and can be found in the driest deserts.

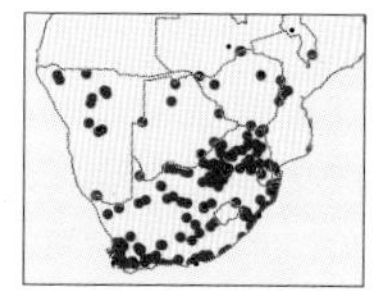

### 1 *Rhodometra sacraria* — Vestal

Forewing 12–14mm. Rests with wings held tent-like against sides of body; 2 distinct seasonal forms; wet-season form forewings creamy yellow with partially pinkish-red costa and oblique stripe; hindwings white; dry-season form browner with brown lines and cell spot. Caterpillar slender, yellow or green. Four species in genus in region. **Biology:** Adults rest on grass stems, active by day and night, abundant at times. Caterpillars feed on various Polygonaceae. Multivoltine throughout the year. **Habitat & distribution:** Probably the most ubiquitous geometrid species; in all habitats across Africa, including Madagascar and La Réunion, also Europe and Asia.

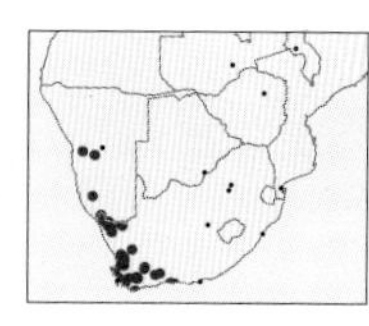

### 2 *Rhodometra participata* — Dimorphic Vestal

Forewing 11–13mm. Very similar to the dry-season form of *R. sacraria* (above), postmedian line always brown and area beyond usually lighter, white cell spot in most specimens with brown striations in median area, a rare purple-red form exists; hindwings white. Caterpillar green with red dorsal stripe. **Biology:** Caterpillars recorded on *Hermbstaedtia* spp. Adults August–April. **Habitat & distribution:** Succulent karoo and dry fynbos habitats; Western Cape, South Africa, north to Namibia.

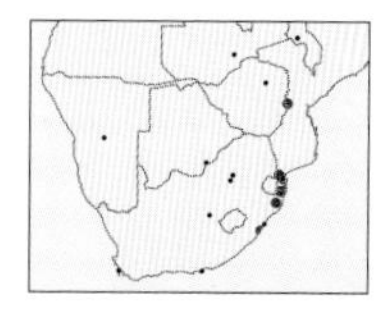

### 3 *Rhodometra satura* — Red Vestal

Forewing 10–12mm. Similar to the wet-season form of *R. sacraria* (above), but forewing ground colour a deep orange-yellow, postmedian line, costa and termen broadly rosy-red, in some individuals forewings are completely red; hindwings brown. **Biology:** Caterpillar host associations unknown. Adults December–February and May–June. **Habitat & distribution:** Coastal bush and sand forest; KwaZulu-Natal, South Africa, north to Mozambique; of conservation concern, not seen south of Maputaland in recent years.

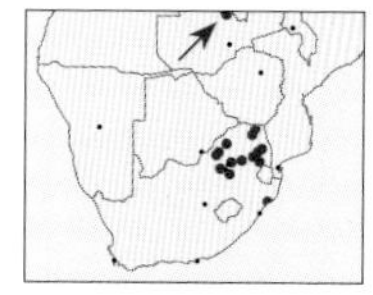

### 4 *Afrophyla vethi* — Rail Vestal

Forewing 14–19mm. Broadly triangular at rest; thorax pale pink to grey; forewings with falcate tips, pinkish to cream, finely spotted with pale brown to pink, crossed by 2 parallel lines, postmedian kinked toward apex with small cell spot; hindwings white with partial postmedian line. Caterpillar cream with variable black dorsal markings and oblique whitish ribs. **Biology:** Caterpillars recorded on Cape Leadwort (*Plumbago auriculata*) and *Psydrax livida* (Rubiaceae). Multivoltine, adults throughout the year. **Habitat & distribution:** Bushveld and forest; KwaZulu-Natal, South Africa, north to tropical Africa, also Madagascar and La Réunion.

1
1
1
2
2
3
4
4

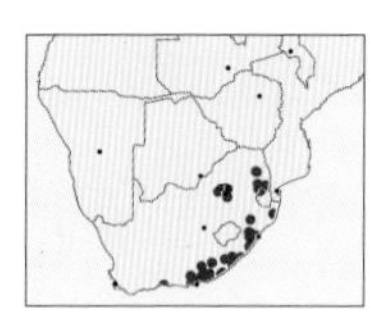

## 1 *Palaeaspilates inoffensa* — Inoffensive Vestal

Forewing 13–17mm. Body and wings pink-cream variably suffused with brown striations; forewings falcate, with a few dark spots, diagnostic straight pinkish-brown postmedian line of variable intensity. Caterpillar variable, green to various shades of brown, occasionally strongly marked with darker dorsal patches. Two species in genus, similar *P. carneata* has the postmedian line kinked and occurs in arid western areas. **Biology:** Caterpillars feed on Cape Leadwort (*Plumbago auriculata*). Multivoltine, adults throughout the year. **Habitat & distribution:** Various relatively moist habitats, often in gardens; Western Cape north to Limpopo, South Africa.

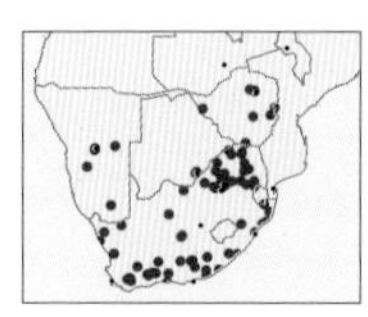

## 2 *Chlorerythra rubriplaga* — Green Vestal

Forewing 11–15mm. Two distinct seasonal forms: wet-season form green with straw margin, forewings with red cilia and yellow and red oblique postmedian line; dry-season form brown or straw with dark line; hindwings lighter, without lines. Caterpillar green with oblique white stripes. One species in genus in region. **Biology:** Caterpillars feed on various thorn trees (*Vachellia* spp.). Multivoltine, adults throughout the year. **Habitat & distribution:** Various habitats where *Vachellia* trees grow; Western Cape, South Africa, north to tropical Africa.

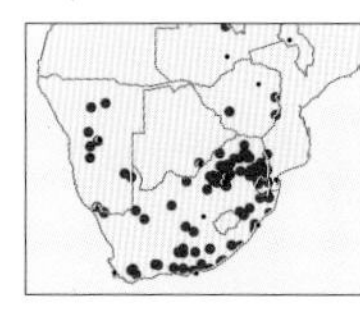

## 3 *Traminda ocellata* — Eyed Vestal

Forewing 10–13mm. Forewings slightly falcate at tip; hindwings with short tail; colour seasonally variable: wet-season form green, wings with yellow to orange border, in some specimens crossed by single yellow median line, postmedian a row small black dots, red circled white cell spot on each wing diagnostic, those on hindwings much larger; dry-season form brown with darker brown-circled wing spots. Caterpillar smooth, green with oblique black-bordered white and or pink lines. Five species in genus in region. **Biology:** Caterpillars feed on thorn trees (*Vachellia* spp.). Multivoltine, adults throughout the year. **Habitat & distribution:** Variety of habitats where thorn trees grow; Western Cape, South Africa, north to tropical Africa.

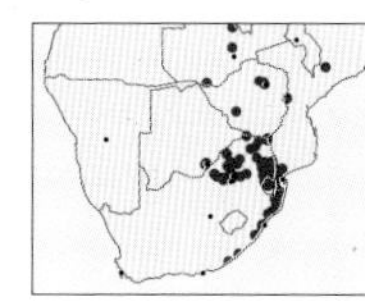

## 4 *Traminda vividaria* — Vivid Vestal

Forewing 11–16mm. Similar seasonal colour forms to *T. ocellata* (above), but slightly larger, forewing margin kinked and distinct pointed tail on hindwings, cell spot only on forewings. Similar *T. obversata* has forewing margin straight and no tail on hindwings. Caterpillar mottled light and dark brown with thorn-like lateral projections. **Biology:** Caterpillars feed on Sicklebush (*Dichrostachys cinerea*) and Sweet Thorn (*Vachellia karroo*). Multivoltine, adults throughout the year. **Habitat & distribution:** Coastal bush and bushveld; Eastern Cape, South Africa, north to tropical Africa, also Madagascar.

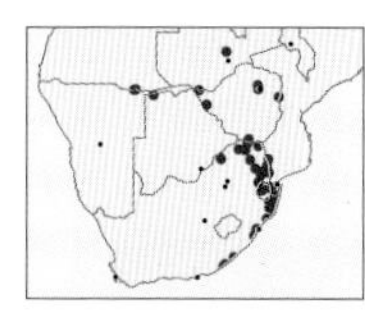

## 5 *Traminda neptunaria* — Neptune's Vestal

Forewing 13–17mm. Varies from pale yellow-green to darker green, never brown; forewings kinked and slightly falcate, with a large white-centred black cell spot and crossed by a dark, yellow-edged line beyond which is a scattered band of black or red dots; hindwings end in a short, pointed tail. Caterpillar green with oblique yellow and white bands and red dorsal spikes. **Biology:** Caterpillars feed on Sicklebush (*Dichrostachys cinerea*) and various straight-thorn trees (*Vachellia* spp.). Multivoltine, adults throughout the year. **Habitat & distribution:** Coastal bush and lowveld bushveld; Eastern Cape, South Africa, north to tropical Africa and Madagascar.

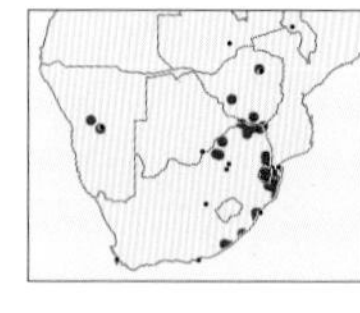

## 6 *Traminda falcata* — Falcate Vestal

Forewing 11–14mm. Forewings falcate, body and wings bone yellow finely striated with brown; forewings with clear cell spot, both wings with dark line ending before costa, absent in some individuals, sometimes short row of black spots. **Biology:** Caterpillar host associations unknown. Multivoltine, adults throughout the year. **Habitat & distribution:** Coastal bush and lowveld bushveld; Eastern Cape, South Africa, north to tropical Africa.

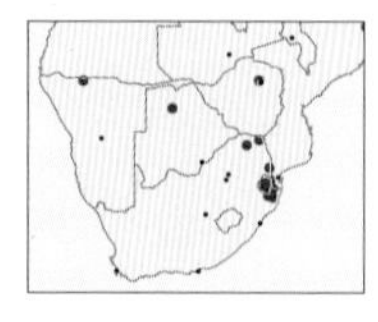

## 7 *Pseudosterrha acuta* — Straight Vestal

Forewing 9–11mm. Similar in colour to *Traminda falcata* (above), but smaller and forewings not falcate, only slightly pointed at apex in female; distinct subterminal row of black dots on both wings ending on postmedian band near costa. Caterpillar green with white lateral line and oblique pink lines. **Biology:** Caterpillars recorded on Cork-bark Thorn (*Vachellia davyi*). Multivoltine, adults throughout the year. **Habitat & distribution:** Lowveld bushveld; KwaZulu-Natal, South Africa, north to Kenya.

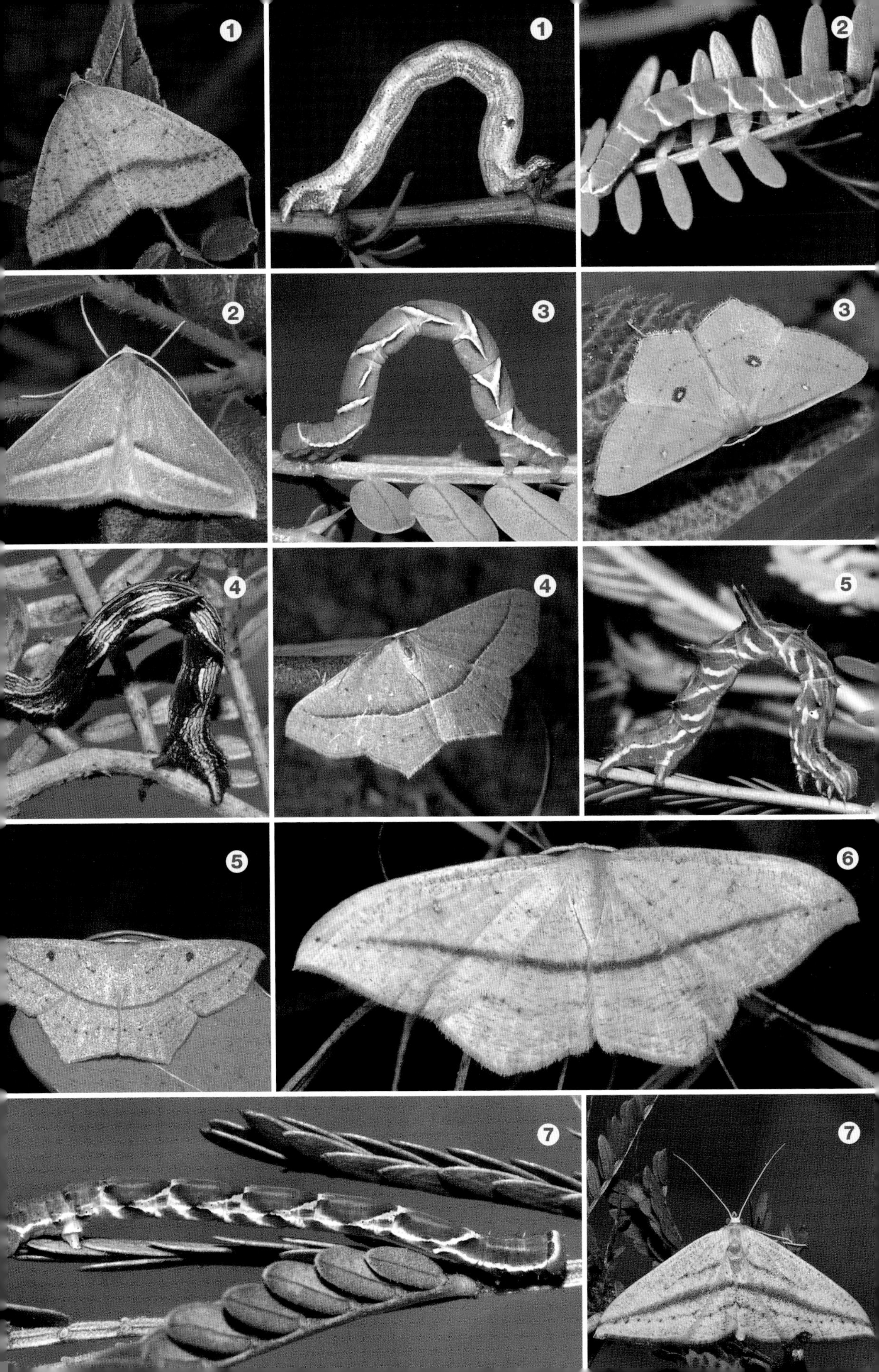

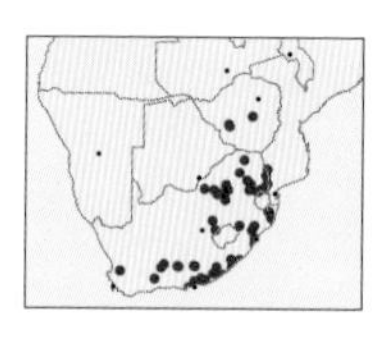

## 1 *Anthemoctena textilis* — Textile Pink

Forewing 7–9mm. Forewings cream with pink margins and strongly marked with curved pink lines of varying widths; hindwings cream with pink margin and faint pink lines, small but distinct cell spot. One species in genus in region. Can be confused with *Acidaliastis curvilinea* (p.226), which has fewer lines and no cell spot. **Biology:** Caterpillar host associations unknown. Adults throughout the year. **Habitat & distribution:** Various grassland habitats; Western Cape, South Africa, north to Zimbabwe.

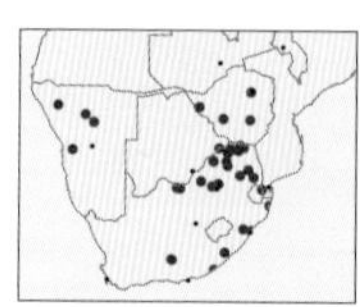

## 2 *Discomiosis crescentifera* — Crescent Wave

Forewing 9–11mm. Wings cream, heavily, but variably, diffused with brown and grey scales, distinctive jagged postmedial line, diagnostic white-ringed black cell spot on each wing. Two species in genus in region, similar *D. arciocentra* has black median band on both wings, prominent on hindwings. **Biology:** Caterpillars feed on *Grewia* spp. (Malvaceae). Adults September–April. **Habitat & distribution:** Relatively dry habitats where *Grewia* grows; Eastern Cape, South Africa, north to tropical Africa.

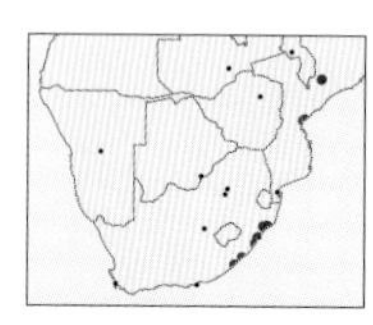

## 3 *Chrysocraspeda leighata* — Storm Wave

Forewing 7–10mm. Very similar to *Eois grataria* (p.158), but smaller and lacks the grid-like markings on wings; variable forms, wings of typical form heavily mottled dark reddish-brown, with conspicuous orange border and fringe of cilia, sometimes mottled brown with yellow border, some forms with large yellow forewing tip; hindwings angled to a point. Caterpillar long, slender, uniform green. Four species in genus in region. **Biology:** Caterpillars recorded on *Combretum bracteosum* (Combretaceae). Adults throughout the year. **Habitat & distribution:** Humid tropical forest and coastal bush; Eastern Cape north to tropical Africa.

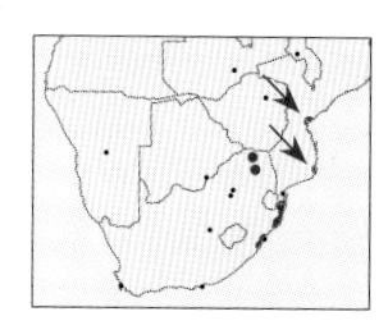

## 4 *Cyclophora unocula* — Eyed Wave

Forewing 12–14mm. Mottled reddish-brown with diagnostic rows of small black dots, representing lines, across both wings and along wing margins, a wide black median band present sometimes; a larger black-ringed white cell spot, almost completely black in some individuals, on hindwings. Four species in genus in region. **Biology:** Caterpillar host associations unknown. Adults throughout the year. **Habitat & distribution:** Coastal bush and lowland tropical forest; KwaZulu-Natal, South Africa, north to Kenya.

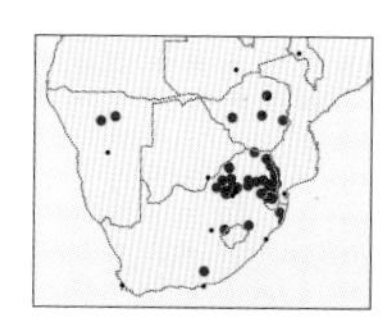

## 5 *Idaea umbricosta* — Dark-edged Rivulet

Forewing 5–6mm. Rests with elongate wings extended at right angles to body similar to some *Eupithecia* (p.158); wings cream, forewings with diagnostic broad diffuse brown band on costa crossed by darker bars. Caterpillar cream with dark triangular dorsal patches. Diverse genus with 41 species in region. **Biology:** Caterpillars feed on wilted leaves of herbaceous plants. Multivoltine, adults throughout the year. **Habitat & distribution:** Arid areas; Eastern Cape, South Africa, north to Malawi.

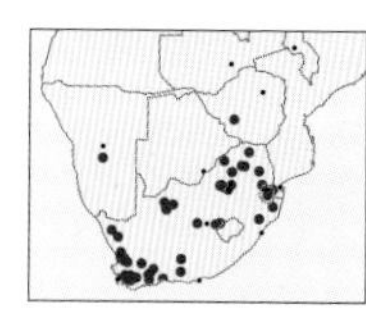

## 6 *Idaea torrida* — Torrid Rivulet

Forewing 6–9mm. Wings cream with orange borders, heavily mottled and lined with black, antemedial and postmedian lines crenulated, black subterminal patches present in some specimens. Caterpillar white, black ventrally. **Biology:** Caterpillars feed on wilted leaves of herbaceous plants. Adults September–April. **Habitat & distribution:** Arid areas; Western Cape, South Africa, north to Angola.

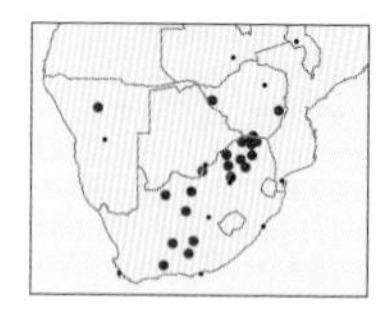

## 7 *Idaea lilliputaria* — Lilliput Rivulet

Forewing 6–7mm. Wings creamy white with faint dark cell spot, basal line strongly developed curving strongly in near costa diagnostic, postmedian line faint in some individuals, with kink near costa; hindwings with strong subterminal line and small cell spot. Caterpillar cream with brown zigzag markings. **Biology:** Caterpillars feed on wilted leaves of herbaceous plants. Adults November–March. **Habitat & distribution:** Arid habitats; Western Cape, South Africa, north to tropical Africa, also Madagascar.

## 8 *Idaea sublimbaria* — Bordered Rivulet

Forewing 5–7mm. Distinctive; both wings creamy white variably suffused with purplish-brown scales, strongly bordered by orange and purplish-grey band, cilia orange, lines parallel and crenulated. **Biology:** Caterpillar host associations unknown. Adults November–March. **Habitat & distribution:** Forest and riverine bush; Western Cape, South Africa, north to Malawi.

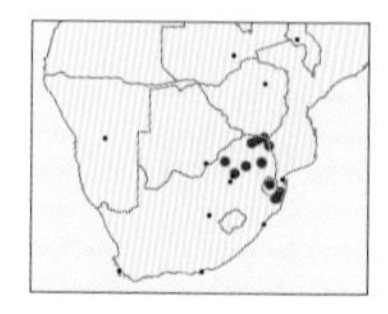

## 9 *Idaea nasifera* — Grey Rivulet

Forewing 7–9mm. Body and wings light grey variably suffused with darker grey scales, diagnostic postmedian line sharply kinked before costa, black cell spot on all wings. **Biology:** Caterpillar host associations unknown. Adults November–March. **Habitat & distribution:** Bushveld; KwaZulu-Natal north to Limpopo, South Africa.

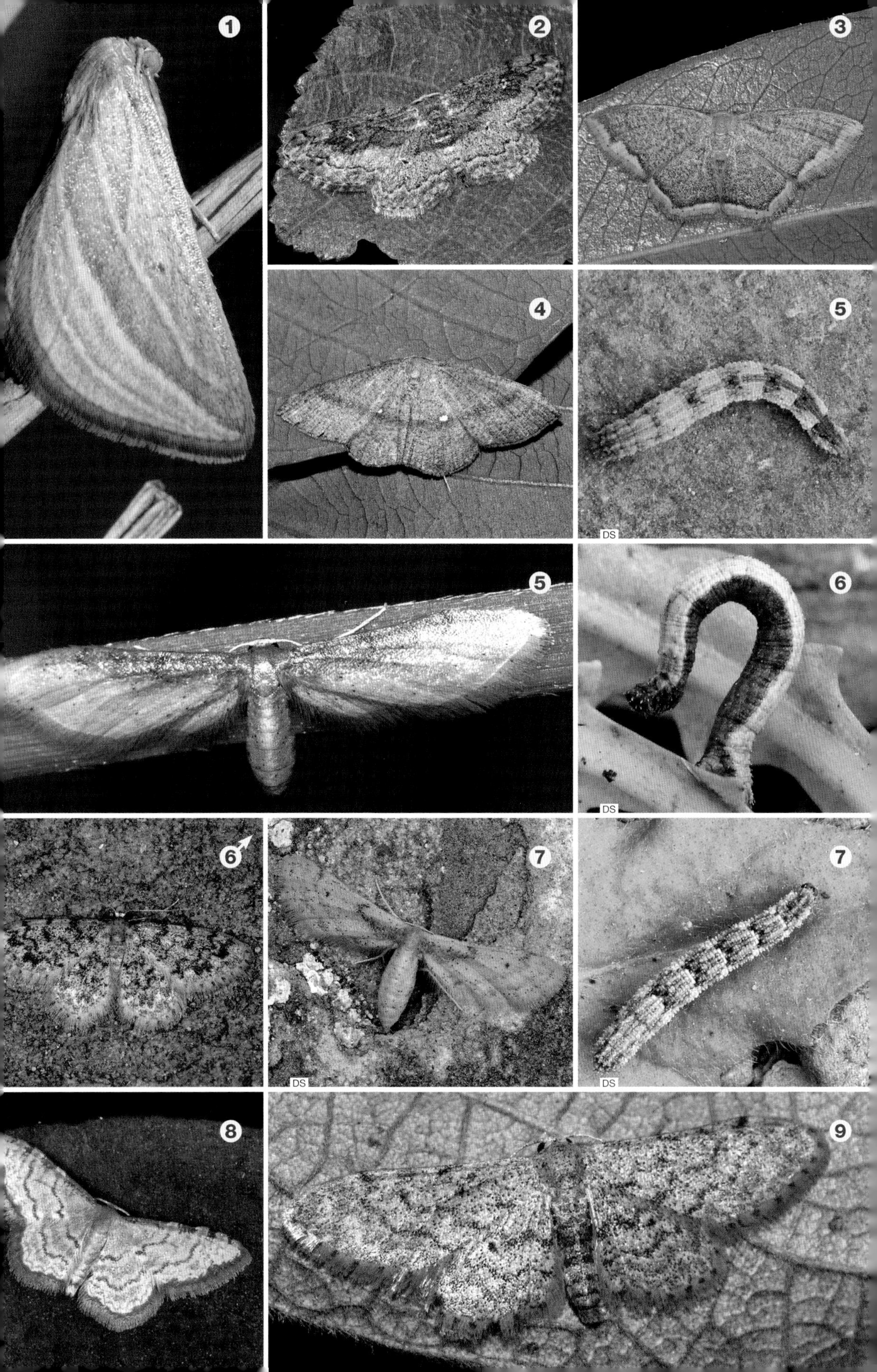
1
2
3
4
5
DS
5
6
DS
6
7
DS
7
DS
8
9

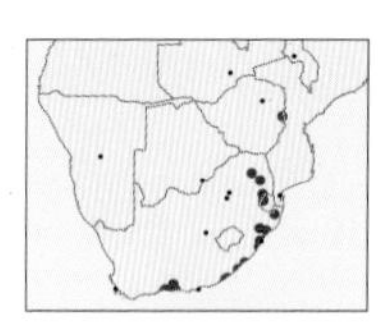

### 1 *Idaea basicostalis* — Dotted Rivulet

Forewing 5–7mm. Body and wings cream variably suffused with purplish-brown scales, diagnostic postmedian line black before costa and before inner margin in some specimens, black cell spot on all wings, margin punctuated with black dots. **Biology:** Caterpillar host associations unknown. Adults November–March. **Habitat & distribution:** Forest and coastal bush; southern Cape north to Zimbabwe.

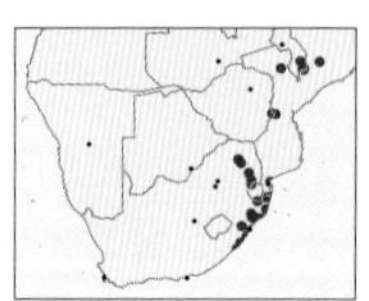

### 2 *Aletis libyssa* — Monarch Looper

Forewing 24–33mm. Convincing mimic of African Monarch butterfly (*Danaus chrysippus*). Body black with variable white spots; wing colour differs between and within populations, from ivory white to dark orange or red basally, broad black distal band enclosing white spots. Caterpillar variable, white to cream to grey, sometimes becoming orange at both ends, variably marked with black spots and bands. Species boundaries in this group not well understood, treated here as one variable species in genus in region. **Biology:** Caterpillars feed on various Wild Loquat trees (*Oxyanthus speciosus*) and *Oxyanthus* vines. Adults strictly diurnal, exude distasteful fluid. Multivoltine, adults throughout the year. **Habitat & distribution:** Forest and coastal bush where *Oxyanthus* spp. grow; Eastern Cape, South Africa, north to tropical East Africa.

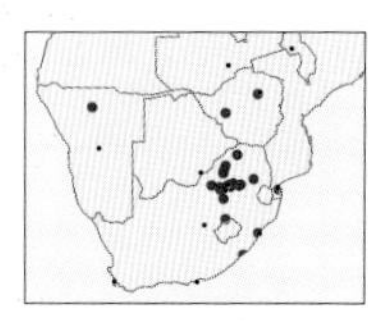

### 3 *Epicosymbia perstrigulata* — Tiled Wave

Forewing 10–12mm. Wings held flat and outspread at rest; body and wings cream variably overlaid by red, sometimes obliterating darker lines and distal half divided into cells, row of small dots along wing margins. Three species in genus in region. **Biology:** Caterpillars recorded on *Pygmaeothamnus zeyheri* (Rubiaceae). Adults September–April. **Habitat & distribution:** Grassland; Eastern Cape, South Africa, north to Namibia and Zimbabwe.

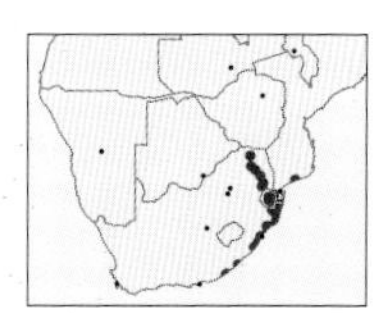

### 4 *Epicosymbia nitidata* — Irregular Wave

Forewing 12–16mm. Wings held flat and outspread at rest; body and wings light brown, both wings similarly marked, median area darker brown, diagnostic postmedian and subterminal lines irregularly jagged. **Biology:** Caterpillars recorded on *Canthium ciliatum* (Rubiaceae) and *Apodytes dimidiata* (Icacinaceae). Adults throughout the year. **Habitat & distribution:** Lowland forest and riverine bush; Eastern Cape north to Limpopo, South Africa, and Mozambique.

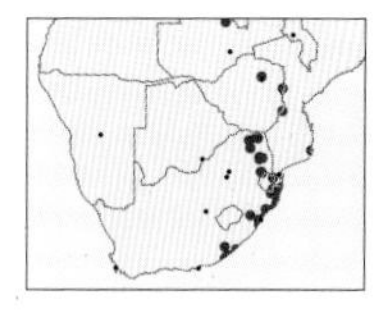

### 5 *Problepsis digammata* — Orange Pendant

Forewing 14–19mm. Wings white with faint grey postmedian and subterminal lines, diagnostically shaped silver pendant mark on forewings with top bar extending well beyond usually open ring; silver-edged yellow band on hindwings. Caterpillar slender, light greenish-grey. Nine species in genus in region. **Biology:** Caterpillars feed on Olive (*Olea europaea*) and Star Jasmine (*Jasminum multiflorum*). Multivoltine, adults throughout the year. **Habitat & distribution:** Forests and lowland riverine bush; Eastern Cape, South Africa, north to tropical Africa.

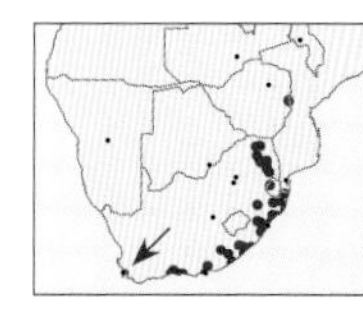

### 6 *Problepsis aegretta* — Silver Pendant

Forewing 14–17mm. Very similar to *P. digammata* (above), but silver pendant mark on forewings forms a ring and top bar does not extend beyond ring; silver-edged band on hindwings. Caterpillar slender, light greenish-brown. Similar *P. mozambica* has pendant round and grey. **Biology:** Caterpillars recorded on *Ligustrum lucidum* (Oleaceae). Multivoltine, adults throughout the year. **Habitat & distribution:** Moist forests; Western Cape, South Africa, north to Zimbabwe.

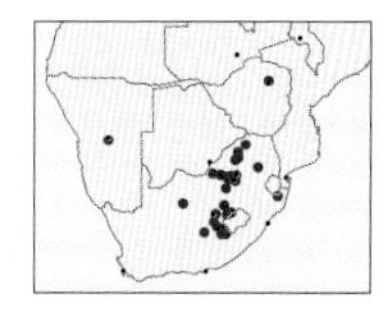

### 7 *Problepsis ctenophora* — Absent Pendant

Forewing 14–19mm. Similar to other *Problepsis* spp. (above), but without silver pendant mark on forewings; ground colour cream in many individuals. Caterpillar slender olive brown with lighter marks, prefect mimic of host plant twigs. Similar *P. centrophora* has a small silver pendant on all wings. **Biology:** Caterpillars feed on Olive (*Olea europaea*). Multivoltine, adults throughout the year. **Habitat & distribution:** Grassland and bushveld where olive trees grow; Eastern Cape, South Africa, north to Kenya.

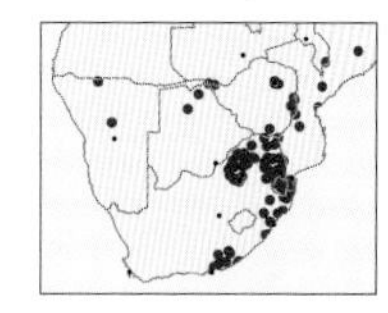

### 8 *Problepsis vestalis* — Vestal Birdling

Forewing 9–14mm. Distinctive species; body and wings white, postmedian line prominent sharply wavy, subterminal area patterned with fine brown and grey patches, each wing with distinctive central large irregular silver and brown blotch. Similar *P. prouti* lacks the central brown and silver blotches. **Biology:** Caterpillar host associations unknown. Adults throughout the year, mimics bird droppings. **Habitat & distribution:** Bushveld and grassland; Eastern Cape, South Africa, north to Malawi.

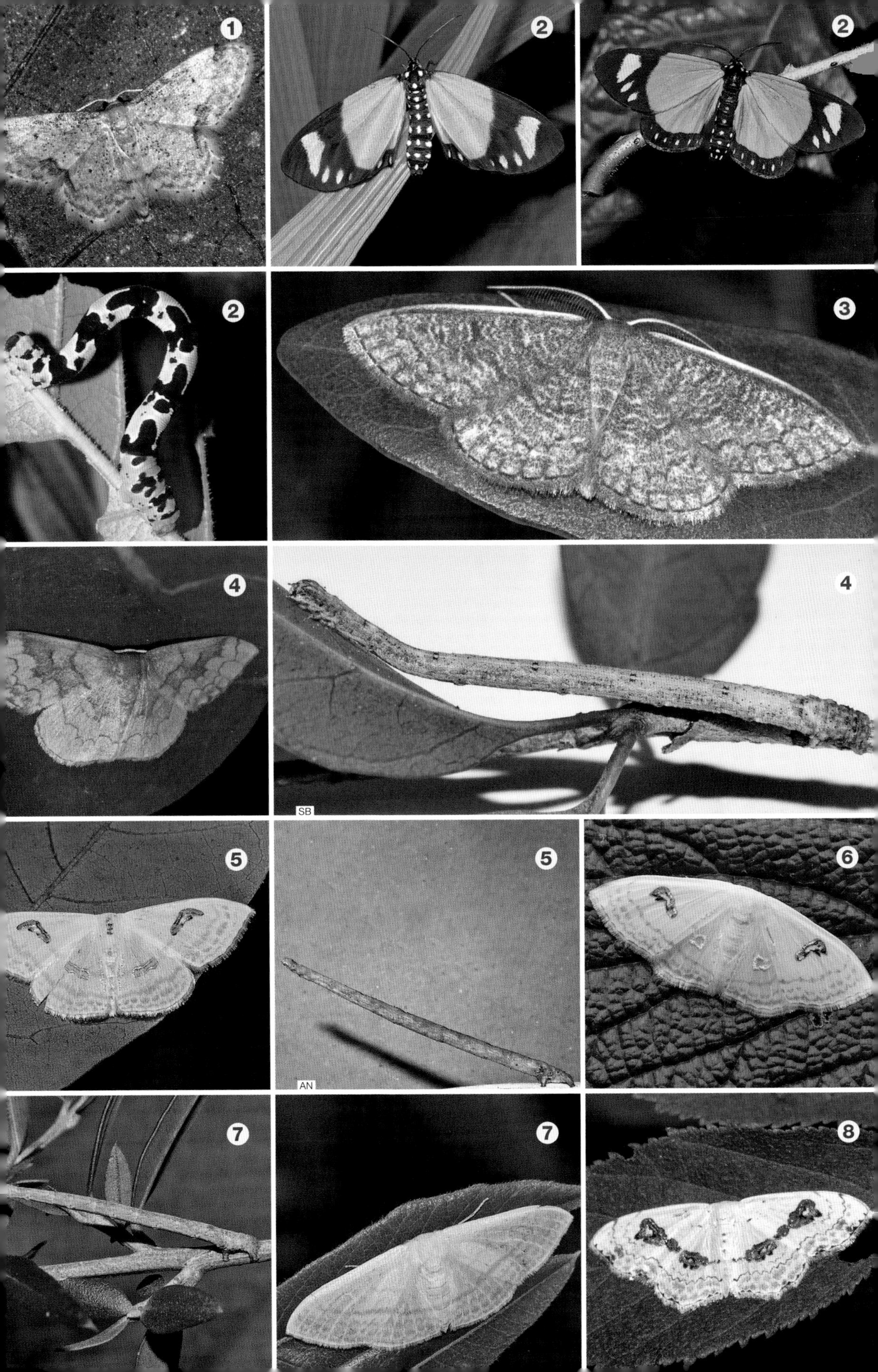

1
2
2
2
3
4
4
SB
5
5
AN
6
7
7
8

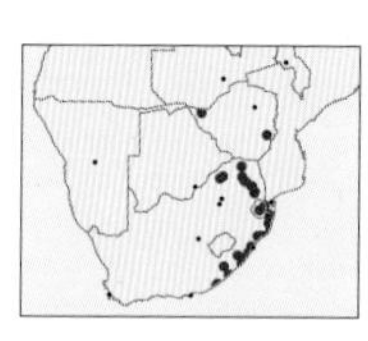

**1 *Somatina sedata*** Sedated Birdling

Forewing 13–16mm. Similar to *Problepsis ctenophora* (p.150), but wings bright white and broader with black cell spot and black broken marginal line, variable fine dusting of black or brown scales. Four species in genus in region. **Biology:** Caterpillar host associations unknown. **Habitat & distribution:** Forest and coastal bush; Eastern Cape, South Africa, north to Zimbabwe.

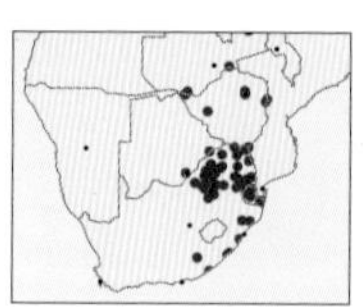

**2 *Scopula caesaria*** Red-bordered Wave

Forewing 9–11mm. Wings creamy brown with scattered small brown spots and numerous wavy red lines, becoming densely packed toward margins. Caterpillar slender, brown with fine reddish crenulated lines. Very diverse genus with 118 species in region, some undescribed. **Biology:** Caterpillars feed on Bloutee (*Oocephala staehelinoides*). Multivoltine, adults throughout the year, peaking in late summer. **Habitat & distribution:** Bushveld, grassland and forest; Eastern Cape, South Africa, north to tropical Africa, Indian Ocean islands, Middle East, Asia and Australia.

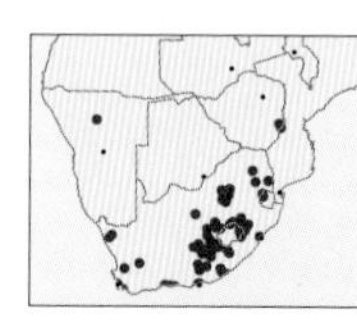

**3 *Scopula gazellaria*** Gazelle Wave

Forewing 8–10mm. Off-white, wings each with black hyphen-like cell spot, series of small black marginal dots, wings crossed by transverse brown wavy lines, diagnostic postmedian line often dark chocolate brown, other lines often absent. **Biology:** Caterpillar host associations unknown. Adults throughout the year. **Habitat & distribution:** In fynbos and grassland; Western Cape, South Africa, north to Zimbabwe.

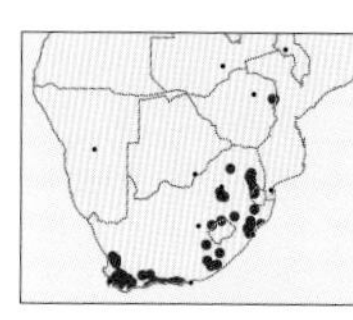

**4 *Scopula inscriptata*** Inscribed Wave

Forewing 8–11mm. Body white with brown crossbands; wings white, patterned with black cell spot and numerous wavy brown bands, chequered cilia and broken marginal line. **Biology:** Caterpillar host associations unknown. Adults September–April. **Habitat & distribution:** In fynbos and moist grassland; Western Cape, South Africa, north to Zimbabwe.

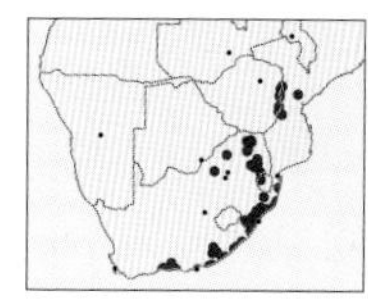

**5 *Scopula internata*** Intern Wave

Forewing 10–13mm. Wings held flat and outstretched; body cream; forewings cream to light pink with small black cell spot on all wings, diagnostic wavy postmedian line punctuated by small black dots on crest, a pink to brown blotch on inner margin diagnostic when present, distal margin with broken black line; hindwings similar but with angular margin and no blotches. Caterpillar slender, mottled orange and brown. **Biology:** Caterpillars polyphagous on leaves of plants growing in undergrowth. Multivoltine, adults throughout the year. **Habitat & distribution:** Undergrowth of moist forest and closed-canopy riverine vegetation; southern Cape, South Africa, north to tropical Africa.

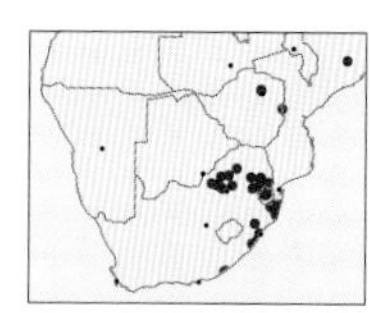

**6 *Scopula nemorivagata*** Pastel Wave

Forewing 11–13mm. Similar to *S. internata* (above) but ground colour pastel pink to white fading to creamy white in dried specimens, distinctive postmedian line red with intermittent black arrow markings, black dorsal spots on body, variable brown to grey terminal blotches often present. Similar *S. fimbrilineata* is cream and has wings crenulated, falcate and more pronounced tail on hindwings. **Biology:** Caterpillar host associations unknown. Adults September–May, peaking in February. **Habitat & distribution:** Bushveld and forest; Eastern Cape, South Africa, north to Zimbabwe.

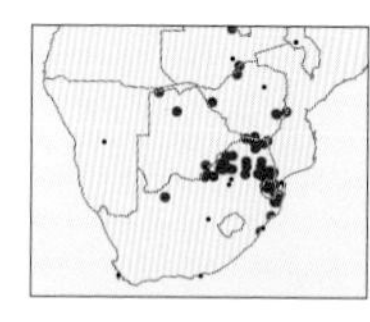

**7 *Scopula rufinubes*** Penny Wave

Forewing 9–11mm. Wings broad, ground colour cream to pink, wavy lines parallel, evenly curved toward costa of forewings; antemedial line curves sharply around black cell spot on hindwings, almost encircling it, variable purplish-brown blotches and shading in subterminal area of both wings, marginal line cream. **Biology:** Caterpillar host associations unknown. Adults throughout the year. **Habitat & distribution:** Bushveld; KwaZulu-Natal, South Africa, north to Ethiopia.

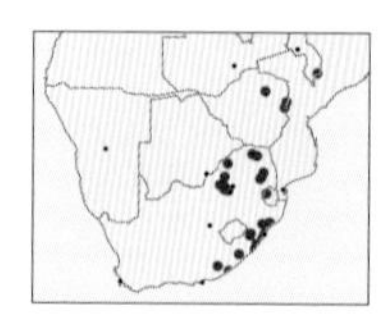

**8 *Scopula latitans*** Bone Wave

Forewing 12–14mm. Similar to *S. minorata* (p.154), but larger, wings broader and lines tapering toward costa, ground colour bone yellow, finely dusted with black, antemedial line can be a broad brown band in some individuals. Caterpillar brown, not as slender as most *Scopula* spp. Several very similar *Scopula* spp. in region that can only be reliably separated with barcoding or detailed examination. **Biology:** Caterpillars recorded on *Isoglossa woodii* (Acanthaceae). Multivoltine, adults throughout the year, easily disturbed by day in undergrowth. **Habitat & distribution:** Forest, riverine vegetation and closed-canopy woodland; Eastern Cape, South Africa, north to tropical Africa.

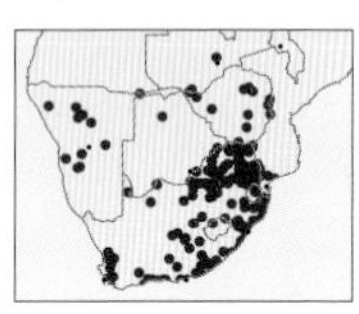

### 1 *Scopula minorata* — Minora Wave

Forewing 9–11mm. Wings held flat and outstretched; body and wings cream, variably dusted with black scales, each wing with black cell spot and crossed by about 5 weakly defined, wavy parallel lines, a line of black dots along distal margins. Caterpillar long and slender, cream with numerous darker fine longitudinal lines. Barcoding indicates that this is a complex of species, treated here as one. **Biology:** Caterpillars polyphagous on low-growing plants. Multivoltine, adults throughout the year. **Habitat & distribution:** Wide variety of habitats, including transformed ones; widespread across Africa, including Madagascar, Mauritius, La Réunion, Comoros and Seychelles to southern Europe and reaching Southeast Asia.

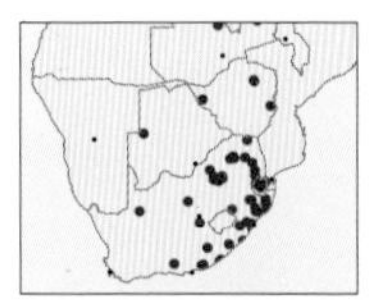

### 2 *Scopula serena* — Serene Wave

Forewing 6–8mm. Similar to *S. minorata* (above), but smaller; wings more elongate and wavy lines evenly spaced. Caterpillar dark mottled grey. **Biology:** Caterpillars recorded on *Selago densiflora* (Scrophulariaceae). Multivoltine, adults throughout the year, often active at dusk. **Habitat & distribution:** Wide variety of habitats, including transformed ones; widespread across Africa.

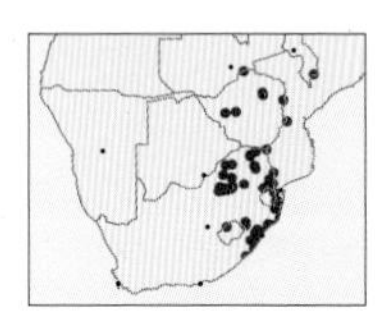

### 3 *Scopula sublobata* — Garden Wave

Forewing 9–11mm. Similar to *S. serena* (above), but larger; distinctive row of black spots on body, wings even more elongate and pointed, postmedian line and median band tapering to a point near apex, heavily dusted with dark brown. Caterpillar dark mottled grey. **Biology:** Caterpillars polyphagous on low-growing plants. Multivoltine, adults throughout the year. **Habitat & distribution:** Moist habitats, well established in gardens; Eastern Cape, South Africa, north to Malawi.

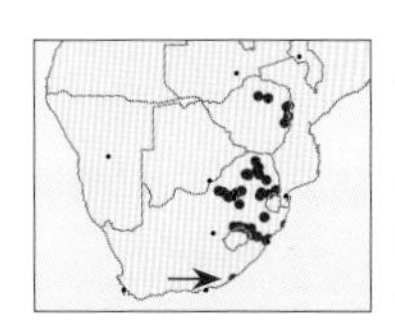

### 4 *Scopula erinaria* — Straight Wave

Forewing 10–12mm. Similar to *S. sublobata* (above), but lines straight and strongly dotted with black triangular marks; ground colour light cream sparsely dotted with black scales. Similar *S. donovani* has postmedian line dots kinked toward costa, *S. bigeminata* lacks the black dots on lines. **Biology:** Caterpillar host associations unknown. Adults throughout the year. **Habitat & distribution:** Relatively moist grassland and bushveld; Eastern Cape, South Africa, north to tropical Africa.

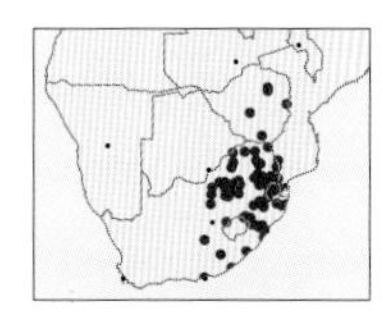

### 5 *Scopula deserta* — Deserted Wave

Forewing 10–12mm. Distinct summer and winter forms; summer-form body and wings ground colour creamy with pinkish-brown shading, cilia pink, postmedian line dotted, curving gently toward costa; winter form dark brick red. Similar *S. dissonans* is grey, never pink or reddish. **Biology:** Caterpillar host associations unknown. Summer brood adults nocturnal, winter brood adults fly in full sunshine late afternoon but are also attracted to light. Adults throughout the year. **Habitat & distribution:** In relatively moist grassland and bushveld; Eastern Cape, South Africa, north to tropical Africa.

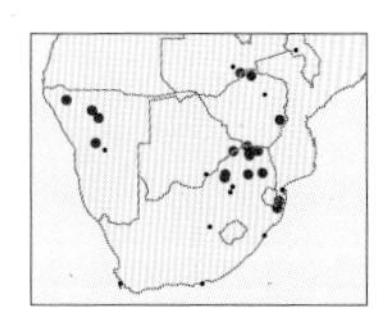

### 6 *Scopula penricei* — Penrice's Wave

Forewing 10–14mm. Body cream; slightly falcate forewings white basally with small cell spot, crossed by indistinct grey proximal lines, darker postmedian line diagnostically bent outward and then sharply inward toward costa, terminal area variably marked with zigzag lines and darker triangular black blotches; hindwings similarly coloured, with short tail. Caterpillar smooth and green. **Biology:** Caterpillars recorded on *Barleria senensis* (Acanthaceae). Bivoltine, adults August–December and March–May. **Habitat & distribution:** Bushveld; KwaZulu-Natal, South Africa, north to Zambia and Angola.

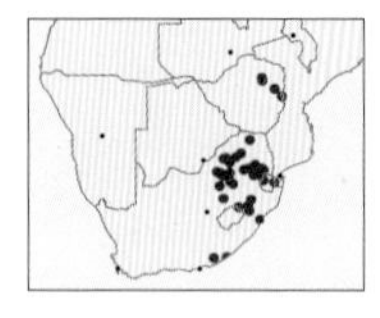

### 7 *Scopula sanguinisecta* — Blotch Wave

Forewing 10–14mm. Similar to *S. penricei* (above); body and wings white; wings each with tiny black cell spot, crossed by a few fine brown lines, forewings with postmedian line gently curving toward costa, not bent outward, and then sharply inward toward costa, 2 (rarely 1) distinctive irregular brown subterminal blotches. **Biology:** Caterpillar host associations unknown. Adults throughout the year. **Habitat & distribution:** Grassland and bushveld; Eastern Cape, South Africa, north to Kenya, also Madagascar.

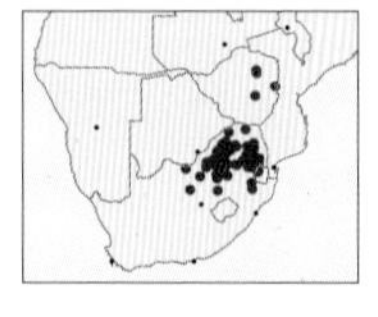

### 8 *Scopula sinnaria* — Fawn Wave

Forewing 9–13mm. Distinctive species; varies from cream with red-brown speckles to reddish-brown; wings crossed by 3 wavy red-brown lines and with brown marginal line, crosslines along veins create 'cells' near margin. **Biology:** Caterpillar host associations unknown. Adults September–May, peaking in February. **Habitat & distribution:** Dry grassland and bushveld; Free State, South Africa, north to Kenya.

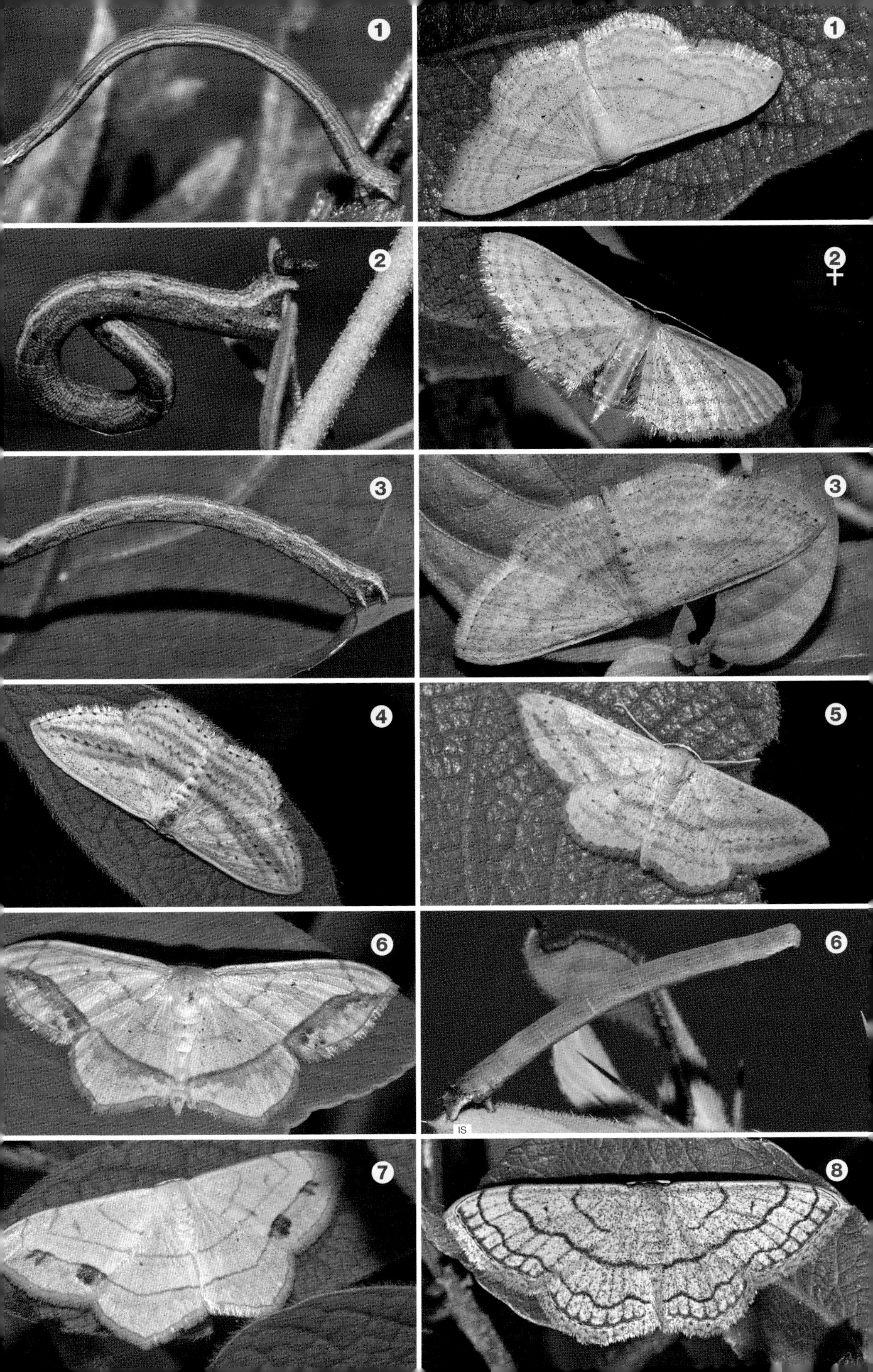

1
1
2
2
♀
3
3
4
5
6
6
IS
7
8

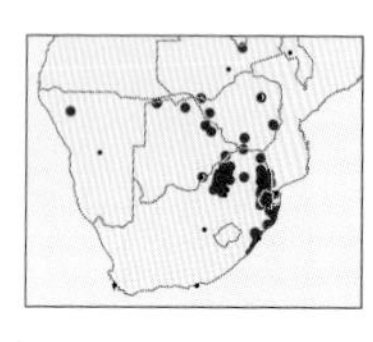

**1 *Scopula rufisalsa*** Brick Wave

Forewing 12–14mm. Forewing slightly falcate, ground colour white, heavily shaded with brick red and variably dusted with dark brown scales, crenulated postmedian line diagnostic, black cell spot on hindwings. **Biology:** Caterpillar host associations unknown. Adults throughout the year, peaking March–April. **Habitat & distribution:** Bushveld; KwaZulu-Natal, South Africa, north to Kenya.

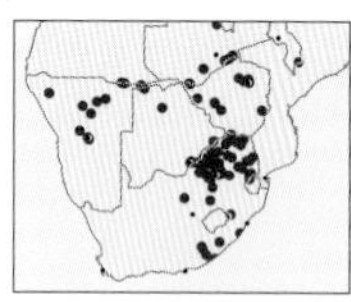

**2 *Scopula sincera*** Clean Wave

Forewing 7–9mm. Distinctive; body and wings bright white with no dusting, black cell spot on all wings, wavy lines light brown to orange, tiny black marginal spots. Similar *S. quadrifasciata* is cream and has broader lines covering most of wing surface. **Biology:** Caterpillar host associations unknown. Adults throughout the year, peaking in March. **Habitat & distribution:** Grassland and bushveld; Eastern Cape, South Africa, north to Zambia and Angola.

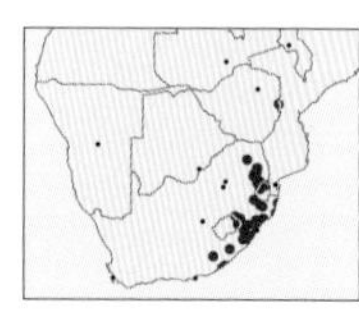

**3 *Scopula opperta*** Oppertune Wave

Forewing 10–14mm. Body and wings pure white, variably dusted with brownish-grey scales, lines dark brown, cell spot on all wings. Caterpillar green with tiny lateral spots. **Biology:** Caterpillars polyphagous on undergrowth plants, including ferns. Multivoltine, adults throughout the year, peaking in March and November. **Habitat & distribution:** Humid forests; Eastern Cape, South Africa, north to Zimbabwe.

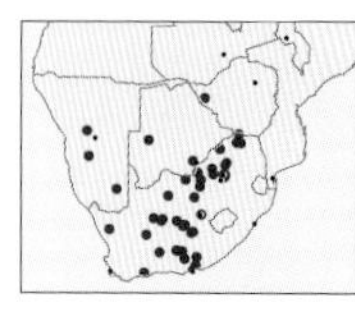

**4 *Scopula picta*** Tent Wave

Forewing 8–11mm. Unlike other waves, wings held tent-like over body; can be confused with *Conchylia decorata* (p.160); wings white, forewings variably suffused with grey, black cell spot, oblique postmedian line orange to grey, black marginal spots. **Biology:** Caterpillar host associations unknown. Adults throughout the year. **Habitat & distribution:** Arid karoo, grassland and bushveld; Western Cape north to Zimbabwe.

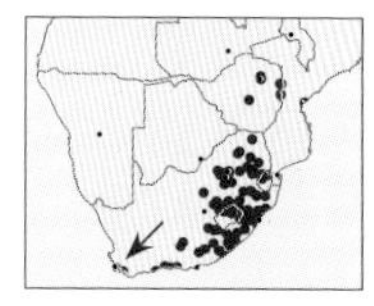

**5 *Scopula punctilineata*** Punctuated Wave

Forewing 8–10mm. Two seasonal forms: body and wings white, slightly dusted with grey scales in wet-season form, heavily in dry-season form with some individuals almost black; wings similar, wavy postmedian line punctuated with black spots on crests, prominent brown to black median band. **Biology:** Caterpillar host associations unknown. Adults throughout the year. **Habitat & distribution:** Moist fynbos, grassland and bushveld; Western Cape, South Africa, north to Zimbabwe.

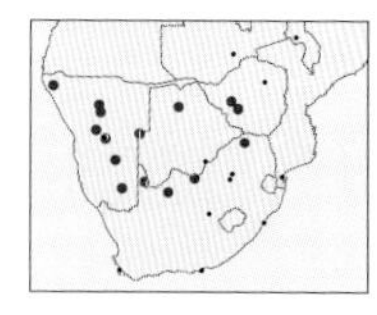

**6 *Scopula phyletis*** Kalahari Wave

Forewing 10–12mm. Very variable species; ground colour bone white, sometimes lacking markings on wings, or finely dusted with black, light grey to black lines, extensive black blotches in subterminal area in some, lines distinctive when present. **Biology:** Caterpillar host associations unknown. Adults throughout the year. **Habitat & distribution:** Arid sandy bushveld in and around the Kalahari basin; Northern Cape, South Africa, north to Namibia, Botswana and Zimbabwe.

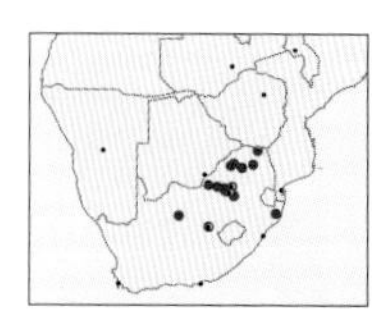

**7 *Zygophyxia roseocincta*** Rosy Wave

Forewing 7–8mm. Distinctive species; rests with elongate forewings held at right angles to body, similar to *Idaea umbricosta* (p.148); body pink, both wings orange with broad pink border; forewings with basal and median areas variably shaded with pink, completely pink in some individuals. **Biology:** Caterpillar host associations unknown. Adults November–April. **Habitat & distribution:** Grassland and bushveld, Free State, South Africa, north to Tanzania.

## Subfamily Larentiinae Carpets

Over 200 regional species. Most are nocturnal. Intricate, carpet-like patterns on the wings. Caterpillars usually rather short and stout for Geometridae, lack appendages and have rounded heads. Most are adapted to moist temperate conditions and can therefore be the dominant geometrid group on temperate mountains, but some are adapted to arid areas. Many are cryptic, host specialists, but some seem to be more polyphagous, with generalist cryptic colouration. Members of the tribe Eupitheciini are often flower feeders.

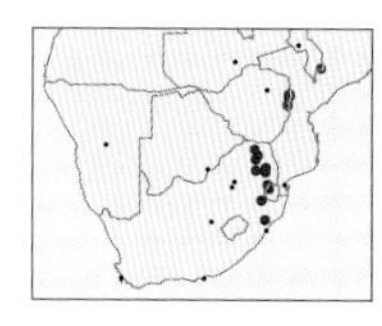

**8 *Asthenotricha pycnoconia*** Brown Shell

Forewing 12–15mm. Pale grey-brown, basal antemedial and postmedian lines yellow and red, other lines indistinct, black cell spot, black and cream lines on veins in terminal area; female with apex of forewings slightly falcate; male with large black androconial patch on forewings. Four species in genus in region. **Biology:** Caterpillar host associations unknown. Adults throughout the year. **Habitat & distribution:** Moist Afromontane forests; KwaZulu-Natal, South Africa, north to tropical Africa.

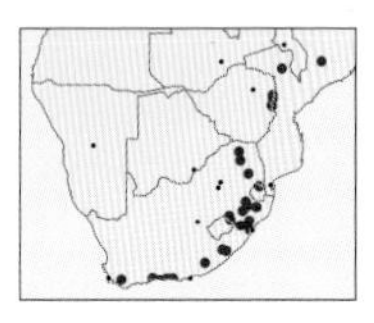

## 1 ***Asthenotricha inutilis*** Kinked Shell

Forewing 11–13mm. Mottled light to reddish-brown, black lines kinked on veins not yellow and red, other lines distinct, tiny black cell spot, no black and cream lines on veins in terminal area; male without large black androconial patch on forewings. Caterpillar short and stout, speckled brown. Similar *A. dentatissima* is bright sulphur yellow. **Biology:** Caterpillar host associations unknown. Adults October–May. **Habitat & distribution:** Moist forests; Western Cape, South Africa, north to tropical Africa.

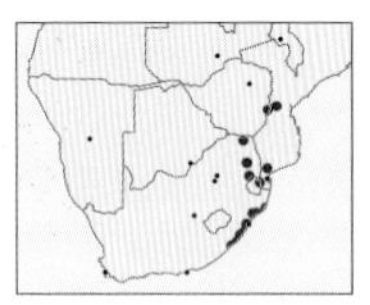

## 2 ***Eois grataria*** Grid Eois

Forewing 10–12mm. Antennae bipectinate in both sexes; wings similar, yellow ground colour and cilia with multiple red or brown lines consisting of halfmoon shapes that appear grid-like, variably suffused with brown or purple shading, black cell spot on forewings. Can be confused with *Chrysocraspeda leighata* (Sterrhinae, p.148). Caterpillar green with red lateral and white-ringed black dorsal spots. Two species in genus in region. **Biology:** Caterpillars feed on *Croton sylvaticus* (Euphorbiaceae). Multivoltine, adults throughout the year. **Habitat & distribution:** Warm humid forests; Eastern Cape, South Africa, north to tropical Africa, also Indian subregion.

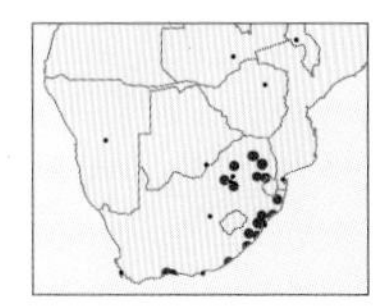

## 3 ***Chloroclystis grisea*** Zigzag Shell

Forewing 6–9mm. Body light brown with darker crossbands; all wings light mottled brown crossed by several black zigzag lines, postmedian line most distinct, interspersed with variable green shading, costa of male straight, not arched. Caterpillar green and fairly plump. Similar *C. consobrina* has postmedian line even with no green shading. Five species in genus in region. **Biology:** Caterpillars feed on flowers of various trees. Multivoltine, adults throughout the year. **Habitat & distribution:** Forests and wooded kloofs; southern Cape, South Africa, north to Kenya.

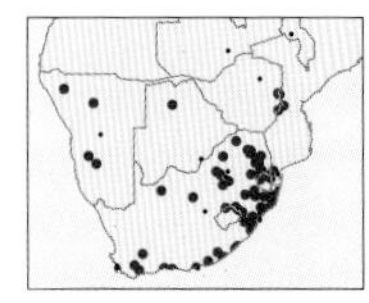

## 4 ***Chloroclystis muscosa*** Lobed Shell

Forewing 8–9mm. Ground colour light green, which fades in dried specimens, with variable density of mottled grey or black lines and blotches; a diagnostic androconial lobe about one-third along arched costa in males. Caterpillar variable; colour adapts to that of the flowers on which it feeds, from grey to light green-brown, plain or with darker crossbands on each segment. **Biology:** Caterpillars feed on flowers of a wide variety of plants. Multivoltine, adults throughout the year. **Habitat & distribution:** Variety of habitats, one of the few Larentiines that can also be found in arid areas; Western Cape, South Africa, north to tropical Africa.

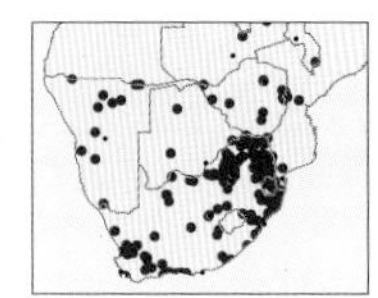

## 5 ***Pasiphila lita*** Lita Shell

Forewing 5–8mm. Very similar to *Chloroclystis muscosa* (above); male with arched costa but lacking distinct lobe; thorax black with some lighter bands, wings ground colour white, not green, densely patterned with dark brown and black patches and bands. Caterpillar variable, colour adapts to that of flowers it feeds on from grey to yellow and pink, plain or mottled. Two species in genus in region. **Biology:** Caterpillars polyphagous on flowers, often on Asteraceae. Multivoltine, adults throughout the year. **Habitat & distribution:** Often abundant in variety of habitats, another of the few Larentiines that can also occur in arid areas; Western Cape, South Africa, north to tropical Africa.

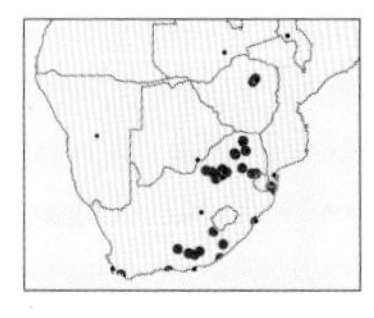

## 6 ***Eupithecia gradatilinea*** Twindash Pug

Forewing 9–11mm. Forewings elongate and held at right angles to body; body and wings light brown with darker mottling, diagnostic dash cell mark and dash further on costa, distinct grate-like postmedian line on hindwing upper side and undersides of wings. Caterpillar brown or green with dorsal projections. Very diverse genus containing hundreds of species globally, 44 described species in region; individual species difficult to identify without detailed examination. **Biology:** Caterpillars feed on Sweet Thorn (*Vachellia karroo*). Bivoltine, adults September–March. **Habitat & distribution:** Relatively moist habitats where *Vachellia* trees grow; Western Cape, South Africa, north to Kenya.

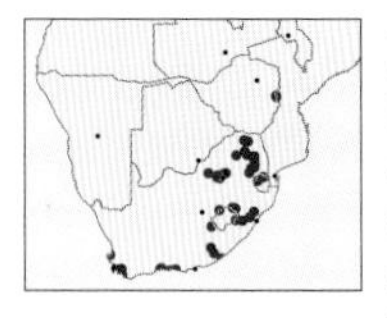

## 7 ***Eupithecia rubiginifera*** Rose Pug

Forewing 9–11mm. Similar to *E. gradatilinea* (above), but cell spot round and with distinctive white dot on forewing margin and large subterminal rose pink to green clear patch. Caterpillar dark brown to grey with dorsal processes. **Biology:** Caterpillars polyphagous on low-growing plants. Multivoltine, adults throughout the year. **Habitat & distribution:** Relatively moist habitats, including gardens; Western Cape, South Africa, north to Zimbabwe.

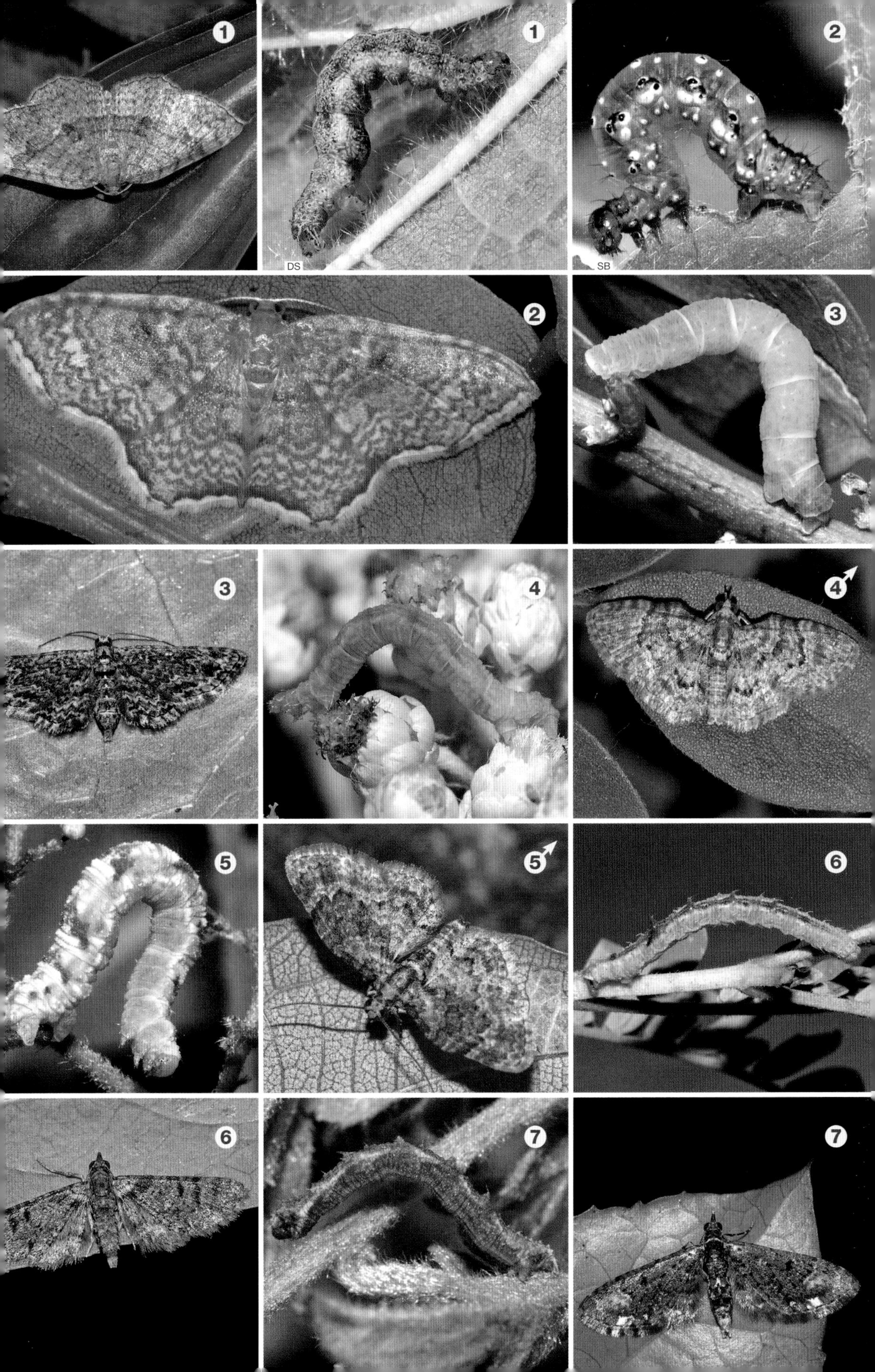
1
1
DS
2
SB
2
3
3
4
4
5
5
6
6
7
7

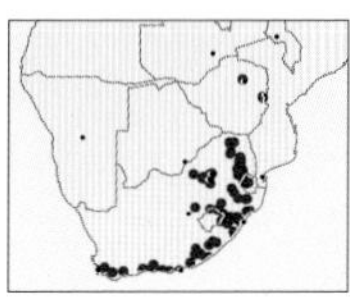

### 1 *Eupithecia infelix* — Copper Pug

Forewing 8–10mm. Forewings elongate, held outstretched and away from abdomen, body and wings coppery brown, variably shaded with darker brown, distinctive white-ringed black cell spot, dark reniform beyond cell in most specimens; hindwings plain brown with weak cell spot. Caterpillar light to dark brown with distinctive pale lateral markings. **Biology:** Caterpillars polyphagous on various low-growing plants in undergrowth, including ferns. Multivoltine, adults throughout the year. **Habitat & distribution:** Closed-canopy forest and bush, including gardens; Western Cape, South Africa, north to Tanzania.

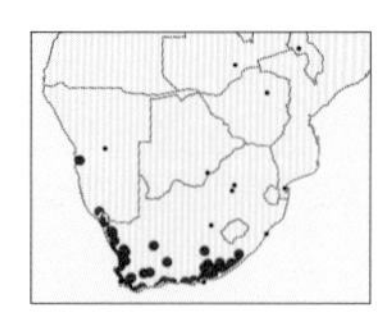

### 2 *Eupithecia inconclusaria* — Desert Pug

Forewing 10–13mm. Distinctive; straw ground colour with many brown lines over forewings and thorax, small cell spot and dark reniform beyond on postmedian line, median area often without markings, hindwings plain straw with darker border. **Biology:** Caterpillars reported to feed on Lamiaceae. Adults throughout the year, primarily September–December. **Habitat & distribution:** Adapted to arid habitats, in Karoo biomes, desert and other arid areas; Cape provinces, South Africa, and Namibia.

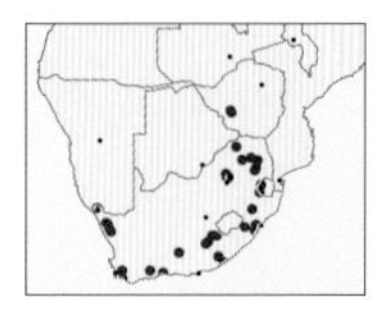

### 3 *Eupithecia sagittata* — Plain Pug

Forewing 10–13mm. Distinctive; plain light grey ground colour, antemedial and postmedian lines undulating almost parallel, comma-shaped black cell spot, broken black marginal line, dark spot in curve of inner margin; hindwings with diagnostic zigzag black patterns. **Biology:** Caterpillar host associations unknown. Adults throughout the year. **Habitat & distribution:** Variety of habitats; Western Cape, South Africa, north to Zimbabwe.

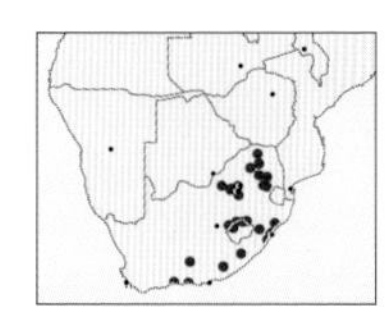

### 4 *Eupithecia infectaria* — Large Pug

Forewing 13–15mm. Largest species in genus; body and wings shiny purplish-brown, fading to dull brown in dried specimens, largish black cell spot in forewings, black and white dots on margin and on veins in some specimens, distinctive white subterminal line. Caterpillar fluffy green and cream, resembling shoots and underside of host plant leaf. **Biology:** Caterpillars feed on new leaves of Sage Wood (*Buddleja salviifolia*). Bivoltine, adults August–December and February–April. **Habitat & distribution:** Forest and wooded areas where host plant grows; southern Cape to Limpopo, South Africa.

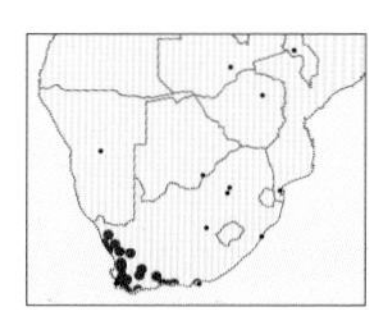

### 5 *Conchylia niditula* — Layered Silver

Forewing 13–15mm. At rest, wings steeply inclined against body; forewings silvery white with 4 longitudinal dark brown bands, one on terminal margin plain, others dusted with golden-brown scales, one along costa, the 2 central bands meeting distally, lower one running along inner margin with no silvery white scales below and sharply angling toward apex diagnostic; hindwings cream with brown fringe. Several very similar species; *C. rhabdocampa* has silvery white area between inner margin and lower band; *C. ditissimaria* lower band ending on inner margin, not kinked inward. Fifteen described species in genus in region. **Biology:** Caterpillar host associations unknown. Adults August–October and April. **Habitat & distribution:** Fynbos and renosterveld habitats; Cape provinces, South Africa.

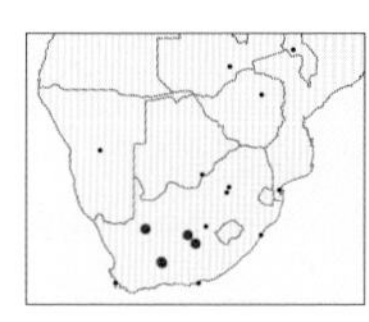

### 6 *Conchylia gamma* — Gamma Silver

Forewing 9–12mm. Differs from *C. niditula* (above) in that the forewing lines begin on inner margin, not base, merging near cell, forming a gamma sign. **Biology:** Caterpillar host associations unknown. Adults in April. **Habitat & distribution:** Nama karoo habitats; Cape provinces and Free State, South Africa.

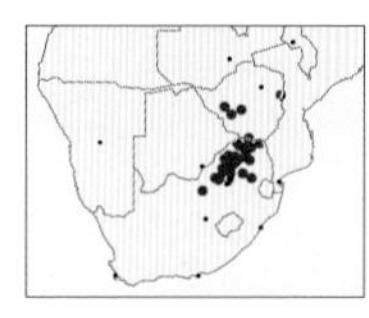

### 7 *Conchylia pactolaria* — Two-band Silver

Forewing 9–10mm. Similar to *C. niditula* (above) in colour and posture, but smaller and forewings crossed by 2 diagnostically wavy, black-bordered orange bands. **Biology:** Caterpillar host associations unknown. Adults October–March. **Habitat & distribution:** Grassland and bushveld; North West province to Zimbabwe.

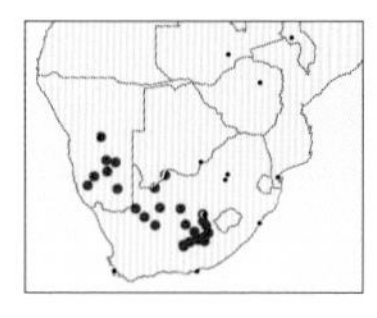

### 8 *Conchylia decorata* — Decorated Silver

Forewing 8–11mm. Can be confused with *Scopula picta* (p.156); antemedial and postmedian bands variably decorated with brick red-brown and grey scales, antemedial band almost absent in some individuals, cilia chequered brown and white; hindwings white shading to brown distally. **Biology:** Caterpillar host associations unknown. Adults August–April. **Habitat & distribution:** Grassy karoo and arid bushveld; Eastern Cape, South Africa, north to Namibia.

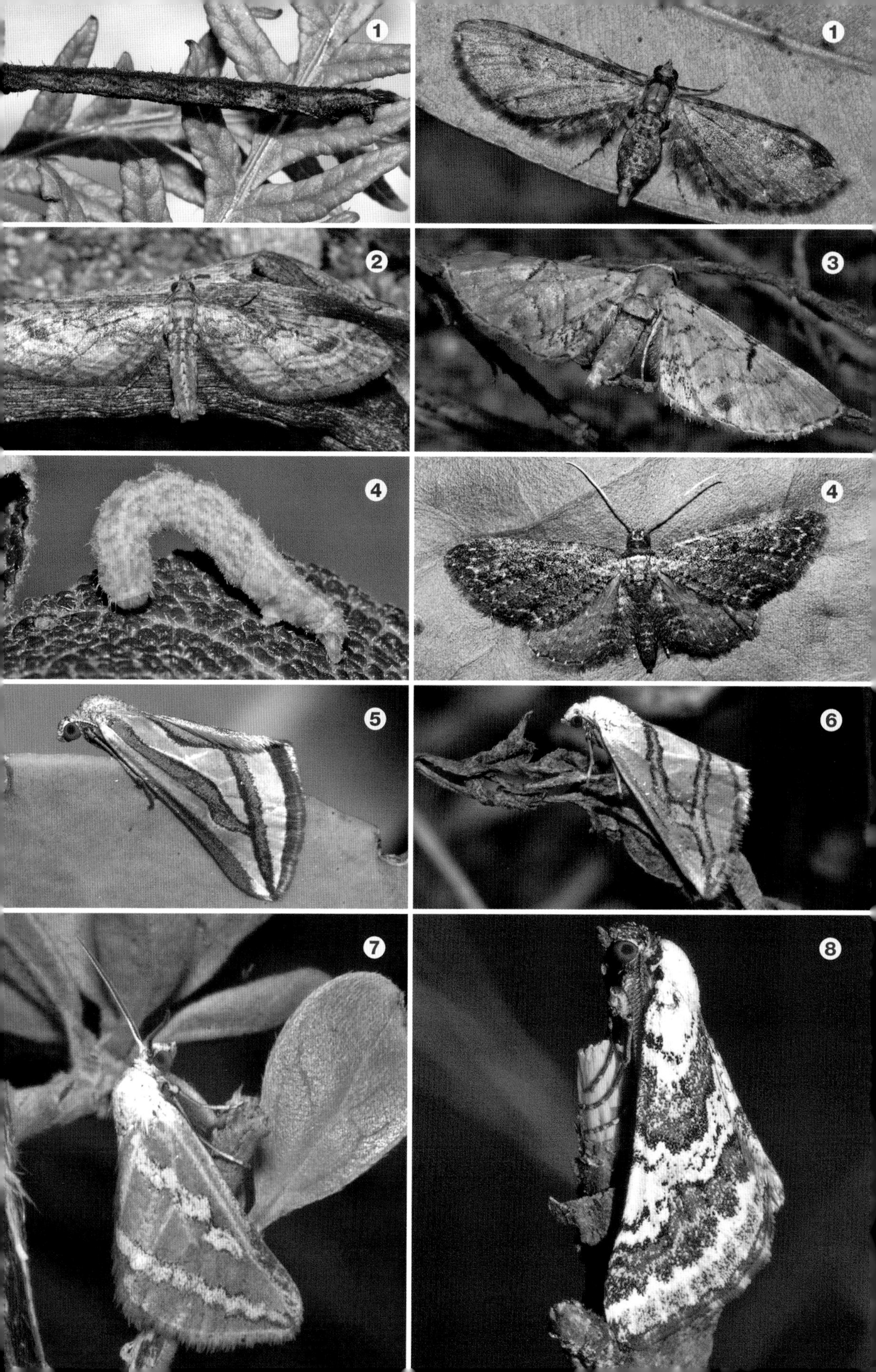
1
1
2
3
4
4
5
6
7
8

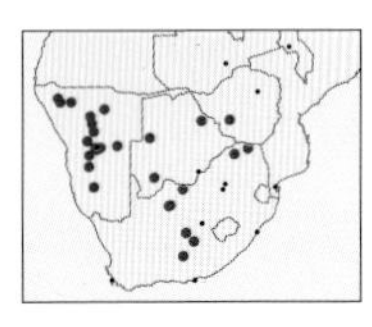

### 1 *Conchylia canescens* Grey Silver

Forewing 8–11mm. Forewings heavily diffused with grey scales, except for thin areas around dark brown antemedial and postmedian bands, which do not reach costa; hindwings grey-brown. Similar *C. sesquifascia* is diffused with brown scales, and bands are darker brown. **Biology:** Caterpillar host associations unknown. Adults December–April. **Habitat & distribution:** Grassy karoo and arid bushveld; Eastern Cape, South Africa, north to Namibia and Zimbabwe.

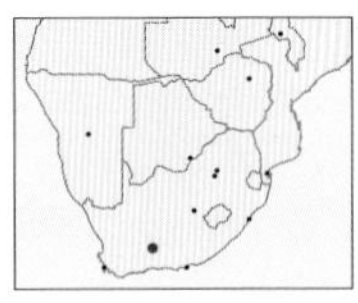

### 2 *Conchylia irene* Blackline Silver

Forewing 13mm. Body and forewings pure white except for straight, black antemedial and postmedian lines not reaching costa; hindwings white, shading to brown distally. Similar *C. actena* has postmedian line reaching costa at apex and antemedial line curved ending in arrow point. **Biology:** Caterpillar host associations unknown. Adults April. **Habitat & distribution:** Upper Karoo Hardeveld; known only from the Nuweveld Mountains near Beaufort West, South Africa.

### 3 *Melanthia ustiplaga* Triangle Shell

Forewing 10–13mm. Thorax dark brown, abdomen lighter brown; forewings light brown with dark brown base, diagnostic triangular white-lined black patch on costa and white dotted subterminal line, hindwings with several black lines of varying intensity. Caterpillar light green with several thin, darker green longitudinal lines. **Biology:** Caterpillars feed on Traveller's Joy (*Clematis brachiata*). Adults August–April. **Habitat & distribution:** Moist forests and wooded kloofs; Western Cape to Limpopo, South Africa.

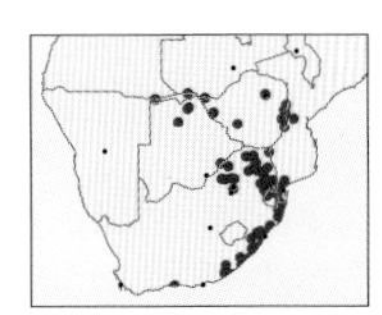

### 4 *Disclisioprocta natalata* Natal Dingy Shell

Forewing 14–19mm. Body and wings light brown, densely mottled and marked with darker brown to almost black, crenulated white postmedian and antemedial lines variably obscured, subterminal white triangular marks diagnostic. Caterpillar light to mottled brown. One species in genus in region. **Biology:** Caterpillars feed on Great Bougainvillea (*Bougainvillea spectabilis*) and Four O'clock Flower (*Mirabilis jalapa*), both Nyctaginaceae. Multivoltine, adults throughout the year. **Habitat & distribution:** Forest and riverine vegetation, including gardens; southern Cape, South Africa, north to tropical Africa, also Madagascar, Comoros and Mauritius.

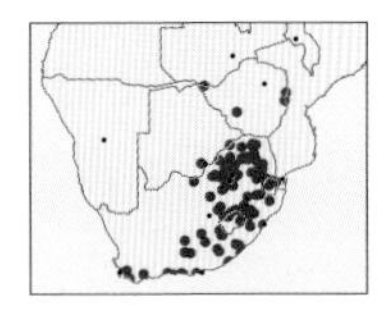

### 5 *Horisme obscurata* Obscure Shell

Forewing 12–14mm. Wings held flat and widely extended; body and wings light brown, both densely patterned with darker brown patches and transverse zigzag lines. Caterpillar pale yellow to brown with short black setae. Six species in genus in region. **Biology:** Caterpillars feed on *Clematis* spp. (Ranunculaceae). **Habitat & distribution:** Variety of moist and not too arid habitats where *Clematis* spp. grow; Western Cape, South Africa, north to tropical Africa.

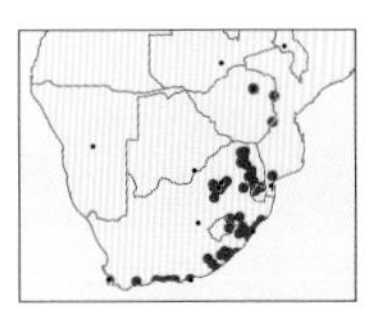

### 6 *Horisme minuata* Pale-bordered Shell

Forewing 11–15mm. Wings held flat and widely extended; head and prothorax dark brown, rest of thorax cream, abdomen light brown with dark dorsal spots; forewings dark brown with wide pale straight band across costa and another posteriorly; hindwings with numerous transverse brown lines. **Biology:** Caterpillars recorded on Traveller's Joy (*Clematis brachiata*). **Habitat & distribution:** Variety of relatively moist habitats where *Clematis* spp. grow; Western Cape, South Africa, north to tropical Africa.

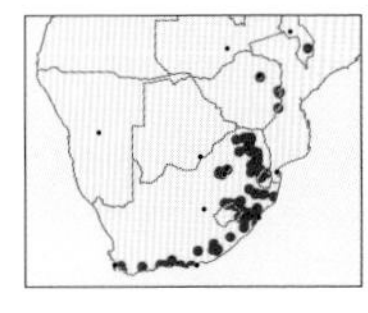

### 7 *Piercia bryophilaria* Mossy Carpet

Forewing 8–11mm. Body and forewings blue-green, usually with numerous black zigzag bars and spots, resembling lichen, occasionally these are fewer in number, often with some orange spots in basal and terminal areas; hindwings uniform white. Twenty-five similar species in genus in region, often difficult to identify without detailed examination. **Biology:** Caterpillars recorded from African violets (*Saintpaulia* spp.). Adults throughout the year. **Habitat & distribution:** Moist forests and riverine vegetation; Western Cape, South Africa, north to tropical Africa.

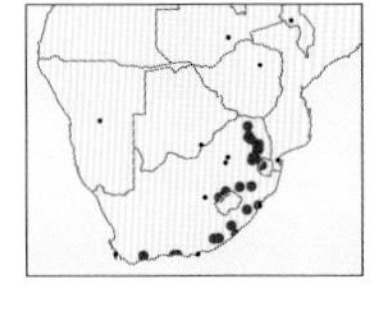

### 8 *Piercia nimipunctata* Triangle Carpet

Forewing 6–8mm. Similar to *P. bryophilaria* (above), but smaller and lines indistinct, diagnostic large triangular black patch on costa of forewings; hindwings grey with faint postmedian line and small cell spot. **Biology:** Caterpillar host associations unknown. Adults September–May, peaking November and February. **Habitat & distribution:** Afromontane forests; Western Cape, South Africa, north to Limpopo, South Africa.

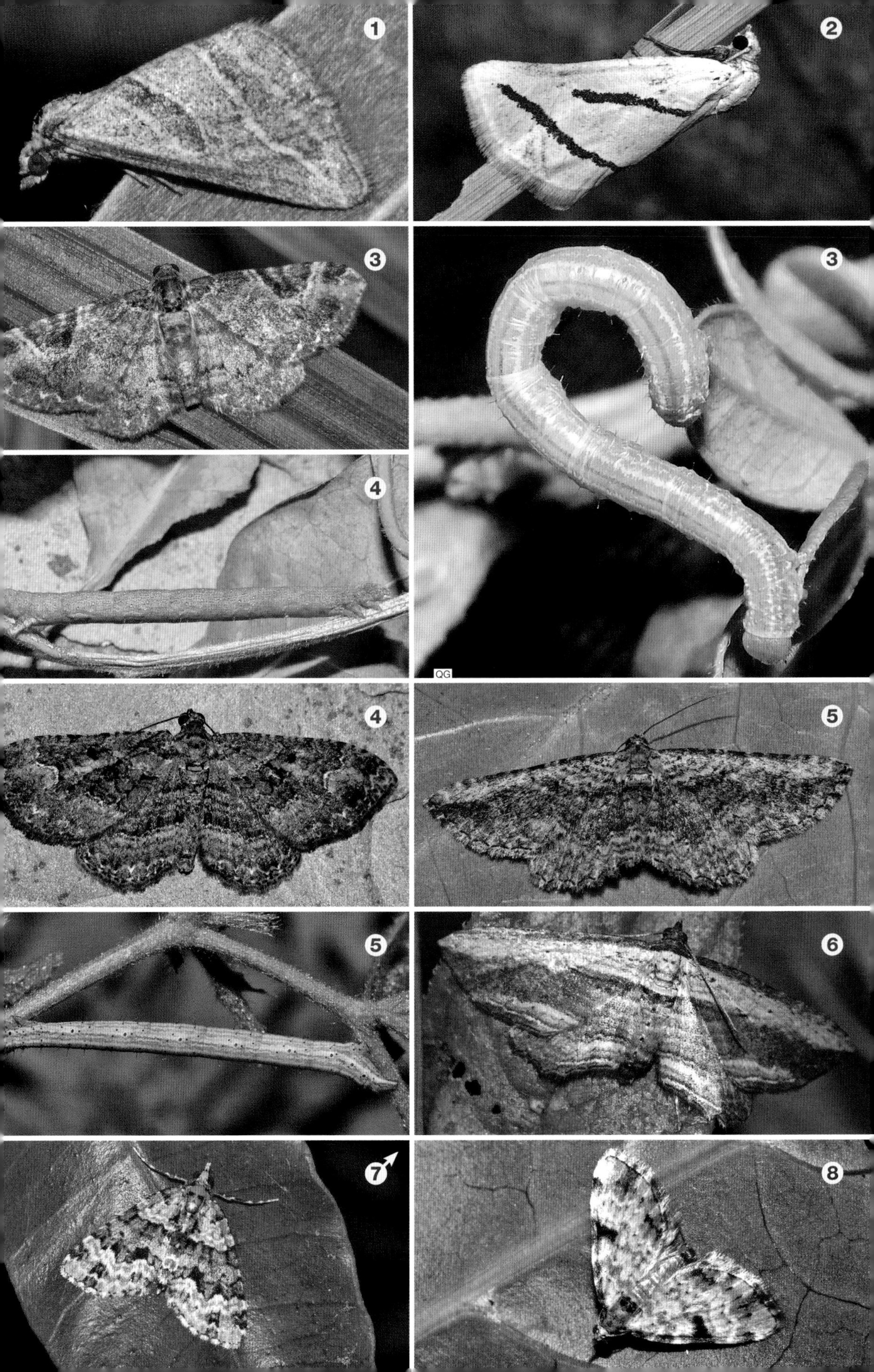
1
2
3
3
4
QG
4
5
5
6
7
8

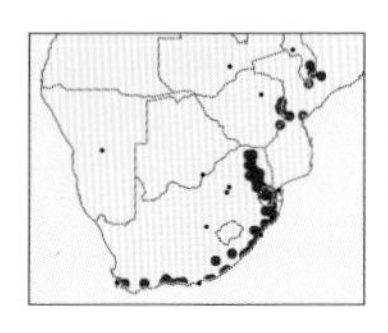

### 1 *Piercia prasinaria* — Bark Carpet

Forewing 9–11mm. Body and forewings dull olive grey, basal, antemedial and postmedian lines double, postmedian sharply kinked, variable dark terminal patches, small cell spot, median area rusty brown in some specimens; hindwings straw to light brown with diagnostic broad dark marginal band. Similar *P. cidariata* lacks the broad marginal band on hindwing, *P. ciliata* has chequered hindwing cilia. **Biology:** Caterpillar host associations unknown. Adults throughout the year, peaking in summer. **Habitat & distribution:** Moist forest and bush; Western Cape, South Africa, north to tropical Africa.

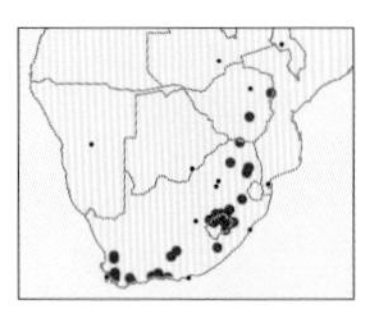

### 2 *Piercia artifex* — Orange-line Carpet

Forewing 7–9mm. Body and forewings mottled brown, antemedial and postmedian lines white edged with bright orange that fades in dried specimens, partial white subterminal line from costa, large, distinctive cell spot; hindwings white with dull lines. **Biology:** Caterpillar host associations unknown. Adults throughout the year, peaking in summer. **Habitat & distribution:** Fynbos and high-altitude grasslands with fynbos elements; Western Cape, South Africa, north to Zimbabwe.

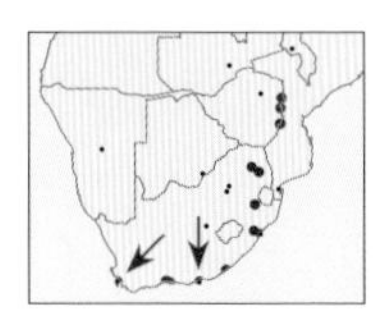

### 3 *Episteira confusidentata* — Confused Longcarpet

Forewing 15–17mm. Broadly oval at rest; elongate forewings slightly pointed; body light green to brown; wings cream, patterned with numerous wavy light and dark green, variable brown shading in basal and median areas, no black spots; hindwings rose-brown. Caterpillar plump and light green with some black spots and brown dorsal stripe on anal claspers. One described species in genus in region. **Biology:** Caterpillars feed on Red Currant (*Searsia chirindensis*). Bivoltine, adults September–November and January–April. **Habitat & distribution:** Moist forests; Western Cape, South Africa, north to Cameroon and Kenya.

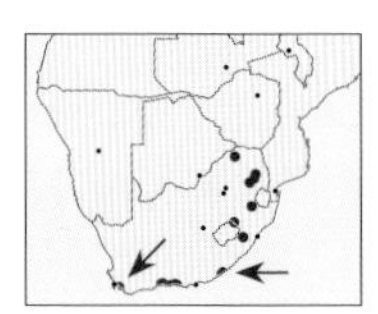

### 4 *Protosteira spectabilis* — Spectacled Longcarpet

Forewing 18–20mm. Elongate forewings, wings sharply pointed; body and forewings white to pale brown, densely patterned with green, postmedian and subterminal lines jagged with sharp black points, a line of small black dots along ends of veins; male hindwings pale red with basal androconial fold and large black patch of hair which releases pheromones, resembling a pair of spectacles in set specimens. One described species in genus in region. **Biology:** Caterpillars recorded on Real Yellowwood (*Podocarpus latifolius*). Adults September–May. **Habitat & distribution:** Moist forests; Western Cape, South Africa, north to tropical Africa, also Madagascar and Comoros.

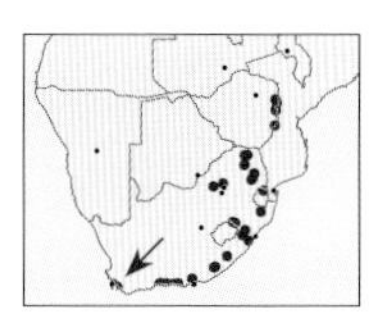

### 5 *Lobidiopteryx eumares* — Rounded Longcarpet

Forewing 16–19mm. Profile triangular at rest; elongate forewings with rounded tips, finely patterned with dense pink, green and brown spots and bands, resembling lichen or bark, random black spots, distinctive white spot on inner margin of wings over abdomen when at rest; hindwings plain cream. One described species in genus in region. **Biology:** Caterpillar host associations unknown. Adults October–March. **Habitat & distribution:** Moist forests and kloofs; Western Cape, South Africa, north to Zimbabwe.

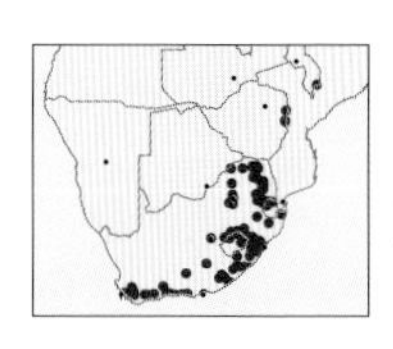

### 6 *Xanthorhoe poseata* — Barred Carpet

Forewing 10–12mm. Colouration variable, mottled brownish, pinkish or greenish but with typical broad dark central band, distal portion may be pale, female with some green on wing; hindwings white with faint grey markings. Eight species in genus in region. **Biology:** Caterpillars recorded from Heliotrope (*Helitropium arborescens*), lettuce and Cherry Plum (*Prunus cerasifera*). Multivoltine, adults throughout the year. **Habitat & distribution:** Variety of moist habitats, including gardens; Western Cape, South Africa, north to tropical Africa.

### 7 *Xanthorhoe exorista* — Double-barred Carpet

Forewing 11–14mm. Broadly triangular at rest; sexually dimorphic; male with white ground colour and median area dark, shape of postmedian line diagnostic, female similar to *X. poseata* (above) but ground colour brick red, never green. Caterpillar brown. **Biology:** Caterpillars polyphagous on herbaceous and tree species. Multivoltine, adults throughout the year. **Habitat & distribution:** Better adapted to more arid conditions than other species in genus, variety of habitats, including gardens; Western Cape, South Africa, north to tropical Africa.

1
2
3
3
4
5
6 ♀
7 ♂

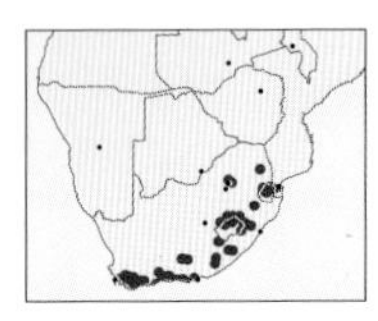

**1 *Haplolabida inaequata*** Intricate Carpet

Forewing 10–14mm. Broadly triangular at rest; forewings marked with complex pattern of transverse zigzag bands in shades of brown, one dominant median cream crossbar of varying width, cell spot comma-shaped, diagnostic postmedian line gently undulating; hindwings tan with faint kinked antemedial line. Caterpillar light brown or green with darker brown and white lateral bars. Two described species in genus in region, similar *H. coaequata* has postmedian line sharply crenulated. **Biology:** Caterpillars feed on various heaths (*Erica* spp.). Multivoltine, adults throughout the year. **Habitat & distribution:** Fynbos and grasslands where *Erica* spp. grow; Western Cape north to Mpumalanga, South Africa.

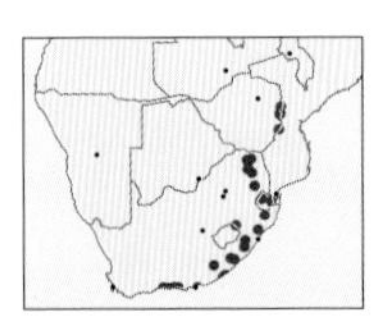

**2 *Gonanticlea meridionata*** Southern Carpet

Forewing 15–17mm. Broadly triangular at rest; similar to *H. inaequata* (above), but much larger, postmedian line jagged, not undulating, and small cell spot; hindwings unmarked, dark brown and crenulated. One species in genus in region. **Biology:** Caterpillars feed on *Clematis brachiata* (Ranunculaceae). Adults September–May. **Habitat & distribution:** Moist forests; Western Cape, South Africa, north to tropical Africa.

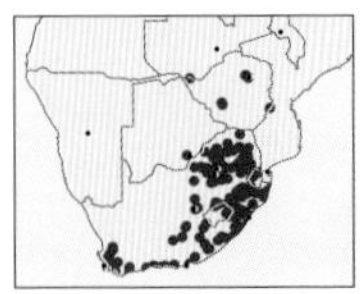

**3 *Mimoclystia pudicata*** Diurnal Carpet

Forewing 10–13mm. Ground colour of both sides of wings orange, forewings marked with numerous horizontal wavy brown lines of varying widths, hues and intensity; some individuals in Cape populations have extensive pink shading on forewing obliterating lines. Caterpillar a slender looper, pale below, pale brown to bright red above. Six described species in region. **Biology:** Caterpillars feed on Tremble Tops (*Kohautia amatymbica*). The only Larentiine in region where adults are diurnal, but also attracted to light. Bivoltine, adults throughout the year, peaking in summer months. **Habitat & distribution:** Variety of relatively moist habitats; Western Cape, South Africa, north to tropical Africa.

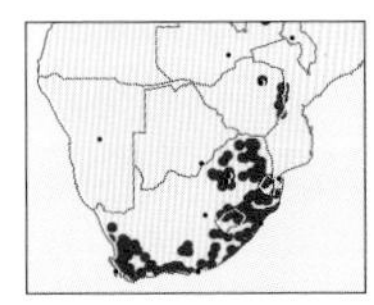

**4 *Mimoclystia explanata*** Explanatory Carpet

Forewing 11–14mm. Broadly triangular at rest; forewings divided into broader dark and lighter bands, variably shaded from grey to reddish-brown; hindwings similar but more lightly marked. Caterpillar plump, variably mottled black and brown. **Biology:** Caterpillars polyphagous on plants in undergrowth. Multivoltine, adults throughout the year. **Habitat & distribution:** Variety of relatively moist habitats; Western Cape, South Africa, north to Malawi.

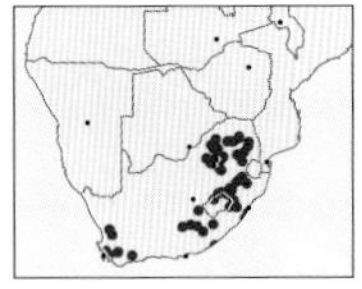

**5 *Mimoclystia undulosata*** Undulating Carpet

Forewing 10–12mm. Can be confused with *Epirrhoe edelsteni* (below) but occurs in moist habitats; diagnostic white to cream band both sides of median area and forewing pattern continues on inner part of hindwings, multiple lines crenulated, median area reddish-brown; basal part of hindwing uniform brown with light lines. Similar *M. tepescens* has extensive white basal and terminal areas. **Biology:** Caterpillar host associations unknown. Multivoltine, adults throughout the year. **Habitat & distribution:** Moist grassland and fynbos habitats; Western Cape north to Limpopo, South Africa.

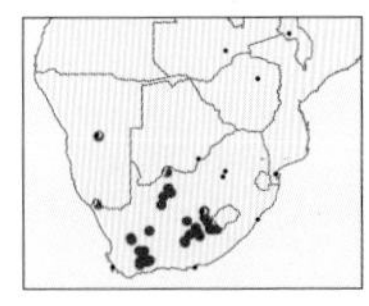

**6 *Epirrhoe edelsteni*** Brick Carpet

Forewing 12–14mm. Can be confused with *Mimoclystia undulosata* (above), but occurs in arid karoo habitats; no white to cream band around median area and hindwings plain grey, multiple lines crenulated, forewing dark brown variably dusted with brick red scales; hindwings uniform brown with light lines. Three species in genus in region. **Biology:** Caterpillar host associations unknown. Adults August–April. **Habitat & distribution:** Arid karoo, renosterveld and bushveld; Western Cape, South Africa, north to Namibia.

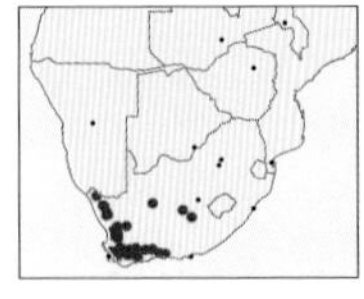

**7 *Euphyia distinctata*** Distinct Carpet

Forewing 9–11mm. Forewings straw-coloured with orange and black crenulated lines, diagnostic median area silvery grey, narrow at inner margin becoming sharply broader midway to costa; hindwings plain straw becoming greyer toward apex. One species in genus in region. **Biology:** Caterpillar host associations unknown. Adults August–November and March, primarily in spring. **Habitat & distribution:** Succulent karoo habitats and along Orange River valley; Cape provinces, South Africa.

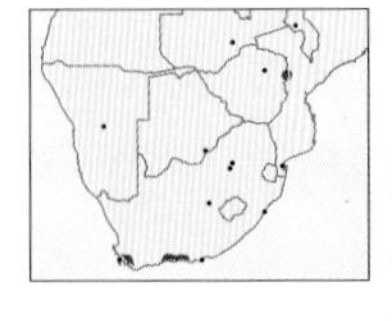

**8 *Parortholitha subrectaria*** Wrench Carpet

Forewing 12–15mm. Broadly triangular at rest; unmistakably marked, forewings light brown, crossed by 2 broad bands graded to darker brown and edged with wavy gold lines, distal band encloses dark dot. Three species in genus in region. **Biology:** Caterpillar host associations unknown. Adults throughout the year, common in habitat. **Habitat & distribution:** Fynbos and fynbos remnant vegetation, apparent disjunct populations; Western Cape, southern Cape and Zimbabwe.

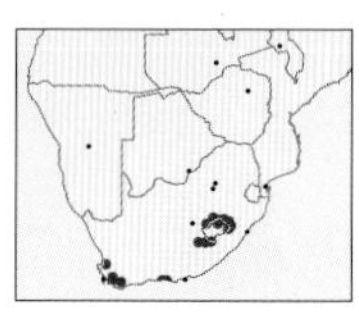

**1 *Parortholitha moerdyki*** Grey Carpet

Forewing 14–17mm. Broadly triangular at rest; a large *Parortholitha*, wings and body grey, forewings with full complement of distinctive white lines of variably intensity, black cell spot; hindwings light grey with faint markings. **Biology:** Caterpillar host associations unknown. Adults October–April. **Habitat & distribution:** Moist fynbos and Afroalpine vegetation; Western Cape north to KwaZulu-Natal, South Africa, and Lesotho.

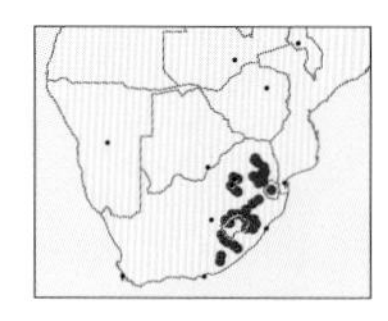

**2 *Parortholitha horismodes*** Copper Carpet

Forewing 14–16mm. Broadly triangular at rest; wings and body with coppery brown sheen, lines parallel and diagnostically kinked; hindwings light brown with many fine lines. **Biology:** Caterpillars feed on *Leucosidea sericea* (Rosaceae). Adults October–April. **Habitat & distribution:** Cool moist grasslands where host plant grows; Eastern Cape north to Limpopo, South Africa.

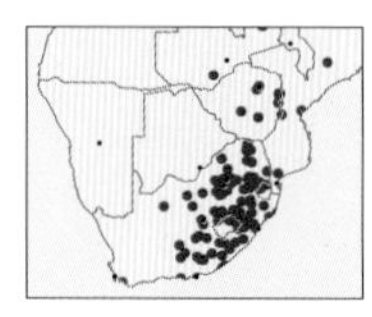

**3 *Pseudolarentia megalaria*** Large Spot Rivulet

Forewing 12–16mm. Triangular at rest; wings off-white, variably dusted with brownish-grey scales, antemedial and postmedian bands distinctive, large brown cell spot, dashed brown line along fringed margin; hindwings cream with fine brown marginal line. One species in genus in region. **Biology:** Caterpillar host associations unknown. Adults throughout the year, can be abundant in summer months. **Habitat & distribution:** Various grassland habitats; relatively moist eastern parts of the region, north to Ethiopia.

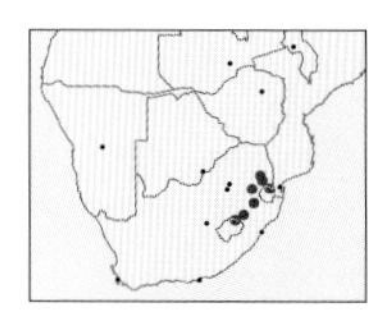

**4 *Perizoma lamprammodes*** Stratified Carpet

Forewing 13–14mm. Broadly triangular at rest; forewings cream with brown costa, marked by numerous distinctive straight oblique brown lines of varying thickness, black cell spot; hindwings with similar lines but curved sharply before inner margin. Three species in genus in region. **Biology:** Host associations unknown. **Habitat & distribution:** High-altitude moist grasslands; Lesotho north to Mpumalanga.

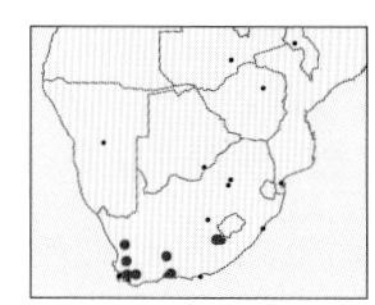

**5 *Perizoma alumna*** Ringed Carpet

Forewing 15–19mm. Broadly triangular at rest; forewings white with variable dense grey shading, median and basal area brown in some individuals, cilia chequered, distinctive round black-ringed white median spot near costa; hindwings uniform grey with fine lines. Caterpillar brown with oblique white lateral marks and on prolegs. **Biology:** Caterpillars feed on *Cliffortia ilicifolia* (Rosaceae). **Habitat & distribution:** Fynbos and renosterveld; Cape provinces, South Africa.

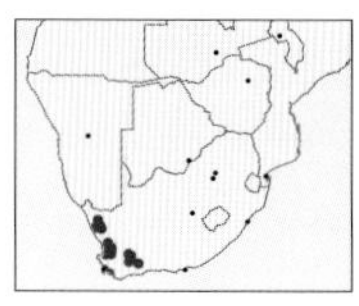

**6 *Entephria petrogenes*** Spotted Carpet

Forewing 10–13mm. Broadly triangular at rest; ground colour light grey, antemedial, basal and postmedian lines broad, black, overlaid with golden scales in some individuals, diagnostic white spots beyond postmedian, cilia light brown and black chequered; hindwings dark grey with fine lines. Seven species in genus in region. **Biology:** Caterpillar host associations unknown. Adults August–October and April. **Habitat & distribution:** Dry fynbos and renosterveld; Western Cape and Northern Cape, South Africa.

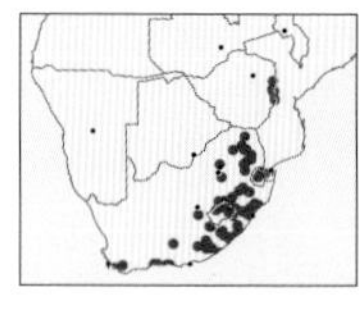

**7 *Scotopteryx bitrita*** Gilt Carpet

Forewing 12–15mm. Outline triangular when at rest; forewings black at base, grey distally, dominated by 2 large black patches rimmed with distinctive gold edging, postmedian line evenly curved in male, sharply curved in female; hindwings cream with brown border. Three species in genus in region. **Biology:** Caterpillar host associations unknown. Adults throughout the year. **Habitat & distribution:** Cool moist fynbos and grassland habitats; Western Cape, South Africa, north to Malawi.

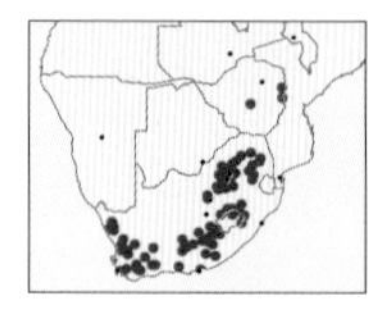

**8 *Scotopteryx deversa*** Diverse Rivulet

Forewing 10–14mm. Broadly triangular at rest; forewings strikingly marked with horizontal brown to black bands, a dominant central pale median band flanked by broad, wavy dark brown to black bands, shape of postmedian line diagnostic; hindwings pale grey to brown with weak pale lines. Ten species in genus in region. **Biology:** Caterpillar host associations unknown. Adults throughout the year. **Habitat & distribution:** Grassland and karroid vegetation; Western Cape, South Africa, north to Zimbabwe.

1
2
3
4
5
5
6
6
7
8

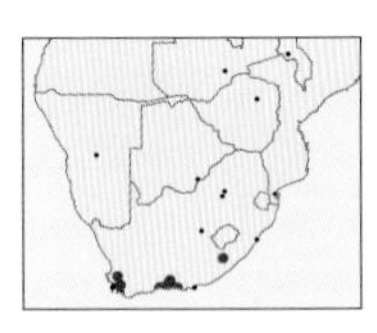

### 1 *Scotopteryx crenulimargo* — Crenulated Rivulet

Forewing 14–16mm. Similar to *S. deversa* (p.168), but shape of postmedian line less extreme and median area variable from dark brown to light brown, but without pale median band; 2–4 triangular white subterminal marks diagnostic, wing margin crenulated. Caterpillar mottled brown with diamond-shaped dorsal marks. **Biology:** Caterpillars feed on *Cliffortia odorata* (Rosaceae). Adults throughout the year, peaking in March. **Habitat & distribution:** Damp places and wetlands where host plant grows; Western Cape and Eastern Cape.

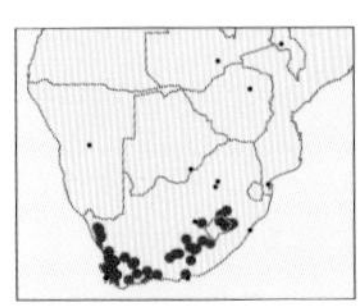

### 2 *Scotopteryx cryptospilata* — Blemished Rivulet

Forewing 14–17mm. Broadly triangular at rest; pale grey forewings crossed by 2 diagnostic slightly wavy, closely adjacent brown lines; hindwings plain brown with faint paler wavy lines; populations regionally different, probably a complex of species. Similar *S. ferridotata* has brown colour and postmedian line sharply kinked. **Biology:** Caterpillar host associations unknown. Adults throughout the year. **Habitat & distribution:** Variety of habitats, grassland, fynbos, renosterveld and succulent karoo; Western Cape to KwaZulu-Natal and Lesotho.

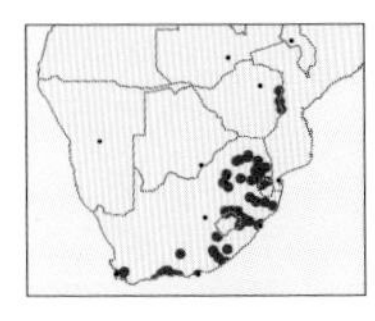

### 3 *Scotopteryx nictitaria* — Straightline Rivulet

Forewing 15–18mm. Broadly triangular at rest; distinguished from other *Scotopteryx* spp. by the straight postmedian line; forewing ground colour shaded by brown of varying intensity; hindwings light grey becoming darker distally. Caterpillar cream with numerous orange to black thin wavy lines. **Biology:** Caterpillar recorded on *Senna italica* (Fabaceae). Adults throughout the year. **Habitat & distribution:** Moist grassland and fynbos; Western Cape, South Africa, north to Kenya.

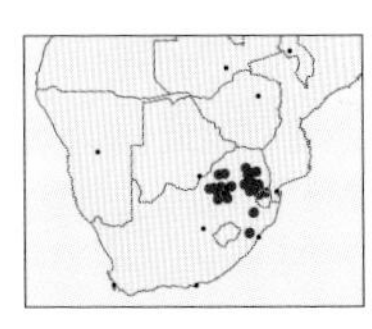

### 4 *Scotopteryx cryptocycla* — Wavyline Rivulet

Forewing 14–17mm. Similar to *S. nictitaria* (above), but brown and postmedian line wavy, as are all other lines on forewings, large black forewing cell spot present in most individuals on upper side, always on underside; hindwings straw-coloured becoming darker brown distally. **Biology:** Caterpillar host associations unknown. Adults throughout the year. **Habitat & distribution:** Highveld grassland; KwaZulu-Natal north to Limpopo, South Africa.

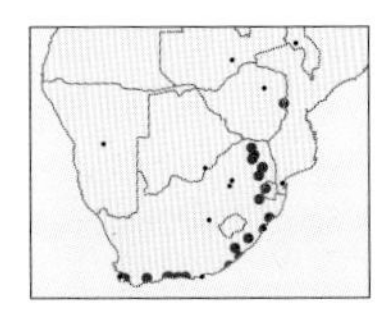

### 5 *Collix foraminata* — Perforated Shell

Forewing 14–17mm. Female light brown, male dark brown, distinct black cell spot on forewings consists of tufts of erect scales in male; hindwing margin crenulated, underside distinctly marked with 2 brown bands and cell spot on all wings. Caterpillar colouration matches that of host plant, green to brown to purple. One species in genus in region. **Biology:** Caterpillars polyphagous on trees. Multivoltine, adults throughout the year. **Habitat & distribution:** Moist forests; Western Cape, South Africa, north to tropical Africa, also Madagascar and Comoros.

## Subfamily Desmobathrinae

A small subfamily with 9 species and 4 genera in region. Slender bodied with long legs and antennae, and fragile wings. Caterpillars in region are generally associated with Sapotaceae.

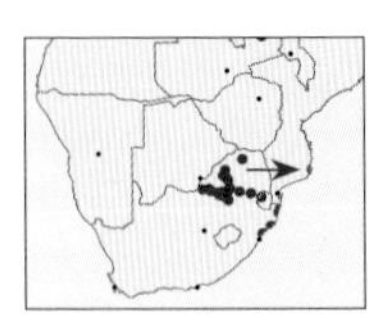

### 6 *Conolophia conscitaria* — Partaker

Forewing 18–27mm. Wings held outstretched at rest; body and wings pinkish-brown to greyish-brown, postmedian lines straight, sometimes spotted with small dark dots, yellowish-brown with graded brown edge on both wings; form *maculata* has a single larger brown spot in lower terminal area of forewing. Caterpillar brown, finely striated with white lines, black-ringed white setal bases. Three species in genus in region, similar *C. aemula* smaller, straw-coloured and heavily dusted with grey scales. **Biology:** Caterpillars feed on Common Red Milkwood (*Mimusops zeyheri*). Bivoltine, adults October–May, peaking in summer. **Habitat & distribution:** Bushveld and forest margins where host plant grows; KwaZulu-Natal, South Africa, north to tropical Africa, Madagascar and La Réunion.

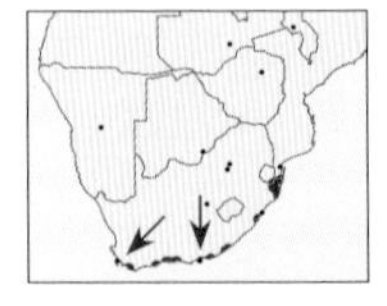

### 7 *Panagropsis equitaria* — Coastal Lesser Partaker

Forewing 12–15mm. Broadly triangular in outline; body and forewings reddish-brown variably dusted with dark brown scales, antemedial and postmedian lines curved inward, containing small black dots diagnostic, black cell dot and sometimes larger black spots in terminal area. Caterpillar smooth, green dorsally, paler underside. Two species in genus in region. **Biology:** Caterpillars feed on White Milkwood (*Sideroxylon inerme*) and Common Red Milkwood (*Mimusops obovata*). Multivoltine, adults throughout the year. **Habitat & distribution:** Coastal habitats where host plants grow; Western Cape, South Africa, north to Mozambique.

1
1
2
3
3
4
5♀
5
6
6
7
7

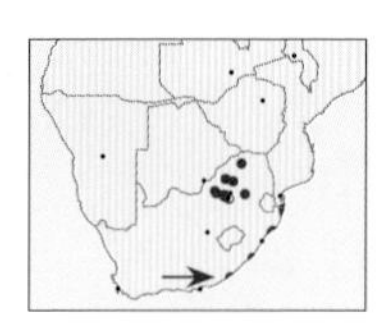

**1 *Panagropsis muricolor*** Brown Lesser Partaker

Forewing 12–16mm. Similar to *Conolophia conscitaria* (p.170), but much smaller and cell spot well developed; body and wings uniformly light brown with diagnostic straight, mostly dark brown postmedian line, spots small and few when present. Caterpillar a looper variably mottled brown, grey and green. **Biology:** Caterpillars feed on Common Red Milkwood (*Mimusops zeyheri*) and Transvaal Milkplum (*Englerophytum magalismontanum*) utilising new leaves before they turn hard. Multivoltine, adults throughout the year. **Habitat & distribution:** Various habitats where host plants grow; Eastern Cape, South Africa, north to Kenya and Sudan.

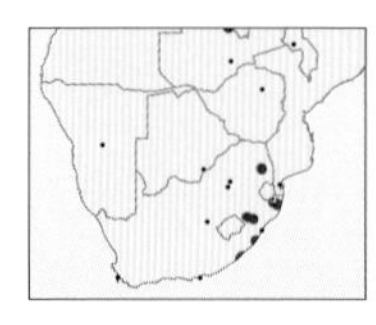

**2 *Ozola pulverulenta*** Falcate Partaker

Forewing 7–10mm. Forewing sharply falcate; head, body and all wings white, sparsely or densely shaded with brownish scales and dusted with black scales, antemedial line sharply curved toward costa, postmedian line undulating and curved inward. One species in genus in region. **Biology:** Caterpillar host associations unknown. **Habitat & distribution:** Lowland bushveld and coastal bush; Eastern Cape, South Africa, north to Tanzania.

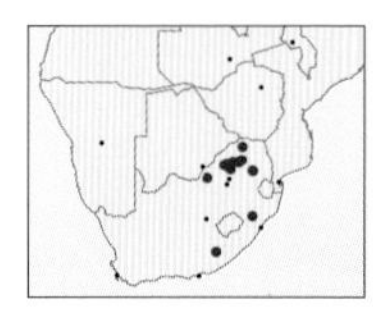

**3 *Barrama impunctata*** White Pinhole

Forewing 8–10mm. Broadly triangular at rest; body and wings hyaline white, finely dusted with loose brown scales that are lost in older individuals, tiny black cell spot on all wings, row of black spots along termen. Caterpillar not described. One species in genus. **Biology:** Caterpillars reported feeding on Fingerleaf (*Vitex obovata*). Adults October–April. **Habitat & distribution:** Highland bushveld; Eastern Cape north to Limpopo, South Africa.

## Subfamily Ennominae

The largest geometrid subfamily; very diverse, with over 900 species recorded from the region. Regional ennomines vary from small (forewing 6mm) to very large (forewing 37mm) and adults are generally more robust than in other subfamilies. Most are cryptic and nocturnal, but some are colourful and diurnal. Caterpillars feed on the leaves of many different plant families and can be found in most habitats. Some are generalists, but many are specialists that are restricted in distribution, making them of conservation concern. Caterpillars feed in the open, mostly at night; most pupate in the soil or in the leaf litter in loosely constructed cocoons. Ennomines form a substantial part of the biomass in many habitats and can be abundant at times, making them an important link in the food chain. Most can be placed in well-defined tribes.

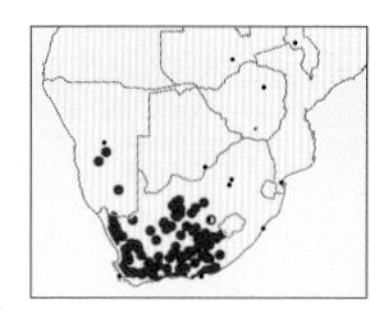

**4 *Acanthovalva focularia*** Orange Thornyvalve

Forewing 10–12mm. Male antennae strongly pectinate, filiform in female; body and forewings light brown with darker transverse bands, sometimes broken, a median band usually the widest and darkest; hindwings distinctively orange with black markings, underside orange to yellow. Four species in genus in region. **Biology:** Caterpillar host associations unknown. Females diurnal, not attracted to light, males nocturnal attracted to light. Adults throughout the year, often abundant in rainy season. **Habitat & distribution:** Karoo biomes, as well as fynbos and grassland that contain some karroid vegetation; western parts of South Africa and Namibia.

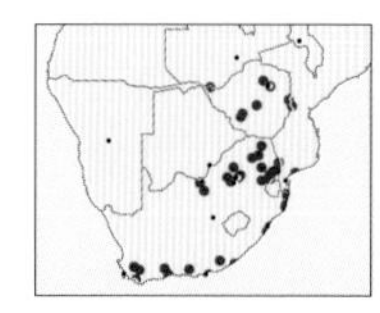

**5 *Acanthovalva bilineata*** Two-line Thornyvalve

Forewing 7–10mm. Thorax mottled black, abdomen brownish-orange; wings dark grey dorsally heavily marked with black stipples and bands, undersides orange with brown speckles and 2 black bands. Caterpillar stout, black with narrow dorsal and wide lateral white and orange stripes. Similar *A. capensis* has white ground colour with grey markings and is restricted to the west coast of South Africa. **Biology:** Caterpillars feed on besembossies (*Thesium* spp.). Bivoltine, adults October–May, never abundant. **Habitat & distribution:** Localised but widespread in variety of habitats; Western Cape, South Africa, north to Kenya.

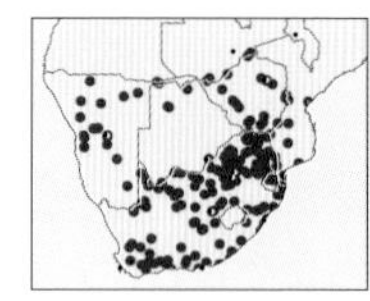

**6 *Acanthovalva inconspicuaria*** Inconspicuous Thornyvalve

Forewing 7–11mm. Body grey; wings light grey with very variable brown and black speckling; forewings crossed by 3 indistinctly defined brown bands and with row of tiny black dots along margin, forewing uniform light grey with no markings in extreme forms, underside slightly darker (but not orange as in *A. bilineata* above). Caterpillar green with white lateral stripe, sometimes row of 5 black spots along sides. **Biology:** Caterpillars feed primarily on *Limeum* spp., also Sierkooltjie (*Corbichonia decumbens*) and Black Thorn (*Vachellia nilotica*). Multivoltine, adults throughout the year, can be abundant in late summer, easily disturbed by day. **Habitat & distribution:** Variety of habitats, individuals in more arid habitats much smaller; Western Cape north through Africa, southern Europe and Near East.

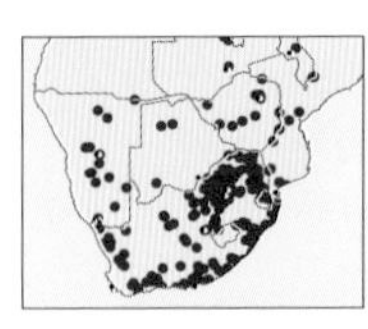

## 1 *Chiasmia brongusaria* — Variable Chiasmia

Forewing 14–17mm. Very variable; yellowish-brown to dark brown, wings crossed by several broken dark lines or no lines, distal section of forewings sometimes with darker brown patch incorporating one large and several smaller brown spots, postmedian line slightly curved outward and strongly curved back to costa near apex diagnostic. Individuals from western populations tend to be darker brown. Caterpillar brown, grey or yellow-green, with diagnostic tiny black dots, often with yellow transverse bands. Eighty species in genus in region, which can be abundant being an important link in the food chain. **Biology:** Caterpillars feed on Sweet Thorn (*Vachellia karroo*), and other *Vachellia* spp. Multivoltine, adults throughout the year, peaking in summer, abundant, often disturbed by day under trees. **Habitat & distribution:** Habitats where *Vachellia* trees grow, including gardens; Western Cape, South Africa, north to East Africa.

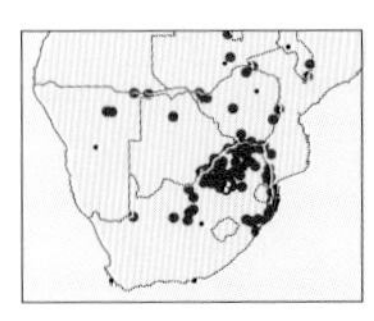

## 2 *Chiasmia furcata* — Forked Chiasmia

Forewing 9–13mm. Thorax light brown, becoming paler toward abdomen; forewings light brown speckled with darker scales and crossed by several wide darker bands of varying intensity, some complete and other partial. Caterpillar variably grey to dark brown, often with red blotches. **Biology:** Caterpillars feed primarily on Umbrella Thorn (*Vachellia tortilis*). Multivoltine, adults throughout the year. **Habitat & distribution:** Bushveld where umbrella thorns grow; KwaZulu-Natal and Northern Cape, South Africa, north to Angola and Zambia.

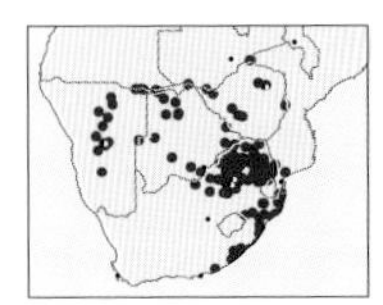

## 3 *Chiasmia multistrigata* — Multistriped Chiasmia

Forewing 9–15mm (smaller individuals from arid areas or dry seasons). Body grey-brown crossed by darker bands; wing ground colour white variably shaded with grey, very dark in some individuals, heavily marked by numerous wavy dark lines, postmedian line crenulated and sharply angled before costa, distal margins blotched with dark brown patches; a scalloped black line loops along hind margin. Similar *C. boarmioides* has forewing postmedian line evenly curved before costa. Caterpillar light green with diagnostic yellowish lateral line and bands across dorsal area between segments. **Biology:** Caterpillars feed on hook thorns (*Senegalia* spp.). Multivoltine, adults throughout the year, peaking in summer months. **Habitat & distribution:** Bushveld where hook thorns grow; Eastern Cape, South Africa, north to Kenya.

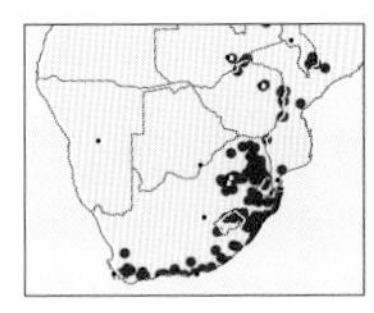

## 4 *Chiasmia simplicilinea* — Simple-lined Chiasmia

Forewing 13–16mm. Body grey to light brown; forewings grey to light brown with 1 straight black line horizontally between wing tips, usually dark patch behind the line and black cell dot; hindwings with 2 transverse lines, broader basal one and thinner distal one black, black cell spot, hind margin with short tail. Melanistic forms with wings fully or partially dark brown are common. Similar *C. rectistriaria* has lines double and hindwing line less curved. Caterpillar olive green with yellow and white lateral line, diagnostic 3 black spots on each segment. **Biology:** Caterpillars commonly feed on Australian wattles (*Acacia* spp.), native host plants unknown. Adults drink fluid from the eyes and dung of mammals. Multivoltine, adults throughout the year, peaking in summer. **Habitat & distribution:** Moist grassland, forest, bushveld and habitats infested with alien wattles; Western Cape, South Africa, north to tropical Africa.

## 5 *Chiasmia natalensis* — Angled Forest Chiasmia

Forewing 11–13mm. Body and wings creamy yellow; wings crossed by thin, wavy yellow and brown lines, cell spot on each wing; falcate forewings diagnostic, 4 small triangular black spots on subterminal line diagnostic; margin of hindwings irregular, with distinct short tail. Similar *C. paucimacula* does not have forewing falcate. Caterpillar light green with yellow head capsule and narrow lateral stripe. **Biology:** Caterpillars feed on Flat-crown Albizia (*Albizia adianthifolia*). Multivoltine, adults throughout the year. **Habitat & distribution:** Coastal bush and lowland forest; KwaZulu-Natal, South Africa, north to Kenya.

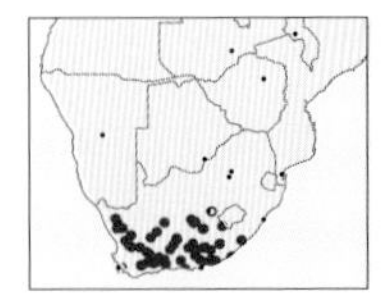

## 6 *Chiasmia observata* — Karroo Chiasmia

Forewing 11–14mm. Body light brown; forewings light brown, marked with 3 tapering partial transverse lines, antemedial and medial lines connected at inner margin diagnostic, area between bands distinctly paler; area beyond distal band distinctly darker; hindwings paler. Similar *C. subcurvaria*, which occurs from Eastern Cape to tropical Africa, has antemedial and basal lines apart at inner margin. Caterpillar plump, green with cream lateral band. **Biology:** Caterpillars feed on Sweet Thorn (*Vachellia karroo*). Well adapted to arid environments, can complete life cycle in 3 weeks, survives drought in pupal stage. Adults appear any time of the year after good rains. **Habitat & distribution:** Arid habitats where the host plant grows; Western Cape north to Free State, South Africa.

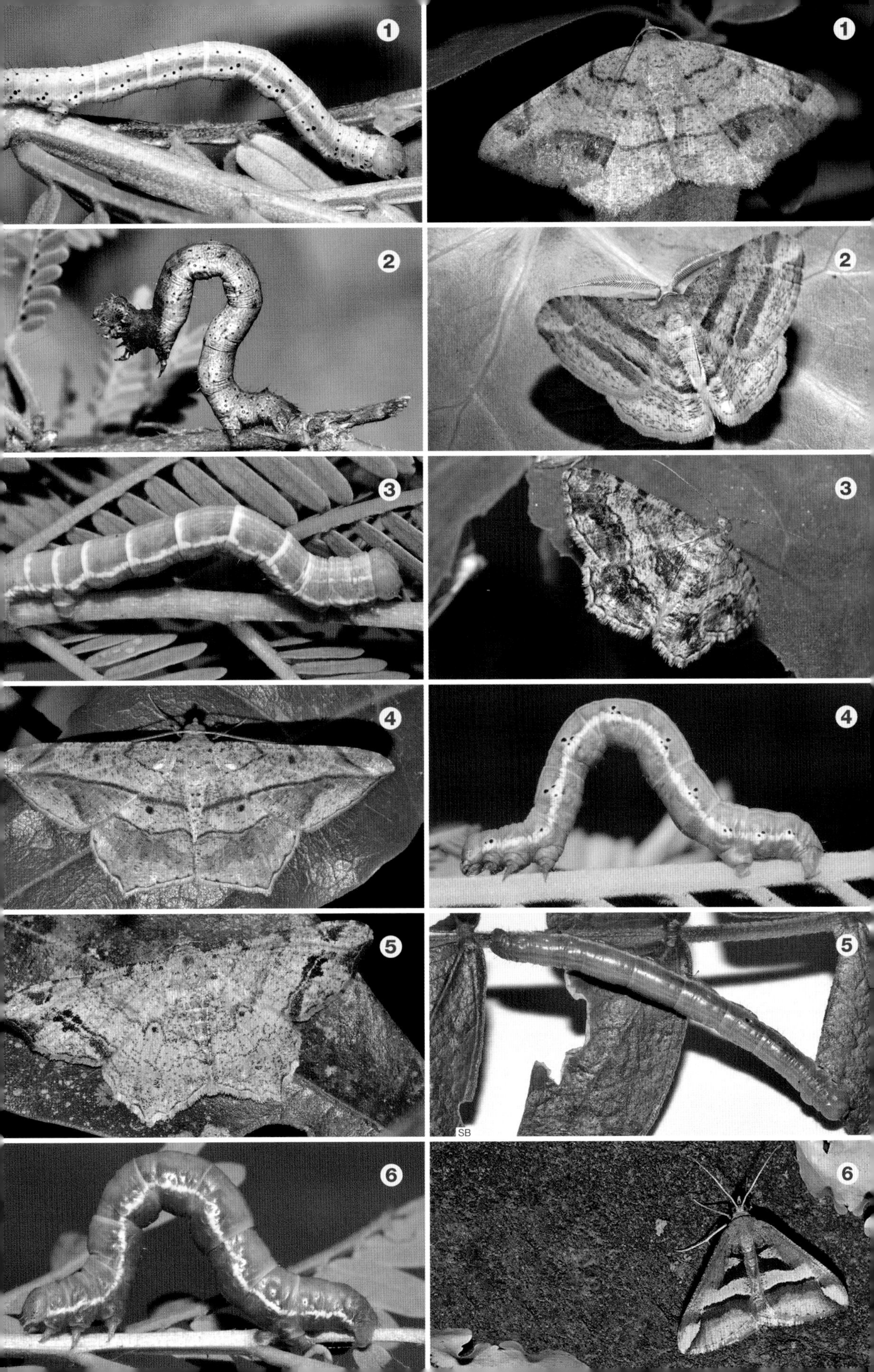
1
1
2
2
3
3
4
4
5
5
SB
6
6

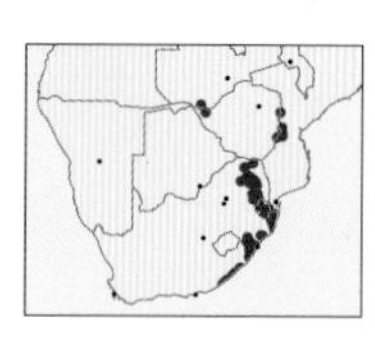

## 1 *Chiasmia umbrata* — Shadowed Chiasmia

Forewing 12–14mm. Cream to brown and heavily marked with dark brown or black, but mostly with broad lighter band across centre of both wings; margins of hindwings scalloped, elongate in male; caterpillars mottled pinkish-brown. **Biology:** Caterpillars feed on various genera of native thorn trees. Multivoltine, adults throughout the year. **Habitat & distribution:** In frost-free moist forest and bush; Eastern Cape, South Africa, north to tropical Africa, also Madagascar.

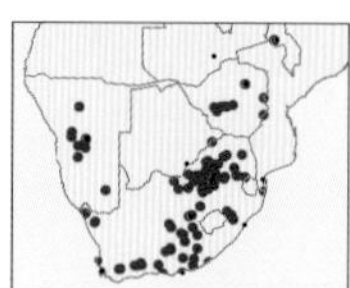

## 2 *Chiasmia turbulentata* — Turbulent Chiasmia

Forewing 13–15mm. At rest, wings folded above body like a butterfly; upperside cream with broad dusty brown margin to wings, more commonly seen underside broadly striped in brown and cream with central black spot, no curved median band across. Similar *C. procidata* is less well marked on underside and has diagnostic curved median band across hindwing. Caterpillar very variable; green with white banding, sometimes dark spots, alternatively dark mottled cream and black. **Biology:** Caterpillars feed on Sweet Thorn (*Vachellia karroo*) and Paperbark Thorn (*V. sieberiana*). Multivoltine, adults throughout the year. **Habitat & distribution:** Various habitats where *Vachellia* trees grow, including transformed habitats; Western Cape, South Africa, north to tropical Africa.

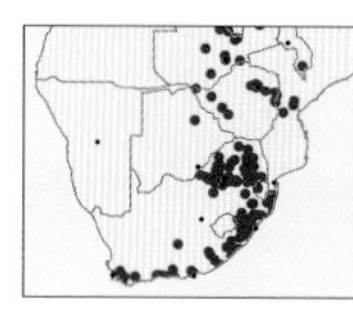

## 3 *Chiasmia streniata* — Seasonal Chiasmia

Forewing 13–16mm. Wet-season form very dark, dry-season form *amandata* much lighter and well-marked; body and wings grey, finely speckled with black dots, orange postmedian line straight over both wings, round bend before costa, variably rusty shading in terminal area, underside white to orange variably sprinkled with brown striations. Caterpillar white dorsally with numerous brown longitudinal lines, laterally yellow with conspicuous black and white dots on each segment. **Biology:** Caterpillars feed on hook thorns (*Senegalia* spp.), also Australian wattles (*Acacia* spp.). Multivoltine, adults throughout the year, seasonal peaks in summer and winter. **Habitat & distribution:** Moist habitats where hook thorns or wattles grow, expanded its range substantially following wattle infestations; Western Cape, South Africa, north to tropical Africa, also Madagascar.

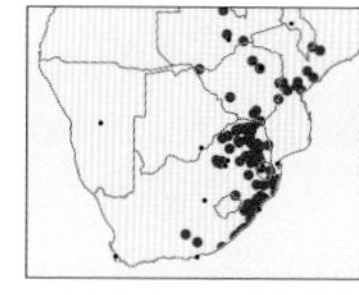

## 4 *Chiasmia confuscata* — Confusing Chiasmia

Forewing 10–14mm. Body and wings dark to light grey, brown in some populations, black cell spot and straight postmedian line, other lines indistinct, variable rusty terminal shading in females, underside white with black speckles and brown postmedian band of varying width. Similar *C. sororcula* has underside yellow and *S. feraliata* is much larger and has terminal area of forewing underside uniform brown with one white dot. Caterpillar green with distinctive yellow lateral and white dorsolateral lines. **Biology:** Caterpillars feed primarily on Flame Thorn (*Senegalia ataxacantha*), on which they can swarm at times. Multivoltine, adults throughout the year, most abundant in February, when they may come to light sources in their thousands. **Habitat & distribution:** Dry and wet forests where their host tree grows; Eastern Cape, South Africa, north to tropical Africa, also Madagascar.

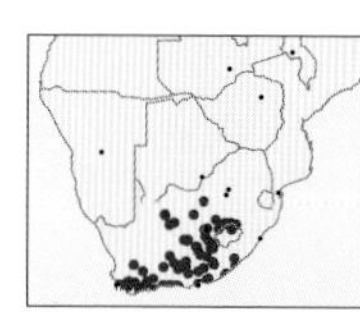

## 5 *Chiasmia semitecta* — Southern Grass Chiasmia

Forewing 10–13mm. Ground colour light grey, variably sprinkled with dark grey, postmedian line evenly curved, prominent elongate black cell spot, black spots beyond postmedian line often present. Several similar species; *C. johnstoni* has forewing postmedian line straight before curving near costa. Caterpillar pinkish-grey with distinct lateral spots. **Biology:** Caterpillars recorded feeding at night on *Helichrysum cymosum* (Asteraceae), and probably other daisies. Multivoltine, adults throughout the year. **Habitat & distribution:** Fynbos, karroid and grassland habitats that contain karroid daisies; Western Cape north to North West, South Africa.

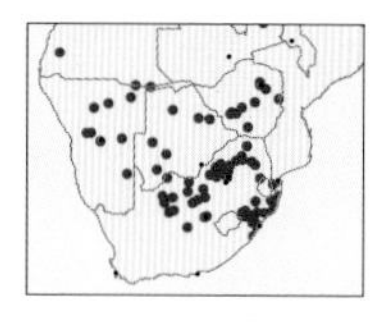

## 6 *Chiasmia marmorata* — Ivory Chiasmia

Forewing 12–15mm. Ground colour creamy white lightly dusted with brown, postmedian line straight, curved sharply inward near costa, small black marks beyond postmedian line, no cell spot, marks and lines often almost absent in some individuals. Similar *C. arenosa* has forewing postmedian line straight, not curving near costa. Caterpillar green with several cream longitudinal lines and conspicuous black spots. **Biology:** Caterpillars feed on Paperbark Thorn (*Vachellia sieberiana*). Multivoltine, adults throughout the year. **Habitat & distribution:** Bushveld; Eastern Cape, South Africa, north to Angola.

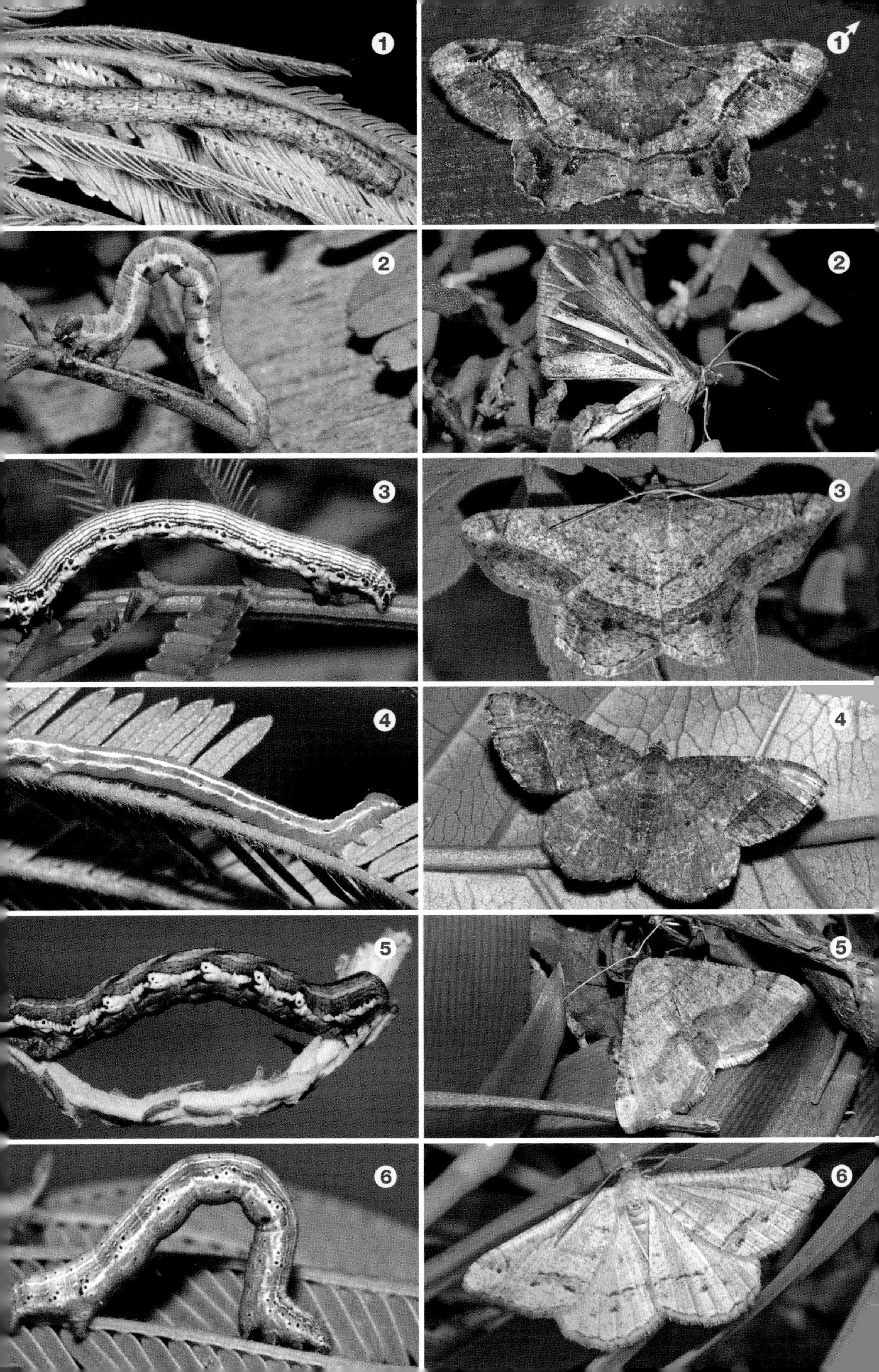
1
1♂
2
2
3
3
4
4
5
5
6
6

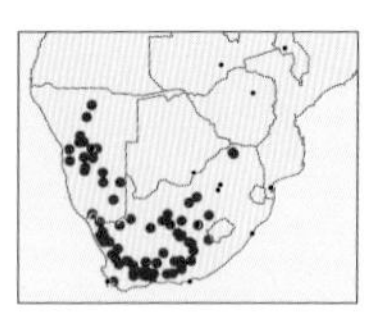

### 1 *Chiasmia inaequilinea* — Thorn Chiasmia

Forewing 10–13mm. Ground colour creamy white, distinctly shaded with brown and black scales between sinuous lines, shading beyond postmedian parallel; hindwings less well marked. Similar *C. grimmia* has fine grey sprinkling of scales and has shading beyond postmedian line black lined and sharply pointed. Caterpillar white, mimicking the white thorns of host plant. **Biology:** Caterpillars feed on Sweet Thorn (*Vachellia karroo*). Multivoltine, adapted to arid conditions, can complete larval stage in 3 weeks, adults throughout the year after rain. **Habitat & distribution:** Arid habitats where host plant grows; Western Cape, South Africa, north to Namibia.

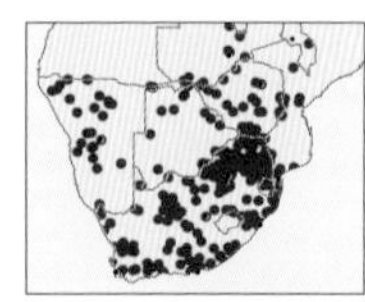

### 2 *Isturgia deerraria* — Thorn Peahen

Forewing 10–14mm. Distinct wet- and dry-season forms but also many intermediates; wet-season form creamy grey sprinkled with distinct brown spots, both wings crossed by 2 brown lines, postmedian line straight, terminal area distinctly darker; dry-season form has lines indistinct or absent, sparse sprinkling of spots or almost uniformly fawn with brown marks in inner margin and postmedian line. Caterpillar also variable, may be green, brown, red or mottled. Eleven species in genus in region; very similar *I. arizeloides* is browner and has denser sprinkling of brown spots. **Biology:** Caterpillars feed on various genera of native thorn trees (Acacieae). Multivoltine, adults throughout the year, nocturnal but often disturbed by day. **Habitat & distribution:** Abundant in habitats that support Acacieae; across Africa and Middle East, also Malta, Spain and Portugal.

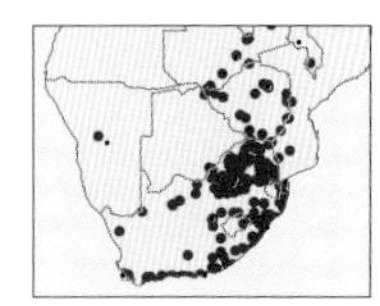

### 3 *Isturgia catalaunaria* — Catalonia Peahen

Forewing 12–16mm. Almost like a large *Acanthovalva inconspicuaria* (p.172); body and wings greyish; summer form densely speckled with brown, postmedian line dark brown sometimes with distinct dark dots; winter form larger, less well marked, almost uniform grey in some individuals. Caterpillar green with numerous longitudinal yellow lines and white lateral band. **Biology:** Caterpillars feed on various herbaceous Fabaceae. Multivoltine, adults throughout the year. **Habitat & distribution:** Various habitats with herbaceous Fabaceae, primarily in the less arid east of the region but also on hills and mountains in the west; Western Cape, South Africa, north to southern Europe.

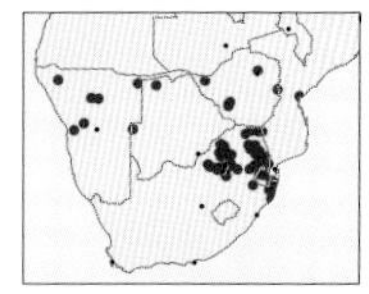

### 4 *Isturgia supergressa* — Plain Peahen

Forewing 15–18mm. Plain cream speckled with brown or grey; diagnostic forewing postmedian line straight but curved slightly inward, 1 prominent black spot in terminal area of each wing in some individuals. Similar *I. exospilata* has the postmedian line slightly curved outward. Caterpillar olive green with numerous crinkled longitudinal white lines. **Biology:** Caterpillars feed on Cork Bush (*Mundulea sericea*). Multivoltine, adults throughout the year, peaking in summer months. **Habitat & distribution:** Bushveld and grassland where the host plant grows, KwaZulu-Natal, South Africa, north to Kenya.

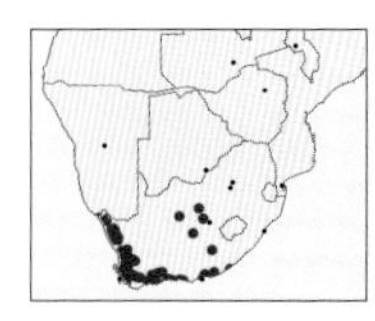

### 5 *Isturgia exerraria* — Cape Peahen

Forewing 11–12mm. Light grey speckled with brown and sometimes orange scales, basal and postmedian lines black, lines well developed near inner margin, postmedian line kinked inward before costa, often a dark postmedian band on forewings, prominent cell spot on all wings, sometimes obliterated when median line is developed. Caterpillar grey-green with dark lateral spots. **Biology:** Caterpillar recorded on *Aspalathus spinosa* (Fabaceae). Multivoltine, adults throughout the year. **Habitat & distribution:** Primarily in succulent karoo biome, but also adjacent dry fynbos and grassland; Western Cape, Eastern Cape, Northern Cape and southern Free State, South Africa.

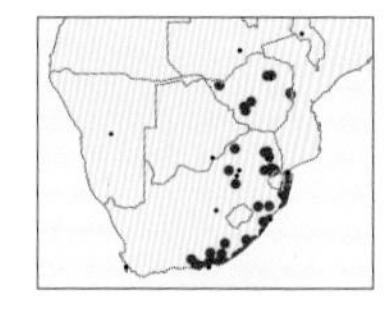

### 6 *Isturgia spissata* — Eastern Peahen

Forewing 12–13mm. Similar to *I. exerraria* (above), larger and postmedian line straight, not kinked, median line when present kinked outward. Caterpillar grey-brown with dark lateral spots. **Biology:** Caterpillars feed on *Vachellia* spp. (Fabaceae). Multivoltine, adults throughout the year. **Habitat & distribution:** Relatively moist habitats where *Vachellia* trees grow; eastern parts of South Africa north to Tanzania.

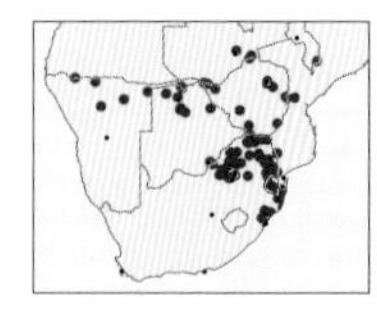

### 7 *Plateoplia acrobelia* — Swollen Thorn

Forewing 10–12mm. Pale yellow to orange, variably dusted with brown, forewings crossed by 3 parallel, orange-brown lines, hindwings by 2 such lines. Unusual caterpillar, stout and markedly swollen at anterior end, colour variable, green to purplish-brown, plain or mottled. One species in genus. **Biology:** Caterpillars feed on sour plums (*Ximenia* spp.). Multivoltine, adults throughout the year, peaking in summer months. **Habitat & distribution:** Bushveld and forest where *Ximenia* spp. grow; KwaZulu-Natal, South Africa, north to Angola and Kenya.

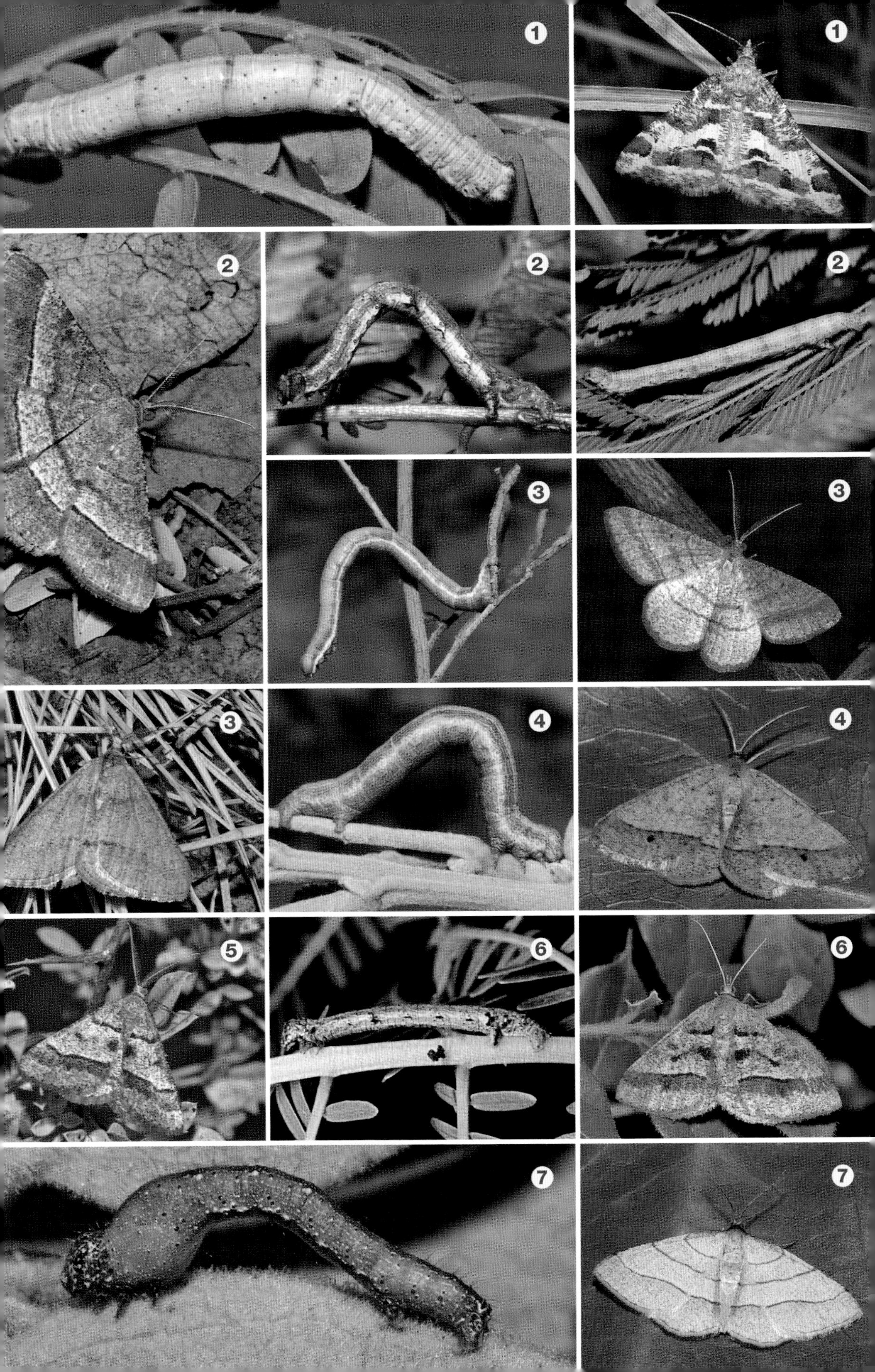
1
1
2
2
2
3
3
3
4
4
5
6
6
7
7

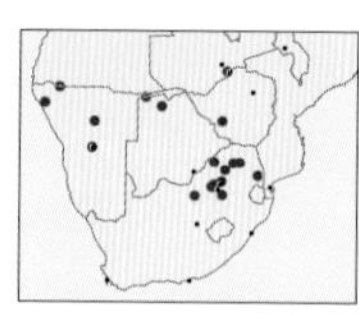

## 1 *Platypepla persubtilis* — Rounded Thorn

Forewing 8–10mm. Wings broadly rounded at rest; forewings fairly uniform pinkish-brown to orange, margin convex with falcate tip, 2 black spots on costa, additional central spot and more black speckles in female; hindwings plain cream, speckled at inner margin. Caterpillar plump and green with red head and legs, and variable red dorsal line and bands. Nine similar species in genus in region. **Biology:** Caterpillars feed on Mistletoe (*Agelanthus natalitius*). Bivoltine, adults September–March. **Habitat & distribution:** Grassland and bushveld; Free State, South Africa, north to Namibia, Botswana and Zimbabwe, with outlying record from Kenya.

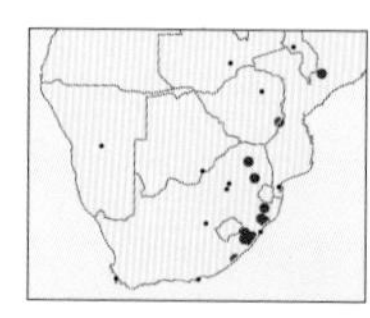

## 2 *Chelotephrina acorema* — Rectangle Thorn

Forewing 13–14mm. Wings widely extended at rest; body pale to dark brown with distinctly darker last thoracic segment; forewings pale brown with diagnostic rectangular paler patch in median area on costa, cream in male and orange in female, some darker spots on both wings. **Biology:** Caterpillar host associations unknown. **Habitat & distribution:** Moist forests; Eastern Cape north to Zimbabwe and Malawi.

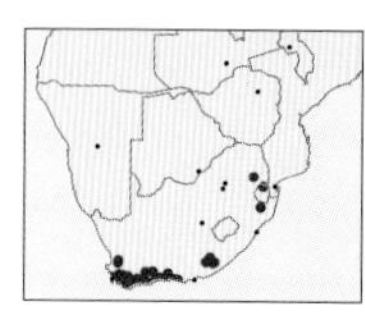

## 3 *Acrasia crinita* — Crossbar Acrasia

Forewing 13–15mm. Body strongly setose, wings held folded against body, similar to Noctuidae; sexually dimorphic; female forewing colouration extremely variable, ranging from brown to yellow or orange, variably shaded in these colours or plain, hindwings white basally, grey distally; male forewings uniform brown, hindwings straw-coloured, crinkled with androconial hairs on upperside. Caterpillar bright green with row of distinctive reddish 'X' marks along dorsal surface, white and yellow lateral spots, head with yellow and red line. Several very similar species; allopatric on isolated mountain ranges where heaths grow, along Great Escarpment up to Tanzania. Fourteen species in genus in region. **Biology:** Caterpillars feed on heaths (*Erica* spp.). Multivoltine, adults throughout the year. **Habitat & distribution:** Fynbos and fynbos remnant vegetation where heaths grow; Western Cape to Mpumalanga, South Africa.

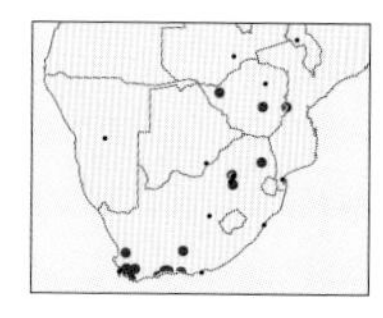

## 4 *Acrasia grandis* — Large Acrasia

Forewing 16–17mm. Similar to *Acrasia crinita* group (above), but larger and forewings more rounded with slightly falcate tip and visible curved postmedian line; females yellow variably dusted with reddish scales. Caterpillar stout, green with yellow tints, yellow head with dark face. **Biology:** Caterpillars feed on proteas (*Protea* spp.). Multivoltine, adults throughout the year. **Habitat & distribution:** Various habitats where *Protea* spp. grow; Western Cape, South Africa, north to Zimbabwe.

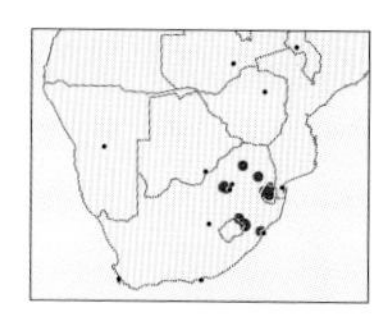

## 5 *Acrasia ava* — Grey Acrasia

Forewing 16–17mm. Forewings grey with purplish tint in some specimens, postmedian line curved and punctuated with black dots in some specimens, wing margins slightly crenulated; male hindwings lack androconial setae and are not crinkled. Caterpillar blue-green, usually with black-dotted dorsal line, black and white marks on face. **Biology:** Caterpillars feed on the Sugarbush (*Protea caffra*). Bivoltine, adults September–April. **Habitat & distribution:** Grasslands where *P. caffra* grows; KwaZulu-Natal north to Limpopo, South Africa.

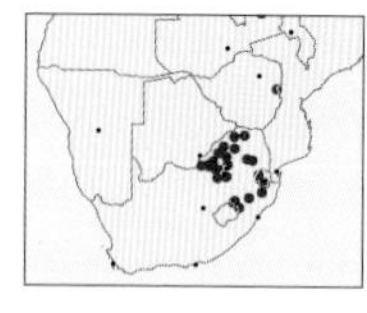

## 6 *Terpnostola ardescens* — Chameleon Thorn

Forewing 11–15mm. At rest, wings held roof-like against body; orange-red with oblique pale line (sometimes 2 lines) across forewings, sometimes a row of brown dots along the line and/or dots between lines and circular cell spot, area beyond line usually paler; hindwings plain pale orange. Caterpillar variable; can change colour to blend in with colour of leaf that it is feeding on, plain pink or plain green or with pink dorsal band or spots. Seven similar species in genus in region. **Biology:** Caterpillars feed on various *Protea* and *Faurea* spp. (Proteaceae). Bivoltine, adults September–April. **Habitat & distribution:** Grassland, bushveld and forest where Proteaceae grow; KwaZulu-Natal, South Africa, north to Zimbabwe.

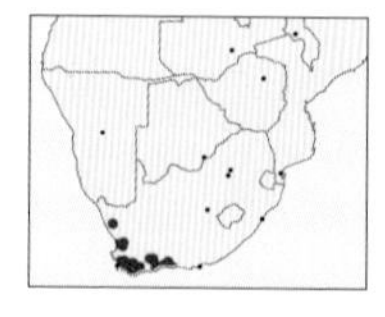

## 7 *Mauna filia* — Conebush Thorn

Forewing 14–15mm. Wings held roof-like against body; thorax strongly setose, sexually dimorphic; male forewings rounded, uniform grey or sometimes reddish-brown in median area with jagged postmedian line; female forewings straight, pointed toward apex, brick red with lighter terminal shading in some individuals; hindwings of both sexes white basally, becoming grey distally. Caterpillar green or brown, sometimes green with pink dorsal marks. One species in genus. **Biology:** Caterpillars feed on various conebushes (*Leucadendron* spp.), also shifted to *Pinus*. Bivoltine, adults August–May. **Habitat & distribution:** Cape fynbos region where conebushes grow; Western Cape and Northern Cape, South Africa.

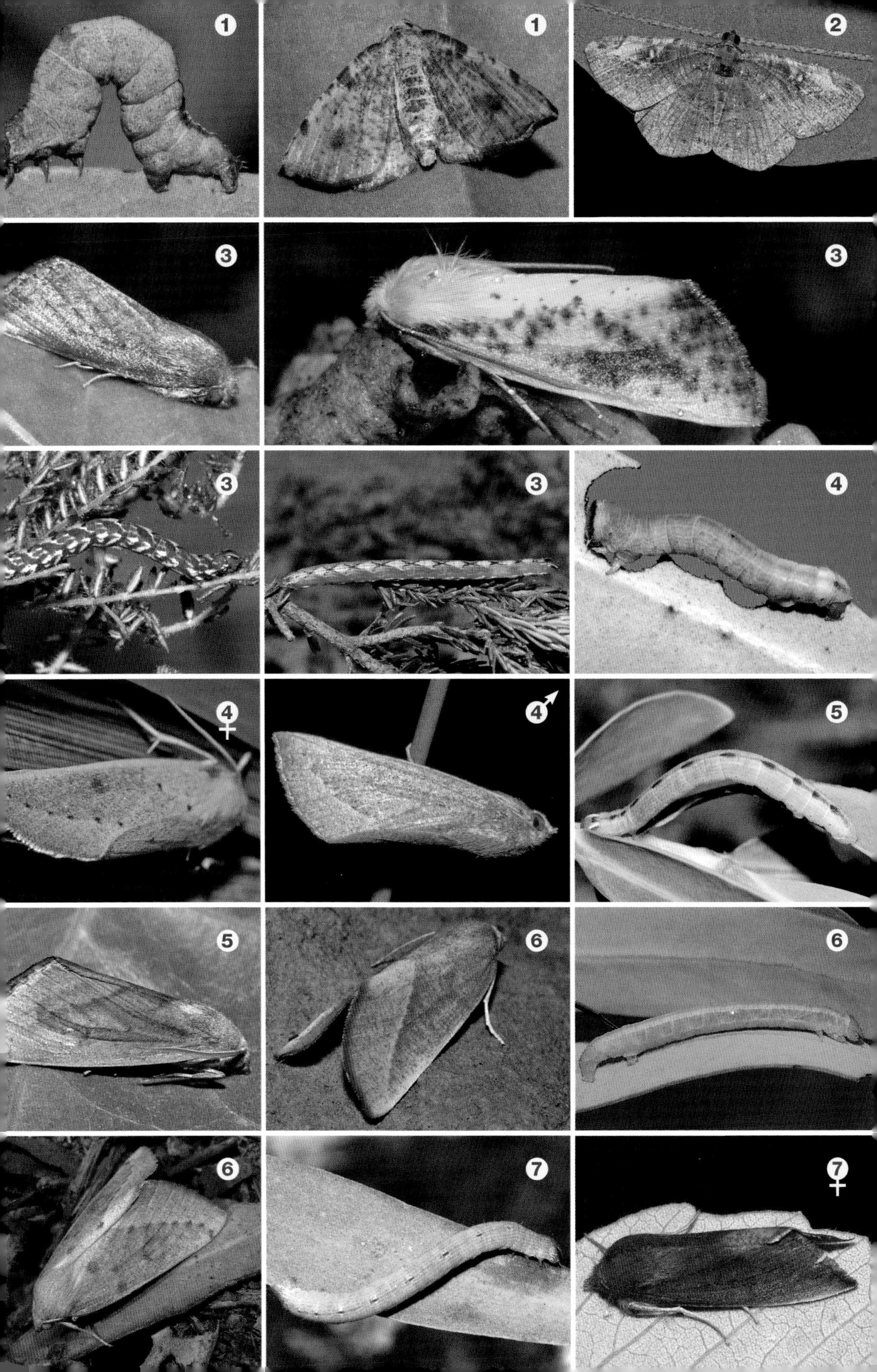
1
1
2
3
3
3
3
4
4♀
4♂
5
5
6
6
6
7
7♀

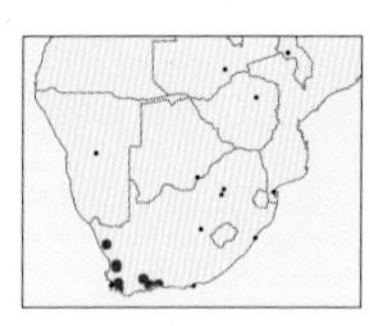

## 1 *Illa nefanda* — Southern Illa

Forewing 17–18mm. Body stout, thorax strongly setose, dark brown, abdomen light brown; forewings darkish brown, tips bend outward, crossed by 2 diagnostically shaped thin wavy black lines, area between lines may be darker nut brown (or graded browns); hindwings cream basally, grading to light brown. Three species in genus in region. **Biology:** Caterpillars feed on *Protea* and *Leucospermum*. Probably univoltine, adults June–September. **Habitat & distribution:** Protea fynbos; Western Cape and Northern Cape, South Africa.

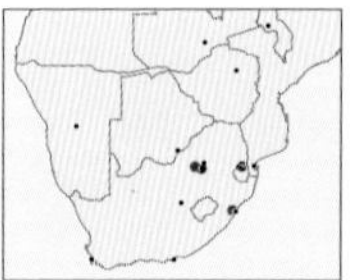

## 2 *Illa reprobata* — Northern Illa

Forewing 13–14mm. Very similar in shape to *I. nefanda* (above), but forewings fairly uniform medium brown in proximal half, and light brown in distal half, shape of lines diagnostic; hindwings white basally, grading to light brown. Caterpillar green or brown with fine pale longitudinal lines and bands, reddish-brown in final instar. **Biology:** Caterpillars feed on Dwarf Savanna Sugarbush (*Protea welwitschii*). Bivoltine, adults October–November and March, seldom attracted to light. **Habitat & distribution:** *Protea* grassland; KwaZulu-Natal and Gauteng, South Africa, and Eswatini.

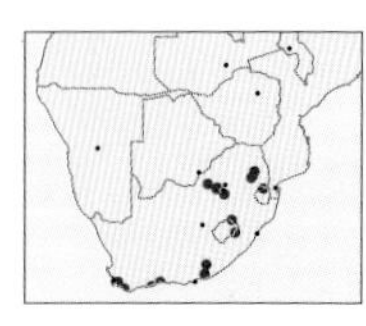

## 3 *Eutelostolmus pictifimbria* — Punctuated Thorn

Forewing 14–15mm. Forewings falcate, body and wings light grey, forewings heavily dusted with coppery brown scales, postmedian line punctuated by black dots, costa straw-coloured, cilia grey. Caterpillar dark brown to black with broad, reddish-brown lateral band. One species in genus. Similar *Dispatha pelopasta* has broader wings, punctuations white and black, cilia white and costa red. **Biology:** Caterpillars feed on Sugarbush (*Protea caffra*). Univoltine, adults August–April. **Habitat & distribution:** Fynbos and *Protea* grassland; Western Cape north to Mpumalanga, South Africa, and Eswatini.

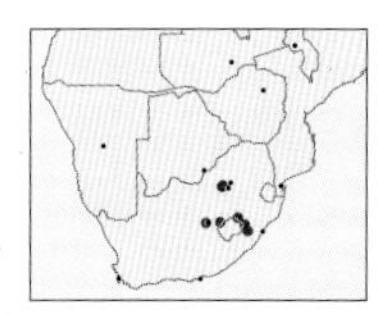

## 4 *Asemoprepes homales* — Pink Protea Thorn

Forewing 16–18mm. Broadly triangular at rest; body and forewings pink-brown; forewings falcate at tip with conspicuous pale-centred brown ring and usually indistinct darker oblique band; hindwings white basally, grading to pink or brown distally, also sometimes with oblique line. One species in genus. **Biology:** Caterpillars feed on Common Sugarbush (*Protea caffra*). Bivoltine, adults July–October and December–March. **Habitat & distribution:** Highveld grassland where sugarbushes grow; KwaZulu-Natal north to Gauteng, South Africa.

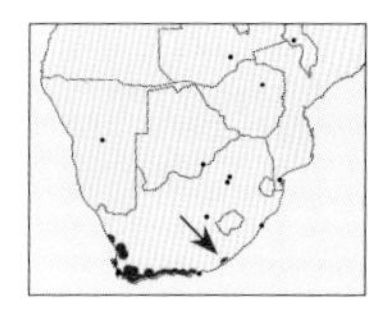

## 5 *Argyrophora arcualis* — Silver Studs

Forewing 10–13mm. Thorax brown; abdomen white; forewings darker brown than in *A. trofonia* (below), also with white reticulated patterning, but distinctive pattern of alternating white and brown bars along distal margin; hindwings plain white with long white cilia. Nine species in genus in region. **Biology:** Caterpillar host associations unknown. Adults throughout the year. **Habitat & distribution:** Fynbos and coastal bush; Western Cape and Eastern Cape, South Africa.

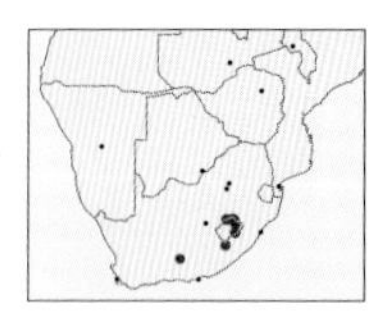

## 6 *Argyrophora leucochrysa* — High Studs

Forewing 13–14mm. Almost entirely white; forewings with distinctive pattern of golden-brown bands; hindwings white to light grey. Caterpillar long and slender, variably mottled in white, yellow, brown and black. **Biology:** Caterpillars feed on *Helichrysum trilineatum* (Asteraceae). Univoltine, adults January–February. **Habitat & distribution:** High Afroalpine vegetation to above 3,000m; Eastern Cape, South Africa, north to Lesotho.

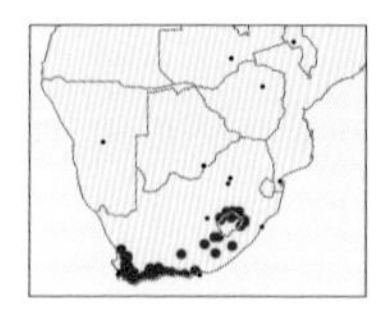

## 7 *Argyrophora trofonia* — Gold Lattice

Forewing 11–14mm. Slender with brown thorax and white abdomen; wings held roof-like against body; forewings orange-brown, patterned with numerous silver spots and dashes; hindwings white. Caterpillar dark brown with broken white lines. Several similar species, some undescribed, allopatric on high mountains where remnant colonies of *Erica* persist north to Tanzania. Can also be confused with the unrelated *Triphassa argentea* (see Pyralidae, p.110). **Biology:** Caterpillars feed on various heaths (*Erica* spp.). Multivoltine, adults throughout the year. **Habitat & distribution:** Fynbos and grassland where heaths grow; Western Cape to KwaZulu-Natal and Free State.

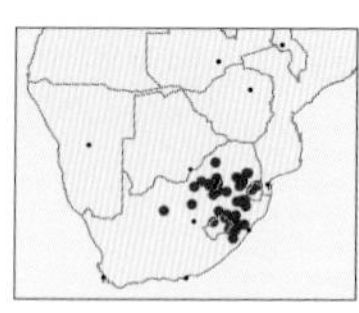

### 1 *Argyrophora variabilis* — Variable Studs

Forewing 10–12mm. Almost entirely white; forewings variably dusted with grey scales, thin dark grey band from apex to centre straight, not ending in darker blotch, diagnostic, other grey bands between veins often present; hindwings pure white. Caterpillar long and slender, pale green with white lateral line. Very similar *Microligia* spp. can be separated by having dark band more curved and ending in a darker blotch in cell and caterpillars being grey. **Biology:** Caterpillars feed on Daisy Tea Bush (*Athrixia elata*) and Marotole (*Helichrysum rugulosum*). Multivoltine, adults throughout the year, peaking in summer. Grassland; Eastern Cape north to Limpopo, South Africa.

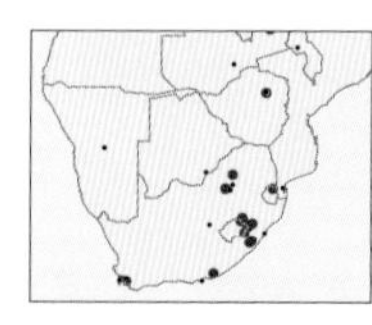

### 2 *Microligia luteitincta* — Lemon Streak

Forewing 14–15mm. Slender, larger than other species in genus; forewings completely lemon yellow, mostly in females, partially so in males, with rest of wing shaded in cream and light grey, grey postmedian line from apex almost to inner margin; hindwings plain white. Eight species in genus in region. **Biology:** Caterpillars feed on *Protea caffra*. Adults appear sporadically throughout the year. **Habitat & distribution:** Grassland; Western Cape, South Africa, north to Zimbabwe.

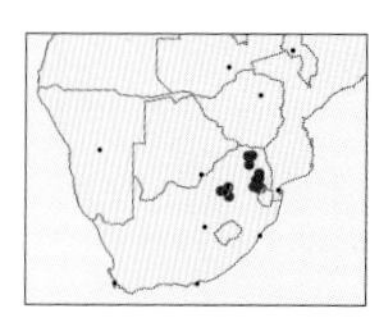

### 3 *Microligia pseudodolosa* — Grey Streak

Forewing 12–13mm. At rest, wings held steeply angled against body; white, fairly densely speckled with grey; forewings with dark band from apex, curved and ending in darker blotch in cell, diagnostic for group to separate from *Argyrophora variabilis* (above); hindwings white. Caterpillar elongate, brown to dark grey. Several very similar allopatric species best separated by distribution: *M. dolosa* Western Cape to coastal KwaZulu-Natal; *M. confinis* along Great Escarpment KwaZulu-Natal and Free State to Eswatini; *M. sepentrionalis* Eastern Highlands, Zimbabwe. **Biology:** Caterpillars feed on Daisy Tea Bush (*Athrixia elata*). Bivoltine, adults September–April. **Habitat & distribution:** Grassland; Gauteng to Limpopo, South Africa.

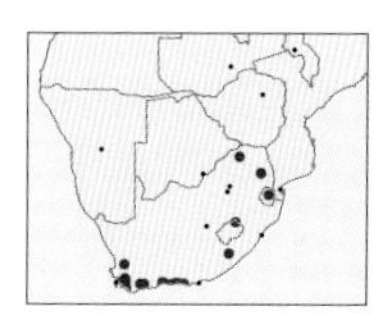

### 4 *Larentioides cacothemon* — Cedar Thorn

Forewing 13–14mm. Antennae filiform in both sexes; females darker in colour, narrow winged, both wings silvery grey variably diffused with light or dark brown scales, lines well developed, double postmedian lines undulating but parallel, diagnostic. Caterpillar green with longitudinal unbroken white lines. Two species (one undescribed) in genus in region. **Biology:** Caterpillars feed on Mountain Cedar (*Widdringtonia nodiflora*, Cupressaceae). Multivoltine, adults throughout the year. **Habitat & distribution:** Mountain grassland and fynbos where Mountain Cedars grow; Western Cape north to Limpopo, South Africa, and Eswatini.

### 5 *Callioratis abraxas* — Dimorphic Tiger

Forewing ♂ 23–28mm, ♀ 27–38mm. Sexually dimorphic and with aposematic colouration; females orange with diagnostic black and white antemedial and postmedian bands merging on inner margin, creating a 'V'-shape; males have forewing bands much broader with white areas much increased almost obliterating orange ground colour, male hindwing upper side and forewing underside form a distinctive androconial organ when at rest. *Callioratis* caterpillars have an extra set of functional prolegs, unusual for Geometridae, orange below, black and yellow banded above, white above in early instars. Four species in genus in region. **Biology:** Caterpillars feed initially on cycads (*Encephalartos* spp.) growing in the shade, later instars polyphagous on a variety of forest trees. Adults diurnal, males form a lek on a tree emerging from canopy, females visit lek for mating but mostly fly near forest floor. Univoltine in south of range, more broods further north, adults throughout the year, primarily in April. **Habitat & distribution:** Moist forests with cycads growing in understorey; Eastern Cape north to KwaZulu-Natal, South Africa, and Eswatini.

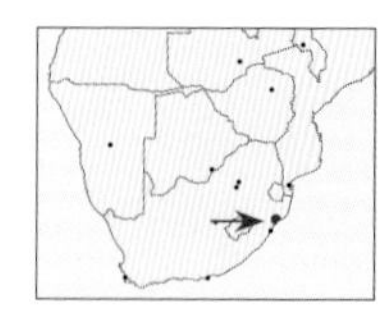

### 6 *Callioratis millari* — Millar's Tiger

Forewing ♂ 29–32mm, ♀ 30–35mm. Sexes alike, thorax black with orange margins, abdomen yellow-orange with black bands; wings reddish-orange with antemedial and postmedian black bands not merging on inner margin, marginal spots, fewer on hindwings; males lack an androconial organ. Caterpillar with yellow and black longitudinal bands. **Biology:** Caterpillars feed initially on Stanger's Cycad (*Stangeria eriopus*) growing in open grassland, polyphagous in later instars. Univoltine, adults diurnal, not forming leks, March–May. **Habitat & distribution:** Grassland at altitudes around 700m that harbour *Stangeria* cycads; formerly in Kloof/Gillitts area where it has become extinct, now critically endangered, only in small patch of grassland near Eshowe, KwaZulu-Natal.

1
1
2♂
3
3
4
4♀
5
5♀
5♂
6
6

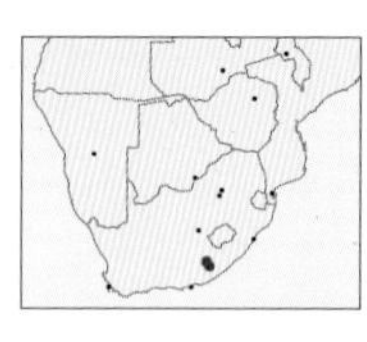

## 1 *Callioratis mayeri* — Mayer's Tiger

Forewing ♂ 30–34mm, ♀ 36–42mm. Similarly patterned to *C. millari* (p.184), but has pale cream to yellow forewings with black-edged dark stripes and bright orange to red hindwings; males with androconial organ in fold of hindwing inner margin. Caterpillars black above and red below. **Biology:** Caterpillars feed on the cycad *Encephalartos friderici-guilielmi.* Univoltine, adults diurnal, males form leks on cliff faces, March–April. **Habitat & distribution:** On higher hills with cliffs where host cycad grows; Eastern Cape, South Africa.

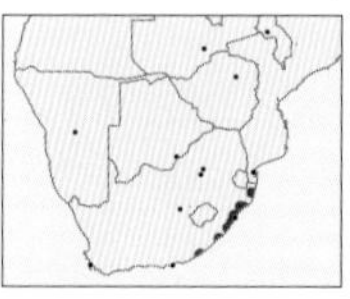

## 2 *Veniliodes pantheraria* — Forest Tigerlet

Forewing 14–15mm. Body and wings orange to yellow; forewings crossed by 4 or 5 black bands from costa variably covered with orange or yellow scales, often with conjoined spots; hindwings plain at base, bearing variable scattered black spots and larger black patch distally. Caterpillar orange with cream lines and black setal bases. Four species in genus in region; similar *V. inflammata* has only 3 black bands from costa and is a grassland species. **Biology:** Caterpillars feed initially on cycads (mostly *Encephalartos villosus*) growing in shade, later instars polyphagous on forest trees. Univoltine, adults diurnal, flying in forests and on forest edge, April–June. **Habitat & distribution:** Moist forests with cycads growing in understorey; Eastern Cape north to KwaZulu-Natal, South Africa.

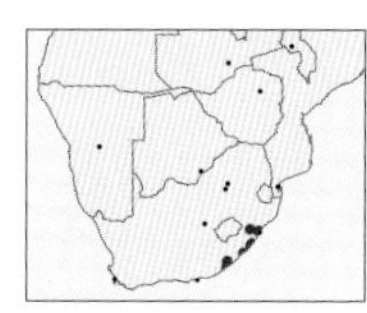

## 3 *Veniliodes setinata* — Grassland Tigerlet

Forewing 11–12mm. Sexually dimorphic; male forewings partially or completely charcoal black, hindwings orange with variable black marginal dots and larger black patch distally; female forewings with a variable lattice of black bands over a yellow to white ground, hindwings as in male; in some populations the male resembles the female. Caterpillar mottled orange and yellow with black spots. **Biology:** Caterpillars feed initially by tunneling into leaf and petiole of Stanger's Cycad (*Stangeria eriopus*), from which they exit in third instar to feed on petals of various flowers. Bivoltine, adults throughout the year, diurnal, flying fast and low over grassland. **Habitat & distribution:** Grasslands where host plant grows; Eastern Cape to KwaZulu-Natal, South Africa.

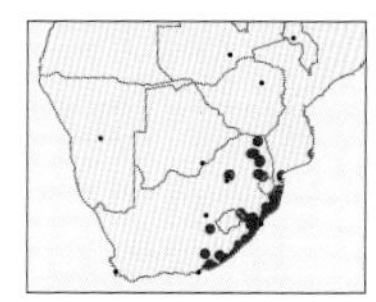

## 4 *Zerenopsis lepida* — Leopard Magpie

Forewing 22–23mm. Orange, both wings covered in large round black spots variable in size and quantity. Caterpillar yellow to orange, sometimes almost red, with black spots, northern populations with black longitudinal stripes. Four species in genus in region. **Biology:** Caterpillars initially obligatory feeders on any cycad, including exotic species, polyphagous after third instar. Multivoltine, adults throughout the year. **Habitat & distribution:** Variety of habitats where cycads grow, including gardens; Eastern Cape north to Limpopo, South Africa, and Mozambique.

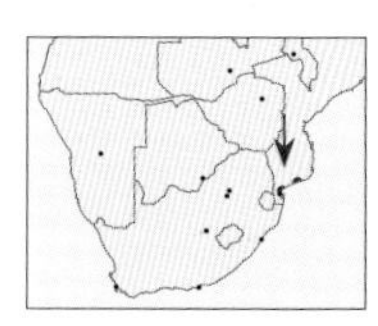

## 5 *Zerenopsis moi* — Arrow Leopard

Forewing 18–20mm. Sexually dimorphic; female yellow with black border, wings rounded, abdomen variable yellow and black; male orange with black borders, enlarged hindwings folded back to form androconial organ inside fold, making hindwings appear pointed as an arrow. Caterpillar orange-red with large black marks and long white setae. **Biology:** Caterpillars initially feed on Maputaland Cycad (*Encephalartos ferox*), polyphagous after third instar. Bivoltine, adults August–September and December. **Habitat & distribution:** Coastal dunes where host plant grows; Mozambique.

## 6 *Zerenopsis geometrina* — Geometric Leopard

Forewing 15–16mm. Body black with orange crossbars in some specimens; wings orange with smaller and less defined black dots but usually more numerous than in *Z. lepida* (above); hindwings orange with sparse black dots, male hindwings folded back slightly to form androconial organ. Caterpillar orange-yellow or white with numerous black spots and long white setae. Similar *Z. meraca* is larger with narrower wings, lighter orange and has very small black dots. **Biology:** Caterpillars initially feed on various cycads, polyphagous after third instar. Bivoltine, adults May–October and December–January. **Habitat & distribution:** Coastal forests where cycads grow; Eastern Cape north to KwaZulu-Natal, South Africa.

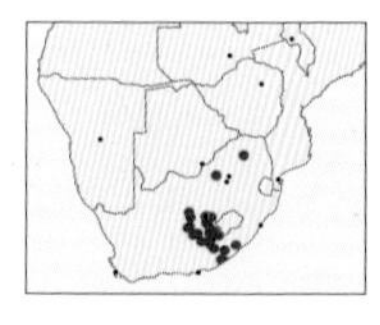

## 7 *Hebdomophruda curvilinea* — Straw Glider

Forewing 14–15mm. At rest, elongate wings held out laterally to form a 'T'-shape reminiscent of gliders; body and wings straw-coloured; forewings with brown speckling near costa, postmedian line curved to apex; hindwings with some speckling. Nineteen species in genus in region. **Biology:** Caterpillar host associations unknown. Adults March–April. **Habitat & distribution:** Grassy karoo; Eastern Cape to Free State, isolated populations in North West province and Limpopo, South Africa.

1
1♂
2
2
3
3
AA
3♂
4
4
5
5♀
6
SB
6
7

**1 *Hebdomophruda crenilinea*** Crenulated Glider

Forewing 14–16mm. Similar in shape to *H. curvilinea* (p.186), but grey with crenulated double postmedian line and median line variably shaded with black; hindwings grey with faint lines. Several very similar species, which can only be reliably separated by examining genitalia or by barcoding. **Biology:** Caterpillar host associations unknown. Adults April–May. **Habitat & distribution:** Upper grassy karoo; Eastern Cape, Northern Cape and Free State, South Africa.

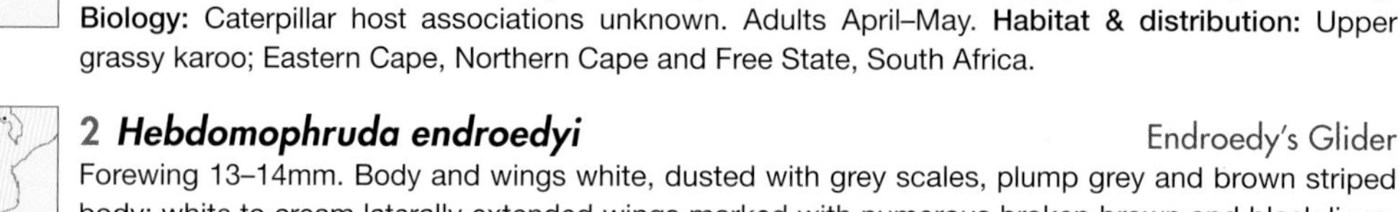

**2 *Hebdomophruda endroedyi*** Endroedy's Glider

Forewing 13–14mm. Body and wings white, dusted with grey scales, plump grey and brown striped body; white to cream laterally extended wings marked with numerous broken brown and black lines, postmedian line dotted near apex, commonly a distinct black forewing cell spot. Several very similar species; *H. imitatrix* is smaller and darker with postmedian line less dotted. **Biology:** Caterpillar host associations unknown. Adults July–August. **Habitat & distribution:** Strandveld and adjacent lower elevation renosterveld; Western Cape, South Africa.

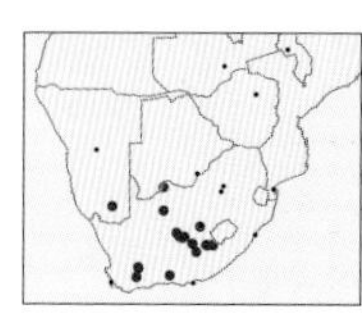

**3 *Hebdomophruda sculpta*** Sculpted Glider

Forewing 12–14mm. Body and wings light brown, dusted with grey scales; wings slightly crenulated, distinctive intricate sculpted pattern of oval and curved lines on forewings; hindwings pale grey with thin black border. **Biology:** Caterpillar host associations unknown. Adults October and March–May. **Habitat & distribution:** Nama karoo, arid grassland and Kalahari; Western Cape, South Africa, north to Namibia and Botswana.

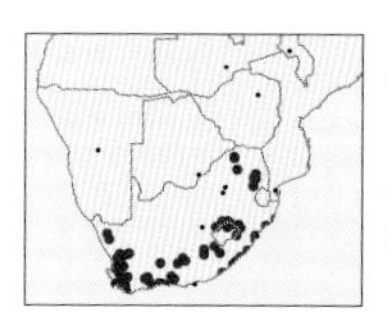

**4 *Pseudomaenas alcidata*** Silver Line Geometric

Forewing 13–15mm. Wings pitched roof-like against body; forewings deep brown, marked by about 3 complete and 3 partial tapering white bands of varying thickness that radiate toward apex; hindwings cream, slightly darker at anterior margin. Caterpillar slender, mottled brown, with orange and red patches. Thirty-three species in genus in region. Barcoding indicates that *P. alcidata* is a complex of similar species. **Biology:** Caterpillars feed on Gonna bushes (*Passerina* spp.). Multivoltine, adults throughout the year. **Habitat & distribution:** Fynbos and fynbos remnant habitats where *Passerina* spp. grow; Western Cape north to Limpopo, South Africa, and Lesotho.

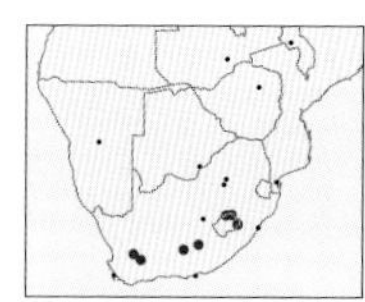

**5 *Pseudomaenas eumetrorrhabda*** Bent Line Geometric

Forewing 16–18mm. Similar to *P. alcidata* (above), but larger with wings more slender, brown with white lines, but these are angled distally to run parallel to hind margin; hindwings white. **Biology:** Caterpillar host associations unknown. Adults December–March. **Habitat & distribution:** Renosterveld and Afroalpine vegetation at high altitudes; along Great Escarpment, Northern Cape to Lesotho.

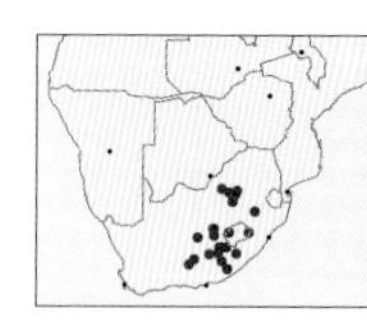

**6 *Pseudomaenas margarita*** Broad Line Geometric

Forewing 13–14mm. Similar to *P. eumetrorrhabda* (above), but much smaller with broader wings that are held flat over body at rest; white lines broader, taking up much of the forewing surface, with light brown border. **Biology:** Caterpillar host associations unknown. Adults February–April. **Habitat & distribution:** Highveld grassland, renosterveld and Afroalpine vegetation; Eastern Cape north to Gauteng, South Africa.

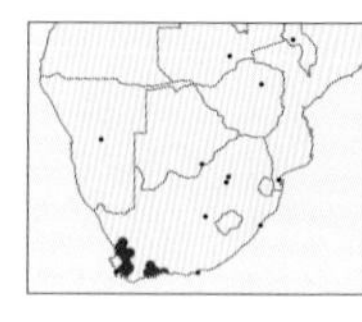

**7 *Pseudomaenas leucograpta*** Triangular Geometric

Forewing 12–13mm. Body reddish-brown, wings held roof-like against body; forewings brown with reddish margins and thick, bent white line along posterior and distal margin, a shorter one basally on anterior margin and oblique one centrally, together forming broken triangular perimeter around wing; hindwings light brown with dark border. **Biology:** Caterpillar host associations unknown. Adults throughout the year. **Habitat & distribution:** Mountain fynbos; Western Cape, South Africa.

**8 *Pseudomaenas oncodogramma*** Space Geometric

Forewing 13–15mm. Wings held flat in broadly triangular outline; forewings falcate, pale brownish-pink, marked with black geometric shapes, a triangle near base, distal ellipses and diagnostic triangular crescent with flattened point; hindwings plain straw-coloured. Several similar species; *P. dukei* has forewings rounded and distal triangular crescent rounded; *P. krooni* has distal triangle arrow-shaped; *P. nigrosema* has wings rounded and distal triangle sickle-shaped. **Biology:** Caterpillar host associations unknown. Adults February–March. **Habitat & distribution:** Cool, moist grasslands; KwaZulu-Natal north to Mpumalanga, South Africa, and Eswatini.

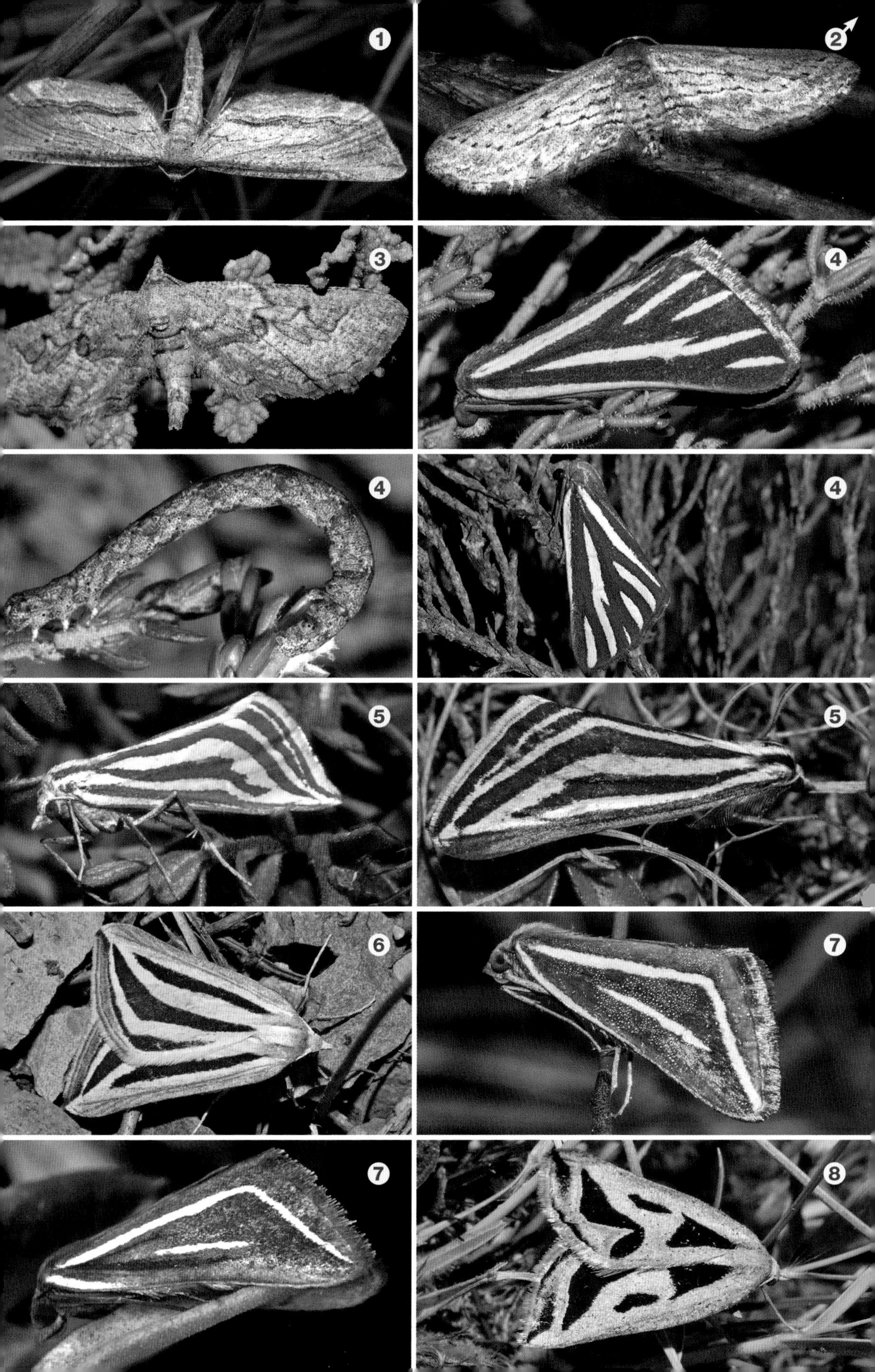
1
2♂
3
4
4
4
5
5
6
7
7
8

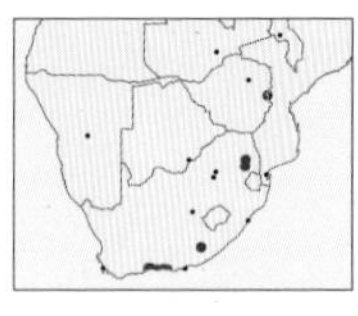

**1 *Pseudomaenas anguinata*** Face Geometric

Forewing 14–15mm. Appears to have a face on wings when at rest; has similar colouring and geometric design to *P. oncodogramma* (p.188), but smaller black patches, comprising triangle, large cell spot and distal narrow band; hindwings with small black cell spot. **Biology:** Caterpillar host associations unknown. Adults February–April. **Habitat & distribution:** Cool, moist fynbos, and in isolated mountain grasslands with fynbos remnant vegetation; southern Cape, Eastern Cape and Mpumalanga, and Zimbabwe.

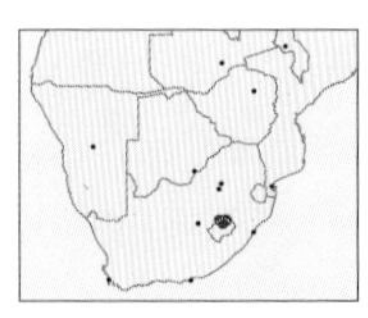

**2 *Pseudomaenas staudei*** Black-lined Geometric

Forewing 14–17mm. Geometrically marked; forewings off-white finely dusted with black scales, grey margin, strongly marked with 3 large, curved and tapering black lines, terminal line often appears smudged; hindwings white. Similar *P. bivirgata* has a black cell spot and basal line in opposite direction. **Biology:** Caterpillar host associations unknown. Adults January–February. **Habitat & distribution:** Afroalpine vegetation above 2,500m; Lesotho.

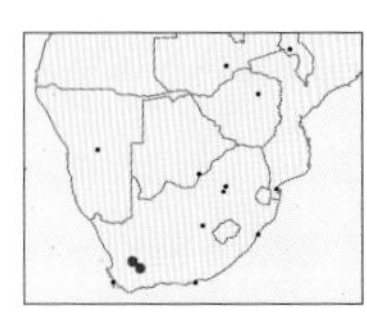

**3 *Pseudomaenas euglyphica*** Hieroglyphic Geometric

Forewing 13–15mm. Geometrically marked; forewings brick red with intricate pattern of triangles, lines, marks and curved band reminiscent of hieroglyphics; hindwings uniform brown. Seven similar intricately patterned species described, each with unique pattern, allopatric on Cape mountains. **Biology:** Caterpillar host associations unknown. Adults in March. **Habitat & distribution:** Renosterveld on Roggeveld escarpment; Northern Cape.

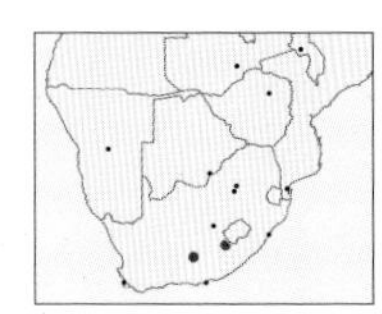

**4 *Pseudomaenas euzonaria*** Belted Geometric

Forewing 13–14mm. Body and wings white; forewing costa brown, dark brown basal line straight, postmedian and terminal lines undulating with alternating dark brown marks and orange streaks like a belt, hindwings with broad brown border. Similar *P. turneri* has much reduced gap between basal and postmedian line and deeper kink in postmedian line. **Biology:** Caterpillar host associations unknown. Adults in March. **Habitat & distribution:** High-altitude renosterveld on Great Escarpment; Eastern Cape.

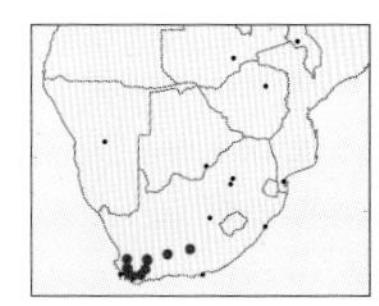

**5 *Pseudomaenas honiballi*** Tyre Geometric

Forewing 11–13mm. Body, wings light grey, dusted with dark brown scales, cilia white, forewings partially shaded with brown, unique intricate pattern of black triangular and octagonal marks, black cell spot; hindwings grey with small cell spot. **Biology:** Caterpillar host associations unknown. Adults October–November and March–May. **Habitat & distribution:** Mountain renosterveld; Western Cape and Eastern Cape.

**6 *Idiodes saxaria*** Fern Thorn

Forewing 15–16mm. Antennae filiform in both sexes; broad winged with forewing falcate, uniformly light to dark brown with antemedial and postmedian lines punctuated by black streaks, sometimes edged with yellow in males, full lines in females. Caterpillar yellow to orange with longitudinal brown lines. Two species in genus in region. **Biology:** Caterpillars feed on various ferns. Multivoltine, adults throughout the year, peaking October and March. **Habitat & distribution:** Cool moist forests and kloofs where ferns grow; Western Cape and Eastern Cape, South Africa.

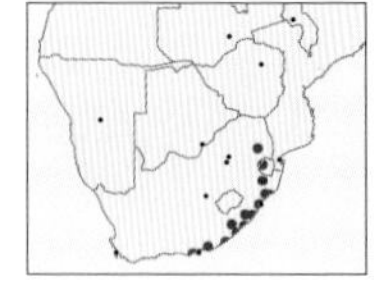

**7 *Pareclipsis oxyptera*** Pointed Thinwing

Forewing 12–14mm. Male dark brown, female light pink-cream to grey; forewings slightly falcate with sharply pointed tip, wings appear thin with minute flat scales, postmedian line straight with one kink, not curved in near costa, small cell spot. Eight described species in genus in region, in need of new genus name, as not closely related to European *Pareclipsis* spp. **Biology:** Caterpillar host associations unknown. Adults April–October, peaking in winter. **Habitat & distribution:** Moist forests; Eastern Cape north to Mpumalanga, South Africa.

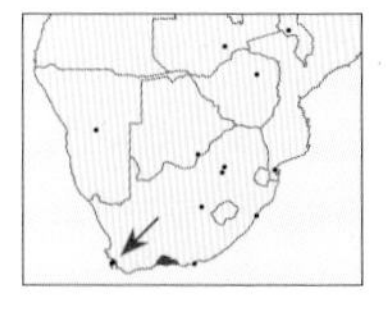

**8 *Pareclipsis incerta*** Darkband Thinwing

Forewing 14–15mm. Wings and body uniform grey with sprinkling of dark brown spots, postmedian a distinctive broad, dark brown band, lacks pointed wing tip and has row of subterminal black dots. **Biology:** Caterpillar host associations unknown. Adults March–May. **Habitat & distribution:** Moist fynbos habitats; Western Cape, South Africa.

1
2
3♂
4
5
6
6
7♂
8

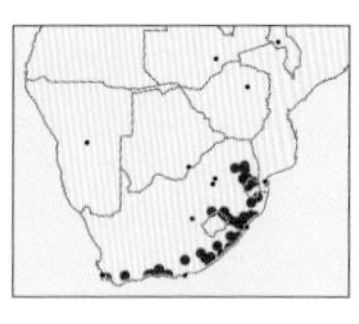

### 1 *Pareclipsis punctata* — Spotted Thinwing

Forewing 12–14mm. Cream to pink, lines made up of small black dots, postmedian line curved in near costa, large black cell spot, wings variably dusted with brown and black spots. Similar *P. distolochorda* has postmedian line with 2 solid lines, not punctuated. **Biology:** Caterpillars recorded on the Green Cliff Brake fern (*Cheilanthes viridis*). Adults throughout the year. **Habitat & distribution:** Moist forests; Western Cape north to Mpumalanga, South Africa.

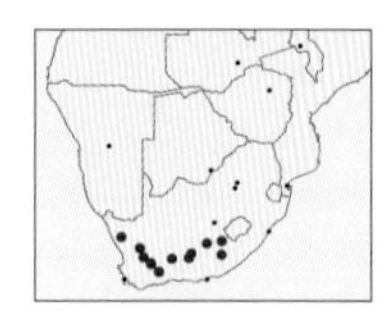

### 2 *Aethiopodes noctuodes* — Black Thorn

Forewing 16–17mm. Wings narrow for genus; forewings charcoal with variable cream spots, black antemedial and postmedian lines with sharp jagged shape diagnostic, a round black ring in cell; hindwings much paler, with indistinct black cell spot and single wavy crossline. Ten species in genus, all in region. **Biology:** Caterpillars feed on *Euryops trifidus* (Asteraceae). Adults October–April. **Habitat & distribution:** Nama karoo habitats; Western Cape, Northern Cape and Eastern Cape, South Africa.

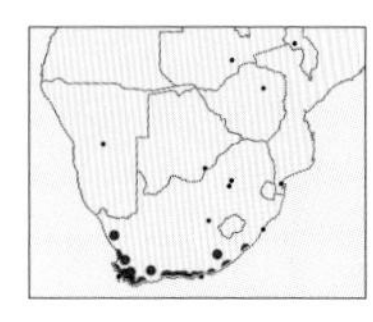

### 3 *Aethiopodes erebaria* — Bietou Thorn

Forewing 17–19mm. Thorax strongly setose; body and forewings light brown to dark brown, almost black in some females, diagnostic forewing basal line deeply indented, postmedian line variably undulating, distal margin diagnostically doubly indented; hindwings cream with very indistinct cell spot. Caterpillar stout, light to dark brown, distinctive white lateral warts. Several similar related species; *A. stictoneura* has basal line evenly curved; *A. paliscia* has postmedian line crenulated; *A. staudei* has distal margin evenly curved. Can also be confused with several *Odontopera* spp. **Biology:** Caterpillars feed on Bietou (*Osteospermum moniliferum*). Multivoltine, adults throughout the year. **Habitat & distribution:** Various moist habitats where bietou plants grow; Western Cape, Northern Cape and Eastern Cape, South Africa.

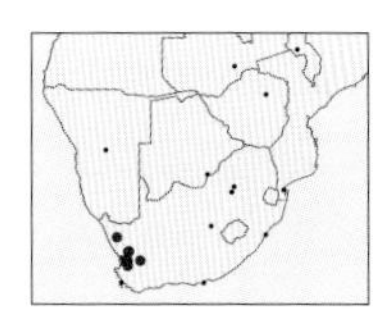

### 4 *Aethiopodes medioumbrata* — Halfdark Thorn

Forewing 15mm. Body light brown with strongly setose thorax; forewings straw-coloured variably sprinkled with black scales, patterned with darker median area crossed by orange veins and outlined by dark wavy lines, distal part much lighter, other dark spots may be present or absent; hindwings white with indistinct cell spot and distal dusting of black scales. **Biology:** Caterpillar host associations unknown. Adults March–May. **Habitat & distribution:** Renosterveld and succulent karoo; Western Cape and Northern Cape, South Africa.

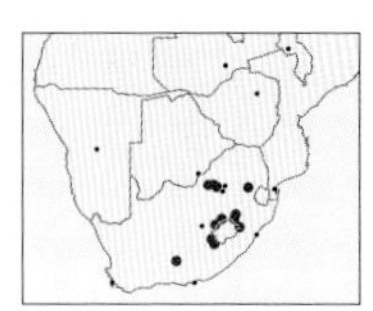

### 5 *Odontopera impeyi* — Yellow-edged Thorn

Forewing 18–19mm. Thorax dark, strongly setose, abdomen brown; forewings crossed by inner curved and outer wavy lines, dark brown to black basally, much lighter beyond outer line, costa yellow; hindwings cream, crossed by only wavy line. Caterpillar mottled in shades of brown, double black lateral lines, white around spiracles. Three species in genus in region. **Biology:** Caterpillars feed on Rock Tannin-bush (*Osyris lanceolata*). Adults October–May. **Habitat & distribution:** Bushveld and kloofs where host plant grows; Eastern Cape north to Mpumalanga, South Africa.

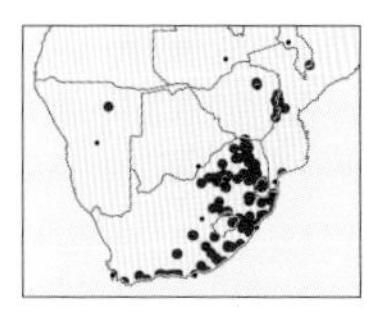

### 6 *Psilocerea pulverosa* — Powdered Thorn

Forewing 17–21mm. Body and wings light brown, apex of forewings slightly falcate, angled point in centre of both fore- and hindwings in both sexes; forewings with small black cell spot and oblique brown and cream line, dusting of black distally in some specimens; hindwings similar. Caterpillar stout, light brown with darker brown transverse bands. Four species in genus in region. **Biology:** Caterpillars feed on Traveller's Joy (*Clematis brachiata*). Multivoltine, adults throughout the year. **Habitat & distribution:** Variety of relatively moist habitats where host plant grows; Western Cape, South Africa, north to tropical Africa.

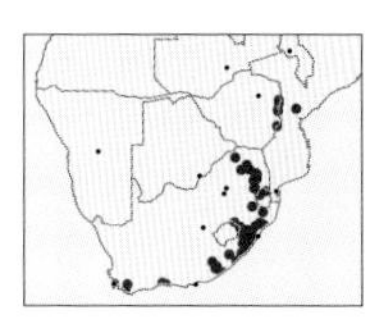

### 7 *Psilocerea immitata* — Imitating Thorn

Forewing 20–22mm. Pale pink to grey; forewings falcate and hindwings extended into distinct short, pointed tails, angled point in centre of both fore- and hindwings in female only; forewings with small black cell spot and darker tip; hindwings with distinct to indistinct darker central patch; both wings crossed by distinctive brown and yellow or orange double postmedian line. **Biology:** Caterpillars feed on Wild Peach (*Kiggelaria africana*) and Lightning Bush (*Clutia pulchella*). Multivoltine, adults throughout the year. **Habitat & distribution:** Moist forests; Western Cape, South Africa, north to Zimbabwe.

1♂
2
3
3
4♂
5
5
6
6♀
7
7♂

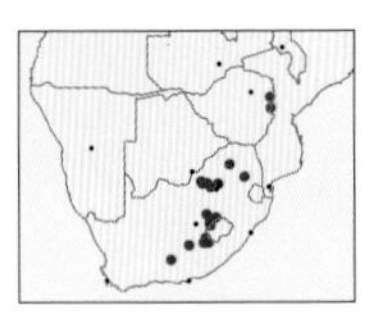

**1 *Xanthisthisa niveifrons*** White-headed Thorn

Forewing 17–22mm. Head and frons white; round to triangular at rest; males yellow, females orange-brown; forewings falcate and with wave in hind margin, wings crossed by oblique dark postmedian line that kinks sharply just before leading margin, sometimes black spots beyond line. Caterpillar green or brown with conspicuous dorsal and lateral processes. Two species in genus in region. Bivoltine, adults September–March. **Biology:** Caterpillars feed on Rock Tannin-bush (*Osyris lanceolata*), also moved onto exotic cypress and pine trees (*Cupressus* and *Pinus*). **Habitat & distribution:** Bushveld and kloofs where host plants grow; Eastern Cape, South Africa, north to Malawi.

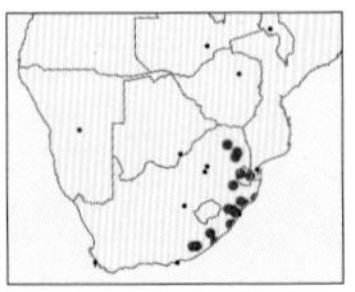

**2 *Anacleora extremaria*** Extreme Bark

Forewing 17–19mm. Body cream with brown band across first abdominal segment; forewings cream with brown stippling and several irregular broken brown bands, postmedian line straight, curved sharply near costa, often a darker central brown spot, small brown dots along wavy distal margin; hindwings similar, margins distinctly toothed. Two species in genus in region; *A. pulverosa* has wings more rounded, margin not toothed and postmedian line evenly curved. **Biology:** Caterpillar host associations unknown. Adults October–December and March–June. **Habitat & distribution:** Moist forests; Eastern Cape north to Limpopo, South Africa.

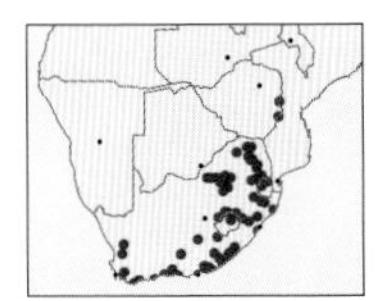

**3 *Aphilopota patulata*** Dotted Umber

Forewing 15–17mm. Broadly triangular outline at rest; body and wings light to medium brown, both wings sometimes crossed by median cloudy darker brown band, median area as rest of wing or much lighter, diagnostic postmedian line dotted and evenly curved, larger irregular black cell spot. Caterpillar plain greyish-brown with small black dots and small yellow lateral marks, dorsal projection posteriorly. Fourteen species in genus in region. **Biology:** Caterpillars feed on various Celastraceae. Multivoltine, adults throughout the year. **Habitat & distribution:** Variety of habitats where Celastraceae grow; Western Cape, South Africa, north to Zimbabwe.

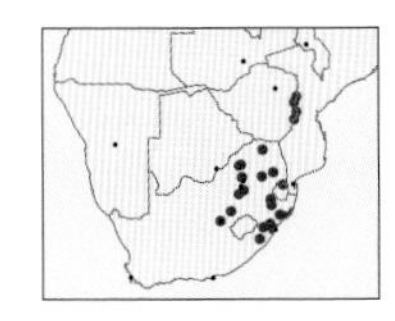

**4 *Aphilopota interpellans*** Curved Umber

Forewing 15–17mm. Brown; forewings crossed by 2 wavy unbroken lines with prominent cell spot, evenly outward curved shape of postmedian line diagnostic; hindwings similar with single outer line. **Biology:** Caterpillars feed on Yellow Carpet Bean (*Rhynchosia totta*). **Habitat & distribution:** Grassland and bushveld; KwaZulu-Natal, South Africa, north to Zimbabwe and Angola.

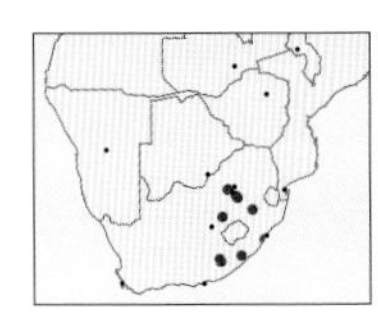

**5 *Aphilopota iphia*** Mosaic Umber

Forewing 17–20mm. Brown with purplish tint; forewing antemedial and postmedian lines diagnostically variably indented, sometimes touching, broken cream terminal band across both wings. Caterpillar intricately marked in mosaic-like pattern of white, pink, black and yellows. **Biology:** Caterpillars feed on *Gymnosporia tenuispina*, probably also other spikethorns. Bivoltine, adults August–September and December–March. **Habitat & distribution:** Grassland; Eastern Cape north to Gauteng, South Africa.

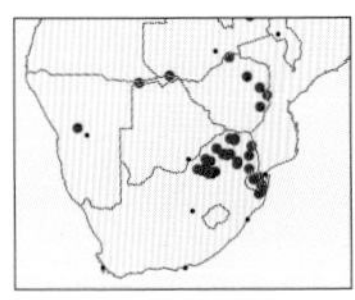

**6 *Aphilopota decepta*** Deceptive Umber

Forewing 18–20mm. Wings held flat and widespread at rest; sooty grey, wings with antemedial and postmedian lines undulating but straight, thin, black and not reaching margins; margins of wings toothed, especially hindwings. Caterpillar a dark grey looper covered in tiny black spots and with irregular lateral yellow patches. **Biology:** Caterpillars feed on spikethorns (*Gymnosporia* spp.). Multivoltine, adults September–May. **Habitat & distribution:** Bushveld; KwaZulu-Natal, South Africa, north to Zambia.

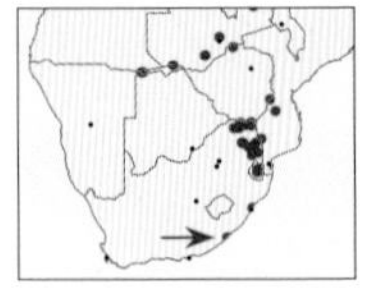

**7 *Racotis breijeri*** Scalloped Satin

Forewing 23–30mm. Females amongst the largest Geometridae in Africa; wings held widely extended at rest; body and wings light grey, both wings with diagnostic scalloped margins, heavily stippled with black and traversed by broken wavy dark lines, a distinctive black-ringed white cell spot and wavy white submarginal line on each wing. Four species in genus in region. **Biology:** Caterpillar host associations unknown. Adults throughout the year. **Habitat & distribution:** Lowland bush and forest; Eastern Cape, South Africa, north to Kenya.

1
1
2
3
3
4
5
5
6
6
7

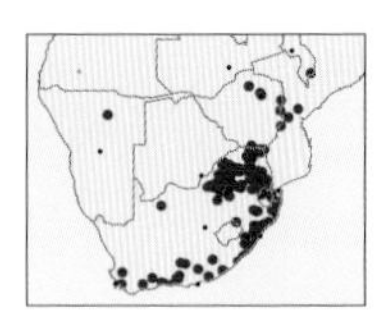

### 1 *Ascotis reciprocaria* — Common Bark

Forewing 18–22mm. Cream to grey, body and wings crossed by wavy black lines, a distinct black-ringed cell spot on each wing, median area heavily suffused with charcoal scales in some individuals; underside lacks terminal dark patches on hindwing underside. Caterpillar light or mottled green to brown with row of 4 dorsal and lateral projections on 5th segment. **Biology:** Caterpillars polyphagous, but obligatory on a plant species after feeding on it for a while. Multivoltine, adults throughout the year. **Habitat & distribution:** Variety of habitats that are not too arid, including transformed habitats; Western Cape north across Africa, Madagascar and Comoros.

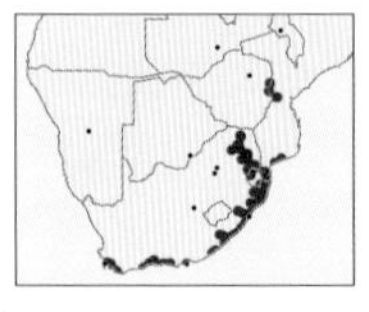

### 2 *Cleora tulbaghata* — Tulbagh Bark

Forewing 16–18mm. Very similar to *Ascotis reciprocaria* (above), but lacks distinct black-ringed cell spots and underside is marked with dark terminal patches on all wings, postmedian line undulating, some individuals have median area dark grey, lacks orange scales on wings. Caterpillar light to dark brown with 1 small pair of projections on 5th segment. Sixteen species in genus, some are best separated by genitalia comparison or barcoding, caterpillars are easier to separate. **Biology:** Caterpillars polyphagous on forest plants. Multivoltine, adults throughout the year. **Habitat & distribution:** Moist forests and kloofs; Western Cape, South Africa, north to tropical Africa.

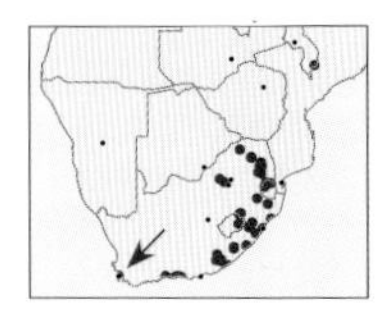

### 3 *Cleora betularia* — Peppered Bark

Forewing 18–20mm. Darker than other *Cleora* spp.; pinkish-brown heavily peppered with darker markings, postmedian line diagnostically jagged, area between lines may be distinctly darker, or distinctly lighter, dark median band in some individuals; hindwing margin not crenulate. Caterpillar pale brown with large dorsal projections on 5th segment. **Biology:** Caterpillars polyphagous on forest plants. Multivoltine, adults throughout the year. **Habitat & distribution:** Forests and kloofs; Western Cape, South Africa, north to Malawi.

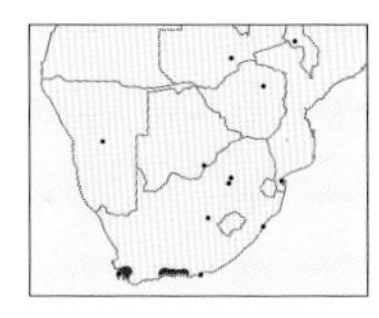

### 4 *Cleora flavivenata* — Yellow-veined Bark

Forewing 17–20mm. Often confused with other *Cleora* spp.; body and wings pale with alternating bands of rusty brown or grey and black lines, postmedian line less regularly jagged than in *C. betularia* (above) and with veins yellow outside of median area, margin of hindwing crenulate, elongate cell spot when present. Caterpillars mottled yellow and green, projection on 5th segment minute or absent. **Biology:** Caterpillars polyphagous on trees. Multivoltine, adults throughout the year. **Habitat & distribution:** In kloofs, in cool, moist mountain fynbos habitats and high forests; Western Cape, South Africa.

### 5 *Cleora munda* — Munda Bark

Forewing 16–18mm. Similar to *C. flavivenata* (above), but smaller, wings narrower and postmedian line in hindwing strongly curved inward with more rounded cell spot. Caterpillar brown to green or mottled, well-developed projections on 5th segment. **Biology:** Caterpillars polyphagous on forest trees. Multivoltine, adults throughout the year. **Habitat & distribution:** Warmer moist forests; Eastern Cape, South Africa, north to tropical Africa, isolated populations in deep river valleys, southern Cape.

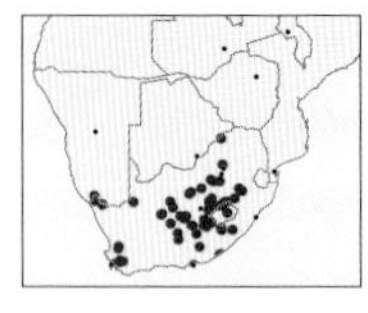

### 6 *Cleora oligodranes* — Brown Bark

Forewing 14–16mm. Unlike other *Cleora* spp.; body and wings straw-coloured, variably shaded with dark brown bands, antemedial and postmedian black lines straight and close to each other. **Biology:** Caterpillar host associations unknown. Adults September–March, peaking in December. **Habitat & distribution:** The only *Cleora* sp. adapted to arid and cold conditions, in river valleys and kloofs; Western Cape north to Limpopo, South Africa.

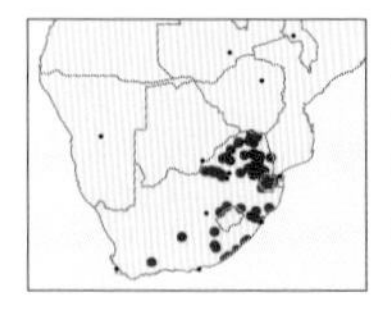

### 7 *Hypomecis ectropodes* — Flat-face Bark

Forewing 11–12mm. Much smaller than similar *Cleora* spp. (above); wings rounded, white dusted with grey and brown scales, antemedial and postmedian lines jagged, row of terminal black spots, underside of wings white and unmarked. See also similar *Parectropis spoliataria* (p.198). Caterpillar light green with purple mark on second proleg and unique anteriorly flattened brown head. Seven described species, several undescribed, in diverse genus in region. **Biology:** Caterpillars feed on *Maytenus albata* (Celastraceae). Bivoltine, adults September–April. **Habitat & distribution:** Bushveld, forest and kloofs; Western Cape north to Limpopo, South Africa, also reported from Zimbabwe.

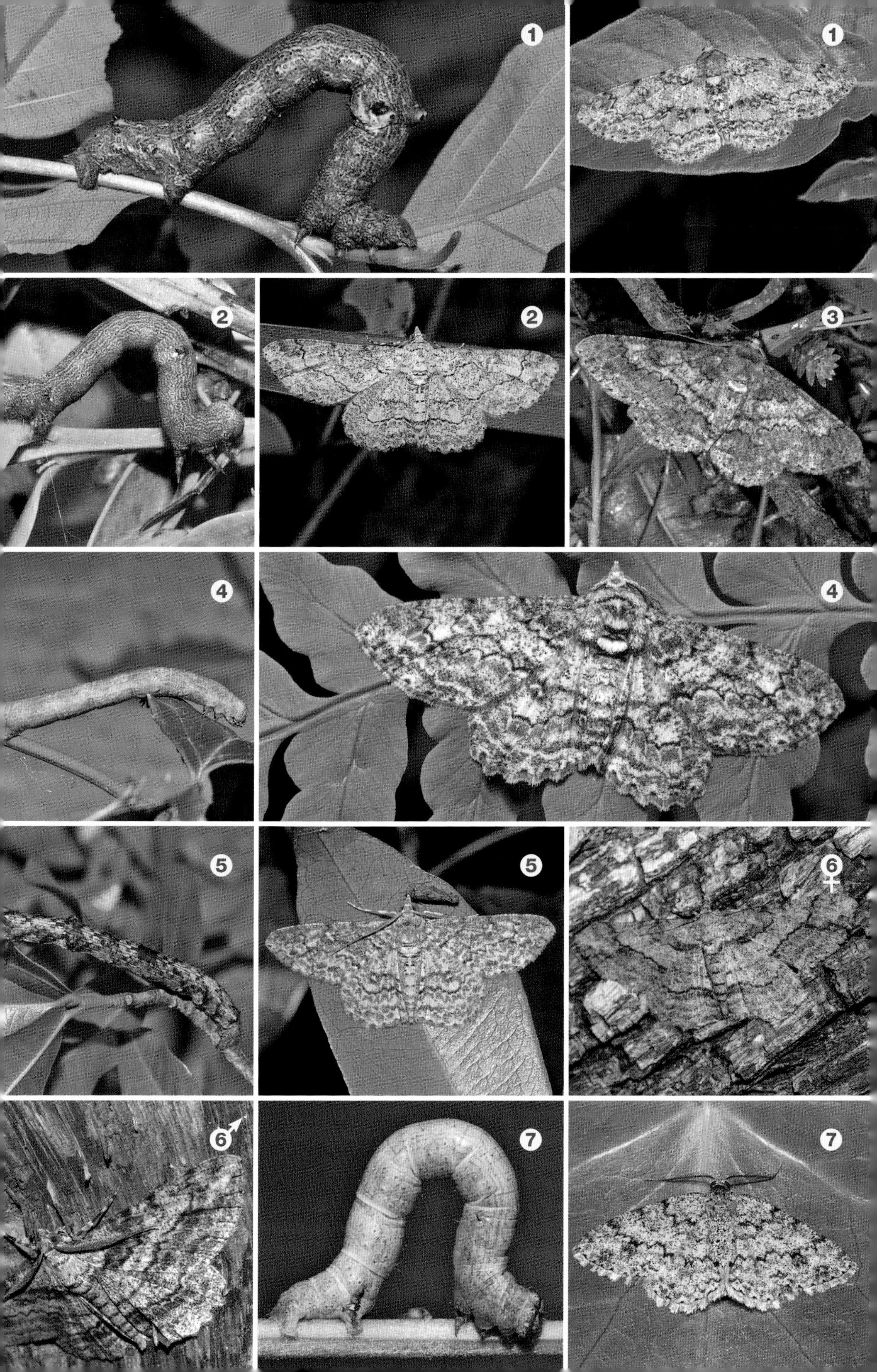
1
1
2
2
3
4
4
5
5
6 ♀
6 ♂
7
7

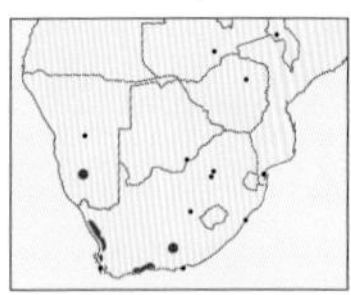

## 1 *Hypomecis gladstonei* — Copperline Bark

Forewing 10–13mm. Male antennae heavily bipectinate; relatively narrow winged, body and wings light brown with distinctive alternating black and coppery brown lines over all wings. Caterpillar mottled brown, perfectly blending in with twigs of host plant. **Biology:** Caterpillars feed on gonna bushes (*Passerina* spp.). Bivoltine, adults September–March. **Habitat & distribution:** Primarily in near coastal vegetation where gonna bushes grow; Western Cape and Northern Cape, but also reported on isolated mountains Eastern Cape and Namibia.

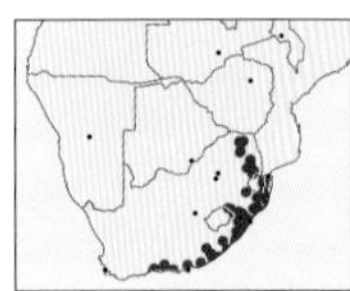

## 2 *Parectropis spoliataria* — Confusing Bark

Forewing 11–13mm. Very similar to *Hypomecis ectropodes* (p.196), best separated by the wing underside, which is heavily marked with grey shading, distinctive darker mark on inner margin in median area; at rest, wings held outspread, body cream with brown crossbands; both wings cream, crossed by numerous variable broken bands of brown or yellowish spots or dots. Ten species in genus in region. **Biology:** Caterpillars reported to feed on lichens. Multivoltine, adults throughout the year. **Habitat & distribution:** Humid forest and bush; Western Cape north to Limpopo, South Africa.

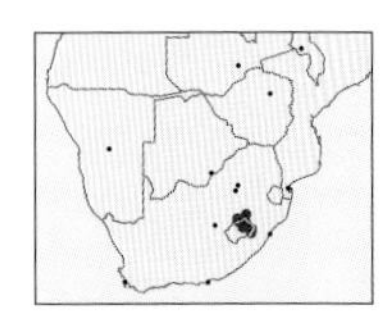

## 3 *Parectropis alticolaria* — Highland Bark

Forewings 15–18mm. Broadly triangular at rest, wing tips rounded; ground colour white, forewings densely suffused with grey and black speckles, crossed by several indistinct darker bands; hindwings paler. **Biology:** Caterpillar host associations unknown. Adults November–February. **Habitat & distribution:** High-altitude grasslands; Lesotho and South Africa.

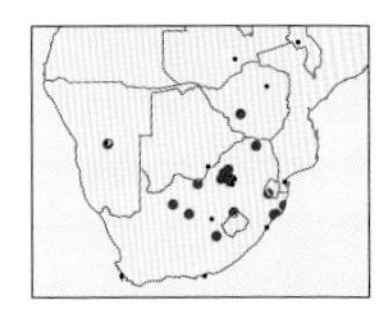

## 4 *Parectropis obliquilinea* — Oblique-lined Bark

Forewings 13–14mm. Narrow winged with wings open at rest; ground colour grey, wings densely suffused with dark brown, 2 parallel oblique black lines on all wings diagnostic. Caterpillar creamy yellow with dense speckling, mimicking underside of host plant leaf. **Biology:** Caterpillars feed on Camphorbush (*Tarchonanthus camphoratus*). Adults November–February. **Habitat & distribution:** Habitats where camphorbushes grow; Eastern Cape, South Africa, north to Zimbabwe and Namibia.

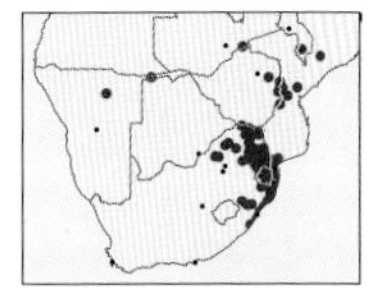

## 5 *Colocleora proximaria* — Pale Dusty Brindle

Forewing ♂ 15–16mm, ♀ 20–23mm. White with varying densities of fine grey speckling; female darker, wings crossed by 3 or 4 black dotted lines. Caterpillar brown or greenish, often becoming pink at both ends. Five species in genus in region; similar *C. grisea* has lines continuous. **Biology:** Caterpillars polyphagous on trees. Multivoltine, adults throughout the year, peaking in summer months. **Habitat & distribution:** Forest and bushveld; Eastern Cape, South Africa, north to tropical Africa.

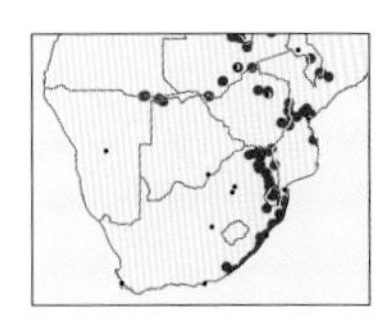

## 6 *Colocleora divisaria* — Diverse Brindle

Forewing ♂ 21–24mm, ♀ 26–31mm. Very variable; typical form has wings mottled brown, crossed by 2 strong black lines, forewing postmedian line diagnostically sharply pointed, median area either a darker shade of brown, of ground colour or much paler, often obliterating lines. Similar *C. simulatrix* has postmedian line not as sharply pointed and is uniform brown. Caterpillar slender, brown and white mottled loopers. **Biology:** Caterpillars polyphagous on trees. Multivoltine, adults throughout the year. **Habitat & distribution:** Warm humid forest and bush; Eastern Cape, South Africa, north to tropical Africa.

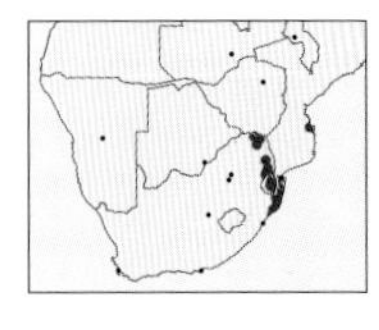

## 7 *Cerurographa bistonica* — Marbled Bark

Forewing 15–17mm. Body stout with dark thorax and paler, banded abdomen; forewings white with dramatic marbled patterning of dark brown lines and patches and thin, black, wavy marginal line; hindwings pale pink, weakly marked, margin wavy. One species in genus in region. **Biology:** Caterpillars feed on Spikethorn (*Gymnosporia buxifolia*). Multivoltine, adults throughout the year. **Habitat & distribution:** Lowland bushveld; KwaZulu-Natal, South Africa, north to Mozambique.

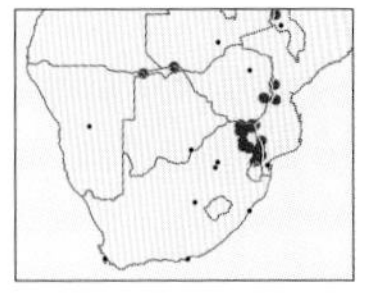

## 8 *Phthonandria pinguis* — Sooty Bark

Forewing 15–17mm. Very broadly triangular at rest; base colour of body and wings white, both very heavily marked with dark brown and black bands and spots, distinctive light mark in centre of terminal area in most specimens; hindwings much lighter, almost white in some specimens, grading to black toward margin. One species in genus in region. **Biology:** Caterpillar host associations unknown. Adults September–April. **Habitat & distribution:** Lowland bushveld; Mpumalanga, South Africa, north to Angola and Malawi.

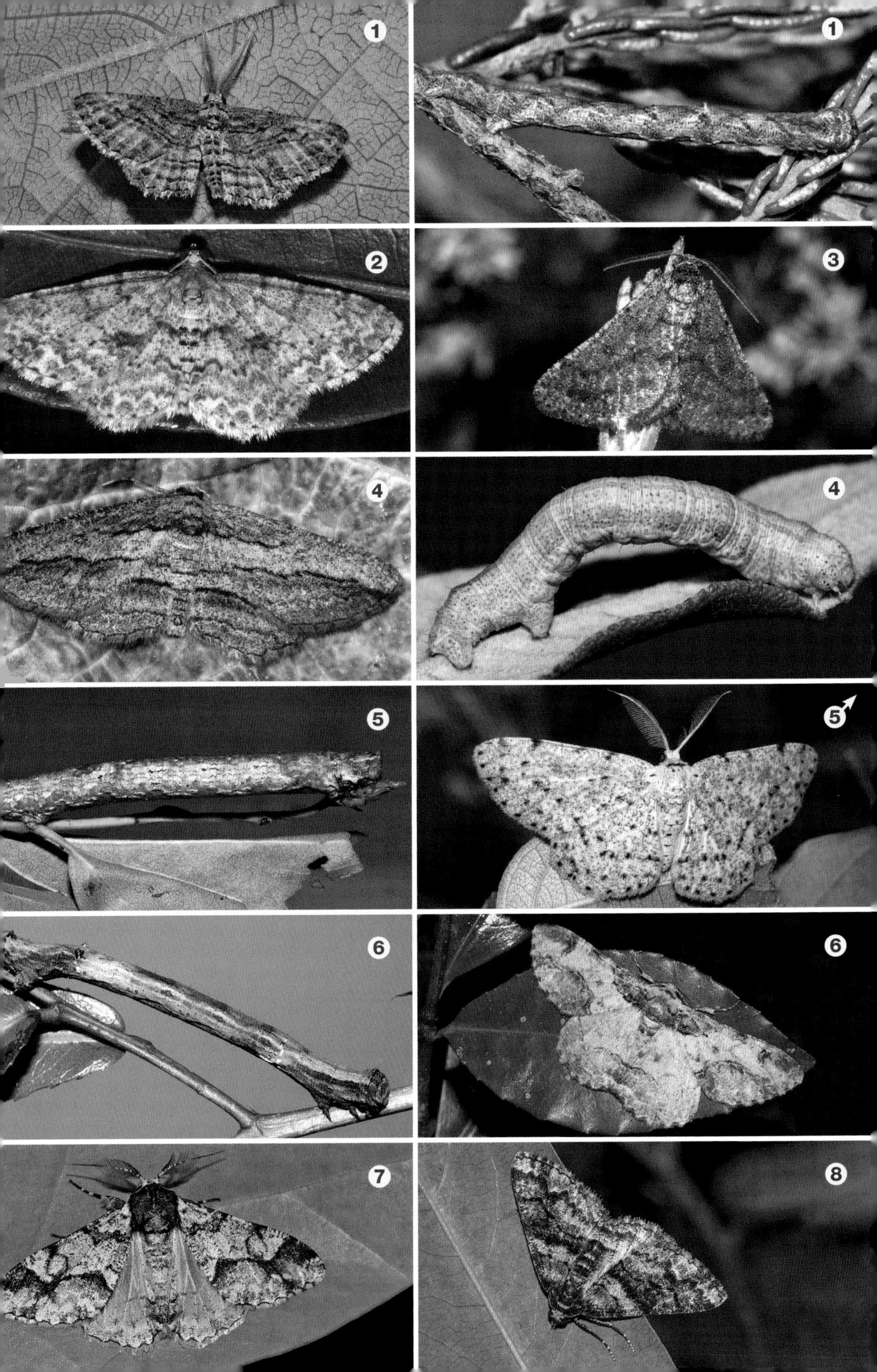
1
1
2
3
4
4
5
5♂
6
6
7
8

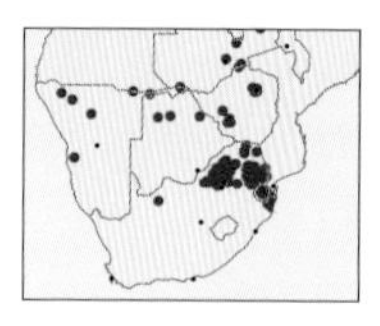

## 1 *Eulycia grisea* — Grey Pennant

Forewing 17–21mm. Elongate at rest, wings held out laterally to form a 'T'-shape (similar to unrelated *Hebdomophruda* spp. pp.186, 188); grey with distinctive brown marking in terminal area, black postmedian line sharply kinked outward, black cell spot on all wings. Similar *E. extorris* has postmedian line straight and dotted. Caterpillar brown with conspicuous orange and grey lateral marks on each segment. Four species in genus in region. **Biology:** Posture thought to mimic a dead twig. Caterpillars feed on bushwillows (*Combretum* spp.). Bivoltine, adults September–April, peaking November and February. **Habitat & distribution:** *Combretum* bushveld; KwaZulu-Natal, South Africa, north to tropical Africa.

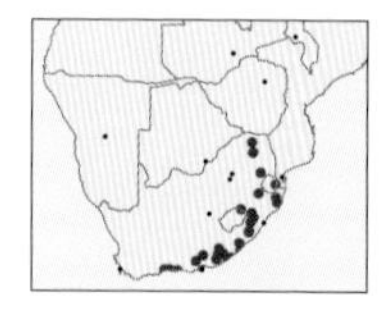

## 2 *Eulycia accentuata* — Black-lined Pennant

Forewing 14–17mm. At rest, wings held flat and outstretched; body plump, white (sometimes grey) crossed by black lines; wings white to grey, marked with several thin black lines that form a lattice-like pattern along hind margin, diagnostic postmedian line deeply indented inward before costa; hindwings with black cell spot. Similar *E. subpunctata* has fewer lines, postmedian line undulating, not deeply indented before costa and occurs in bushveld. Caterpillar brown with red spots laterally and variable white lateral shading, several small pairs of dorsal spikes. **Biology:** Caterpillars feed on various Celastraceae. Multivoltine, adults throughout the year. **Habitat & distribution:** Forest and moist woodland; Western Cape north to Limpopo, South Africa.

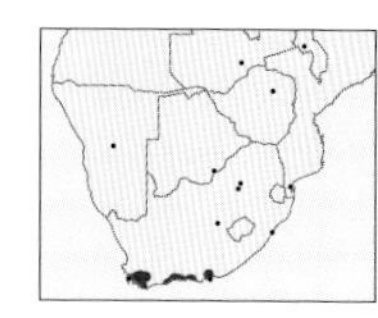

## 3 *Loxopora dentilineata* — Dotted Bark

Forewing 12–15mm. Sexually dimorphic; male straw-coloured, dotted with black dots of various sizes, postmedian line straight with triangular black dots on both wings and grey shading beyond, row of round terminal dots; female completely covered in dense black dots almost obliterating lines in some specimens. Caterpillar very colourful with alternating lines, spots and bands of pink, orange, yellow, black and cream. One described species in genus in region, several similar undescribed species on mountains in remnant fynbos vegetation north to Malawi. **Biology:** Caterpillars feed on *Berzelia lanuginosa* (Bruniaceae). Adults throughout the year. **Habitat & distribution:** Fynbos; Western Cape and Eastern Cape, South Africa.

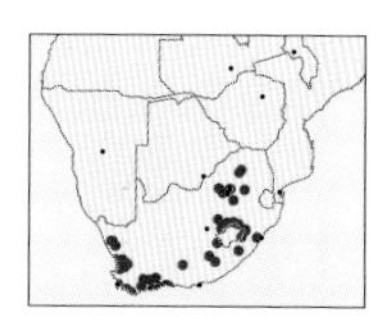

## 4 *Omphalucha crenulata* — Crenulate Brindle

Forewing 15–17mm. Body stout, thorax setose and dark brown, abdomen paler; forewings grey, heavily speckled, series of orange and brown patches in terminal area, median area lighter, black antemedial line semicircular, postmedian line with large kink in forewing and diagnostically crenulated wing margins. Caterpillar smooth, brown with longitudinal double lines, pink head. Eleven species in genus in region. Similar *O. maturnaria* brown, not grey, and wing margins only slightly crenulated. **Biology:** Caterpillars feed on various *Searsia* spp. Multivoltine where conditions allow, adults throughout the year. **Habitat & distribution:** Variety of habitats where *Searsia* spp. grow; Western Cape north to Limpopo, South Africa.

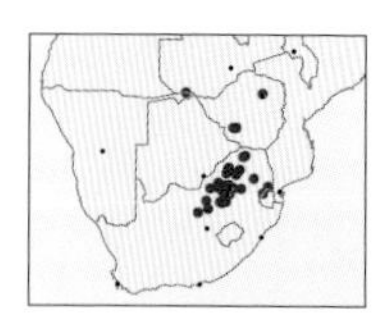

## 5 *Omphalucha albosignata* — Clouded Brindle

Forewing 15–19mm. Similar to *O. crenulata* (above), but hindwings partially white (hidden when at rest) and wing margins only slightly crenulated, grey-brown; forewings crossed by 2 thin black lines, antemedial line rounded, postmedian line sharply pointed. Caterpillar slender, mottled brown. **Biology:** Caterpillars feed on various *Searsia* spp. Bivoltine, adults September–March, peaking October and February. **Habitat & distribution:** Bushveld; Free State, South Africa, north to Zimbabwe and Namibia.

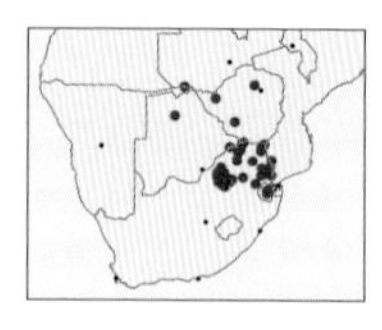

## 6 *Omphalucha indeflexa* — Unflexed Brindle

Forewing 14–15mm. Similar to *O. albosignata* (above), but browner, not grey, antemedial and postmedian lines much straighter on both wings, and hindwings not partially white. Caterpillar light brown with yellowish markings, black projections on last segment. **Biology:** Caterpillars feed on bushwillows (*Combretum* spp.). **Habitat & distribution:** *Combretum* bushveld; Eswatini north to Zimbabwe and Namibia.

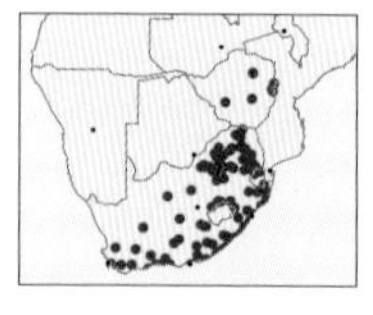

## 7 *Xylopteryx arcuata* — Cryptic Bark

Forewing 12–16mm. Extremely cryptic; broadly oval outline at rest; forewings heavily mottled with grey to brown or black, sometimes with some white in median area; hindwings cream with wide brown distal margin. Caterpillar a plump, light brown looper with conspicuous diamond-shaped dorsal marks. Five species in genus in region. **Biology:** Caterpillars feed on various *Searsia* spp. Multivoltine, adults throughout the year, peaking in summer. **Habitat & distribution:** Variety of habitats where *Searsia* spp. grow; Western Cape, South Africa, north to Kenya.

1
1
2
2
3
3♂
4
4♀
5
5
6
6
7
7

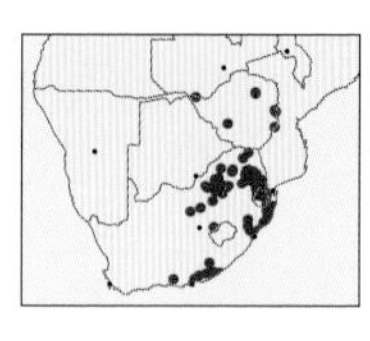

### 1 *Xylopteryx prasinaria* — Lichen Bark

Forewing 12–16mm. Wings held flat and outstretched; body greenish to grey; wings heavily mottled in green, purple and brown, possibly a lichen mimic, postmedian line strongly indented inward and diagnostic curve before costa. Caterpillar a plump, pinkish, knobbled looper. **Biology:** Caterpillars feed on Spikethorn (*Gymnosporia buxifolia*) growing in the shade. **Habitat & distribution:** Forest and riverine bush; Eastern Cape, South Africa, north to Kenya.

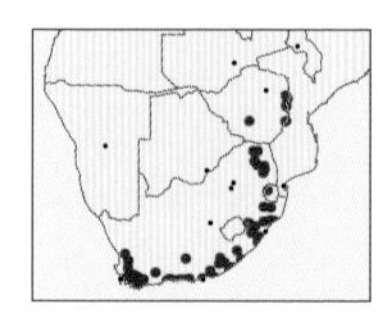

### 2 *Xylopteryx protearia* — Patchwork Bark

Forewing 18–20mm. Extremely variable in colour from pink-grey to dark brown, typically with an oval brown cell spot, which is crossed by 2 jagged black lines, shape of postmedian line, when visible, diagnostic, median area may be almost white or darker than rest of wings. Caterpillar plump and brown with conspicuous white dorsal patch. **Biology:** Caterpillars feed on various Celastraceae spp. Multivoltine, adults throughout the year. **Habitat & distribution:** Forests and moist mountain fynbos; Western Cape, South Africa, north to tropical Africa.

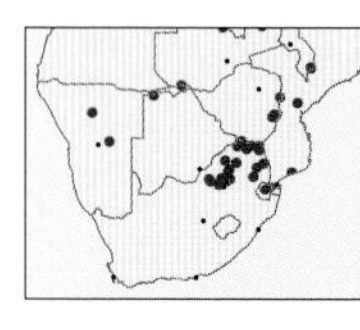

### 3 *Zeuctoboarmia hyrax* — Grey Bark

Forewing 22–25mm. Ground colour off-white; both wings with oval black-ringed, grey-centred stigma, wings vary from almost white, through dusted with black, to heavily marked in black (especially distally), wings crossed by series of thin, jagged black lines, shape of postmedian line diagnostic. Caterpillar mottled orange. Five species in genus in region. **Biology:** Caterpillars recorded on Alfalfa (*Medicago sativa*) and Peruvian Pepper Tree (*Schinus molle*). Multivoltine, adults throughout the year. **Habitat & distribution:** Bushveld; Eswatini north to tropical Africa.

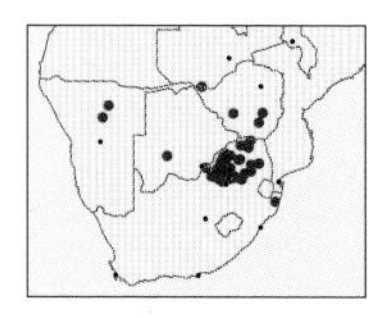

### 4 *Zeuctoboarmia cataimena* — Croton Bark

Forewing 17–20mm. Bipectinate antennae well developed; broad winged with long cilia, ground colour off-white, body and wings heavily and evenly dusted with black and brown scales. Caterpillar perfectly camouflaged to blend in with the leaves and twigs of its host plant. **Biology:** Caterpillars feed on Lavender Feverberry (*Croton gratissimus*). Multivoltine, adults throughout the year. **Habitat & distribution:** Bushveld on hills where lavender feverberries grow; KwaZulu-Natal, South Africa, north to Zimbabwe and Namibia.

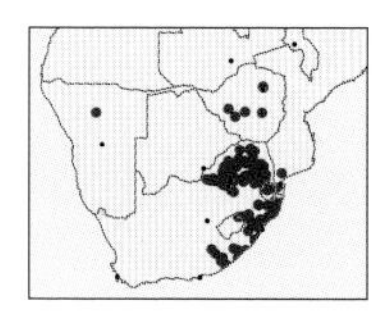

### 5 *Cabera strigata* — Marbled Cabera

Forewing 10–13mm. Pinkish-brown, wings slightly to heavily spotted with black, more so in females, in males crossed by 2 or 3 indistinct orange to black lines; small black cell spot on forewings and short black stripe leading to apex. Caterpillar slender, green to dark brown, some mottled and other striped in these colours. Six described species in genus in region. **Biology:** Caterpillars, like all regional Caberini, feed on Rhamnaceae, primarily on Buffalo Thorn (*Ziziphus mucronata*), also Soap Creeper (*Helinus integrifolius*). Multivoltine, adults throughout the year, peaking in summer. **Habitat & distribution:** Bushveld and forest; Eastern Cape, South Africa, north to Zimbabwe and Namibia.

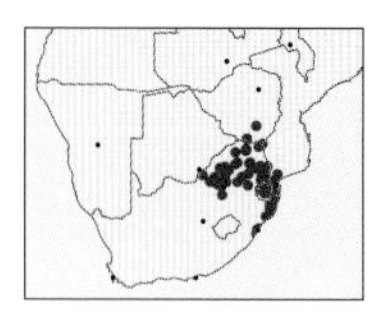

### 6 *Cabera elatina* — Raised Cabera

Forewing 8–10mm. Thorax dark brown; abdomen mottled green and brown; diagnostic base of forewings almost black extending along costa, distal portions white to light brown, heavily marked with black; hindwings mottled black and brown. Caterpillars slender, light green or brown, sometimes with white lateral line, or tinged with orange at both ends. Similar *C. aquaemontana* is uniform brown, never white, with double postmedian line and lacks the black shading along costa. **Biology:** Caterpillars feed on Buffalo Thorn (*Ziziphus mucronata*). Bivoltine, adults September–April. **Habitat & distribution:** Bushveld; KwaZulu-Natal, South Africa, north to Kenya.

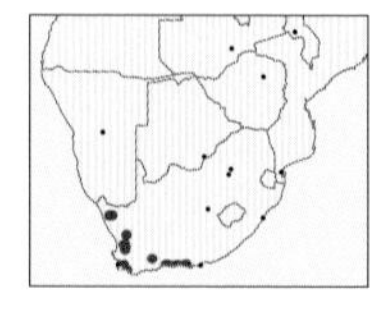

### 7 *Cabera pseudognophos* — Cape Cabera

Forewing 12mm. Wings and body silvery grey, variably sprinkled with black and brown scales, diagnostic jagged light brown to black lines on both wings, small black cell spot. Similar *C. neodora* has lines straight. Caterpillar green with 2 distinctive white lines. **Biology:** Caterpillars feed on *Trichocephalus stipularis* (Rhamnaceae). Bivoltine, adults throughout the year, peaking September and March. **Habitat & distribution:** Fynbos and sandveld; Northern Cape, Western Cape and Eastern Cape, South Africa.

1
2
2
2
2
2
3
3
4
4
5
5♂
6
6
7
7♀

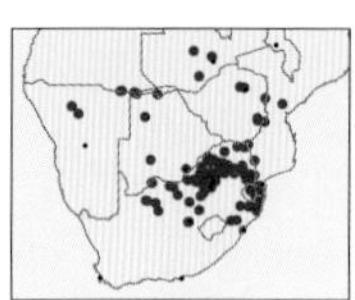

**1 *Lhommeia subapicata*** Banded Scallop

Forewing 16–19mm. Very broadly triangular at rest; distinct seasonal forms, wet-season form yellow and dry-season form straw; wings lightly spotted with brown (more so on underside), diagnostic large subapical mark on costa and postmedian pink-brown band, stronger on hindwings, antemedial line absent. Caterpillar mottled brown with lateral projection on 5th segment. Similar *L. biskraria* is often green and has subapical mark and lines thinner, and a developed antemedial line with a median band. Two described species in genus in region. **Biology:** Caterpillars feed on Buffalo Thorn (*Ziziphus mucronata*). Multivoltine, adults throughout the year. **Habitat & distribution:** Bushveld and grassland where host tree grows; Free State, South Africa, north to tropical Africa.

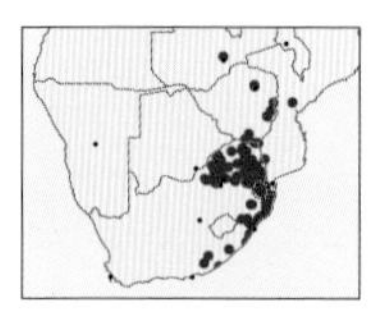

**2 *Erastria leucicolor*** Chain Scallop

Forewing 13–15mm. Variable forms, ground colour ranging from yellow to green to brown; both wings speckled with brown scales and with a wide light brown transverse band, underside of wings with diagnostic chain-like row of red spots. Caterpillar brown, slender and stick-like. Three species in genus in region. **Biology:** Caterpillars feed on Buffalo Thorn (*Ziziphus mucronata*). Multivoltine, adults throughout the year. **Habitat & distribution:** Bushveld and forest; Eastern Cape, South Africa, north to tropical Africa, also Madagascar, Seychelles.

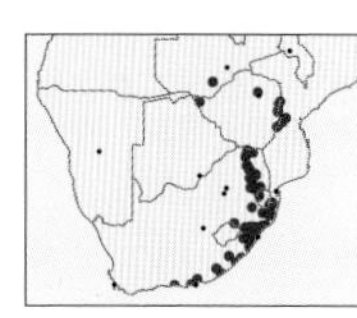

**3 *Erastria madecassaria*** Variable Scallop

Forewing 17–20mm. Colour variable, from yellow-brown to green to dark brown, sometimes red-brown; forewings falcate, crossed by 2 or 3 wavy brown lines, area between lines may be lighter, outer margin of wing often darker (also on underside); hindwings with distinctly undulating and crenulate margin. **Biology:** Caterpillar host associations unknown in region, elsewhere feeds on Chinese Date (*Ziziphus jujuba*) and Indian Jujube (*Z. mauritiana*). Multivoltine, adults throughout the year. **Habitat & distribution:** Moist forests; Eastern Cape, South Africa, north to tropical Africa, also Madagascar and Indian Ocean islands.

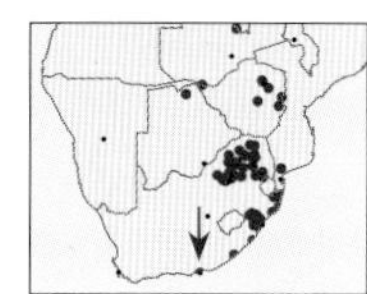

**4 *Coenina poecilaria*** Wisp Wing

Forewing 13–20mm. Wings orange to pinkish-grey with rectangular white patches, black dots and white hindwing inner margin; typical Gonodontini posture when at rest, forewings held rolled up and extended forward and hindwings held well back, with large gap between wings. Caterpillar cream with dark lateral lines and finger-like dorsal projections. Two species in genus in region. **Biology:** Caterpillars polyphagous on herbaceous plants. Multivoltine, adults throughout the year; females diurnal, flying low over grassland ovipositing on many plants; males nocturnal, attracted to light. **Habitat & distribution:** Grasslands and bushveld; Eastern Cape, South Africa, north to Malawi.

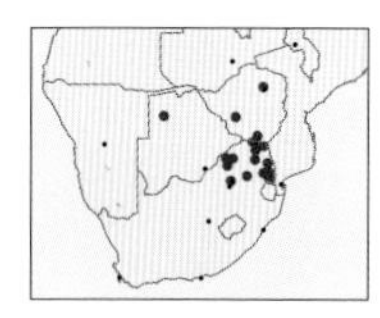

**5 *Coenina dentataria*** Toothed Wisp Wing

Forewing 16–18mm. Similar to *C. poecilaria* (above), but light cream-yellow wings with a few dark blotches and lines; hind margin of forewings extended into a lobe and of hindwings crenulate. **Biology:** Caterpillars polyphagous. Adults nocturnal, October–March. **Habitat & distribution:** Bushveld; KwaZulu-Natal, South Africa, north to tropical Africa.

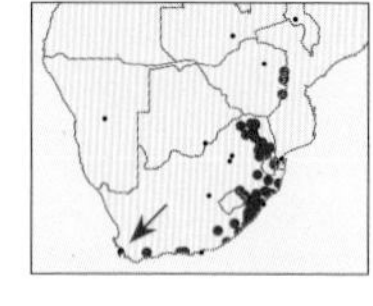

**6 *Psilocladia obliquata*** Tailed Wisp Wing

Forewing 16–19mm. Males creamy yellow with brown markings; hindwings crossed by parallel brown lines and with short apical tail. Females silvery-white with violet markings and more distinct crenulation on hindwing. One species in genus in region. **Biology:** Caterpillar host associations unknown. Multivoltine, adults throughout the year. **Habitat & distribution:** Moist forests; Western Cape, South Africa, north to tropical Africa.

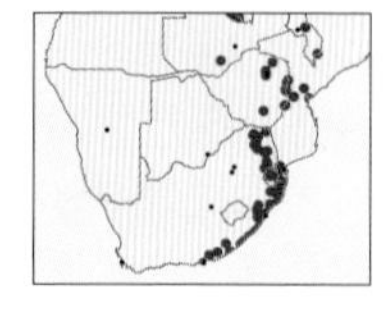

**7 *Xenimpia erosa*** Ragged Wisp Wing

Forewing ♂ 13–16mm, ♀ 17–20mm. Strongly sexually dimorphic; male wings yellow crossed by parallel brown lines, hindwings slightly scalloped; female forewings reddish-brown, both wings extremely scalloped, giving the moth a ragged appearance. Bizarre plump caterpillar brown with pronounced finger-like dorsal lobes. Four described species in genus in region. **Biology:** Caterpillars polyphagous on trees. Multivoltine, adults throughout the year. **Habitat & distribution:** Moist forest and bush; Eastern Cape, South Africa, north to tropical Africa.

1
1
2
2
3
3
4
4 ♀
5
6 ♀
7 ♂
IS
7

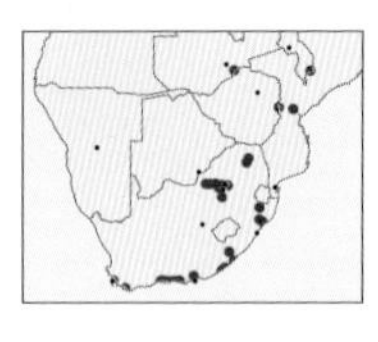

### 1 *Xenimpia maculosata* — Variable Wisp Wing

Forewing ♂ 15–17mm, ♀ 19–20mm. Similar distinctive posture to *X. erosa* (p.204), but more falcate and with hindwing edge flat not scalloped; variable in colour and markings pink-brown, often with fine black dusting, blotches and a cream band across the forewings. Plump brown caterpillars have finger-like projections. **Biology:** Caterpillars recorded on Forest Elder (*Nuxia floribunda*). Multivoltine, adults throughout the year, peaking in summer. **Habitat & distribution:** Forest and riverine vegetation; Western Cape, South Africa, north to Zambia.

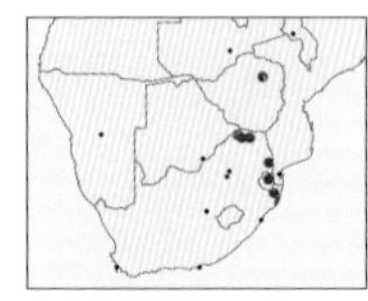

### 2 *Sesquialtera ridicula* — Small Wisp Wing

Forewing 10–13mm. Smaller and narrower winged than other species in tribe Gonodontini; wings grey and brown with black cell spot in forewings; hindwings with diagnostic roughly triangular black patch in median area and margin sharply toothed to form several small tails. Two species in genus in region; *S. ramecourti* lacks the black mark in hindwing. **Biology:** Caterpillar host associations unknown. Multivoltine, adults throughout the year, peaking in summer. **Habitat & distribution:** Bushveld; KwaZulu-Natal, South Africa, north to Somalia.

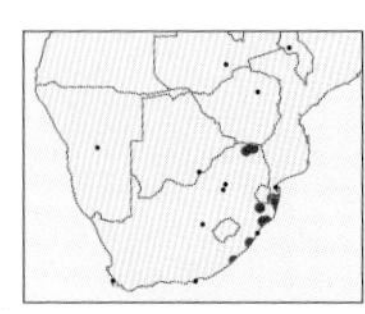

### 3 *Miantochora venerata* — Unwisp Wing

Forewing 23mm. At rest, wings held widely extended and flat with forewings not folded up, unlike other species in tribe Gonodontini; body and wings yellow to light brown or grey in some individuals; forewings with falcate tips, speckled with black and with 2 larger black spots, and crossed by straight, pale double postmedian line; hindwings extended to points, crossed by stronger double postmedian line, starting pale then becoming dark brown. One species in genus in region. **Biology:** Caterpillar host associations unknown. **Habitat & distribution:** Moist forests; Eastern Cape, South Africa, north to tropical Africa.

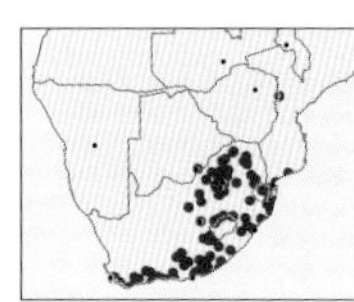

### 4 *Sicyodes cambogiaria* — Dimorphic Tent Elegant

Forewing 13–15mm. Wings held broadly angled alongside body, distal margins of forewings strongly concave, giving a distinctive tent profile when viewed from above; ground colour in male usually bright yellow but greenish in some specimens; sexes dimorphic; forewings of female with oval brown spot centrally and broad dark brown distal band; forewings of male crossed by 2 narrow diagonal brown lines. Caterpillar green, sometimes with reddish dorsal band. Eleven similar species in genus in region all feeding on Celastraceae. Very similar *S. olivescens* is primarily olive green, caterpillar is grey, and is found in more arid habitats; *S. biviaria* has conspicuous black dots on margin and is found in moist forests. **Biology:** Caterpillars feed on various *Gymnosporia* spp. Multivoltine, adults throughout the year. **Habitat & distribution:** Relatively moist bushveld and kloofs, absent from arid habitats and moist forests; Western Cape, South Africa, north to Kenya.

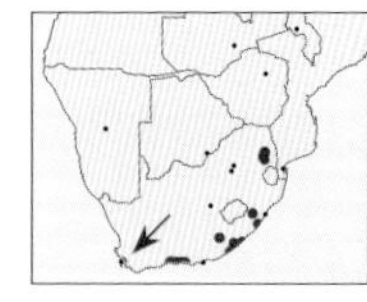

### 5 *Sicyodes algoaria* — Variable Tent Elegant

Forewing 14–16mm. Larger than similar species; sexually dimorphic; forewings of male olive green, some with large black dots, postmedian line straight, often indistinct; forewings of female vary from yellow to red and permutations in between. Caterpillar slender, light green with 2 dorsal and 1 set of lateral red processes. **Biology:** Caterpillars feed on Candlewood (*Pterocelastrus tricuspidatus*). Multivoltine, adults throughout the year. **Habitat & distribution:** Moist forests and coastal bush; Western Cape, north to Limpopo, South Africa.

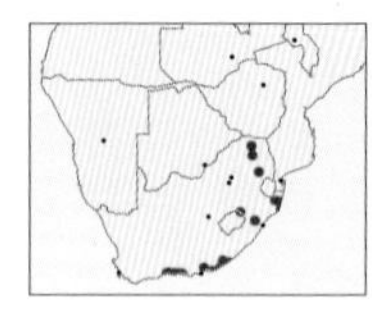

### 6 *Sicyodes ocellata* — Eye Tent Elegant

Forewing 13–14mm. Colour variable, from green through pink to purplish-brown, diagnostic eyed cell spot on forewings, lines narrow and punctuated by small black dots. Caterpillar slender, green, spotted with numerous brown dots and a few blotches, 2 lateral processes on 5th segment, where caterpillar is kinked. **Biology:** Caterpillars feed on *Mystroxylon aethiopicum* (Celastraceae). Multivoltine, adults throughout the year. **Habitat & distribution:** Moist forests and coastal bush; Western Cape north to Limpopo, South Africa.

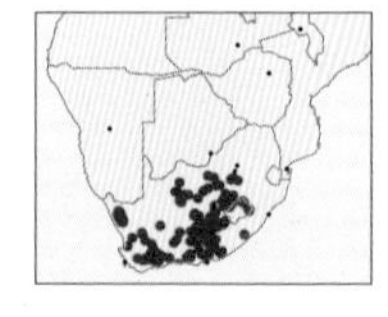

### 7 *Drepanogynis bifasciata* — Sargasso Elegant

Forewing 13–16mm. Thorax green, abdomen brown; forewings green, marked with 2 wide, meandering white lines (edged with brown in some specimens) and white ring between, margins with white flashes, sometimes brown between these; hindwings white basally with central brown cell spot, wide light brown margin often chequered brown and white. Nineteen species in genus in region. **Biology:** Caterpillar host associations unknown. **Habitat & distribution:** Karoo biomes and grassland containing some karroid vegetation; Western Cape north to North West, South Africa.

1
2
3
4
4
5
5
6
6
7

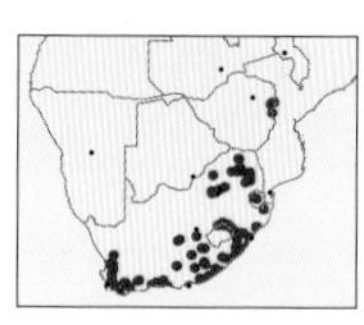

**1 *Drepanogynis mixtaria*** Mixed Elegant

Forewing 13–16mm. Wings held widely outstretched when at rest; female with forewings falcate and hindwings toothed, those of male less crenulate; body and wings cream to pale reddish-brown, wings crossed by 2 wavy brown lines, area between lines usually darker, often marbled and with white markings. Caterpillar brown. Several similar species, some undescribed; *D. serrifasciaria* has postmedian line straight and evenly crenulated; *D. insolens* has postmedian line sharply curved outward. **Biology:** Caterpillars polyphagous on trees. Adults December–June, peaking in March. **Habitat & distribution:** Variety of habitats; Western Cape, South Africa, north to Zimbabwe.

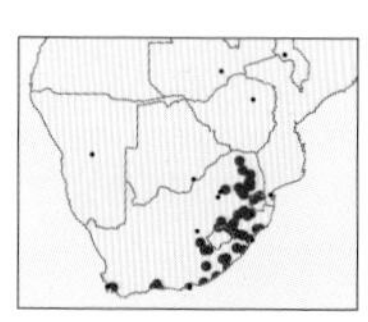

**2 *Drepanogynis determinata*** Robust Elegant

Forewing 19–20mm. Robust species with very setose body, colour varies from green to brown; both wing margins diagnostically crenulated, more so in female, black cell spot in forewings with 2 well-defined lines, median area darker in most specimens. Caterpillar resembles a brown thorn with a pair of processes on first and last segments. Similar *D. monas* has margins not crenulated. **Biology:** Caterpillars polyphagous on trees. Univoltine, adults March–May. **Habitat & distribution:** Moist forest and grassland; Western Cape north to Limpopo, South Africa.

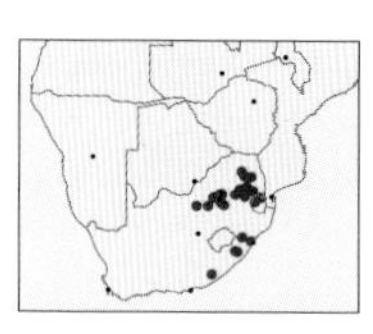

**3 *Drepanogynis gloriola*** Glorious Elegant

Forewing 14–15mm. Wings held tight against body and over perch when at rest; frons and head pink, thorax and abdomen white; forewings a shiny golden colour with broad, white-edged pink postmedian band ending on apex, margins pink; hindwings grey with pink cilia. Similar *D. soni* has postmedian band broader, forewing area at apex more extensively pink, and occurs in fynbos in Western Cape. **Biology:** Caterpillar host associations unknown. Adults September–March. **Habitat & distribution:** Grassland; Eastern Cape north to Limpopo, South Africa.

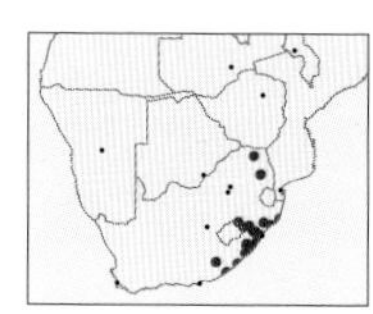

**4 *Drepanogynis valida*** Fourspot Elegant

Forewing 14–17mm. Wings held flat and open when at rest; sexually dimorphic; males ivory cream, females yellow to orange; large black cell spot on forewings, smaller one on hindwing, thin and straight lines diagnostic, sparse sprinkling of black scales over wings. Similar *D. curvifascia* has lines thicker and prominently black, and postmedian curved outward; *D. angustifascia* has lines dull brown and slightly curved or absent. **Biology:** Caterpillar host associations unknown. Univoltine, adults March–May. **Habitat & distribution:** Afromontane forest; Eastern Cape north to Limpopo, South Africa.

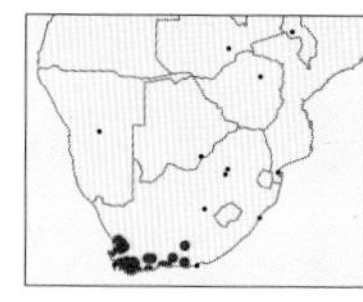

**5 *Drepanogynis sinuata*** Mustache Elegant

Forewing 16–17mm. Wings held flat and broadly triangular when at rest; ground colour straw to grey with variable sprinkling of brown scales, black cell spot, antemedial line jagged, postmedian line curved outward, terminal area beyond line darkly shaded, wing margins crenulated. Caterpillar slender, grey with dark blotches. Similar *D. agrypna* has postmedian line straight; *D. nigrobrunnea* is dark brown and black. **Biology:** Caterpillars recorded on *Searsia pterota* (Anacardiaceae). Univoltine, adults August–October. **Habitat & distribution:** Fynbos and coastal bush; Western Cape and Eastern Cape.

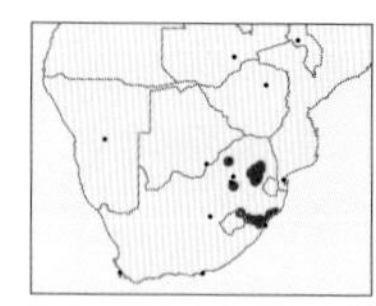

**6 *Drepanogynis fuscimargo*** Margin Elegant

Forewing 16–17mm. Wings held flat and open when at rest; ground colour yellow, postmedian line regularly crenulated, area beyond brick red forming a broad band, variable pinkish dots along costa, antemedial line broken, margins crenulated. Similar *D. angustimargo* has postmedian line less crenulated, band much narrower and margins not crenulated. **Biology:** Caterpillar host associations unknown. Probably bivoltine, adults October–November and January–February. **Habitat & distribution:** Grassland; KwaZulu-Natal north to Limpopo, South Africa.

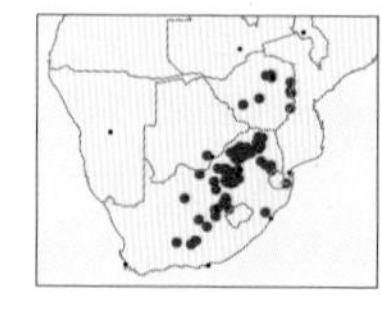

**7 *Apleroneura tripartita*** Partition Elegant

Forewing 14–16mm. Can be confused with *Drepanogynis fuscimargo* species group (above), but postmedian line straight, not crenulated; at rest, wings broadly spread, with distal teeth and divided into 3 sections, basal and distal sections dark brown and central one yellow with small cell spot, wings variably sprinkled with black dots or striations. Caterpillar straw to light brown with pale longitudinal lines, blending in with stem of host plant. Twenty-seven species in genus in region. **Biology:** Caterpillars feed on Bushveld Asparagus (*Asparagus laricinus*). Bivoltine, adults September–April. **Habitat & distribution:** Arid areas, Nama karoo, grassland and bushveld; Eastern Cape, South Africa, north to Zimbabwe.

1
1
2
2♂
3
4♀
5
6
7
7

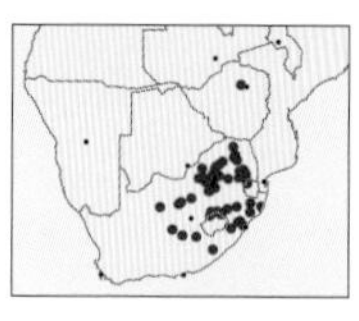

### 1 *Apleroneura epione* — Web Elegant

Forewing 11–12mm. Smaller than other species in genus; ground colour white with brown lines and veins brown with light striations on both wings giving a web-like appearance; basal and terminal areas dark brown in form *fuscomarginata*. Caterpillar green with white and yellow longitudinal lines, blending in with leaves of host plant. **Biology:** Caterpillars feed on *Asparagus* spp. Bivoltine, adults October–March. **Habitat & distribution:** Arid areas, Nama karoo, grassland and bushveld; Eastern Cape, South Africa, north to Zimbabwe.

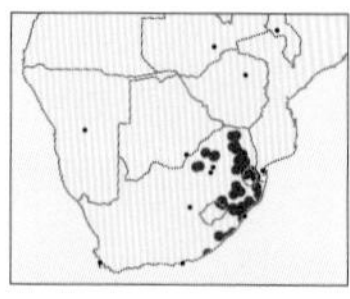

### 2 *Apleroneura admiranda* — Marble Elegant

Forewing 12–14mm. Distinctive species with sharply pointed wings; ground colour in male shiny white with golden-brown shading and lines giving it a marbled appearance, black cell spot; female with deep red to pink shading. **Biology:** Caterpillars recorded on *Syncolostemon foliosus* (Lamiaceae). Bivoltine, adults October–March. **Habitat & distribution:** Grassland; Eastern Cape north to Limpopo, South Africa.

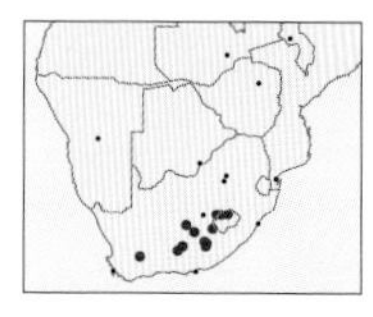

### 3 *Apleroneura hypoplea* — Sandy Elegant

Forewing 13–14mm. Ground colour light brown; the smooth, curved shape of postmedian line and general sandy suffusion over all wings diagnostic for this species group, small elongate cell spot. Similar *A. chromatina* has postmedian line straight; *A. fumosa* has postmedian line straight and cell spot round; *A. nephelochora* is much larger with ground colour distinctly coppery. **Biology:** Caterpillar host associations unknown. Bivoltine, adults October–March. **Habitat & distribution:** Nama karoo and Afroalpine vegetation; Western Cape, South Africa, north to Lesotho.

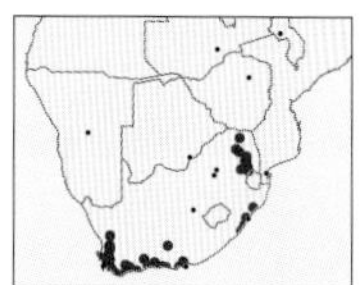

### 4 *Derrioides villaria* — Curved Elegant

Forewing 16–18mm. Thorax brown and densely covered in rich reddish setae; abdomen reddish-brown; wings dark pink to red brown, each with dark cell spot, forewings crossed by faint antemedial and more distinct oblique postmedian paired yellowish and brown lines, diagnostically slightly crenulated. Seventeen species in genus in region. Similar *D. latipennis* has line straighter and not crenulated. **Biology:** Caterpillars recorded on Butterspoon Tree (*Cunonia capensis*) and Gonna (*Passerina vulgaris*). Multivoltine, adults throughout the year. **Habitat & distribution:** Fynbos and grassland where remnant fynbos elements occur; apparently 3 isolated populations: Western Cape and Eastern Cape; KwaZulu-Natal coastal region; eastern escarpment of Mpumalanga and Limpopo, South Africa.

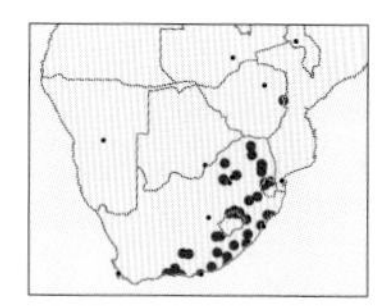

### 5 *Derrioides arcuifera* — Forest Elegant

Forewing 17–20mm. Sexually dimorphic; colouration variable; males have shades of green on forewings, conspicous half round mark on costa, wing pattern is distinctive and cannot be mistaken for any other *Derrioides*; females larger, dark reddish-grey and wing tip more falcate. Genetic diversity suggests that isolated populations may be different species. **Biology:** Caterpillars recorded on *Halleria lucida* (Stilbaceae). Univoltine, adults January–May, peaking in March. **Habitat & distribution:** Forests and kloofs; Western Cape, South Africa, north to Zimbabwe.

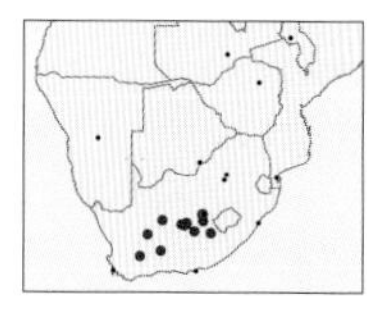

### 6 *Derrioides strigulosa* — Banded Ground Elegant

Forewing 16–17mm. Thorax setose, dark brown; forewings straw-coloured with diagnostic but variable broad, wavy, dark brown band across upper side, which bears a central cell spot and is bordered by darker lines punctuated with tiny white dots. **Biology:** Caterpillar host associations unknown. Probably univoltine, adults January–June, peaking in March/April. **Habitat & distribution:** Nama karoo and grassland; Western Cape north to Free State, South Africa.

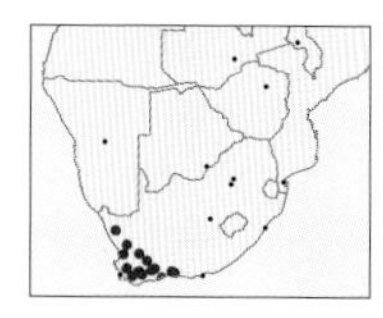

### 7 *Derrioides miltophyris* — Ochre Ground Elegant

Forewing 14–15mm. Several species in species group, often sympatric, the ochreous suffusion on the forewings and relatively wider median area are diagnostic; small black cell spot, some specimens with reddish shading on forewing. Similar *D. dentatilinea* has postmedian line interspersed with white marks; *D. griseisparsa* is uniform grey-brown; *D. oinophora* has median area much narrower and white; *D. rosea* has ringed cell spot and sprinkled with rosy red scales. **Biology:** Caterpillar host associations unknown. Probably univoltine, adults January–June, peaking in March/April. **Habitat & distribution:** Succulent karoo and dry fynbos; Northern Cape and Western Cape.

1
1
2♀
3
4
5
6
7

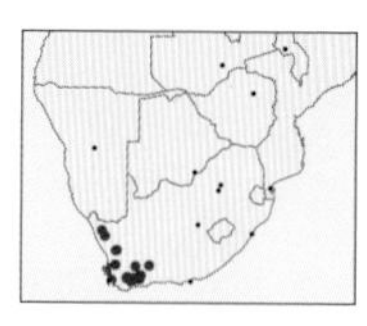

**1 *Derrioides cnephaeogramma*** Punctuated Elegant

Forewing 15–17mm. Several species in species group; punctuated postmedian line diagnostic, colour varies from deep red to olive green, variably sprinkled with black scales, distinct black cell spot. Caterpillar brick red with processes on 5th and last segments. Allopatric, but similar *D. rufaria* lacks punctuations on postmedian line; *D. hypenissa* lacks punctuations and has forewing distinctly falcate. **Biology:** Caterpillars reared on Karee (*Searsia lancea*). Probably univoltine, adults March–August, peaking in April. **Habitat & distribution:** Strandveld, succulent karoo and dry fynbos; Northern Cape and Western Cape, South Africa.

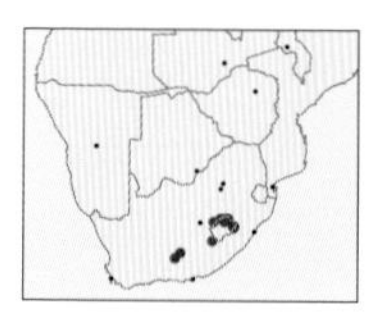

**2 *Derrioides hilaris*** Yellow Berg Elegant

Forewing 16–17mm. Wings held roof-like over body; body and wings bright yellow covered by bright pink dots and marks of varying density, some individuals almost completely pink; hindwings cream, variably shaded with grey. **Biology:** Nocturnal adults perfectly camouflaged when at rest by day between the numerous yellow flowers in habitat. Probably univoltine, adults December–March. **Habitat & distribution:** High-altitude Afroalpine vegetation; Eastern Cape to Lesotho.

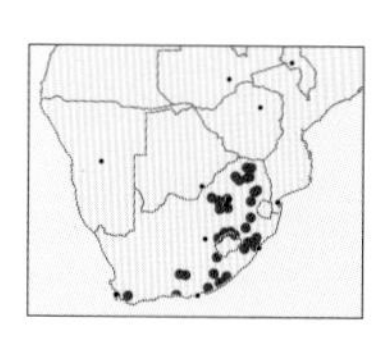

**3 *Aspilatopsis orthobates*** Pink Pastel Elegant

Forewing 15–16mm. Pastel pink to golden pink with straight brown and white postmedian line ending on apex of wing. Twenty-three very similar species described in genus in region, forewings in various hues of pastel colours; difficult to separate species based on wing patterns alone, best separated by careful examination of structures. **Biology:** Caterpillar host associations unknown. Adults August–April. **Habitat & distribution:** Fynbos and grassland; Western Cape to Limpopo, South Africa.

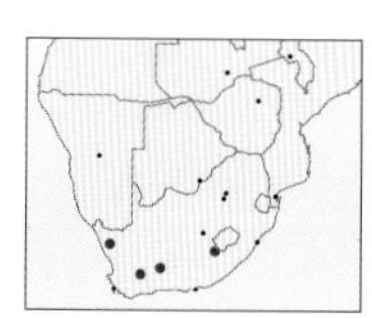

**4 *Axiodes smaragdaria*** Emerald Bold Elegant

Forewing 15–16mm. Bold, dark green median area between sharply pointed white antemedial and acutely angled postmedian lines, pinkish tint on costa and inner angle of both wings but not extensively pink, cilia uniform gold; hindwings light green with sharply curved line. Similar *A. smaragdarioides* has purple suffusion on both wings and chequered cilia. Sixty species in genus in region, fairly uniform in shape. **Biology:** Caterpillar host associations unknown. Probably univoltine, adults March–April. **Habitat & distribution:** High-altitude renosterveld along Great Escarpment; Northern Cape and Eastern Cape.

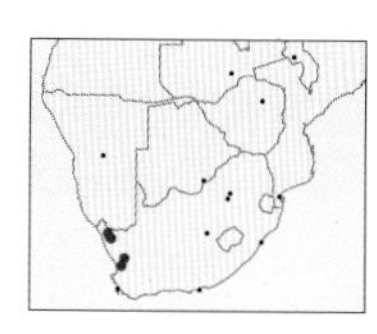

**5 *Axiodes rhodampyx*** Angled Bold Elegant

Forewing 14–15mm. Sexually dimorphic; male head and abdomen rosy, thorax and wings green, forewing median area much darker green, postmedian line with diagnostic sharp angle before costa; female as male but thorax, basal and terminal areas brick red. Similar *A. asteiochlora* has postmedian line evenly curved. **Biology:** Caterpillar host associations unknown. Probably univoltine, adults July–August. **Habitat & distribution:** Dry sandy fynbos and succulent karoo; Western Cape and Northern Cape, South Africa.

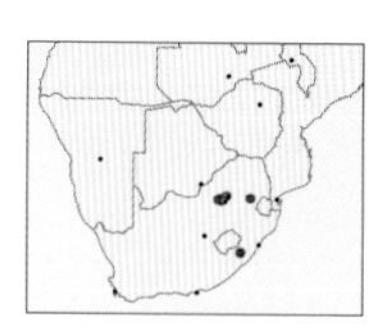

**6 *Axiodes commutata*** Large Winter Elegant

Forewing 18–19mm. Sexually dimorphic; male light grey, body densely setose, median area variably shaded with dark brown, basal and terminal areas lighter, antemedial line broken or absent, postmedian line undulating, hindwings light grey; female dark brown to copper, postmedian line straighter, antennae filiform. Similar *A. synclinia* has postmedian line straight and antemedial line sharply pointed outward. **Biology:** Caterpillars feed primarily on Besembos (*Thesium utile*), sometimes also herbaceous Asteraceae. Univoltine, a true winter moth, adults active during cold fronts in winter, May–August. **Habitat & distribution:** Highveld grassland; KwaZulu-Natal north to Gauteng, South Africa.

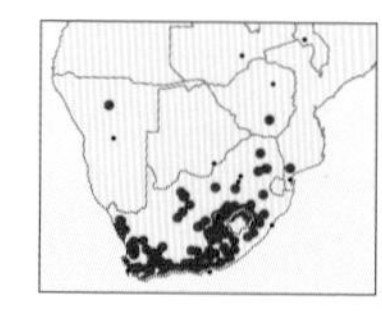

**7 *Axiodes dochmoleuca*** Variable Grey Elegant

Forewing 13–14mm. Very variable, body densely setose, wings grey, variably sprinkled with white, antemedial line sharply pointed outward, postmedian line slightly curved, lightly crenulated, basal and terminal area shaded with light brown in some individuals. Caterpillar light brown with dense rows of dark brown crenulated lines. Several very similar species that can only be reliably separated through careful examination of specimens, taxonomy still unresolved. **Biology:** Caterpillars feed on *Felicia filifolia* (Asteraceae). Adults throughout the year. **Habitat & distribution:** Primarily Karoo biomes but also in fynbos and grassland, primarily in more arid habitats; Western Cape, South Africa, north to Zimbabwe.

1
M
1
2
3
4
5♂
6♂
7
7

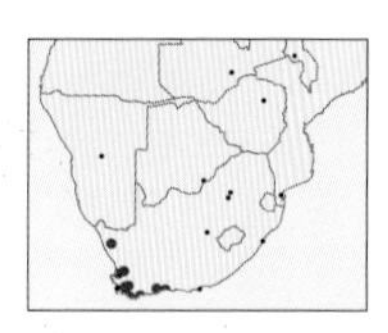

### 1 *Axiodes albilinea* — Feather Elegant

Forewing 14mm. Male with well-developed bipectinate antennae, body setose; forewings light grey with variable patches of brown shading, lines white, diagnostic triangular black patch from base on inner margin; hindwings brown, sometimes with darker patches. Similar *A. bipartita* lacks black triangular marking in basal area. **Biology:** Caterpillar host associations unknown. Adults June–September. **Habitat & distribution:** Fynbos and renosterveld; Western Cape and Northern Cape.

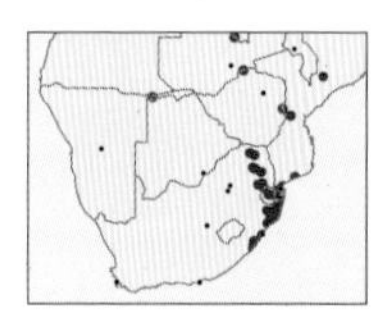

### 2 *Euexia percnopus* — Variegated Elegant

Forewing 15–19mm. Stout with strongly setose thorax; male has abdomen upturned at rest, displaying tuft of setae; colour very variable, ranging from grey to brown or reddish, sometimes green or permutations of these colours, forewing antemedial line diagnostically jagged, postmedian line straight, both lines crenulated in hindwing, median area sometimes darker; females larger with extreme falcate wing tip. One species in genus in region. **Biology:** Caterpillars recorded on Forest False Spikethorn (*Putterlickia verrucosa*) and Golden Dewdrop (*Duranta erecta*). Multivoltine, adults throughout the year. **Habitat & distribution:** Moist forest and bush; Eastern Cape, South Africa, north to Kenya.

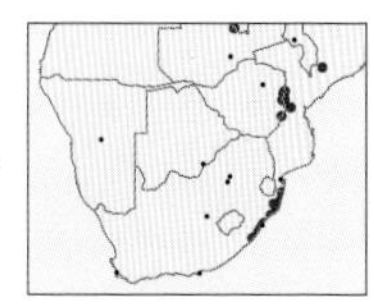

### 3 *Plegapteryx anomalus* — Swift Elegant

Forewing 20–24mm. Similar to *Euexia percnopus* (above), but green upper side, larger and wings elongate and 'swift-like' with distinct white-dotted black eyespot; underside of wings yellow. Caterpillar long, slender, light brown with single process on anal clasper. **Biology:** Caterpillars recorded on *Keetia gueinzii* (Rubiaceae). Adults throughout the year. **Habitat & distribution:** Moist forest and bush; Eastern Cape, South Africa, north to tropical Africa.

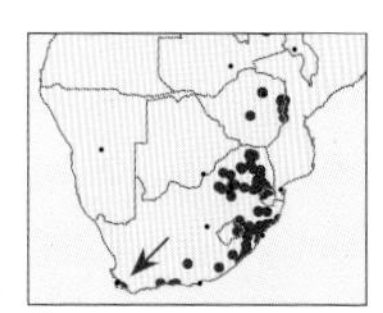

### 4 *Epigynopteryx maeviaria* — Oblique Barred

Forewing 13–18mm. Broadly triangular at rest; male forewings slightly hooded apically, light straw colour, with tiny central black dot and oblique brown bar, some brown dots behind bar; hindwings similar, but lighter; female yellow with postmedian band darker. **Biology:** Caterpillar host associations unknown. Adults throughout the year. **Habitat & distribution:** Moist fynbos and grassland; Western Cape, South Africa, north to tropical Africa.

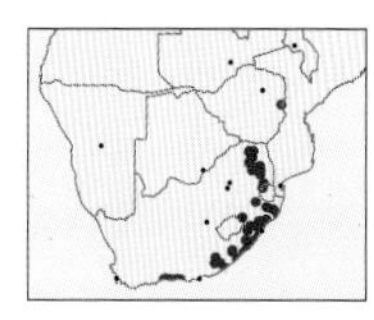

### 5 *Omizodes ocellata* — Ocellate Thorn

Forewing 19–22mm. Male with strongly pectinate antennae; thorax strongly setose, forewings slightly falcate, wing colour very variable, may be yellow, green or green-yellow, female red-brown, a few larger brown or black spots, often arranged in bands across wing, spots encircled with white/pale blue rings near apex of forewings, some small brown sprinkling of round dots (not striations). Caterpillar with black and yellow longitudinal lines, broad yellow lateral band diagnostic, head grey. Two species in genus in region. **Biology:** Caterpillars feed on Wild Pomegranate (*Burchellia bubalina*). Probably bivoltine, adults September–April. **Habitat & distribution:** Moist forests where host plant grows; Western Cape north to Limpopo, South Africa.

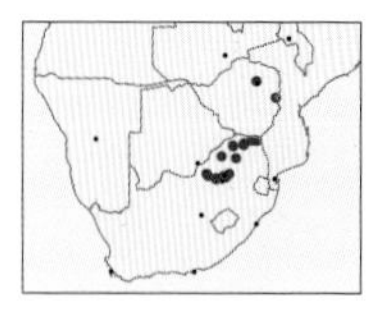

### 6 *Omizodes complanata* — Flattened Thorn

Forewing 17–20mm. Often confused with *O. ocellata* (above); red-brown to green, dry-season forms more plain than wet-season forms; wings less rounded, with 2 bands of white-ringed black spots sometimes visible, wings with diagnostic fine brown striations. Caterpillar purplish with fine yellow lines but lacks broad yellow lateral band. **Biology:** Caterpillars feed on Wild Gardenia (*Rothmannia capensis*). Bivoltine, adults October–February. **Habitat & distribution:** Kloofs and ridges where host plant grows; Gauteng, South Africa, north to Zimbabwe.

### 7 *Phoenicocampa terinata* — Flame Line

Forewing 13–18mm. Body and forewings cream, forewings with yellow costa, and finely dusted with brown scales, completely brown in some specimens, thin bright red lines, sometimes partially black; margins of forewings and hindwings bright yellow, all wings with a black cell spot; hindwings uniform yellow. Two species in genus in region. Similar *P. rubrifasciata* has forewings more pointed, yellow without brown dusting and thick red lines. **Biology:** Caterpillar host associations unknown. Adults September–December and April. **Habitat & distribution:** Coastal forest and lowland bush; KwaZulu-Natal, South Africa, north to southern Mozambique.

1
2
3
3♂
SB
4
5
5♂
6
6
7♂

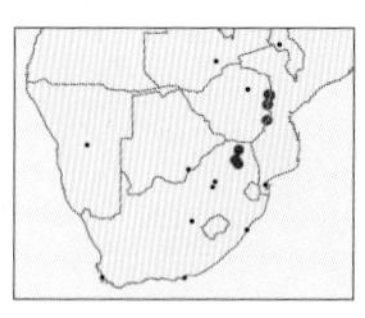

## 1 *Pitthea lacunata* — Southern Highflier

Forewing 16–19mm. Body black above, yellow below; antennae of males strongly pectinate; upper- and underside of wings similar; forewings black, crossed by 2 wide, bright yellow bands touching outer wing margins; hindwings similar, but with only 1 band. Two species in genus in region; *P. trifasciata* has bands white, shorter, not touching outer wing margins. **Biology:** Caterpillars recorded on *Keetia gueinzii* (Rubiaceae). Adults fly high in sunshine in forests, also attracted to light at night. **Habitat & distribution:** Afromontane forests; Limpopo, South Africa, and Eastern Highlands, Zimbabwe.

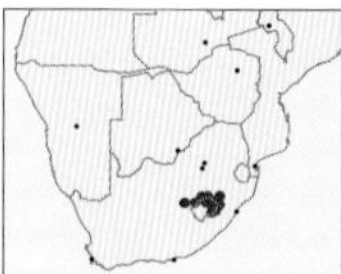

## 2 *Gnophos spinicosta* — Mountain Gnophos

Forewing 11–12mm. Wings held flat and extended laterally; body and wings light grey with dense brown and yellow stipples forming several indistinct bands across wings; indistinct black ring in centre of each wing and tiny black spots along distal margins. Five described species in genus in region. Similar allopatric *G. delagardei* from Cape provinces has same markings but is less well marked. **Biology:** Caterpillar host associations unknown. **Habitat & distribution:** High-altitude grasslands around Maluti Mountains; Free State, Lesotho and KwaZulu-Natal.

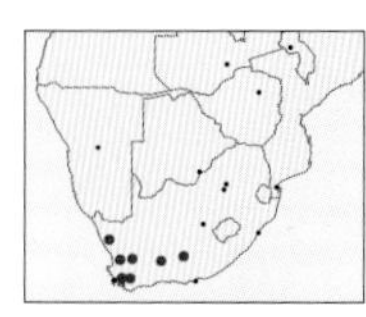

## 3 *Gnophos rubricimixta* — Red Gnophos

Forewing 10–13mm. Wings held flat and extended laterally; body and wings straw-coloured, heavily sprinkled with brick red and brown scales, lines indistinct, round cell spot on each wing. Several similar undescribed species. **Biology:** Caterpillar host associations unknown. **Habitat & distribution:** Renosterveld and karroid vegetation types; Northern Cape, Western Cape and Eastern Cape.

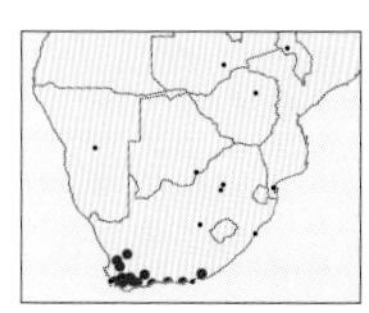

## 4 *Hypotephrina confertaria* — Renosterbos Camo

Forewing 12–13mm. Broadly oval-triangular at rest; thorax with 2 black spots, body and forewings densely mottled in brown, white, orange and black, postmedian line slightly kinked near inner margin, then curved and kinked inward before costa but not reaching costa. Caterpillar mottled brown, cream and black with broken white dorsal line, blending in perfectly with twigs of host plant. Nine similar species in genus in region, very well camouflaged by their disruptive patterns to blend into leaf litter. **Biology:** Caterpillars feed on Renosterbos (*Dicerothamnus rhinocerotis*). Multivoltine, adults throughout the year. **Habitat & distribution:** Renosterveld; Northern Cape, Western Cape and Eastern Cape, South Africa.

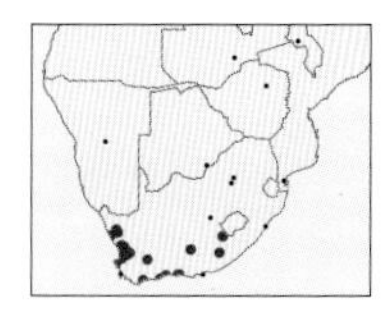

## 5 *Hypotephrina serrimargo* — Evencurve Camo

Forewing 12–14mm. Very similar to *H. confertaria* (above), distinguished by the postmedian line that is evenly curved and not kinked near costa. **Biology:** Caterpillar host associations unknown. Adults September–April. **Habitat & distribution:** Fynbos and renosterveld, as well as remnant patches of these vegetation types on mountains; Western Cape to eastern Free State, South Africa.

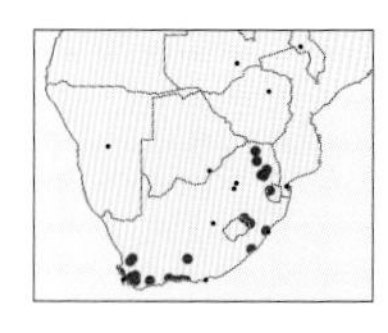

## 6 *Hypotephrina exmotaria* — Hourglass Camo

Forewing 12–14mm. Differs from other *Hypotephrina* spp. by the evenly curved antemedial and postmedian lines with orange median area resembling an 'hourglass', and by dark shading of basal and terminal areas. **Biology:** Caterpillars feed on *Anthanasia trifurcata* (Asteraceae). Adults September–May. **Habitat & distribution:** Fynbos and grasslands containing remnant patches of fynbos; Western Cape to Limpopo, South Africa.

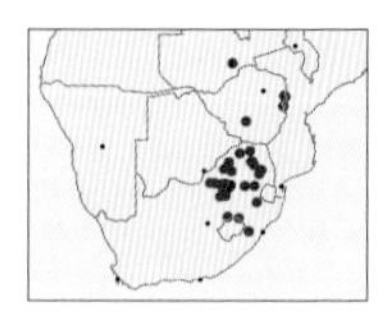

## 7 *Menophra jansei* — Janse's Menophra

Forewing 16–20mm. Ground colour cream, wings crossed by several horizontal black lines and heavily patterned with yellowish-brown stippling and patches, straight lines on both wings diagnostic, margins of both wings crenulate. Caterpillar slender, grey to brown, sometimes mottled with black. Five species in genus in region; similar *M. caeca* has ground colour light grey with uniform brown striations. **Biology:** Caterpillars polyphagous on trees and shrubs growing in shade. Multivoltine, adults throughout the year where conditions allow. **Habitat & distribution:** Forest and thick riverine bush, also in gardens with similar habitat; KwaZulu-Natal, South Africa, north to Zimbabwe.

## 8 *Menophra serrataria* — Serrated Menophra

Forewing 14–16mm. Similar in colouration to *M. jansei* (above), but with more strongly scalloped margins to both wings and postmedian line curved diverging from antemedial line; hindwing line crenulated and evenly curved; an extreme form has basal and terminal areas light brown to almost white. Similar *M. contemptaria* has hindwing line acutely angled. Caterpillar pale green-grey to brown. **Biology:** Caterpillars polyphagous on forest trees. Multivoltine, adults throughout the year. **Habitat & distribution:** Moist forests; Eastern Cape, South Africa, north to Zimbabwe.

1
2
3
4
4
5
6
7
7
8
8

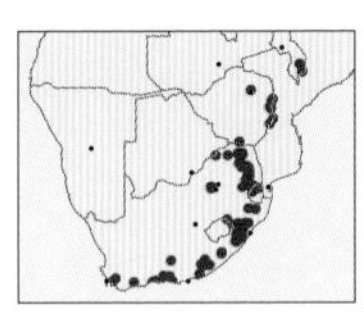

## 1 *Oedicentra albipennis* — Marbled White

Forewing 15–18mm. Forewings white to pink, heavily marbled with brown and black lines and patches, a dominant broad white or paler band across centre of wings bounded by wavy black lines; hindwings white with grey marginal bands. Caterpillar mottled dark grey to brown with swellings on dorsal surface, kinked in 2 places to imitate broken twig. One species in genus in region. **Biology:** Caterpillars polyphagous on forest trees, obligatory on host after feeding there for some time. Multivoltine, adults throughout the year. **Habitat & distribution:** Forest and thick riverine bush; Western Cape, South Africa, north to tropical Africa.

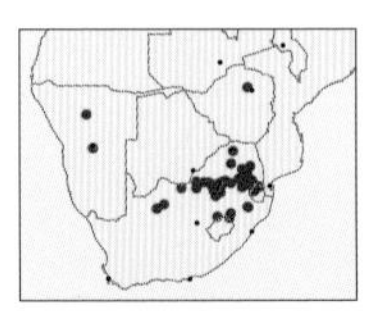

## 2 *Heterostegane rectistriga* — Straight-line Yellow Wave

Forewing 10–11mm. Wings held flat and extended laterally; ground colour creamy yellow, stippled with brown spots, both wings crossed by usually 4 straight transverse lines, outer one thickest, sometimes only this outer line distinct. Caterpillar pale green with lateral cream stripe and numerous distinctive dorsal and lateral ripples. Five species in genus in region. Similar *H. auranticollis* is smaller and has lines orange-brown and indistinct. **Biology:** Caterpillars feed on Elephant's Root (*Elephantorrhiza elephantina*). Bivoltine, adults September–March. **Habitat & distribution:** Grassland where host plant grows; KwaZulu-Natal, South Africa, north to Zimbabwe and Namibia.

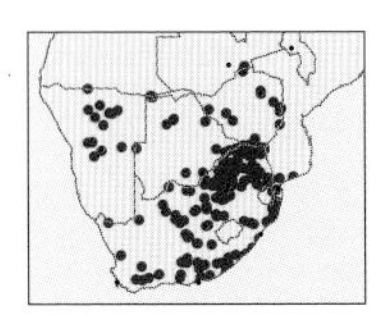

## 3 *Heterostegane aridata* — Cell Wave

Forewing 6–9mm. Wings held flat and widely extended; body pale orange-yellow; forewings pale orange-cream crossed by 2 or 3 wavy orange lines which are well developed in wet-season forms and almost absent in dry-season forms; sometimes small black cell spot, often median line extends to termen in 2 places creating terminal cells; hindwings similar. Caterpillar similar to *H. rectistriga* (above), but has white bands between segments. **Biology:** Caterpillars feed on thorn trees (*Vachellia* spp.). Multivoltine, adults August–May, peaking in summer. **Habitat & distribution:** Habitats where *Vachellia* trees grow, including gardens; Western Cape, South Africa, north to Kenya.

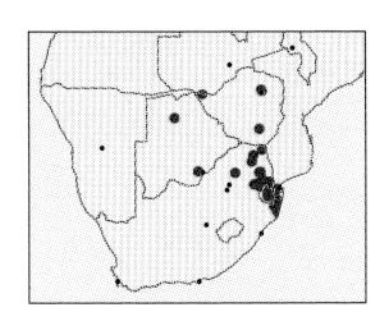

## 4 *Heterostegane bifasciata* — Red-spot Wave

Forewing 6–8mm. Ground colour cream, wings variably diffused with brown scales, brick red spots in terminal area, lines parallel, sharply crenulated; underside of wings distinctive orange and black. **Biology:** Caterpillar host associations unknown. Adults October–March. **Habitat & distribution:** Thick bush and forest; KwaZulu-Natal, South Africa, north to Zimbabwe and Botswana.

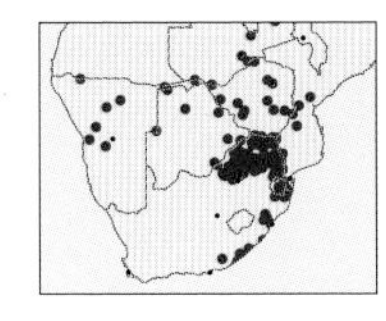

## 5 *Orbamia obliqua* — Four Spot Brindle

Forewings 8–10mm. Creamy grey with variable density of brown mottling and several distinct to poorly formed wavy black lines crossing wings; grey-centreed black-ringed cell spot on all wings, rings also visible as strong black dots on yellowish underside; diagnostic postmedian line is sharply bent back before costa. Similar *O. octomaculata* has postmedian line evenly curved before costa and is larger. Caterpillars slender, brown loopers. **Biology:** Caterpillars feed on bushwillows (*Combretum* spp.). **Habitat & distribution:** *Combretum* bushveld; Eastern Cape, South Africa, north to Zimbabwe.

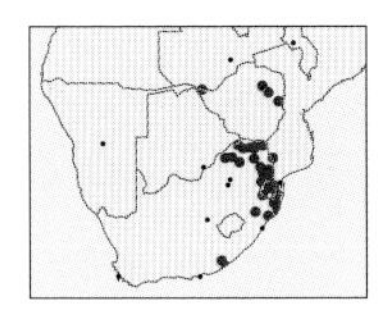

## 6 *Zamarada adiposata* — Fat Zamarada

Forewing 10–11mm. Body reddish-brown; both wings largely transparent with iridescent golden sheen and broad, irregular reddish-brown marginal band. Caterpillar bright green. Forty-nine species in genus in region, most species characterised by having largely transparent wings with dark margins. **Biology:** Caterpillars feed on Weeping Boer-bean (*Schotia brachypetala*). Multivoltine, adults throughout the year. **Habitat & distribution:** Subtropical bush and forest, Eastern Cape, South Africa, north to Mozambique and Zimbabwe.

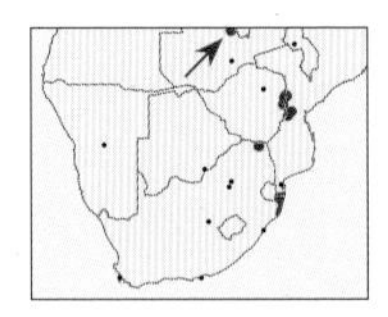

## 7 *Zamarada dentigera* — Toothed Zamarada

Forewing 11–13mm. Prominent round black cell spot in greenish transparent portion of each wing, marginal band wide and brown and marked with zigzag lines, marginal setae chequered light and dark brown. **Biology:** Caterpillar host associations unknown. Adults throughout the year. **Habitat & distribution:** Tropical bush and forest; KwaZulu-Natal, South Africa, north to tropical Africa.

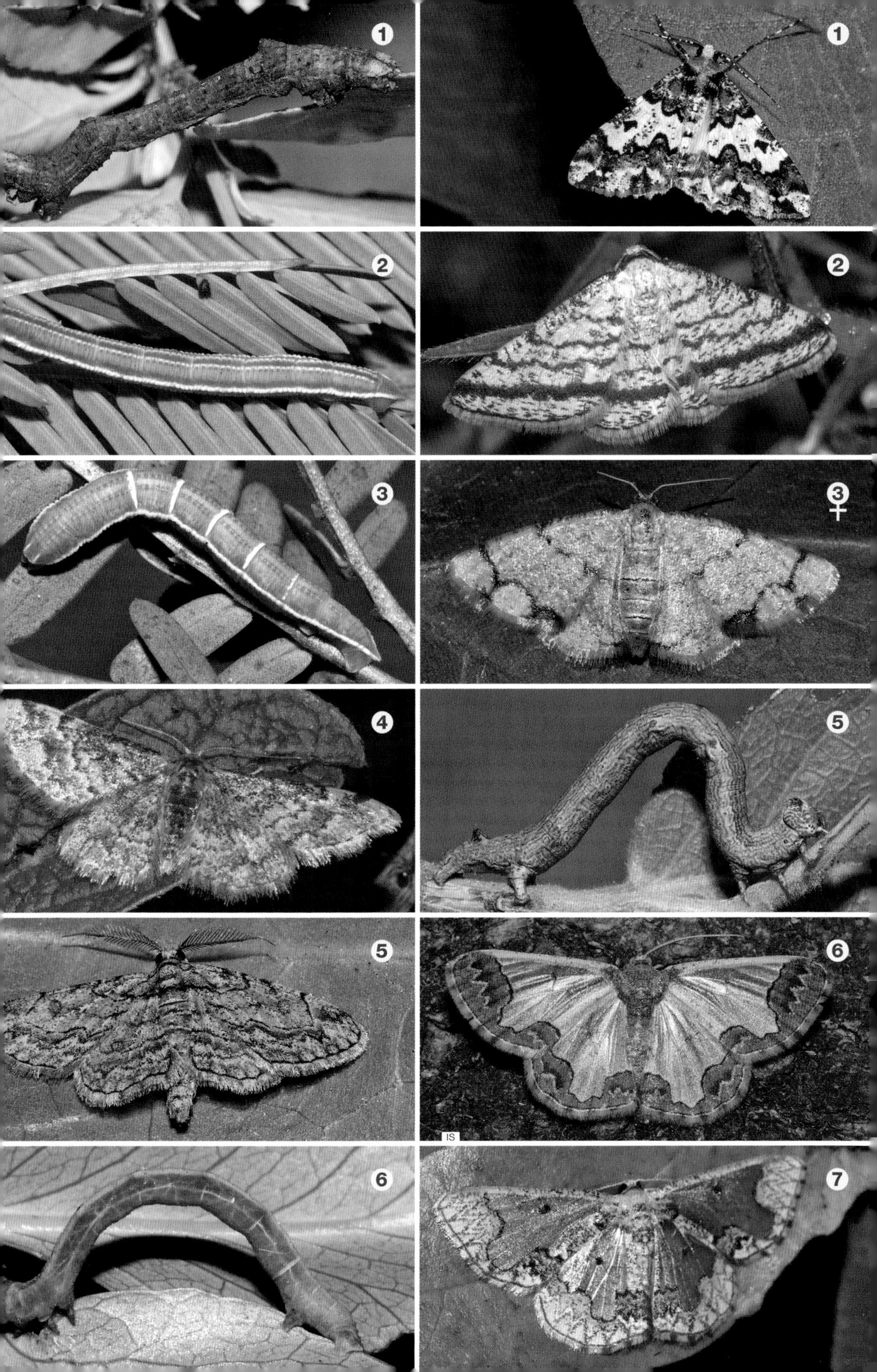
1
1
2
2
3
3♀
4
5
5
6
6
IS
7

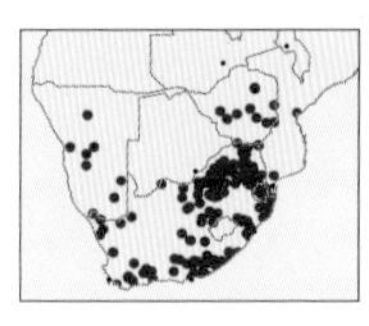

## 1 *Zamarada pulverosa* — Dusty Zamarada

Forewing 11–13mm. Pink-brown with a black-ringed eyespot on forewings; median area purplish, transparent, often heavily dusted in brown scales that can cover the whole wing; broad light brown marginal band often marked with darker brown blotches, postmedian line variably and unevenly crenulated, the broad marginal band on underside of wing uniform dark brown. Caterpillar green with white lateral line and bands. Very similar *Z. metallicata* much darker, almost black in some specimens and has postmedian line evenly toothed. **Biology:** Caterpillars feed on Sweet Thorn (*Vachellia karroo*). Multivoltine, adults throughout the year where conditions allow. **Habitat & distribution:** Habitats where *Vachellia* trees grow, including gardens; Western Cape, South Africa, north to Kenya.

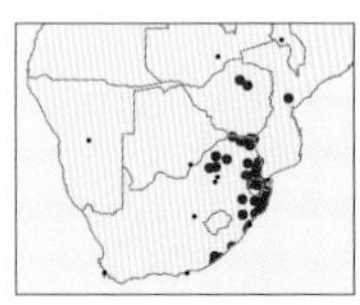

## 2 *Zamarada deceptrix* — Deceptive Zamarada

Forewing 9–10mm. Similar to *Z. pulverosa* (above), but lighter, and terminal area with cream patches in both wings; broad marginal band on underside of hindwing is diagnostic – the inner half cream, the outer half dark brown. Caterpillar also similar to that of *Z. pulverosa*, but has cream bands between segments and diagnostic black dorsal patch on head. **Biology:** Caterpillars feed on Robust Thorn (*Vachellia robusta*). Multivoltine, adults throughout the year where conditions allow. **Habitat & distribution:** Habitats where *Vachellia* trees grow; Eastern Cape, South Africa, north to Kenya.

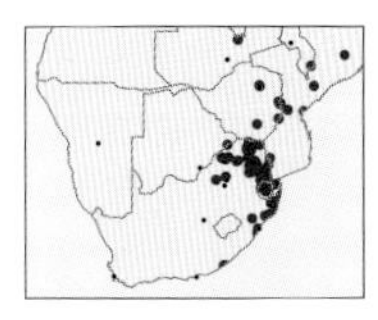

## 3 *Zamarada plana* — Flat Zamarada

Forewing 11–14mm. Body pink-brown; transparent section of wings golden yellow or yellow-green, opaque border brown, marked with short sections of dark brown and cream zigzag lines and black triangular marks. **Biology:** Caterpillars feed on Weeping Boer-bean (*Schotia brachypetala*). Bivoltine, adults August–May, peaking November and March. **Habitat & distribution:** Forest and bushveld where *Schotia* trees grow; Eastern Cape, South Africa, north to tropical Africa.

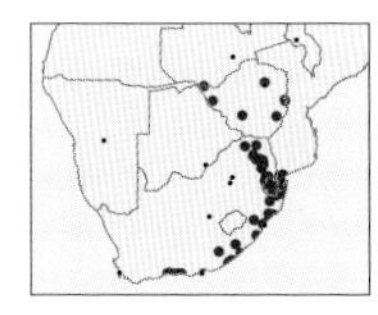

## 4 *Zamarada transvisaria* — Ochna Zamarada

Forewing 11–14mm. Body brown sometimes with yellow dorsal spots on abdomen; wings variable in colour, transparent median section metallic green, often suffused with brown striations; a small black cell spot; broad, wavy, distal opaque band usually reddish-brown. Caterpillar elongate, green-brown to grey sometimes with white lateral spots blending in well with twigs of host plant. **Biology:** Caterpillars feed on forest *Ochna* spp. Multivoltine, adults throughout the year. **Habitat & distribution:** Moist forests; Western Cape, South Africa, north to tropical Africa.

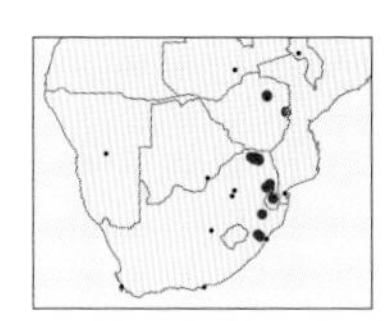

## 5 *Zamarada ordinaria* — Lichen Zamarada

Forewing 11–12mm. Body and wings mottled in various shades of white, light brown and black, perfectly blending in with the lichen-covered bark of its host plant; postmedian line evenly curved with no prominent outward projection as in other similar *Zamarada* spp. in region. **Biology:** Caterpillars feed on Paperbark Thorn (*Vachellia sieberiana*). Multivoltine adults throughout the year. **Habitat & distribution:** Grassland and bushveld where host plant grows; KwaZulu-Natal, South Africa, north to Zimbabwe.

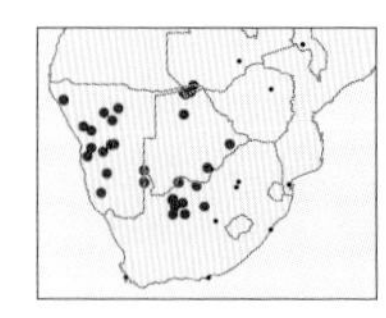

## 6 *Zamarada ascaphes* — Sandveld Zamarada

Forewing 9–11mm. Wings narrower in this group than in most *Zamarada* spp.; body and wings light grey finely striated with thin, black lines with a sprinkling of silvery scales in fresh specimens, oblique postmedian line straight and not curved toward costa, terminal area with partial black shading. Similar *Z. differens* has postmedian line curved toward costa and is browner; *Z. jansei* has terminal band completely dark, almost black. **Biology:** Caterpillar host associations unknown. Adults August–April. **Habitat & distribution:** Kalahari sandveld and desert; Northern Cape to Namibia.

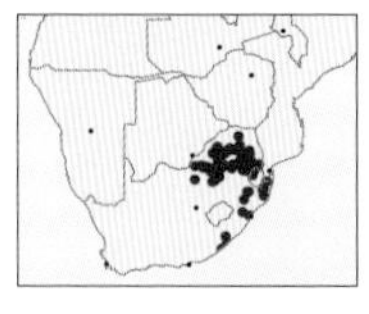

## 7 *Ligdia batesii* — Bates Carpet

Forewing 10–11mm. Very distinctive; black thorax, white abdomen; wings white to cream with brown basal section extending along front margin, small dot behind this and wide straight subterminal brown band and marginal patches. Caterpillar a slender, green looper with white lateral band and crossbars. Three species in genus in region. **Biology:** Caterpillars feed on Hook Thorn (*Senegalia caffra*). **Habitat & distribution:** Bushveld and grassland where host tree grows; Eastern Cape north to Limpopo, South Africa.

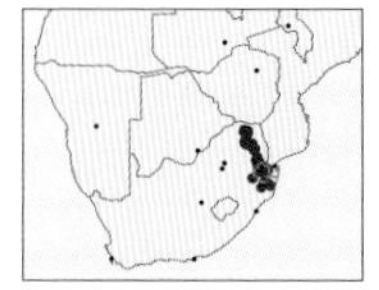

## 8 *Ligdia pectinicornis* — Peppered Carpet

Forewing 10–11mm. Differs from *L. batesii* (above) with subterminal band being broken, with yellow marks and further away from margins, terminal area and partially median area peppered with brown. Similar *L. interrupta* has postmedian band curved. **Biology:** Caterpillar host associations unknown. **Habitat & distribution:** Forests; KwaZulu-Natal north to Limpopo, South Africa.

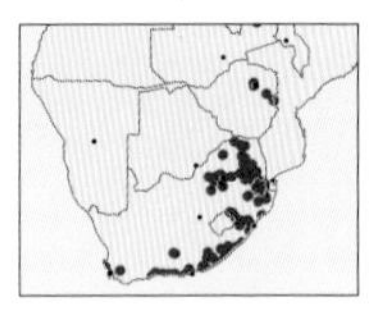

### 1 *Obolcola petronaria* — Lined Carpet

Forewing 9–12mm. Broadly triangular at rest; wings vary from cream to light brown, circular eyespot; antemedial, median and postmedian lines straight, thin, orange dotted with black; hindwings similar but lighter. Caterpillar green with variable red, white and black spots and small dorsal processes. Nine species in genus in region. Similar *O. aliena* has solid black lines and medial area often shaded; *O. pulverea* is dark grey with lines indistinct; *O. pallida* is white with broad black lines. **Biology:** Caterpillars feed on Transvaal Kooboo Berry (*Mystroxylon aethiopicum*). Multivoltine, adults throughout the year. **Habitat & distribution:** Forest and bushveld where host plant grows; Western Cape, South Africa, north to Kenya.

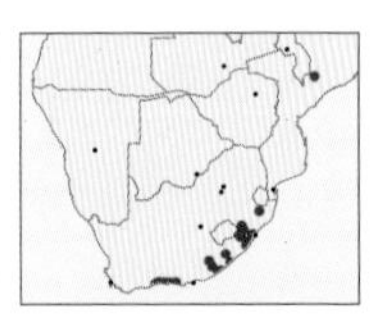

### 2 *Obolcola decisa* — Decisive Carpet

Forewing 9–11mm. Broadly triangular at rest; similar to *O. petronaria* group (above), but postmedian line crenulated and distinctively curved outward, conspicuous black marks in terminal and basal areas. **Biology:** Caterpillar host associations unknown. Adults throughout the year. **Habitat & distribution:** Moist forests; Western Cape, South Africa, north to Mozambique.

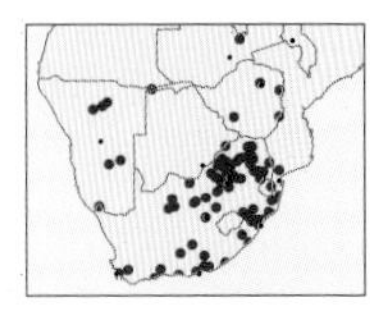

### 3 *Nassinia caffraria* — Threaded Looper

Forewing 14–18mm. Rests with elongate wings held outstretched diagonally away from body; head and thorax white with orange and black spots; forewings cream with 2 rows of orange dots, each with black centre; hindwings paler cream with fewer and smaller spots. Three species in genus in region. **Biology:** Caterpillar host associations unknown. Adults often active by day but also attracted to light. **Habitat & distribution:** Variety of habitats; Western Cape, South Africa, north to tropical Africa.

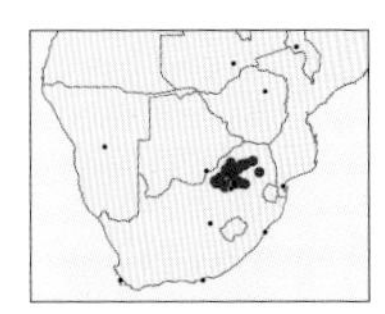

### 4 *Nassinia pretoria* — Orange-threaded Looper

Forewing 15–18mm. Similar to *N. caffraria* (above); forewings darker almost pinkish and less rounded; hindwings orange and dots on abdomen prominently ringed with orange. Caterpillar similar to *Aphilopota* caterpillars (p.194), swollen thorax, dark brown with numerous tiny white marks, yellow and black lateral spots. Similar *N. aurantiaca* is smaller, has all wings orange and is restricted to Maputaland. **Biology:** Caterpillars feed on Bell Spikethorn (*Gymnosporia tenuispina*). Bivoltine, adults diurnal but also attracted to light, September–April. **Habitat & distribution:** Bushveld on rocky ridges where host plant grows; Gauteng, Mpumalanga and Limpopo, South Africa.

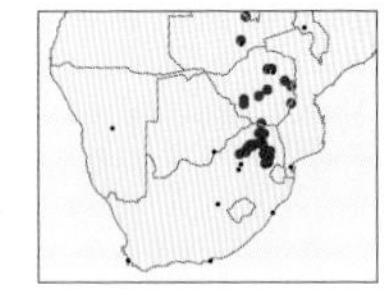

### 5 *Rhodophthitus commaculata* — Pied Magpie

Forewing 18–23mm. Black-spotted white thorax and orange abdomen; forewings white, marked with broad, dark brown to black swathes and white areas marked with black dots, terminal white area diagnostic; hindwings creamy white with scattered brown to black spots. Three species in genus in region; similar *R. atacta* has forewings almost completely covered in black, with no white in terminal area, and hindwings same white as forewings with 2 pectinated grey bands. Could also be confused with the similar unrelated *Rhypopteryx rhodalipha* (Lymantriinae, p.398). **Biology:** Caterpillars reported feeding on *Lantana* spp. Adults October–April. **Habitat & distribution:** Bushveld and forest; Mpumalanga, South Africa, north to Kenya.

### 6 *Oaracta auricincta* — Broad Concentric

Forewing 10–11mm. Orange body and wings, 3 shiny dark brown dorsal projections on abdomen; wings orange with variable red-brown striations, prominent black cell spot, antemedial and postmedian lines with rose gold sheen, diagnostic postmedian line evenly curved. Caterpillar tapering toward head, small dorsal projections on 5th segment, mottled green and brown. Four species in genus in region. **Biology:** Caterpillars feed on sourberries (*Dovyalis* spp.), Salicaceae. Bivoltine, adults October–May. **Habitat & distribution:** Moist forests; Western Cape north to Limpopo, South Africa.

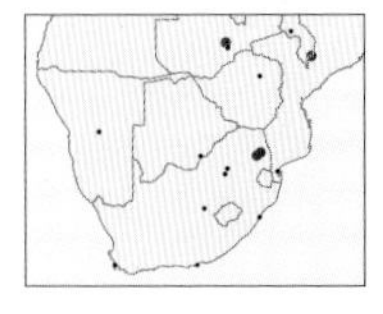

### 7 *Oaracta maculata* — Narrow Concentric

Forewing 10–11mm. Red to orange body and wings, 4 shiny dark brown dorsal projections on abdomen; wings with fingerprint-like concentric patterning of silver to brown scales, large dark brown patch on forewings in some specimens, antemedial and postmedian lines with silver sheen, diagnostic postmedian line pointed and curved back to costa. Caterpillar elongate, small dorsal projections on 5th segment, mottled grey with black or white spots. **Biology:** Caterpillars feed on Snuff-box Tree (*Oncoba spinosa*) and Governor's Plum (*Flacourtia indica*), both Salicaceae. Multivoltine, adults throughout the year. **Habitat & distribution:** Lowland tropical forests; Limpopo, South Africa, north to Kenya, also Madagascar and Indian Ocean islands.

1
1
2♂
3
4
4
5
6
6
7
IS
7

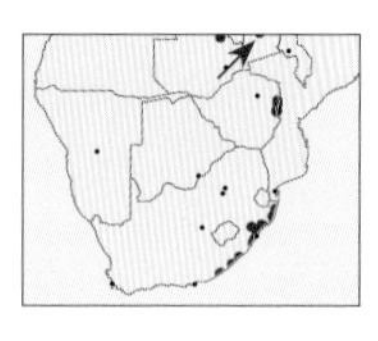

### 1 *Ochroplutodes haturata* — Heteromorph

Forewing 14–16mm. Rests with wings extended flat; forewings slightly falcate, diagnostic white frons and antennae; adults extremely polymorphic; one form is canary yellow with irregular sprinkling of reddish-brown spots and marks; another light brown, heavily irrorated with dark brown scales and 2 distinct yellow rectangular patches on forewings; another is dark olive green with distinctive yellow and black postmedian line; yet another is cream with longitudinal brown striations. Species is best identified by shape of wings and white frons and antennae. One species in genus in region. **Biology:** Caterpillar host associations unknown. Adults throughout the year. **Habitat & distribution:** Tropical forest and coastal bush; Eastern Cape, South Africa, north to Zimbabwe, also Uganda.

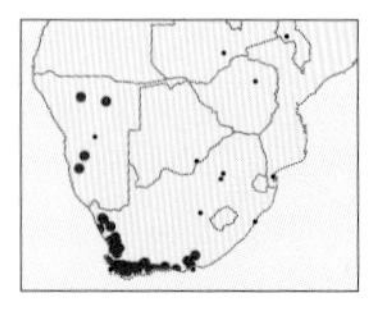

### 2 *Proutiana ferrorubrata* — Iron-red Cape Moth

Forewing 9–12mm. Rusty iron brown to grey, variably sprinkled with brown dots; diagnostic forewing antemedial and postmedian lines straight across wing, area beyond postmedian line darker, line of tiny black terminal dots; hindwings grey-brown, paler toward base. Caterpillar a fairly stout, pale green looper. Two species in genus. **Biology:** Caterpillars feed on Wild Clove Bush (*Montinia caryophyllacea*). Multivoltine when conditions allow, adults throughout the year. **Habitat & distribution:** Dry fynbos, renosterveld and Karoo where host plant grows; Eastern Cape to Namibia.

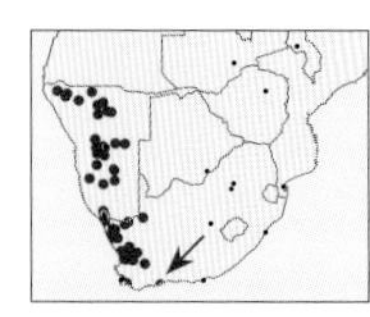

### 3 *Proutiana perconspersa* — Crenulated Cape Moth

Forewing 9–12mm. Differs from *P. ferrorubrata* (above) by the jagged curved lines on forewings. Caterpillar a fairly stout, mottled brown looper. **Biology:** Caterpillars feed on Wild Clove Bush (*Montinia caryophyllacea*). Adults September–March. **Habitat & distribution:** Dry fynbos, renosterveld and karoo where host plant grows; Western Cape to Namibia.

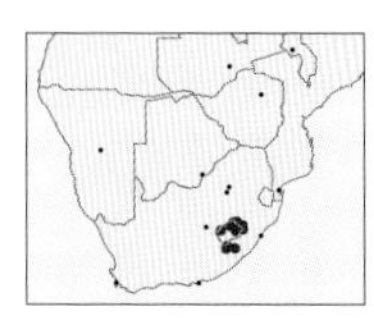

### 4 *Biclavigera uloprora* — Yellow Pennant

Forewing 14mm. Male wings very elongate and held diagonally away from abdomen; body and wings cream, variably densely overlaid with black and yellow-orange mottling, formed into 2 main bands crossing wings; cilia chequered black and yellow. Five described species in genus in region; females in this genus (where known) are wingless. **Biology:** Caterpillar host associations unknown. Adult males fly actively by day but are also attracted to light, January–February. **Habitat & distribution:** High-altitude Afroalpine vegetation; Eastern Cape and Lesotho.

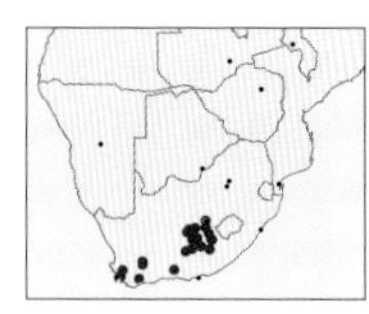

### 5 *Biclavigera praecanaria* — Copper Pennant

Forewing 13mm. Body and wings brownish-grey with veins copper shaded, diagnostic postmedian line evenly curved to costa with outward tooth-like marks and white shading in between, especially in specimens from northern populations; cilia chequered black and cream. **Biology:** Caterpillar host associations unknown. Adults February–May. **Habitat & distribution:** Karoo and grassy karoo habitats; Western Cape to southern Free State.

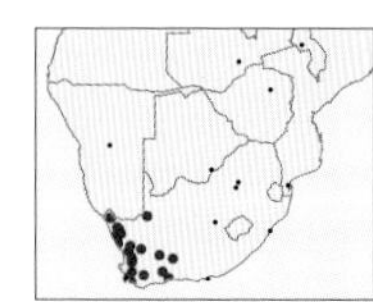

### 6 *Biclavigera deterior* — Sand Pennant

Forewing 12mm. Body and wings of male sandy cream to grey; forewings densely peppered with brown, lines straight when present, often a faint white submarginal line; hindwings plain brown except for inner edge which has same colouration as forewings, diagnostic cilia uniform brown, examples from wetter areas more boldly marked; female wingless. Caterpillar white with numerous brown markings on dorsal and ventral surface, legs red, and red dorsal patches with yellow lateral dots. Similar *B. fontis* has lines when present slightly curved and chequered cilia. **Biology:** Caterpillars feed on Rooibos (*Aspalathus linearis*). Adults March–May. **Habitat & distribution:** Dry fynbos, renosterveld and karoo habitats; Western Cape and Northern Cape.

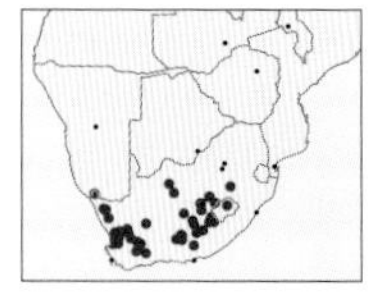

### 7 *Hemixesma anthocrenias* — Peppered Cream

Forewing 11–12mm. Broadly triangular at rest; male body and forewings cream with brown stippling, a distinct dark brown postmedian band across forewings initially curves outward then straightens toward leading margin of wings; female dark brown with cream postmedian band. One described species in genus. **Biology:** Caterpillar host associations unknown. Adults August–April. **Habitat & distribution:** Karoo and arid grassland with karroid vegetation; Western Cape to Namibia.

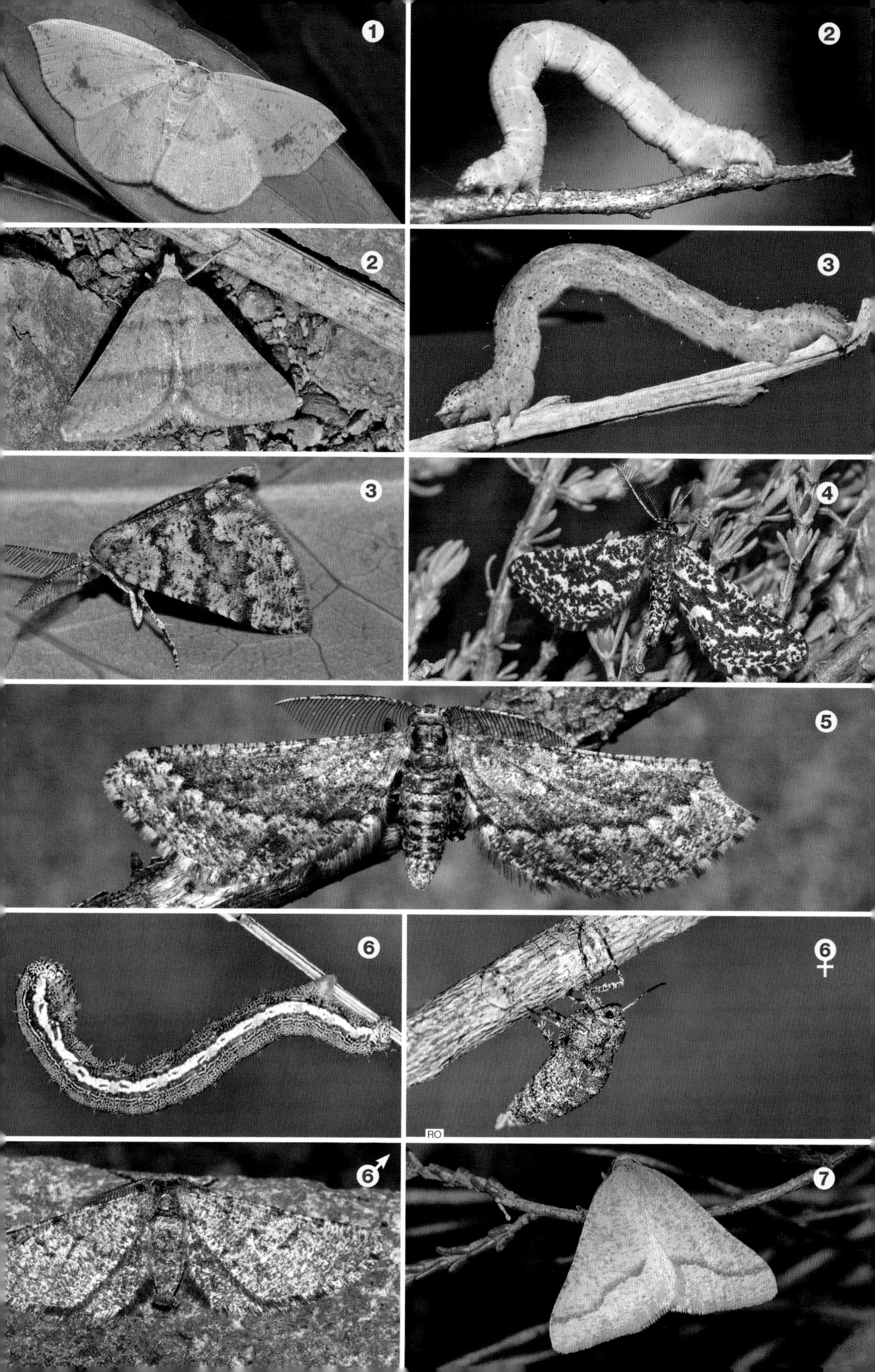
1
2
2
3
3
4
5
6
6 ♀
RO
6 ♂
7

## Subfamily Geometrinae Emeralds

Over 200 species described in this subfamily for the region. Most of the species are nocturnal; their green or lichen colouration allows them to blend into the foliage by day, where they are not readily disturbed. Caterpillars can be identified by having a bifid (deeply incised) head capsule, and are found singly. Geometrinae are cryptic species blending in with their surroundings (with only one exception in the region, *Petovia marginata*, which is aposematic and diurnal). Many are host specialists, cryptically adapted to their host plant, but some groups appear to be more polyphagous. Widespread across most habitats.

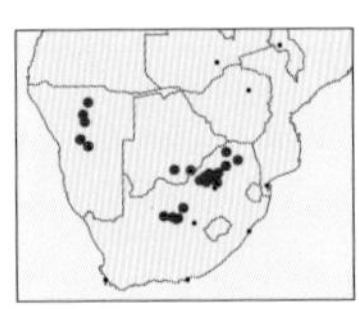

### 1 *Acidaliastis curvilinea* — Straw Emerald

Forewing 7–9mm. Wings held roof-like over body; 2 distinct forms, one straw-coloured, the other green; lines cream, oblique merging near apex of wing, only 1 line visible in some individuals; straw form has rest of forewing lightly dusted with reddish scales, green form densely so with green scales; hindwings white, unmarked in both forms. Three species in genus in region; *A. bicurvifera* has the lines parallel not merging near apex; *A. prophanes* has crenulated lines. **Biology:** Caterpillar host associations unknown. Adults September–March. **Habitat & distribution:** Arid bushveld; Northern Cape, South Africa, north to Namibia.

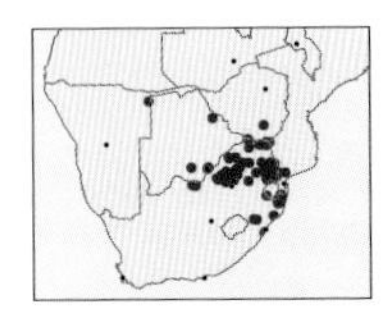

### 2 *Allochrostes biornata* — Ornate Emerald

Forewing 7–10mm. Wings held flat and outstretched; thorax and centre of wings green, abdomen red and white; both wings with wide bright to dull red chequered margin enclosing row of white shading of varying intensity. Marginal band rarely white, never red. Five species in genus in region. **Biology:** Caterpillar host associations unknown. Adults September–May. **Habitat & distribution:** Bushveld; KwaZulu-Natal, South Africa, north to Kenya.

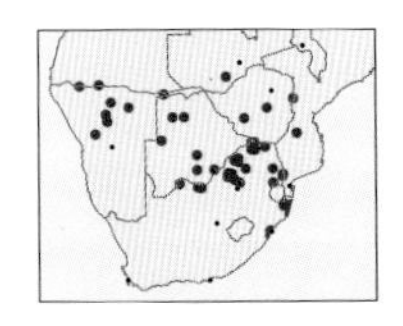

### 3 *Allochrostes impunctata* — Unspotted Emerald

Forewing 8–11mm. Wings held flat and outstretched; thorax green, abdomen green or white; forewings light to medium green with white costa, faint white lines and faint white cell spot when present, a red- and white-spotted band along both wing margins. Similar *A. saliata* has distinct red cell spots on all wings. **Biology:** Caterpillar host associations unknown. **Habitat & distribution:** Bushveld; KwaZulu-Natal, South Africa, north to tropical Africa.

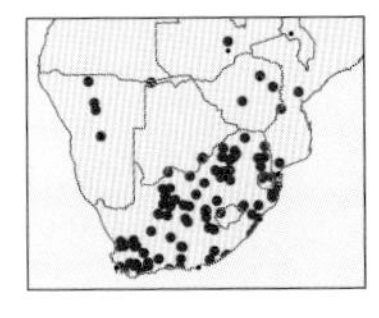

### 4 *Microloxia ruficornis* — Flower Emerald

Forewing ♂ 8–10mm, ♀ 12–14mm. Usually broadly triangular at rest; body green; forewings green, finely dusted with white and with yellowish margin, pale lines slightly curved; hindwings with single line; frons with reddish margins. Caterpillar light green with paler lateral markings. **Biology:** Caterpillars feed primarily on daisy flowers (Asteraceae), rarely on other flowers. Multivoltine, adults throughout the year. **Habitat & distribution:** Habitats where daisies grow, including severely disturbed places; across Africa, also Madagascar and Middle East.

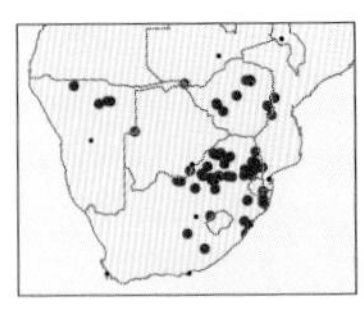

### 5 *Mixocera albistrigata* — Cream-lined Emerald

Forewing 8–10mm. Head and antennae pale orange, body and wings green to pale blue, sometimes olive, white dorsal line down centre of thorax and abdomen; forewings with white margin, postmedian line broad, slightly curved across both wings, antemedial line thinner on forewing but present, even if faint. Five species in genus in region; *M. frustratoria* has lines straight and lacks antemedial line; *M. xanthostephana* has only a straight postmedian line, only on forewings. **Biology:** Caterpillar host associations unknown. Adults October–April. **Habitat & distribution:** Bushveld and grassland; Eastern Cape, South Africa, north to tropical Africa.

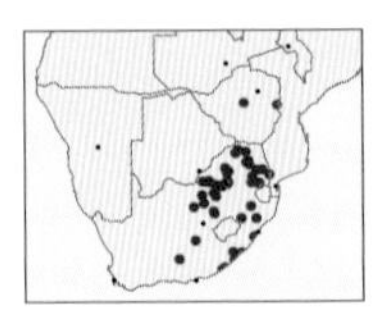

### 6 *Syndromodes cellulata* — White-edged Emerald

Forewing 10–11mm. Broadly triangular at rest; head and antennae white, body and wings pale green; forewings with thin white line along costa, small white cell spot, crossed by single oblique white postmedian line, cilia green and white. Five species in genus in region. **Biology:** Caterpillars feed on Sweet Thorn (*Vachellia karroo*). Bivoltine, adults September–May. **Habitat & distribution:** Grassland and bushveld where host tree grows; Eastern Cape, South Africa, north to Zimbabwe.

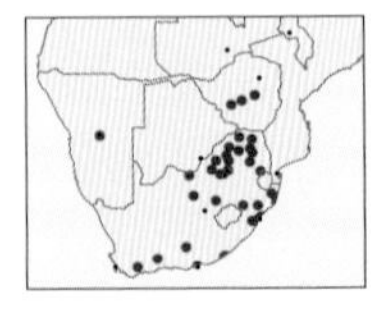

### 7 *Syndromodes invenusta* — Dotted Line Emerald

Forewing 9–10mm. Similar to *S. cellulata* (above), but postmedian line is dotted with white spots and cilia white from base. Similar *S. prasinops* has no markings on wings. Caterpillar a slender green looper with extreme bifid head resembling new shoot of host plant. **Biology:** Caterpillars feed on Sweet Thorn (*Vachellia karroo*). Bivoltine, adults September–March. **Habitat & distribution:** Where host tree grows; Western Cape, South Africa, north to tropical Africa.

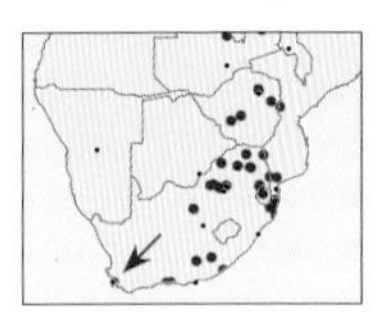

**1 *Omphacodes punctilineata*** Dotted Margin Emerald

Forewing 10–13mm. Head and antennae white, white dorsal spots on abdomen; wings green with white dotted lines, row of distinct white marginal dots at end of veins, costa white, cilia chequered, dark and light. Caterpillar green with serrated dorsolateral ribbing. Three species in genus in region; similar *O. delicata* lacks the white marginal dots; *O. vivida* has plain green cilia and no marginal line. **Biology:** Caterpillars feed on thorn trees (*Vachellia* spp.). Multivoltine where conditions allow, adults throughout the year. **Habitat & distribution:** Various habitats where *Vachellia* trees grow; Western Cape, South Africa, north to Zimbabwe.

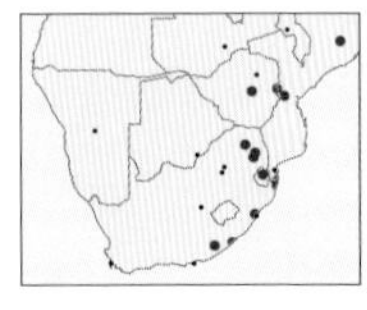

**2 *Rhodesia viridalbata*** Frosted Emerald

Forewing 10–13mm. Rests with wings held flat and outstretched; thorax green, abdomen pale; both wings green with fine, jagged white lines and distinctive large white blotches; forewings with speckled white leading margin. Three species in genus in region; *R. alboviridata* has white blotches much reduced; *R. depompata* lacks white blotches. **Biology:** Caterpillars feed on Simple-spined Num-num (*Carissa edulis*) and Mountain Ebony (*Bauhinia variegata*). Bivoltine, adults September–December and March–May. **Habitat & distribution:** Moist forests; Eastern Cape, South Africa, north to tropical Africa, also Madagascar and Comoros.

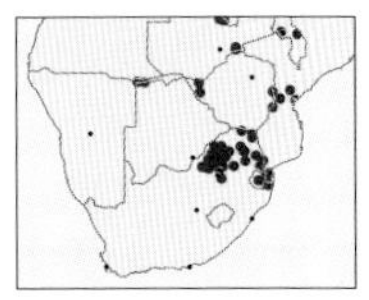

**3 *Antharmostes papilio*** Butterfly Emerald

Forewing 18–20mm. Thorax green, abdomen brown; outstretched wings green with slightly falcate apex and broad brown irregular posterior band, closely resembling withered leaves; hindwings with short, pointed tail. Caterpillar a uniform light green or brown looper with short 'tail' and brown bifid head. Two species in genus in region. **Biology:** Caterpillars feed on various bushwillows (*Combretum* spp.). Bivoltine, adults August–May. **Habitat & distribution:** *Combretum* bushveld; KwaZulu-Natal north to tropical Africa.

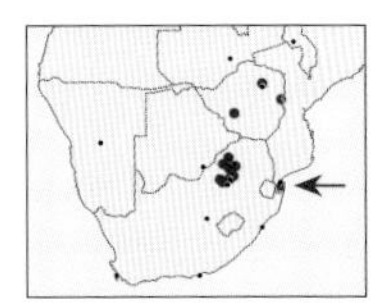

**4 *Paragathia albimarginata*** Paved Emerald

Forewing 11–14mm. Unmistakable; thorax green and abdomen black; wings vivid green, edged and crossed by curved and white-edged black bands; hindwings with short tail. One species in genus in region. **Biology:** Caterpillar host associations unknown. **Habitat & distribution:** Bushveld and dry forest; Gauteng, South Africa, north to tropical Africa.

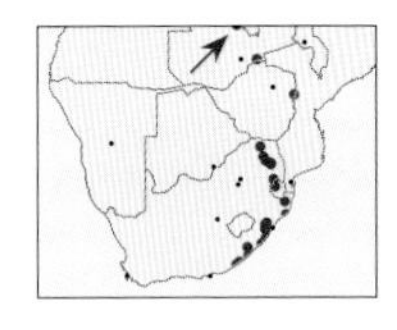

**5 *Lophostola atridisca*** Dark Dotted Emerald

Forewing 12–15mm. Thorax green, abdomen brown; wings dull green crossed by fine broken white lines, each with white-ringed brown cell spot. Caterpillar slender, green, sometimes with ventral red spots. One described species in genus in region. **Biology:** Caterpillars feed on both *Senegalia* and *Vachellia* thorn trees. Multivoltine, adults throughout the year. **Habitat & distribution:** Forest and moist bush; Eastern Cape, South Africa, north to Kenya.

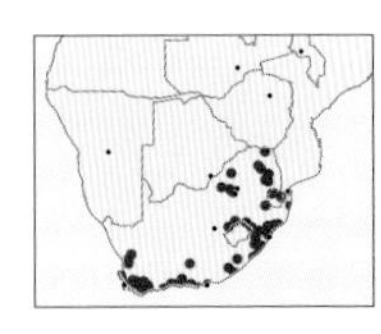

**6 *Adicocrita koranata*** Circlet Kora

Forewing 16–17mm. Thorax green, antennae white, triangular section of brown or white abdomen visible at rest; wings green, crossed by sparse lines of small black dots, and with red marginal line, white costa and cilia; diagnostic margin of hindwings partially crenulate. Caterpillar green to pink with finely flattened setae. **Biology:** Caterpillars recorded on White Pear (*Apodytes dimidiata*), and Glossy Currant (*Searsia lucida*). Multivoltine, adults August–April. **Habitat & distribution:** Forest, kloof and riverine bush; Western Cape north to Limpopo, South Africa also reported from Zimbabwe.

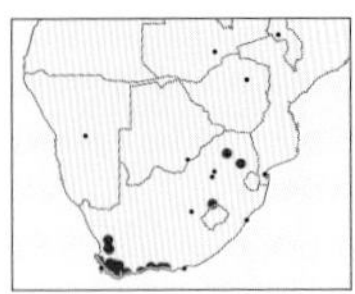

**7 *Adicocrita discerpta*** Framed Kora

Forewing 14–16mm. At rest, wings held flat and outstretched; wings framed by orange and purplish lines, which fade in dried specimens, distinctive red cell spot; hindwings green with brown margin and red cell spot. Caterpillar green with variable line of dorsal red markings. **Biology:** Caterpillars recorded on Bietou (*Osteospermum moniliferum*) and *Penaea acutifolia* (Penaeaceae). Multivoltine, adults throughout the year. **Habitat & distribution:** Moist fynbos in fynbos region and on high mountains where remnant fynbos vegetation persists; Western Cape north to Limpopo, South Africa.

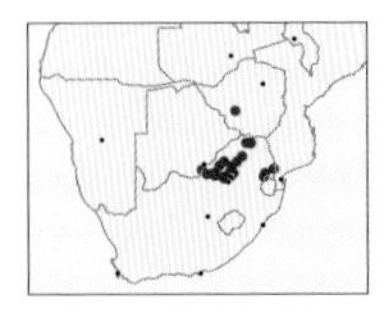

**8 *Centrochria deprensa*** Depressed Kora

Forewing 11–14mm. At rest, wings held flat and outstretched; wings light green, framed by orange and purplish margin with variable silver marginal blotches, red-lined silver cell spot on all wings, faint postmedian line. Caterpillar green with dark dorsal line near anal claspers. Two species in genus in region. **Biology:** Caterpillars feed on Mountain Karee (*Searsia leptodictya*). Multivoltine, adults throughout the year. **Habitat & distribution:** Bushveld where host tree grows; Gauteng, South Africa, north to Zimbabwe.

1
1
2♂
3
3
4
5
5♂
6
6
7
7
8
8

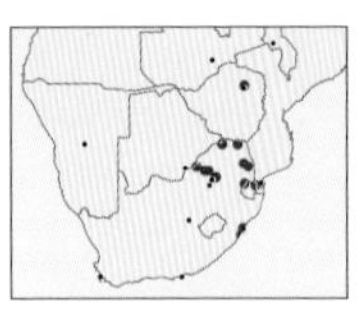

**1 *Celidomphax rubrimaculata*** Blotched Emerald

Forewing about 15mm. Thorax yellow, abdomen banded with brown; wings in females light brown, green in males, both wings marked with small brown cell dots and with 2 large brown blotches on margin, resembling diseased leaf. Caterpillar light green or brown with darker blotches, sometimes red dorsally. **Biology:** Caterpillars feed on bushwillows (*Combretum* spp.). **Habitat & distribution:** Bushveld; KwaZulu-Natal north to tropical Africa.

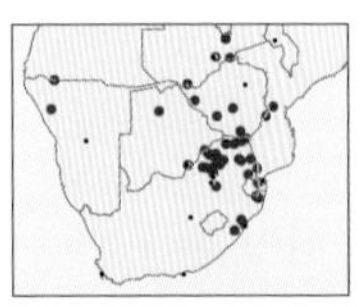

**2 *Celidomphax quadrimacula*** Spot Emerald

Forewing 15–16mm. Thorax green, abdomen red-brown, anterior segments visible at rest; wings light green with yellow margins and yellow veins, 2 very faint small brown spots on each wing. **Biology:** Caterpillars feed on Small-leaved Guarri (*Euclea undulata*). **Habitat & distribution:** Bushveld; KwaZulu-Natal north to Zambia.

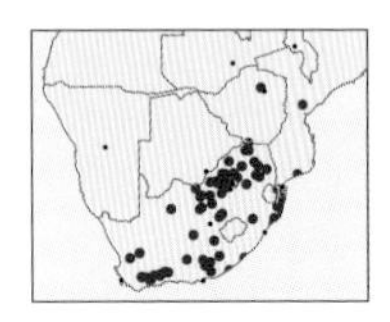

**3 *Chlorocoma clopia*** Confused Lattice

Forewing 8–11mm. Antennae white, body and wings bright green with white costa; wings covered in a lattice of white streaks, therefore often mistaken for a small *Prasinocyma immaculata* (see below) but has no cell spot or white marks on inner margin. Caterpillar smooth green with white to yellow lateral line, red lateral spots and red or yellow dorsal arrow marks. Four species in genus in region. **Biology:** Caterpillars feed on *Vachellia* thorn trees. Multivoltine, adults throughout the year, peaking in summer. **Habitat & distribution:** Various habitats where host trees grow; Western Cape north to Zimbabwe.

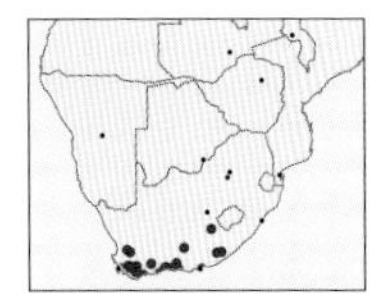

**4 *Chlorocoma didita*** Fullstop Emerald

Forewing 9mm. Uniform pale green with yellowish costa and small black cell spot on each wing. Caterpillar green with dorsal and lateral oblique white and red markings, sometimes whitish with oblique red stripes, or pink; individual can change colour when moving to different *Erica* spp. Almost identical *C. eucela* replaces *C. didita* in northern South African populations where *Erica* spp. grow. **Biology:** Caterpillars feed on various heaths (*Erica* spp.). Multivoltine, adults throughout the year. **Habitat & distribution:** Fynbos and grassland where *Erica* spp. grow; Western Cape, Northern Cape and Eastern Cape.

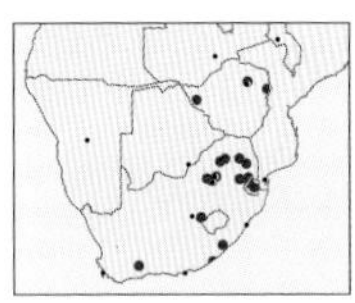

**5 *Paraprasina discolor*** Triangular Emerald

Forewing 17–22mm. Similar to some *Prasinocyma* spp. (below), but forewing narrowly triangular; male with only 2 spurs and large body; head, costa and antennae white, body and wings pale green, wings with thin white striations in some individuals, plain in others. Caterpillar thick, blue-green with cream lateral line and rows of fine white dots. One species in genus in region. **Biology:** Caterpillars feed on cabbage trees (*Cussonia* spp.). Bivoltine, adults September–February. **Habitat & distribution:** Kloofs and rocky ridges where host trees grow; Western Cape north to Zimbabwe.

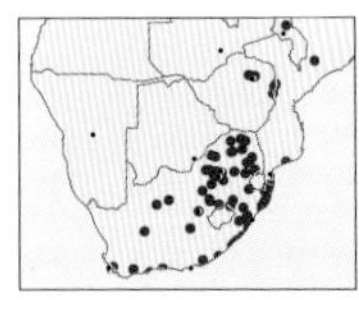

**6 *Prasinocyma immaculata*** Immaculate Lattice

Forewing 13–15mm. Similar to *Chlorocoma clopia* (above), but larger, body and wings pale transparent green to bluish-green with fine white lattice-like striations, a tiny black cell spot on all wings diagnostic. Caterpillar pale mottled green, covered in stubby white setae. Eighteen similar species in genus in region, which are often confused. Very similar *P. magica* has fewer and larger striations and larger cell spots. **Biology:** Caterpillars feed on Taaibos (*Searsia pyroides*). Multivoltine, adults throughout the year. **Habitat & distribution:** Open woodland and savanna; South Africa north to Sudan and Ethiopia.

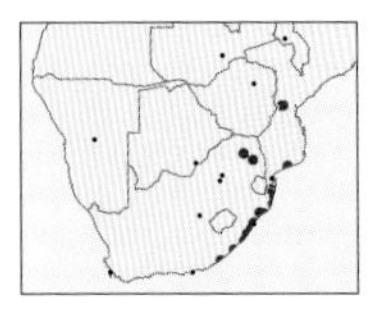

**7 *Prasinocyma oculata*** Eyed Lattice

Forewing 15–18mm. Wings held open and flat on substrate; antennae bright white; body pale green, wings blue-green with white striations, a single red-brown-ringed dark cell spot on each wing; distinctive white spot halfway on inner margin of forewings. Similar *P. vermicularia* is bluer and has cell spots much reduced. **Biology:** Caterpillar host associations unknown. Adults throughout the year. **Habitat & distribution:** Moist forest and coastal bush; Eastern Cape north to tropical Africa.

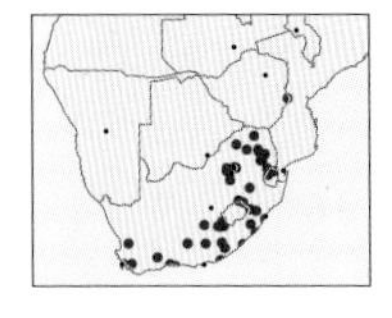

**8 *Prasinocyma chloroprosopa*** White-spot Lattice

Forewing 10–13mm. Head and antennae white, body and wings uniform bluish-green with no striations or other markings except for small white spots on antemedial line area in some individuals. Caterpillar green with red-spotted cream lateral band, red or white dorsal spots on some individuals, covered in stubby white setae. **Biology:** Caterpillars feed on flowers of *Searsia* trees and shrubs. Caterpillars develop quickly, within weeks, when host plants flower, adults throughout the year. **Habitat & distribution:** Fynbos, grassland and bushveld where *Searsia* spp. grow; Western Cape, South Africa, north to Tanzania.

1 ♀
1 ♂
IS
1
2
3
3
4
4
5
5
6
6
7
8
8

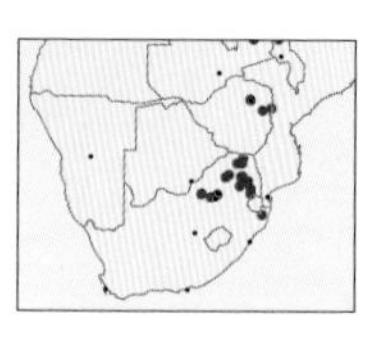

### 1 *Prasinocyma pictifimbria* — Coloured-border Lattice

Forewing 11–14mm. Lighter green than other similar species, with white lattice streaks, distinctive margin coloured in red/white/black spots with golden cilia in between, costa white to golden, head and antennae white. **Biology:** Caterpillars recorded on Currant Resin Tree (*Ozoroa sphaerocarpa*) and *Monotes glaber*. Adults throughout the year. **Habitat & distribution:** Bushveld; KwaZulu-Natal, South Africa, north to Kenya.

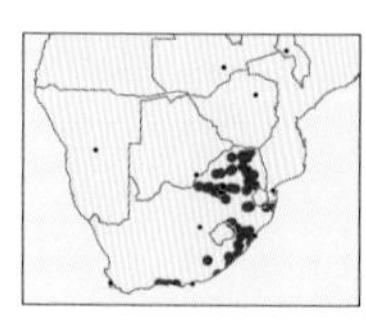

### 2 *Heterorachis despoliata* — Red-edged Emerald

Forewing 11–13mm. Head, antennae and legs orange-red, thorax green, abdomen red; wings plain green, lightly irrorated with white scales, except for thin yellow and purple marginal band, cilia purple. Caterpillar slender, light green with white lateral line and short tail. Thirteen species in genus in region. **Biology:** Caterpillars feed on Rock Alder (*Afrocanthium mundianum*). Bivoltine, adults September–April. **Habitat & distribution:** Forest, riverine vegetation and rocky ridges where host plant grows; Western Cape north to Limpopo, South Africa.

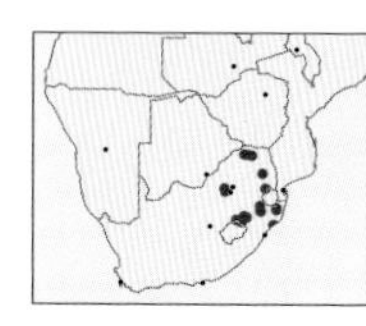

### 3 *Heterorachis roseifimbria* — Rosy-fringed Emerald

Forewing 11–13mm. Similar to *H. despoliata* (above), but wings not finely irrorated with white, yellow marginal band wavy with distinctive wider wave on inner margin. Caterpillar green, dorsoventrally flattened and indented between segments. Similar *H. devocata* has marginal band purple, broader and deeply indented. **Biology:** Caterpillars recorded on Hairy Turkey-berry (*Canthium ciliatum*) and Sand Apple (*Pygmaeothamnus zeyheri*), both Rubiaceae. Multivoltine, adults throughout the year. **Habitat & distribution:** Grassland on rocky ridges; KwaZulu-Natal north to Limpopo, South Africa.

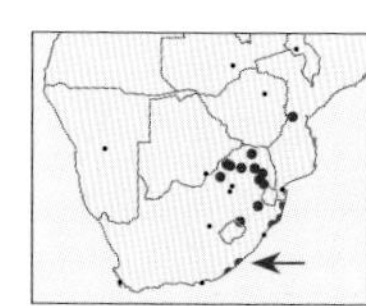

### 4 *Heterorachis simplicissima* — Simple Emerald

Forewing 13–16mm. Several similar unmarked species in genus; uniform green except for pale costa and antennae. Caterpillar mottled brown and yellow, warty, with a pair of dorsal projections on middle segments. Similar *H. pervirides* has costa and antennae red and scales around cell darker; *H. disconotata* is smaller with rounder wings. **Biology:** Caterpillars feed on Rock Alder (*Afrocanthium mundianum*). Bivoltine, adults October–March. **Habitat & distribution:** Forest, riverine vegetation and rocky ridges where host plant grows; Eastern Cape, South Africa, north to Mozambique.

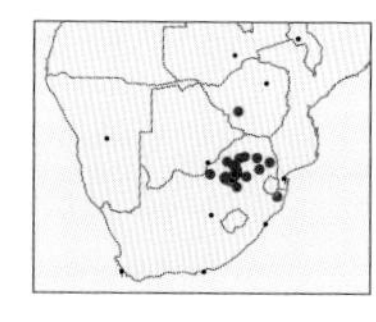

### 5 *Omphax idonea* — Smooth Omphax

Forewing 13–17mm. Wings held open flat against surface; thorax green, abdomen brown, wings rich plain green, costa cream, cilia cream and pink, margins not crenulated, underside golden. Caterpillar green, with white lateral and dorsal lines. Ten species in genus in region; similar *O. plantaria* is darker green, slightly crenulated brown margins, underside green to straw. **Biology:** Caterpillars feed on African Medlar (*Vangueria infausta*). Bivoltine, adults September–February. **Habitat & distribution:** Rocky ridges in bushveld and forest where host plant grows; KwaZulu-Natal, South Africa, north to Zimbabwe.

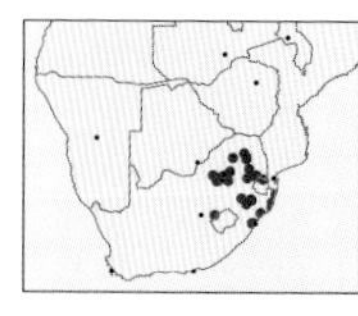

### 6 *Omphax bacoti* — Confused Omphax

Forewing 13–17mm. Wings folded over body, thorax green, abdomen cream; forewings plain green, costa cream, cilia cream with slight pinkish tint in some individuals; hindwings plain cream with green tint on anal margin. Can be confused with *Rhadinomphax divincta* (below), but much larger. **Biology:** Caterpillar host associations unknown. Bivoltine, adults September–February. **Habitat & distribution:** Grassland; KwaZulu-Natal north to Limpopo, South Africa.

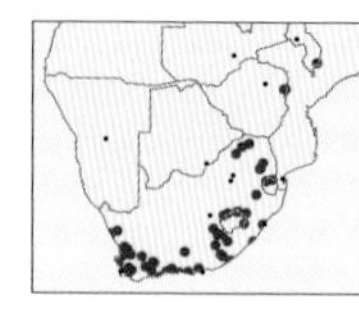

### 7 *Rhadinomphax divincta* — Two-phase Emerald

Forewing 11–14mm. Rests with wings held steeply roof-like against body; head and legs brown, thorax green, abdomen white; forewings plain vivid green with faint orange costa, diagnostic plain green cilia; hindwings white with faint green margin. Four species in genus in region. **Biology:** Caterpillars feed on *Cliffortia repens* (Rosaceae). Adults August–May. **Habitat & distribution:** Fynbos and moist grasslands with fynbos elements; Western Cape, South Africa, north to Malawi.

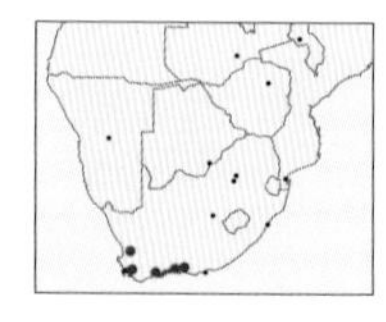

### 8 *Rhadinomphax pudicata* — Pink-edged Emerald

Forewing 9–11mm. Rests with wings held steeply roof-like against body; head and legs red, thorax green, abdomen brown; forewings plain vivid green, costa cream, red toward apex, diagnostic red cilia, no cell spot; hindwings white with faint pink margin. Similar *R. sanguinipuncta* has distinct red or brown cell spot. **Biology:** Caterpillars recorded on *Aspalathus* (Fabaceae). Bivoltine, adults October–March. **Habitat & distribution:** Various fynbos habitats; Western Cape and Eastern Cape, South Africa.

1
2
2
3
3
4
4
5
5
6
7
7
8
8

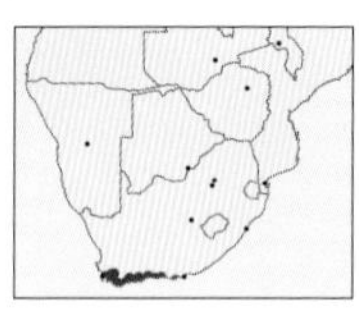

### 1 *Rhadinomphax trimeni* — White-eye Emerald

Forewing 12–14mm. Rests with wings held more open than other species in genus; head orange, legs white, thorax and wings yellowish-green, costa red-edged orange, row of marginal white spots and red/white cilia, distinctive red- and black-ringed white cell spot; hindwings white with faint red margin. **Biology:** Caterpillar host associations unknown. Adults October–January. **Habitat & distribution:** Moist fynbos; Western Cape and Eastern Cape, South Africa.

### 2 *Prosomphax callista* — Cirrus Emerald

Forewing 14–16mm. At rest, wings held roof-like against body; antennae white, head and legs orange, body green; forewings green with narrow orange costa, spotted with diffuse silvery white dots and streaks, which are almost absent in some populations. Two species in genus in region. **Biology:** Caterpillar host associations unknown. Adults throughout the year. **Habitat & distribution:** Fynbos, renosterveld and moist grasslands; Western Cape north to Gauteng, South Africa.

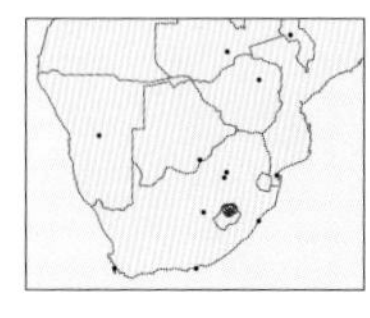

### 3 *Prosomphax horitropha* — Clouded Emerald

Forewing 17–19mm. Similar in colouring and posture to *P. callista* (above), but larger, with much more distinct silvery white markings and is allopatric. **Biology:** Caterpillars feed on Drakensberg Gonna (*Passerina drakensbergensis*). Probably bivoltine, adults January–April. **Habitat & distribution:** Afroalpine heathland above 2,500m; Lesotho.

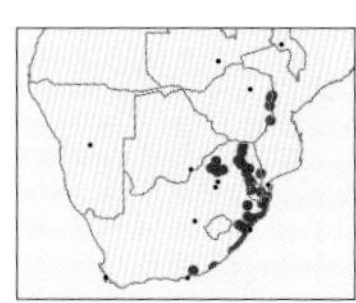

### 4 *Thalassodes quadraria* — Squared Emerald

Forewing 14–17mm. Antennae and top of head white; body and wings bluish-green; wings semi-transparent with small pale scribbling and indistinct wavy lines, margins white with yellow to pink shading; hindwings squared at apex. Caterpillar smooth, elongate, green with red dorsal line in some specimens. One species in genus in region. **Biology:** Caterpillars polyphagous on dicotyledonous trees. Multivoltine, adults throughout the year. **Habitat & distribution:** Frost-free forest or riverine vegetation; Eastern Cape, South Africa, north to tropical Africa, including Indian Ocean islands, also India and Sri Lanka.

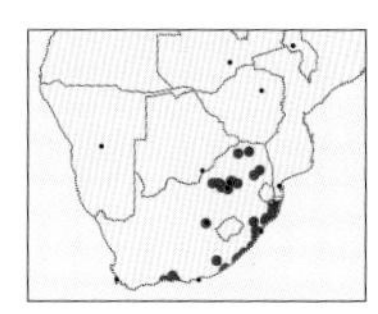

### 5 *Victoria mirabilis* — Forest Victoria

Forewing 17–21mm. Thorax and dorsal abdomen brown, becoming white posteriorly; wings banded in green and white, with scalloped distal margins, red cell spot, diagnostic forewing antemedial band curved inward, postmedian band often incomplete and shaded with pink. Caterpillar green to brown with dorsal process. Four described species in genus in region; similar *V. fuscithorax* has white bands broader and antemedial band straight, occurs in tropical bushveld. **Biology:** Caterpillars feed on various mistletoes (Loranthaceae). Multivoltine, adults throughout the year. **Habitat & distribution:** Forests and highveld kloofs; southern Cape north to Limpopo, South Africa.

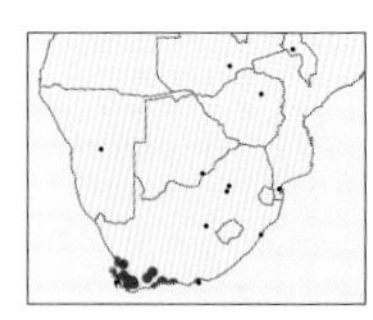

### 6 *Victoria albipicta* — Cape Victoria

Forewing 17–19mm. Similar to *V. mirabilis* (above), but forewings mostly green with white antemedial bar straight and thin, postmedian bar very reduced or absent in some individuals; hindwings white with pale green border. Caterpillar pale green. **Biology:** Caterpillars feed on Cape Mistletoe (*Viscum capense*). Multivoltine, adults throughout the year. **Habitat & distribution:** Moist fynbos habitats; Western Cape and Eastern Cape, South Africa.

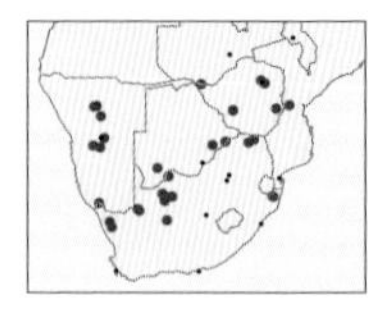

### 7 *Victoria triplaga* — Desert Victoria

Forewing 11–16mm. Body and wings uniform light green to yellow or a mixture of these colours, in some individuals; a brown- and red-ringed white cell spot on each wing, wing margins orange, sprinkled with reddish spots and variable scalloped blotches. **Biology:** Caterpillars feed on various mistletoes (Loranthaceae). Adults September–March. **Habitat & distribution:** Arid habitats; Northern Cape, South Africa, north to southern Sahara.

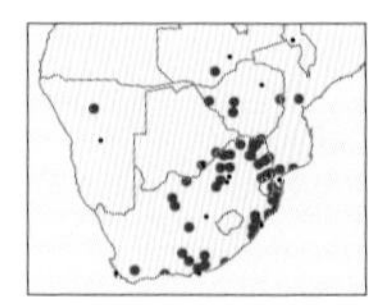

### 8 *Chlorissa albistrigulata* — White-streaked Chlorissa

Forewing 10–11mm. Body and wings light blue-green, marked with numerous fine wavy white streaks, costa yellow, postmedian line crenulated on both wings. Caterpillar slender, light to dark green or brown. Eight species in genus in region. **Biology:** Caterpillars polyphagous, often on flowers. Multivoltine, adults throughout the year. **Habitat & distribution:** Variety of habitats, often in disturbed areas; Western Cape, South Africa, north to tropical Africa, also Comoros.

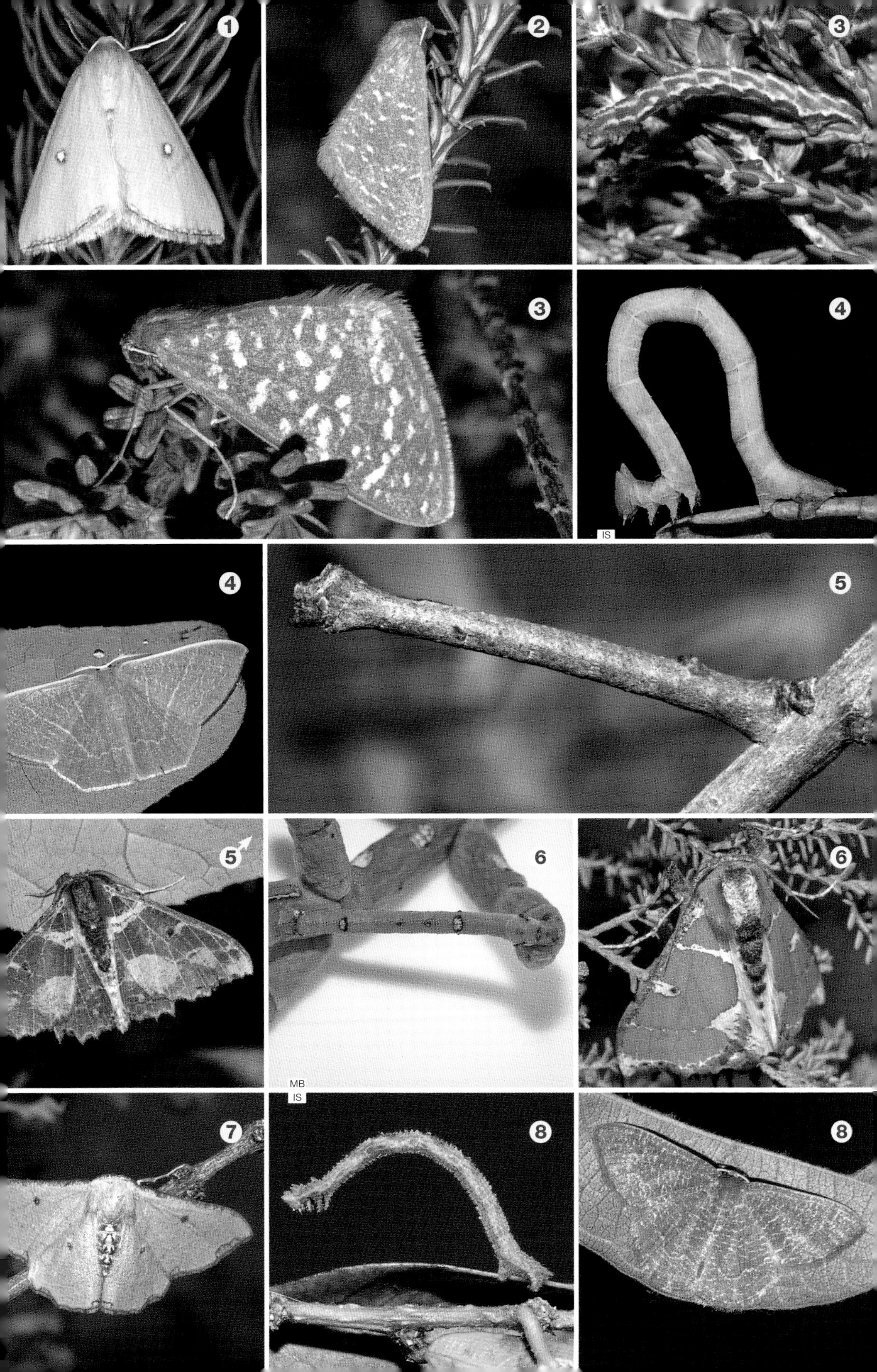

1
2
3
3
4
IS
4
5
5
6
6
MB
IS
7
8
8

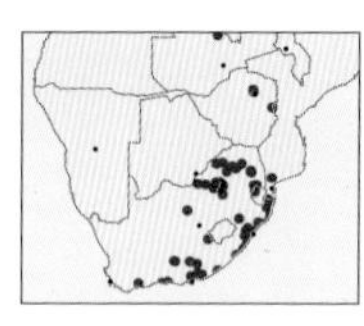

**1 *Chlorissa attenuata*** Attenuated Chlorissa

Forewing 10–11mm. Thorax green, abdomen brown or green with brown spots; wings green without white streaks, held flat and widespread, darker green and white postmedian line diagnostically evenly curved, small darker green-brown cell spots. Similar *C. articulicornis* has lines straight on both wings; *C. dorsicristata* has lines crenulated and brownish. **Biology:** Caterpillar host associations unknown. Multivoltine, adults throughout the year. **Habitat & distribution:** Forest and riverine vegetation; Western Cape, South Africa, north to tropical Africa, also Comoros.

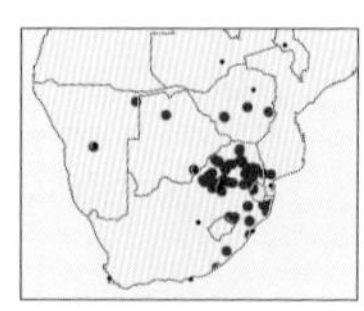

**2 *Neromia rubripunctilla*** Red-dot Neromia

Forewing ♂ 10–12mm, ♀ 14–16mm. Body and wings plain green, sparsely covered in scales, paler costa, indistinct transverse white lines and single diagnostic small red cell spot; less distinct dot on hindwings (sometimes not visible). Caterpillar variable, green to red-brown, sometimes with white markings, extreme bifid head, covered in short stubby setae. Ten species in genus in region; similar *N. strigulosa* is a slightly different green and lacks red cell spot. **Biology:** Caterpillars feed on hook thorns (*Senegalia* spp.). Bivoltine, adults September–March. **Habitat & distribution:** Forest and bushveld where *Senegalia* spp. grow; Eastern Cape, South Africa, north to Zimbabwe, Namibia, Botswana.

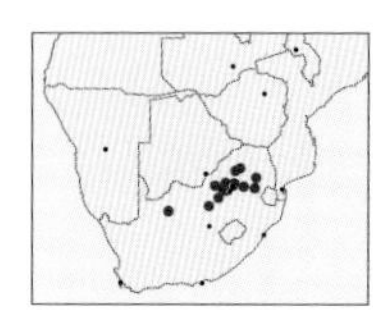

**3 *Neromia cohaerens*** Coherent Neromia

Forewing 9–13mm. Head and antennae white, legs pink, body green; forewings plain green densely covered in scales with pink costa, no cell spot. Caterpillar smooth, plain green, very slender with extreme bifid head, perfect mimic of host plant. Similar *N. activa* has a faint crenulated postmedian line across both wings. **Biology:** Caterpillars feed on Elephant's Root (*Elephantorrhiza elephantina*). Bivoltine, adults September–February. **Habitat & distribution:** Grassland where host plant grows; Northern Cape to Limpopo, South Africa.

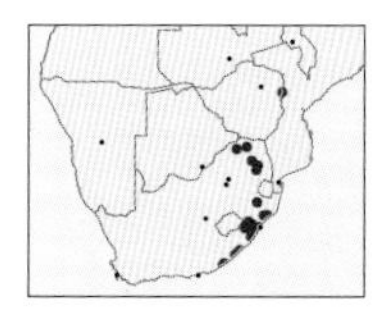

**4 *Neromia impostura*** Eyed Neromia

Forewing 10–15mm. Forewings finely dusted with green, with brown and yellow marginal line, and cilia chequered white and brown, lines jagged, distinct relatively dark white-ringed cell spot on both wings. **Biology:** Caterpillar host associations unknown. Adults throughout the year. **Habitat & distribution:** Moist forests; Eastern Cape, South Africa, north to Zimbawe.

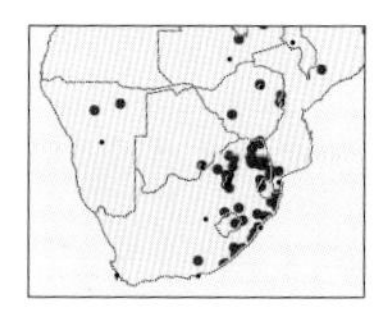

**5 *Comibaena leucospilata*** Singed Concealer

Forewing 12–16mm. Thorax green, abdomen orange with white dorsal spots; wings green with white costa, reddish hind margin and red and white chequered cilia, each wing with small black cell spot, diagnostic postmedian line curved inward before costa. Bizarre caterpillar brown, adorning itself with debris from its surroundings. Four species in genus in region; similar *C. coryphata* has postmedian line straight to costa. **Biology:** Caterpillars polyphagous, often on flowers. Multivoltine, adults throughout the year. **Habitat & distribution:** Grassland, bushveld and forest; Eastern Cape, South Africa, north to tropical Africa.

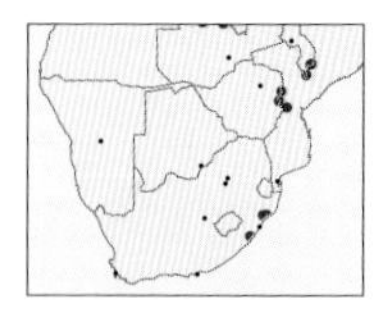

**6 *Comibaena barnsi*** Blotched Concealer

Forewing 16–17mm. Similar to *C. leucospilata* (above), but larger with reddish cell spots and conspicuous large reddish-brown marginal blotch covering both wings. **Biology:** Caterpillar host associations unknown. Adults throughout the year. **Habitat & distribution:** Tropical forest; KwaZulu-Natal, South Africa, north to tropical Africa.

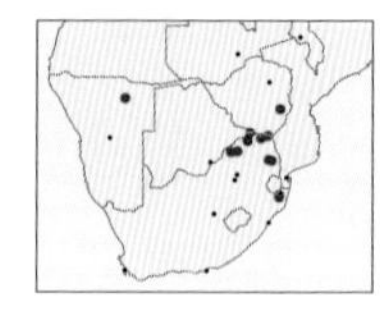

**7 *Microbaena pulchra*** Spectacled Concealer

Forewing 9–12mm. Thorax green, abdomen white to pink; forewings emerald green with tiny black cell spot and orange-edged white patches, including 2 large semicircular patches on hind margin, thin dashed brown line along distal edge. Caterpillar brownish-orange, camouflaged with large pieces of leaves carried on back. One species in genus in region. **Biology:** Caterpillars recorded on Broad-leaved Resin Tree (*Ozoroa obovata*). Adults September–April. **Habitat & distribution:** Bushveld; KwaZulu-Natal, South Africa, north to Ethiopia and Israel.

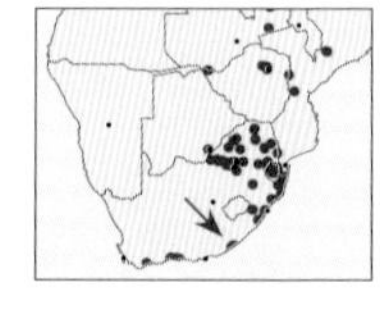

**8 *Comostolopsis stillata*** Pink-laced Concealer

Forewing 6–8mm. Body and wings blue-green, wings edged with fine red line and fringe of yellow seta, bearing 2 rows of cream-ringed red dots, outer row may be linked by cream band. Caterpillar green with reddish head, concealing itself with surrounding debris. Five species in genus in region. **Biology:** Caterpillars polyphagous on flowers of various trees. Multivoltine, adults throughout the year, peaking in summer. **Habitat & distribution:** Forest and riverine bush, including gardens; Western Cape, South Africa, north to tropical Africa, including Madagascar and Seychelles.

1
2
2
3
3
4
5
IS
5
6
7
8

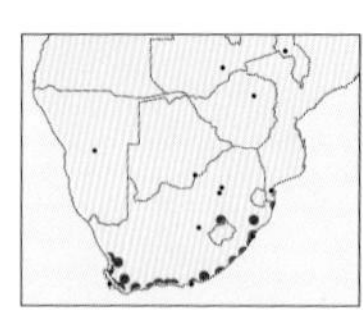

**1 *Comostolopsis germana*** Bietou Tip Concealer

Forewing 6–8mm. Body and wings light green; forewings with postmedian line white, crenulated, with brown spots in some individuals, both wings with small, reddish cell spot. Caterpillar rotund, light green with white dorsal line and indistinct white scribblings. Similar *C. apicata* has postmedian line curved but not crenulated and bigger red cell spots. **Biology:** Caterpillars monophagous on Bietou (*Osteospermum moniliferum*), concealing themselves in new shoots on which they feed. Introduced as biocontrol agent in Australia, where Bietou is an invasive weed. Multivoltine, adults throughout the year. **Habitat & distribution:** Humid habitats where host plant grows; Western Cape to KwaZulu-Natal, introduced to Australia.

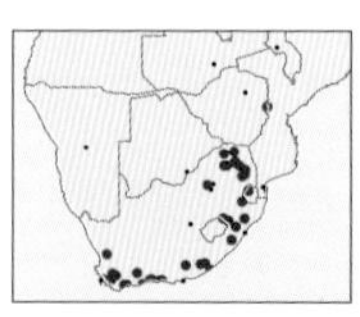

**2 *Comostolopsis capensis*** Blue Concealer

Forewing 9–11mm. Similar to *C. germana* (above), but larger, body and wings blue-green and cell spot black. Caterpillar stout, light green, with purple head and legs. **Biology:** Caterpillars feed on Cape Sumach (*Colpoon compressum*). Adults August–May. **Habitat & distribution:** Cool, moist fynbos, forests and grasslands; Western Cape, South Africa, north to Zimbabwe.

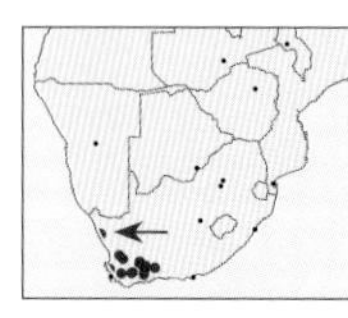

**3 *Eucrostes rufociliaria*** Rosy Eucrostes

Forewing 10mm. Usually rests with wings angled roof-like against body; thorax and costa rosy orange, fading to green posteriorly; hindwings rosy orange, both wings with purple cilia. Three species in genus in region. **Biology:** Caterpillar host associations unknown. Adults July–April. **Habitat & distribution:** Succulent karoo habitats; Western Cape and Northern Cape.

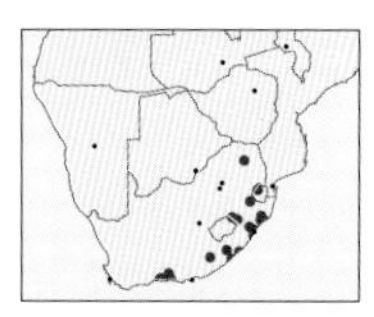

**4 *Dolosis illacerata*** Dotted Dolosis

Forewing 14–16mm. Thorax green, abdomen light brown, with dark spots; wings yellowish-green to bright green, large brown and black cell spot, lines wavy with black and white spots; hindwing margin produced into 2 triangular points. Caterpillar green, stout, surface crinkled, with thin lateral line. One species in genus. **Biology:** Caterpillars feed on Turkey-berry (*Canthium inerme*). Bivoltine, adults November–April. **Habitat & distribution:** Moist forests; Western Cape north to Limpopo, South Africa.

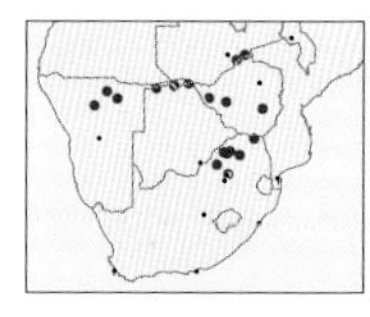

**5 *Mimandria cataractae*** Eyed Duster

Forewing 16 –18mm. Wings held flat and extended laterally; body and wings white with sparse to thick dusting of green to grey scales; both wings crossed by zigzag black postmedian lines, a variable row of small red dots beyond postmedian line, absent in some individuals. One species in genus in region. **Biology:** Caterpillar host associations unknown. Adults November–March. **Habitat & distribution:** Bushveld; Gauteng, South Africa, north to Kenya.

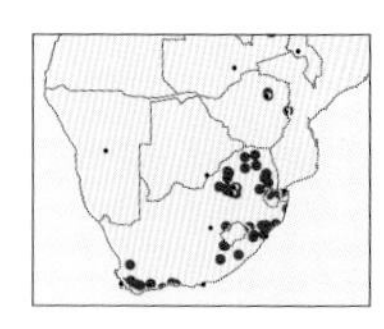

**6 *Pingasa distensaria*** Scribbled Duster

Forewing 17–22mm. White, body and wings speckled with brown scales, crenulated antemedial line and jagged postmedian line diagnostic. Caterpillar fairly stout, typical for tribe Comostolini with distinctive lateral ribbing, green ventrally with lateral yellow to orange line, dorsal parts in variety of shades of green to red. **Biology:** Caterpillars polyphagous on trees. Multivoltine, adults throughout the year, peaking in summer. **Habitat & distribution:** Forest, wooded kloofs and riverine vegetation, also gardens; Western Cape, South Africa, north to tropical Africa.

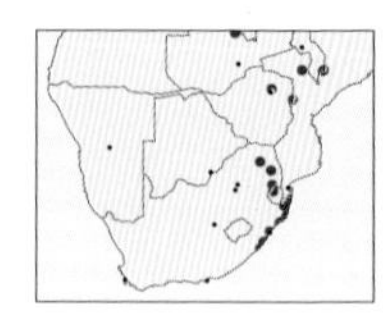

**7 *Pingasa ruginaria*** Yellow Duster

Forewing 18–21mm. Wings held flat and laterally extended; body green, wings finely dusted with green, variable reddish scales in postmedian area, postmedian line gently undulating apart from minute outward kinks in some specimens, basal part of underside of wings diagnostically deep yellow. Similar *P. rhadamaria* has wings on underside pure white; *P. lahayei* is smaller with uniform dark grey dusting. Caterpillar distinctive, with large round lateral and dorsal projections on each segment. **Biology:** Caterpillars recorded on Taaibos (*Searsia pyroides*), probably polyphagous. Multivoltine, adults throughout the year. **Habitat & distribution:** Coastal bush and moist forests; Eastern Cape, South Africa, north to tropical Africa.

1
1
JuB
2
3
4
4
5
6
6
7
7

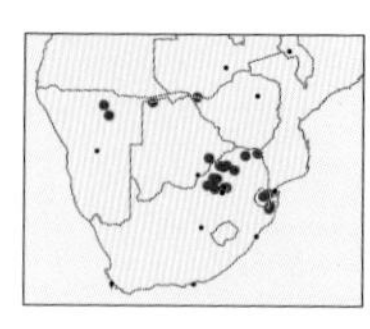

### 1 *Mictoschema swierstrai* — Grey Duster

Forewing ♂11–12mm, ♀ 18–22mm. Wings held tent-like over body; body and wings shades of grey, undulatingly kinked postmedian line diagnostic, distinct black marginal line. Caterpillars perfectly adapted to host plant, with dorsal side green with white 'stipules' and ventral side cream with numerous reddish dots; in earlier instars it is completely cream with reddish dots, mimicking twigs. Two species in genus in region; *M. tuckeri* has median area much darker and postmedian line straighter but still curved. **Biology:** Caterpillars feed on Lavender Feverberry (*Croton gratissimus*). Bivoltine, adults October–April. **Habitat & distribution:** Coastal bush and moist forests; KwaZulu-Natal, South Africa, north to tropical Africa.

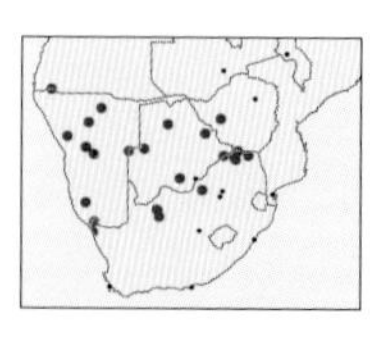

### 2 *Holoterpna errata* — Arid Duster

Forewing ♂8–10mm, ♀ 13–14mm. Similar to *M. swierstrai* (above), wings held in a peculiar way with large gap between forewings and hindwings; white wings densely dusted with grey, lines indistinct with postmedian almost straight, no distinct black marginal line. **Biology:** Caterpillar host associations unknown. Adults November–March. **Habitat & distribution:** Arid sandy habitats; Northern Cape, South Africa, north to Namibia and Zimbabwe.

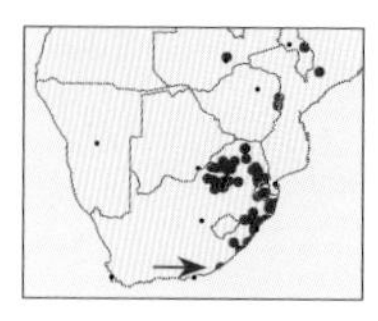

### 3 *Petovia marginata* — Clouded Orange

Forewing 15–23mm. Highly variable species previously described under several different names; body orange with black crossbars; wings typically orange with wide black margins and prominent black veins, but veins may be weakly marked or absent; forewings often cream or brownish. Caterpillar white, orange at both ends, with black lateral and dorsal lines and crossed by black bars. **Biology:** Caterpillars gregarious on various African medlars (*Vangueria* spp.). The only fully diurnal species in the subfamily in region. Univoltine, adults October–December in region. **Habitat & distribution:** Bushveld and forest where medlars grow; Eastern Cape, South Africa, north to tropical Africa.

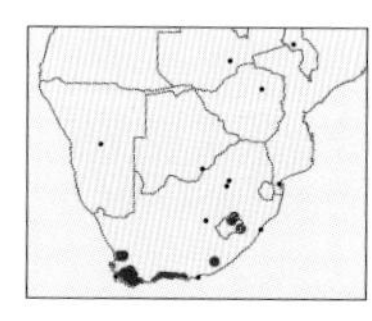

### 4 *Argyrographa moderata* — Rivulet Emerald

Forewing 10–12mm. Wings held roof-like alongside the body; forewings vivid green crossed by 4 broad, wavy white bands; hindwings white with green dots along margin. One species in genus in region. **Biology:** Caterpillar host associations unknown. **Habitat & distribution:** Moist fynbos in the fynbos region and on high mountains where remnant fynbos vegetation persists; Western Cape north to KwaZulu-Natal, South Africa.

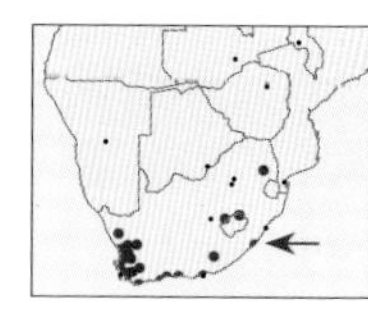

### 5 *Dichroma equestralis* — Horse Head

Forewing 14–15mm. Unmistakable; bright green wings held steeply sloped against body, wings marked with jagged white 'V'-shaped markings and lines, Lesotho populations have lines in median area broken up. One species in genus. **Biology:** Caterpillars feed on gonna bushes (*Passerina* spp.). Adults August–January. **Habitat & distribution:** Fynbos and grasslands containing remnant fynbos vegetation; Western Cape north to Mpumalanga, South Africa.

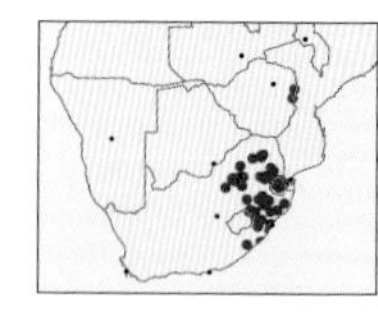

### 6 *Acollesis fraudulenta* — Fraudulent Emerald

Forewing 10–14mm. Body and wings mottled bluish-green; forewings with white antemedial and postmedian lines slightly crenulated and curve slightly inward, no cell spots. Three species in genus in region; similar *A. umbrata* has small black cell spots and straight lines; *A. terminata* is yellowish-green with straight postmedian line. **Biology:** Caterpillar host associations unknown. Adults November–March. **Habitat & distribution:** Grassland habitats; Eastern Cape, South Africa, north to Zimbabwe.

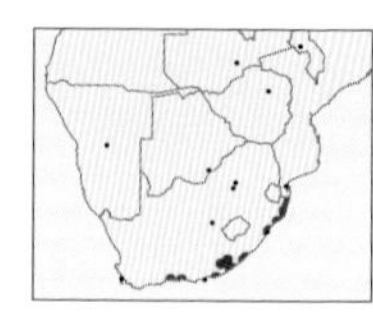

### 7 *Lasiochlora diducta* — Eye Emerald

Forewing 13–16mm. Wings held flat and open when at rest; body and wings uniform green, lines represented by small white dots, margin slightly crenulated with purplish-brown cilia, small cream cell spot on each wing and a large 'eye-like' brown and white spot at base of forewing postmedian line. Two species in genus in region. **Biology:** Caterpillars recorded on Cat-thorn (*Scutia myrtina*, Rhamnaceae) and Climbing Flat Bean (*Dalbergia obovata*, Fabaceae). Multivoltine, adults throughout the year. **Habitat & distribution:** Forest and coastal bush; southern Cape north to KwaZulu-Natal, South Africa.

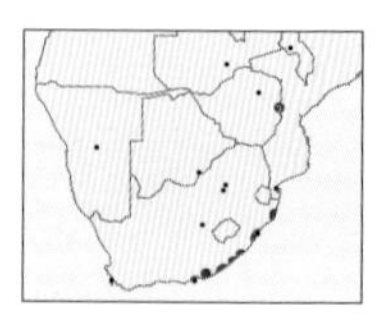

### 8 *Lasiochlora bicolor* — Two-tone Emerald

Forewing 13–16mm. Wings held closed and roof-like over body; head and costa white; forewings plain green with white and black dotted lines; hindwings uniform rosy pink with no markings. Subspecies in north of region has additional markings on forewing. **Biology:** Caterpillars recorded on *Scutia myrtina* (Rhamnaceae). Adults November–January. **Habitat & distribution:** Forest and coastal bush; Eastern Cape, South Africa, north to Kenya.

1
1
1
2
3
3
SW
4
5
6
7
8♂

## FAMILY BOMBYCIDAE Silk Moths

**Medium-sized, robust, hairy moths with prominent comb-like antennae in both sexes. Wings broadly triangular. About 160 species globally, most in Asia, probably only 5 in southern Africa. Best-known representative the Domesticated Silkworm, *Bombyx mori*.**

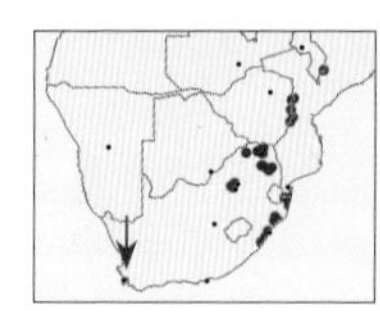

### 1 *Racinoa pallicornis* Small Silk Moth

Forewing ♂ 11mm, ♀ 16mm. At rest, wings held out laterally with hindwings folded beneath forewings and abdomen twisted to one side; body and wings pale to reddish-brown; forewings apically falcate, with fine wavy outer line; hindwings round with pronounced apical lobe. Caterpillar reddish- to greyish-brown, whitish dorsally, with short fleshy tail; at rest, raises front part of body off substrate, resembling a broken twig. A number of similar species in region, 2 described, some undescribed. **Biology:** Caterpillars feed on both wild and cultivated fig trees (*Ficus* spp.) and spin a flat yellow silk cocoon on the underside of leaves. Multivoltine, found throughout the year. **Habitat & distribution:** Widespread in the wet warmer southern and eastern parts of the subcontinent, from Eastern Cape north into Zimbabwe.

## FAMILY BRAHMAEIDAE Brahmin Moths

**Small family, restricted to forests, which include some species that exhibit the most complicated and detailed wing patterns amongst any moths. Caterpillars have long, paired hairy spines near each end of the body. Only 18 species globally, 4 in the region.**

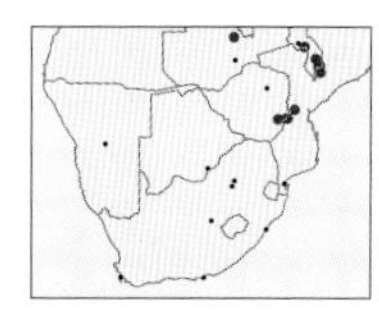

### 2 *Dactyloceras neumayeri* Neumayer's Brahmin

Forewing 46–67mm. Body brown; wings mainly brown toward base, with complicated pattern of wavy white and brown lines and row of eyespots along margin of forewing. Only species in genus in region. **Biology:** Host associations unknown. **Habitat & distribution:** Afromontane forests; Zimbabwe, Mozambique, north through East Africa to Ethiopia.

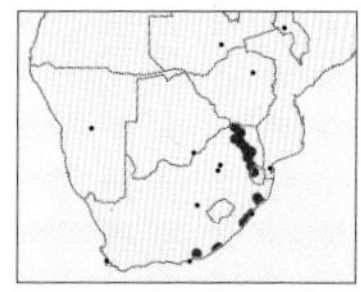

### 3 *Spiramiopsis comma* Comma Moth

Forewing 28–32mm. Black thorax and white abdomen; forewings black basally, silver to brown distally, with large, black 'comma' centrally. Caterpillar mottled pinkish-brown, with 2 pairs of paired, long, hairy projections anteriorly, and long, bent posterior horn. One species in genus. **Biology:** Caterpillars feed on the forest creepers *Secamone alpinii* and *S. gerrardii*. Possibly bivoltine, adults recorded July–March. **Habitat & distribution:** Afromontane forests and coastal bush; Eastern Cape, South Africa, to Mozambique.

## FAMILY EUPTEROTIDAE Monkey Moths

**Medium to large (forewing 11–70mm) with hairy, plump bodies and broad, rounded wings densely covered with elongate scales and fringed with setae. Most are off-white to various shades of brown. Most have slow, flapping flight and are nocturnal. Adults are a favourite food of bushbabies and other monkeys, hence the common name. Caterpillars usually densely covered with long hairs. Eupterotidae are associated with a wide variety of host plant families but individual species mostly specialise on a particular group. About 325 described species globally, 81 in the region.**

### Subfamily Striphnopteryginae

Largely restricted to Africa, with some New Guinean and Australian species. Defined on some features (the uncus) of the male genitalia.

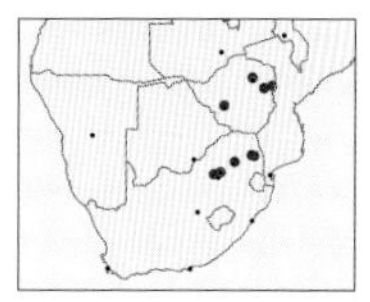

### 4 *Janomima westwoodi* Inquisitive Monkey

Forewing 39–52mm. Fawn in colour; forewings crossed by several pairs of fine dark lines in both sexes; densely irrorated with black scales in males; hindwings with large black patch at base. Caterpillar black, with long, recumbent tufts of black and white hairs, conspicuous 'exclamation mark' on headshield. Two species in genus in region. **Biology:** Caterpillars polyphagous; often seen on the ground moving between hosts. Bivoltine, adults August–November and January–February. **Habitat & distribution:** Savanna habitats; from Gauteng, South Africa, north to tropical Africa.

1
1
IS
2
3
4
♀
3

**1** ***Lichenopteryx despecta*** Despised Monkey

Forewing 19–25mm. Body and wings grey; wings speckled with darker brown or black and crossed by several indistinct wavy lines. Caterpillar greenish with red and yellow marks on each segment; long tufts of setae on each segment and long anterior 'horns'. Only species in genus in region. **Biology:** Caterpillars recorded on Kei Apple (*Dovyalis caffra*) and *Dalbergia obovata*. Probably multivoltine, adults September–May. **Habitat & distribution:** Lowland forest and woodland; from southern KwaZulu-Natal, South Africa, north to Tanzania.

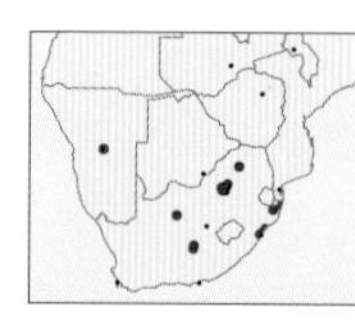

**2** ***Marmaroplegma paragarda*** Speared Monkey

Forewing ♂ 23–25mm, ♀ 30–32mm. Easily recognised by bold pattern of white-edged light brown 'spear points' on forewing; hindwings orange-brown with brown marginal triangles. Caterpillar unknown. Three species in genus in region. **Biology:** Host associations unknown. Univoltine, adults September–October. **Habitat & distribution:** Grasslands; in South Africa, Mozambique and Zimbabwe.

**3** ***Phyllalia patens*** Clay Monkey

Forewing 25–35mm. Plain pinkish-brown with shining scales and slightly darker, hairy body, males with strongly pectinate antennae. Caterpillar covered in dense dark brown irritant hairs. Seven species in genus in region. **Biology:** Caterpillars feed on grasses. Multivoltine. **Habitat & distribution:** Wet grasslands and fynbos, from the coast to over 3,000m, seemingly absent from dry areas; endemic to South Africa.

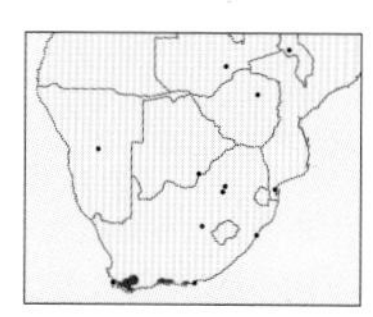

**4** ***Phyllalia thunbergii*** Thunberg's Monkey

Forewing 33–38mm. Plain brown with zigzag dark brown median and postmedian lines. **Biology:** Host associations unknown. Adults present during warmer months, October–April. **Habitat & distribution:** Restricted to the fynbos region; Western Cape and Eastern Cape.

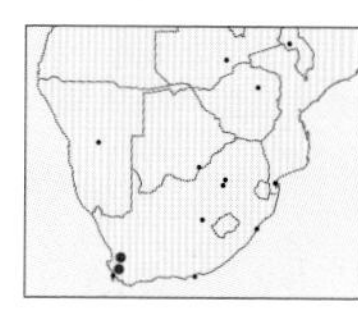

**5** ***Phyllalia alboradiata*** Fynbos Monkey

Forewing 27mm. Plain brown, distinguished by prominent branching white veins on both wings. Caterpillar covered in irritant hairs. **Biology:** Caterpillar host associations unknown. Adults recorded in April. **Habitat & distribution:** Fynbos habitats; Western Cape, South Africa.

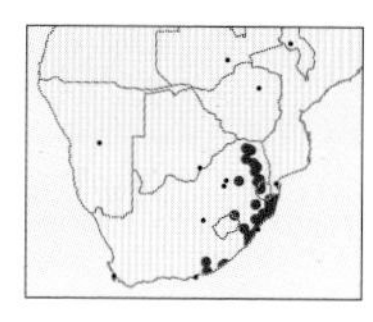

**6** ***Poloma angulata*** Angular Monkey

Forewing ♂ 21–28mm, ♀ 33mm. Male with strongly pectinate antennae, and dark brown, hairy thorax; forewings pinkish-brown basally, darker centrally and distally, thin pale median lines; hindwings pinkish. Female with shorter pectinate antennae, pinkish-brown, medial lines visible. Caterpillar mottled bluish-grey with white spots and oblique stripes, and red prolegs. Three species in genus in region. **Biology:** Caterpillars feed on *Canthium* spp. (Rubiaceae). Multivoltine, adults August–April. **Habitat & distribution:** Forests and coastal bush where *Canthium* is present; Eastern Cape to Limpopo, South Africa.

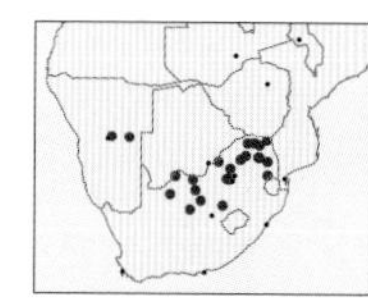

**7** ***Phiala incana*** White Monkey

Forewing 17–22mm. Head and abdomen orange, legs orange and black striped; wings white with variable scattered black scales, often grouped into diffuse bands, thorax with long white hairs. Caterpillar with alternating erect tufts of black and white hairs. Often confused, many similar species, some undescribed, some sympatric, in need of taxonomic revision. Nineteen described species in genus in region. **Biology:** Caterpillars feed on *Asparagus* spp. Multivoltine, 8 weeks to complete life cycle, adults September–April. **Habitat & distribution:** Dry grassland and savanna; probably restricted to southern Africa.

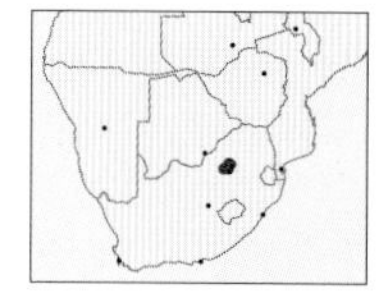

**8** ***Phiala flavipennis*** Diurnal Phiala

Forewing 20–22mm. Body and upper side of forewings white, underside of all wings dusty orange; forewings with dotted black postmedian and terminal lines in male, irregular dusting of black scales. Final instar caterpillar with short tufts, black and white striped dorsally with cream lateral bands. **Biology:** Caterpillars feed on various grasses. Females oviposit spiral of eggs around a grass stem or blade. Adults diurnal, males fly fast above grass, only in early morning; females fly only in late afternoon in sunshine to oviposit. Appears orange when in flight, thought to provide aposematic protection against diurnal predators. Univoltine, adults late November to early December. **Habitat & distribution:** Highveld grassland; Gauteng, South Africa. Probably under-recorded due to unusual flight times and not visiting light sources.

1
1
2
3
RO
3
4
5
6
6
7
7
8
8

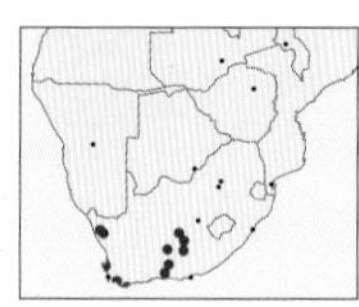

### 1 *Rhabdosia vaninia* — Zebra Monkey

Forewing 20–22mm. Head, legs and abdomen brownish-orange, thorax with long white hairs with 2 black bands; wings white, dusted with black or dark brown scales between veins; underside of both wings uniform with fine brown dusting. Inland specimens are browner. Similar *R. patagiata* has the zebra pattern repeated on the underside of wings. Caterpillars dark brown with longitudinal white bands. Two species in genus in region. **Biology:** Caterpillars feed on *Asparagus rubicundus*. Early instars gregarious in silken nests spun on host plant, final instar caterpillars found singly. **Habitat & distribution:** Various vegetation types where *Asparagus* grows; Western Cape, Northern Cape and Eastern Cape.

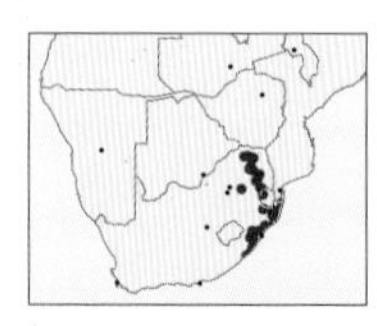

### 2 *Striphnopteryx edulis* — Edible Monkey

Forewing 25–53m. Ground colour cream, variably irrorated with various shades of brown, both wings patterned with multiple wavy lines. Caterpillar orange with dense upright tufts of pale pink hairs. One species in genus in region. **Biology:** Caterpillars polyphagous on dicotyledonous plants. Multivoltine, adults September–May, often abundant in forests in autumn. **Habitat & distribution:** Coastal bush and forests; from Pondoland north into Mozambique and Zimbabwe.

## Subfamily Janinae

Restricted to Africa, and separated from other subfamilies by unique form of the male genitalia.

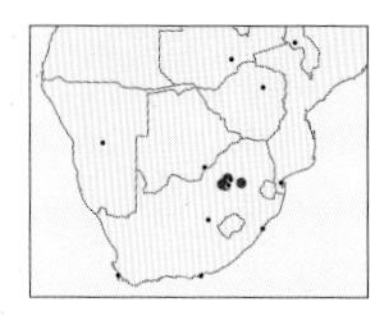

### 3 *Hemijana subrosea* — Blushing Monkey

Forewing 20–26mm. Body and wings pinkish-red; forewing surface irrorated with grey scales, less so in female, crossed by several thin wavy black lines. Caterpillar has 2 forms, one with narrow longitudinal black, red and yellow stripes and sparce tufts of long hairs, the other bluish-grey with prominent dorsoventral black and red longitudinal bands. Three species in genus in region. **Biology:** Caterpillars feed on Sand Apple (*Pygmaeothamnus zeyheri*). Bivoltine, adults September–February. **Habitat & distribution:** Highveld grassland; Gauteng and Mpumalanga, also Angola and Zimbabwe.

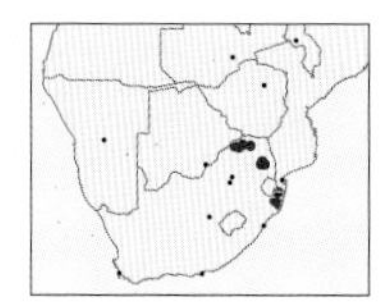

### 4 *Poloma variegata* — Variable Monkey

Forewing 21–25mm. Highly variable, from almost entirely brown to bright pink or red; body typically strongly setose and red; forewings brown to pink, with dark patch crossed by white zigzag lines near base and straight white line and diffuse darker patch distally; hindwings pink. Caterpillar silvery white with long tufts of setae dorsally. **Biology:** Caterpillars feed on False Turkeyberry (*Plectroniella armata*) and Porcupine Bush (*Pyrostria histrix*). Bivoltine, adults September–December and February–March. **Habitat & distribution:** Lowveld savanna; South Africa and Mozambique.

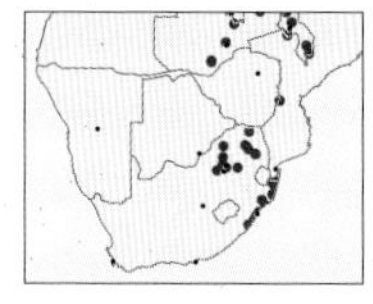

### 5 *Jana transvaalica* — Banded Monkey

Forewing 45–57mm. Thorax with black collar; forewings light brown (melanistic forms also occur) with dark spots near base, a row of tiny dots next to thin postmedian line, followed by a dark, elongate patch; hindwings crossed by 2 black bands. Two species in genus in region. **Biology:** Caterpillars feed on *Afrocanthium mundianum* and *Psydrax obovata* (Rubiaceae). Bivoltine, adults September–May. **Habitat & distribution:** Forests and savanna; Pondoland north into tropical Africa.

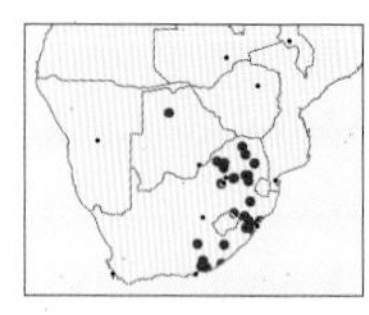

### 6 *Tantaliana tantalus* — King Monkey

Forewing 47–63mm. Similar to *J. transvaalica* (above); thorax black; forewings light brown (melanistic forms also occur) heavily irrorated with brown to black scales, postmedian line deeply toothed, dark patch around cell spot elongate; hindwings crossed by 2 black bands, basal band curved. **Biology:** Caterpillars feed on various Boraginaceae and Oleaceae. Bivoltine, adults July–May, peaking in spring and late summer. **Habitat & distribution:** Forests and grassland; Eastern Cape to Zimbabwe.

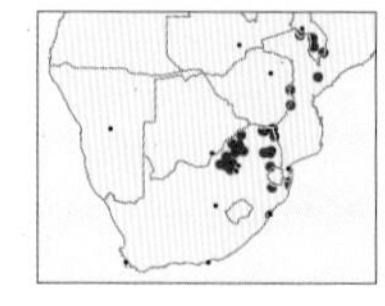

### 7 *Stenoglene obtusa* — Obtuse Monkey

Forewing ♂ 18–23mm, ♀ 26–28mm. Variable, yellow to orange or red, abdomen often flexed to one side; forewings crossed by 1 or 2 broken darker lines; hindwings orange. Hairy caterpillar bluish-grey with black dots, cream dorsal and yellow-orange lateral lines, reddish-brown head. Eight species in genus in region. **Biology:** Caterpillars feed on monkey orange trees (*Strychnos* spp.), at times partially defoliating trees, heavily parasitised by Tachinidae flies. Multivoltine, adults October–May. **Habitat & distribution:** Common and widespread where *Strychnos* grows; KwaZulu-Natal, South Africa, north to tropical Africa.

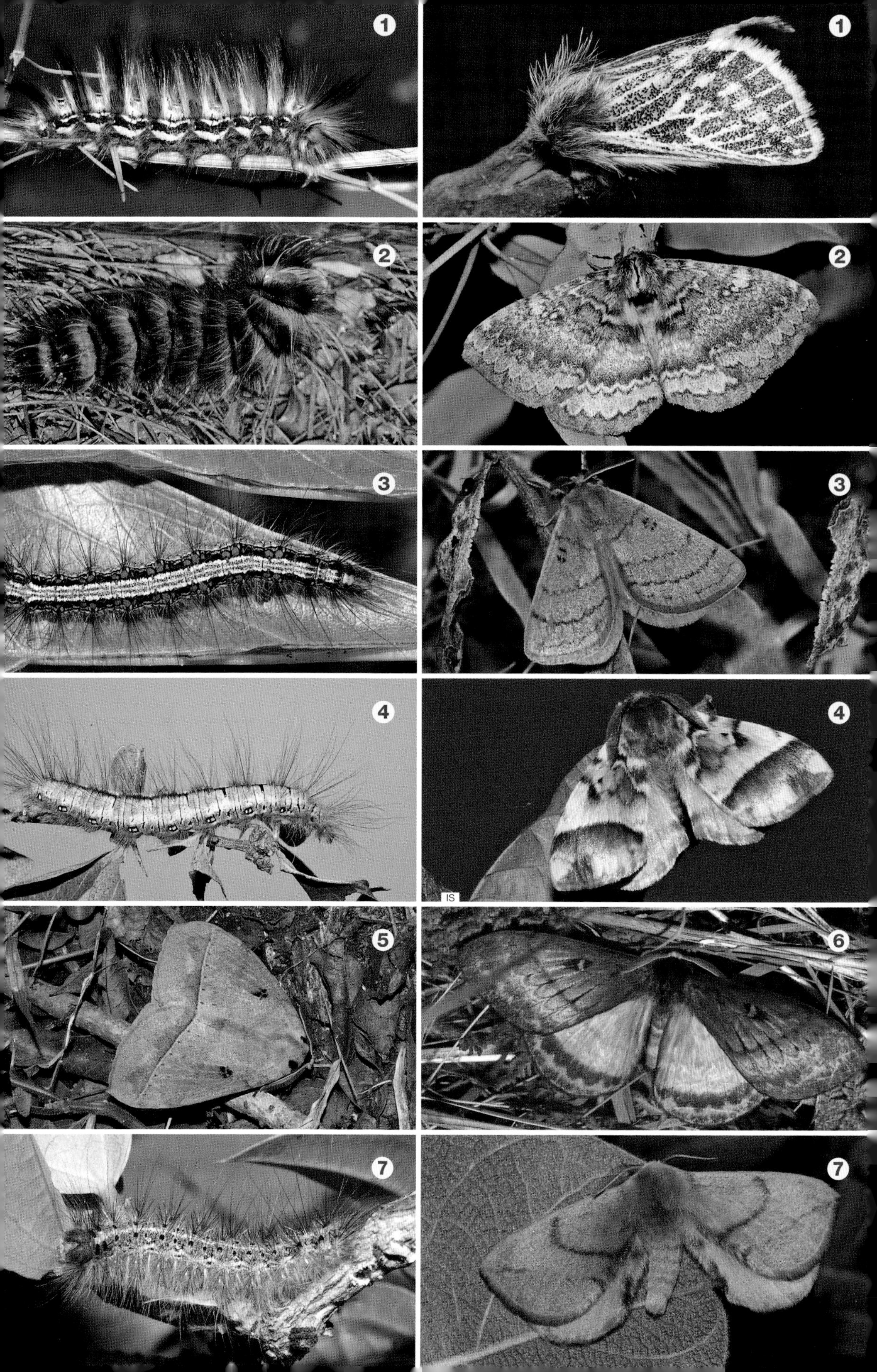
1
1
2
2
3
3
4
4
IS
5
6
7
7

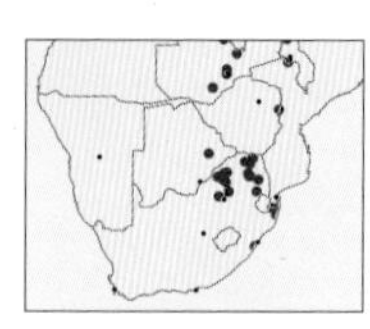

### 1 *Stenoglene roseus* — Rosy Monkey

Forewing ♂ 18–20mm, ♀ 23–24mm. Body and wings deep yellow; forewings variably shaded with rosy pink or absent, distinct brick red postmedian line, faint in females; hindwings plain yellow. **Biology:** Caterpillar host associations unknown. Adults October–May. **Habitat & distribution:** Bushveld and woodland; KwaZulu-Natal, South Africa, north to tropical Africa.

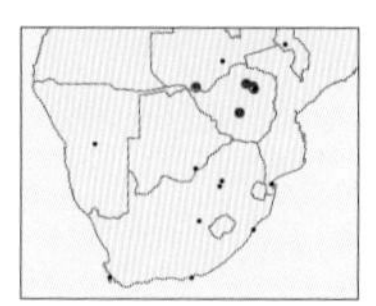

### 2 *Hibrildes norax* — Veined Monarch

Forewing 27–35mm. Variable in colour and sexually dimorphic. Orange; wings rounded, males white, females yellow to orange, with thick brown veins, sometimes black tip to forewings and black band along hindwing margin. Two species in genus in region. **Biology:** Caterpillar host associations unknown. Adults diurnal and nocturnal, with slow 'butterfly-like' flight. Probably bivoltine, adults September and February. **Habitat & distribution:** Coastal and inland forests; KwaZulu-Natal, South Africa and Zimbabwe north to tropical Africa (previously occurred near Durban but not recorded in South Africa since 1960).

## FAMILY LASIOCAMPIDAE Eggar and Lappet Moths

**Medium to large, nocturnal moths attracted to light, with very stout, setose bodies. Antennae usually strongly bipectinate in both sexes; proboscis absent and labial palps often projecting forward, giving 'snouted' appearance. Forewings elongate and relatively narrow; hindwings brown or grey, rounded, and at rest, often project beyond forewings. Often strongly sexually dimorphic; females larger and bulkier than males and rarely wingless. Caterpillars generally flattened, with tufts of short, sometimes brightly coloured hairs and bristles, often extending into long lateral tufts. Setae highly urticating. Some caterpillars are gregarious in the early instars and may build communal webs. Caterpillars of many species congregate on tree trunks by day in a large mass, sometimes reacting to a sudden noise by jumping. Caterpillars disperse up the tree at night to feed on foliage. Produce tough cocoons, sometimes attached to host plant, and can remain dormant to survive droughts for years, others pupate in the soil. Cocoons of some species used in silk industry. Eggars and lappets can be found in almost all vegetated habitats. Some 2,000 species globally, with approximately 227 in region.**

### Subfamily Lasiocampinae

Antennae are less than half the forewing length, the labial palps are usually short and held upright, and there is no small accessory vein in the costal region of the forewing.

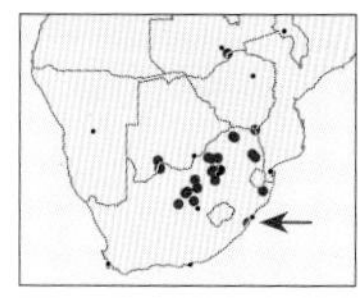

### 3 *Anadiasa affinis* — Related Eggarlet

Forewing ♂ 10–11mm, ♀ 16mm. Body brown, strongly setose; forewings variably banded in shades of brown; median area a darker swathe, with distinctive white patch on inner margin; hindwings white with diagnostic black triangular mark at anal angle. Caterpillar brown with darker dorsal crossbars on anterior segments and blue head, setae brown. Six species in genus in region. **Biology:** Caterpillars feed on Sweet Thorn (*Vachellia karroo*). Multivoltine, adults in warmer months September–April. **Habitat & distribution:** Dry thorn bushveld; South Africa to Malawi.

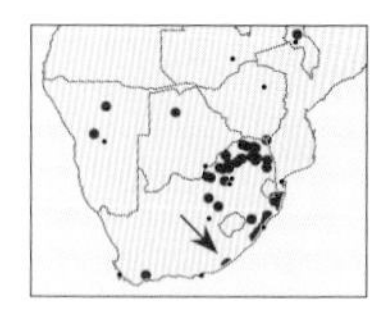

### 4 *Anadiasa punctifascia* — Chestnut Eggarlet

Forewing ♂ 10–11mm, ♀ 16mm. Body light to dark chestnut brown, strongly setose; forewings variable brown, with pattern of light radiating veins; hindwings brown with setose margins. Caterpillar grey with dark crossbars anteriorly lacking long forward-directed setal tufts behind head, setae white, head grey. Possibly a complex of species. **Biology:** Caterpillars polyphagous, feeding primarily on various Fabaceae. Multivoltine, adults in warmer months August–April. **Habitat & distribution:** Primarily in forest and bushveld habitats, but also suburban gardens; across southern and East Africa.

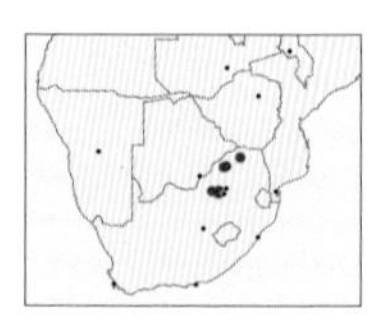

### 5 *Anadiasa jansei* — Janse's Eggarlet

Forewing ♂ 10mm, ♀ 13mm. Light to darker brown setose body; male forewing straw-coloured with dark brown lines and cell spot, female darker grey-brown. Caterpillar with short setae, light brown with dorsoventral light bands. *A. swierstrae* similar but allopatric in Northern Cape. **Biology:** Caterpillars feed on Elephant's Root (*Elephantorrhiza elephantina*). Early instars gregarious in silken nest on host plant, final instars single. Bivoltine, adults October–March. **Habitat & distribution:** Grasslands at higher elevations; Gauteng, North West province and Limpopo, South Africa.

1
2
3
3
4
4
5
5

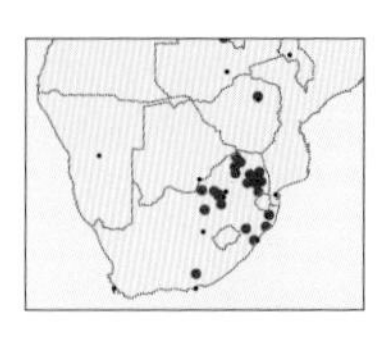

**1 *Beralade perobliqua*** Oblique Eggarlet

Forewing 15–20mm. Sexes similar; white, except for orange front margin and 2 (rarely 1) oblique row of black dots on forewing; legs banded orange and black; form *fumosa*, prevalent in Eastern Cape, light brown with similar pattern. Caterpillar brown and white, with tufts of long setae on some segments. Four similar species in genus in region. **Biology:** Caterpillars feed on Sweet Thorn (*Vachellia karroo*). Adults November–March. **Habitat & distribution:** Widespread across eastern Africa from South Africa to Zambia, outlying record from Eritrea.

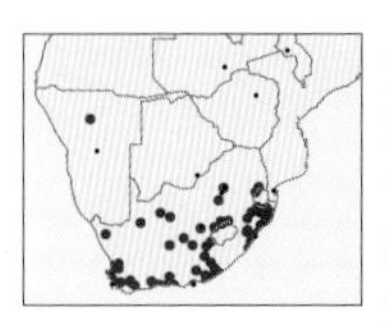

**2 *Bombycomorpha bifascia*** Barred Eggarlet

Forewing ♂ 15mm, ♀ 20mm. Off-white with 2 pale brown bands across forewings (sometimes one in female) and black cell spot. Caterpillar black with narrow red or yellow longitudinal markings and dense yellow setae. Three species in genus. **Biology:** As with all *Bombycomorpha* species, caterpillars feed on Anacardiaceae, primarily *Searsia* spp. but also *Ozoroa* spp. and introduced pepper trees (*Schinus* spp.). Eggs laid in clusters around a twig, cluster covered in a fine layer of silk; caterpillars gregarious; pupation in shallow soil. *Searsia* trees defoliated by caterpillars tend to grow more prolifically the following season than other trees. Multivoltine, adults throughout the year. **Habitat & distribution:** Sympatric with other *Bombycomorpha* spp. in some places; widespread where *Searsia* trees are common, including transformed habitats, but seemingly absent from northern bushveld; South Africa; subsp. *borealis* from East Africa.

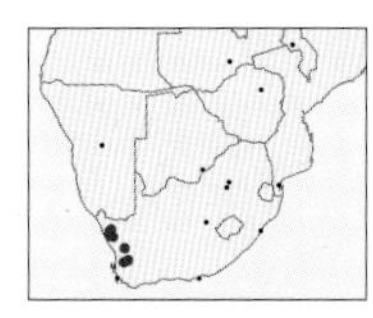

**3 *Bombycomorpha dukei*** West Coast Eggarlet

Forewing ♂ 17mm, ♀ 22mm. White and similar to *B. bifascia* (above), but has more distinct, darker brown lines across forewing, cell spot tiny. Caterpillar similar to *B. bifascia*, but setae tend to be denser and orange. **Biology:** Life history similar to *B. bifascia*. Probably univoltine, adults February–May. **Habitat & distribution:** Endemic to the west coast region of South Africa.

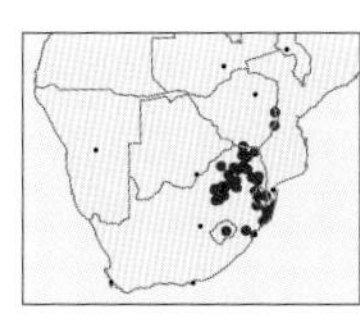

**4 *Bombycomorpha pallida*** Pallid Eggarlet

Forewing ♂ 13–15mm, ♀ 20mm. All white, no lines, prominent dark cell spot on forewing. Caterpillar similar to *B. bifascia*, but yellow setae are white in some specimens and tend to be less dense and shorter. **Biology:** Life history similar to *B. bifascia*. Multivoltine, adults August–May. **Habitat & distribution:** Various habitats; in northeastern southern Africa, South Africa, Mozambique and Zimbabwe.

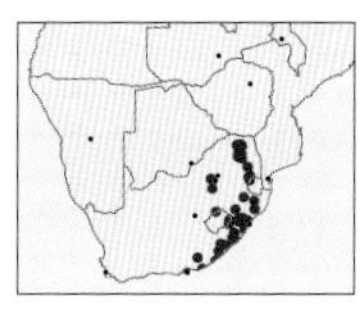

**5 *Bombycopsis bipars*** Divided Eggar

Forewing ♂ 13–19mm, ♀ 21–28mm. Setose with very prominent 'snout'; forewings light to dark brown, with dark longitudinal bar, white cell spot, lighter oblique lines, subterminal band broken with several crossbars is diagnostic; hindwings plain pale yellowish. Caterpillar reddish-brown with 2 dark dorsoventral longitudinal stripes and conspicuous lateral tufts, pale setae. Confused with many similar species in the past. Nineteen species in genus in region. **Biology:** Caterpillars feed on *Halleria lucida* (Stilbaceae) and various Asteraceae. Multivoltine, adults throughout the year. **Habitat & distribution:** Wet forests and riverine vegetation; Eastern Cape to Limpopo, South Africa.

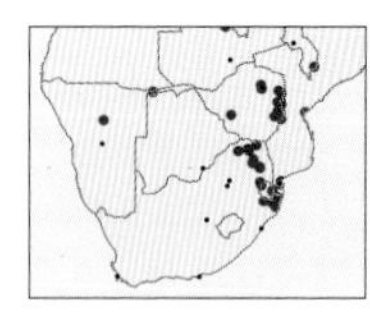

**6 *Bombycopsis nigrovittata*** Blackstreak Eggar

Forewing ♂ 14–20mm, ♀ 25–32mm. Squared-off outer margin of the wing and weak claw-like markings on the distal margin distinguish this and similar species from other *Bombycopsis* groups. Caterpillar dull brown with tufts of brown setae. Several very similar species from further south in South Africa. **Biology:** Caterpillars polyphagous. Multivoltine, adults throughout the year. **Habitat & distribution:** Forest and bushveld; Zululand, South Africa, north to DRC and Uganda.

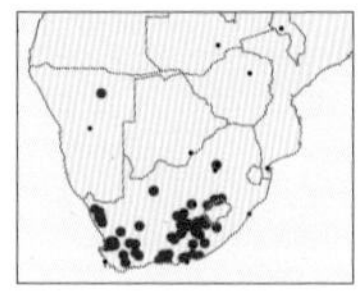

**7 *Bombycopsis tephrobaphes*** Karoo Eggar

Forewing ♂ 12–17mm, ♀ 16–22mm. Body cream; strongly setose forewings light to dark grey with orange veins and row of small black dots along distal margin; distinctive central black band with white cell spot. Greyish colouration diagnostic. Caterpillar has dark reddish body with several black, longitudinal dorsoventral lines, weakly covered in grey setae. **Biology:** Caterpillars feed on African Sheepbush (*Pentzia incana*) and other Asteraceae. Multivoltine, adults throughout the year. **Habitat & distribution:** Widespread and common in karroid habitats; South Africa.

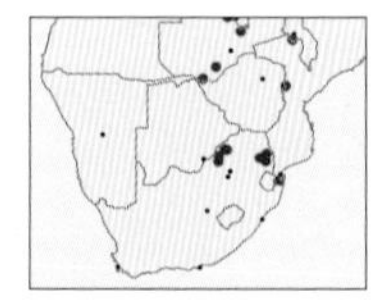

**8 *Gastroplakaeis meridionalis*** Dusted Eggar

Forewing 24–25mm. Body white speckled with reddish-brown scales, densely at base and distally a conspicuous black cell spot. Caterpillar unknown. Monotypic genus. **Biology:** Host plants unknown. Probably bivoltine, adults September–December and February–March. **Habitat & distribution:** Lowland bushveld and woodland; Maputaland north to southern DRC.

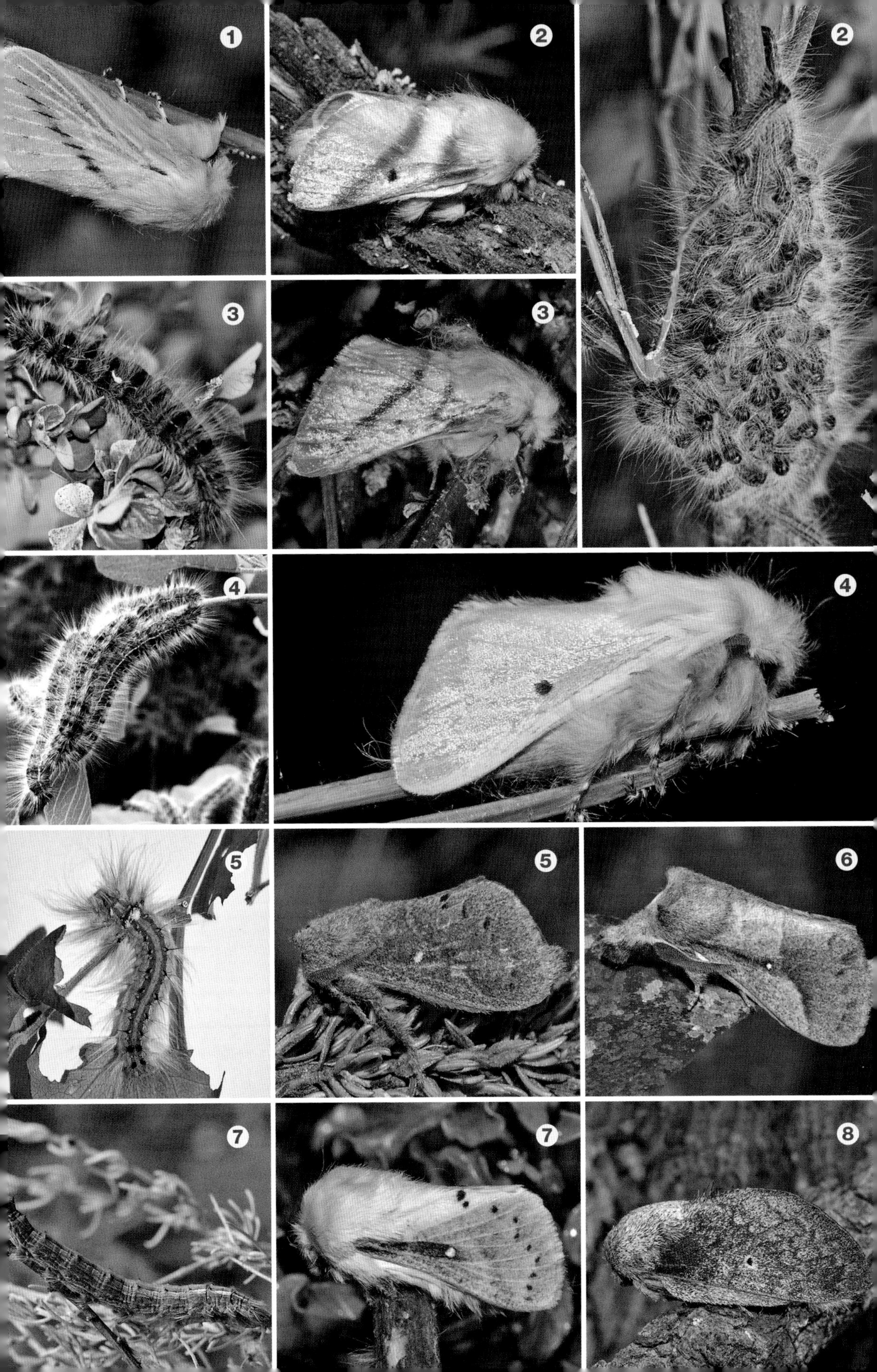

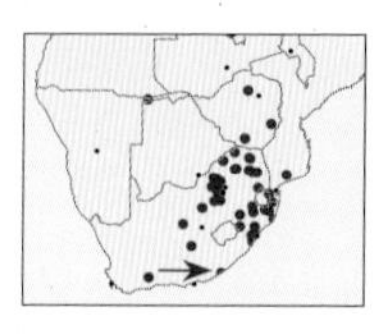

### 1 *Braura truncata* — Truncate Lappet

Forewing ♂ 19–24mm, ♀ 35–40mm. Thorax dark brown with lighter longitudinal bands; forewings light to dark brown with wavy paler lines along margin; hindwings plain, male brown, female cream. Caterpillar grey-brown with black spots along dorsal surface. Three very similar species (others not as white) in genus in region. **Biology:** Caterpillars feed on Sweet Thorn (*Vachellia karroo*), Paperbark Thorn (*V. sieberiana*), Black Wattle (*Acacia mearnsii*) and Mexican Weeping Pine (*Pinus patula*). Multivoltine, adults August–April. **Habitat & distribution:** Various habitats where thorn trees grow, expanding its range due to wattle infestation; South Africa north to Angola and Tanzania.

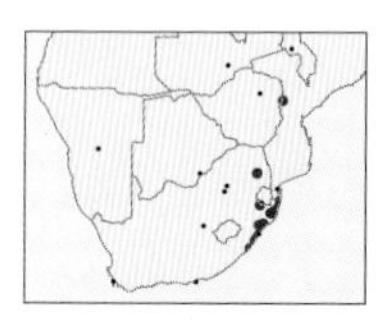

### 2 *Catalebeda cuneilinea* — Toothed Cream Spot

Forewing 33–58mm. Thorax reddish-brown with diagnostic white dorsal spot posteriorly; forewings reddish-brown, crossed by 3 darker wavy lines; hindwings plain brown. Caterpillar dark brown dusted with short white setae with lighter red or orange dorsal crossbars and very long lateral tufts of white setae, longer anteriorly. Three species in genus in region. **Biology:** Caterpillars feed on *Dalbergia* spp. Multivoltine, adults throughout the year. **Habitat & distribution:** Forests where *Dalbergia* trees grow; South Africa, Mozambique, Zimbabwe to tropical Africa.

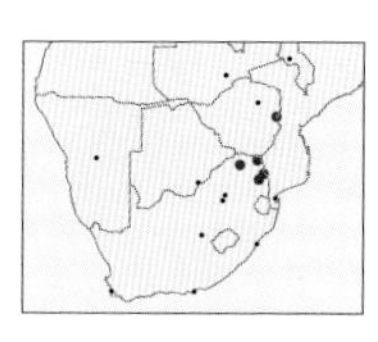

### 3 *Dollmania cuprea* — Purple Eggar

Forewing ♂ 24–28mm, ♀ 42mm. Thorax and forewings purplish to reddish-brown crossed by darker wavy bands, with crescent-shaped cell dot ('stigma'); hindwings pale reddish-brown. Caterpillar, early instars yellow with black bands; final instar black, red and cream, with blue head and prominent lateral tufts of white setae. Four similar species in genus. **Biology:** Caterpillars feed on fig trees (*Ficus* spp.), Sjambok Pod (*Cassia abbreviata*) and Pink Shower (*C. javanica*). Multivoltine, adults October–March. **Habitat & distribution:** Lowland bushveld and woodland; Limpopo, South Africa.

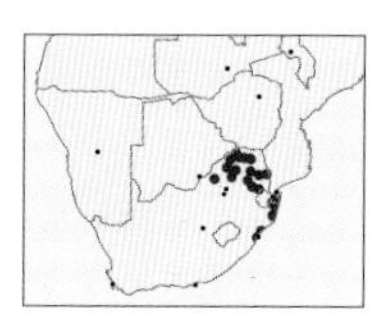

### 4 *Eucraera salammbo* — Salammbo

Forewing 20–29mm. Thorax and forewings orange-yellow; abdomen in male black; wings crossed by angled white lines, with a small black cell spot; scalloped margins of hindwings project forward beyond forewing when at rest, bearing androconial organ in males. Caterpillar black with complex white or pink markings and tufts of long white or black setae. *E. koellikerii* similar, darker with more pronounced cell spot. Three species in genus in region. **Biology:** Caterpillars feed on Marula (*Sclerocarya birrea*), once recorded on *Eucalyptus*. Multivoltine, adults October–April. **Habitat & distribution:** Bushveld and subtropical forest where marula trees grow; from South Africa through East Africa to Ethiopia.

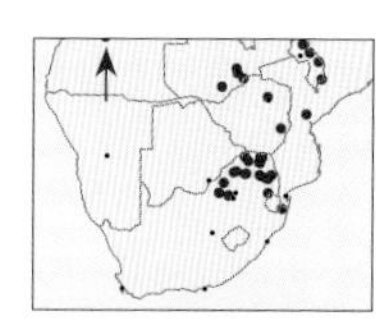

### 5 *Eucraera gemmata* — Budded Lappet

Forewing 18–25mm. Similar to *E. salammbo* (above) but smaller and paler creamy olive with banded orange and black abdomen. Multicoloured caterpillar ornamented with long tufts of black and white setae. **Biology:** Caterpillars feed on Fabaceae and Anacardiaceae. Multivoltine, adults October–April. **Habitat & distribution:** Bushveld and woodland; from Maputoland north to tropical Africa.

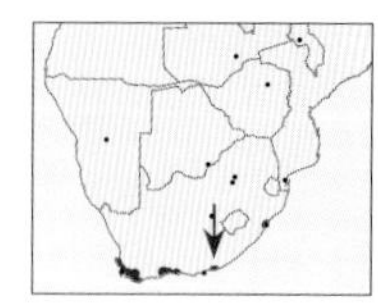

### 6 *Eutricha capensis* — Cape Lappet

Forewing ♂ 22–26mm, ♀ 36–40mm. Body and wings fawn, median and postmedian lines black edged with white, crenulated, median line curved inward at costa; wings more rounded in female; hindwings fawn, reddish-brown at base. Caterpillar with black dorsal band edged by white and long lateral tufts of ginger setae, mauve marking near head. Often confused with other species in the genus. *Eutricha* spp. separated easier with caterpillars than adults. Five species in genus in region. **Biology:** Caterpillars conspicuous, aggregate on trunks of trees during the day. Polyphagous on trees from many families. Multivoltine, adults throughout the year. **Habitat & distribution:** Mesic habitats, including those transformed; restricted to the fynbos biome, Western Cape and Eastern Cape.

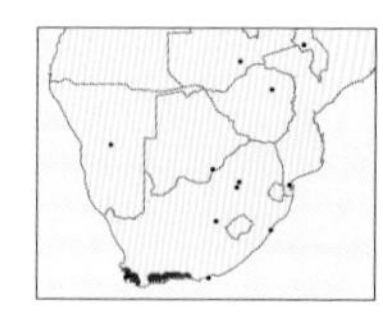

### 7 *Eutricha bifascia* — Fynbos Lappet

Forewing ♂ 24–26mm, ♀ 36–42mm. Similar to *E. capensis*; darker reddish-brown; forewing median area dark brown, median and postmedian lines black not crenulated. Caterpillar ginger with 4 black dorsoventral bands, purple and ginger setae but white setae absent. **Biology:** Unlike the tree-feeding *E. capensis*, caterpillars found singly feeding on various fynbos bushes and herbs, including *Erica*, *Osteospermum* and *Passerina* spp. Multivoltine, adults September–February. **Habitat & distribution:** Fynbos; Western Cape and Eastern Cape.

1
1♂
2
2
3
IS
3♂
4
5
5
6
RO
MB
6♀
7♀
7

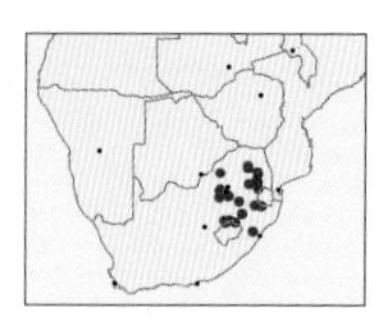

## 1 *Eutricha obscura* — Highveld Lappet

Forewing ♂ 23–26mm ♀ 36–41mm. Female similar to *E. capensis* (p.252); fawn-coloured, forewing median area same as ground colour; forewings of male densely covered in dark scales, sometimes obscuring lines, lines black, not white edged. Caterpillar similar to *E. capensis* but dorsal setae white and dorsoventral bands black. **Biology:** Like *E. capensis*, congregate on tree trunks by day, polyphagous, feeding on the foliage of many different tree species at night. Multivoltine, adults September–February. **Habitat & distribution:** Forest and riverine bush, including transformed habitats; Highveld areas, Free State, KwaZulu-Natal, Gauteng, Mpumalanga and Limpopo, South Africa.

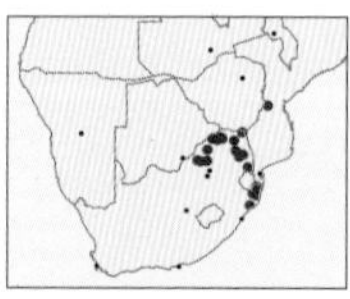

## 2 *Eutricha morosa* — Lowveld Lappet

Forewing ♂ 25–27mm, ♀ 32–36mm. Similar to *E. obscura* (above), but wing white, heavily irrorated with darker brown or charcoal scales, sometimes almost covering the ground colour, lines thin, wavy white or dark. Caterpillar light grey to very dark with white markings and long white or yellow setae. **Biology:** As with other *Eutricha* spp., caterpillars feed on a wide variety of trees. Multivoltine, adults August–May. **Habitat & distribution:** Lowveld bushveld and woodland; Zululand, South Africa, north to tropical Africa.

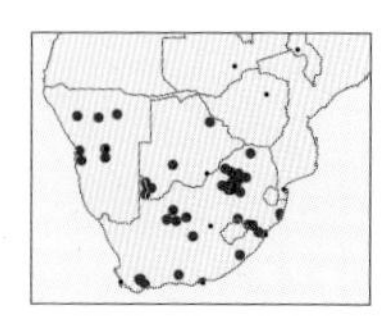

## 3 *Gonometa postica* — African Silk Moth

Forewing ♂ 20–22mm, ♀ 39–45mm. Adults sexually dimorphic; male with red-brown abdomen, purple-brown forewings and small brown hindwings with transparent margin; female much larger, with yellow abdomen, forewings purple-brown with white crossbands, hindwings with yellow base and dark brown margin. Caterpillar, early instars yellow and black with white setae, final instar black and white with diagonal bands and tufts of long irritant setae. Four species in genus in region. **Biology:** Caterpillars feed primarily on *Vachellia* trees in South Africa, but also on a wide range of plants, including *Brachystegia*, *Elephantorrhiza*, *Pinus radiata* and *Julbernardia*. Cocoons produce high-quality silk harvested by local peoples for centuries. Commercial exploitation is hindered by calcium oxalate crystals on the cocoon surfaces. Dried cocoons traditionally used as ankle rattles. Ingestion of cocoons can cause death of cattle, antelope and other ruminants, as the silk blocks the rumen, causing starvation. Univoltine, with long resting period as pupa, adults throughout the year after good rains. **Habitat & distribution:** Wherever thorn trees grow; South Africa north to Kenya and CAR.

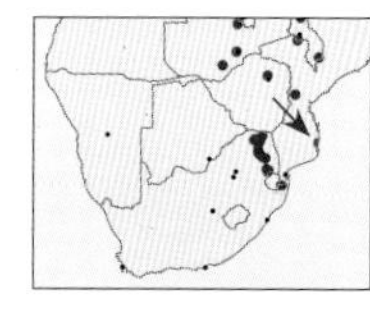

## 4 *Grammodora nigrolineata* — Black-lined Eggar

Forewing 20–30mm. Body and wings cream; forewings patterned with fine longitudinal black lines and a few red lines; hindwings uniform cream. Monotypic genus. **Biology:** Caterpillars polyphagous on trees. Adults September–April. **Habitat & distribution:** Found in lowveld bushveld, forest and woodland; Maputoland, South Africa, north to DRC and Tanzania.

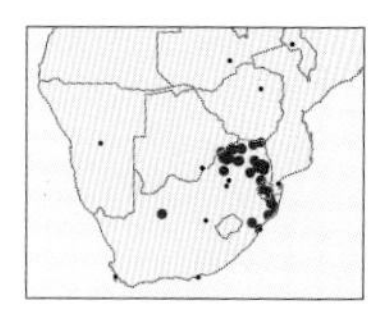

## 5 *Henometa clarki* — Clark's Eggar

Forewing 20–28mm. Body and forewings red-brown, sometimes with violet sheen, forewings crossed by 3 darker wavy lines, conspicuous shiny tympanal organ at base of upper side of forewing; hindwings pale reddish. Caterpillar grey with dark dorsal band, short red setae and long tufts of white and black setae. Two species in genus in region. **Biology:** Caterpillars feed on Weeping Boer-bean (*Schotia brachypetala*), also recorded on Sicklebush (*Dichrostachys cinerea*). Multivoltine, adults August–April. **Habitat & distribution:** Lowveld bushveld areas; South Africa, Botswana, Namibia and Zimbabwe.

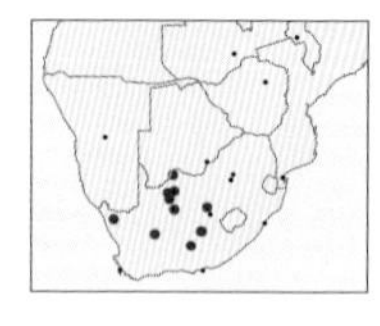

## 6 *Napta straminea* — Ivory Lappet

Forewing ♂ 14–17mm, ♀ 25–26mm. Body, wing undersides and hindwing upper side white; upper side of forewing plain ivory, forewing costa and underside variably irrorated with dark brown scales in some specimens, otherwise no markings. Antennae black with brown pectinations. Six species in genus, one in region. **Biology:** Host association unknown. Possibly bivoltine, adults September–April. **Habitat & distribution:** Nama karoo and Kalahari; South Africa, probably adjacent Botswana and Namibia.

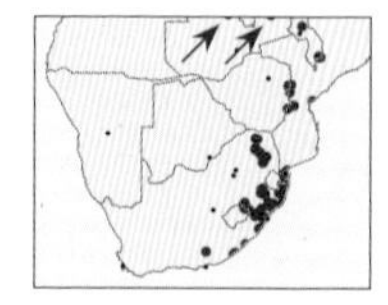

## 7 *Leipoxais acharis* — Related Lappet

Forewing ♂ 14–16mm, ♀ 22–26mm. Has unusual broad pentagonal outline at rest; male dark reddish-brown with forewing slightly rounded, dark postmedian line sharply indented with distinctive white cell spot; female fawn to light brown, with forewing falcate, weaker postmedian indentation and cell spot smaller; hindwings uniform brown. Caterpillar white (sometimes grey) dorsally, brown to black laterally with orange dots, dense tufts of white setae at both ends. Three described species and several undescribed regional species in genus. **Biology:** Caterpillars polyphagous on woody and herbaceous plants, gregarious in early instars. Multivoltine, adults throughout the year. **Habitat & distribution:** Confined to high-rainfall areas in forest, bush and gardens; Eastern Cape north into tropical Africa.

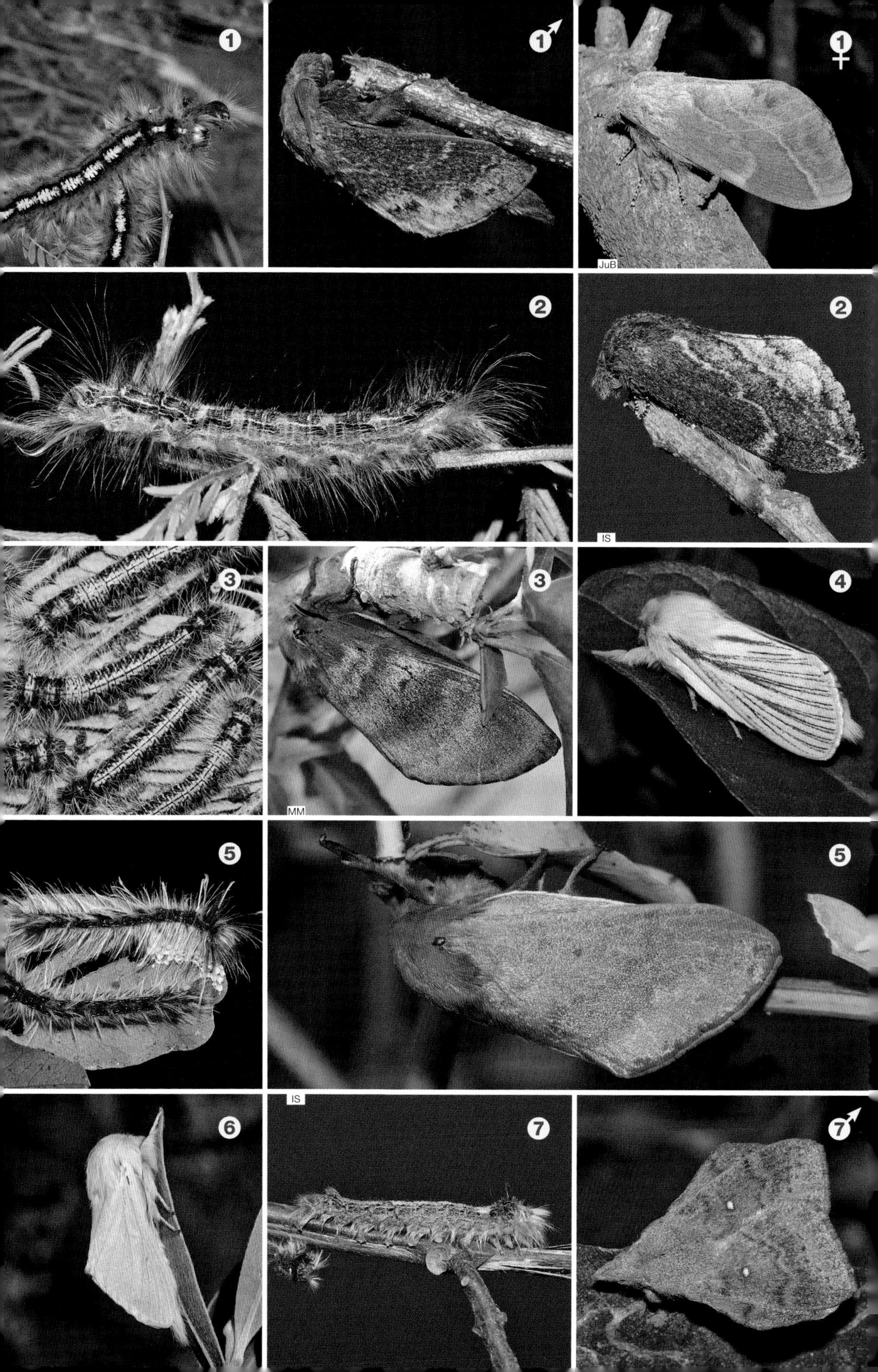
1
1♂
1♀
JuB
2
2
IS
3
3
MM
4
5
5
IS
6
7
7♂

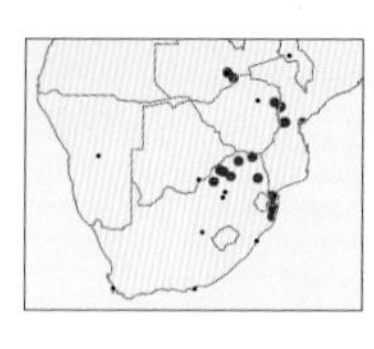

## 1 *Soligena juna*
Fawny Lappet

Forewing ♂ 14–16mm, ♀ 22–26mm. Body and wings fawn to rose-brown; male forewing slightly falcate, female wings heavily scalloped, 2 parallel yellow lines across forewing; small brown cell spot; curved brown line across hindwing. Two species in genus in region. **Biology:** Caterpillars feed on Silver Cluster-leaf (*Terminalia sericea*). Multivoltine, September–April. **Habitat & distribution:** Sandy bushveld where host plant grows; Zululand, South Africa, north to Kenya.

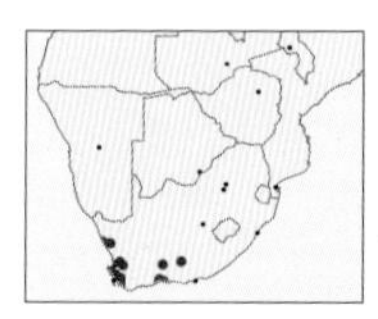

## 2 *Mesocelis monticola*
Mountain White Spot

Forewing ♂ 15–17mm. Strongly sexually dimorphic; male winged, with golden body and dark brown to black wings, each with a central white translucent window; female wingless, with golden-yellow body. Caterpillar black, with longitudinal white lines, covered in long white and black setae. Two black lines across orange head capsule diagnostic for *Mesocelis*. Two species in genus, both in region. **Biology:** Caterpillars polyphagous, often seen on ground walking and feeding on different plants. Multivoltine, adults throughout the year. **Habitat & distribution:** Wet fynbos and grassland; mainly Western Cape, but also on mountains with fynbos remnants in Nyanga, Zimbabwe, and Mozambique.

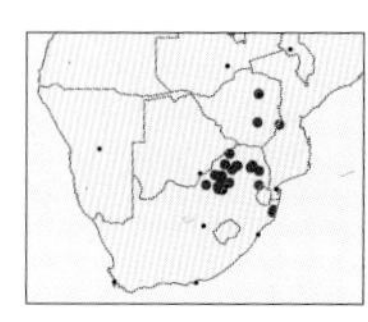

## 3 *Metajana marshalli*
Marshall's Lappet

Forewing ♂ 21–25mm, ♀ 26–32mm. Thorax pale grey, brown in front; forewings mottled brown and crossed by 2 jagged dark lines; hindwings pinkish, basal half darker. Caterpillar, early instars bluish with white dorsal band; final instar yellow and black, sometimes red, with long lateral tufts of yellow, red or white setae. Two species in genus in region; similar *M. wahlbergi* allopatric, Namibia north, more species further north. **Biology:** Caterpillars feed on *Combretum* spp. Bivoltine, adults August–March. **Habitat & distribution:** Bushveld; Zululand, South Africa, north to Tanzania.

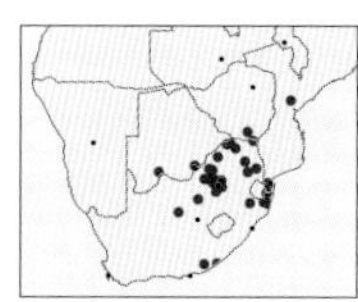

## 4 *Odontocheilopteryx myxa*
Hairy Eggarlet

Forewing ♂ 12–15mm, ♀ 20–25mm. Thorax white; forewings with complex patchwork of cream, greys and browns, with central white spot and unusual hairy lobe on posterior margin (projecting upward when at rest); hindwings cream to white, cilia not chequered. Caterpillar orange with red and black dorsal bars, long setae in tufts protruding laterally. Similar *O. obscura*, sympatric, has hindwing cilia chequered. Three species in genus in region. **Biology:** Caterpillars feed primarily on thorn trees (*Vachellia* and *Senegalia*), also use various Australian wattles (*Acacia* spp.) and recorded on *Eriosema*, all Fabaceae. Multivoltine, adults throughout the year. **Habitat & distribution:** Bushveld; widely distributed across Africa from South Africa to Ethiopia, also Madagascar.

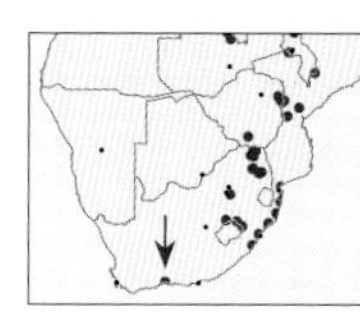

## 5 *Pallastica pallens*
Hawthorn Lappet

Forewing ♂ 20–26mm, ♀ 32–48mm. Sexually dimorphic; male body and forewing reddish-brown, forewing crossed by 2 double lines of black dots; abdomen of female banded black and brown, forewings yellowish-brown crossed by 2 converging pale bands. Caterpillar brown with 2 transverse red and black bristles on anterior segments. Four species in genus in region. **Biology:** Caterpillars polyphagous on many tree species. Multivoltine, adults throughout the year. **Habitat & distribution:** Forest and riverine bush; southern Cape, South Africa, north to tropical Africa.

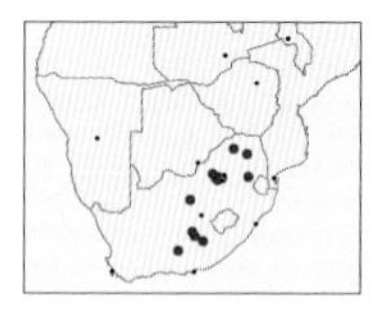

## 6 *Philotherma rennei*
Curdled Eggar

Forewing ♂ 35–38mm, ♀ 38–41mm. Colour of body and forewings variable, straw to reddish-brown, variably irrorated with black scales, more so in females; postmedian line strongly curved, cell spot black with white dot, distal margin with small black spots. Caterpillar brown with white lateral bars merging ventrally, densely covered in long brown setae. Six species in genus in region. Easily confused with many similar species further north, allopatric in different habitats. **Biology:** Caterpillars polyphagous on trees and shrubs, often seen walking over ground and roads, enter buildings for shelter during winter. Possibly univoltine, adults September–March. **Habitat & distribution:** Karroid and highveld grasslands; Graaff-Reinet to Polokwane, South Africa.

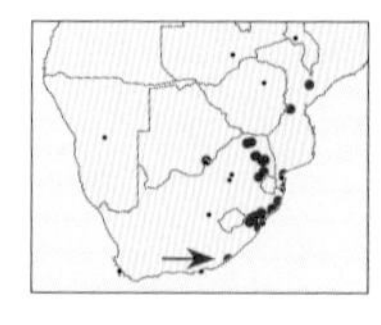

## 7 *Philotherma rosa*
Rose Eggar

Forewing ♂ 40–43mm, ♀ 48–51mm. Very similar to *P. rennei* (above), but larger with postmedian line straight and curved near costa; females often heavily irrorated with black scales. Caterpillar dark with pale lateral bars and covered in long, reddish-brown setae. Similar to *P. aniera* from Namibia. **Biology:** Caterpillars polyphagous on trees and shrubs. Possibly univoltine, adults August–December. **Habitat & distribution:** High-rainfall areas in forest, coastal bush and gardens; Eastern Cape to Tanzania.

1
2
2♂
MM
3
3
4
4
5♂
6
6
7
PW
7♂
7♀

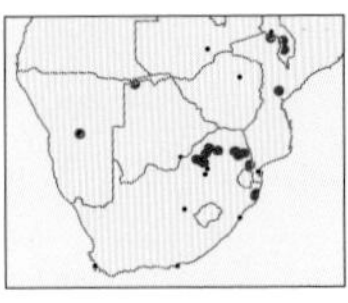

### 1 *Rhinobombyx cuneata* — Zigzag Eggarlet

Forewing ♂ 15–16mm, ♀ 23–25mm. Male body and wings dark brown, black cell spot with white centre, postmedian line black, strongly zigzagged; hindwing uniform brown, strongly scalloped front margin projects from beneath forewing when at rest; female larger, fawn-coloured, cell spot black; hindwing lighter with dark median line. Caterpillar with black, yellow and blue dorsal patterning and strong lateral tufts of black and white setae. Monotypic genus. **Biology:** Caterpillars feed on bushwillows (*Combretum*). Multivoltine, adults August–May. **Habitat & distribution:** Lowveld *Combretum* bushveld; Zululand, South Africa, north to Tanzania.

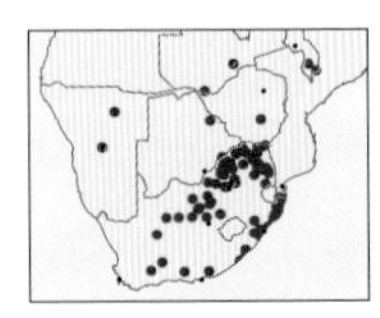

### 2 *Sena prompta* — Interjection

Forewing ♂ 11–15mm, ♀ 16–19mm. Ground colour white, variably irrorated with pale grey to brown scales; forewings crossed by 3 straight diagonal lines and with characteristic central white dash like 'exclamation mark'; hindwings white to pale brown. Caterpillar black and white banded with orange spots and few very long tufts of both black and white setae in early instars; final instar with many blue, orange, yellow and white lines and spots. Nine species in genus in region. **Biology:** Caterpillars feed primarily on thorn trees (*Vachellia* spp.), as well as Sicklebush (*Dichrostachys cinerea*) and Veld Violet (*Ruellia cordata*). Multivoltine, adults throughout the year. **Habitat & distribution:** Wide range of semi-arid vegetation types across southern and eastern Africa, north to Somalia.

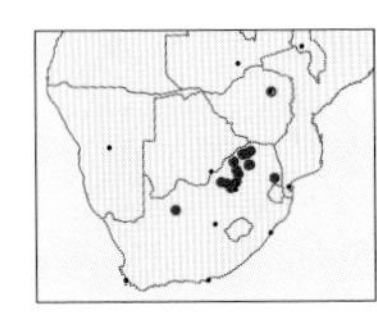

### 3 *Sena donaldsoni* — Donaldson's Eggarlet

Forewing ♂ 14–16mm, ♀ 21–24mm. Similar to *S. prompta* (above), but with white dash on forewing longer and bent into more of a 'V'-shape extending to base, broken in some individuals, only submarginal line present. Caterpillar dull yellow and brown. **Biology:** Caterpillars recorded on Yellow Thatching Grass (*Hyparrhenia dissoluta*). Probably multivoltine, adults September–March. **Habitat & distribution:** Thorn bushveld; South Africa north to Ethiopia.

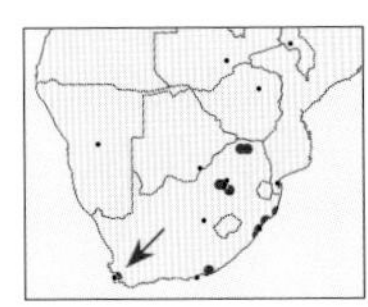

### 4 *Stoermeriana aculeata* — Saddleback

Forewing ♂ 22–26mm, ♀ 29–32mm). Thorax cream with dark shoulders; forewing light brown with large white oval patches on posterior margin, these meeting to form a white 'saddle' when perched, oblique postmedian line curved outward, black cell spot distinct; hindwings uniform light brown. Caterpillar grey with brown oval markings dorsally and 2 dark brown bands behind head. Six species in genus in region. **Biology:** Caterpillars feed on Forest Natal Mahogany (*Trichilia dregeana*), also Umbrella Tree (*Maesopsis eminii*) and pines (*Pinus leiophylla* and *P. patula*). Probably multivoltine, adults throughout the year. **Habitat & distribution:** Widely recorded in different habitats; South Africa and Kenya.

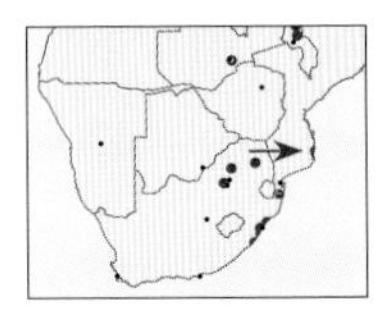

### 5 *Stoermeriana cuneata* — Dark Saddleback

Forewing ♂ 21–24mm, ♀ 29mm. Similar to *S. aculeata* but darker reddish-brown, oblique postmedian line curved inward and cell spot indistinct. Caterpillar light brown, darker dorsally, lateral setae white tipped. **Biology:** Caterpillars feed on Nana-berry (*Searsia dentata*), probably also on other trees. Apparently multivoltine, adults throughout the year. **Habitat & distribution:** In widely different habitats; from Pondoland in South Africa north to Mozambique and Zambia.

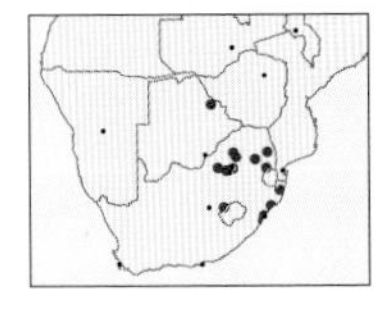

### 6 *Streblote carinata* — Carinate Eggar

Forewing ♂ 15–18mm, ♀ 27–32mm. Strongly sexually dimorphic; females bulkier and with broader wings; forewings of male indented and all wings with straight outer margins (convex in female); thorax grey dorsally, separated by white band from reddish sides; forewing reddish-brown with a white streak at base; fading distally where crossed by white line; hindwings cream. Caterpillar light grey with dark dorsal band and setal tufts behind head. Eleven species in genus in region. **Biology:** Caterpillars polyphagous on trees and shrubs. Multivoltine, adults throughout the year. **Habitat & distribution:** In widely different habitats; KwaZulu-Natal,, South Africa, north to Botswana, Zimbabwe and Tanzania.

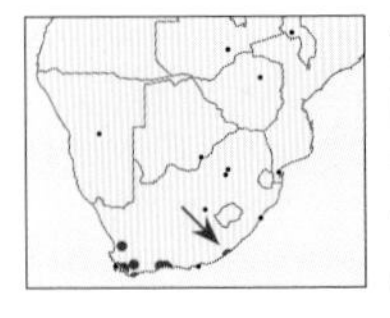

### 7 *Streblote cristata* — Crested Eggar

Forewing ♂ 21mm, ♀ 26mm. Red, variably irrorated with golden scales; sinuous white lines on forewings; hindwings deep red-brown. Caterpillar variable, cream to greenish-grey, sometimes blue, with pale longitudinal stripes and large orange or red setal clumps behind head. Similar *S. jansei* allopatric in highveld grasslands. **Biology:** Caterpillars polyphagous on woody plants. Multivoltine, adults August–May. **Habitat & distribution:** Fynbos; Western Cape and Eastern Cape, South Africa.

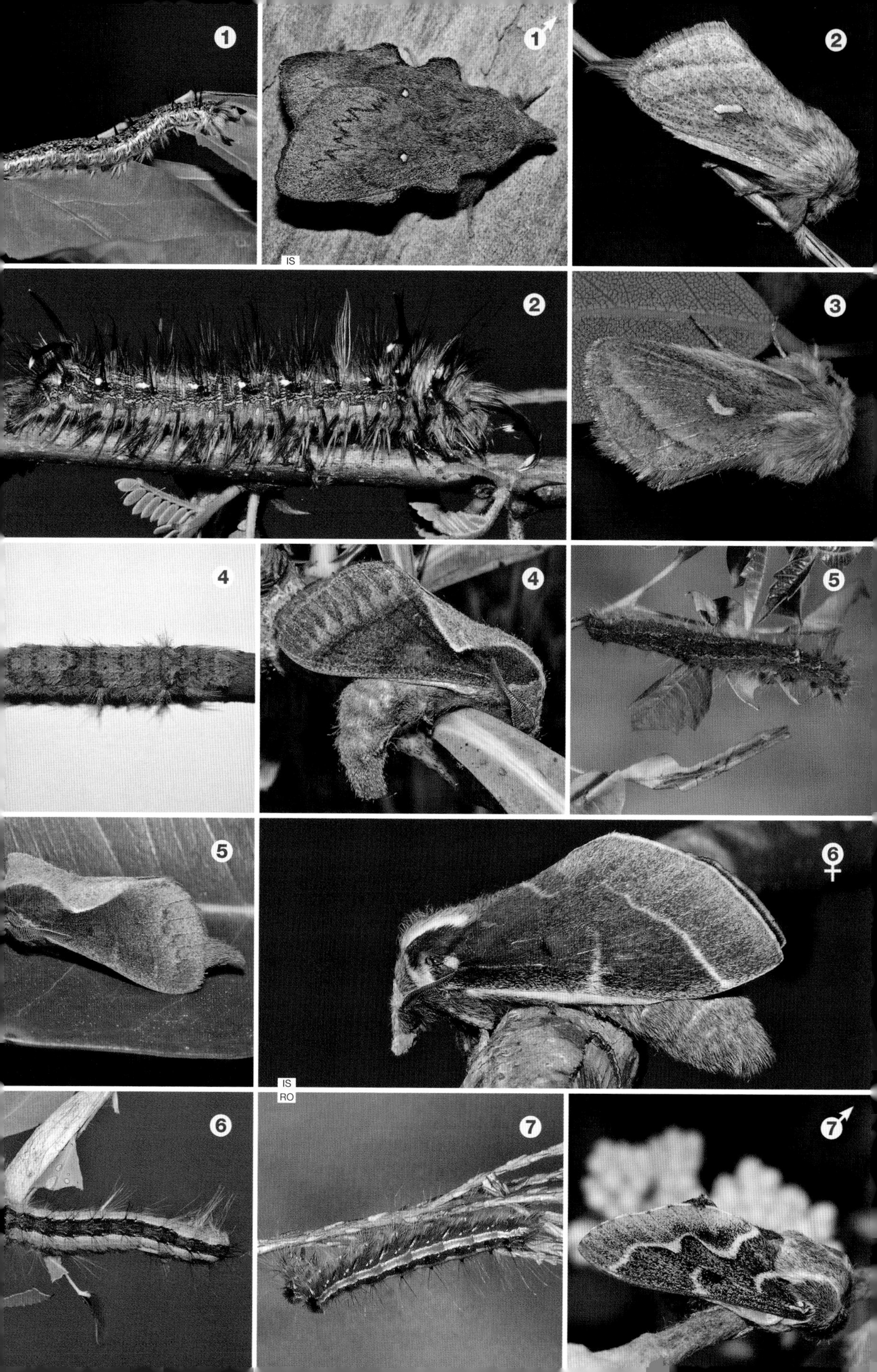
1
1♂
IS
2
2
3
4
4
5
5
6♀
IS
RO
6
7
7♂

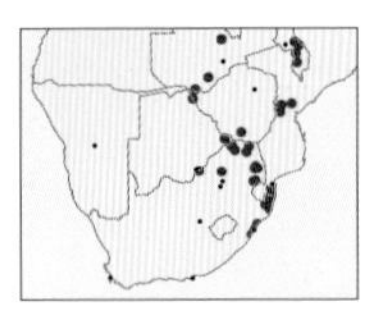

### 1 *Trichopisthia monteiroi* — Lemon Lappet

Forewing ♂ 24–27mm, ♀ 38–42mm. Cream to yellow; forewing straight, median lines brown, cell spot orange; sinuous fore-margin of hindwings bear black setae and project forward from beneath forewings at rest. Caterpillar black with pairs of transverse white bars dorsally and long white setae. Two species in genus in region. **Biology:** Caterpillars feed on Marula (*Sclerocarya birrea*). Multivoltine, adults October–April. **Habitat & distribution:** Coastal bush and bushveld where marula trees grow; southern KwaZulu-Natal coast, South Africa, north to DRC and Tanzania.

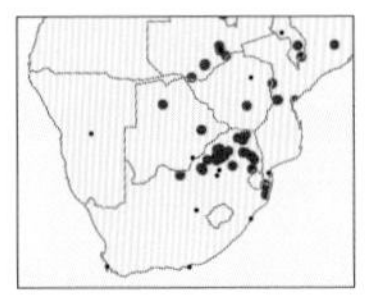

### 2 *Trichopisthia igneotincta* — Fiery Lappet

Forewing ♂ 19–21mm, ♀ 25–27mm. Extremely variable patterns of yellow, orange, red and dark brown on forewing, median lines prominent or absent; forewing colour repeats on otherwise yellow hindwings on the part that is visible when at rest. Caterpillar light grey to blue, with black or brown and yellow or orange transverse bands and long prominent tufts of white setae at both ends. **Biology:** Caterpillars feed on Silver Cluster-leaf (*Terminalia sericea*). Multivoltine, adults September–April, abundant at times when different forms can be found together. **Habitat & distribution:** Sandy bushveld where silver Cluster-leafs grow; Zululand, South Africa, north to DRC and Tanzania.

## FAMILY SATURNIIDAE Emperor Moths

**Includes some of the largest and best-known moths. Adults have small heads and vestigial mouthparts and do not feed, living for only a week or so. Adults live off their fat reserves stored in their heavy bodies. They are covered with hair-like scales and the large, often lobed wings are normally marked with prominent ringed 'eyes' with a clear centre spot. Males use enlarged, comb-like antennae to detect pheromones produced by females up to several kilometres distant. Caterpillars large, cylindrical and often covered by spines or hairs; some are cryptic, others very colourful. Some species are significant defoliators of trees; a few, such as the well-known mopane worm, are exploited for human food. Some species pupate in a silken cocoon, others underground. Most species in the region are strictly nocturnal, with females generally attracted to light in the early evening and males after midnight. Some are diurnal and are protected by aposematic colouration (*Pseudaphelia* and *Eochroa*); in others the males fly rapidly in the early morning or late afternoon but their females are nocturnal (*Ludia* and *Vegetia*). Most have a slow, flapping flight. Their life cycle is relatively long, univoltine or bivoltine, with some trivoltine in subtropical areas. Adults have short specific flight periods; some swarm in good seasons. These moths are most numerous in forest and bushveld, but can be found in other habitats. About 2,300 described species globally, 81 in the region.**

### Subfamily Saturniinae

Most species have eyespots, lunate markings or translucent panels in their wings. They are short-lived, nocturnal and generally large to very large. Oblong eggs laid in clusters, pupal cocoon of yellow silk.

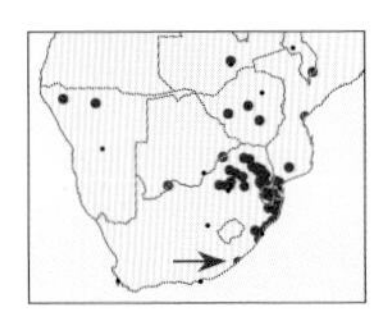

### 3 *Argema mimosae* — African Moon Moth

Forewing 50–62mm. Body yellowish, wings green; forewings with brown leading edge; hindwings with long twisted tails; both wings with prominent eyespot. Caterpillar green, expanded body segments with long dorsal spiny projections. Cocoon silvery, pitted with small holes. Two species in genus, one in region. **Biology:** Caterpillars feed primarily on Marula (*Sclerocarya birrea*), but recorded also on Corkwood (*Commifera mollis*), Duiker-berry (*Excoecaria africana*), walnut and Tamboti (*Spirostachys africana*). Small holes in cocoon possibly mimic parasitoid emergence holes. Bivoltine, adults peak in October and February but occasionally in other months. **Habitat & distribution:** Coastal bush and bushveld where marula trees grow; from KwaZulu-Natal, South Africa, north to Angola and Ethiopia.

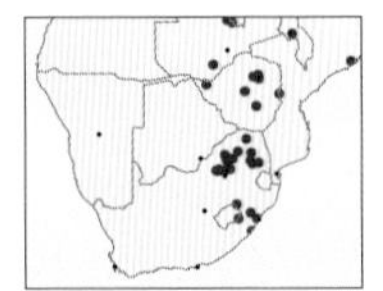

### 4 *Cinabra hyperbius* — Banded Emperor

Forewing 43–64mm. Several distinct regional forms: KwaZulu-Natal and Mpumalanga populations uniform red; Gauteng populations uniform yellow; Waterberg and Zimbabwe populations, both forms, including intermediates. Forewing postmedian and basal lines grey, tiny clear eyespot; hindwings with large dark grey and black eyespot, postmedian line black and strong grey and violet marginal shading in some specimens. Caterpillar yellow-green with small black spines, red lateral triangles, black ventral dots. One species in genus in region; genetic information indicates probable complex of species. **Biology:** Caterpillars feed on Proteaceae in South Africa, also on miombo trees (*Brachystegia* and *Julbernardia* spp.) in Zimbabwe and further north. Bivoltine or univoltine, adults October–April. **Habitat & distribution:** Moist grassland, forest and woodland where host plants occur. Several allopatric populations from KwaZulu-Natal, South Africa, north to tropical Africa.

1
1
2
IS
2
3
AC
3
3
AC
4
4

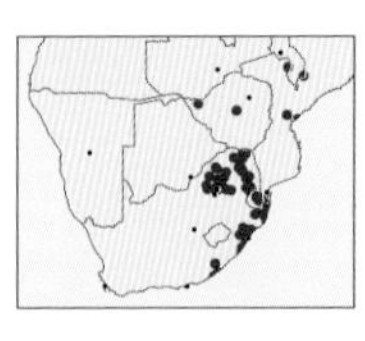

### 1 *Epiphora mythimnia* — White-ringed Atlas Moth

Forewing 46–72mm. Thorax with white band posteriorly; wings reddish-brown with marginal dots and lines, small eyespot near apex; both wings with very large, clear windows, oval in hindwing, teardrop in forewing, each entirely circled in yellow and black. Caterpillar initially black and white, becoming greenish with rows of curved yellow spikes, blue-tipped anteriorly. Five regional species in genus. Similar *E. scribonia* has the clear window oval in both wings and feeds on *Croton* (*Euphorbiaceae*). **Biology:** Caterpillars feed primarily on Buffalo Thorn (*Ziziphus mucronata*), sometimes on Soap Bush (*Helinus integrifolius*), both Rhamnaceae. Probably bivoltine, adults August–April. **Habitat & distribution:** Widespread in moist eastern bushveld and forest; Eastern Cape, South Africa, north to East Africa.

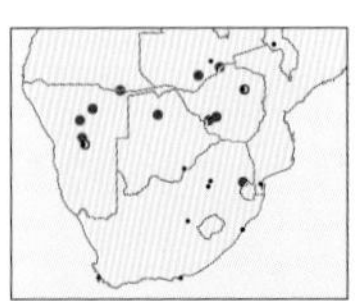

### 2 *Epiphora bauhiniae* — Southern Atlas Moth

Forewing 52–63mm. Similar to *E. mythimnia* (above) but basal part of both wings white. Caterpillar light green with short, sharp white or blue-tipped red spines. **Biology:** Caterpillars feed on *Ziziphus* spp. Probably bivoltine, adults November–April. **Habitat & distribution:** Bushveld; Namibia, Botswana and Zimbabwe, one record from South Africa, north to tropical Africa.

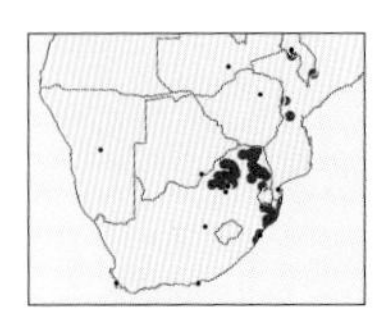

### 3 *Aurivillius fusca* — Cat's Eye Emperor

Forewing 53–72mm. Two distinct colour forms, brick red and yellow, both with pinkish-red shading in terminal areas of all wings; lines zigzag, small red-circled eyespot on forewings, larger on hindwings. Caterpillar green with pale vertical bands and purplish dorsal spikes. One species in genus in region. **Biology:** Caterpillars feed on Weeping Wattle (*Peltophorum africanum*) and Flat-crown (*Albizia adianthifolia*), also on African Locust Bean (*Parkia biglobosa*) north of region. Probably bivoltine, adults September–May. **Habitat & distribution:** Coastal bush and bushveld; Pondoland, South Africa, north to Tanzania.

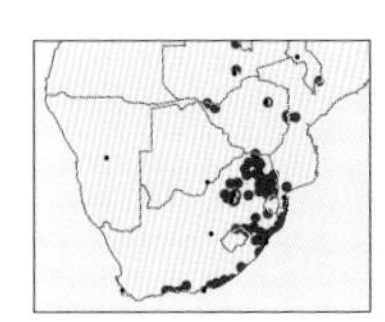

### 4 *Bunaea caffra* — Common Emperor

Forewing 58–95mm. Body red-brown; wings brown and pink, traversed by sharp brown line; clear variable rectangular window on forewing, larger orange eyespot on hindwing ringed with black and white. Caterpillar dramatically patterned in black and red, with rows of conspicuous white dorsal spines, in some individuals final instar completely red apart from white spines. One species in genus in region. **Biology:** Caterpillars feed primarily on cabbage trees (*Cussonia* spp.) and Cape Ash (*Ekebergia capensis*) in the region. Bivoltine, adults July–April, peaking in the summer months. **Habitat & distribution:** Moist wooded habitats, including gardens; southern Cape, South Africa, north to tropical Africa.

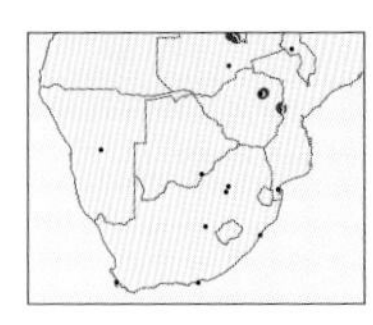

### 5 *Bunaeoides macrothyris* — Large-eyed Emperor

Forewing 75–95mm. Body brown with orange collar; wings with well-demarcated brown bases; forewings white anteriorly, light brown posteriorly, with large clear red- and white-ringed eyespot; hindwings darker with large red eyespot circled in white. Caterpillar light blue-green with branched spines. **Biology:** Caterpillars feed mostly on Zebrawood (*Brachystegia spiciformis*) and *Julbernardia*. Adults in January. **Habitat & distribution:** Miombo woodland; Zimbabwe north to East Africa.

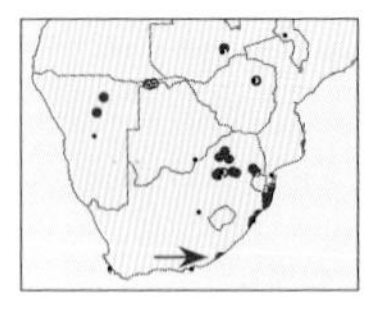

### 6 *Cirina forda* — Pallid Emperor

Forewing 45–56mm. Uniform pale fawn or pinkish and thinly scaled; hindwings pointed posteriorly in male, rounded in female, with inconspicuous grey-ringed eyespot. Caterpillar yellowish with transverse black bands and short, white hairs. One species in genus in region. **Biology:** Caterpillars feed on wide variety of woody plants, often on Wild Syringa (*Burkea africana*). **Habitat & distribution:** Bushveld and coastal bush, Eastern Cape, South Africa, north to tropical Africa.

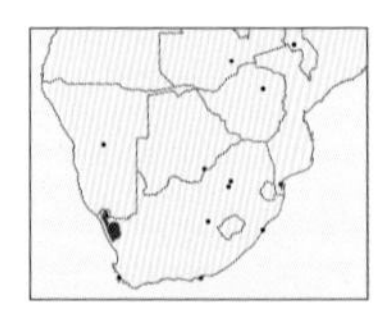

### 7 *Eochroa trimenii* — Roseate Emperor

Forewing 33mm. Unmistakable, 2 distinct forms: normally wings bright pink with black and orange marginal bands and large, dark eyespots; in other form pink is replaced by yellow. Caterpillar cream with transverse black-spined bands. Monotypic genus. **Biology:** Caterpillars feed on *Melianthus* spp. Diurnal, presumed to be aposematic. Univoltine, adults April–May. **Habitat & distribution:** Namaqualand stony hills to Richtersveld and adjacent coastal plain where host plants grow; Northern Cape, South Africa.

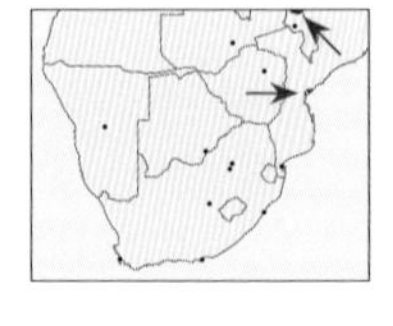

### 8 *Decachorda pomona* — Roundwing

Forewing 23–28mm. Four similar species in genus in region; unlike any other saturniid, all wings markedly rounded, small to tiny clear eyespots. *D. pomona* in coastal bush and forests, orange with violet postmedian line; *D. fulvia* like *D. pomona*, but brick red with yellow eyespot, in forests; *D. rosea* orange or pink, in miombo woodland, Zimbabwe northward. **Biology:** Caterpillars feed on various grasses. **Habitat & distribution:** Mozambique coast, north to tropical Africa.

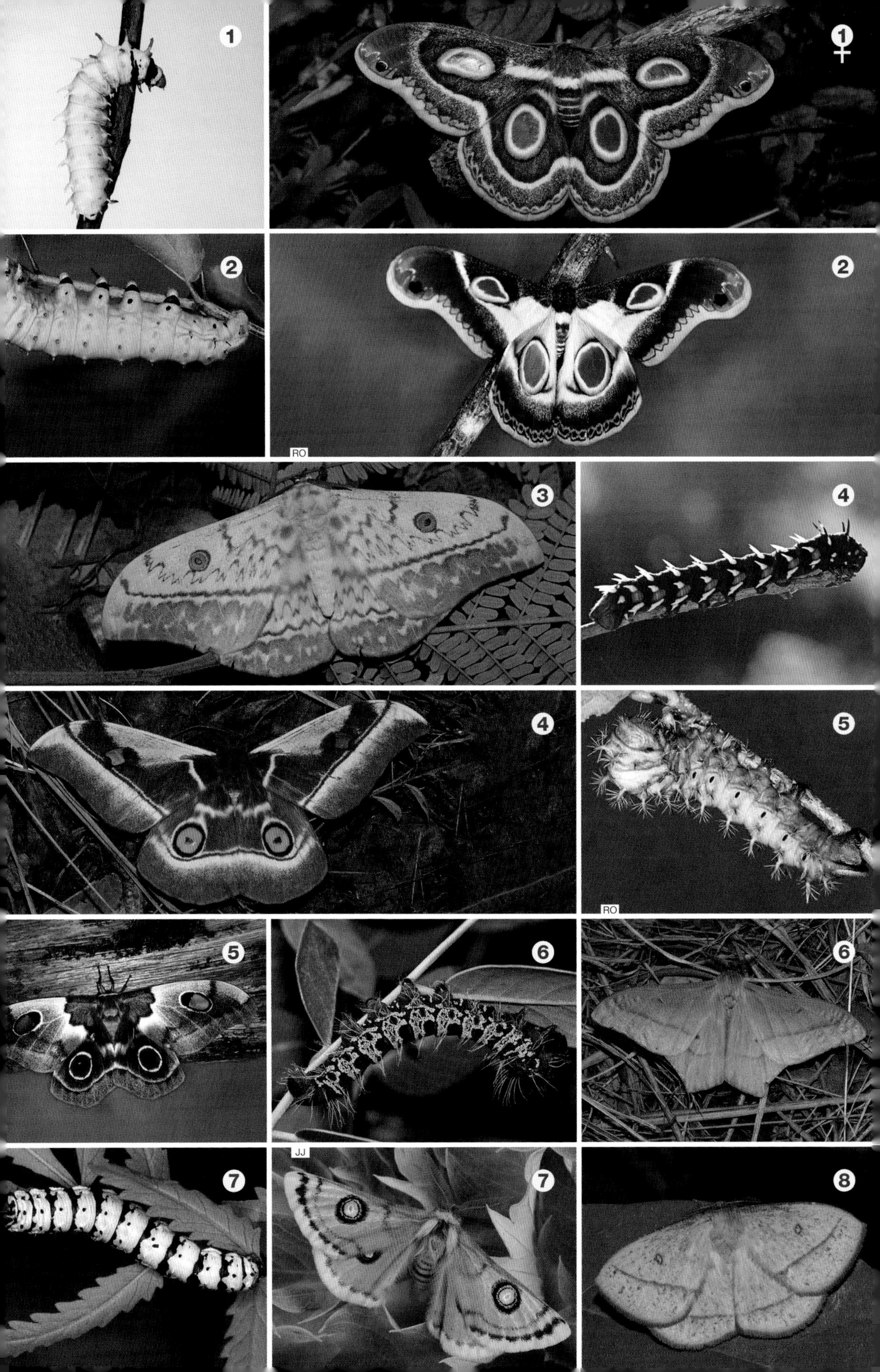

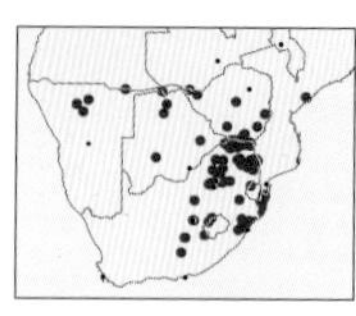

**1 *Gonimbrasia belina*** Mopane Moth

Forewing 58–66mm. Variable, from greyish-green through yellow and orange to reddish-brown; thorax with white collar, both wings bisected by double white and black lines, postmedian straight, median with slight kink which is not repeatedly dentated, forewing eyespot variable, small, ringed in orange, black and sometimes white; hindwing eyespot larger. Caterpillar covered with multicoloured round scales, with hairy dorsal spines. Ten species in genus in region. **Biology:** Caterpillars polyphagous on variety of trees, common on Mopane (*Colophospermum mopane*); may defoliate trees. Caterpillars used extensively as food and one of region's most economically important insects. Traditionally eviscerated and dried, eaten dry, fried or incorporated into stews. Bivoltine, adults recorded throughout the year in subtropical areas but most often October–March. **Habitat & distribution:** Widespread in bushveld and grassland habitats, From Orange River valley north to tropical Africa.

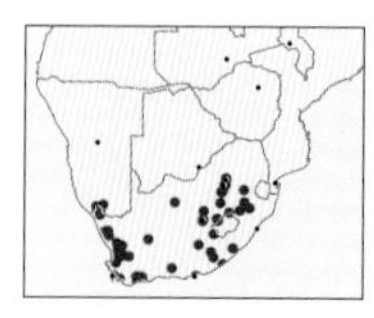

**2 *Gonimbrasia tyrrhea*** Zigzag Emperor

Forewing 60–65mm. Brown; zigzag white line near base of forewings, both wings with wavy black, white-edged line near margins and large brown and black eyespot. Caterpillar black and setose, with white, red and yellow scales. **Biology:** Caterpillars polyphagous on trees and shrubs, including exotic trees. Univoltine, adults mainly September–December in summer-rainfall areas and March–May in winter-rainfall areas. **Habitat & distribution:** Arid Cape vegetation types and highveld grassland; South Africa and southern Namibia.

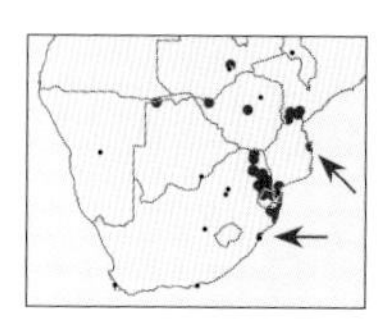

**3 *Gonimbrasia zambesina*** Zambezi Emperor

Forewing 63–75mm. Body brown to olive green; forewings olive green centrally with whitish margins, eyespot small; hindwings pink-based, eyespots very large. Caterpillar black, covered in dense green and yellow scales, and black or red spiny tubercles. **Biology:** Caterpillars feed primarily on *Diospyros* spp. but occasionally on other trees. Bivoltine, adults throughout the year, peaking in summer months. **Habitat & distribution:** Lowveld bushveld and woodland; Zululand, South Africa, north to Kenya.

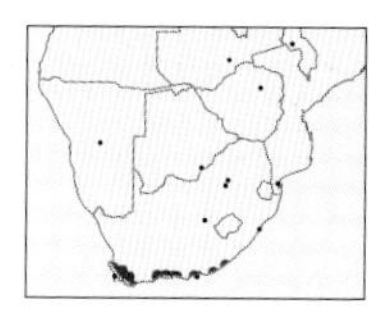

**4 *Nudaurelia cytherea*** Pine Emperor

Forewing 60–75mm. Body and wing colour variable brown to orange, both wings dusted with white and with large oval eyespots with yellow, black and pink or white rings. Caterpillar huge (up to 120mm), dark red, with bands of large blue and yellow scales and short, branched spines. Six species in genus in region. **Biology:** Caterpillar polyphagous, primarily on *Protea* spp., but also Cape Beech (*Rapanea melanophloeos*), *Searsia* and *Watsonia* spp.; also on exotic tree species, such as pines, various *Eucalyptus* spp., apple, guava, quince, etc. Although reputedly edible, not popular as food item. Bivoltine, adults throughout the year but peaks in October–November and March–May. **Habitat & distribution:** Mountain and coastal fynbos; absent in native forests but sometimes abundant in commercial pine and *Eucalyptus* plantations; Western Cape and Eastern Cape.

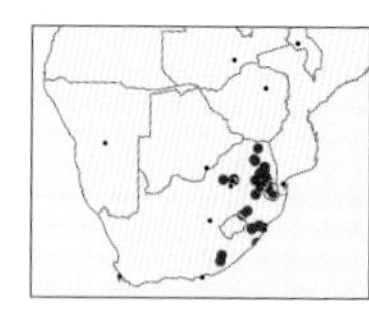

**5 *Nudaurelia clarki*** Protea Emperor

Forewing 60–75mm. Similar to *N. cytherea* (above), but body and wings yellow to orange, less dusted; eyespots round, not as oval and postmedian line straighter and not as kinked below hindwing eyespot; caterpillar similar. **Biology:** Caterpillars feed primarily on *Protea* spp., also exotic pines. Univoltine in mountains, adults February–April, possibly bivoltine at lower elevations near coast, adults October–May. **Habitat & distribution:** Primarily in mountain grassland where *Protea* spp. grow, also *Protea* grassland near coast; Eastern Cape mountains north to Soutpansberg, South Africa.

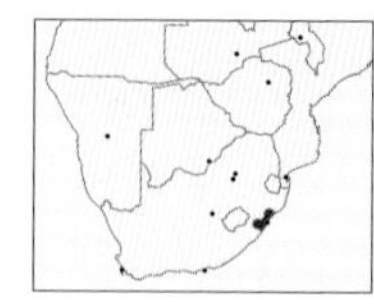

**6 *Nudaurelia gueinzii*** Variable Emperor

Forewing 56–76mm. Orange body and wings; forewings with small yellow eyespot ringed with dark red; hindwing eyespot large, yellow, ringed with black and pink. Caterpillar similar to *N. cytherea* (above). **Biology:** Caterpillars polyphagous on various tree species, including introduced species. Bivoltine, adults August–February. **Habitat & distribution:** Moist forests; KwaZulu-Natal, South Africa, also reported from Zimbabwe and further north to tropical Africa.

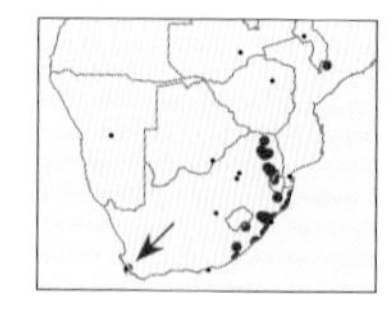

**7 *Nudaurelia wahlbergii*** Wahlberg's Emperor

Forewing 44–66mm. Body yellow-orange; wings yellow to orange with pink/mauve banding; eyespots on forewing smaller than on hindwing, circled with yellow and pink. Caterpillar black with red or orange spiky projections. **Biology:** Caterpillars polyphagous on many tree species. Multivoltine, adults throughout the year. **Habitat & distribution:** Forest and coastal bush; Eastern Cape, South Africa, north to tropical Africa.

1
1
2
2
AM
3
3
IS
4
5
6
6
LP
7
7

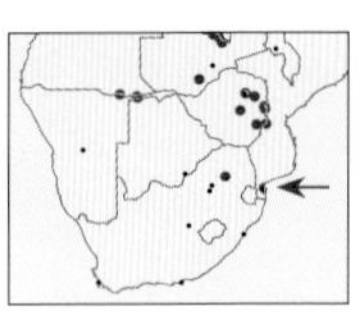

## 1 *Imbrasia ertli* — Diverse Emperor

Forewing 55–66mm. Pale to dark brown; forewing eyespot small and clear, that on hindwing large, with black, pink and white rings; hindwings of male pointed, females rounded, prominent white lines and spots on female forewings. Caterpillar black with red head, earlier instars with long white hairs. One species in genus in our region. **Biology:** Caterpillars feed on miombo trees (*Julbernardia* and *Brachystegia* spp.) also on other trees during rearing; gregarious in all instars, descending at maturity to pupate. In DRC they are collected as a valuable seasonal food resource ('chini'). Probably bivoltine, adults October–May. **Habitat & distribution:** Miombo woodland; Zimbabwe north to tropical Africa.

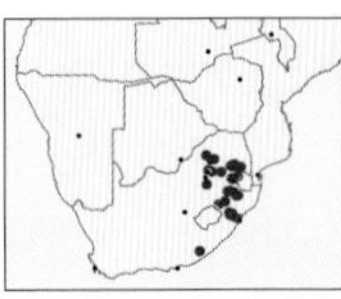

## 2 *Bunaeopsis arabella* — Injured Emperor

Forewing 46–68mm. Ground colour yellow, blood red around base; lines black with grey shading, crenulated, broken toward apex; well-developed eyespots on all wings, shaded with charcoal between veins in terminal area; underside of wings more extensively shaded. Caterpillar black with cream lateral line and cream around base of short black spines. Seven species in genus in region. **Biology:** Caterpillars feed on various grasses. Univoltine, adults December–January. **Habitat & distribution:** Grassland; Eastern Cape north to Limpopo, South Africa.

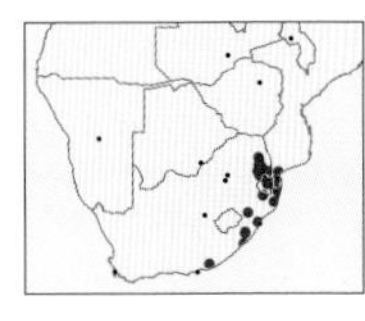

## 3 *Melanocera menippe* — Chestnut Emperor

Forewing 61–70mm. Easily recognised by the mostly red wings, with yellow to brown margins and large black, white-ringed eyespots on both wings; white medial line extends onto costal margin. Caterpillar deep red with black bands and spines. Two species in genus in region. Very similar *M. dargei* is a Maputaland endemic, has basal area in forewing almost black, median line does not extend onto costal margin, reddish colour less extensive. **Biology:** Caterpillars feed primarily on *Ochna* spp., sometimes on other forest trees. Probably bivoltine, adults September–February. **Habitat & distribution:** Coastal bush and forest; Eastern Cape, South Africa, north to Kenya.

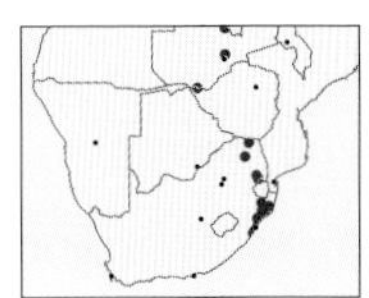

## 4 *Lobobunaea angasana* — Saturnine Emperor

Forewing 66–94mm. Largest moth in South Africa; body brown with distinctive white collar anteriorly; forewings falcate, variable shades of brown, with minute clear spot; hindwings brown with large black eyespot ringed in red and pink. Caterpillar green, swollen, with bulbous ridges and white lateral line. Two species in genus in region. **Biology:** Caterpillar polyphagous on trees belonging to at least 5 families. Bivoltine, adults August–November and January–February. **Habitat & distribution:** Moist Afrotropical forests; Pondoland, South Africa, north to tropical Africa.

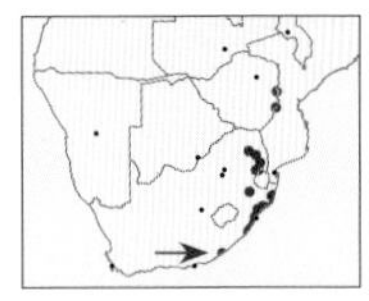

## 5 *Pseudobunaea tyrrhena* — Cat's Paw Emperor

Forewing 54–75mm. Thorax orange with white anterior collar; forewings falcate, orange to brick red with darker grey-brown median area defined by zigzag lines and grey margin, eyespot tiny, white; hindwings orange, eyespot large, black with white centre, conspicuous dark brown 'cat's paw' mark on underside of all wings. Caterpillar green, squat and with ridged swellings on each segment, pale yellow lateral line. Two species in genus in region. **Biology:** Caterpillars polyphagous, mostly on white stinkwoods (*Celtis* spp.), Pigeon Wood (*Trema orientalis*) and Cape Ash (*Ekebergia capensis*). Multivoltine, adults throughout the year. **Habitat & distribution:** Coastal bush and forest; Eastern Cape, South Africa, north to tropical Africa.

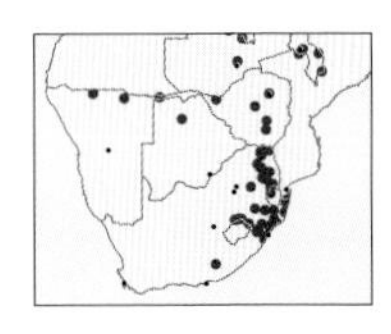

## 6 *Pseudobunaea irius* — Irian Emperor

Forewing 60–72mm. Variable in colour, from pale fawn of violet-grey wings to yellow or red, brown lines prominent to almost absent, variably crenulated rather than zigzag; brown claw markings on underside reduced, often absent on forewing. Caterpillar has white dorsal spots. **Biology:** Caterpillars polyphagous on many tree species, including exotics such as poplar. Bivoltine in colder areas, multivoltine in subtropical areas, adults July–March, peaking November and March. **Habitat & distribution:** Many different habitats; Eastern Cape, South Africa, north to tropical Africa.

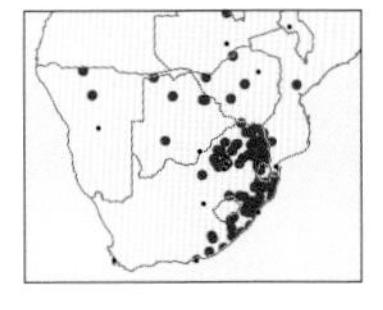

## 7 *Gynanisa maja* — Speckled Emperor

Forewing 55–75mm. Males with very large, brown bipectinate antennae; forewings with complex pattern of brown and white lines and mottling and small clear eyespot; hindwings with large orange and pink-ringed eyespot. Caterpillar green to yellow with red and white lateral line and dense white spotting; segments ridged, bearing yellow-tipped pearly spines. Twenty-two species in genus, 2 in region. Very similar *G. ata* (Zimbabwe north) is larger with male forewings slightly falcate. **Biology:** Caterpillars feed on various Fabaceae trees, including introduced wattles, allowing range expansion. Bivoltine, adults throughout the year, peaking in summer months. **Habitat & distribution:** Wooded habitats, including grasslands with wattle infestations; Eastern Cape, South Africa, north to tropical Africa.

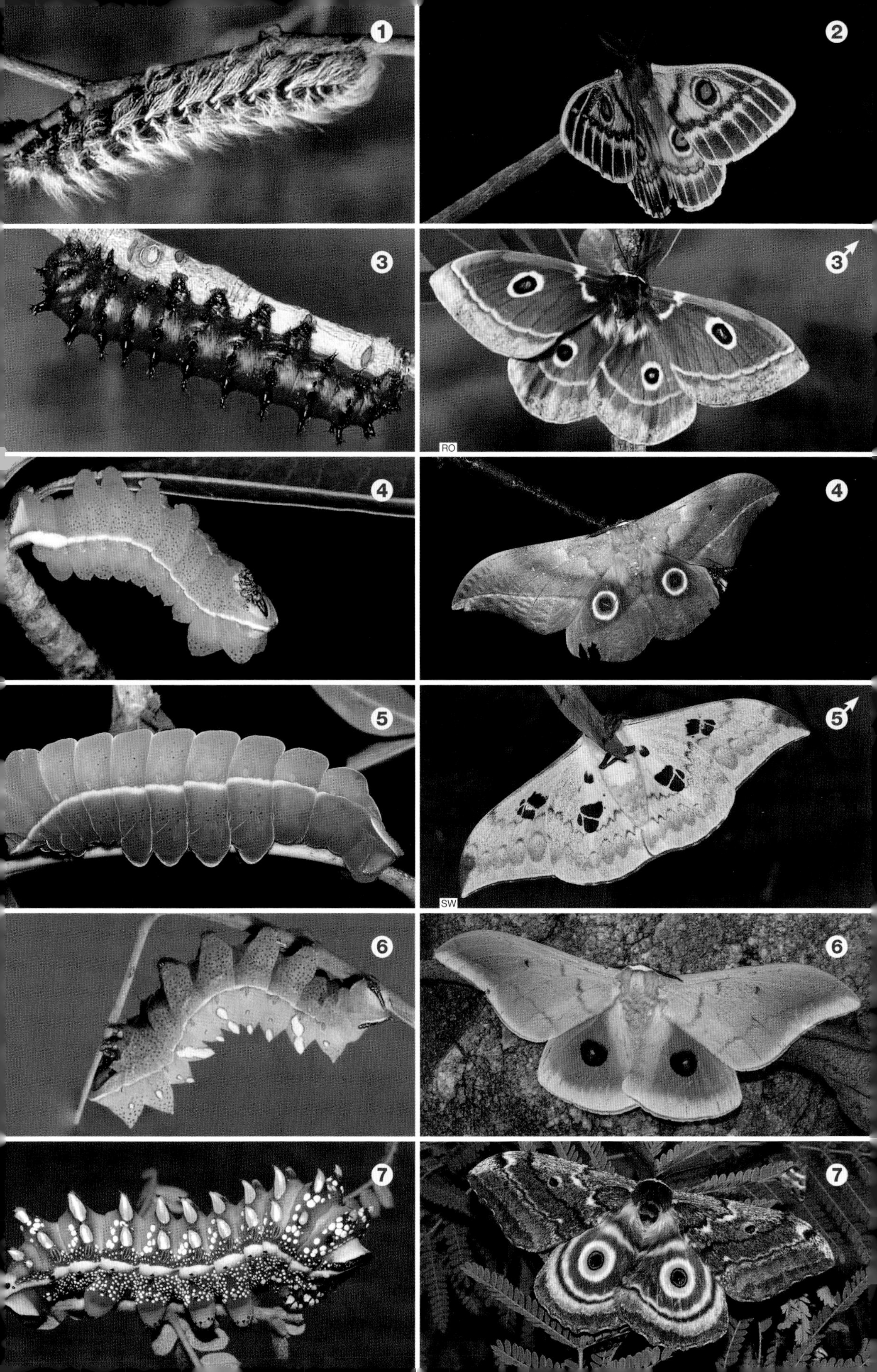
1
2
3
3♂
RO
4
4
5
5♂
SW
6
6
7
7

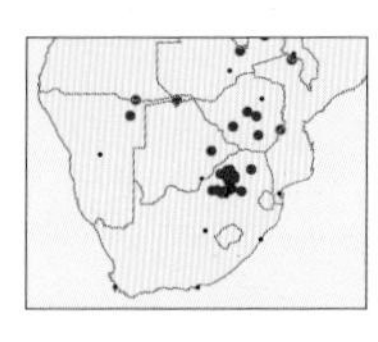

### 1 *Rohaniella pygmaea* — Pygmy Emperor

Forewing 36–43mm. Thorax reddish-brown with white collar; forewings pink with faint lines and tiny central clear spot; hindwings orange with pink borders and large, dark eyespot. Caterpillar plump and sluggish, green with red and white lateral line. One regional species in genus. **Biology:** Caterpillars feed on Wild Syringa (*Burkea africana*). Bivoltine, adults September–March. **Habitat & distribution:** Bushveld where wild syringas grow; Gauteng, South Africa, north to tropical Africa.

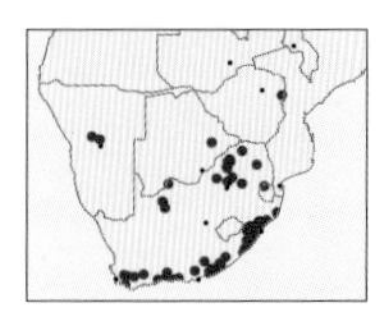

### 2 *Heniocha apollonia* — Southern Marbled Emperor

Forewing 31–47mm. Thorax white with brown central patch; forewings white, patterned with fawn to brown patches and lines, pink-tipped, with large eyespot; hindwings paler with smaller eyespot. Caterpillar green with transverse row of white, spiny tubercles on each segment. Five species in genus in region. **Biology:** Caterpillars feed on Sweet Thorn (*Vachellia karroo*) and introduced wattles (*Acacia* spp.). Bivoltine, adults October–March, occasionally in other months in warmer areas. **Habitat & distribution:** Many different habitats, absent in habitats experiencing severe frost, probably extended its range in response to wattle expansions; Western Cape, South Africa, north to Zambia.

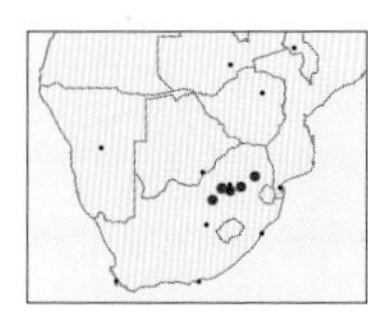

### 3 *Heniocha flavida* — Highveld Marbled Emperor

Forewing 38–46mm. Similar to *H. apollonia* (above), but wings more rounded, ground colour generally more yellow, often completely yellow, and brown lines on the forewings are further apart. Caterpillar pale green with white lateral line and white spinose projections on each segment. **Biology:** Caterpillars feed on Elephant's Root (*Elephantorrhiza elephantina*). Univoltine, adults October–January. **Habitat & distribution:** Highveld grasslands that experience severe frost where elephants roots grow; seems restricted to the central highveld of South Africa.

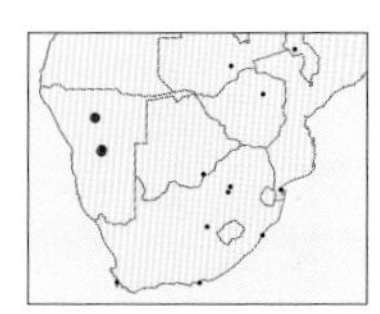

### 4 *Heniocha distincta* — Namibia Marbled Emperor

Forewing 34–38mm. Similar to *H. apollonia* (above), but brown patch on thorax much reduced or absent; wings heavily irrorated with grey scales covering the entire wing area except outer bar and oblique basal bar. **Biology:** Caterpillars feed on Elephant's Root (*Elephantorrhiza elephantina*). Univoltine, adults December–March. **Habitat & distribution:** Habitats where elephants roots grow; seems restricted to central plateaux in Namibia.

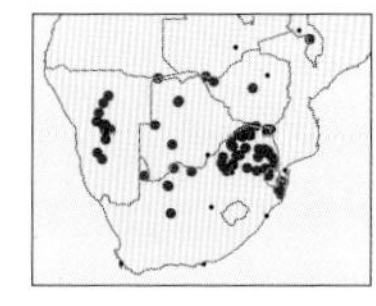

### 5 *Heniocha dyops* — Marbled Emperor

Forewing 36–58mm. Similar to *H. apollonia* (above), but thorax pure white; forewings with yellow to orange scalloped line along distal margins, forewing eyespot touching postmedian shading, 2 basal lines diverging; eyespots on hindwings absent or reduced to grey dot. Caterpillar green with yellow lateral stripe and marked purple-tipped dorsal spines. **Biology:** Caterpillars feed on various *Senegalia* spp. Univoltine in west to bivoltine or trivoltine in east of range, adults December–February in west where it can swarm in good seasons, July–May in east of range. **Habitat & distribution:** Moist and arid bushveld where *Senegalia* trees grow; Orange River valley, north to East Africa.

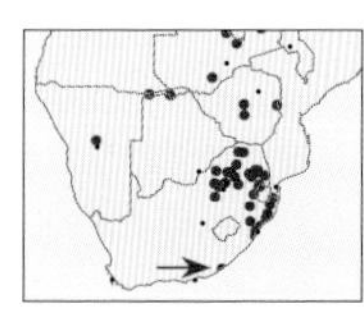

### 6 *Heniocha marnois* — Grey-bordered Marbled Emperor

Forewing 46–60mm. Paler version of *H. dyops* (above) with large white areas between the brown lines on forewing, forewing eyespot not touching postmedian shaded areas, 2 basal lines parallel; eyespots on hindwings absent or reduced to grey dot. Caterpillar pale green with pale yellow vertical and horizontal bands, and white dorsal spikes on each segment. **Biology:** Caterpillars feed on various *Senegalia* spp. Bivoltine or trivoltine, adults July–April. **Habitat & distribution:** Moist bushveld where *Senegalia* trees grow, absent from western arid bushveld; East London, South Africa, north to tropical Africa.

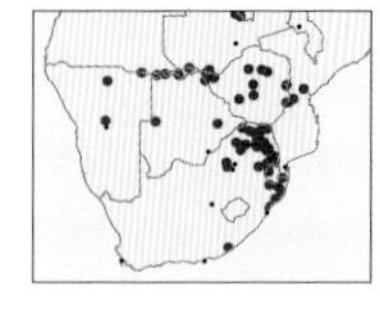

### 7 *Campimoptilum kuntzei* — Lunar Prince

Forewing 24–35mm. Variably mottled in brown to red shades; eyespot reduced to small, black-edged crescent on forewing, tiny or obsolete on hindwing. Caterpillar variable, usually white or cream with rows of black spots and feathery hairs, but some green. One species in genus in region. **Biology:** Caterpillars feed on a variety of Fabaceae trees and shrubs. Bivoltine, adults October–May. **Habitat & distribution:** Moist and dry forest and bushveld; KwaZulu-Natal, South Africa, north to tropical Africa.

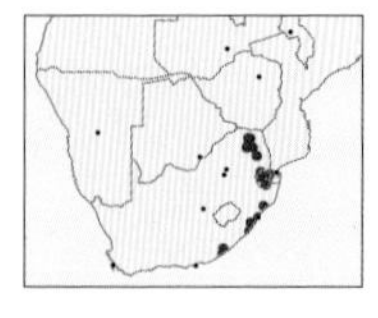

### 8 *Ludia goniata* — Black Prince

Forewing ♂ 24mm, ♀ 29mm. Similar to *L. delegorguei* (p.270) but darker with a distinct kink or double tooth in middle of margin of hindwings. **Biology:** Caterpillars feed on Coastal Silver Oak (*Brachylaena discolor*). Bivoltine, adults August–April. **Habitat & distribution:** Coastal bush and forests; Eastern Cape, South Africa, north to Tanzania.

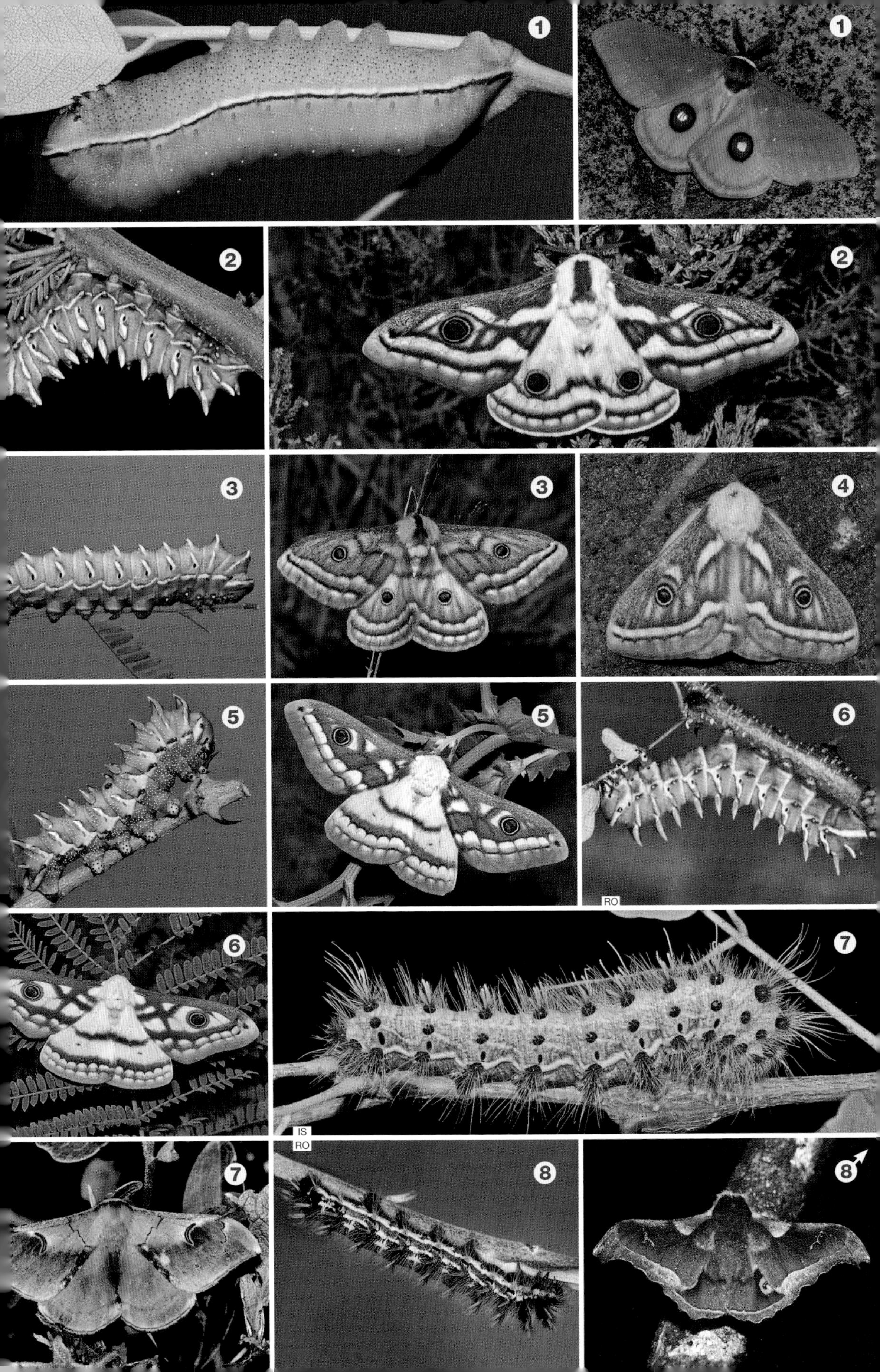

1
1
2
2
3
3
4
5
5
6
RO
6
7
IS
RO
7
8
8

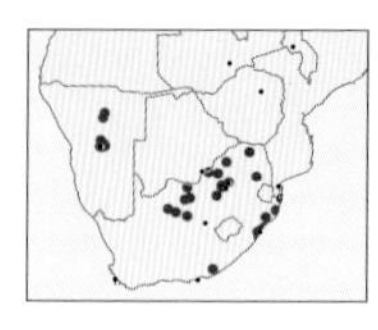

## 1 *Ludia delegorguei* — Delegorgue's Prince

Forewing ♂ 23–25mm, ♀ 26–37mm. Males smaller than females; body brown to charcoal; forewings (falcate in male), with pale perimeter around broad dark centre, and with small '3'-shaped clear spot; hindwings red at base, brown distally; eyespot with broad outer black ring. Caterpillar pale yellow with variable black side-bars and spots and long white hairs. Five similar species in genus in region. **Biology:** Caterpillars feed on various woody Asteraceae. Females nocturnal, attracted to light, males fly by day early in the morning. Bivoltine, adults August–April. **Habitat & distribution:** Grassland and bushveld; Eastern Cape to tropical Africa.

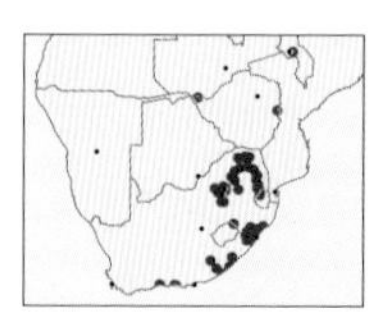

## 2 *Holocerina smilax* — Variable Prince

Forewing ♂ 25–30mm, ♀ 35–41mm. Males markedly smaller than females with strongly falcate (sickle-shaped) forewings; body and centres of wings dark red-brown, outer margins paler orange-brown; unusual large, irregular clear spot on forewings, minute one on hindwings. Caterpillar with long setae and distinct forms, some dramatically striped black and white, other forms irregularly mottled with red. Four very similar species in genus in area, 3 species Zimbabwe northwards. **Biology:** Caterpillars feed primarily on Tree Fuchsia (*Halleria lucida*), sometimes Cape Ash (*Ekebergia capensis*); accepts many other plants in captivity. Multivoltine, adults August–June. **Habitat & distribution:** Moist forests, on hills and riverine thickets where host plants grow; southern Cape, South Africa, north to tropical Africa.

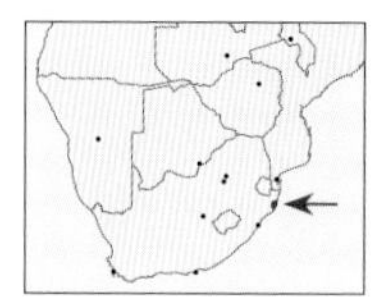

## 3 *Micragone cana* — Pale Prince

Forewing 24–30mm. Body and wings dull brown with no eyespots or clear spots on wings; forewings falcate in male only; hindwings pink at base, dark markings on inner margin and tip. Caterpillar black with red head, covered in long irritant white hairs. **Biology:** Caterpillars feed on Water Berry (*Syzygium cordatum*). **Habitat & distribution:** Widespread across South, East and Central Africa.

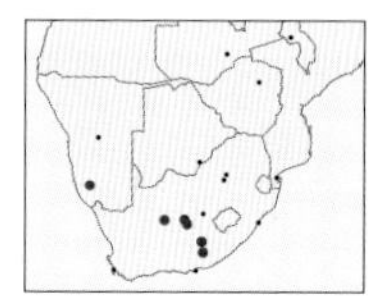

## 4 *Vegetia dewitzi* — Dewitz's Princeling

Forewing 17–30mm. Females larger than males; thorax brown with yellow collar (and sometimes longitudinal bands); forewings brown, speckled yellow and with yellow bands, margin yellow, eyespot an almost complete clear ring; hindwings red at base, brown distally; eyespot similar to those in forewings, encircled with yellow then black. Caterpillar black with yellow and white markings. Four species in genus, all in region. **Biology:** Caterpillars feed on Wild Rosemary (*Eriocephalus ericoides*). Univoltine, adults May–October. Males diurnal with rapid flight, females nocturnal, attracted to light. **Habitat & distribution:** Upper grassy karoo; Eastern Cape, Northern Cape and southern Namibia.

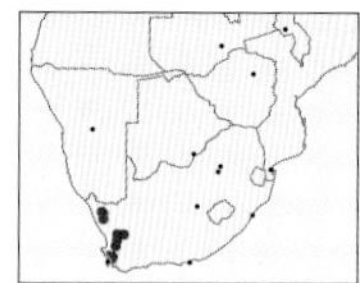

## 5 *Vegetia ducalis* — Ducal Princeling

Forewing 20–26mm. Dark brown to black body; forewings dark brown with wavy white bands and broken white to yellow margin; eyespot clear, '3'-shaped; hindwing orange-red in male, pink at base in female, with yellow eyespot. Caterpillar black with orange markings and dense tufts of white hairs. **Biology:** Caterpillars feed on Wild Rosemary (*Eriocephalus africanus*), and other *Eriocephalus* spp. Univoltine, adults April–September. Males diurnal with rapid flight, females nocturnal, attracted to light. **Habitat & distribution:** Succulent karoo; Western Cape to Northern Cape.

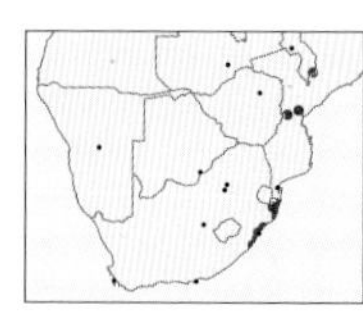

## 6 *Tagoropsis flavinata* — Gold Marbled Emperor

Forewing 40–46mm. Can be confused with *Pselaphelia flavivitta* (p.272). Male body and wings pale to deepish yellow, female (depicted) richer yellow-orange; wings with transverse wavy lines; small eyespot in forewings only (distinguishing it from *P. flavivitta*). Caterpillar black, marked with white reticulating lines and spiny red projections. Two species in genus in region. **Biology:** Caterpillars feed on *Allophylus* spp. Multivoltine, adults throughout the year. **Habitat & distribution:** Coastal bush and forest; Pondoland, South Africa, north to tropical Africa.

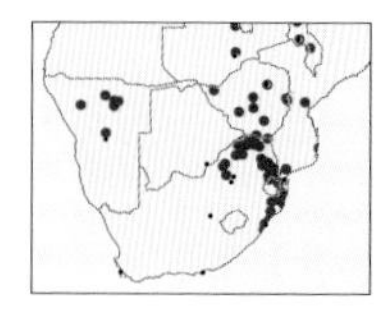

## 7 *Usta terpsichore* — Cavorting Emperor

Forewing 33–65mm. Thorax light to dark brown; wings white, heavily dusted with brown, central part filled with brown; large eyespot brown-centred, ringed with yellow and black; hindwings plain white with brown marginal band and similar eyespot. Caterpillar black with orange patches and small spiny projections. Two species in genus in region. **Biology:** Caterpillars feed on various plants including Forest Corkwood (*Commiphora woodii*), Chinaberry (*Melia azedarach*), Pepper Tree (*Schinus molle*), Marula (*Sclerocarya birrea*) and *Commiphora* spp. Multivoltine, adults throughout the year. **Habitat & distribution:** Variety of lowland forest and bushveld; KwaZulu-Natal, South Africa, north to tropical Africa; Namibian population seemingly isolated.

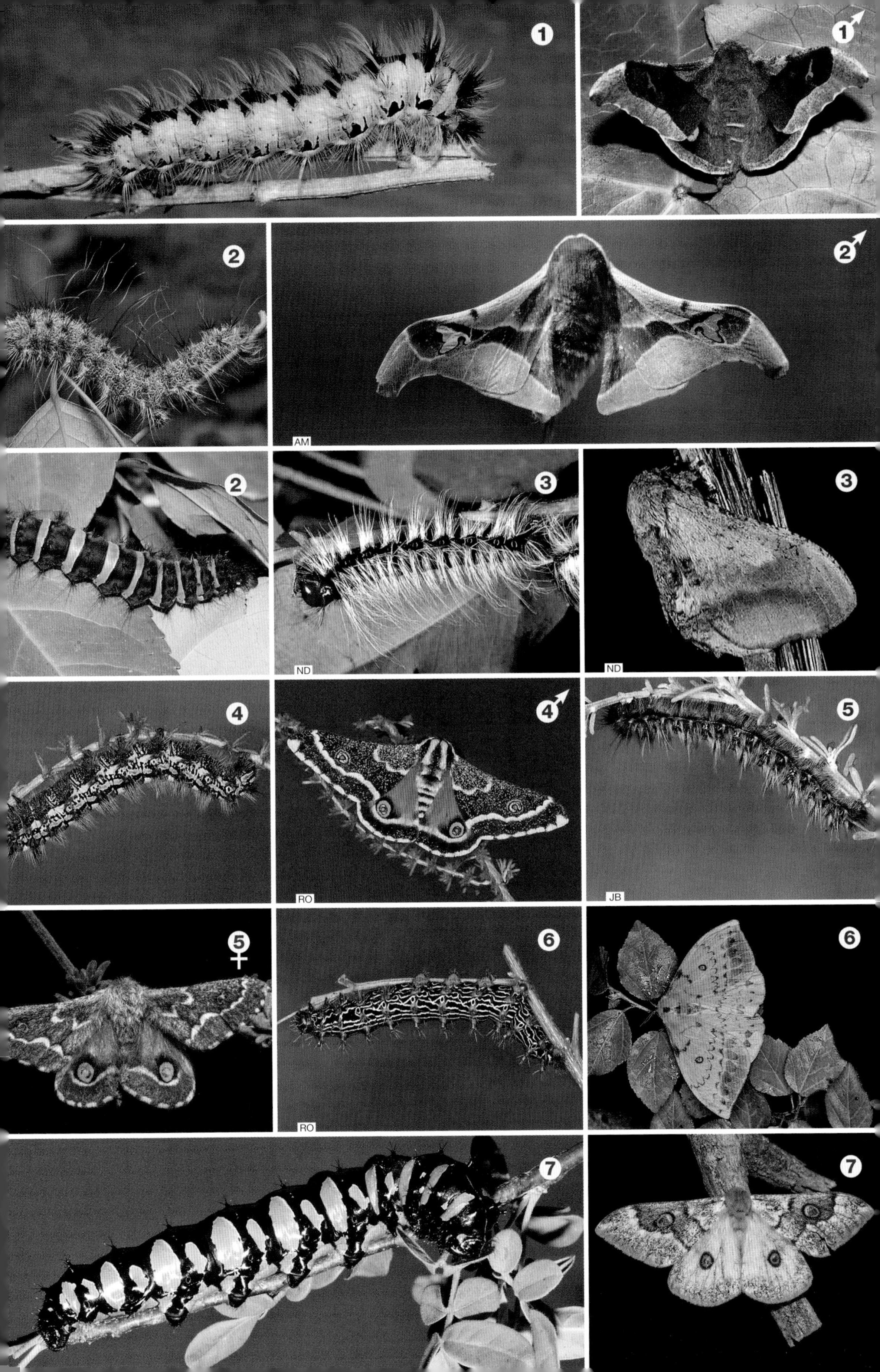
1
1♂
2
2♂
AM
2
3
ND
3
ND
4
4♂
RO
5
JB
5♀
6
RO
6
7
7

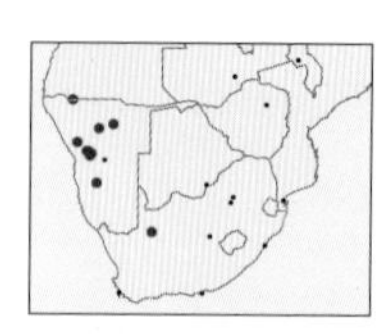

**1 *Usta wallengrenii*** — Wallengren's Emperor

Forewing 45mm. Pale yellowish thorax and wings densely speckled with dark grey. Caterpillar black with narrow yellow bands and spiky black projections. **Biology:** Caterpillars feed on various *Commiphora* spp. Adults November–March. **Habitat & distribution:** In desert and semidesert; Namibia and southern Angola, one record from Northern Cape.

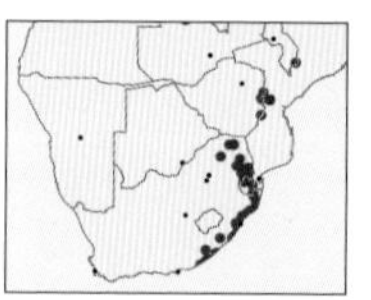

**2 *Urota sinope*** — Tailed Emperor

Forewing 43–52mm. Body and forewings orange to brown, forewings crossed by 2 distinct white bands, each enclosing brown line; hindwings of male with distinctive short, angular tails; hindwings of females kinked, but no tail; eyespots reduced to small white spots. Caterpillar distinctive, yellow with black transverse bands and white setae. Two species in genus in region. **Biology:** Caterpillars feed primarily on coral trees (*Erythrina* spp.) in our region. Multivoltine, adults throughout the year. **Habitat & distribution:** Forests and coastal bush where coral trees grow; Eastern Cape, South Africa, north to tropical Africa.

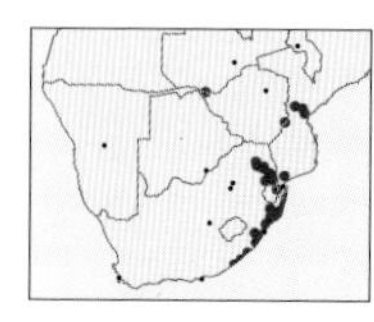

**3 *Pselaphelia flavivitta*** — Leaf Emperor

Forewing 40–48mm. Male ground colour rich yellow, wings slightly irrorated with brown scales, straight postmedian and wavy median brown lines, small white tip to forewings; all wings with small clear eyespot ringed by orange, black and white; female ground colour pale yellow, wings richly irrorated with brown and white scales. Caterpillar smooth, green with red lateral line, tapering at both ends to resemble folded leaf. Only species in genus in region. **Biology:** Caterpillars feed primarily on *Trichilia* spp., also recorded on wild ginger (*Afromomum* spp.). Probably multivoltine, adults August–May, peaking in summer. **Habitat & distribution:** Subtropical forest and thick bush where *Trichilia* spp. grow; Eastern Cape, South Africa, north to Kenya; isolated population in the remnant forest at Victoria Falls.

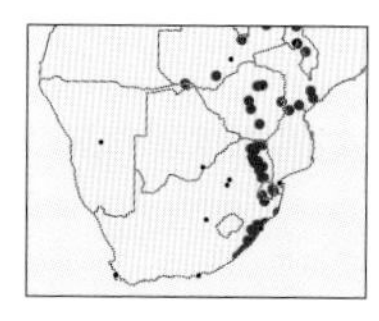

**4 *Pseudaphelia apollinaris*** — Apollo Moth

Forewing 28–42mm. One form thorax yellow, abdomen white with row of black dots dorsally; wings translucent white, outer margins grey with yellow spots; 2 eyespots yellow/orange with black ring, those on hindwings minute; second form (KwaZulu-Natal and Eastern Cape) wings translucent yellow with one forewing eyespot. Caterpillar banded black and white with orange lateral stripe; distinctive black horn at rear of body. One species in genus in region. **Biology:** Caterpillars feed on honeysuckle trees (*Turraea* spp.), accepts other trees in captivity. Adults diurnal, aposematic, fly slowly in the afternoon, attracted to light sometimes. Multivoltine, adults throughout the year, peaking in summer. **Habitat & distribution:** Frost-free coastal bush, moist bushveld and forest where honeysuckle grows; Eastern Cape to Ethiopia.

## FAMILY SPHINGIDAE Hawkmoths

**Distinctive medium to large, robust, streamlined moths with spindle-shaped bodies, narrow, pointed forewings, small hindwings and filiform antennae usually hidden under wings at rest; outline usually triangular when at rest. Renowned for their rapid, sustained flight and ability to hover when feeding. Most nocturnal, attracted to light, but many active around dusk or dawn, some diurnal. Hawkmoths are important pollinators, developing specialised mutualistic relationships with many plants. To prevent pollination by other generalist pollinators, flowers of many plant species have exceptionally long narrow tubes specialised for hawkmoths which have an extremely long proboscis, many times the length of the body. Many are active migrants, multiple species participating annually in mass nocturnal migrations crossing mountain ranges and sometimes even oceans. As a result, widely distributed, with adults often encountered far away from their actual caterpillar habitats. Caterpillars have distinctive dorsal abdominal horn on last segment; most smooth and cryptic, some have large anterior eyespots. Pupae of some have unusual free proboscis. About 1,450 species globally, 111 in region.**

### Subfamily Sphinginae

Lack patch of short sensory hairs on inner surface of labial palps (tasting organs of mouthparts); male genitalia symmetrical. Caterpillars with paired, lateral lines running obliquely along each body segment, body surface finely granulose, and proboscis of pupa looped away from body.

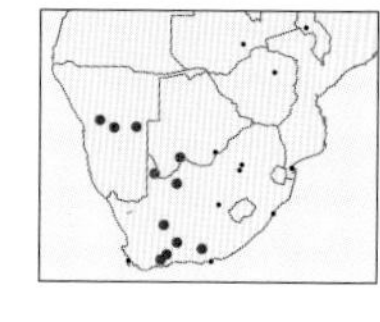

**5 *Hoplistopus penricei*** — Cribage Hawk

Forewing 20–26mm. Body grey, thorax with black 'collar', abdomen with lateral black spots on each segment; wings variable black striations on plain pale to dark grey; hindwings paler, cilia white. One species in genus in region. **Biology:** Caterpillars feed on Trumpet Thorn (*Catophractes alexandri*). Adults present October–March. **Habitat & distribution:** Arid areas; Cape provinces, South Africa, north to Botswana, Namibia and Angola.

1
1♂
RO
2
2♀
3
3♂
3♀
4
SW
4
5
5

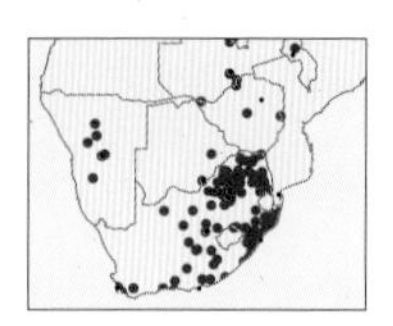

**1 *Acherontia atropos*** Death's Head Hawkmoth

Forewing 40–60mm. Dark mottled brown with diagnostic skull-like marking on thorax; abdomen and hindwings with orange or yellow bars; proboscis short. Caterpillars, earlier instars green with short spines, final instar yellow, body with diagonal yellow and blue or slate bars and dark spots; abdominal horn curved, very spiky. Pupa underground, glossy brown, proboscis fused to body. One species in genus in region. **Biology:** Caterpillars polyphagous on woody and herbaceous plants, often seen on ground moving between plants. Bivoltine or trivoltine in warmer areas, adults throughout the year peaking in warmer months. Adults well known for habit of raiding beehives for honey, and ability to produce squeaking sounds when handled. **Habitat & distribution:** Widespread in natural and transformed habitats; migratory species through Africa, Middle East and Europe, rarely also Asia.

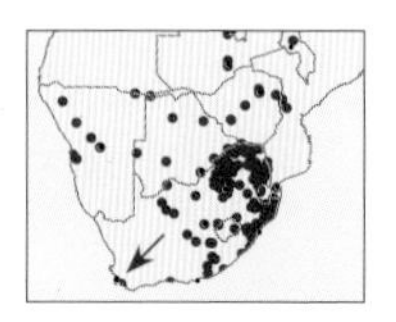

**2 *Agrius convolvuli*** Convolvulus Hawk

Forewing 40–50mm. Variably mottled grey and black; characteristic pink and black bars on sides of abdomen. Caterpillar green or brown, sometimes orange-brown, typically with dark diagonal bars laterally above row of black dots; abdominal horn long, spiny, backward curving. Pupa in soil, shiny brown, proboscis free and looping out from head. Two regional species in genus. **Biology:** Caterpillars primarily on Convolvulaceae, sometimes Lamiaceae and other plants. Adults use exceptionally long proboscis to feed on tubular flowers; important pollinator of baobab trees. Multivoltine, adults throughout the year, primarily October–March, flies from dusk to dawn. **Habitat & distribution:** Widespread in natural and transformed habitats; active migrant through Africa, Middle East, Europe, western Asia, Australia and New Zealand.

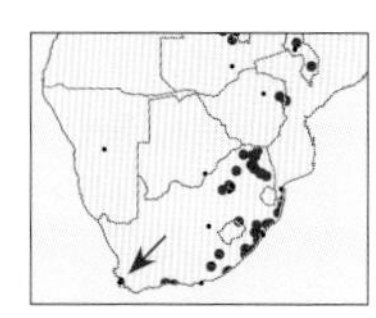

**3 *Coelonia fulvinotata*** Fulvous Hawkmoth

Forewing 55mm. Mottled brown and black, resembling *Acherontia atropos* (above), but without skull-like thoracic markings, instead 2 paler pinkish spots at rear of thorax, hindwings yellow at base; proboscis very long. Caterpillars variable, green with oblique violet stripes, yellow with oblique green stipes or brownish; thoracic segments swollen and wrinkled; abdominal horn long, papulate and curled. Pupa brown, with free, arched proboscis loop. One regional species in genus. **Biology:** Caterpillars polyphagous on woody and herbaceous plants. Probably bivoltine, adults September–May. **Habitat & distribution:** Moist forest, bush and gardens; Western Cape north through Africa and Indian Ocean islands.

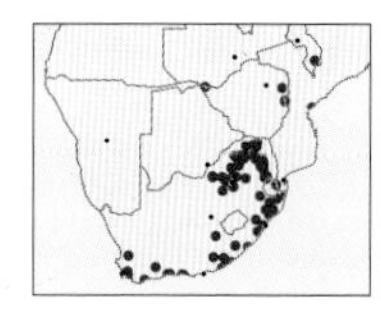

**4 *Macropoliana natalensis*** Natal Sphinx Hawkmoth

Forewing 55–62mm. Pale grey with black streaks on forewings and black and yellow band encircling thorax; hindwings brown. Caterpillar blue and green with oblique pale green or mauve stripes on each segment, or brown, abdominal horn small, folded inward. Two similar regional species in genus, possibly a complex of species. **Biology:** Caterpillars feed on olive trees (*Olea* spp.). Adults recorded throughout the year. **Habitat & distribution:** Where olive trees grow; Western Cape to Ethiopia and into West Africa.

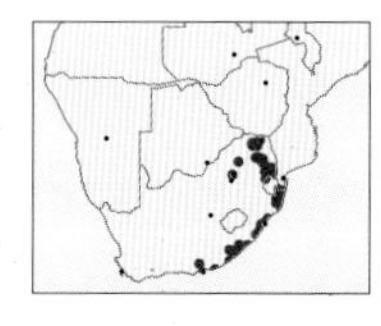

**5 *Oligographa juniperi*** Juniper Hawk

Forewing 28–30mm. Grey with thin black line along thorax and fine black markings on forewings; hindwings dark brown. Caterpillar light green with oblique whitish stripes, abdominal horn straight, red-tipped. One species in genus in region. **Biology:** Caterpillars feed on Cape Honeysuckle (*Tecomaria capensis*) and other Bignoniaceae. Bivoltine, adults throughout the year, peaking November and February. **Habitat & distribution:** Coastal bush and forests; Eastern Cape, South Africa, north to Mozambique.

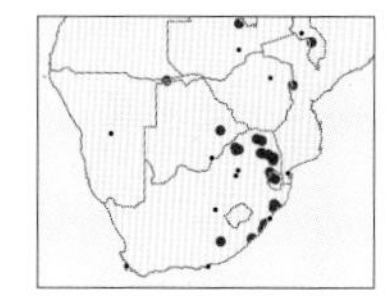

**6 *Platysphinx piabilis*** Measly Hawkmoth

Forewing 59–62mm. Male body and forewings light orange with numerous small dark markings; hindwings orange densely dotted with red, large basal black-ringed eyespot; female forewings densely covered in brown to grey scales; hindwings almost completely red in some individuals; proboscis short and stout. Caterpillar pale green, bluish or brown with paler oblique stripes; abdominal horn yellow, short and curved. One species in genus in region. **Biology:** Caterpillars feed on various Faboideae trees. Probably multivoltine, adults August–May. **Habitat & distribution:** Frost-free forest, bushveld and woodlands; Eastern Cape, South Africa, north to tropical Africa.

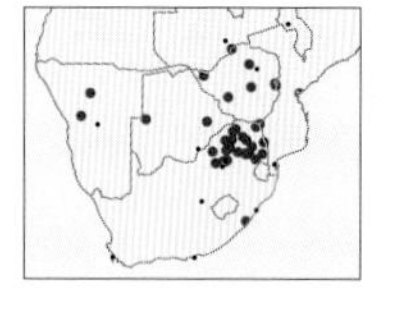

**7 *Praedora marshalli*** Marshall's Hawk

Forewing 18–24mm. Body grey, thorax pale, abdomen with central black line; forewings attractively marbled with wavy transverse black lines, 2 short black longitudinal lines; hindwings grey. Caterpillar light green with white lateral line and oblique white bars, abdominal horn thin and straight. Three similar regional species in genus. **Biology:** Caterpillars feed on Tinderwood (*Clerodendrum ternatum*). Bivoltine, adults October–April. **Habitat & distribution:** Bushveld; Gauteng, South Africa, north to Angola and Somalia.

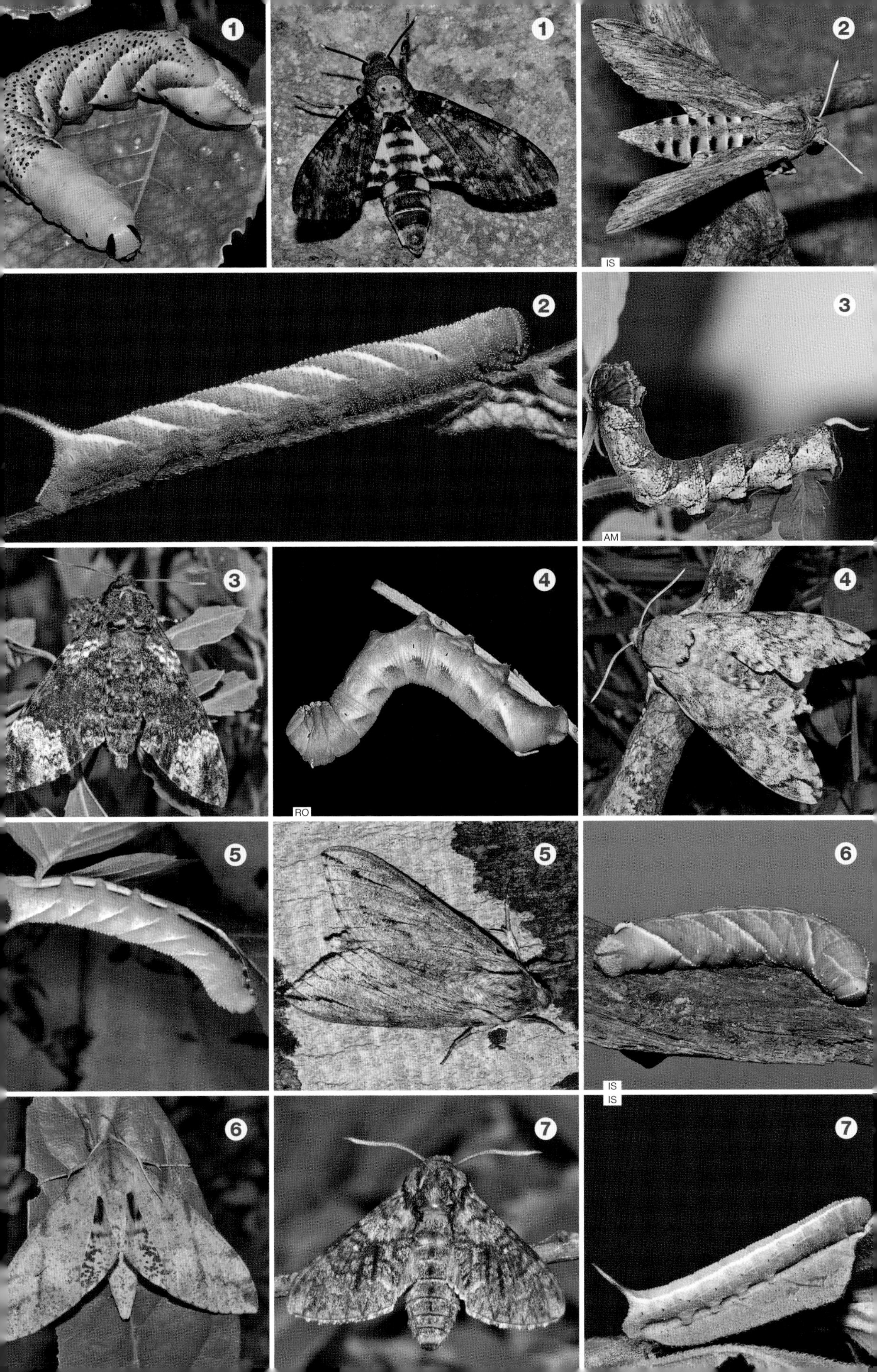
1
1
2
IS
2
3
AM
3
4
RO
4
5
5
6
IS
IS
6
7
7

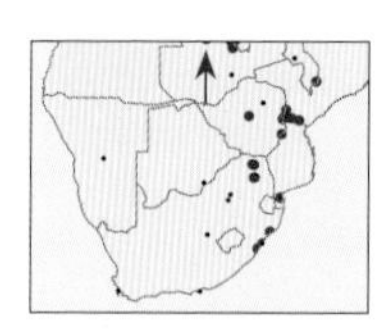

### 1 *Xanthopan morganii* — Morgan's Hawkmoth

Forewing 55–68mm. Thorax dull brown to green-grey, abdomen with conspicuous yellow and black banding laterally; forewings grey to brown to black with wavy black lines or bars; hindwings dark brown with conspicuous bright yellow basal patches. One regional species in genus. Its sister species, *X. praedicta* from Madagascar, is famous for its exceptionally long proboscis (3.25 × body length), being the sole pollinator of Darwin's orchid, a relationship Darwin predicted before the moth was discovered. Also one of the main pollinators of endemic baobabs (*Adansonia* spp.) and various orchid species. **Biology:** Caterpillars feed on various Annonaceae. Adults throughout the year. **Habitat & distribution:** Moist forests; KwaZulu-Natal north across sub-Saharan Africa.

## Subfamily Smerinthinae

Occurs largely in Africa and Asia. Hindwings often have an eyespot; proboscis of pupa reduced in size and fused to body of pupa (no free loop). Caterpillar has oblique stripes, as well as longitudinal stripes running length of body.

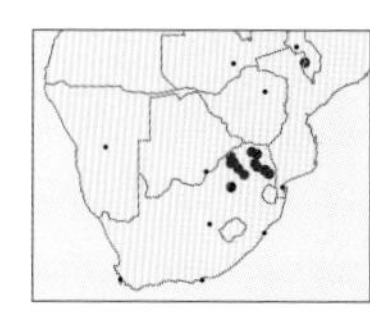

### 2 *Afroclanis calcareus* — Ruby Hawkmoth

Forewing 34–38mm. Pink to reddish-brown with darker thorax; forewings with darker wing tip; hindwings pink-orange. Caterpillar green, with oblique white stripes and dark lateral spot on each segment; abdominal horn red. Two species in genus in region. Similar *A. neavi* has dark median line across wing; Zimbabwe northward. **Biology:** Caterpillars feed on Coastal Golden-leaf (*Bridelia macrantha*) and Round-leaved Bloodwood (*Pterocarpus rotundifolius*). Bivoltine, adults October–March. **Habitat & distribution:** Bushveld and woodland; northern South Africa to DRC and Tanzania.

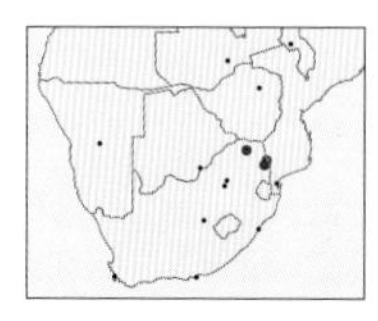

### 3 *Batocnema africanus* — Harlequin Hawk

Forewing 35–43mm. Flesh-coloured, with characteristic olive-green markings on thorax, wings and first abdominal segment; proboscis short. Caterpillar green with oblique bluish stripes and long, slender abdominal horn. One species in genus in region. **Biology:** Caterpillars feed primarily on Marula (*Scelerocarya birrea*), also on mango. Bivoltine, adults October–March. **Habitat & distribution:** Bushveld and coastal bush where marula trees grow; KwaZulu-Natal, South Africa, north to Kenya.

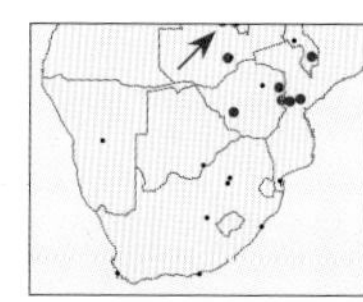

### 4 *Falcatula falcatus* — Falcate Hawkmoth

Forewing 28–41mm. Body and forewings pale grey; forewings with small black spot at base and darker transverse bands curving inward near costa (compare *Rufoclanis numosae*, p.248); hindwings yellowish with dark lines. Caterpillar yellow-green with paired dorsal spots, last pair extending as blue patch to reach yellow abdominal horn. One species in genus in region. **Biology:** Caterpillars feed on Marula (*Sclerocarya birrea*) and coral trees (*Erythrina* spp.). Adults recorded throughout the year. **Habitat & distribution:** Miombo woodland and forests; Zimbabwe into West and East Africa.

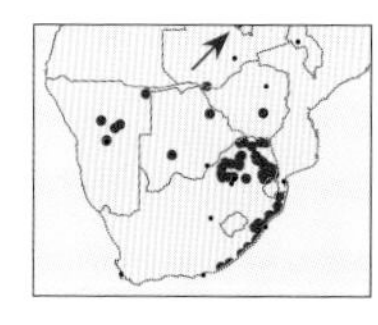

### 5 *Lophostethus dumolinii* — Arrow Sphinx

Forewing 65–85mm. Largest hawkmoth in southern Africa; scalloped margins of forewings and contrasting brown and arrow-shaped white markings diagnostic. Caterpillar huge, pale green, unlike other Sphingidae, with black spines on each segment; abdominal horn black, spiny. One species in genus in region. **Biology:** Caterpillars feed primarily on wide variety of woody Malvaceae. Bivoltine, univoltine in arid areas, adults October–May. **Habitat & distribution:** Various habitats free from severe frost, where woody Malvaceae occur; Eastern Cape, South Africa, north to tropical Africa and beyond.

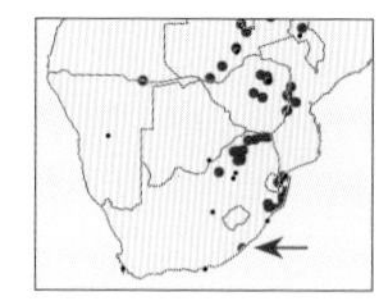

### 6 *Andriasa contraria* — Contrary Andriasa

Forewing 35–37mm. Body and wings light pinkish-brown; forewings with small black basal spot, tiny orange cell spot and a few diffuse darker blotches distally, crossed by jagged fine lines and with row of tiny dots near distal margin; hindwings orange. Two very similar species in genus in region. *Neopolyptychus convexus* is also similar but darker, having larger orange cell spot. **Biology:** Caterpillars feed on Zebrawood (*Brachystegia spiciformis*), Miombo (*Julbernardia globiflora*) and *Lasianthus* spp. Adults throughout the year. **Habitat & distribution:** Forest woodland and kloofs; KwaZulu-Natal, South Africa, north to tropical Africa.

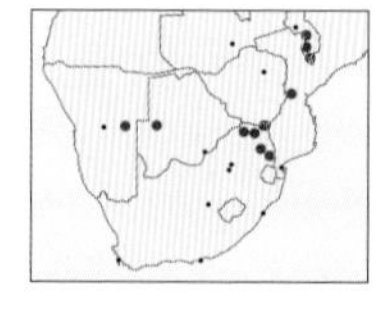

### 7 *Neoclanis basalis* — Wounded Hawk

Forewing 15–20mm. Antennae bright yellow, body and forewings pale yellow; hindwings with bright red basal patch. Caterpillar pale green with oblique white stripes, abdominal horn white and curved. One species in genus in region. **Biology:** Caterpillars feed on Apple Leaf Tree (*Philenoptera violacea*). Bivoltine, adults October–March, peaking in October and February. **Habitat & distribution:** Dry bushveld and woodland; northern South Africa into Central and East Africa.

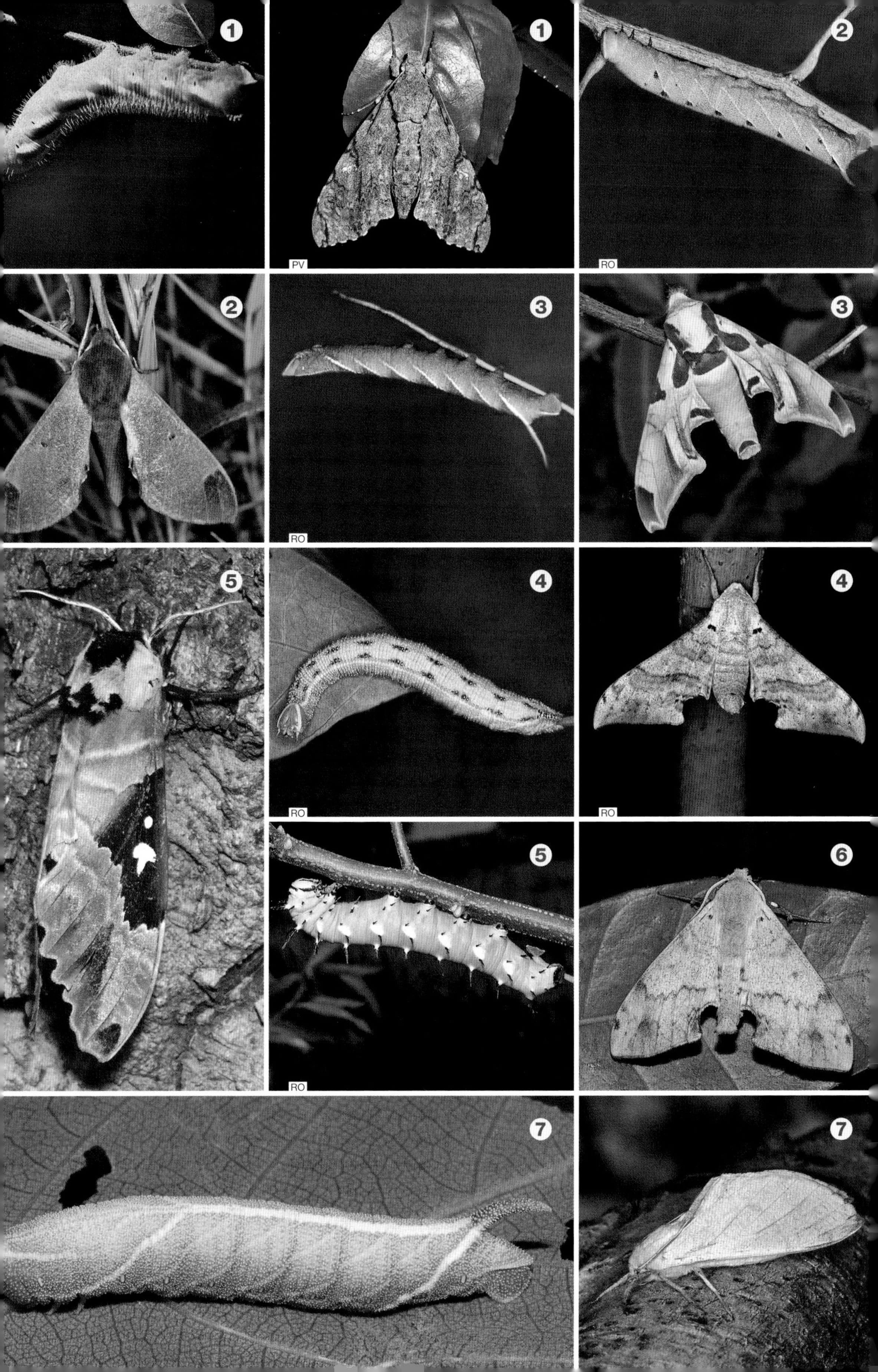
1
1
PV
2
RO
2
3
RO
3
5
4
RO
4
RO
5
RO
6
7
7

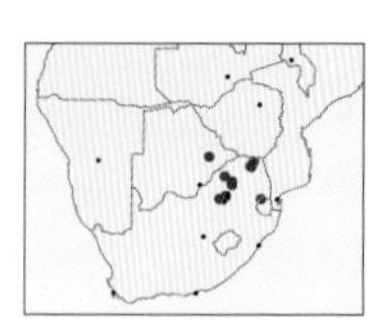

### 1 *Phylloxiphia punctum* — One-spot Redwing

Forewing 28–37mm. Antennae white, body and forewings fairly uniform reddish-brown with a single black dot near apex of forewing, postmedian line indistinct but curved (compare *Theretra capensis*, p.284); hindwings graded from cream to pink to brown distally, male hindwings sharply pointed. Caterpillar bluish-green with faint oblique yellow stripes or pinkish-brown with reddish stripes. Three species in genus in region. **Biology:** Caterpillars feed on *Ochna* spp. Bivoltine, adults September–March, peaking October and March. **Habitat & distribution:** Bushveld on ridges where *Ochna* trees grow; Gauteng, South Africa, north to Botswana, Namibia and tropical Africa.

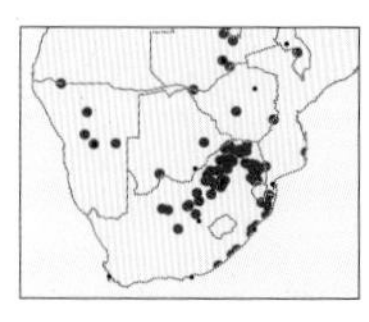

### 2 *Polyptychus grayii* — Sword Hawk

Forewing 38–46mm. Proboscis short; body grey or pinkish; forewings falcate and toothed with curved terminal dark shading, 1 or 2 small black spots at base and series of darker crossbars; hindwings darker. Caterpillar green, rough and granular, with diagonal yellow bars; abdominal horn papulate, short, stout and curved. One species in genus. **Biology:** Caterpillars feed on various woody Boraginaceae. Multivoltine, adults throughout the year. **Habitat & distribution:** Forest, coastal bush and bushveld where Boraginaceae trees grow; Eastern Cape, South Africa, north to East Africa and Sudan.

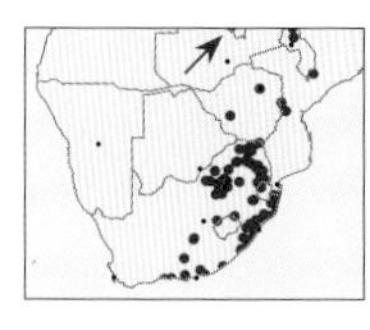

### 3 *Pseudoclanis postica* — Mulberry Hawkmoth

Forewing 32–47mm. Body and forewings olive to light brown, sometimes pinkish, transverse lines wavy; hindwings orange with dark basal patch and distal bar; proboscis very short. Caterpillar green with pale oblique stripes, those on last segment more pronounced and leading to long abdominal horn. Three species in genus in area, *P. molitrix* plain olive on all wings with straight transverse lines. **Biology:** Caterpillars polyphagous on trees, mostly Cannabaceae and Moraceae. Multivoltine, adults throughout the year. **Habitat & distribution:** Common in habitats that support host trees, including transformed habitats, absent from arid west; southern Cape, South Africa, north to tropical Africa.

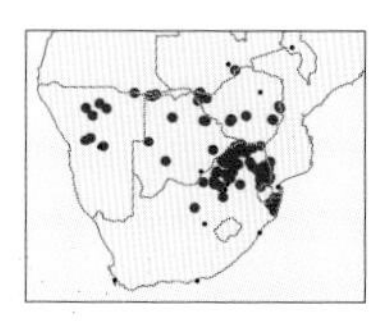

### 4 *Rufoclanis numosae* — Wavy Rufoclanis

Forewing 24–29mm. Body and forewings dark grey to pinkish grey; forewings divided by wavy dark median and subterminal lines, postmedian line straight, behind which wing is much darker, hind margin of all wings waved. Caterpillar green with yellow to pink lateral markings, very long, abdominal horn often red. Three species in genus in area. **Biology:** Caterpillars feed on raisin trees (*Grewia* spp.). Bivoltine, adults October–April. **Habitat & distribution:** Dry bushveld, Zululand, South Africa, north to East Africa.

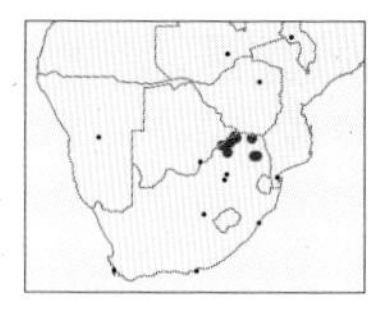

### 5 *Rufoclanis jansei* — Jansen's Hawk

Forewing 30mm. Forewings shorter than *R. numosae* (above), with similar wavy black lines but postmedian line crenulated, ground colour not darker posteriorly and hindwings pink. Caterpillar green with oblique white stripe to abdominal horn. **Biology:** Caterpillars feed on Star Chestnut (*Sterculia rogersii*). Bivoltine, adults October–February. **Habitat & distribution:** Dry bushveld; Limpopo to Zimbabwe and Tanzania.

## Subfamily Macroglossinae

Patch of small, short sensory hairs on inner surface of labial palps. Caterpillar has eyespots on the abdomen rather than on the head; short, stumpy tail at end of body.

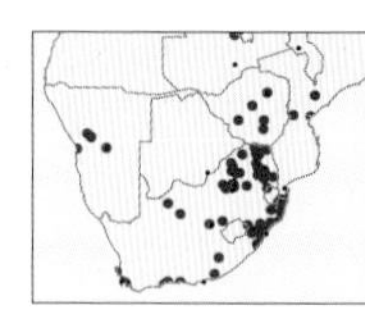

### 6 *Daphnis nerii* — Oleander Hawkmoth

Forewing 55–58mm. Unmistakable, richly marbled in browns and green with white and pink bands. Caterpillar green or brown with white horizontal band and spots, distinctive blue and white eyespot near head; spiracles black; orange abdominal horn short and spiny. One species in genus in region. **Biology:** Caterpillars feed primarily on various Apocynaceae, including the widely planted exotic Oleander (*Nerium oleander*). Multivoltine, adults throughout the year. **Habitat & distribution:** Forest, bushveld and transformed habitats where host plants grow; an active migrant that can be found far away from suitable caterpillar habitat; widely distributed across Africa, Europe and Asia; introduced to Hawaiian Islands to control invasive oleanders and acts as pollinator for endangered flora there.

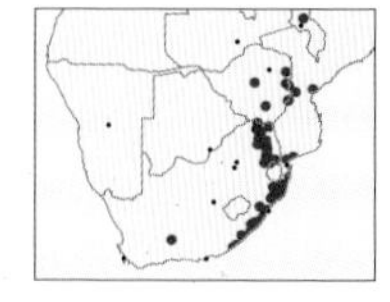

### 7 *Euchloron megaera* — Verdant Hawk

Forewing 42–53mm. Bright green, some small grey and black patches along margins of forewings; hindwings bright orange with black markings. Caterpillar with anterior lateral eyespot. One species in genus in region. **Biology:** Caterpillars feed on various vines of the grape family (Vitaceae). Multivoltine, adults throughout the year. **Habitat & distribution:** Moist frost-free forest and bush; Eastern Cape north across Afrotropics, including Madagascar, Mauritius, La Réunion, also Yemen.

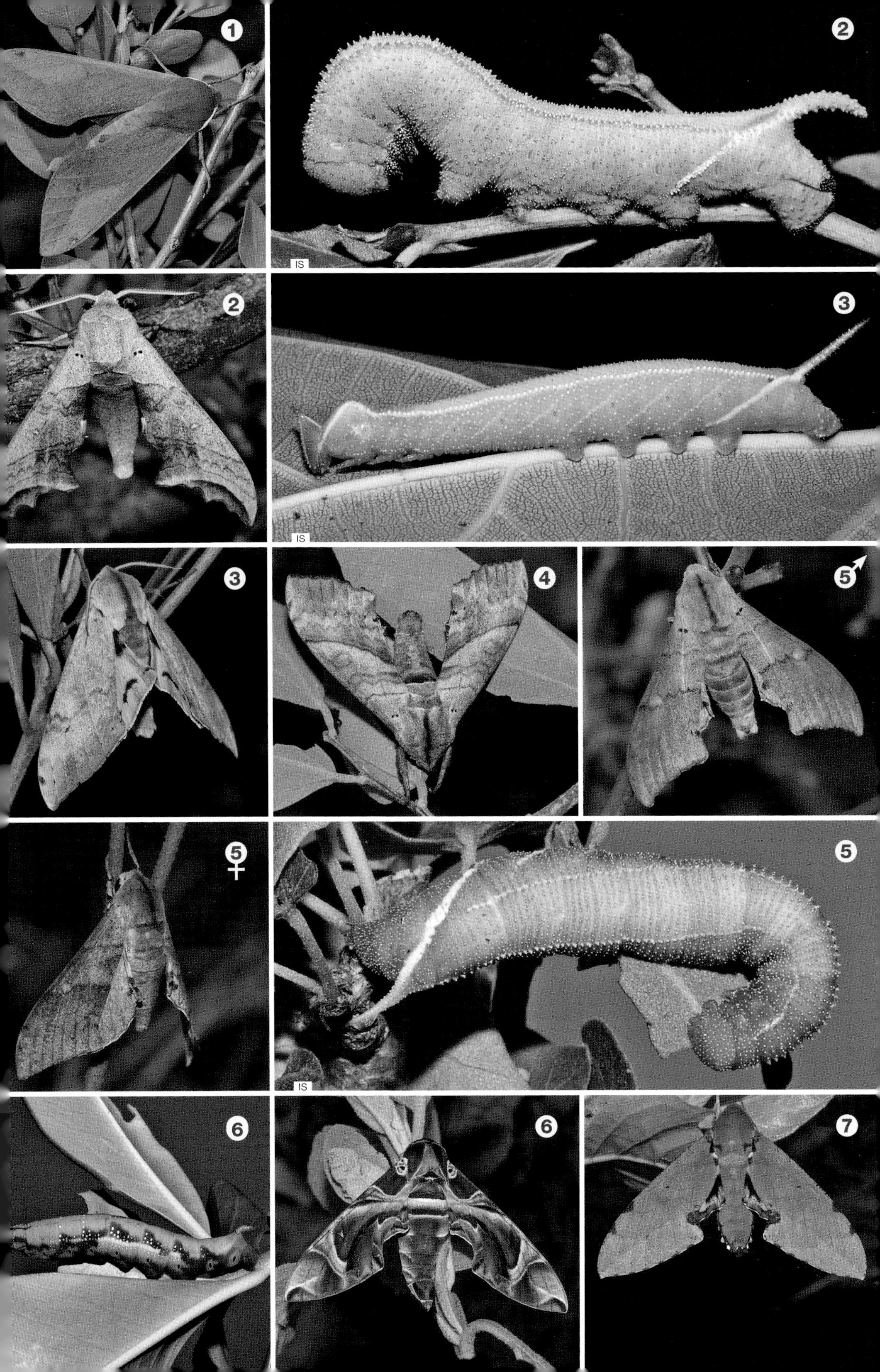
1
2
IS
2
3
IS
3
4
5♂
5♀
5
IS
6
6
7

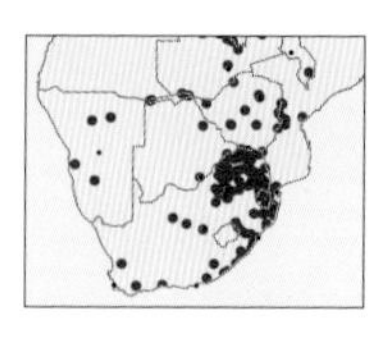

## 1 *Nephele comma* — Comma Nephele

Forewing 32–43mm. Olive green to reddish-brown, abdomen with darker lateral bands; distinctive white comma-shaped mark on forewings (absent in some populations). Caterpillar dark with yellow lateral line and red and white spots, rarely green, short and flat abdominal horn. Ten similar species in genus in region. **Biology:** Caterpillars feed on various Apocynaceae. Multivoltine, adults throughout the year, peaking in summer. **Habitat & distribution:** Forest and bush. Adults common migrants across sub-Saharan Africa and Madagascar, sometimes found far away from suitable caterpillar habitat.

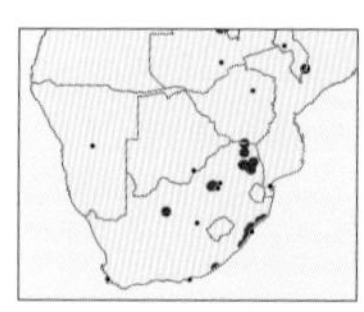

## 2 *Nephele accentifera* — Accented Nephele

Forewing 33–36mm. Abdomen marked in broad black and narrow white stripes and 3 adjacent white spots on forewing that converge to form a 'Y' shape. Caterpillar green to brown with pale lateral stripe that widens abruptly on posterior segments, abdominal horn long, slightly upturned. Similar species *N. peneus* has the 'Y' shape mark broken up. **Biology:** Caterpillars feed on *Ficus* spp. Multivoltine, adults throughout the year, peaking in late summer months. **Habitat & distribution:** Forests and bush where *Ficus* spp. grow; Eastern Cape, South Africa, north to tropical Africa, also Comoros and Madagascar.

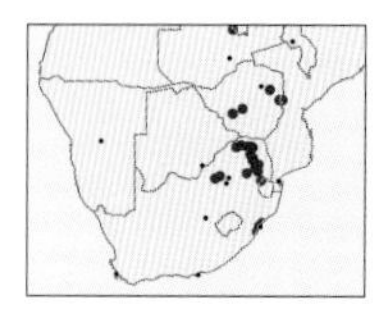

## 3 *Nephele vau* — V Nephele

Forewing 33–35mm. Body and forewings vary from pink to olive with large darker brown distal patch on forewings bearing white 'V' near leading edge; hindwings pink to olive with darker distal patch. Caterpillar green with dark lateral mark on side of segments 5 and 6, abdominal horn long, straight and rough. **Biology:** Caterpillars feed on Simple-spined Num-num (*Carissa edulis*) and Natal Plum (*C. macrocarpa*). Multivoltine, adults throughout the year, peaking in late summer months. **Habitat & distribution:** Forest and bush where *Carissa* spp. grow; KwaZulu-Natal, South Africa, north to tropical Africa, more common in tropical parts, also Saudi Arabia.

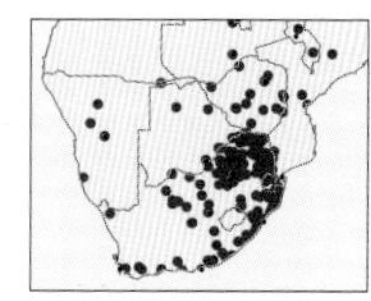

## 4 *Hippotion celerio* — Silver-striped Hawkmoth

Forewing 31–38mm. Brown with silvery streaks and distinct silvery oblique band along forewings; hindwings with black stripes and pink bases. Caterpillar green or brown with pale lateral line, dark yellow-ringed eyespot and long, straight, red abdominal horn with black tip. Six species in genus in region. **Biology:** Caterpillars feed primarily on various Vitaceae and Nyctaginaceae spp., sometimes on other plants. Multivoltine, adults throughout the year, peaking in late summer months. **Habitat & distribution:** Range of habitats; a major migratory species, often making up the bulk of mass hawkmoth migrations in late summer; common across Africa, Europe, Asia and Australasia.

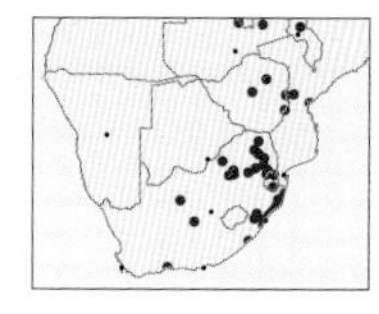

## 5 *Hippotion osiris* — Large Striped Hawkmoth

Forewing 43–49mm. Often confused with *H. celerio*, light brown with black lateral crossbars at base of abdomen (distinguishing it from similar *H. celerio* but not *Hyles livornica*, p.282). Forewings with silvery bands encircling dark central band; hindwings pink, with 2 black crossbars. Caterpillar green or brown with orange head, traversed by broad brown bands, thin wavy lateral line and 2 eyespots; abdominal horn extremely short. Pupa with proboscis enclosed in projecting sheath. **Biology:** Caterpillars polyphagous on a large variety of plants. Multivoltine, adults throughout the year, peaking in late summer months. **Habitat & distribution:** In a range of relatively moist habitats; a migratory species; Western Cape, north across Africa, also Madagascar and Spain.

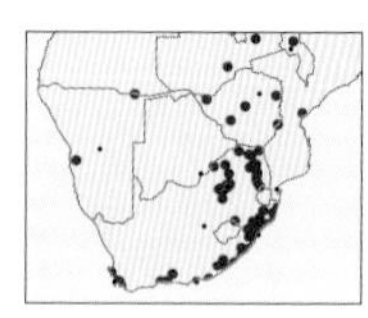

## 6 *Hippotion eson* — Common Striped Hawkmoth

Forewing 37–42mm. Light brown with darker lines running down forewings; hindwings, rose-red with black base and margin. Caterpillar brown (rarely green), with darker oblique bands merging ventrally, 2 yellow eyespots on third and fourth segments, legs reddish; abdominal horn long, pale, curved. Pupa pale with dotted wing pads. **Biology:** Caterpillars feed on Arum Lily (*Zantedeschia aethiopica*) and other Araceae, also on Common Impatiens (*Impatiens walleriana*) and sometimes Vitaceae. Multivoltine, adults throughout the year. **Habitat & distribution:** Moist habitats, including gardens where host plants grow; Western Cape through Africa, Madagascar and Indian Ocean islands.

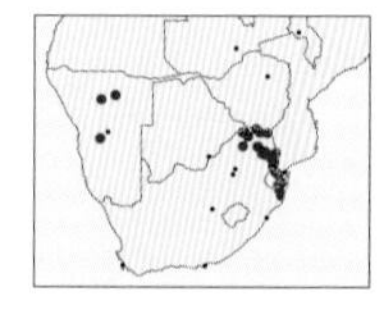

## 7 *Hippotion rosae* — Grey Hippotion

Forewing 32–49mm. Body grey; antennae pink; forewings grey speckled black, small brown dot centrally and diffuse brown patch on incised hind margin; hindwings yellowish. Caterpillar green with white-dotted eyespot on third segment and another eyespot on fourth; abdominal horn dark, pointed. **Biology:** Caterpillars feed on Bushveld Grape (*Cissus rotundifolia*). Adults August–May. **Habitat & distribution:** Frost-free moist forest to arid bushveld; Zululand, South Africa, north to East Africa and Saudi Arabia.

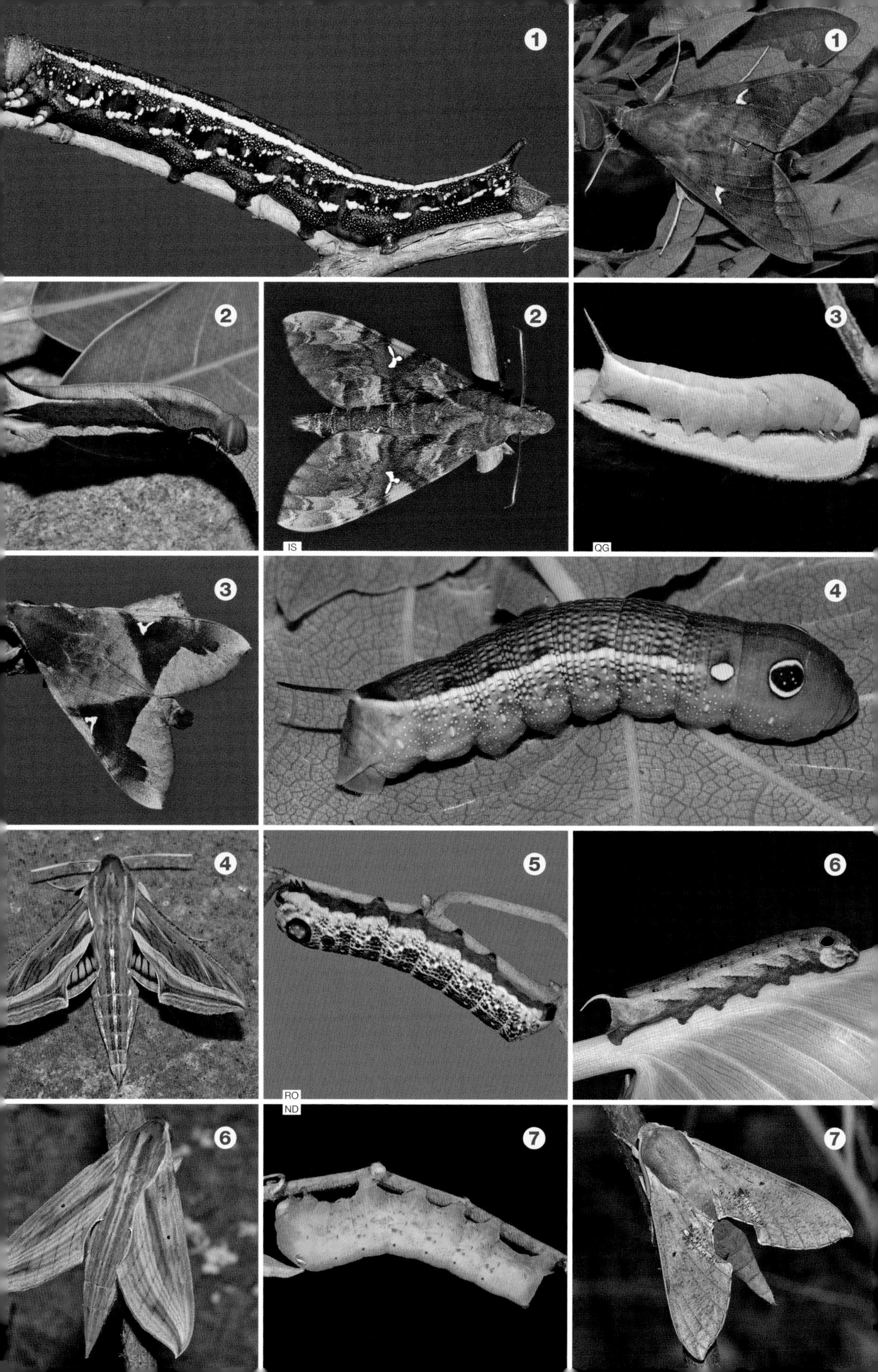
1
1
2
2
IS
3
QG
3
4
4
5
RO
ND
6
6
7
7

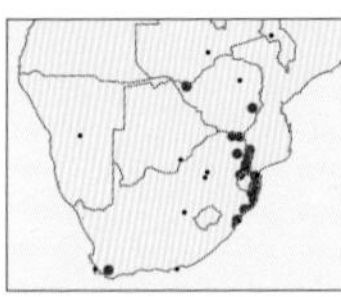

### 1 *Hippotion roseipennis* — Straw Hippotion

Forewing 22–26mm. Similar to some *Theretra capensis* (p.284); light brown, forewings with central irregular brown blotch with oblique postmedian line of small black dots distally, not to apex and pink costa; hindwings uniform dark brown. Caterpillar green with yellow and black eyespot, body dotted with small white spots, spiracles white surrounded by brown shading; abdominal horn dark, white and pointed, on orange conical base. **Biology:** Caterpillars feed on veld grapes (*Cissus* spp.). Probably bivoltine, adults throughout the year, peaking in August and April. **Habitat & distribution:** Lowland bushveld and coastal bush; KwaZulu-Natal, South Africa, north to Ethiopia.

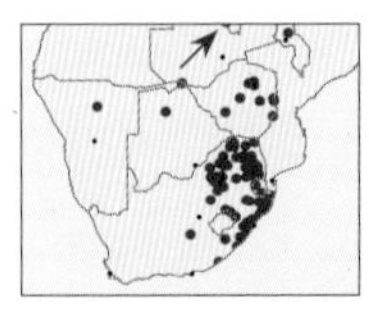

### 2 *Basiothia medea* — Small Verdant Hawk

Forewing 25–32mm. Grass green; hindwings orange with brown terminal shading, white cilia. Caterpillar green or brown with dark eyespot on first abdominal segment, following 7 segments with large yellow and brown triangular patches; abdominal horn straight, on reddish base. Four species in genus in region. **Biology:** Caterpillars feed on various herbaceous Rubiaceae. Multivoltine, adults throughout the year. **Habitat & distribution:** Moist open habitats; northeastern South Africa north across Africa, including Madagascar and Indian Ocean islands; active migrant.

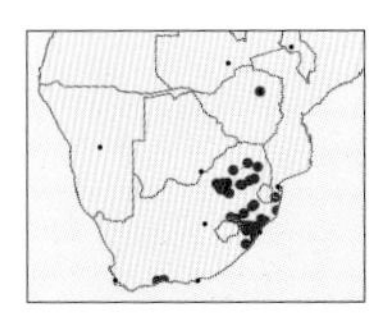

### 3 *Basiothia schenki* — Brown Striped Hawkmoth

Forewing 24–33mm. Brown with single white dorsal line; forewings brown with dark band split by white veins, bordered by whitish band; hindwings red with brown border. Caterpillar charcoal with pale lateral stripes and numerous white dots, 2 pairs of eyespots; abdominal horn dark, straight. **Biology:** Caterpillars feed on Wild Verbena (*Pentanisia angustifolia*). Multivotine, adults August–May. **Habitat & distribution:** Moist grassland; southern Cape, South Africa, north to Zimbabwe and DRC.

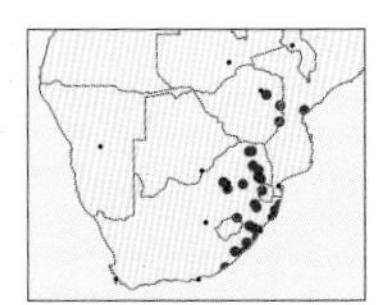

### 4 *Basiothia charis* — Lesser Brown Striped Hawk

Forewing 27mm. Very similar to *B. schenki* (above) and often confused, but double white line down centre of thorax and abdomen is diagnostic. **Biology:** Caterpillars feed on *Gossypium* and *Vernonia* spp. Multivotine, adults August–May. **Habitat & distribution:** Moist grassland; southern Cape, South Africa, north to Sierra Leone and Ethiopia.

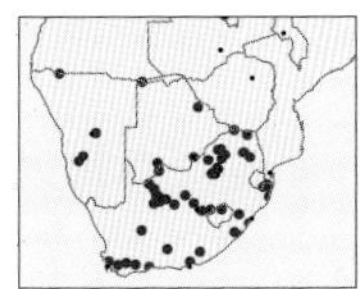

### 5 *Hyles livornica* — Striped Hawk

Forewing 24–33mm. Similar to *Hippotion osiris* (p.280), white scales on veins of forewings diagnostic; thorax brown with white margins, abdomen brown, marked with black and white bands; forewings brown with bold medial white stripe and crossed by fine white lines on veins, hind margin pale; hindwings brown with central pink section and pale margin. Caterpillar yellow with broad longitudinal black and orange lines, orange abdominal horn straight. One species in genus in region. **Biology:** Caterpillars polyphagous on herbaceous and low-growing woody plants belonging to many families. Probably bivoltine, adults throughout the year, peaking in spring and autumn. **Habitat & distribution:** Widespread, mostly in more arid habitats; across Africa, including Madagascar.

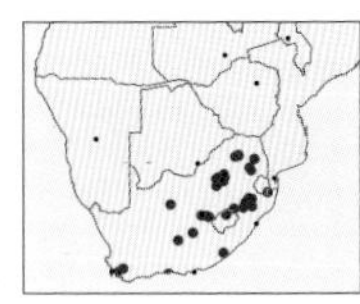

### 6 *Rhodafra opheltes* — Broad-striped Hawk

Forewing 26–32mm. Body greenish-brown, black side patches at base of abdomen; broad yellowish medial band across forewings, broken brown bands anteriorly, continuous band posteriorly; hindwings red with black base and distal band. Caterpillar green with dark dorsal and lateral lines, eyespots on each segment; short abdominal horn. Two species in genus in region. Similar *R. marshalli* lacks yellowish medial band across forewing; Zimbabwe northwards. **Biology:** Caterpillars feed on Wild Scabious (*Scabiosa columbaria*). Bivoltine, adults throughout the year, peaking October and March. **Habitat & distribution:** Grassland and fynbos; Western Cape to Limpopo, South Africa.

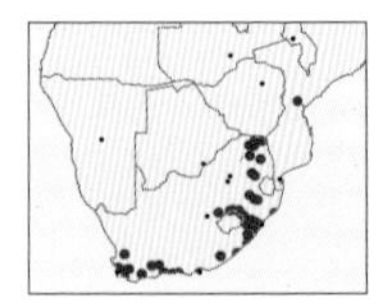

### 7 *Theretra cajus* — Cream-striped Hawk

Forewing 28–30mm. Similar to some *Hippotion* spp. (above); body stout, grey or brown, with pale double dorsal stripe; forewings pointed, dramatically bisected longitudinally by tapering white band between 2 darker bands; hindwings dark brown, not pink, with cream central stripes. Caterpillars green or brown, cream lateral stripe and lateral spots on each segment, large eyespot, abdominal horn slender, curved. **Biology:** Caterpillars feed on various Araceae, including arum lilies. Multivoltine, adults throughout the year, peaking in summer. **Habitat & distribution:** Moist fynbos, grasslands and forests where Araceae grow; Western Cape, South Africa, north to tropical Africa.

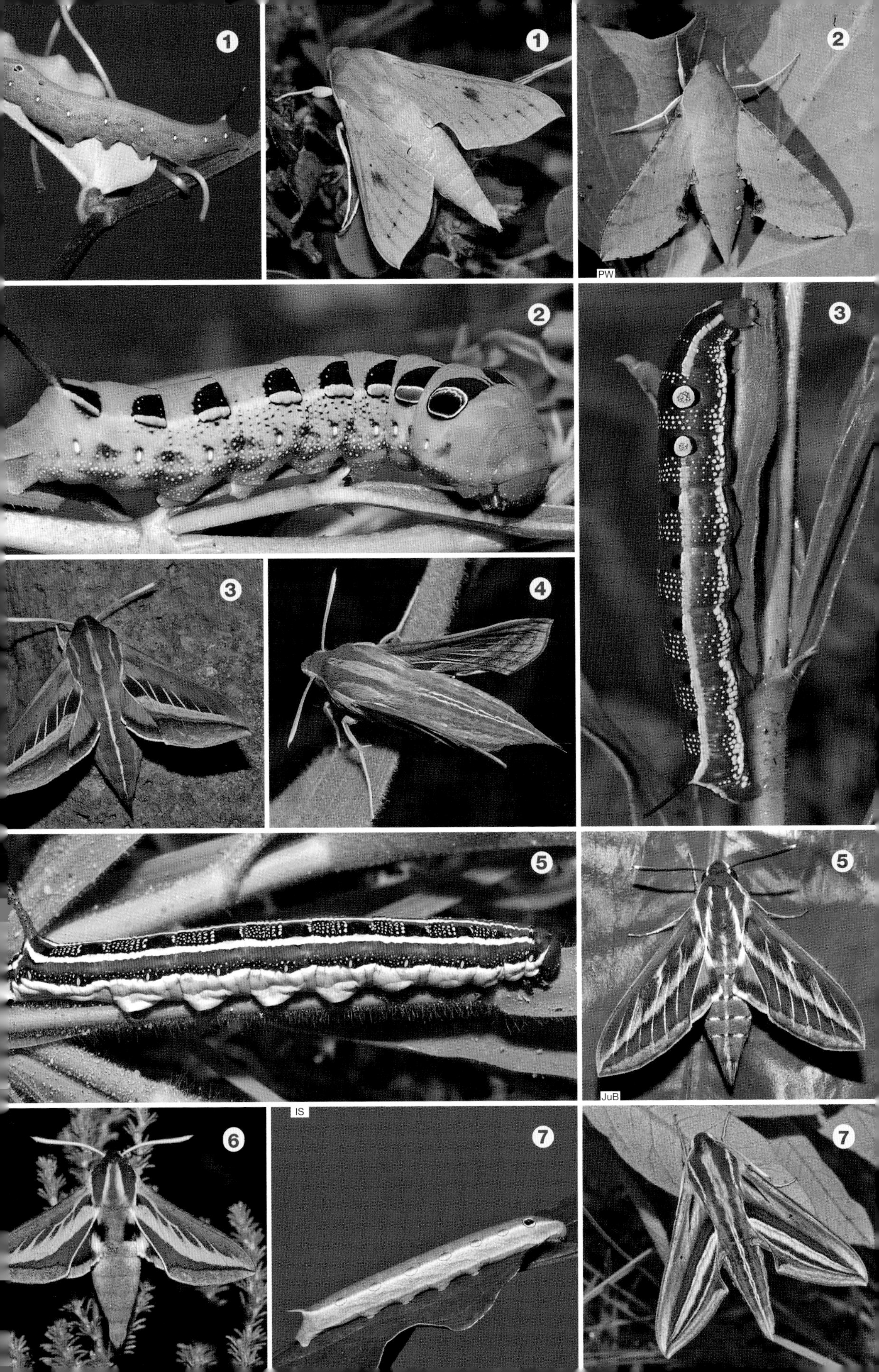
1
1
2
PW
2
3
3
4
5
5
JuB
IS
6
7
7

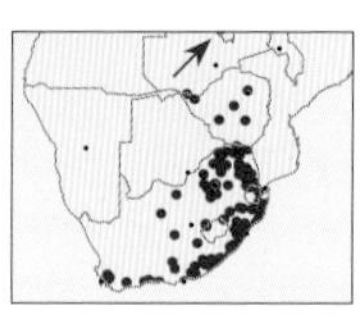

**1 *Theretra capensis*** Grape Hawkmoth

Forewing 41–50mm. Often confused with *Phylloxiphia punctum* (p.278) and *Hippotion roseipennis* (p.282); body and forewings uniformly grey-brown to reddish-brown, distal part of forewing slightly darker, variable poorly defined darker patch laterally, postmedian line straight toward apex; hindwings pink to red with thin brown margin, never brown. Caterpillar green with pale lateral stripe, variably bluish below with faint diagonal stripes, black and white eyespot on thorax, mauve abdominal horn very short and stout. Five species in genus in region. **Biology:** Caterpillars feed on various wild and cultivated grapevines (Vitaceae). Multivoltine, adults throughout the year, peaking in summer months. **Habitat & distribution:** Forest and open habitat; a common species; Western Cape, South Africa, north to East Africa.

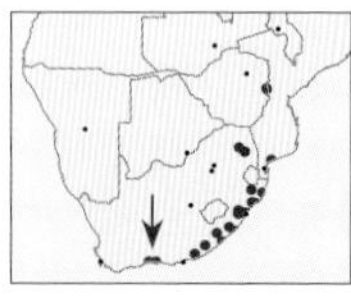

**2 *Theretra orpheus*** Orpheus Hawkmoth

Forewing 22–26mm. Body dark brown; forewings with diffuse longitudinal pale band and poorly defined darker brown to black patches; hindwings dark grey to brown with indistinct black lines. Caterpillar yellow with distinct anterior eyespot and complex pattern of brown wavy markings. **Biology:** Caterpillars feed on orchids, including Leopard Orchid (*Ansellia africana*), *Polystachya ottoniana*, *P. pubescens* and *Aerangis* spp. Multivoltine, adults throughout the year, peaking in late summer. **Habitat & distribution:** Moist forest; southern Cape, South Africa, north to tropical Africa, including Comoros and Madagascar.

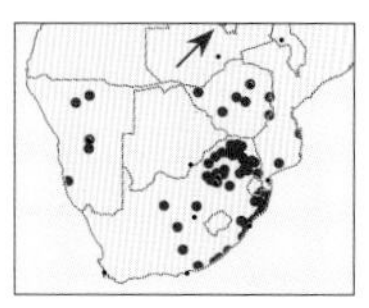

**3 *Cephonodes virescens*** Oriental Bee Hawkmoth

Forewing 25–35mm. Unmistakable, stout yellow to green body, dark abdominal band; clear, black-veined scale-less wings and tuft of hairs at tip of abdomen. Adults covered in brown scales just after emergence but shed scales on first vibration of wings. Caterpillar with raised horny abdominal band behind head and tail, colour variable, brown, green, yellow or black, with strong horizontal stripes and rings around spiracles; abdominal horn long, dark, curved and pointed. One species in genus in region. **Biology:** Caterpillars feed on woody Rubiaceae. Common day-flyers, hovering while inserting long proboscis into flowers, tail fan extended. Multivoltine, adults August–April. **Habitat & distribution:** Variety of vegetation types; across Africa and Madagascar, also Middle East, Asia and Australia; migratory.

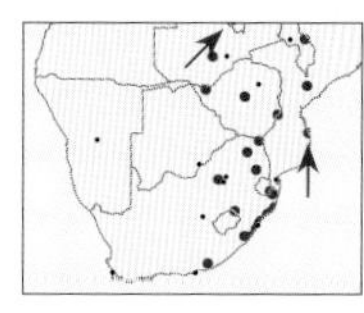

**4 *Leucostrophus alterhirundo*** White-barred Hawk

Forewing 16–19mm. Body robust, dark grey or black, with wide white 'H'-shaped band across abdomen, tip of abdomen with conspicuous fan of black and white setae; both wings uniform dark grey or black, hindwings sometimes with pale anterior patch. Caterpillar green, grey of brown. **Biology:** Adults diurnal. Caterpillars feed on Hard Pear (*Strychnos henningsii*). Multivoltine, adults throughout the year. **Habitat & distribution:** Caterpillars in subtropical and tropical areas; KwaZulu-Natal north across sub-Saharan Africa; adults migrate, sometimes seen far away from their caterpillar habitat.

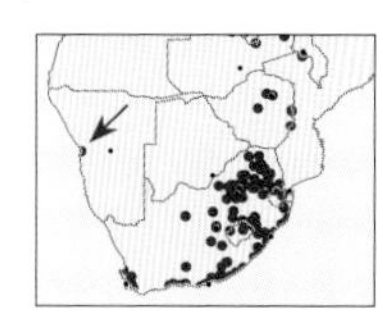

**5 *Macroglossum trochilus*** African Hummingbird Hawkmoth

Forewing 15–19mm. Body and forewings orange-brown; hindwings yellow at base, changing to orange then brown distally; distinctive black and white tail fan opens in flight. Caterpillar green or brown, rarely pink, with narrow longitudinal stripe, long abdominal horn straight and spiky. One species in genus in region. **Biology:** Adults diurnal, hovering like hummingbirds and flitting between tubular flowers such as lavender and salvias. Caterpillars feed on various Rubiaceae. Multivoltine, adults throughout the year, peaking in autumn. **Habitat & distribution:** Common in gardens; widespread through Africa.

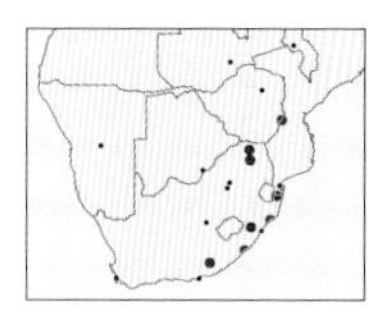

**6 *Odontosida magnifica*** Magnificent Scalloped Hawkmoth

Forewing 21–26mm. Body dark grey; forewings grey, dark triangular section anteriorly containing one white spot, distal margin distinctively scalloped; hindwings with striking yellow basal patch and red and brown distal sections. Caterpillar deep green, marked with blue, abdominal horn bluish. Two species in genus in region. **Biology:** Caterpillars feed on Kei Apple (*Dovyalis caffra*). Probably bivoltine, adults October–May. **Habitat & distribution:** Moist forest; Eastern Cape, South Africa, north to Zimbabwe.

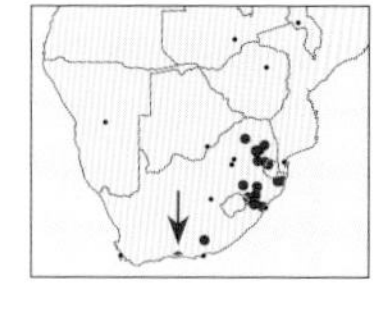

**7 *Odontosida pusillus*** Scalloped Hawkmoth

Forewing about 20–23mm. Slightly smaller than *O. magnifica* (above) and dark grey to reddish-brown; dark lines cross the forewings, which have numerous scallops along distal margins and are held out away from abdomen; hindwings pinkish-brown with dark distal margins, also with scalloped margins. Caterpillar apple green with oblique green and white stripes and reddish spiracles. **Biology:** Caterpillars feed on Sweet Yellow Bells (*Hermannia incana*) and *Triumfetta* spp. (Malvaceae). Bivoltine, adults October–March. **Habitat & distribution:** Grassland; southern Cape to Limpopo, South Africa.

1
1♀
2
2
3
3
4
SBa
5
5
6
7

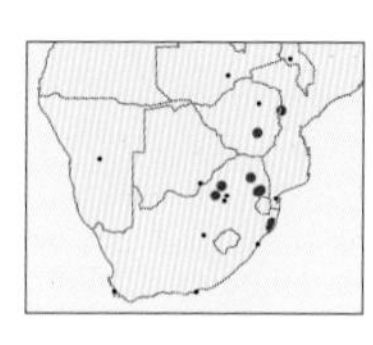

**1 *Sphingonaepiopsis ansorgei*** Ansorge's Hawklet

Forewing 14–17mm. Thorax grey-brown, abdomen squat, reddish-brown; forewings grey basally, crossed by broad diagonal brown band and distal thin wavy brown line, wings held outward to expose abdomen when at rest; hindwings uniform reddish. Caterpillar black with orange legs. Two species in genus in region. **Biology:** Caterpillars feed on various herbaceous Rubiaceae. Bivoltine, adults October–May. **Habitat & distribution:** Open habitats; KwaZulu-Natal, South Africa, north to tropical Africa.

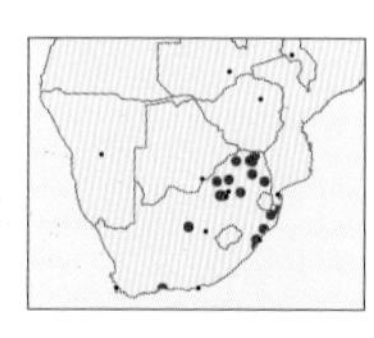

**2 *Sphingonaepiopsis nana*** Dwarf Hawklet

Forewing 12–14mm. Paler grey than *S. ansorgei* (above), with triangular brown patch on forewings crossed by thin white line. Caterpillar slender, green sometimes with pinkish sheen, whitish longitudinal lateral line and very long, slender abdominal horn. **Biology:** Caterpillars feed on various herbaceous Rubiaceae. Bivoltine, adults September–May. **Habitat & distribution:** Widespread across sub-Saharan Africa.

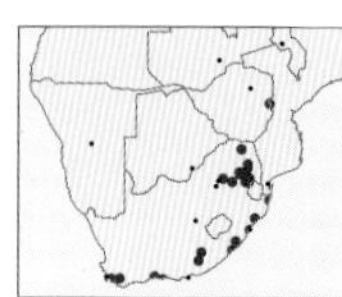

**3 *Temnora namaqua*** Namaqualand Temnora

Forewing 21–23mm. Forewings falcate and strongly incised posteriorly, light to dark brown with darker tips, median bar crenulated, sharpy curved inward near costa; hindwings yellow to orange with darker, sinuous margin. Caterpillar green with oblique white to cream stripes and a thick, red abdominal horn. Seventeen species in genus in region, many very similar and often confused. **Biology:** Caterpillars feed primarily on jakkalsstert (*Anthospermum* spp.) in fynbos habitats, also *Apodytes* and *Strychnos* spp. in adjacent forests. Bivoltine, adults August–May. **Habitat & distribution:** Remnant fynbos and moist Afromontane forests; Western Cape to Zimbabwe.

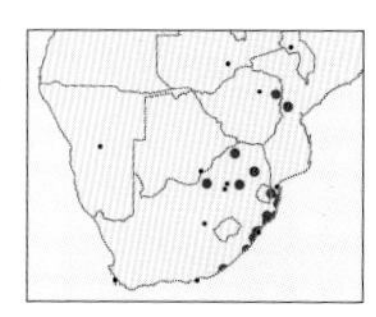

**4 *Temnora marginata*** Marginate Temnora

Forewing 22mm. Body grey to light brown with broad sharply tapering abdomen; large darker triangular marking across distal section of forewings, often with oblique golden band across median area, no cell spot; hindwings orange-brown; both wings with undulating distal margins. Caterpillar light green, oblique white bands across lower lateral parts, abdominal horn blue-green, straight and pointed. **Biology:** Caterpillars feed on Bird-berry (*Psychotria capensis*). Bivoltine, adults October–June. **Habitat & distribution:** Forest and bush; Eastern Cape, South Africa, north to Kenya, also Comoros.

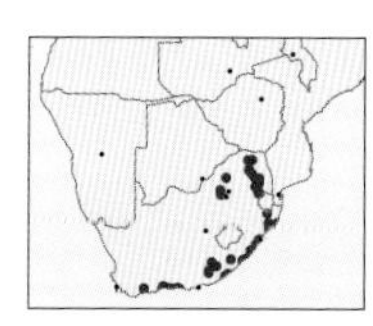

**5 *Temnora plagiata*** Brown Spot Temnora

Forewing 19–26mm. Grey-brown to pinkish; forewings with dark oblique bar near base and large, distinctive, rather quadrangular dark-brown block more distally; hindwings pinkish-brown with dark line near distal margin. Caterpillar dull green with yellow dorsolateral stripes. **Biology:** Caterpillars feed on White Pear (*Apodytes dimidiata*). Bivoltine, adults September–June, peaking in summer months. **Habitat & distribution:** Forest and bush; Western Cape, South Africa, north to tropical Africa.

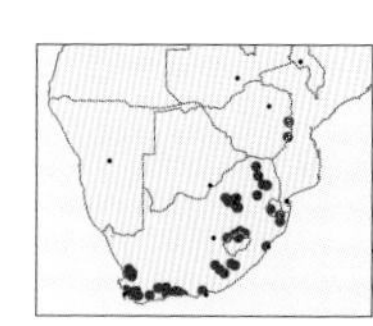

**6 *Temnora pylas*** Barred Yellow-winged Temnora

Forewing 18–20mm. Body and forewings brown, forewings with median dark, semicircular bar and intricate pattern markings toward brown wing tip; hindwings with yellow bases becoming orange then brown distally. Caterpillar green, marked with lateral yellow bars and rows of large yellow dorsal dots. **Biology:** Caterpillars feed primarily on *Anthospermum* spp. (Rubiaceae), also on *Leucas milanjiana* (Lamiaceae). Bivoltine, adults August–May, primarily in summer months. **Habitat & distribution:** Fynbos and moist grassland habitats where host plants occur; Western Cape, South Africa, north to Zimbabwe.

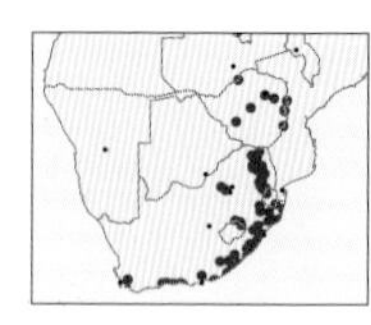

**7 *Temnora pseudopylas*** Lobed Temnora

Forewing 19–22mm. Body and forewings light to dark brown, forewings with large triangular darker patch near apex and distinctively scalloped distal margin; hindwings yellow with wide brown distal band. **Biology:** Caterpillars feed on Red Star Cluster (*Pentas lanceolata*), *P. bussei*, Jakkalsstert (*Anthospermum aethiopicum*) and *Leucas milanjiana*. Probably bivoltine, adults throughout the year. **Habitat & distribution:** Wide variety of habitats; Western Cape, South Africa, north to Ethiopia.

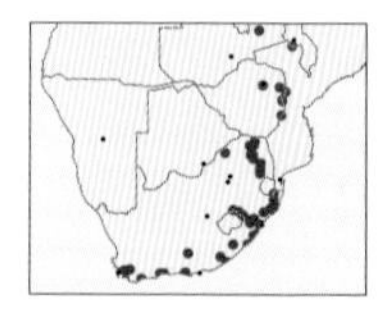

**8 *Temnora pylades*** Yellow-winged Temnora

Forewing 17–20mm. Similar to *T. pseudopylas* (above), but with less strongly scalloped margin to forewings. Caterpillar grey-green with distinctive oblique cream shading on anal segments, yellow eyespot on thoracic segment. **Biology:** Caterpillars feed on Jakkalsstert (*Anthospermum aethiopicum*), *Leucas milanjiana* and *Pentas lanceolata*. Bivoltine, adults throughout the year, peaking October and March. **Habitat & distribution:** Fynbos and moist grassland habitats where host plants occur; Western Cape, South Africa, north to Kenya.

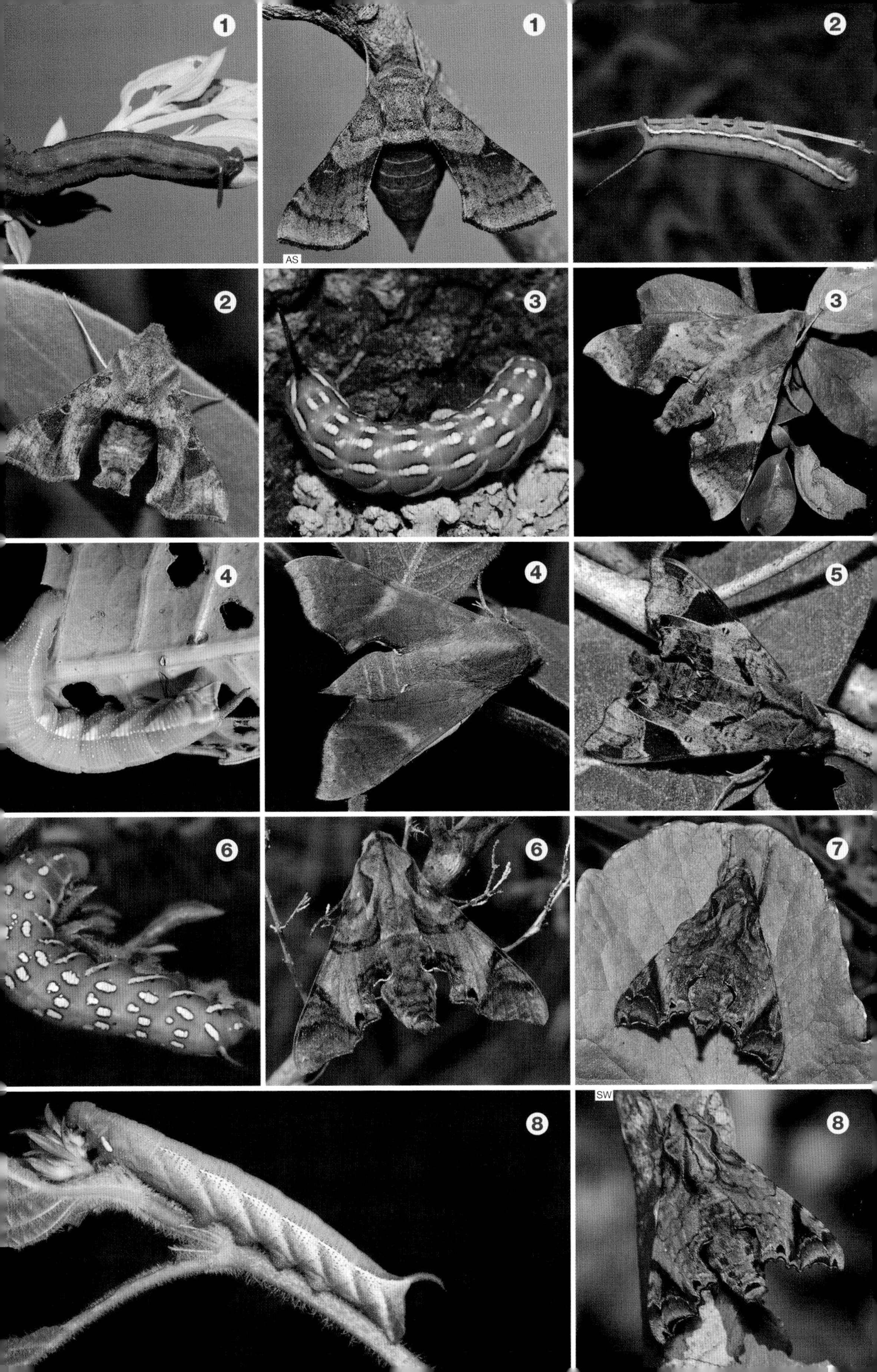

1
1
AS
2
2
3
3
4
4
5
6
6
7
8
SW
8

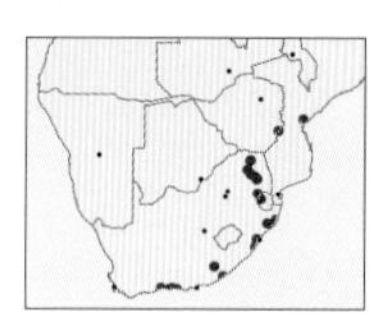

### 1 *Temnora zantus* — Grey Tip Temnora

Forewing 24–28mm. Body and wings dark brown, lines weak or absent, grey to cream wing tip, heavily scalloped wings. Caterpillar green with white spots, sometimes larger laterally. **Biology:** Caterpillars feed on Wild Pomegranate (*Burchellia bubalina*), also Hard Pear (*Strychnos henningsii*) and White Pear (*Apodytes dimidiata*). Mutivoltine, adults throughout the year. **Habitat & distribution:** Moist forest habitats where host plants occur; southern Cape, South Africa, north to Kenya.

# FAMILY NOLIDAE Tuft Moths

**Small moths with subdued green, grey or brown mottled colouration. Most have characteristic tufts of raised scales on the forewings, and adults emerge from the pupa via a vertical slit. Caterpillars lack first pair of prolegs and may have a greatly swollen thorax. Most species are specialists on a plant family or genus; regionally, Nolinae are associated with 13 plant families, Chloephorinae and Westermanniinae are primarily associated with Combretaceae, and Eariadinae and Bleninae with Ebenaceae. About 1,400 species described globally, 161 in the region.**

## Subfamily Eligminae

Males have specialised genitalia that hold open the female reproductive system, a system unique within Lepidoptera. The forewing is generally cut off squarely at the tip, and caterpillars have fully developed prolegs on abdominal segments 3–6.

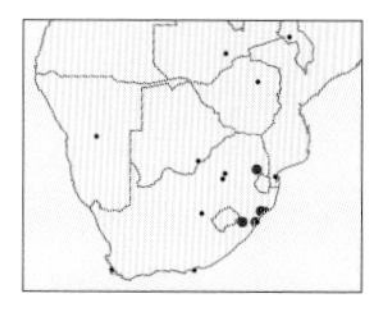

### 2 *Selepa transvalica* — Transvaal Square

Forewing 9–12mm. Body brown; forewings variably blotched bluish-brown crossed by 2 double curved brown lines, a black spot between these; hindwings white, tinted brown at margins. Caterpillar banded black and yellow with very long, white-tipped setae. Four species in genus in region. **Biology:** Caterpillars feed on Leg-ripper (*Smilax anceps*) and *Uapaca* spp. Adults February, June and October. **Habitat & distribution:** Moist forest; KwaZulu-Natal to Malawi.

## Subfamily Nolinae

Small to very small drab moths lacking external ocelli. The caterpillars are covered in short hairs grouped into tufts.

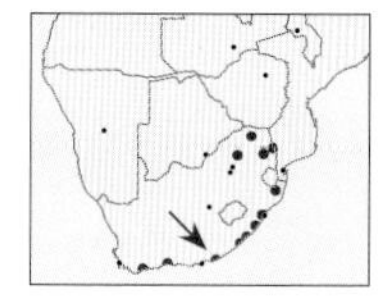

### 3 *Meganola bispermutata* — Shoulder-patch Nolid

Forewing 9–11mm. Body and forewings light mottled grey with a dark patch halfway along fore-margin of wing, and darker patches and lines near distal margin, dotted postmedian line; hindwings grey. Caterpillar black with red spots and orange sides, long brushes of pale lateral setae and 2 large tufts of brown setae dorsally. One of about 30 similar species in region, most mottled grey. **Biology:** Caterpillars feed on Silver-leaf Milkplum (*Englerophytum natalense*). Adults November–December and March–June. **Habitat & distribution:** Variety of frost-free habitats; Western Cape, South Africa, north to Zimbabwe.

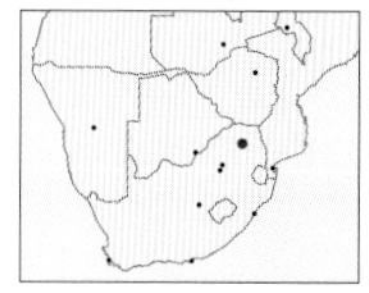

### 4 *Meganola pernivosa* — Bronze-edged Nolid

Forewing 14mm. Shiny white forewings with 2 black patches on costa, a zigzag line between the posterior pair of patches and 2 or 3 thin black lines posterior to this, bronze terminal shading; hindwings light grey. Two similar species in region, many more further north; *M. steniphona* from KwaZulu-Natal has terminal shading black. **Biology:** Caterpillar host associations unknown. Adults fly in November. **Habitat & distribution:** Moist Afromontane forest; known only from Woodbush, Limpopo, South Africa.

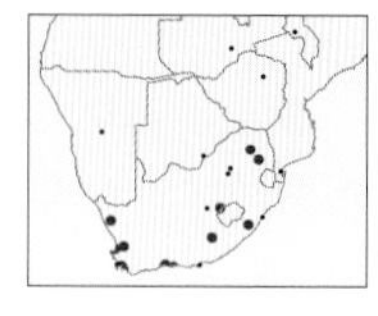

### 5 *Nolidia unipuncta* — One-dot Nolid

Forewing 6–8mm. At rest, strongly triangular in outline, palps projecting forward as pronounced snout; body and forewings ash grey, crossed by series of wavy dark lines and with darker margin, single tiny dark dot on shoulder. Caterpillar pale, translucent pink to green, with rosettes of pale spines, sometimes red dorsal patches. Eight species in genus in region. **Biology:** Caterpillars feed on heathers (*Erica* spp.) and Gonna Bush (*Passerina corymbosa*). Bivoltine, adults August–November and March–April. **Habitat & distribution:** Fynbos; Western Cape north to Limpopo, South Africa.

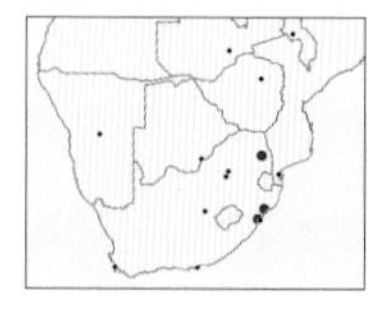

### 6 *Evonima littoralis* — Totem Nolid

Forewing 7mm. At rest, elliptical in outline, with short palps; body and forewings white with patches of grey and brown, tuft of bright brown scales at base of thorax; hindwings grey. Caterpillars pale green with long, white setae. Accumulates shed skins to create 'totem' of old headshields on head. One regional species in genus. **Biology:** Caterpillars feed on Buffalo Thorn (*Ziziphus mucronata*). Probably bivoltine, adults October–April. **Habitat & distribution:** Bushveld and forest; KwaZulu-Natal and Limpopo, South Africa.

1
1
KB
2
2
3
3
4
5
SM
5
6
IS
IS
6
6

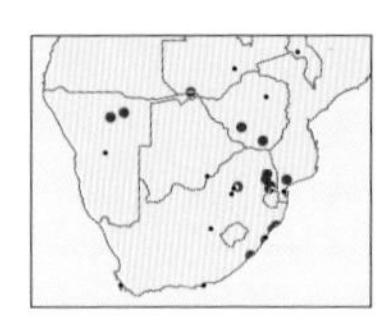

### 1 *Nola cana* — Arches

Forewing 6–8mm. Body and forewings light grey mottled with brown scales, median area of forewing darker brown, postmedian line strongly curved outward, large black cell spot with some shiny scales diagnostic. Caterpillar pale green, white dorsal line, reddish lateral lines, setal bases red, ringed by black on alternating segments. Eighteen species in genus in region. **Biology:** Caterpillars recorded feeding on flowers of *Senegalia ataxacantha.* Probably bivoltine, adults October–March. **Habitat & distribution:** Bushveld and forest; Eastern Cape, South Africa, north to Zimbabwe and Namibia.

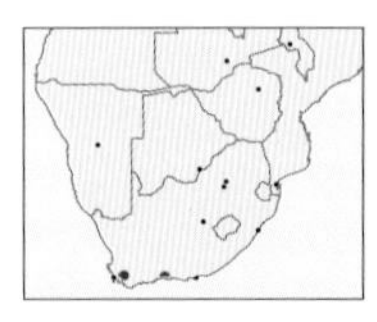

### 2 *Vandamia stellenboschi* — Copper Nolid

Forewing 7mm. At rest, strongly triangular in outline, palps projecting forward; body and forewings coppery brown with parallel, black-edged white lines, larger white patches; hindwings white. Caterpillar pale translucent green, longitudinal white dorsal lines, rosettes of pale spines, long black dorsal and anterior setae. Nine species in genus, endemic to region. **Biology:** Caterpillars feed on *Cliffortia odorata.* Bivoltine, adults October–November and March. **Habitat & distribution:** Fynbos and forest; Western Cape, South Africa.

## Subfamily Chloephorinae

Largely tropical, the adults have scale tufts on the forewing cell (portion of wing surrounded by wing veins). Included are several economically important species that feed on cola, cocoa and walnuts. *Arcyophora* spp. suck lachrymal secretions from cattle and may transmit viruses and bacteria.

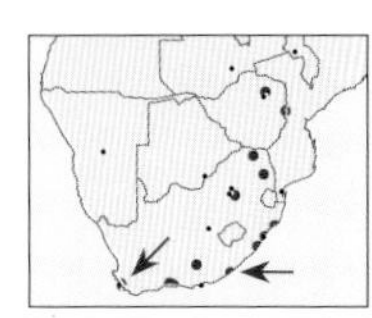

### 3 *Acripia subolivacea* — Scalloped Nolid

Forewing 11–14mm. Highly variable; green, brown or a mix of the colours, thorax with paler medial line; forewings broad, distinctly scalloped, crossed by 2 darker crenulated lines, brown cell spot with central white dot; hindwings pale brown with dark cell spot. One regional species in genus. **Biology:** Caterpillars feed on Crossberry (*Grewia occidentalis*). Probably multivoltine, adults throughout the year. **Habitat & distribution:** Forest and bushveld; Western Cape, South Africa, north to Kenya.

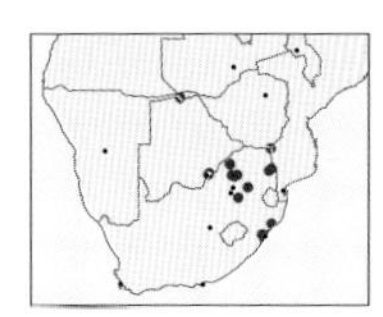

### 4 *Arcyophora longivalvis* — Cattle Eye Moth

Forewing ♂ 16–20mm, ♀ 22–28mm. Sexually dimorphic; body and forewings of male orange-brown, covered in fine grid of wavy broken lines; 2 stronger oblique lines cross wings, sometimes with white spot in postmedian area; female similar but mostly silvery grey; hindwings cream. Caterpillar green with white longitudinal line, tapering posteriorly with elongate stretched out anal claspers, typical for genus. Nine species in genus in region; similar *A. patricula* darker brown with black dots. **Biology:** Caterpillars feed on Red Bushwillow (*Combretum apiculatum*), and Silver Cluster-leaf (*Terminalia sericea*). Adults suck lachrymal fluids from eyes of cattle. Multivoltine, adults August–May. **Habitat & distribution:** Bushveld; KwaZulu-Natal, South Africa, north to tropical Africa.

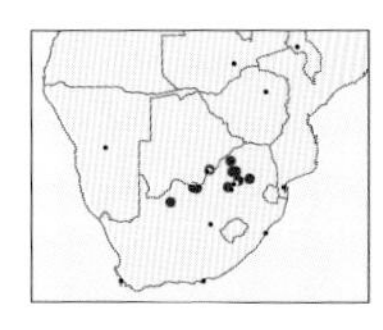

### 5 *Arcyophora piperitella* — Pastel Flag

Forewing ♂ 9mm, ♀ 12mm. Sexually dimorphic; male forewings pastel straw to pink with faint lines, dark black cell spot in some specimens; female shaded pastel pink to grey; hindwings straw to brown. Caterpillar green with white longitudinal line, tapering posteriorly. **Biology:** Caterpillars feed on Silver Cluster-leaf (*Terminalia sericea*). Bivoltine, adults September–March, abundant in February. **Habitat & distribution:** Dry *Terminalia* sandveld; Northern Cape, South Africa, north to tropical Africa.

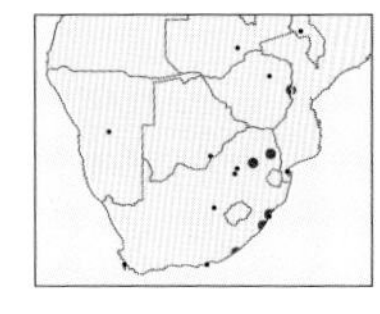

### 6 *Goniocalpe heteromorpha* — Tiptail

Forewing 9–11mm. Wings held steeply to sides and enlarged tip of abdomen bent upward; grey to brown; forewings falcate and broad, crossed by 2 curved white, black or black and white lines, a few small black dots sometimes distally, large cell spot black or white or absent; hindwings cream. Caterpillar green with 2 white longitudinal stripes and forked tail. Two regional species in genus. **Biology:** Caterpillars feed on bushwillows (*Combretum* spp.). Multivoltine, adults throughout the year. **Habitat & distribution:** Forest and riverine bush; Eastern Cape, South Africa, north to Zimbabwe.

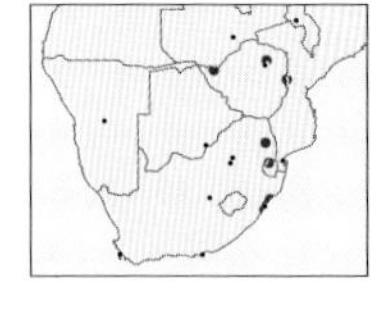

### 7 *Lophocrama phoenicochlora* — Large Vernal

Forewing 10–11mm. Body brown or green; forewings broad, bright green with brown and white chequered costa, sometimes crossed by broad, pale brown band outlined by darker brown lines, occasionally only marginal lines visible; hindwings white or light pink with darker margin. One regional species in genus. **Biology:** Caterpillars feed on various Malvaceae. Multivoltine, adults throughout the year. **Habitat & distribution:** Frost-free forest and bushveld; KwaZulu-Natal, South Africa, north to tropical Africa and Madagascar.

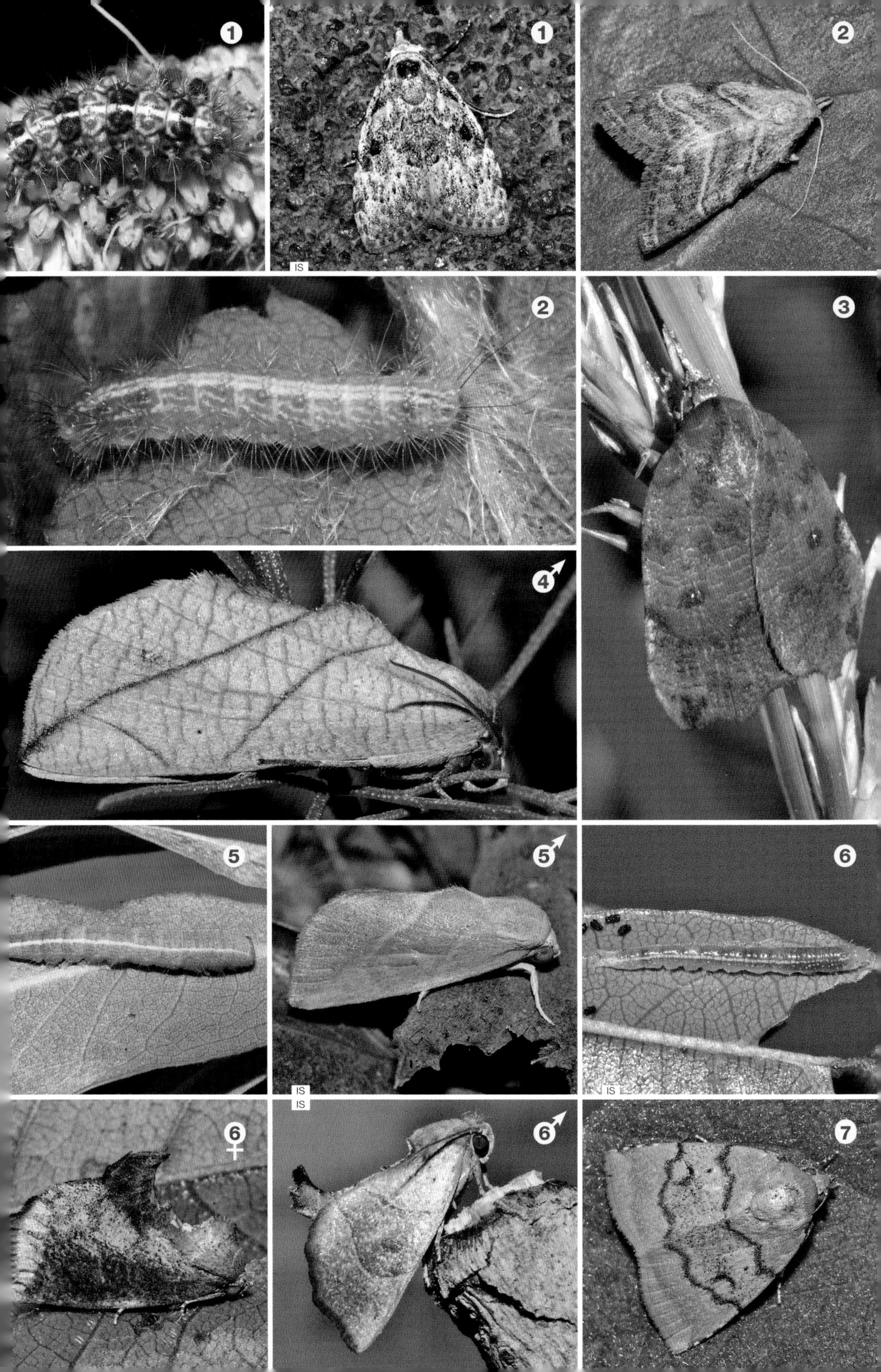
1
1
IS
2
2
3
4♂
5
5♂
IS
6
IS
6♀
IS
6♂
7

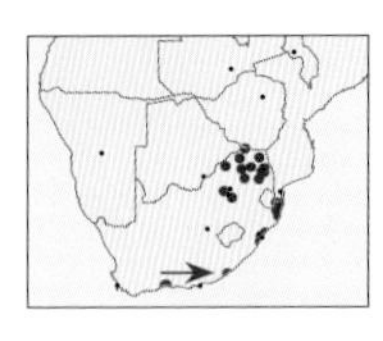

### 1 *Maurilia arcuata* — Arch Drab

Forewing 9–15mm. Body and forewings straw, grey-brown or orange; forewings short, squared off apex, crossed by thin wavy brown to red lines; hindwings off-white. Caterpillar mottled brown to black, greatly swollen thorax. One regional species in genus. **Biology:** Caterpillars polyphagous on trees. Multivoltine, adults throughout the year, peaking February–March. **Habitat & distribution:** Forest, bushveld and coastal bush; Western Cape, South Africa, north to tropical Africa, Madagascar and La Réunion.

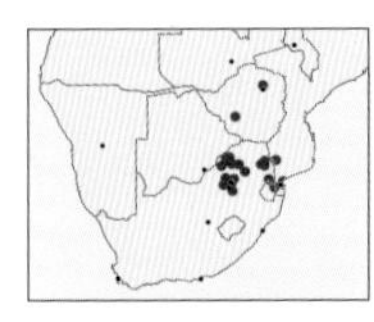

### 2 *Neaxestis rhoda* — Red Flag

Forewing 9–11mm. Body and forewings mottled red-brown, crossed by 2 thin, widely spaced, dark brown lines, postmedian line rounded; hindwings white. Caterpillar mottled green, 2 longitudinal white lines, not markedly swollen anteriorly. Five species in genus in region. **Biology:** Caterpillars feed on Red Bushwillow (*Combretum apiculatum*). Multivoltine, adults August–May. **Habitat & distribution:** *Combretum* bushveld; Gauteng, South Africa, north to Tanzania.

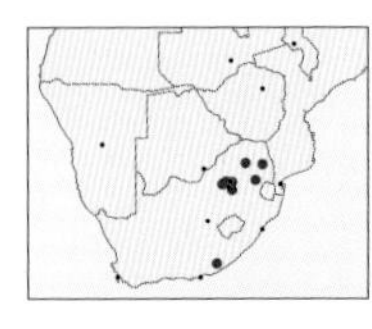

### 3 *Neaxestis acutangula* — Acute Angles

Forewing 11–12mm. Similar to *N. rhoda* (above), but forewings mustard brown and postmedian line pale and bent into an acute angle midway. Caterpillar pale green with white lateral stripe. **Biology:** Caterpillars feed on River Bushwillow (*Combretum erythrophyllum*). Bivoltine, adults September–March. **Habitat & distribution:** Riverine vegetation; Eastern Cape north to Limpopo, South Africa.

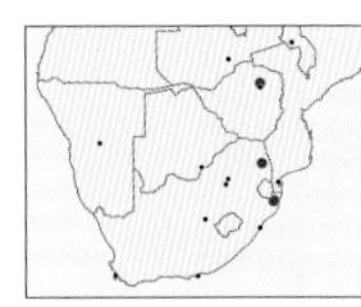

### 4 *Odontestis striata* — Brown V

Forewing 14–15mm. Body brown with dorsal ridged tufts of setae; forewings mottled light and dark brown, a crescent-shaped dark line centrally; hindwings light brown. Caterpillar mottled brown with orange dots. Two species in genus in region. **Biology:** Caterpillars feed on Broom Creeper (*Cocculus hirsutus*). Multivoltine, adults throughout the year. **Habitat & distribution:** Lowland bushveld; KwaZulu-Natal, South Africa, north to tropical Africa.

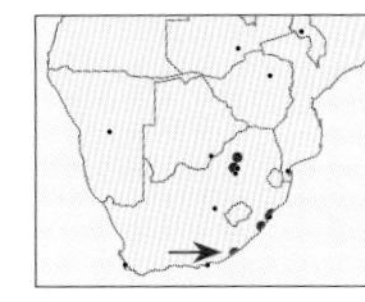

### 5 *Paraxestis rufescens* — Saddled Camel

Forewing 10mm. Light brown with wings steeply inclined, forming sharp dorsal ridge raised into a distinct hump, with darker saddle-like patch behind. Caterpillar pale green. One species in genus in region. **Biology:** Caterpillars feed on bushwillows (*Combretum* spp.). Multivoltine, adults throughout the year. **Habitat & distribution:** Coastal & riverine bush and forest; Eastern Cape north to Limpopo and Tanzania.

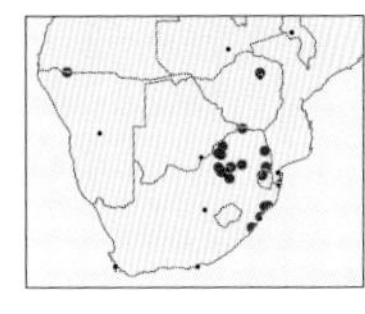

### 6 *Pardasena virgulana* — Grey Square

Forewing 7–8mm. Body dark grey; forewings broad and dark grey, crossed by narrow or thick network of black lines or with variable black patches. Five species in genus in region. **Biology:** Caterpillars polyphagous on flowers and fruit of trees. Multivoltine, adults throughout the year. **Habitat & distribution:** Forest and bushveld; KwaZulu-Natal, South Africa, north to tropical Africa and Indian Ocean islands.

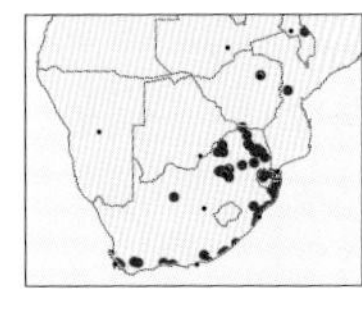

### 7 *Earias biplaga* — Spiny Bollworm

Forewing 9–12mm. Body and forewings mottled greenish-yellow; forewings crossed by wavy yellow lines, area between the lines sometimes brown, hind margin brown; hindwings white with dark margin. Caterpillar spiky, mottled light and dark brown. Four regional species in genus. **Biology:** Caterpillars feed on various Malvaceae. Multivoltine, adults throughout the year. **Habitat & distribution:** Moist habitats, often in gardens; across sub-Saharan Africa, including adjacent Atlantic and Indian Ocean islands.

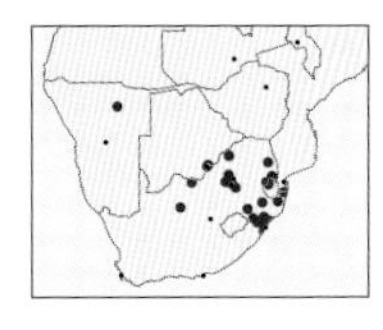

### 8 *Earias cupreoviridis* — Cupreous Bollworm

Forewing 6–9mm. Similar to *E. biplaga* (above), but smaller, lacking lines; forewings with an orange marginal patch, small round brown cell spot and wide orange and brown distal band. Caterpillar similar to that of *E. biplaga*. **Biology:** Caterpillars feed primarily on *Sida* spp., including invasive weeds such as Flannel Weed (*S. cordifolia*) and Arrowleaf Sida (*S. rhombifolia*), helping to control them. Multivoltine, adults throughout the year. **Habitat & distribution:** Open woodland, savanna and bushveld; KwaZulu-Natal north across Africa and Asia.

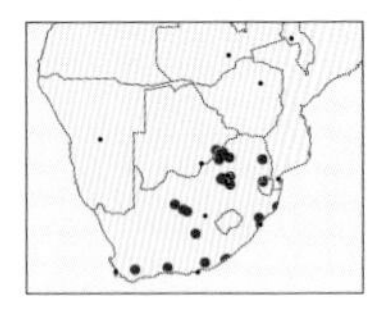

### 9 *Earias insulana* — Insular Bollworm

Forewing 9–11mm. Straw brown, sometimes with greenish tint, 3 zigzag lines across forewings. Often confused with *E. ansorgei*, which is uniform bluish-green with faint or no lines. Caterpillar spiky and mottled brown. **Biology:** Caterpillars feed on wide variety of Malvaceae, including crops such as okra and cotton. Multivoltine, adults throughout the year. **Habitat & distribution:** Variety of habitats, including transformed habitats; across Africa, southern Europe, Middle East, Asia and Australia.

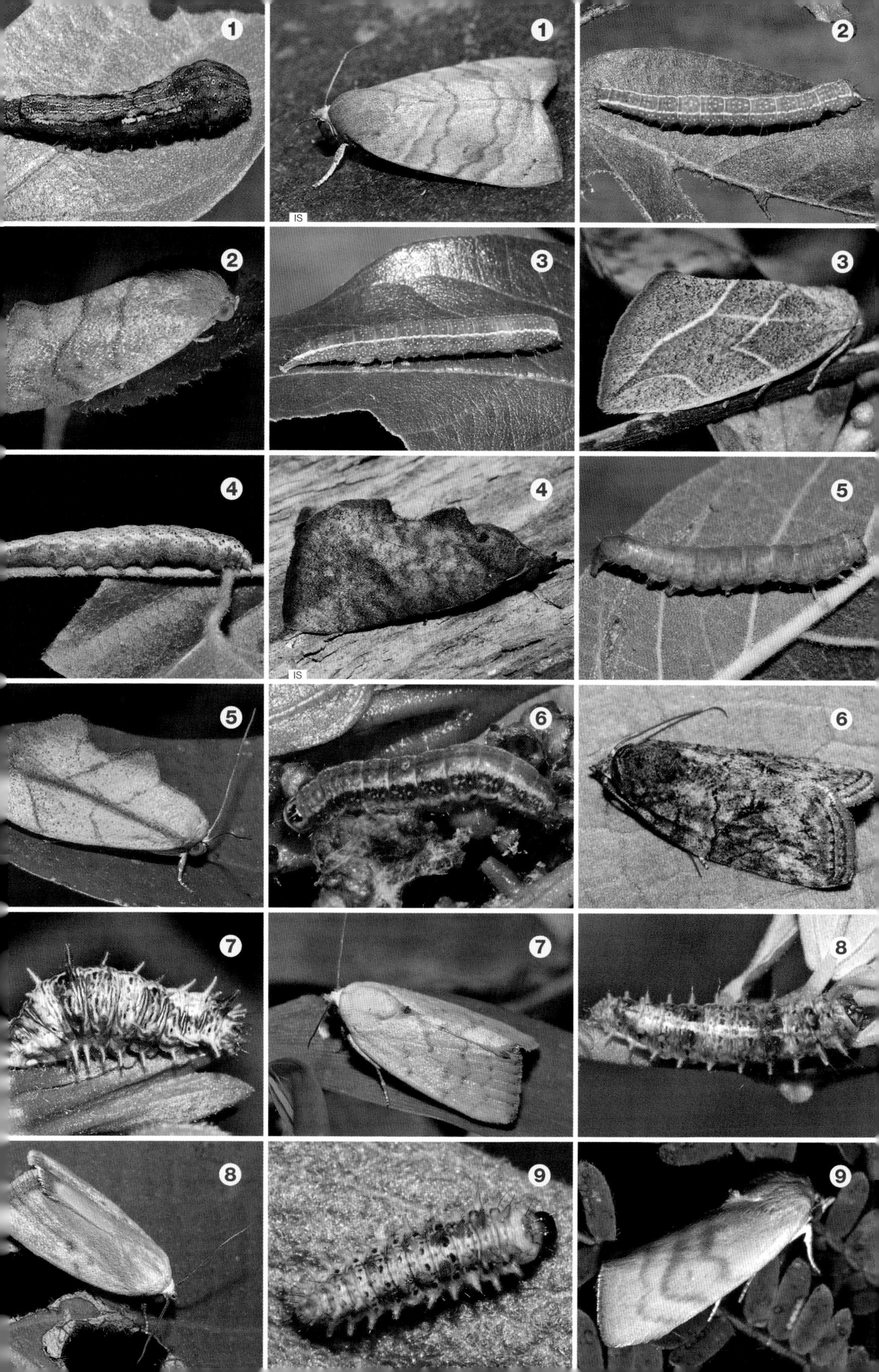

1
1
IS
2
2
3
3
4
4
IS
5
5
6
6
7
7
8
8
9
9

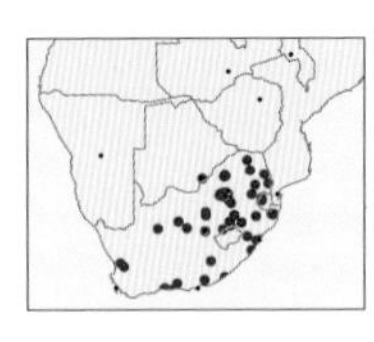

**1 *Blenina squamifera*** Lichen Square

Forewing 10–14mm. Variable; forewings held flat, with squared-off apex; body and forewings covered by complex and variable network of green, grey or brown patches outlined by transverse black lines or bands, closely resembling lichen; hindwings white with dark marginal band. Caterpillar plump anteriorly, tapering toward rear, pale to bright green with longitudinal cream line. Four regional species in genus. **Biology:** Caterpillars feed on Bloubos (*Diospyros lycioides*), also other Ebenaceae. Multivoltine, adults throughout the year, peaking in summer. **Habitat & distribution:** Variety of habitats; Western Cape to tropical Africa.

## Subfamily Westermanniinae

The forewings often silver and satiny, lacking raised scales. The males lack sound-producing organs. Surface of the pupae beaded, and the pupae 'shiver' within their cocoons. There is some host specialisation for *Terminalia* (Combretaceae).

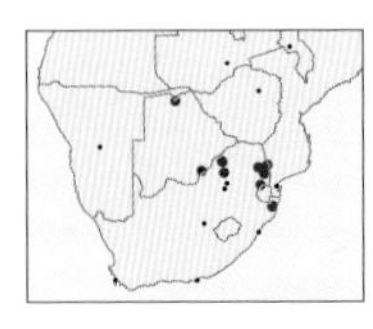

**2 *Negeta luminosa*** Luminous Cloud

Forewing 6–8mm. Body and forewings white; wings with a wide brownish-blue arch-shaped band just before hind margin, sprinkled with bright shiny scales; hindwings white with smoky margin. Caterpillar thin, green with black head capsule. Four species in genus in region. **Biology:** Caterpillars feed in leaf folds on Combretaceae. Multivoltine, adults February–August. **Habitat & distribution:** Bushveld; KwaZulu-Natal, South Africa, north to tropical Africa, Saudi Arabia, Yemen and Madagascar.

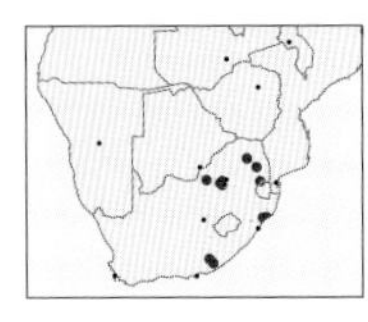

**3 *Westermannia convergens*** Convergent Nolid

Forewing 12–13mm. Thorax light brown; forewings broader posteriorly, 2 converging curved lines divide forewings into 3 sections; hindwings white. Six species in genus in region. **Biology:** Caterpillar host associations unknown. Adults September–April, peaking October and February. **Habitat & distribution:** Forest and riverine bush; Eastern Cape to Limpopo, South Africa.

# FAMILY NOTODONTIDAE Prominents

**Tuft of setae on hind margin of forewings of most species, which projects to side of body at rest. Adults heavy-bodied and dull-coloured mimics of bark, dead leaves or broken twigs. Wings held roof-like against body. May sham death and fall to the ground when disturbed. Caterpillars mostly solitary feeders, but some are often gregarious, such as *Rhenea* and subfamily Thaumetopoinae. The caterpillars are exceptionally diverse in form, some with bizarre shapes and ornamented with dorsal horns, tubercles, bumps or spikes, with dramatic change in appearance between moults. About 3,800 species described globally, with 225 in the region.**

## Subfamily Notodontinae

Antennae in the male are pectinate along their entire length, and the proboscis is short and broad, or absent. The body of the caterpillar is never hairy.

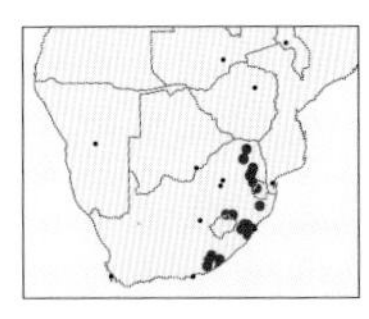

**4 *Cerurella natalensis*** Natal Cerurella

Forewing 17–18mm. Body white with black, mid-dorsal spot on thorax; front legs extremely setose and held extended forward; forewings white with fine wavy grey to black lines and row of black dots along margin diagnostic; hindwings white with yellow veins and black marginal dots. Two species in genus in region. **Biology:** Caterpillars feed on Small-leaved Brown-ironwood (*Homalium rufescens*). Adults August–May. **Habitat & distribution:** Moist forest; Eastern Cape to Limpopo, South Africa.

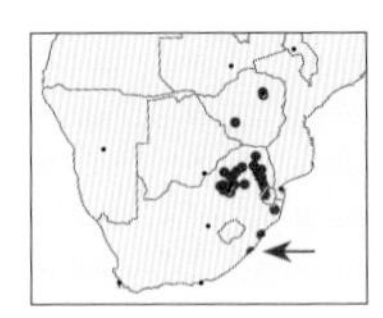

**5 *Hampsonita esmeralda*** Esmeralda's Kitten

Forewing 17–24mm. Lichen mimic; thorax setose, pale grey to bluish or green with black spots and rings; forewings similar, crossed by crenulated black lines and with black dots along distal margin; hindwings white with black dots posteriorly. Caterpillar plump, pale green with brown to reddish blotches around spiracles, 2 long, pointed 'tails' raised in defensive pose when disturbed. One species in genus in region. **Biology:** Caterpillars feed on *Protea* spp. and African Beech (*Faurea saligna*). Bivoltine, adults September–April. **Habitat & distribution:** Where Proteaceae are present; Eastern Cape to tropical Africa.

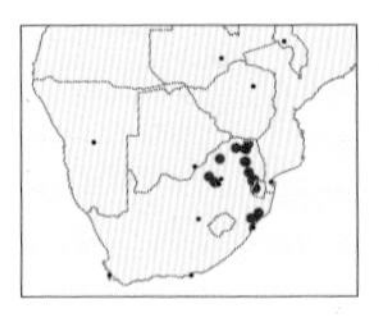

**6 *Atrasana postica*** Dusky Prominent

Forewing 26mm. Body dark grey to black; forewings light grey, variably marked with broken black and brown zigzag lines, a conspicuous curved black band cutting off hind corner; hindwings white with black mark on anal angle. Four similar species in genus in region. **Biology:** Caterpillars reared on Pink Wild Pear (*Dombeya burgessiae*). Probably bivoltine, adults October–April. **Habitat & distribution:** Moist and dry forest and bush; KwaZulu-Natal, South Africa, north to Zimbabwe.

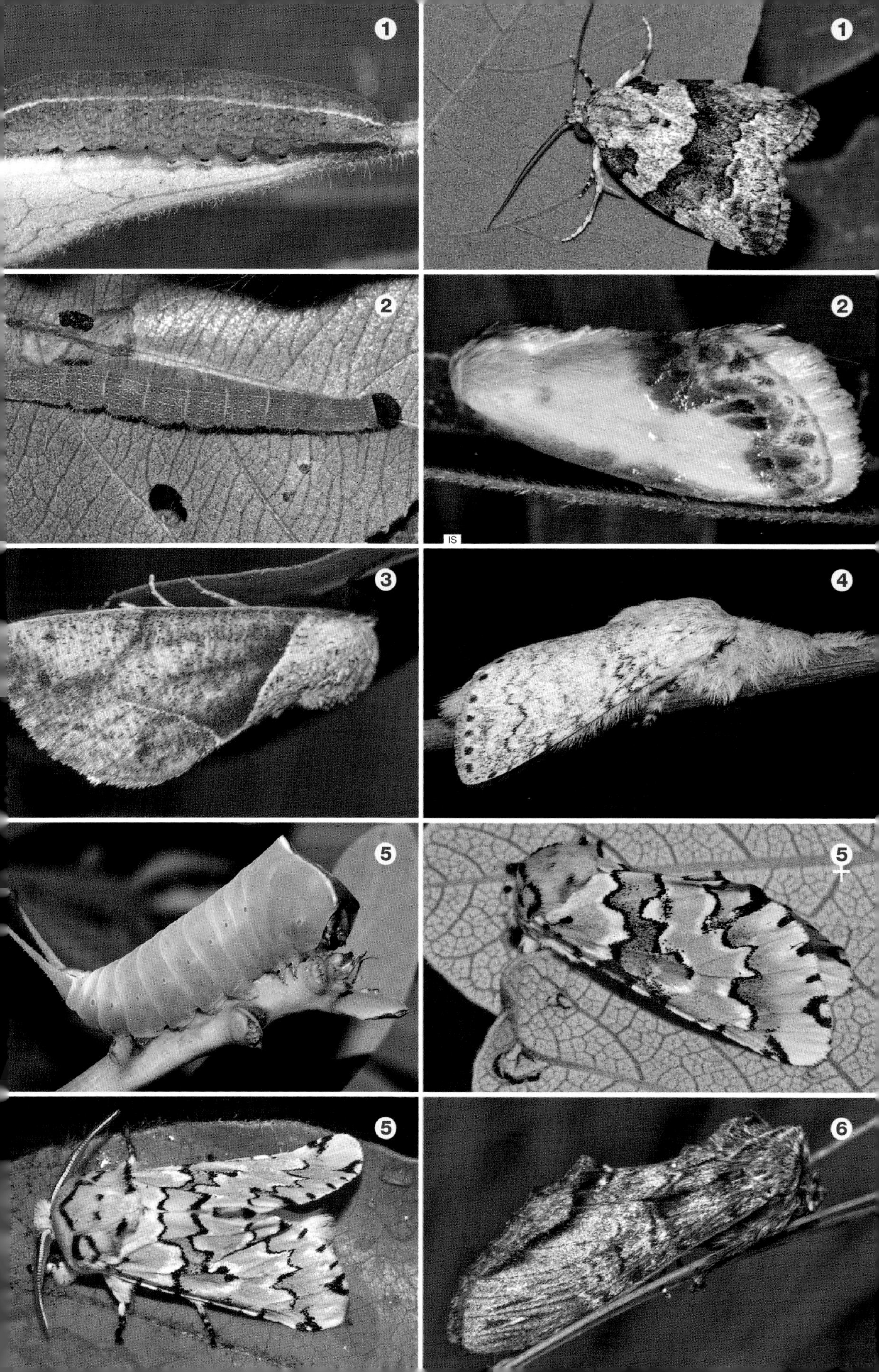
1
1
2
2
IS
3
4
5
5
♀
5
6

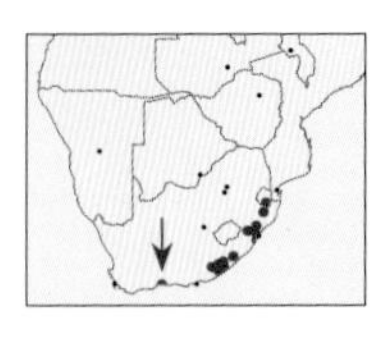

**1 *Atrasana uncifera*** Brown Prominent

Forewing 18mm. Smaller than *A. postica* (p.294), with wings more rounded; body and forewings brown not grey, variably marked with broken black and brown zigzag lines, a conspicuous curved black band cutting off hind corner; hindwings light brown with several light or dark brown bands. **Biology:** Caterpillars feed on various Malvaceae trees. Probably bivoltine, adults September–April. **Habitat & distribution:** Moist forest and coastal bush; southern Cape, South Africa, north to Zimbabwe.

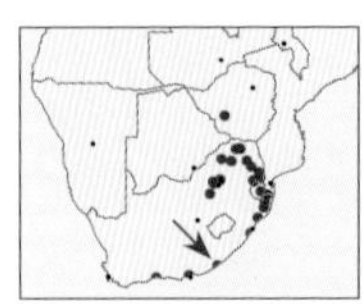

**2 *Notocerura spiritalis*** Spirited Puss

Forewing 22–28mm. Thorax white with 2 transverse black bands, abdomen darker, usually black; legs white with black tips; forewings white with black marginal dots, may be crossed by 2 broad black bands but wings may be almost completely white; hindwings white with black central spot and marginal dots. Caterpillar greenish-brown with large white lateral dots, anterior section raised into bizarre triangular ridge, hind end raised with 2 elongate curled 'tails' when threatened. One species in genus in region. **Biology:** Caterpillars feed on Thorn Pear (*Scolopia zeyheri*) and Governor's Plum (*Flacourtia indica*, Salicaceae). Bivoltine, adults September–March. **Habitat & distribution:** Forest and moist, thick bush; southern Cape, South Africa, north to Zimbabwe, Mozambique, Zambia and Malawi.

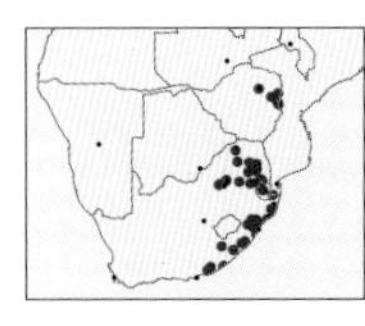

**3 *Pseudorethona albicans*** Lichen Cat

Forewing 12–15mm. Thorax setose, off-white with black markings; forewings white to light brown with network of fine lines, usually darker black median and distal bands and brown distal area; hindwings of male white with marginal black dots, female grey-brown. Caterpillar pale green with white lateral line and long curved 'tails'. One species in genus in region. **Biology:** Caterpillars feed on Wild Apricot (*Dovyalis zeyheri*), Forest Raisin (*Grewia lasiocarpa*) and Wild Mulberry (*Trimeria grandifolia*). Multivoltine, adults throughout the year. **Habitat & distribution:** Forest and moist, thick bush; Eastern Cape, South Africa, north to tropical Africa.

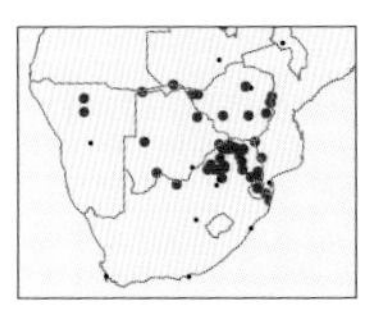

**4 *Afroplitis bergeri*** Squirrel Cat

Forewing ♂ 20–25mm, ♀ 25–30mm. Antennal shaft bright pink, strongly bipectinate, orange; body and forewings grey with black lines, sometimes pinkish; hindwings white basally, brown toward edge. Caterpillar pale green to pink or grey, hind end broadened and upturned. Three regional species in genus, similar *A. dasychirina* has antennae pectinations black, not orange. **Biology:** Caterpillars feed on Silver Cluster-leaf (*Terminalia sericea*). Bivoltine, adults September–April. **Habitat & distribution:** Sandy bushveld; northern KwaZulu-Natal, South Africa, north to Malawi.

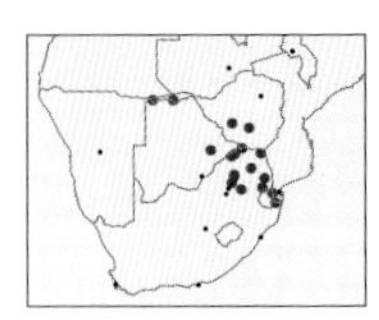

**5 *Amyops ingens*** Giant Prominent

Forewing 30–42mm. Abdomen orange, black markings below and at tip; elongate forewings white with black margins, streaked grey and brown; hindwings white with broad black margins. Caterpillar brown with long, splayed anterior projections, ending in hooked spike, rear end inflated with eyespot, resembling reptile head. Earlier instars mimic bird droppings. Monotypic genus. **Biology:** Adults rest on tree trunks, raise their wings to display aposematic body and hindwings abruptly when disturbed. Caterpillars feed on Silver Cluster-leaf (*Terminalia sericea*). Bivoltine, adults October–February. **Habitat & distribution:** Sandy bushveld; northern KwaZulu-Natal, South Africa, north to tropical Africa.

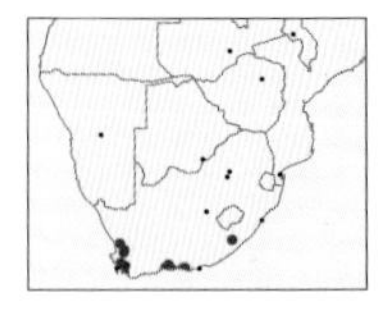

**6 *Catochria catocaloides*** Prominent Underwing

Forewing 26–32mm. Body robust, with long orange antennae, grey thorax and orange abdomen; forewings grey crossed by wavy brown lines, hindwings orange with central wavy black band and brown border. Caterpillar pale green with yellow lateral lines and markings; head pink with white lateral band. Two species in genus in region; similar *C. postflava* allopatric, in protea grassland on north-eastern escarpment, KwaZulu-Natal to Limpopo. **Biology:** Caterpillars feed primarily on Proteaceae (*Protea*, *Leucadendron*), but also on Monterey Pine (*Pinus radiata*) and Hyaena-poison (*Hyaenanche globosa*). Multivoltine, adults recorded throughout the year. **Habitat & distribution:** Fynbos; Western Cape and Eastern Cape, South Africa.

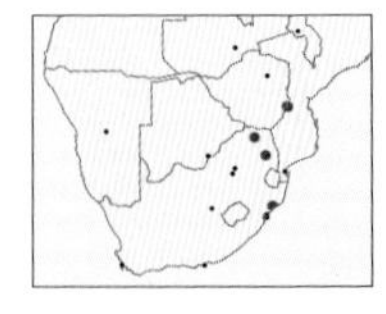

**7 *Chlorocalliope calliope*** Calliope

Forewing 23–24mm. Lichen mimic. Thorax mottled green and brown, prominent dorsal tuft of setae; forewings mottled green with irregular brown lines with distinct broken black lines on termen; hindwings pale brown. One species in genus in region; similar to several *Desmeocraera* spp. (p.298), terminal black markings on forewings diagnostic. **Biology:** Caterpillars feed on Fluted Milkwood (*Chrysophyllum viridifolium*). Bivoltine, adults October–November and March–June. **Habitat & distribution:** Moist forests; Eastern Cape, South Africa, north to Zimbabwe.

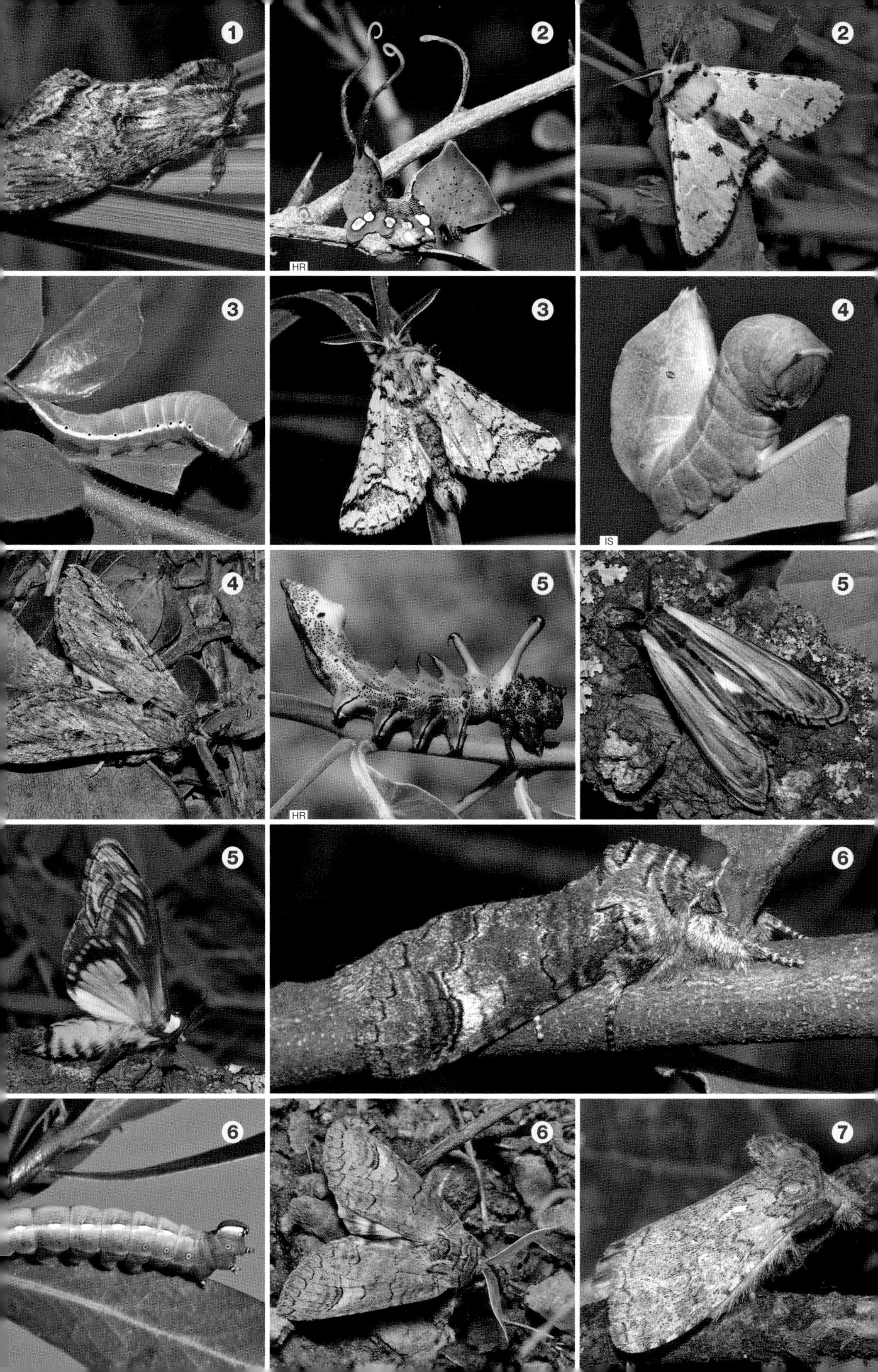
1
2
HR
2
3
3
4
IS
4
5
HR
5
5
6
6
6
7

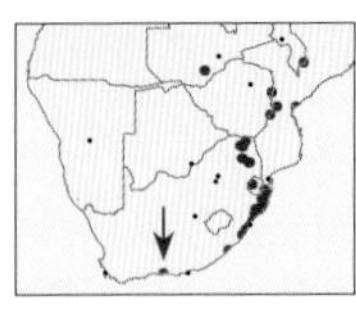

## 1 *Desmeocraera latex* — Olive Prominent

Forewing 20–30mm. Thorax brown and setose, abdomen brown with greenish tip; forewings dull green with brown and black dots, those along hind margin forming a line; hindwings creamy brown with dark distal patch. Caterpillar light brown with darker oblique stripes and pair of laterally projecting 'horns' on thorax, and a white triangle laterally on abdominal segments 7 and 8. Seventeen similar species in genus in region. **Biology:** Caterpillars feed on various milkwoods (*Sideroxylon*, Sapotaceae). Multivoltine, adults throughout the year. **Habitat & distribution:** Mainly along east coast where milkwoods occur; Eastern Cape, South Africa, north to tropical Africa.

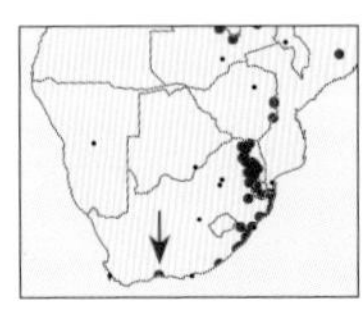

## 2 *Desmeocraera interpellatrix* — Pale-green Desmeocraera

Forewing 16–21mm. Stout; forewings more uniform greenish-grey than in *D. latex* (above), with a few darker markings. Caterpillar pale mottled green with pale lateral line and reddish patches dorsally. **Biology:** Caterpillars feed on Water Berry (*Syzygium cordatum*) and *Eugenia* spp. Multivoltine, adults throughout the year. **Habitat & distribution:** Moist forest and bush; Eastern Cape north to Zimbabwe.

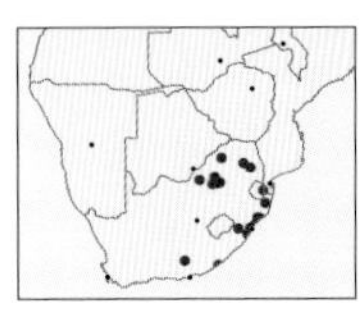

## 3 *Desmeocraera vernalis* — Vernal Prominent

Forewing 18–23mm. Short wings; colour variable but always greyish and not green, thorax and forewings olive grey to light brownish-grey with black markings, often without markings. Caterpillar pale green with white and orange dorsal and oblique yellow lateral lines. **Biology:** Caterpillars feed on various bushwillow trees (*Combretum* spp.). Multivoltine, adults August–April. **Habitat & distribution:** Forest and bushveld where *Combretum* spp. grow; Eastern Cape, South Africa, north to Zimbabwe and Botswana.

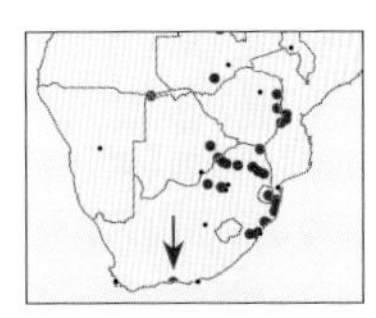

## 4 *Desmeocraera basalis* — Basal Prominent

Forewing 19–25mm. Thorax strongly setose with brown dorsal patch; forewings grey to greenish-grey with diagnostic darker area near base; hindwings white with brown patch along inner margin. **Biology:** Caterpillar host associations unknown. Multivoltine, adults throughout the year. **Habitat & distribution:** Variety of forest and bushveld habitats; Eastern Cape, South Africa, north to tropical Africa.

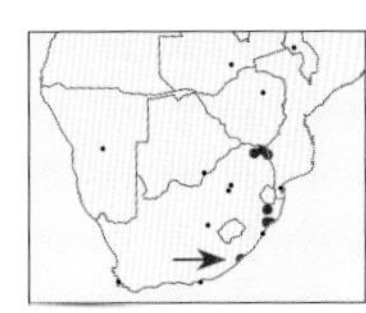

## 5 *Speideliana maget* — Hooded Bark

Forewing 13–16mm. Thorax with forward-projecting tufts of setae, abdominal tuft extending well beyond wings when at rest; forewings brown with bluish scales in median area, subterminal line straight; hindwings uniform brown with faint light brown lines. Two species in genus in region, similar *S. erichkaestneri* has crenulate subterminal line and white cell spot. **Biology:** Caterpillar host associations unknown. Perhaps bivoltine, adults August–October and January–May. **Habitat & distribution:** Lowveld forest and riverine bush; Eastern Cape to Limpopo, South Africa.

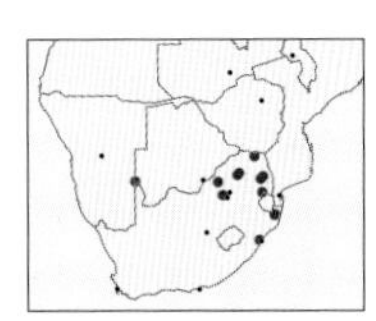

## 6 *Helga cinerea* — Helga's Forktail

Forewing 11–13mm. Thorax with tufts of setae dorsally, prothorax orange; forewings held narrowly against body, mottled cream with 2 narrow sub-basal black bars evenly curved diagnostic, a gold band anteriorly and crescent-shaped spot distally sometimes present. Caterpillar slender, greenish with orange banding and 2 very long posterior forked 'tails'. One species in genus in region. **Biology:** Caterpillars feed on Buffalo Thorn (*Ziziphus mucronata*). Multivoltine, adults throughout the year. **Habitat & distribution:** Bushveld; KwaZulu-Natal, South Africa, north to tropical Africa.

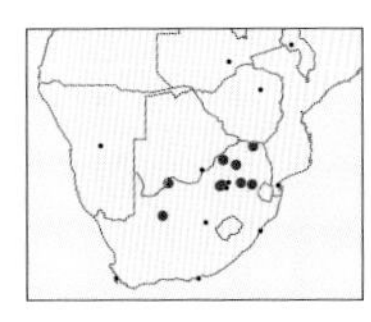

## 7 *Leptolepida rattus* — Grey Spoontail

Forewing 13–17mm. Similar to *Helga cinerea* (above), thorax with dorsal crest; body and forewings dark grey with black shading, 2 narrow sub-basal black bars angled diagnostic; hindwings white with narrow brown border. Caterpillar plump with enlarged and upturned spoon-like 'tail', green with white crossbars and longitudinal lines. Three similar species in genus in region. **Biology:** Caterpillars feed on Hook Thorn (*Senegalia caffra*). Multivoltine, adults September–March. **Habitat & distribution:** Grassland and bushveld; Northern Cape to Limpopo, South Africa, and Namibia.

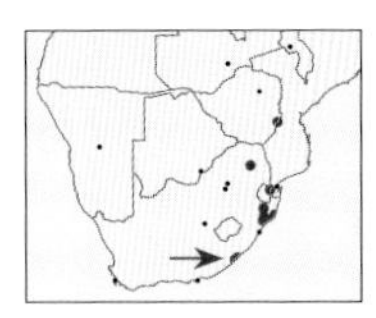

## 8 *Batempa mutabilis* — Horned Bark

Forewing 13–19mm. Thorax with upward-projecting tufts of setae, legs strongly setose; forewings mottled brown with dense pattern of dark, wavy lines; hindwings uniform light brown. Two species in genus in region; similar *B. buceros* has double basal and postmedian lines. **Biology:** Caterpillar host associations unknown. Adults August–May, can be abundant in Zululand forests in autumn. **Habitat & distribution:** Moist forests; Eastern Cape to Zimbabwe.

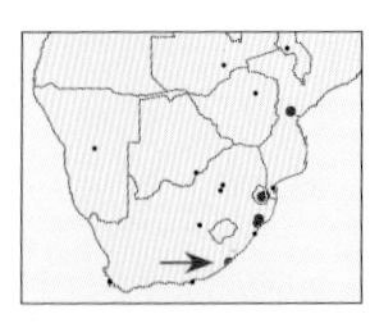

### 1 *Eurystauridia iphis* — Green Bark

Forewing 12–16mm. Similar to *Batempa mutabilis* (p.298); prominent black discal spot, chequered cilia and light brown tuft on thorax diagnostic; forewings and body vary from green to yellowish-brown; hindwings light brown. One species in genus in region. **Biology:** Caterpillar host associations unknown. Adults October–April. **Habitat & distribution:** Moist lowland forests; Eastern Cape to Mozambique.

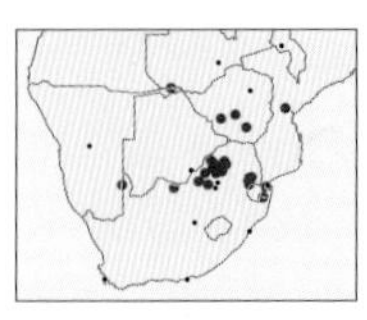

### 2 *Polelassothys plumitarsus* — Feather Foot

Forewing 11–15mm. Body and forewings with complex disruptive pattern of cream, pink, grey, brown and black patches; legs very setose, dorsal crest of setae on thorax and another tuft behind it; hindwings white with dotted border. One species in genus in region. **Biology:** Caterpillars feed on Silver Cluster-leaf (*Terminalia sericea*). Probably bivoltine, adults October–April. **Habitat & distribution:** Sandy bushveld where host plant grows; Zululand, South Africa, north to Malawi.

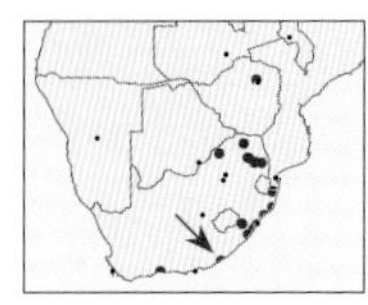

### 3 *Rhenea mediata* — Looped Prominent

Forewing 15–21mm. Body pinkish-brown; forewings darker basally then crossed by 2 black lines that end wide apart at inner margin, area between lines pink; hindwings white in male, tinged with grey in female. Caterpillar yellow-green with black dorsal patches on each segment. Three species in genus in region. **Biology:** Caterpillars feed primarily on Cape Ash (*Ekebergia capensis*), on which *Rhenea* spp. sometimes occur in great numbers defoliating trees, but populations crash soon afterwards, resulting in no permanent harm to trees. Also recorded on Velvet Bushwillow (*Combretum molle*), Wild Plum (*Harpephyllum caffrum*) and *Terminalia* spp. Multivoltine, adults throughout the year. **Habitat & distribution:** Forest and bushveld; southern Cape, South Africa, north to tropical Africa and Madagascar.

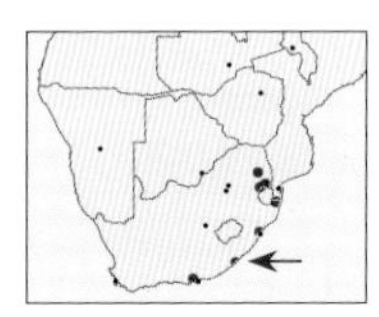

### 4 *Rhenea michii* — Michi's Prominent

Forewing 13–16mm. Similar to *R. mediata* (above), with pink transverse band across forewings divided into 2 distinct oval, pale pink patches separated by darker pink band and 2 black lines ending close together at inner margin. Caterpillar with blue and yellow longitudinal bands, black dots laterally and paired black spots dorsally. **Biology:** Caterpillars feed on Cape Ash (*Ekebergia capensis*). Bivoltine, adults December and March–June. **Habitat & distribution:** Moist forest; Eastern Cape to Zimbabwe.

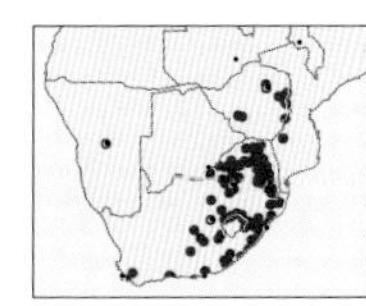

### 5 *Sarimarais bicolor* — Bicoloured Prominent

Forewing 17–24mm. Setose thorax brown with large black dorsal tuft, legs highly setose, abdomen orange-yellow with black markings; forewings plain chocolate brown with faint crosslines; hindwings pale yellow to orange, black spot on anal margin. Caterpillar green with white stripes and small lateral red spots, short hard setae dorsally. Two species in genus in region. **Biology:** Caterpillars feed on *Searsia* spp. Multivoltine, adults throughout the year. **Habitat & distribution:** Various habitats, from coast to over 2,500m, seemingly absent from arid areas; Western Cape, South Africa, north to Zambia.

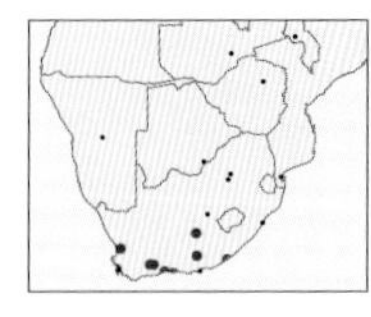

### 6 *Sarimarais peringueyi* — Peringuey's Prominent

Forewing 17–20mm. Similar to *S. bicolor* (above), highly setose, with prominent dorsal tuft on pale grey thorax; forewings pale grey with faint lines, hindwings cream with brown shading distally, no black spot on anal margin. **Biology:** Caterpillars feed on Nana-berry (*Searsia dentata*) and Glossy Currant (*S. lucida*). Bivoltine, adults August–December and April. **Habitat & distribution:** Many different habitats, sympatric with *S. bicolor* in some places; Western Cape, Eastern Cape and Northern Cape, South Africa.

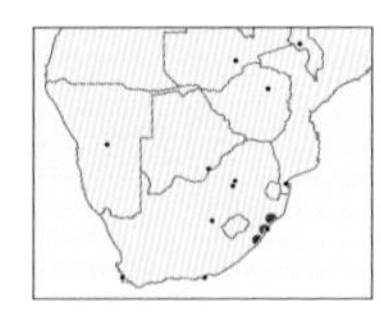

### 7 *Stemmatophalera curvilinea* — Broken Twig

Forewing 18–23mm. Adults resemble a broken twig when at rest; thorax light brown to straw-coloured, rest of body darker brown; forewings very variable, mottled dark greenish-brown at the base, then crossed by curved black line bordered by lighter cream with black and olive markings; hindwings cream with single black patch on distal margin. Caterpillar pale green, becoming white dorsally, with mid-dorsal yellow line. Three very similar species in genus in region, not distinguishable on external characters. **Biology:** Caterpillars polyphagous on tree species. Possibly univoltine, adults March–August. **Habitat & distribution:** Moist forest and coastal bush; Eastern Cape, South Africa, north to tropical Africa.

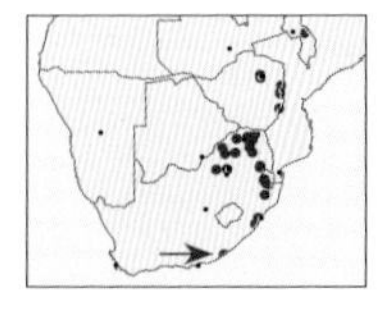

### 8 *Rosinella rosinaria* — Rosy Prominent

Forewing 16–20mm. Thorax, abdomen and forelegs highly setose; body and forewings rosy pink, forewings with black and green lines and spots; 2 distinct black-bordered rosy lines over thorax, rosy and green colours fade in old specimens; hindwings white with black mark on anal angle. One species in genus in region. **Biology:** Caterpillar host associations unknown. Adults August–April. **Habitat & distribution:** Forests and thick bush; Eastern Cape to tropical Africa.

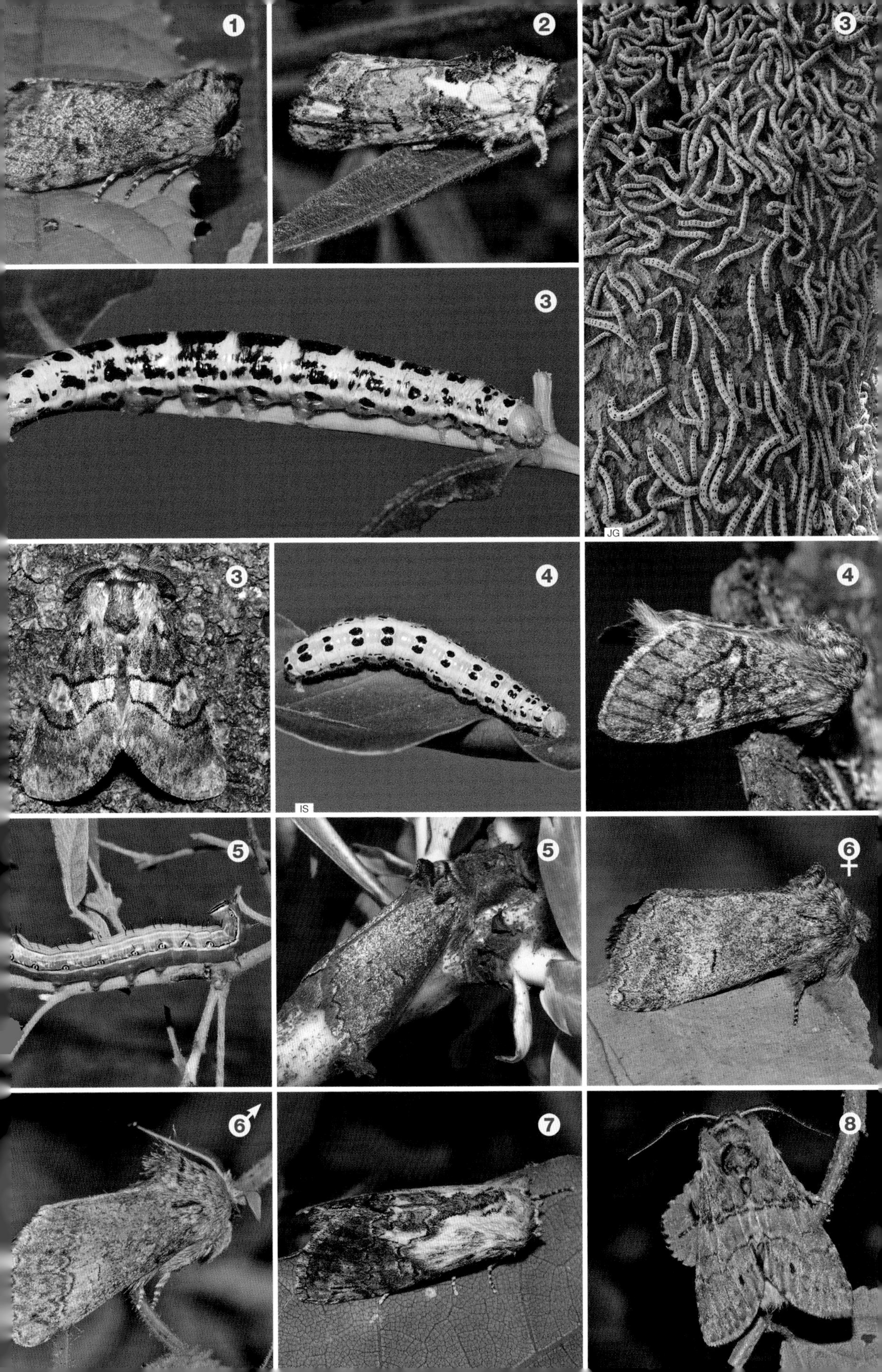
1
2
3
3
JG
3
4
IS
4
5
5
6 ♀
6 ♂
7
8

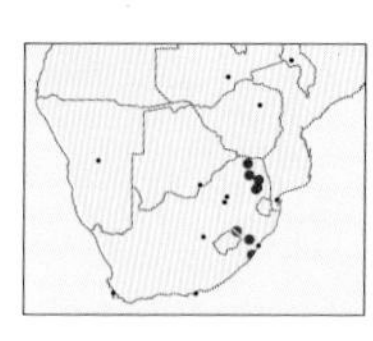

**1 *Zuluana quarta*** Zulu Prominent

Forewing 17–19mm. Thorax and forelegs highly setose; body and forewings mottled brown and cream; forewings crossed by 2 bold brown lines, a crescent-shaped cell spot with oblique black bar toward apex surrounded by white scales; hindwings uniform cream with grey cilia. **Biology:** Caterpillar host associations unknown. **Habitat & distribution:** Moist forests; KwaZulu-Natal to Limpopo, South Africa.

## Subfamily Phalerinae

Antennae of the male are either pectinate or filiform, the tarsal claws are divided into 2 toward their tips. The proboscis is longer than the head but not as long as the thorax.

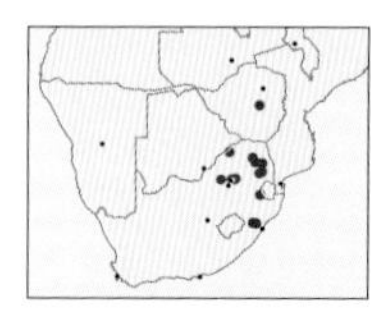

**2 *Antheua aurifodinae*** Gold Digger

Forewing 14–20mm. Thorax yellow, abdomen orange; forewings with a golden sheen, yellow with black and red stripe and 2 black flashes at end of wing; hindwings creamy white. Six species in genus in region. **Biology:** Caterpillar host associations unknown. *Antheua* spp. feed on herbaceous Fabaceae. Adults October–April. **Habitat & distribution:** Grassland; KwaZulu-Natal, South Africa, north to Zimbabwe.

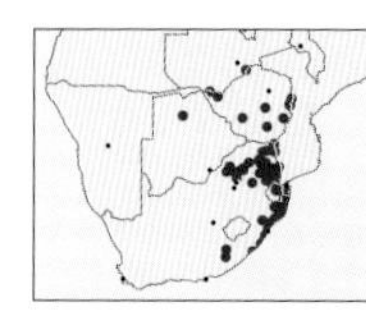

**3 *Antheua tricolor*** Tricoloured Frill

Forewing 18–22mm. Thorax yellow, abdomen dark basally, orange distally; forewings yellow, sprinkled with small black dots centrally; hindwings charcoal. **Biology:** Caterpillar host associations unknown. Adults throughout the year. **Habitat & distribution:** Frost-free grasslands not subjected to heavy frosts; Eastern Cape, South Africa, north to tropical Africa.

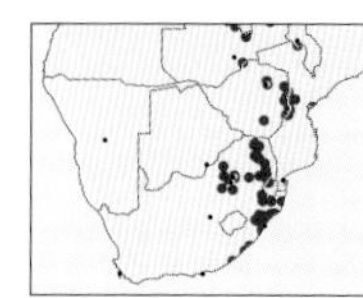

**4 *Antheua simplex*** Burnished Frill

Forewing 18–30mm. Variable, thorax white often with orange bar, abdomen black; forewings ivory with longitudinal grey bands; hindwings white in male, grey in female. Caterpillar black with red head, white tubercles and long white setae. **Biology:** Caterpillars feed on Fabaceae, including tick clovers (*Desmodium* spp.), Monkey Tamarind (*Mucuna pruriens*) and soybean. Multivoltine, adults throughout the year. **Habitat & distribution:** Moist, frost-free grasslands; Eastern Cape, South Africa, north to Zambia.

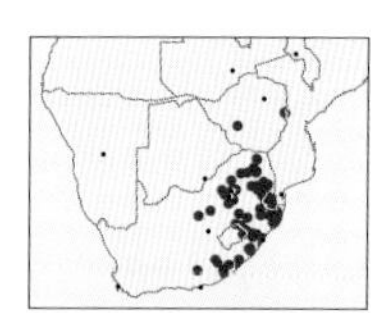

**5 *Polienus capillatus*** Streaked Buff

Forewing 14–16mm. Head and thorax with dense, long white setae; tip of abdomen with tuft of long setae projecting well beyond wings; forewings cream with yellowish and brown shading, denser in females, but always with a strong longitudinal brown line running along centre of wing; hindwings white or light grey. One species in genus in region. **Biology:** Caterpillars feed on various grasses (Poaceae). Bivoltine, adults September–April. **Habitat & distribution:** Moist grasslands; Eastern Cape, South Africa, north to Zimbabwe.

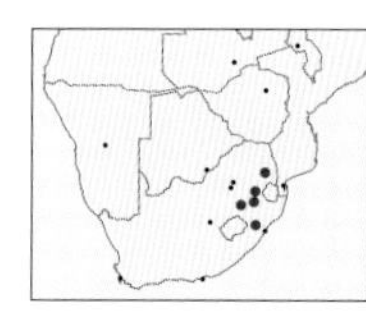

**6 *Phyllaliodes flavida*** Olive Buff

Forewing 19–21mm. Head, thorax and forewings olive green to shiny white, abdomen orange-brown, forewings with variably broken anteromedial and postmedian lines, subterminal line sometimes present; hindwings white with brown basal scales in some specimens; in form *brunnea* thorax and forewings are brown. Three species in genus in region. **Biology:** Caterpillars feed on Creeping Tephrosia (*Tephrosia macropoda*, Fabaceae). Adults November–February. **Habitat & distribution:** Moist highveld grasslands; KwaZulu-Natal to Mpumalanga, South Africa.

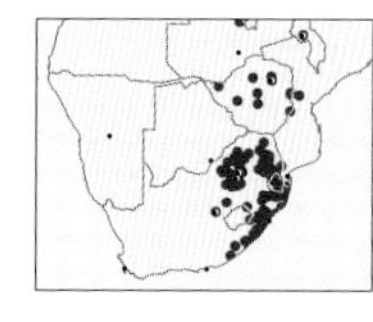

**7 *Rigema ornata*** Ornate Frill

Forewing 16–24mm. Thorax cream to pink crossed by transverse orange band; forewings elongate, cream, with central pink to purple stripe marked with darker veins, also scalloped purple distal edge; hindwings white with marginal dots. Caterpillar yellow with orange lateral band. Two species in genus in region; very similar *R. woerdeni* larger with a prominent brown discal spot on forewing. **Biology:** Caterpillars feed on grasses and sedges, such as Yellow Nut Grass (*Cyperus esculentus*). Bivoltine, adults September–June. **Habitat & distribution:** Moist grasslands; Eastern Cape, South Africa, north to Zambia.

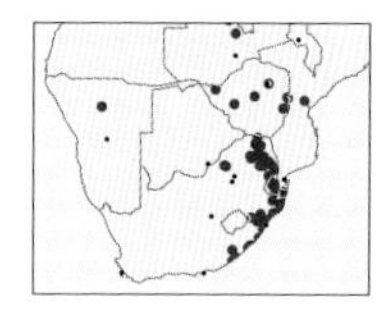

**8 *Phalera imitata*** Imitating Bufftip

Forewing 21–27mm. Head and prothorax orange, abdomen yellow; forewings mottled grey, crossed by fine black lines, a distinctive large, oval, yellow-orange patch at apex; hindwings white and brown. Caterpillar unknown. Two species in genus in region. **Biology:** Adults rest with wings wrapped closely around body, mimicking broken twig. Caterpillars feed on grasses and the twining legume *Rhynchosia viscosa*. Bivoltine, adults October–April. **Habitat & distribution:** Frost-free grasslands; Eastern Cape, South Africa, north to East Africa.

1
2
3
4
5
6
7
7
8

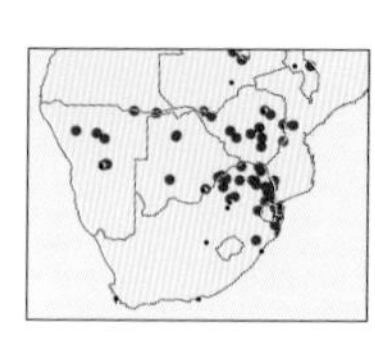

### 1 *Phalera lydenburgi* — Lydenburg's Buff

Forewing 21mm. Forewings plain grey with toothed distal margin with brown crescent mark and small brown patch distally. Caterpillar banded black and white with yellow lateral stripe and 2 long tufts of setae behind head. **Biology:** Caterpillars feed on grasses, including Guinea Grass (*Panicum maximum*) and maize. Bivoltine, adults throughout the year. **Habitat & distribution:** Frost-free grasslands; KwaZulu-Natal, South Africa, north to tropical Africa.

## Subfamily Pygaerinae

Antennae of both sexes pectinate, up to the tip in males. Each facet of the compound eye in adults is fringed with short hairs. There is a protuberance on abdominal segment 8 of the caterpillar.

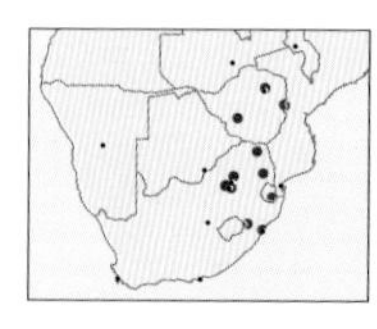

### 2 *Clostera violacearia* — Violet Clostera

Forewing 14–19mm. Thorax dark brown, raised into a conspicuous hump; forewings dark red-brown to greyish-brown crossed by faint broken lines and with violet tinge, especially in lighter distal section in male, orange-brown with purplish markings in female; hindwings white, suffused with brown. Caterpillars light brown with golden dorsal band and small oblique marks. Five species in genus in region. **Biology:** Caterpillars feed on Proteaceae (*Faurea* and *Protea*). Bivoltine, adults throughout the year. **Habitat & distribution:** Various habitats with Proteaceae; KwaZulu-Natal, South Africa, north to Zambia and Malawi.

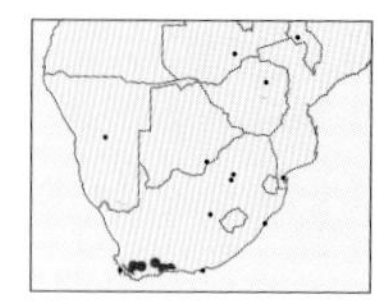

### 3 *Clostera limacodina* — Slug Clostera

Forewing 12–15mm. Greyish-brown clump of setae on thorax; forewings brown with orange basal patch and crossed by oblique white line in male, almost absent in female. Caterpillar setose with red head capsule and longitudinal broken black and white lines and yellow lateral dots. **Biology:** Caterpillars feed on Proteaceae (*Protea* and *Leucadendron*). Multivoltine, adults September–May. **Habitat & distribution:** Protea fynbos; Western Cape, South Africa.

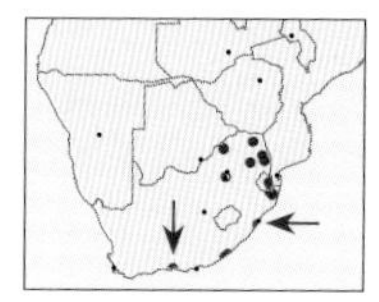

### 4 *Janthinisca joannoui* — Jo's Unicorn

Forewing 17–21mm. Thorax dark anteriorly with spike-like dorsal projection; setose tip of abdomen projects beyond wings; forewings brown with faint round or crescent-shaped spots, crossed by 1 stronger diagonal and other fine, vein-like lines, postmedian line ends well before apex. Caterpillar pale green with irregular red-brown dorsal patches and pair of diverging posterior 'tails'. One species in South Africa, 5 additional similar species further north in genus in region. **Biology:** Caterpillars feed on Weeping Boer-bean (*Schotia brachypetala*), probably also *Schotia afra*. Multivoltine, adults throughout the year. **Habitat & distribution:** Coastal bush and bushveld; Eastern Cape north to Limpopo, South Africa.

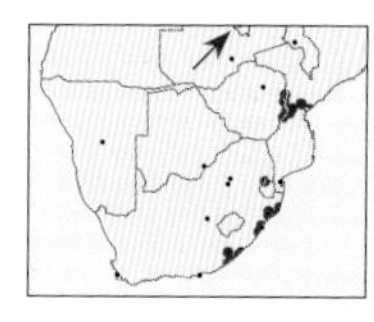

### 5 *Peratodonta heterogyna* — Blackface Unicorn

Forewing 19mm. Head flat and black; thorax with tuft of dark setae projecting dorsally; abdomen and forewings brown; forewings crossed by a fine diagonal dark line within a paler band; meandering pale lines on wing with paler and darker brown between; hindwings cream with darker borders. Caterpillar pale green with irregular brown patches and short, diverging tails. Four very similar species, not distinguishable externally, in genus in region, other 3 species in Zimbabwe and Mozambique northwards. **Biology:** Caterpillars feed on leguminous trees. Multivoltine, adults throughout the year. **Habitat & distribution:** Lowland forests and coastal bush; Eastern Cape, South Africa, north to Zimbabwe.

## Subfamily Scranciinae

The narrow, elongate forewings, long legs and partially long and short bipectinate male antennae are characteristic.

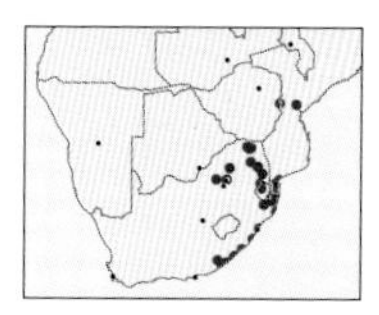

### 6 *Phycitimorpha dasychira* — Zig-zag Phycitimorpha

Forewing 14–16mm. Thorax brown, abdomen pale grey; forewings grey-brown to reddish-brown, crossed by fine, wavy black lines, row of longitudinal black bars near apex; hindwings white to light brown. Caterpillar slender, light brown with pink patches, numerous longitudinal white lines and pair of long, black terminal tails. Easily confused with Phycitinae moths (p.104). One species in genus in region. **Biology:** Caterpillars feed on knobwoods (*Zanthoxylum* spp.). Multivoltine, adults throughout the year. **Habitat & distribution:** Forest and thick bush where knobwoods grow; Eastern Cape, South Africa, north to Zimbabwe.

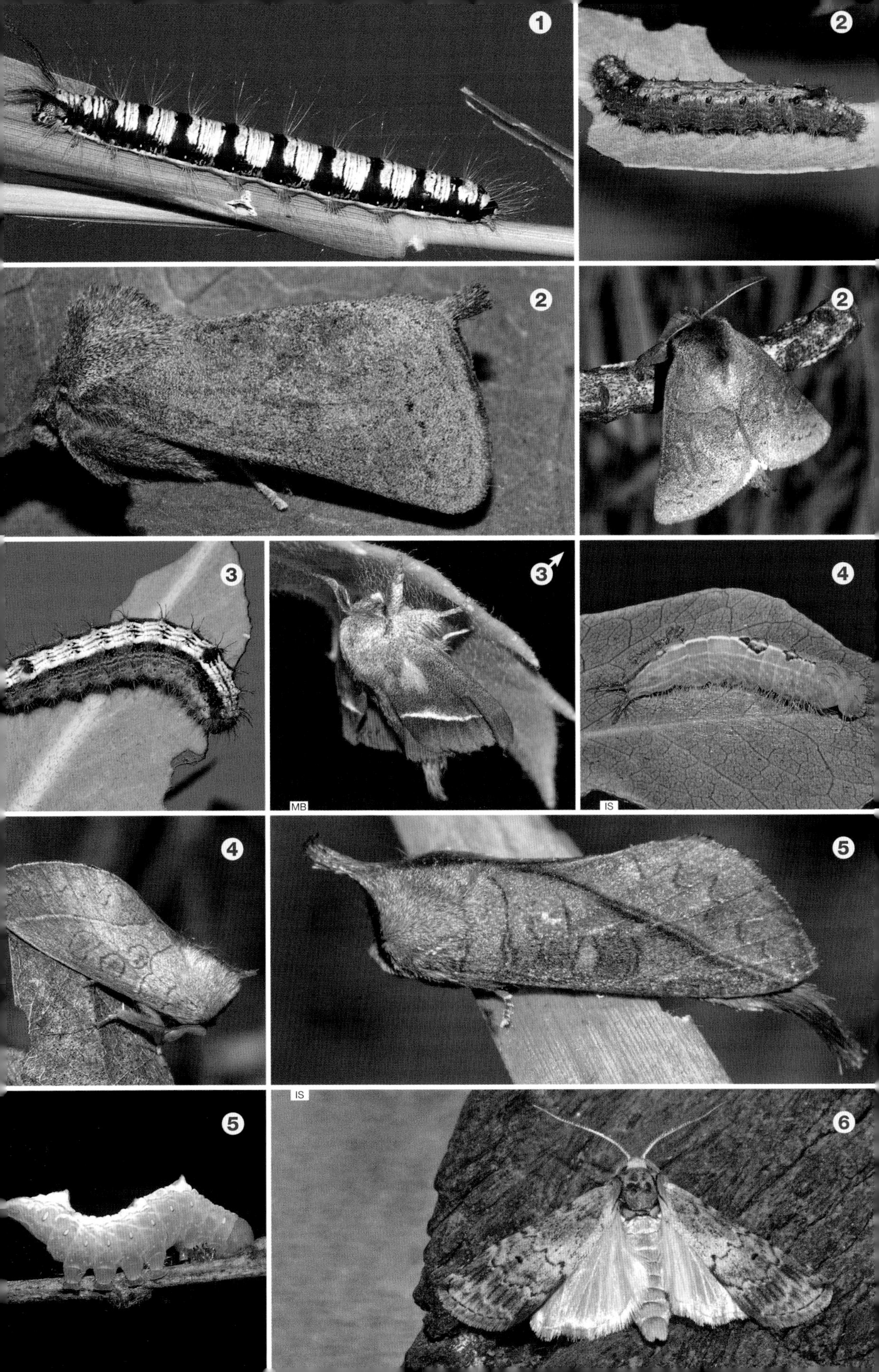

1
2
2
2
3
3♂
MB
4
IS
4
5
IS
5
6

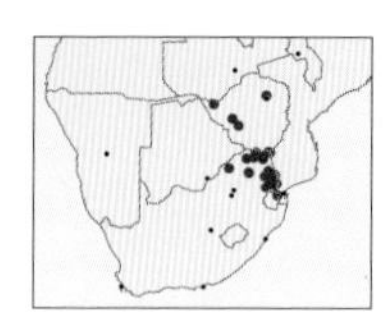

### 1 ***Postscrancia discomma*** Spike Twig

Forewing 15–17mm. Forewings cream with peach tint, 1 large oval black cell spot, also small black spots proximally and longitudinal lines toward apex; hindwings cream basally, becoming sooty distally. Caterpillar slender and brown with marked dorsal black spikes held apart, posterior end raised upward and bearing long terminal tails. One species in genus in region. **Biology:** Caterpillars feed on White Bush-berry (*Flueggea virosa*). Bivoltine, adults September–May, peaking in November and March. **Habitat & distribution:** Bushveld; Eswatini north to Somalia.

## Subfamily Thaumetopoeinae Processionary Moths

The colonial caterpillars leave their silken webs at night en masse when they migrate to nearby host plants. The proboscis is reduced or absent, and the antennae are pectinate along their full length in both sexes. The tarsal claws are simple (ending in a single point).

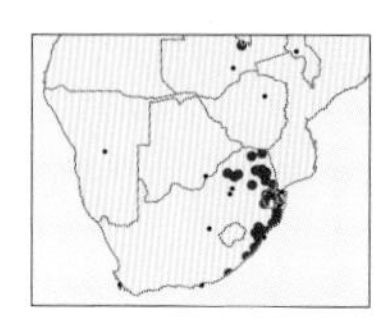

### 2 ***Anaphe reticulata*** Reticulated Bagnet

Forewing 16–26mm. Forewings white, crossed by network pattern of bold dark brown to black lines, an oblique line extending into basal area; hindwings cream. Caterpillar setose and green or yellowish-brown. Two species in genus in region; similar *A. panda* has no oblique line extending into basal area. **Biology:** Caterpillars gregarious and move in a procession. Polyphagous on trees, often on Malvaceae, pupate in a compacted mass of silken cocoons ('bagnet'). Univoltine, adults September–May, peaking in midsummer. **Habitat & distribution:** Bushveld and forest; Eastern Cape, South Africa, north to tropical Africa.

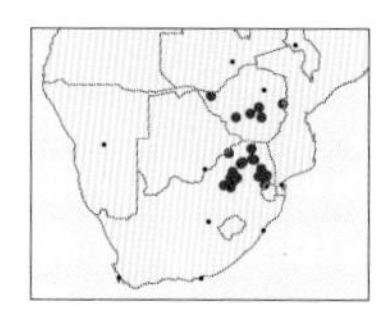

### 3 ***Paraphlebs singularis*** Slug Prominent

Forewing 14mm. Body orange, very setose; forewings short like Limacodidae, primarily orange with 3 broken longitudinal black bands and line of black bars along distal margin, which is fringed with black and orange-chequered setae. Caterpillar black with long white setae, orange spots on legs and prolegs, lateral yellow stripe. One species in genus in region. **Biology:** Caterpillars gregarious but do not have silken retreat, feed on Red Indigo-bush (*Indigofera hilaris*). Bivoltine or univoltine, adults October–April. **Habitat & distribution:** Grassland and bushveld; Gauteng, South Africa, north to Zimbabwe.

# FAMILY EREBIDAE Owlet Moths, Tiger Moths, Tussock Moths, Piercing Moths, Snout Moths and others

**The largest family of moths, Erebidae is one of 6 families within the superfamily Noctuoidea. Although most are fairly large, some of the subfamily Hypenodinae have wingspans as small as 6mm. The family is defined by features of the wing veins (the cubital vein typically splits into 4 branches in both fore- and hindwings viz. quadrifid venation). Some Erebidae (Arctiinae, Lymantriinae, Micronoctuini) were previously treated as families, and Erebidae now consists of 18 subfamilies. Many of the caterpillars are superficially similar to those of true Noctuidae, being arboreal, smooth semi-loopers with reduction of some grasping prolegs, which in more typical caterpillars from other families are present on all abdominal segments. The adaptations for semi-loopers and other loopers (as in Geometridae and Noctuidae) occurred via convergent evolution. In contrast, caterpillars of Tussock Moths (Lymantriinae) and 'woolly bears' of Tiger Moths (Arctiinae) are covered in hairs and bristles, often barbed and urticating, and capable of causing allergic skin reactions. These secondary hairs arise from specialised plates (verrucae). The region has over 1,500 described species among 18 subfamilies.**

## Subfamily Aganainae

This subfamily is most closely related to Litter Moths (Herminiinae). Typically, they include large and brightly coloured tropical or subtropical moths with enlarged and upturned labial palps. The caterpillars usually feed on plants containing cardenolide toxins in their latex (Apocynaceae, Asclepiadaceae, Moraceae), and store (sequester) them, eventually passing them on to the aposematic adults. The caterpillars have the full complement, or only somewhat reduced, number of prolegs. There are 16 species in the region.

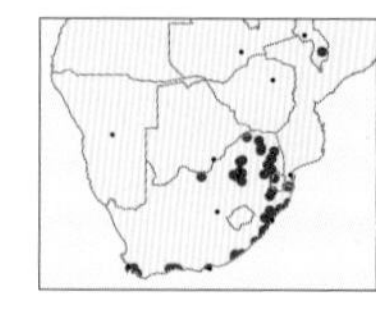

### 4 ***Asota speciosa*** Specious Tiger

Forewing 22–32mm. Extremely variable; white bars with black dots around basal area and 3 dots on thorax diagnostic; base of forewings, thorax and hindwings bright orange or yellow or white, with or without white and black bars and borders, rest of forewing khaki with cream wing veins, or of ground colour. Caterpillar variable; hairy with red or brown head capsule and red posterior parts; body orange to cream to white and banded in black. One species in genus in region. **Biology:** Caterpillars feed on various *Ficus* spp., cut veins supplying latex to leaf before feeding on it. Multivoltine, adults throughout the year, peaking in summer. **Habitat & distribution:** Moist habitats; Western Cape, South Africa, north to tropical Africa.

1
1
IS
1
2
IS
2
3
3
4
4

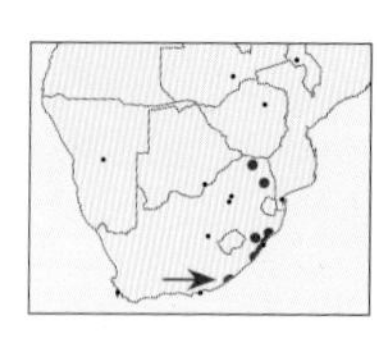

### 1 *Digama aganais* — Grey Agana

Forewing 16–18mm. Brownish-grey with 2 small black dots, female with cream blotch halfway up on costa; hindwings uniform yellow to orange; thorax grey with black dots. Caterpillars fairly squat with long hairs, black sides and head, orange tufts in 2 rows along back. Six species in genus in region. **Biology:** Caterpillars feed on *Carissa* spp. in region. Multivoltine, adults August–April. **Habitat & distribution:** Moist forests; Eastern Cape, South Africa, north to tropical Africa.

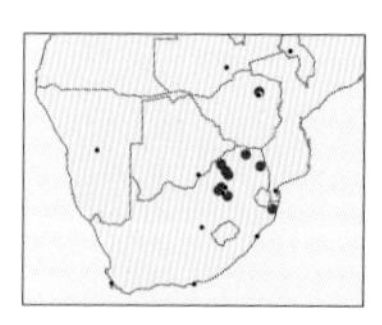

### 2 *Digama meridionalis* — Meridian Digama

Forewing 16–20mm. Forewing white to dark brown or greyish with dark brown wing veins enclosing cells, 2 or 3 black spots, sometimes with dark postmedian blotch; hindwings cream; abdomen yellow with row of central black spots. **Biology:** Caterpillar host associations unknown. Adults October–May. **Habitat & distribution:** Bushveld and wooded kloofs; KwaZulu-Natal, South Africa, north to tropical Africa.

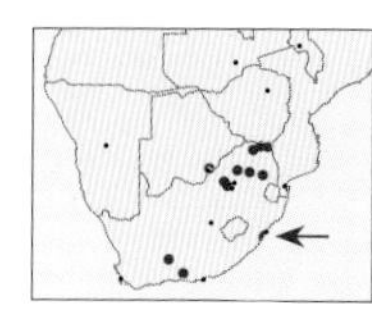

### 3 *Digama ostentata* — Fourspot Agana

Forewing 15–17mm. Superficially similar to *Galtara nepheloptera* (p.372); thorax and forewing grey, 4 diagnostic cream-ringed black spots on forewings, cream postmedian line broken or absent, row of smaller black spots on costa, shading on costa occasionally; hindwings uniform yellow to orange; abdomen orange with dorsal black spots. Easily confused with some *Ethmia* spp. (Depressariidae), which are smaller. **Biology:** Caterpillar host associations unknown. Adults October–May. **Habitat & distribution:** Bushveld and wooded kloofs; Western Cape, South Africa, north to Mozambique.

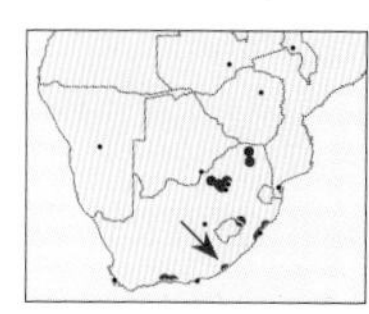

### 4 *Sommeria spilosoma* — Intricate Agana

Forewing 17–21mm. Forewing white, broken up by variable sinuous black bands and spots, terminal area silvery grey; hindwings diagnostic, cream to grey with dark black border; thorax white with black spots, abdomen yellow with central black spots. Caterpillar similar to *D. aganais* (above), but pinkish dorsally and cream to white ventrally. Five species in genus in region; similar *S. sinuosa* has hindwings yellow to orange with no border. **Biology:** Caterpillars feed on various Apocynaceae. Multivoltine, adults throughout the year. **Habitat & distribution:** Forest and wooded kloofs; Western Cape, South Africa, north to Kenya.

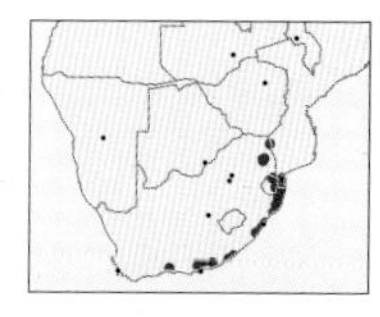

### 5 *Sommeria culta* — Cult Agana

Forewing 15–17mm. Forewing similarly marked to *S. spilosoma* (above), but ground colour grey, variable black patches lined with cream and terminal area distinctively shaded with black between veins; hindwings orange with black spot on apex. Caterpillar white with grey lateral shading and dorsal lines. **Biology:** Caterpillars feed on *Carissa edulis* (Apocynaceae). Multivoltine, adults throughout the year. **Habitat & distribution:** Lowland forest, riverine and coastal bush; Western Cape, South Africa, north to Kenya.

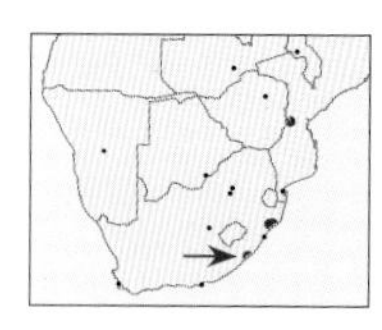

### 6 *Soloe tripunctata* — Dotted Agana

Forewing 17–20mm. Body, fore- and hindwings yellow to orange, forewings with 2 black dots and a thin black border halfway up on costa; hindwings with single black cell spot. Two species in genus in region; similar *S. fumipennis* has ground colour grey. **Biology:** Caterpillar host associations unknown. Adults September and March–April. **Habitat & distribution:** Warm lowland forests; Eastern Cape to Kenya.

## Subfamily Herminiinae Littermoths

Most species feed cryptically on dead rather than live leaves, but also on flowers, senescing leaves, fruit, algae, dead insects, dung, fungi, moss and ferns. Many species live in the undergrowth of moist forests. Caterpillars are dull coloured, slow moving, and covered in short, peg-like hairs. There are 44 regional species.

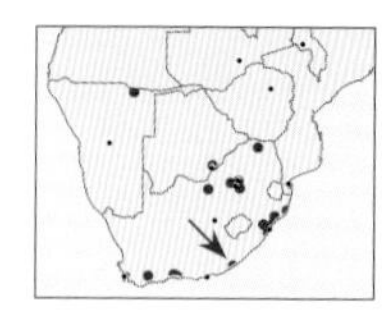

### 7 *Simplicia extinctalis* — Creamline Littermoth

Forewing 10–12mm. Forewings uniformly fawn or buff, with 2 faint basal transverse lines and 1 apical cream line; hindwings off-white with lighter margin. Three species in genus in region. **Biology:** Caterpillars polyphagous on dried plant matter. **Habitat & distribution:** Variety of habitats; Western Cape, South Africa, north to Arabian Peninsula.

1
2
3
4
5
7
1
4
5
IS
6
DS
7

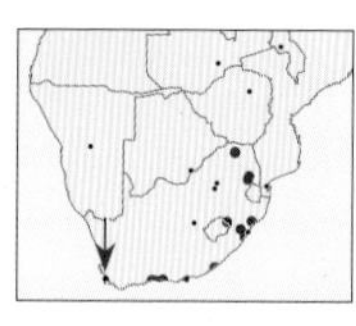

## 1 *Hydrillodes uliginosalis* — Banded Littermoth

Forewing 15–16mm. Similar to *Nodaria nodosalis*, but larger, postmedian and subterminal lines parallel; colouration variable; forewings grey, brown or buff, with median area darker or lighter or same as rest of forewing; hindwings dirty grey. One species in genus in region. **Biology:** Caterpillar host associations unknown. Adults throughout year. **Habitat & distribution:** Moist forests; Western Cape, South Africa, north to tropical Africa.

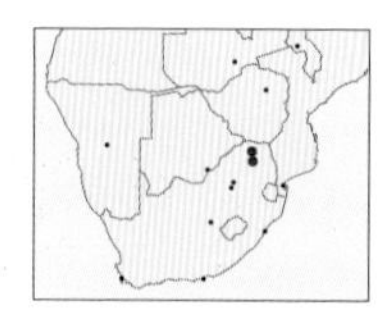

## 2 *Bertula bellaria* — Umber Littermoth

Forewing 12–13mm. Triangular in outline when at rest; ground colour of body and wings straw, variably heavily dusted with umber scales, except in areas adjacent to lines, postmedian and antemedial black, jagged, subterminal straight. Seven species in genus in region. **Biology:** Caterpillar host associations unknown. Adults October and February–March. **Habitat & distribution:** Moist forests; Limpopo, South Africa.

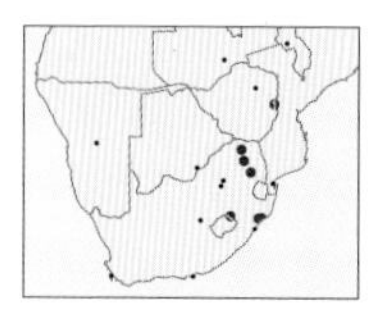

## 3 *Bertula pulvida* — Curved-horn Littermoth

Forewing 13–15mm. Triangular outline at rest; palpi long and curved backward over thorax; body and wings brown, postmedian and antemedial black, crenulated or dotted, cream subterminal slightly curved with kink near apex sometimes with dark shading, comma-shaped cell spot; hindwings lighter brown with curved subterminal and postmedian lines. **Biology:** Caterpillar host associations unknown. Adults September–May. **Habitat & distribution:** Moist forests; KwaZulu-Natal north to tropical Africa.

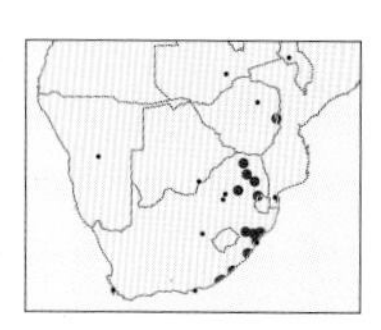

## 4 *Bertula inconspicua* — Porrect-horn Littermoth

Forewing 11–13mm. Triangular outline at rest, palpi long porrect; ground colour of body and wings grey-brown to rusty brown, often darker in median area, shape of cream to orange postmedian and antemedial line diagnostic, subterminal line absent, dash cell spot; hindwings uniform brown. Caterpillar slender, green to brown with sparse long black setae. **Biology:** Caterpillars feed on leaves of *Dalbergia obovata* (Fabaceae). Adults October and February–March. **Habitat & distribution:** Forests and riverine vegetation; Eastern Cape, South Africa, north to Zimbabwe.

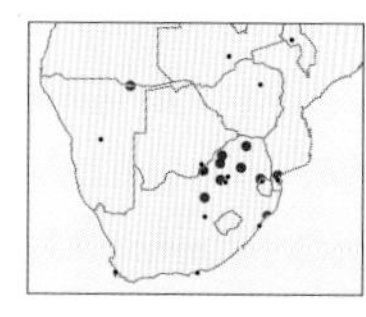

## 5 *Naarda nigripalpis* — Black-palp Littermoth

Forewing 11–14mm. Conspicuous black extended palps; body and wings variably grizzled light and dark brown, lines black and crenulated; forewings with 2 buff patches near costa, one encircling a black cell spot in most specimens, absent in some. Three similar species in genus in region. **Biology:** Caterpillar host associations unknown. Adults November–March. **Habitat & distribution:** Closed-canopy forest, kloofs and riverine vegetation; KwaZulu-Natal, South Africa, north to Arabian Peninsula.

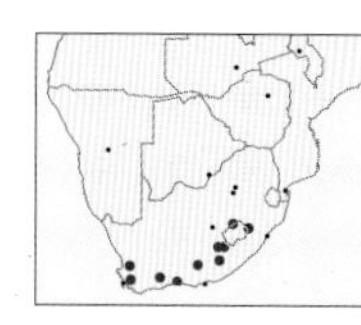

## 6 *Polypogon typomelas* — Triangle Littermoth

Forewing 10–11mm. Palps porrect with last segment turned up; body and wings light grey; shape of forewing postmedian line and black triangular cell spot distinctive, 3 further triangular marks on costa; hindwings white to grey. Sixteen diverse species in genus in region, some adapted to drier habitats, unusual for subfamily. **Biology:** Caterpillar host associations unknown. Adults August–April. **Habitat & distribution:** Fynbos and grassland habitats; Western Cape, South Africa, north to Lesotho.

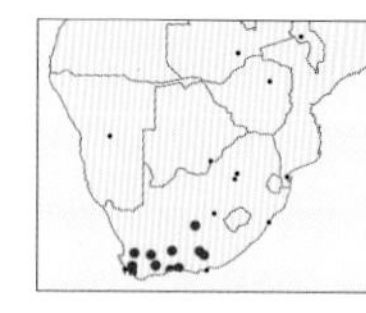

## 7 *Polypogon canofusca* — Cape Littermoth

Forewing 11–16mm. Palps porrect with last segment turned up; covered in black and grey scales, wings with antemedial, postmedian and subterminal lines jagged, cream spots on subterminal; 2 conspicuous cream spots on forewings, often black bordered; hindwings without cream spots. Can be confused with *Plecoptera* spp. (Anobinae p.316), but long upturned palpi diagnostic. **Biology:** Caterpillar host associations unknown. Adults November–April. **Habitat & distribution:** Fynbos and karoo biomes; Cape provinces, South Africa.

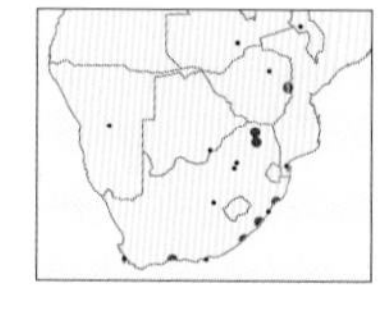

## 8 *Polypogon melanomma* — Concorde Littermoth

Forewing 8–11mm. Palps porrect with last segment turned up; wings narrow, body dark brown, wings pale pinkish-brown, fading in dried specimens, antemedial, postmedian and subterminal lines jagged, black median band in most specimens; 2 conspicuous orange spots on forewings, black bordered or only black; hindwings less well marked. Differs from *P. canofusca* (above) by being smaller and with narrower wings. **Biology:** Caterpillar host associations unknown. Adults November–April. **Habitat & distribution:** Moist forests; Western Cape, South Africa, north to Zimbabwe, another subspecies in Kenya.

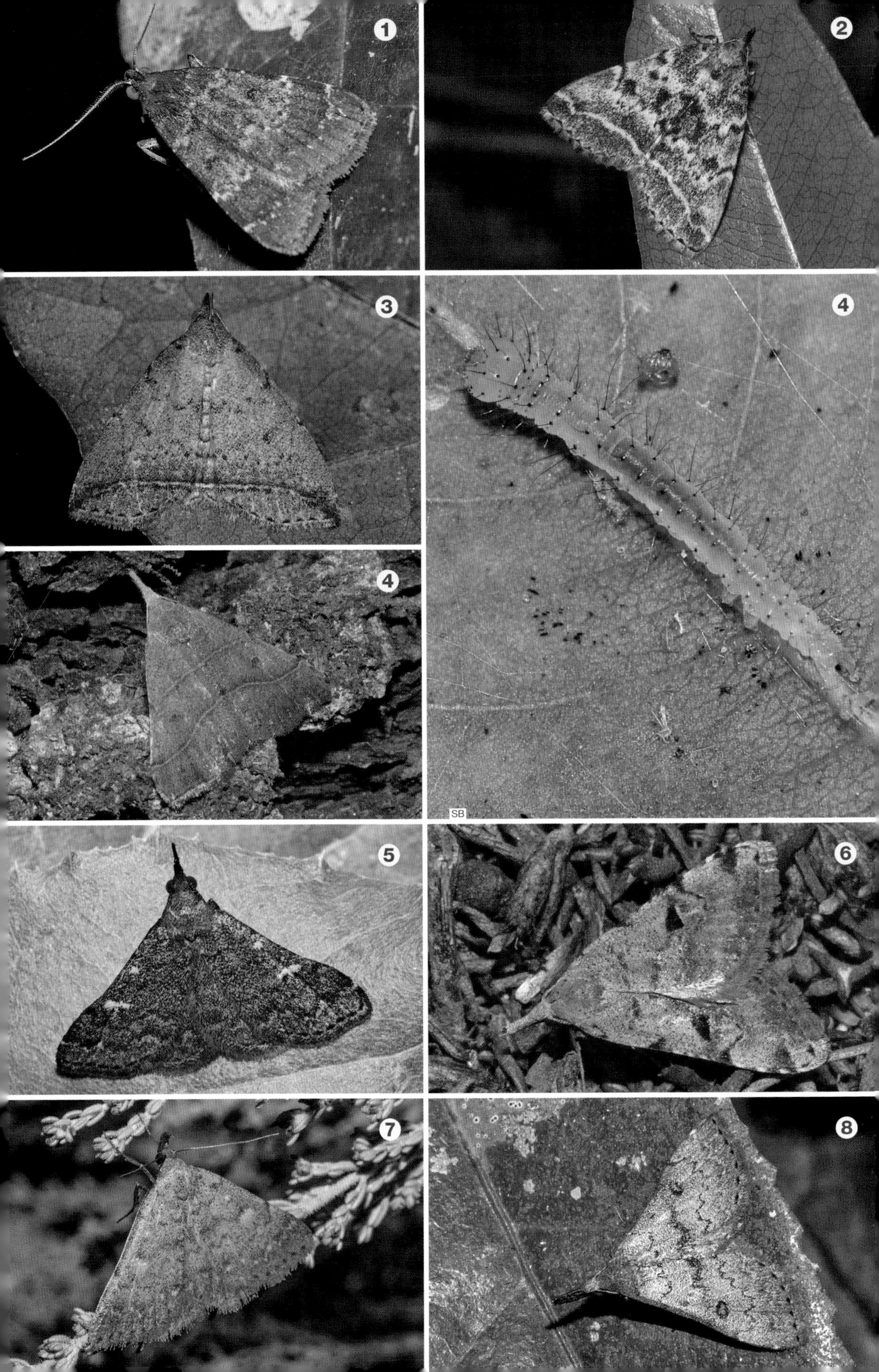
1
2
3
4
4
SB
5
6
7
8

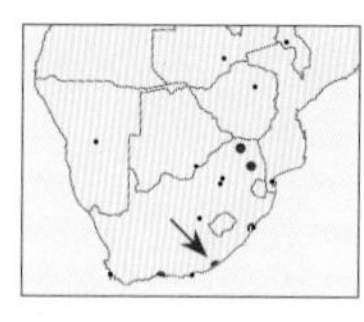

**1 *Ceraptila reniferalis*** Antlered Littermoth

Forewing 14–15mm. Tan to grey; prominent palps; darker form has black transverse bar near base of wing, edged with rufous and with zigzag submarginal line; smooth curved antemedial line with adjacent black in median area and black subterminal markings near apex. Caterpillar smooth, lime green, plump, with white dorsolateral lines running along body. Two species in genus in region, *C. amanialis* lacks black subterminal markings. **Biology:** Caterpillars feed on a Dogbane Vine (*Riocreuxia picta*). Adults August and February–April. **Habitat & distribution:** Moist forests; Western Cape north to Limpopo, South Africa.

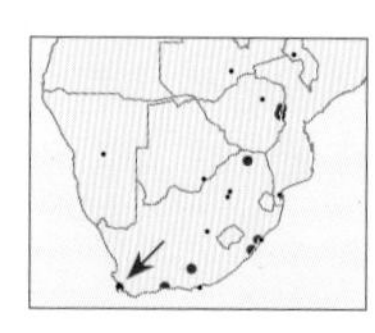

**2 *Aventiola truncataria*** Folded Littermoth

Forewing 7–10mm. Body and wings straw to light pink; prominent palps; conspicuous triangular mark on costa; forewings characteristically folded similar to *Epiplema* spp. (Uraniidae, p.142), but double falcate forewings and long palps diagnostic. One species in genus in region. **Biology:** Caterpillar host associations unknown. **Habitat & distribution:** Forests, wooded kloofs and gardens; Western Cape north to Zimbabwe.

## Subfamily Hypeninae Snout Moths

At rest, they typically have a triangular outline, and a conspicuous 'snout' of enlarged labial palps, similar to Herminiinae. Forty-nine species in region.

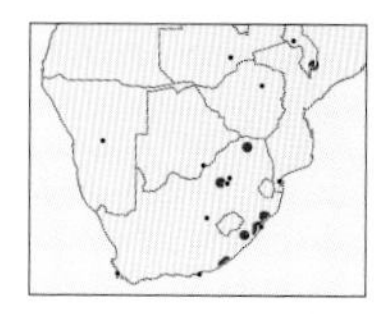

**3 *Hypena erastrialis*** Eyed Snout

Forewing 12–14mm. Variable; forewings dark brown with broad, wavy, tan median band, with diagnostic yellow-bordered ocellus near wing apex; lighter individuals have forewings bicoloured plum basally and tan, ocellus with dark surround; hindwings grey. Caterpillar either black with white longitudinal lines and white-ringed black spots, or yellow with brown dorsum. Thirty-two species in genus in region. **Biology:** Caterpillars feed on the nettle *Didymodoxa caffra*. Multivoltine, adults throughout the year. **Habitat & distribution:** Forest and riverine bush; Eastern Cape to tropical Africa, also Madagascar and La Réunion.

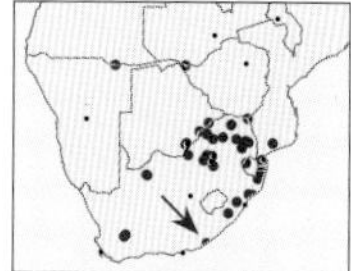

**4 *Hypena simplicalis*** Simple Snout

Forewing 10–13mm. Uniform grey-brown to charcoal ground colour; forewings with postmedian line comprising reddish-brown and yellow lines, similar, fainter antemedial line sometimes present; hindwings uniform charcoal grey. Caterpillar smooth and translucent pale green. **Biology:** Caterpillars recorded feeding on spiderworts (*Commelina*). Multivoltine, adults throughout the year. **Habitat & distribution:** Variety of habitats, including transformed habitats; Western Cape north throughout Africa, Saudi Arabia and Mauritania, also Madagascar and Indian Ocean islands.

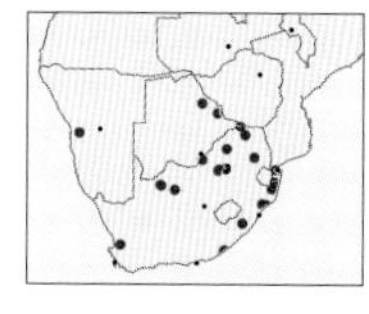

**5 *Hypena commixtalis*** Chalky Snout

Forewing 10–12mm. Forewings divided into tan and brown basal and median area, and brown terminal area, both variably marked with chalky white patches. Similar *H. jussalis* has an oblique grey shading between basal and median area; *H. fumidalis* is larger with wings more pointed. Caterpillar uniform lemon green, with sparse, short setae and indistinct white lateral line. **Biology:** Caterpillars recorded feeding on the yellow everlasting *Helichrysum epapposum*. Multivoltine, adults August–May. **Habitat & distribution:** Variety of habitats; Western Cape north throughout Africa, also Saudi Arabia.

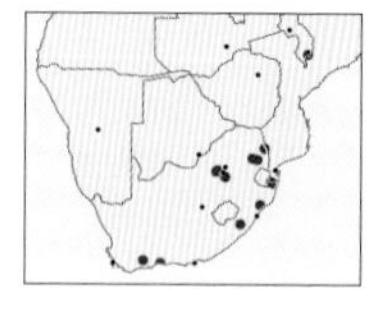

**6 *Hypena lividalis*** Brown Triangle Snout

Forewing 9–12mm. Conspicuous snout of labial palps; forewing with basal and median areas tan or rich brown in an elongate triangle, postmedian oblique, straight with grey shading beyond, small black cell spot; hindwings uniform tan or grey, darkening toward wing margin. Similar *H. laetalis* is larger, wings broader and has postmedian line less oblique. **Biology:** Caterpillars polyphagous on low-growing plants. Multivoltine, adults throughout the year. **Habitat & distribution:** Relatively moist habitats; Western Cape north throughout Africa, occasional migrant in Europe.

**7 *Hypena senialis*** Whiteline Snout

Forewing 13–15mm. Unmistakable; prominent snout; forewings with diagnostic white double postmedian line with marks near apex and white longitudinal line from base with area below black, cell spot white or black; hindwings light to dark brown. **Biology:** Caterpillar host associations unknown. Adults September–March. **Habitat & distribution:** Moist forests; Western Cape, South Africa, north to tropical Africa.

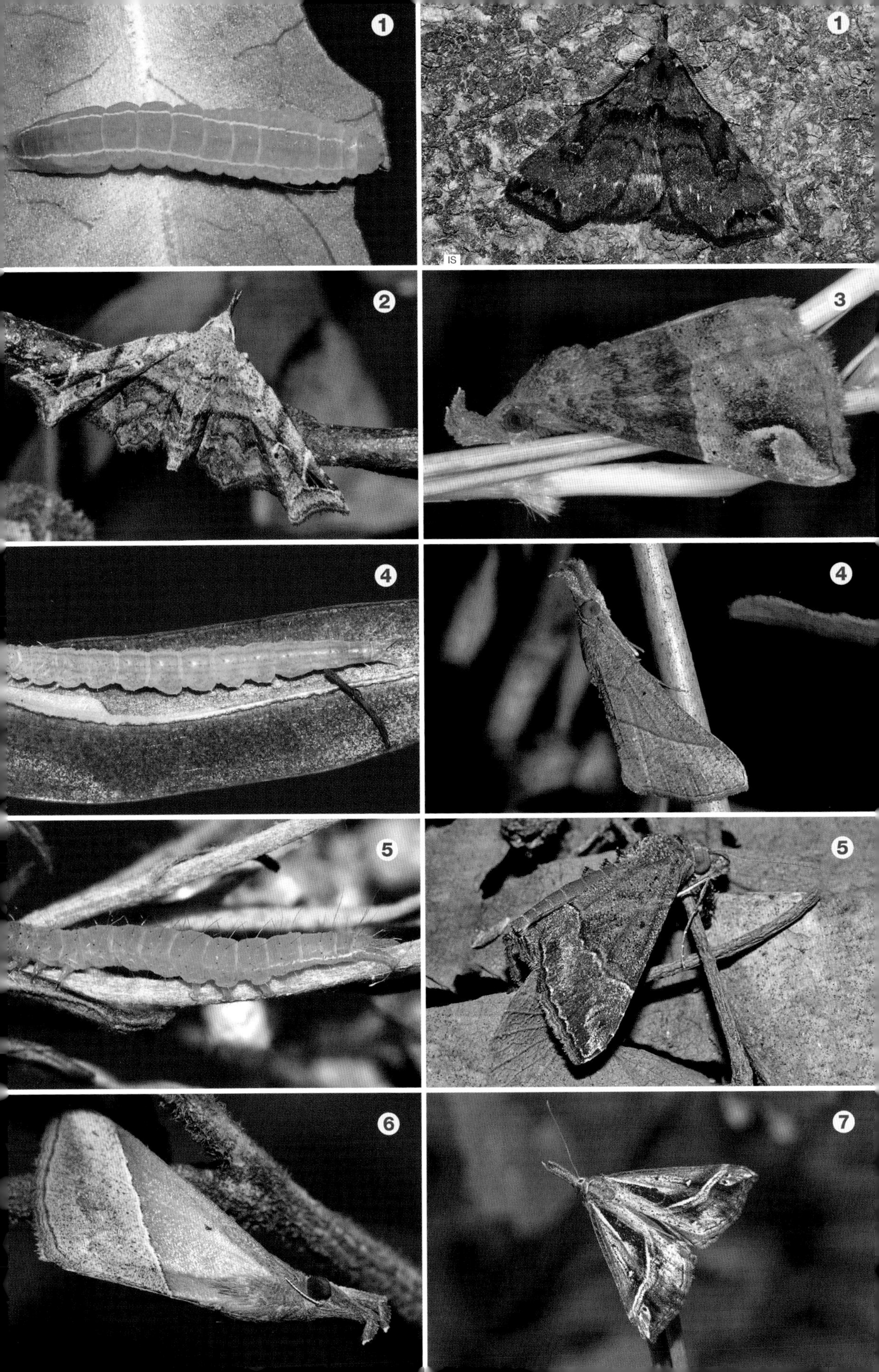
1
1
IS
2
3
4
4
5
5
6
7

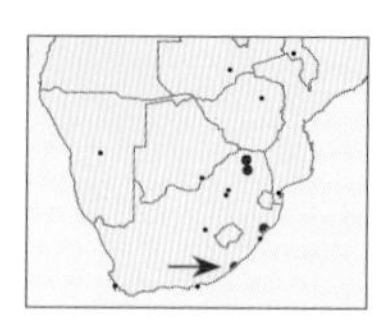

### 1 *Hypena mesomelaena* — Saddle Snout

Forewing 15–16mm. Unmistakable; snout curved upward; body and wings dark brown to charcoal, male with distinctive pale-lined black mark on inner margin in median area resembling a saddle when at rest, absent in female; hindwings light to dark brown. **Biology:** Caterpillar host associations unknown. Adults October–December and March–May. **Habitat & distribution:** Moist forests; Eastern Cape, South Africa, north to tropical Africa.

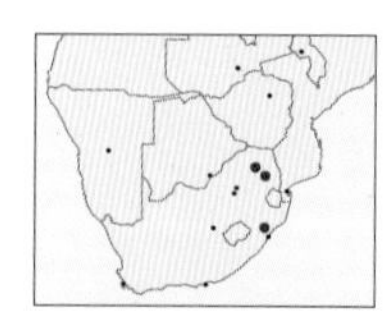

### 2 *Hypena melanistis* — Creamline Snout

Forewing 15–16mm. Unmistakable; snout curved upward; light to dark brown, falcate forewing with broad, curved, cream basal line, postmedian line cream and row of white subterminal spots; hindwings uniform, light to dark brown. **Biology:** Caterpillar host associations unknown. Adults October and March. **Habitat & distribution:** Moist forests; Eastern Cape north to Limpopo, South Africa.

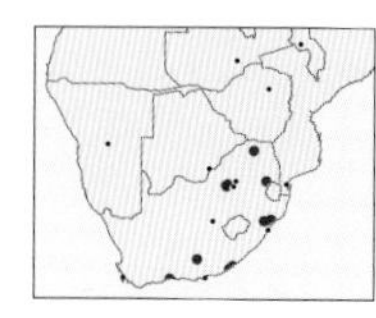

### 3 *Sarmatia interitalis* — Sickle Snout

Forewing 14–15mm. Similar to *Hypena erastrialis* (p.312); snout curved upward; body, basal and median area of forewings dark brown, area beyond postmedian line variable light brown to whitish in some individuals, diagnostic inner half of subterminal line black with dark brown shading extending to wing margin appearing like a sickle when at rest; hindwings uniform brown. One species in genus in region. **Biology:** Caterpillar host associations unknown. Adults throughout the year. **Habitat & distribution:** Forests and riverine vegetation; Western Cape, South Africa, north to tropical Africa.

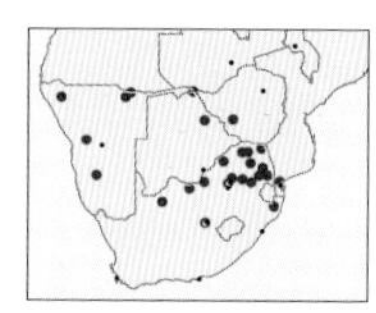

### 4 *Zekelita coniodes* — Whitebase Snout

Forewing 8–10mm. Snout long, curved upward near apex; body and ground colour of forewings white, terminal area variably spotted with black, rusty red and bluish scales, postmedian line slightly curved; hindwings charcoal grey. Similar *Z. anistropha* is larger and has ground colour grey with postmedian line straight. Caterpillar purplish with broad cream lateral band and orange spots. Twelve species in genus in region. **Biology:** Caterpillars feed on *Indigofera filipes* (Fabaceae). Multivoltine, adults November–May. **Habitat & distribution:** Bushveld; Free State, South Africa, north to tropical Africa.

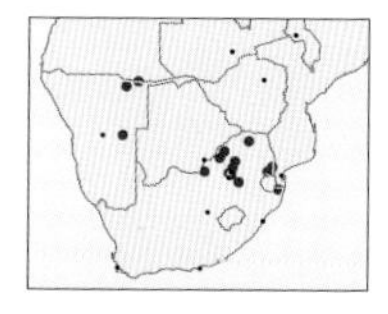

### 5 *Zekelita poecilopa* — Angled Snout

Forewing 8–10mm. Similar to *Z. coniodes* (above), but forewings more greyish, with distinctive postmedian line sharply turned back toward costa and darker orange to black blotch around cell, basal area on inner margin white with some markings. **Biology:** Caterpillar host associations unknown. Adults October–May. **Habitat & distribution:** Bushveld; KwaZulu-Natal, South Africa, north to tropical Africa, also Madagascar.

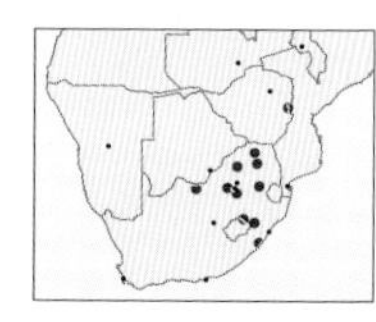

### 6 *Zekelita tinctalis* — Plainline Snout

Forewing 12–13mm. Snout long, curved upward near apex; body and ground colour of forewings brown, basal area shaded with grey, postmedian line straight, white, widening toward apex, slightly falcate; hindwings charcoal grey. **Biology:** Caterpillar host associations unknown. Adults October–May. **Habitat & distribution:** Forest and bushveld; KwaZulu-Natal, South Africa, north to tropical Africa.

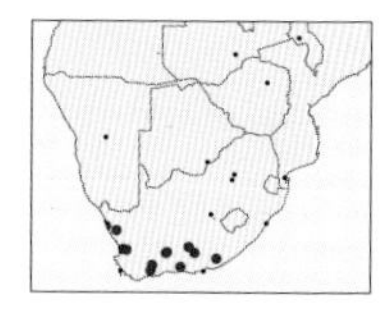

### 7 *Zekelita poliopera* — Cape Snout

Forewing 13–15mm. Snout long, not curved upward near apex; body and ground colour of forewings varies from light brown to black, postmedian line oblique, black near inner margin, sometimes obscured, grey shading on costa to apex diagnostic; hindwings brown to charcoal grey. **Biology:** Caterpillar host associations unknown. Adults throughout the year. **Habitat & distribution:** Dry fynbos, renosterveld and karoo habitats; Cape provinces, South Africa.

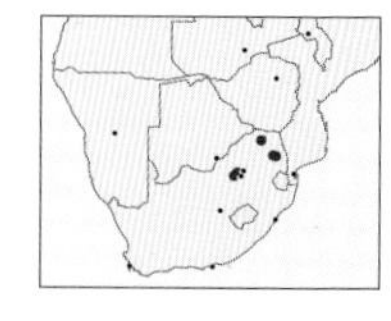

### 8 *Proluta deflexa* — White Bar Snout

Forewing 10–12mm. Grey or brown, with robust, relatively short snout formed from labial palps; forewings with black or brown shading in middle of wing, enclosing short, chalky white bar in cell, postmedian line diagnostically wavy, curved around cell; hindwings dirty yellow, darkening toward apex. Two species in genus in region; similar *P. staudei* is smaller and has postmedian line evenly curved around cell, not wavy. Caterpillar slender, uniform grey or brown, mimics twining host plant. **Biology:** Caterpillars feed on various Malpighiaceae. Bivoltine, adults December–April. **Habitat & distribution:** Grassland and bushveld; Gauteng, South Africa, north to Ethiopia, nominate subspecies in Madagascar.

1
2
3
4
IS
4
5
6
7
8
IS
8

## Subfamily Rivulinae Straws

A small subfamily, long and slender semi-looper caterpillars. They appear to feed largely on low-growing plants. There are 14 species in the region.

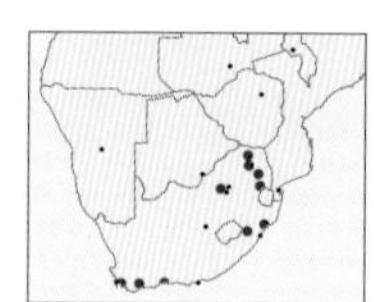

### 1 *Zebeeba meceneronis* — Partridge Mutt-nose

Forewing 8–9mm. Projecting snout of blunt palps; forewings grey, with brown shading toward apex, ending in a distinctive black mark on wing margin, diagnostic complex bar near base of wing, composed of a reddish and a white bar and indistinct black-ringed ocellus at wing border; hindwings white to grey with cell spot and black margin. Caterpillar a smooth, green semi-looper with bold white lateral and dorsal stripes. Ten species in genus in region. **Biology:** Caterpillars feed on *Asparagus*, pupal cocoons covered in the small *Asparagus* leaves, arranged lengthwise. **Habitat & distribution:** Moist forest and fynbos; Western Cape, South Africa, north to Limpopo, South Africa and Tanzania.

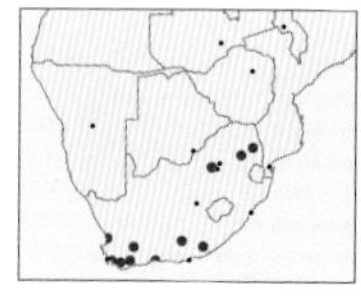

### 2 *Zebeeba mediorufa* — Saddle Mutt-nose

Forewing 7–9mm. Variable; thorax with 3 transverse brown bars; forewings mottled grey and brown with white in some, small cell spot, diagnostic black-edged brown to cream blotch on inner margin in median area, resembling a saddle when at rest, row of dots along apex of wing. Caterpillar a slender, grey semi-looper, with short, sparse bristles and 2 black, dorsal patches near rear of body. **Biology:** Caterpillars feed on various *Asparagus* spp. Multivoltine, adults throughout the year. **Habitat & distribution:** Variety of habitats where host plants grow; Western Cape, South Africa, north to Limpopo, South Africa.

## Subfamily Anobinae

A small subfamily with 38 species in region. In adults, the head is dark, and the light-coloured thorax has a dark collar. Sexual dimorphism is common.

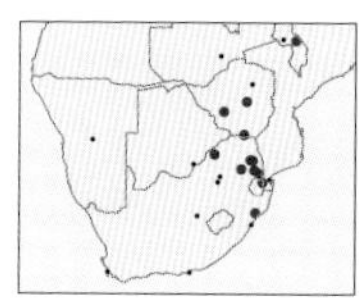

### 3 *Anoba atripuncta* — Dog Knob

Forewing 14–17mm. Forewing creamish-brown shading to dark brown, darker in terminal area, postmedian sharply turned back, diagnostic basal black patch has triangular section on a stalk off inner margin resembling a dog in profile, cell spot small or absent. Eight similar species in genus in region. Caterpillar a pale green semi-looper with vertical cream bands joined at their base by a cream line. **Biology:** Caterpillars feed on Hairy Caterpillar Pod (*Ormocarpum trichocarpum*). Multivoltine, adults throughout the year. **Habitat & distribution:** Lowveld bushveld and forests; KwaZulu-Natal, South Africa, north to tropical Africa.

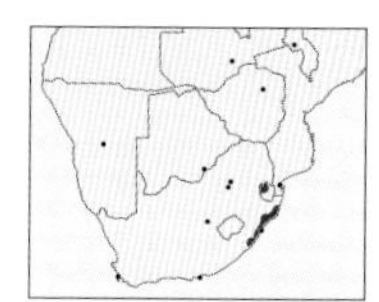

### 4 *Anoba disjuncta* — Triangle Knob

Forewing 14–18mm. Similar to *A. atripuncta* (above), but basal black patch a triangle not stalked from inner margin, often merging with inner margin in some individuals, cell spot large; hindwings dirty grey. **Biology:** Caterpillars feed on *Dalbergia armata*. Multivoltine, adults throughout the year. **Habitat & distribution:** Moist forests; Eastern Cape, South Africa, north to tropical Africa.

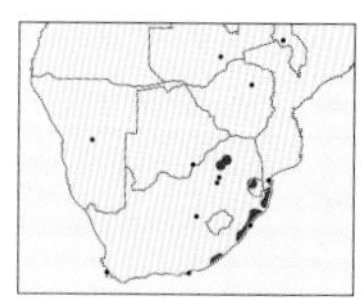

### 5 *Anoba atriplaga* — Dolphin Knob

Forewing 17–19mm. Variable; forewings straw to brown with darker tips, diagnostic black basal patch variable but elongate, mostly merging with postmedian line, resembling a dolphin in profile; hindwings with darker margin. Caterpillar green, with warts and densely covered in white setae. **Biology:** Caterpillars feed on Climbing Flat Bean (*Dalbergia obovata*). Multivoltine, adults throughout the year. **Habitat & distribution:** Forests and wooded kloofs; Eastern Cape, South Africa, north to tropical Africa.

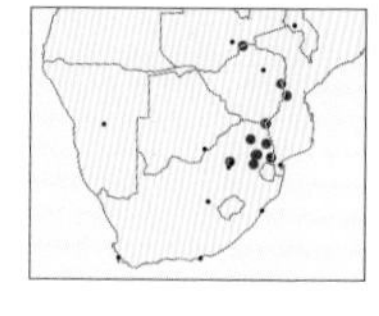

### 6 *Anoba hamifera* — Upright Knob

Forewing 11–12mm. Darker and smaller species fading in old specimens; forewings with postmedian line upright ending in a sharp point before turning back curving to costa, black basal patch thin and upright; hindwings dirty grey with dark border. **Biology:** Caterpillar host associations unknown. Adults throughout the year. **Habitat & distribution:** Bushveld and grassland; Gauteng, South Africa, north to Kenya.

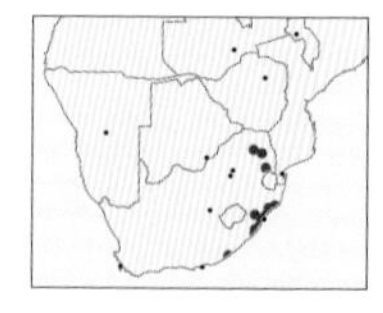

### 7 *Anoba plumipes* — Plumed Knob

Forewing 15–16mm. Forewings pale brown with diagnostic undulating postmedian line, variably lined or spotted with black separating brown wing tips, black basal patch variably triangular on thin stalk; hindwings pale brown. Caterpillar a green semi-looper. **Biology:** Caterpillars feed on Climbing Flat Bean (*Dalbergia obovata*). Adults throughout the year. **Habitat & distribution:** Moist forests; Eastern Cape north to Limpopo, South Africa.

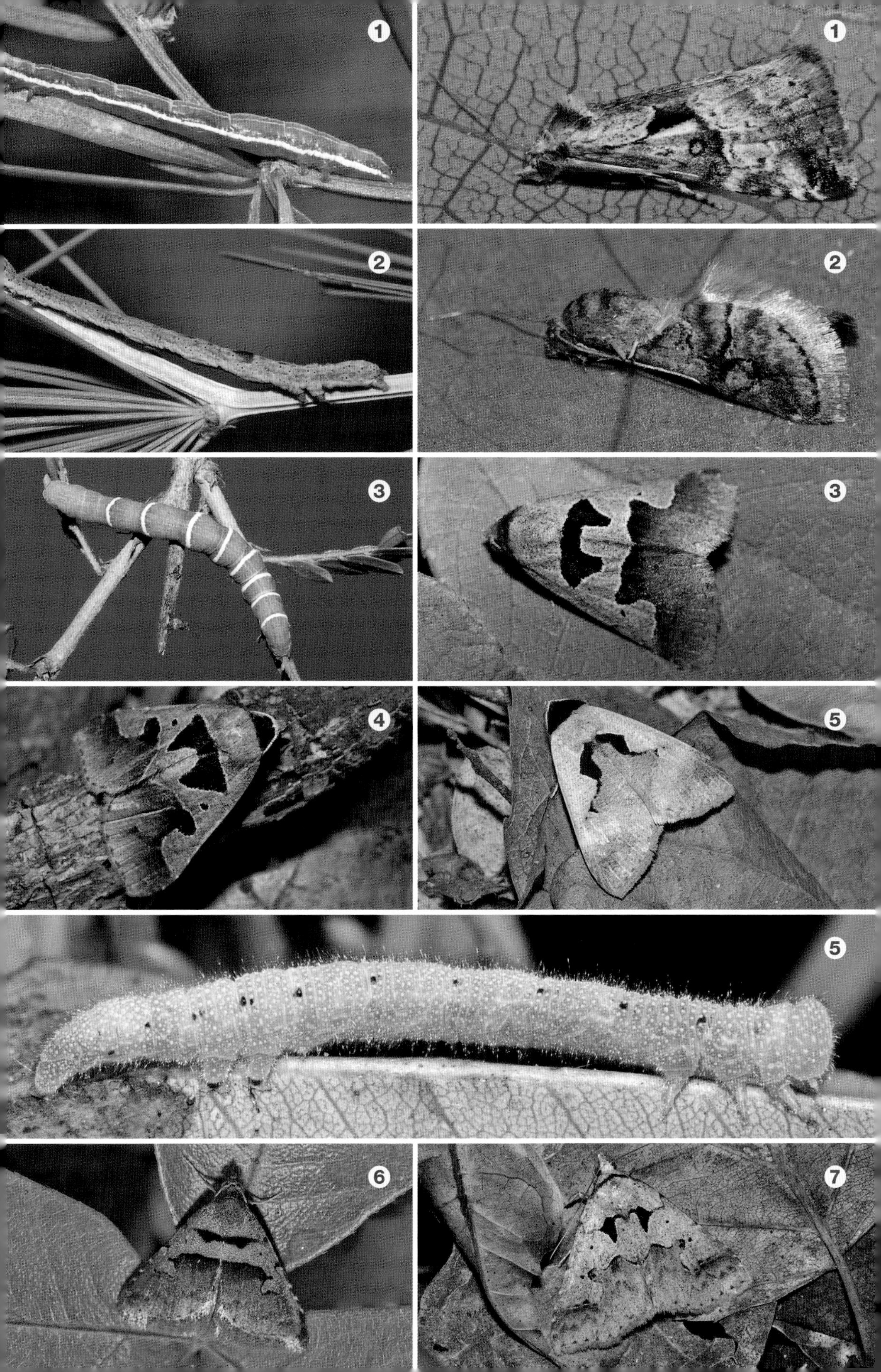
1
1
2
2
3
3
4
5
5
6
7

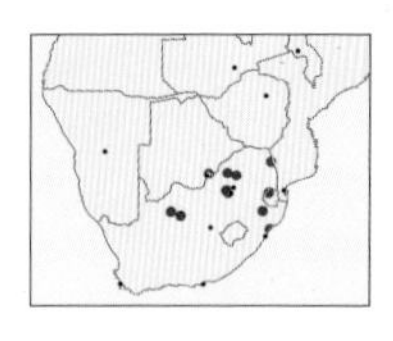

## 1 *Baniana arvorum* — Field Knob

Forewing 8–13mm. Colour variable, reddish-brown; base of forewings cream speckled with black scales, antemedial line crenulated, separated from brown or grey tips by sinuous pale yellow to orange postmedian line; hindwings grey or brown with paler base. Caterpillar slender, pale green with 4 white lines running along length of body. Two species in genus in region. **Biology:** Caterpillars feed on Elephant's Root (*Elephantorrhiza elephantina*). Univoltine, adults September–December. **Habitat & distribution:** Grassland where host plant grows; KwaZulu-Natal, South Africa, north to Malawi.

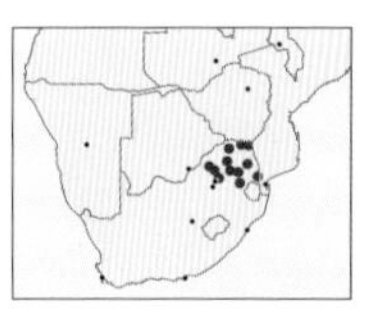

## 2 *Baniana culminifera* — Bicolor Knob

Forewing 8–10mm. Forewings bicoloured, basal and median areas straw-coloured, shaded by grey scales in some specimens, black wavy postmedian line forming dark saddle across body when at rest, terminal area dark brown. Similar to *B. arvorum* (above), but straw ground colour, antemedial line absent and postmedian line black bordered. Caterpillar translucent green with fine hairs and faint lines. Univoltine, adults September–December. **Biology:** Caterpillars feed on *Elephantorrhiza burkei* (Fabaceae). **Habitat & distribution:** Rocky bushveld hills; Gauteng, South Africa, north to Tanzania.

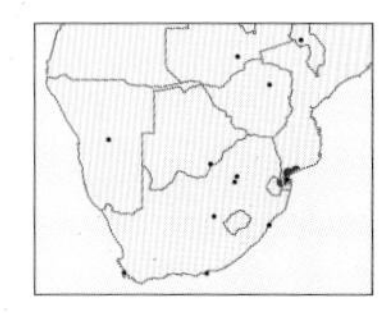

## 3 *Marcipa mediana* — Orange Knob

Forewing 15mm. Head brown; forewings cream or pale yellow with brown to black triangle on costa near tip of wings, females have additional black basal patch and black spot at base of light cream undulating postmedian line; hindwings cream or yellow. One species in genus in region. **Biology:** Caterpillar host associations unknown. **Habitat & distribution:** Coastal bush and tropical forest; Mozambique, South Africa, north to Kenya.

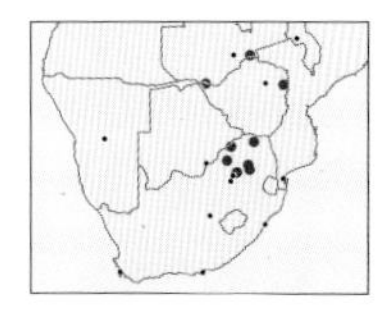

## 4 *Plecoptera sarcistis* — Rosy False-thorn

Forewing 12–14mm. Head and frontal part of thorax black, white between antennae; forewing ground colour plain rosy brown with rosy pink margins, characteristic black blotch beyond black cell spot; hindwings translucent brown with darker terminal shading. **Biology:** Caterpillar host associations unknown. Adults throughout the year. **Habitat & distribution:** Bushveld; Gauteng, South Africa, north to Tanzania.

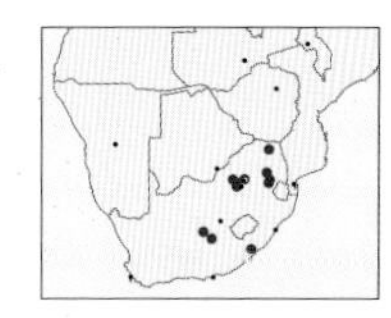

## 5 *Plecoptera arctinotata* — Wavy False-thorn

Forewing 10–12mm. Forewings grey-brown variably dusted with black, large round black cell spot in male, small in female, wavy antemedial and postmedian lines parallel; hindwings grey or brown. Caterpillar pale green, smooth, with 4 cream lines along thorax and diagonal cream lines on abdominal segments. Twenty-one species in genus in region; can be confused with *Acantholipes* spp. (pp.348, 350). **Biology:** Caterpillars feed on Hook Thorn (*Senegalia caffra*). Bivoltine, adults October–March. **Habitat & distribution:** Grassland; Eastern Cape north to Limpopo, South Africa.

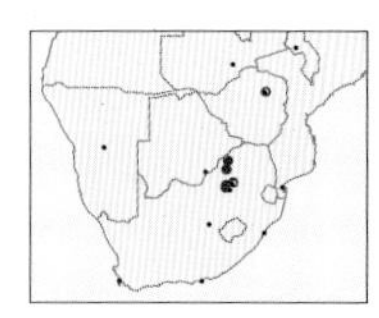

## 6 *Plecoptera rufirena* — Red-eye False-thorn

Forewing 13–15mm. Forewings grey-brown with purplish sheen, cell spot with red inner, wavy basal, antemedial and postmedian lines but terminal line only slightly wavy; hindwings without cell spot. Caterpillar pale greenish-yellow, smooth, with numerous longitudinal lines. Similar *P. misera* smaller, with wavy lines. Also similar to *Polypogon canofusca* (Herminiinae, p.310), but palpi short and lacks 2 orange spots around cell. **Biology:** Caterpillars feed on *Rhynchosia monophylla* (Fabaceae). Bivoltine, adults October–March. **Habitat & distribution:** Grassland and bushveld; Gauteng, South Africa, north to Zimbabwe.

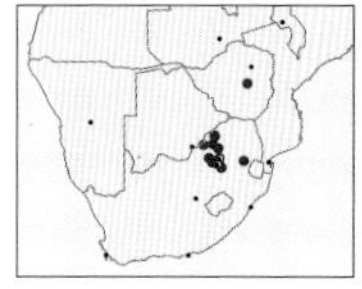

## 7 *Plecoptera annexa* — Annexed False-thorn

Forewing 12–15mm. Forewings grey with violet sheen, wavy basal, antemedial and postmedian lines, more visible in females, comma cell spot, terminal line straight with diagnostic dark, tooth-like outer margin and light grey terminal shading; hindwings light brown, darker beyond curved terminal line. Caterpillar a greenish-yellow semi-looper with numerous red crinkled longitudinal lines. **Biology:** Caterpillars feed on *Indigofera setiflora* (Fabaceae). Bivoltine, adults September–March. **Habitat & distribution:** Grassland and bushveld; northern Free State, South Africa, north to Zimbabwe.

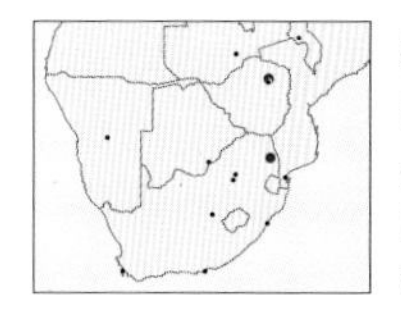

## 8 *Plecoptera flaviceps* — Slate False-thorn

Forewing 12–15mm. Grey to fawn; forewings bicoloured, darker tips, postmedian line with diagnostic curve, antemedial line wavy, comma cell spot; hindwings grey or fawn. Caterpillar grey or green with longitudinal bands. **Biology:** Caterpillars feed on African Blackwood (*Dalbergia melanoxylon*). Adults February–March. **Habitat & distribution:** Bushveld; Limpopo, South Africa, and Zimbabwe.

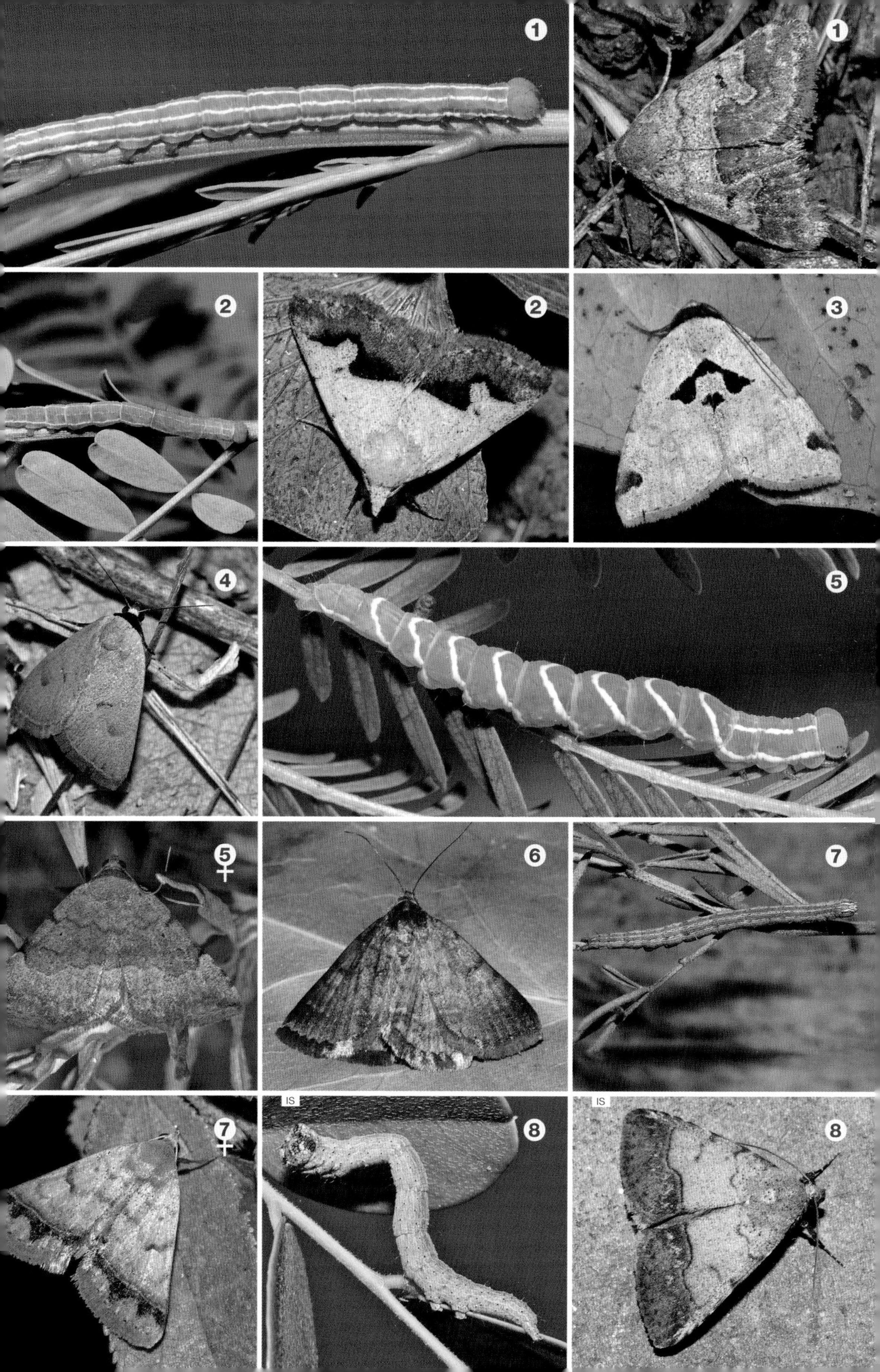
1
1
2
2
3
4
5
5♀
6
7
7♀
IS
8
IS
8

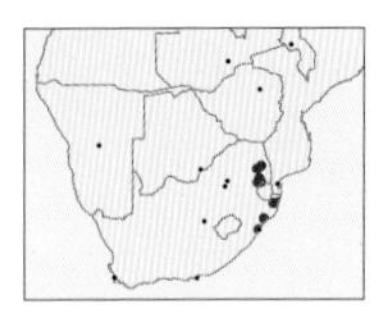

### 1 *Plecoptera laniata* — Straight False-thorn

Forewing 13–15mm. Similar to *P. flaviceps* (p.318), but antemedial and postmedian lines straight, cell spot as 2 separate spots. Caterpillar a brown semi-looper with orange rings around setal bases. **Biology:** Caterpillars recorded on *Dalbergia armata* (Fabaceae) and *Clerodendrum glabrum* (Lamiaceae). Multivoltine, adults throughout the year. **Habitat & distribution:** Forest and riverine bush; KwaZulu-Natal, South Africa, north to Zambia.

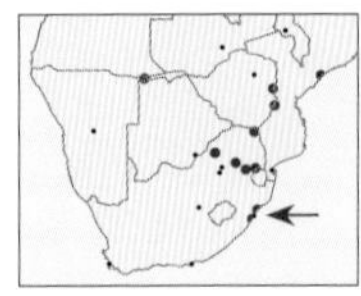

### 2 *Plecoptera stuhlmanni* — Track False-thorn

Forewing 11–12mm. Purplish-grey dusted with golden scales, fading in dried specimens, straight antemedial line, track-like row of black spots beyond postmedian line and black triangle on costa before, diagnostic; hindwings similar to forewings, without cell spot. Similar *P. zonaria* has antemedial line curved and lacks row of black spots. **Biology:** Caterpillar host associations unknown. Adults September–April. **Habitat & distribution:** Lowveld bush; KwaZulu-Natal, South Africa, north to Uganda.

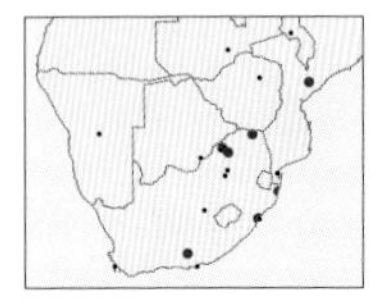

### 3 *Plecoptera poderis* — Variable False-thorn

Forewing 12–14mm. Black frons, head and front of thorax; ground colour variable from light brown to dark brown, lines crenulated, antemedial and postmedian lines often obscure or absent, kidney-shaped cell spot, sometimes surrounded by a ring of grey scales, some white spots on terminal line, variable in number; hindwings translucent brown with darker terminal shading. Caterpillar a blue semi-looper with white lines and between segments. **Biology:** Caterpillars recorded on *Dalbergia obovata* (Fabaceae) and *Clerodendrum glabrum* (Lamiaceae). Multivoltine, adults throughout the year. **Habitat & distribution:** Forest and riverine bush; Eastern Cape, South Africa, north to tropical Africa.

## Subfamily Boletobiinae Concealer Owls

The caterpillars are typically concealed, with very few living in the open as in most other Erebidae. Small nocturnal moths found everywhere, many are well adapted to the most arid environments, some adults easily disturbed by day. There are 183 described species in the region.

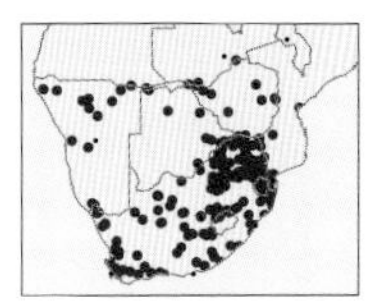

### 4 *Eublemma caffrorum* — Banner Eublemma

Forewing 6–8mm. Variable in colour, but distinctive pattern, with wings forming an anterior yellow triangle separated from a broad brown or violet-pink or straw band; hindwings orange. Caterpillar pale green, with sparse white hairs; often with dark red spots or bands. Large diverse genus with 88 described species in region. **Biology:** Caterpillars feed concealed in leaf fold on various mallows (Malvaceae). Multivoltine, adults throughout the year, easily disturbed by day, abundant at times. **Habitat & distribution:** Almost every habitat where Malvaceae grow, including transformed habitats, where it feeds on weeds; Western Cape, South Africa, north to Namibia, Botswana and Zambia.

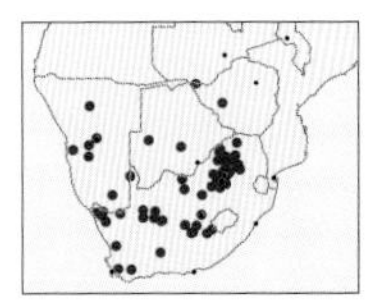

### 5 *Eublemma hemichiasma* — Bicolour Eublemma

Forewing 7–8mm. Forewings bicoloured, basal half white, remainder variable from light- to chocolate brown or ochre, usually with 2 distinct transverse bands with a paler area between them; hindwings white with grey margins. Similar *E. seminivea* can only be separated by examining genitalia or barcoding, but allopatric, found east of Great Escarpment north to Kenya. **Biology:** Caterpillar host associations unknown. Adults throughout the year. **Habitat & distribution:** Variety of relatively arid habitats west of Great Escarpment; Western Cape, South Africa, north to Namibia and Zimbabwe.

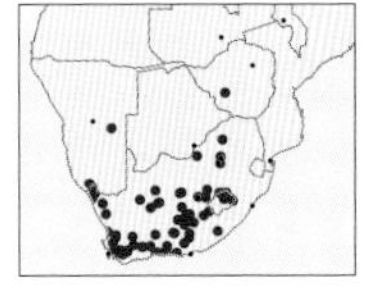

### 6 *Eublemma bipartita* — Divided Eublemma

Forewing 7–9mm. Forewings with basal white band and terminal brown to black band bordered with irregular white stripe. Caterpillar squat, cream, variegated in maroon-red, with white tubercles and fine white hairs. **Biology:** Caterpillars recorded in flowers of Long-stalk Parachute Daisy (*Ursinia paleacea*). Adults throughout the year. **Habitat & distribution:** Variety of habitats where Asteraceae grow, including the most arid areas; Western Cape to Namibia and Zimbabwe.

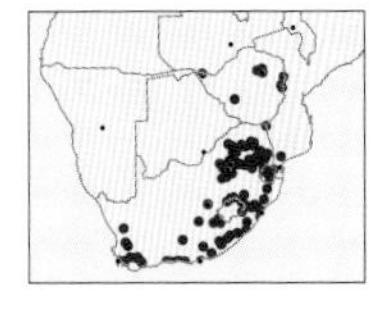

### 7 *Eublemma ornatula* — Ornate Eublemma

Forewing 7–8mm. Forewing ground colour white, shaded with hues of light to dark brown with 3 white transverse bands, shape of postmedian band diagnostic; hindwings white-brown with thin marginal dark stripe. Very similar *E. ceresensis* is smaller and occurs in more arid western areas, but sympatric in some. Caterpillar squat, pale cream-green, with fine white hairs and small black dots. **Biology:** Caterpillars feed inside *Helichrysum* flower heads. Multivoltine, adults throughout the year. **Habitat & distribution:** Fynbos and grassland in the eastern part of the region; Western Cape, South Africa, north to Kenya.

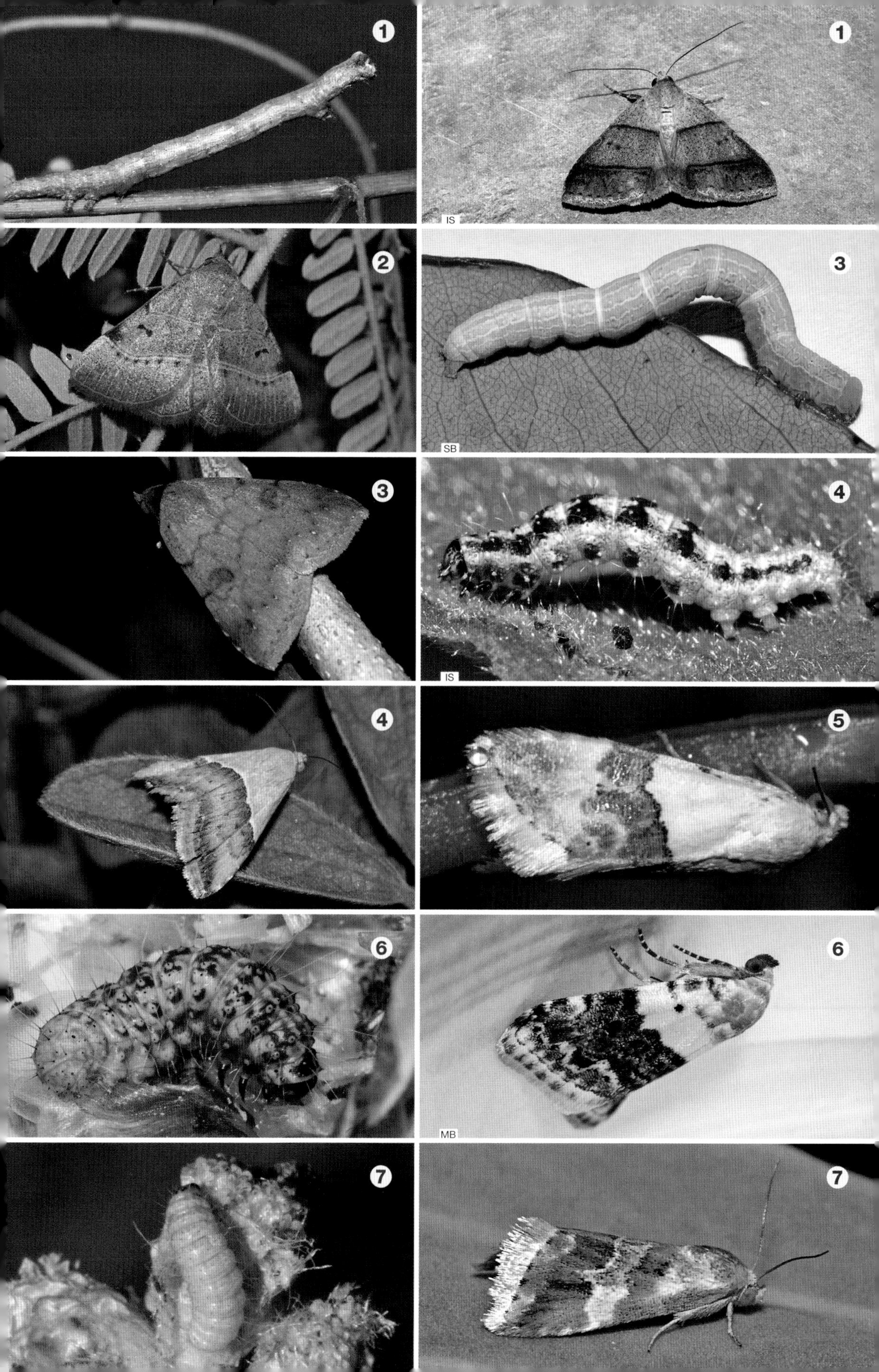
1
1
IS
2
3
SB
3
4
IS
4
5
6
6
MB
7
7

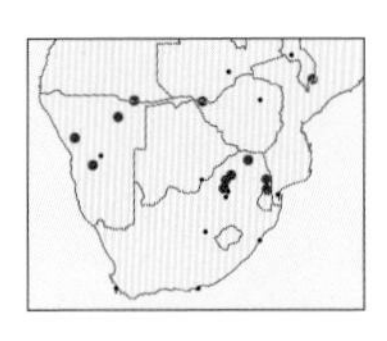

### 1 *Eublemma alexi* — Gem Eublemma

Forewing 6–7mm. Forewings rich chocolate brown, with 2 broad, white transverse bands that form a continuous white stripe and triangle at rest; hindwings dirty grey darkening toward ciliate margin. Caterpillar squat, pale greyish-green, with darker markings on each segment and sparse long hairs. **Biology:** Caterpillars construct shelters of leaves spun together in borages, *Heliotropium* spp. Bivoltine, adults October–March. **Habitat & distribution:** Bushveld; Mpumalanga, South Africa, north to tropical Africa.

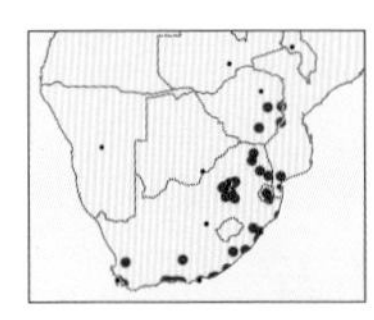

### 2 *Eublemma pennula* — Crested Eublemma

Forewing 6–8mm. Wings and body pale grey-brown; male forewings with diagnostic prominent black crest held up when at rest, absent in females, 2 olive green bands each ending in a small black patch on inner margin; hindwings white with fine grey striations. Caterpillar white, grub-like, with small black head capsule. **Biology:** Caterpillars feed inside flower heads of various Asteraceae. Multivoltine, adults throughout the year, peaking October–March. **Habitat & distribution:** Fynbos and grasslands; Western Cape, South Africa, north to tropical Africa.

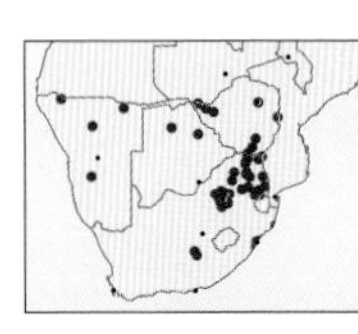

### 3 *Eublemma flavens* — Funnel Eublemma

Forewing 5–7mm. Basal half of forewing cream shading to buff with thin black and white lines forming a diagnostic oblique funnel-shaped median area toward apex; hindwings white to grey. **Biology:** Caterpillar host associations unknown. Adults throughout the year. **Habitat & distribution:** Grassland and bushveld; Eastern Cape, South Africa, north to Zimbabwe, also Ethiopia.

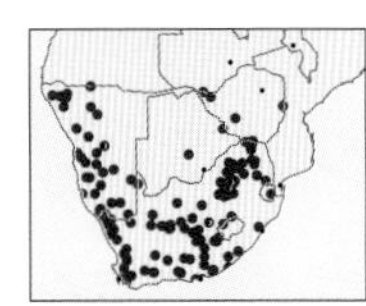

### 4 *Eublemma delicata* — Delicate Eublemma

Forewing 7–8mm. Forewings with white ground colour variably dusted with brown, almost obliterating all white in some individuals, 2 transverse, waved brown and grey bands, diagnostically becoming buff toward wing margin, with distinctive row of shiny black marginal dots; hindwings cream or grey. **Biology:** Caterpillar host associations unknown. Adults throughout the year. **Habitat & distribution:** Drier habitats, can be abundant in arid areas; Western Cape, South Africa, north to Ethiopia.

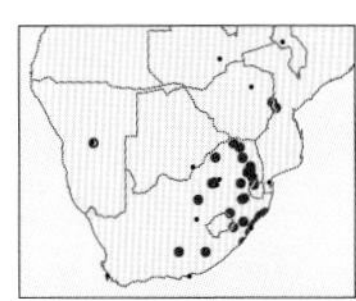

### 5 *Eublemma cochylioides* — Rosy Eublemma

Forewing 6–8mm. Head white, basal half of forewings cream with brown shading at postmedian, terminal area light rose-pink to violet or brown, distinctive cream and black marks along subterminal line; hindwings white to grey. **Biology:** Caterpillars feed primarily on Asteraceae. Adults throughout the year. **Habitat & distribution:** Variety of habitats; throughout the Afrotropical region.

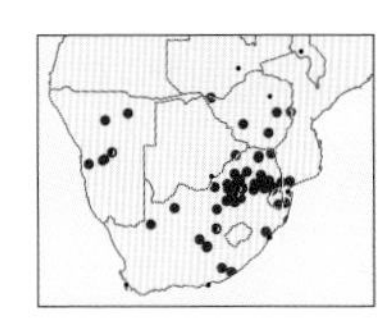

### 6 *Eublemma mesophaea* — Grey Eublemma

Forewing 7–8mm. Largely greyish-brown; forewings with transverse brown and white bands including broad, dark brown basal band; hindwings with faint median brown band. Caterpillar squat, cream with olive green bands running along length of body, which has thin white hairs. **Biology:** Caterpillars feed inside daisy flower heads (Asteraceae). Multivoltine, adults throughout the year. **Habitat & distribution:** Variety of habitats where Asteraceae grow; Eastern Cape, South Africa, north to Yemen.

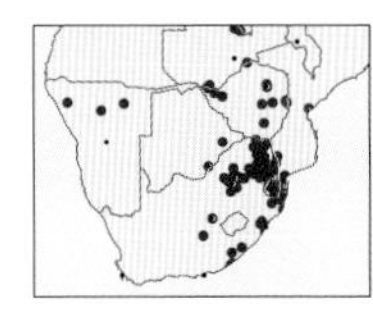

### 7 *Eublemma baccatrix* — Lined Eublemma

Forewing 7–9mm. Forewing colour variable, from tan to dark brown, and covered in diagnostic parallel transverse white lines, wing tips falcate, dark in some individuals; hindwings cream to brown. Caterpillars reddish-brown with long white hairs. **Biology:** Caterpillars feed in leaf fold on morning glory plants (*Ipomoea* spp.). Multivoltine, adults throughout the year. **Habitat & distribution:** Grassland and bushveld; Eastern Cape, South Africa, north to tropical Africa and Indian Ocean islands.

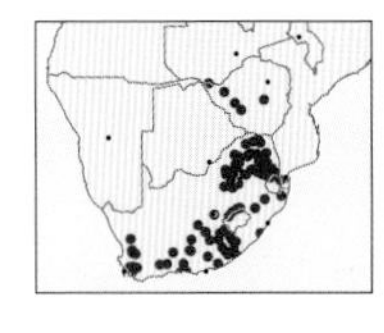

### 8 *Eublemma flaviceps* — Buff Eublemma

Forewing 9–11mm. Forewings buff, dusted with brown scales, and with 3 broad, whitish streaks running length of wing, cream streak on costa curving in toward apex diagnostic; hindwings uniform cream. **Biology:** Caterpillar host associations unknown. Adults throughout the year, peaking in summer months. **Habitat & distribution:** Grassland and fynbos, absent from arid western areas where it is replaced by several similar looking species; Western Cape, South Africa, north to Zimbabwe.

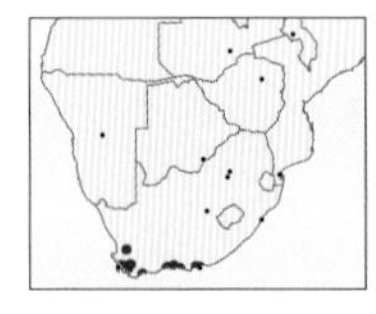

### 9 *Eublemma glaucizona* — Fynbos Eublemma

Forewing 7–9mm. Unmistakable; ground colour brown, straight antemedial line with grey median area, postmedian line strongly curved around distinctive brown and cream patch, terminal area rusty brown; hindwings cream. **Biology:** Caterpillar host associations unknown. Adults throughout the year. **Habitat & distribution:** Rare for this genus, a habitat specialist in fynbos; Western and Eastern Cape, South Africa.

1
1
IS
2
2
3
4
5
6
6
7
7
8
9

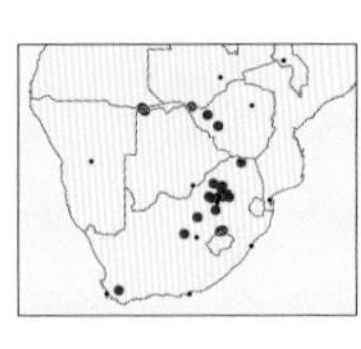

## 1 *Eublemma goniogramma* — Arrow Eublemma

Forewing 9–10mm. Distinctive; forewings deep yellow, variably dusted with brown and red scales, diagnostic, white postmedian line turns back on itself to form an arrow point, white central streak completes the arrow, white dots on margin extend into grey cilia; hindwing colour same as forewing, with chequered marginal cilia. **Biology:** Caterpillar host associations unknown. Adults throughout the year, peaking in summer. **Habitat & distribution:** Grassland and bushveld; Western Cape, South Africa, north to Namibia and Zimbabwe.

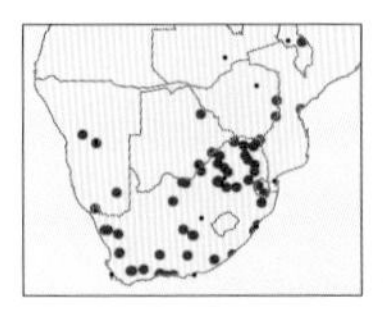

## 2 *Eublemma admota* — Anvil Eublemma

Forewing 10–12mm. Fairly bulky; wings and body variable grey-brown to white, diagnostic anvil-shaped apical part of forewings brown or olive green, separated from rest of wing by sinusoidal white line. Caterpillar plump, black head capsule, pale olive green with darker mottling, and covered in sparse, long white hairs arising from white plates. **Biology:** Caterpillars feed on hairy *Solanum* spp., making shelter from hairs. Multivoltine, adults throughout the year. **Habitat & distribution:** Variety of habitats where *Solanum* spp. grow, including transformed habitats; Western Cape, South Africa, north to Ethiopia.

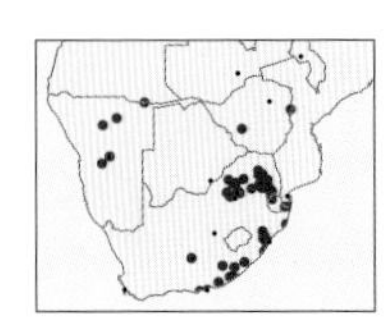

## 3 *Eublemma costimacula* — Scale Eublemma

Forewing 7–12mm. Forewings with basal ochre marking and terminal brown marking, both defined by fine white lines, subapical dark marking at edge of wing; hindwings grey; thorax orange. Caterpillar squat, flesh-coloured, with black head capsule and thoracic shields. **Biology:** Caterpillars feed on various scale (*Duplachionaspis*) and mealy bug insects (Coccoidea), living in shelter of cemented frass and wax filaments of prey. Like *E. scitula* (p.326), of possible use against crop pests. **Habitat & distribution:** Variety of habitats, including gardens; Eastern Cape north throughout the Afrotropical region.

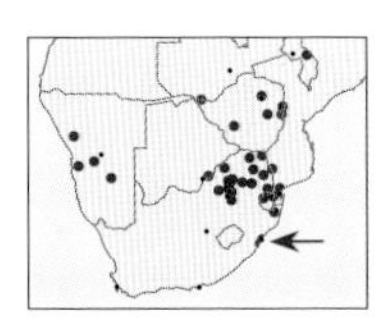

## 4 *Eublemma gayneri* — Sharp Eublemma

Forewing 6–8mm. Colouration very variable, ranging from very light tan through shades of grey, pale green and brown to dark brown; fore- and hindwings similarly marked, shape of straight postmedian line sharply angled back to costa diagnostic, occasionally with dark margins. Caterpillar slender, greenish, with brown prolegs and covered in fine, long hairs. **Biology:** Caterpillars polyphagous inside flowers and fruit, also known to predate mealy bugs (Coccoidea). Multivoltine, adults throughout the year. **Habitat & distribution:** Variety of habitats, including transformed habitats; KwaZulu-Natal north to tropical Africa and Madagascar.

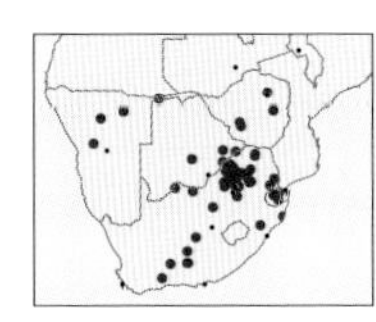

## 5 *Eublemma nigrivitta* — Thorn Eublemma

Forewing 8–10mm. Wings and body brown, wings finely barred in wavy, brown lines with darker band in centre and end of wing; in some forms these bars separate a pale distal region of wing tip. Caterpillar a pale green semi-looper with white stripe running along side of body, darker markings on each segment and body covered in very fine long hairs. Easily confused with *Metachrostis* spp. (p.326), which lack the dark central band. **Biology:** Caterpillars feed on the flowers of Sweet Thorn (*Vachellia karroo*). Multivoltine, adults throughout the year. **Habitat & distribution:** Various habitats where *Vachellia* trees grow; Western Cape, South Africa, north to Malawi.

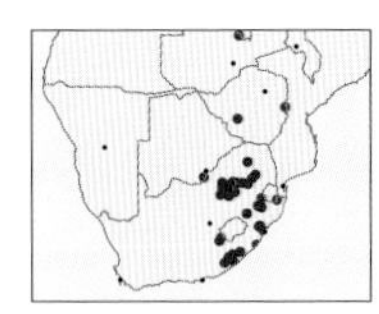

## 6 *Eublemma staudingeri* — Blackstreak Eublemma

Forewing 9–12mm. Ground colour cream to light brown, forewings with diagnostic black longitudinal streaks on veins, ending on white jagged terminal line, cilia cream; hindwings light or dark brown. **Biology:** Caterpillar host associations unknown. Adults throughout the year. **Habitat & distribution:** Mesic arid grassland and bushveld; Eastern Cape, South Africa, north to Zimbabwe, and Yemen.

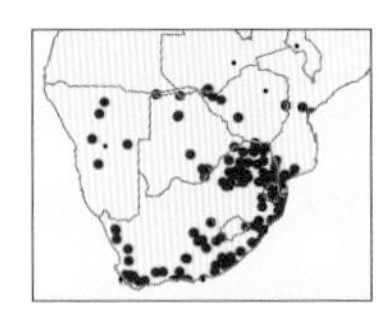

## 7 *Eublemma bolinia* — Earthy Eublemma

Forewing 6–9mm. Variable; forewings chestnut brown, with central pale transverse band bordered by scalloped black line or a white diagonal stripe; hindwings black with thin antemedial white bar and marginal white spots. Similar *E. metacrypta* lacks the 2 distinct white hindwing spots. **Biology:** Caterpillar host associations unknown. Adults expose hindwings when disturbed, possible jumping spider mimic. Adults throughout the year. **Habitat & distribution:** Variety of habitats; Western Cape north to Ethiopia.

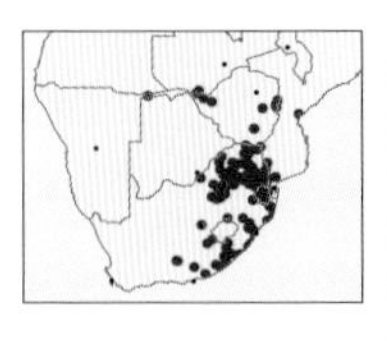

## 8 *Eublemma apicata* — Whitetip Eublemma

Forewing 9–11mm. Forewing ground colour varies from light brown to almost black, black-ringed white cell spot extending in white streaks in some specimens, lines mostly obscured, diagnostic white tip on apex with a smaller one on inner angle; hindwings brown to black with lighter cilia. **Biology:** Caterpillar host associations unknown. Adults throughout the year. **Habitat & distribution:** Less arid grassland and bushveld; Eastern Cape, South Africa, north to Kenya.

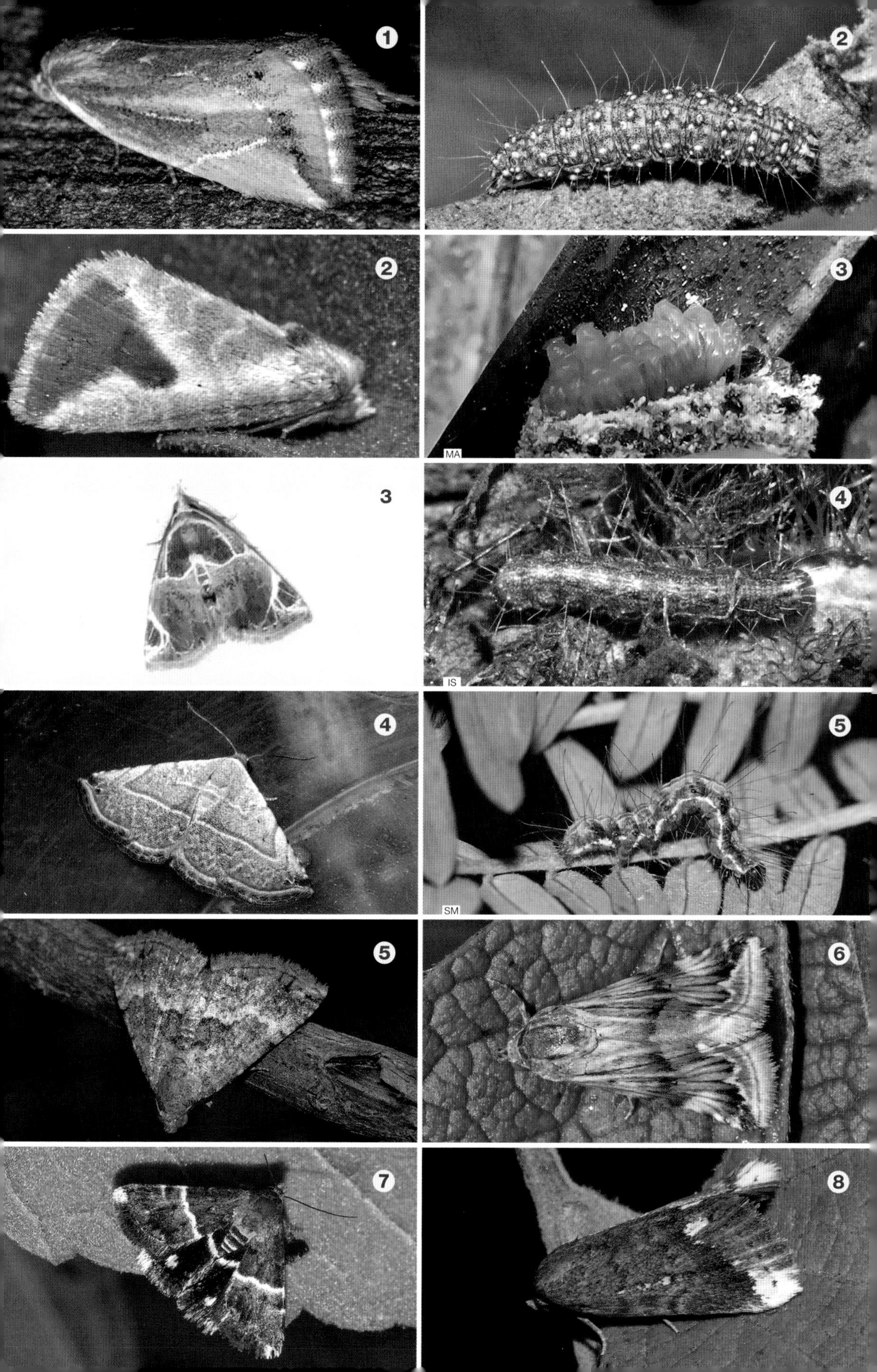
1
2
2
3
MA
3
4
IS
4
5
SM
5
6
7
8

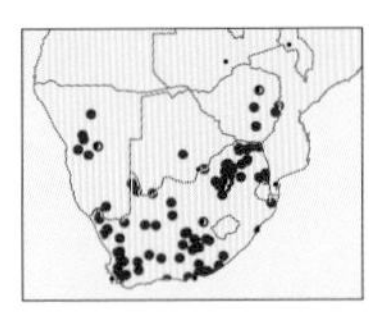

**1 *Eublemma scitula*** Large-eye Eublemma

Forewing 7–9mm. Colouration variable; forewings white or buff with broad light to dark brown, complex band bordered in black and white stripe apically, a large diagnostic eyespot in cell; hindwings white or grey. Caterpillar bright pink-red, with sucker-like prolegs. **Biology:** Caterpillars carnivorous, feeding on various scale insects and mealy bugs, including Red Scale (*Aonidiella aurantii*), wax scales (*Ceroplastes*) and mealy bugs. Caterpillars feed on all stages of their host, with a preference for eggs (female scales are often eaten). Has been considered as a biological control agent. Adults September–March. **Habitat & distribution:** Relatively arid habitats, including deserts; Western Cape to Namibia, nominate subspecies North and East Africa.

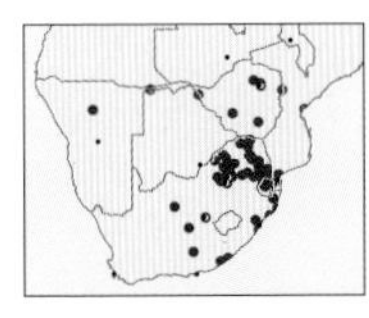

**2 *Honeyana ragusana*** Converging Eye-tip

Forewing 8–9mm. Head, frons and front of thorax brown; forewing white, finely dusted with rows of reddish-brown scales, many lines emanating from inner margin and costa converging on tip of wing, 1 or 2 black eye marks on wing tip, cilia with brown and white lines; hindwings white. Two species in genus in region. **Biology:** Caterpillar host associations unknown. Multivoltine, adults throughout the year. **Habitat & distribution:** Variety of habitats; Eastern Cape north to tropical Africa and Indian Ocean islands.

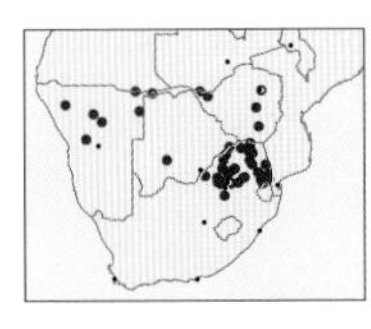

**3 *Honeyana plumbosa*** Plain Eye-tip

Forewing 8–9mm. Similar to *H. ragusana* (above), but with no lines on forewings and cilia plain orange, eyespot on tip similar, ground colour varies from pearly white to pearly grey; hindwings pearly white. **Biology:** Caterpillar host associations unknown. Multivoltine, adults throughout the year. **Habitat & distribution:** Grassland and bushveld; Free State, South Africa, north to Ethiopia.

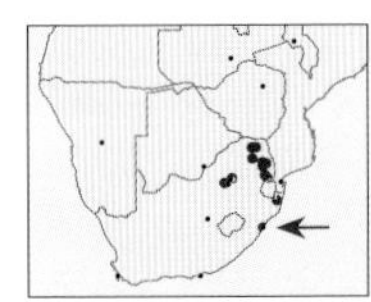

**4 *Metachrostis postrufa*** Dull Widebar

Forewing 8–10mm. Wings brown and grey with complex patterning, but with distinct wide median transverse brownish bar, delineated by thin black lines. Caterpillar white-yellow with long, white hairs and black bands, and lateral and dorsal yellow stripes. Eleven species in genus in region, adults mostly similar but caterpillars distinctive. **Biology:** Caterpillars feed on *Asparagus laricinus*. Multivoltine, adults throughout the year. **Habitat & distribution:** Grassland and bushveld; KwaZulu-Natal, South Africa, north to Ethiopia.

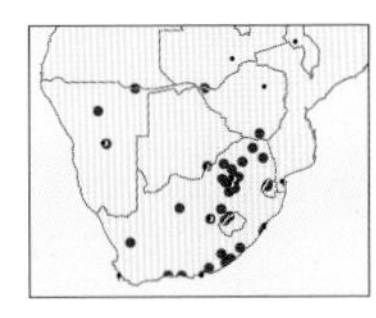

**5 *Metachrostis decora*** Decorous Widebar

Forewing 9–10mm. Similar to *M. postrufa* (above), but marked with pink median shading diagnostically reduced near inner margin. Caterpillar white with yellow prothorax and prolegs, each segment with double black band enclosing yellow stripe and carrying fine short black hairs. **Biology:** Caterpillars feed on *Albuca* spp. (Asparagaceae). Multivoltine, adults throughout the year. **Habitat & distribution:** Variety of habitats; Western Cape, South Africa, north to Saudi Arabia and Madagascar.

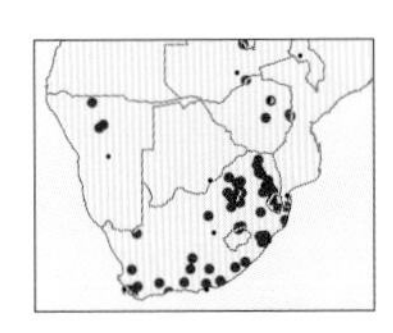

**6 *Metachrostis snelleni*** Dotted Widebar

Forewing 8–9mm. Wings with bluish sheen in live specimens, pale to very dark brown in dried specimens, with complex waved pattern; fawn-coloured terminal band or pink patches conspicuous in dark forms; inner part of hindwings like forewing. Caterpillar pale green-pink, with fine white hairs and yellow markings over spiracles. **Biology:** Caterpillars feed inside bulbs and fruit of various Asparagaceae. Multivoltine, adults throughout the year. **Habitat & distribution:** Variety of habitats where Asparagaceae grow; Western Cape, South Africa, north to Saudi Arabia and Madagascar.

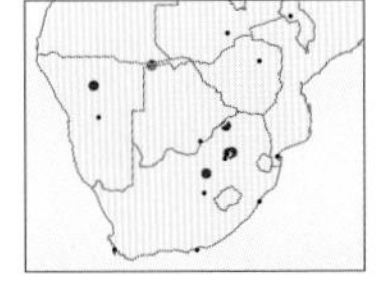

**7 *Metachrostis eupethecica*** Grey Widebar

Forewing 6–8mm. Forewings light grey, lacking shading of bright scales of other species in genus, lines black, zigzag subterminal line, diagnostic orange dots around cell; hindwings grey. **Biology:** Caterpillar host associations unknown. **Habitat & distribution:** Bushveld; Free State, South Africa, north to Namibia.

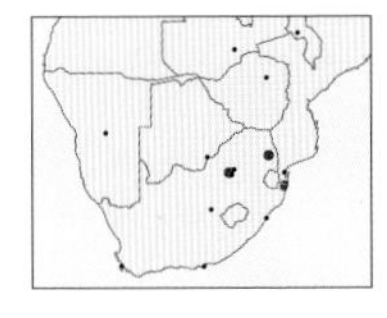

**8 *Corgatha chionocraspis*** Purple Flag

Forewing 7–9mm. Frons white, body and wing ground colour purplish-red, antemedial and postmedian lines white, area between postmedian and subterminal lines variably sprinkled with white scales, distinct cell spot on all wings in males. Nine species in genus in region, adults hold wings open when at rest. **Biology:** Caterpillar host associations unknown. Adults October–May. **Habitat & distribution:** Coastal bush, forests and moist kloofs; KwaZulu-Natal, South Africa, north to Uganda.

1
2
3
4
4
5
5
IS
6
IS
6
IS
7
8

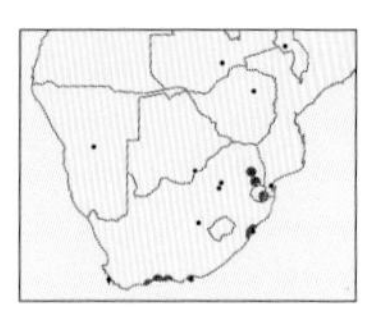

### 1 *Corgatha producta* — Variable Flag

Forewing 7–9mm. Forewings falcate, pinkish-orange through to shades of brown to grey, but shape of lines and bands constant, black subterminal spot only in some individuals, cilia white; hindwings same as forewings. **Biology:** Caterpillar host associations unknown. Adults September–May. **Habitat & distribution:** Forests and riverine bush; Western Cape, South Africa, north to Zimbabwe.

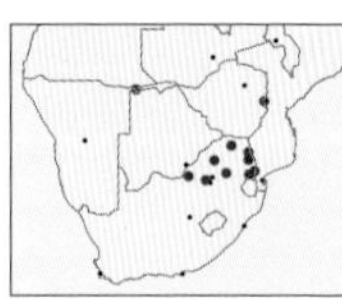

### 2 *Corgatha micropolia* — Grey Flag

Forewing 5–6mm. Body and wings light grey, sparsely dusted with black scales, thin purple margin, wider on costa in median area, wavy antemedial and postmedian lines barely visible in some individuals; hindwings same as forewings. **Biology:** Caterpillar host associations unknown. Adults September–March. **Habitat & distribution:** Bushveld; Gauteng, South Africa, north to Zimbabwe and Namibia.

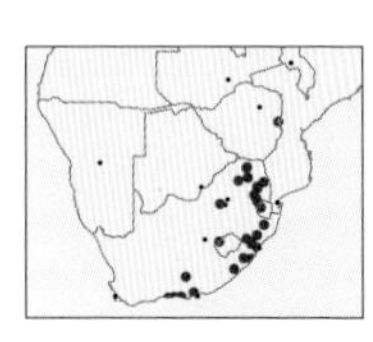

### 3 *Corgatha albigrisea* — Whiteline Flag

Forewing 8–9mm. White, wings variably mottled grey and white, antemedial line black, white jagged postmedian line and white subterminal line (sometimes as spots) diagnostic; hindwings same as forewings. **Biology:** Caterpillar host associations unknown. Adults October–May, peaking in late summer. **Habitat & distribution:** Moist grasslands and woodland; Western Cape, South Africa, north to Zimbabwe.

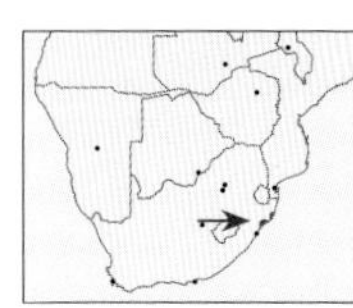

### 4 *Corgatha wojtusiaki* — Wotju Flag

Forewing 9–10mm. Body and wings uniformly fawn-coloured, wings with faint transverse, waved lines and a subterminal line of small white dots. Caterpillar thin, off-white with fine longitudinal streaking; head and rear of body light brown, with bizarre, hair-like projections ending in a spatulate knob. **Biology:** Caterpillar host associations unknown. **Habitat & distribution:** Tropical coastal bush; Zululand, South Africa, also Nigeria, São Tomé and Príncipe.

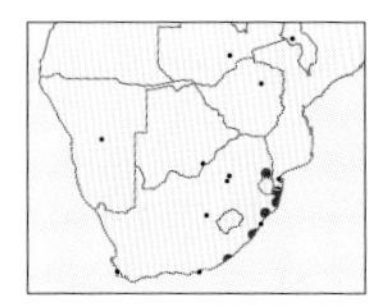

### 5 *Cautatha drepanoidea* — Curved Flag

Forewing 8–9mm. Similar to *Corgatha producta* (above), but shape of cream and orange postmedian and antemedial lines and brown cilia diagnostic, ground colour variable from orange-brown through to dark brown; hindwings same as forewings. Two species in genus in region. **Biology:** Caterpillar host associations unknown. Adults September–May. **Habitat & distribution:** Coastal bush and sand forest; Eastern Cape north to Mpumalanga, South Africa.

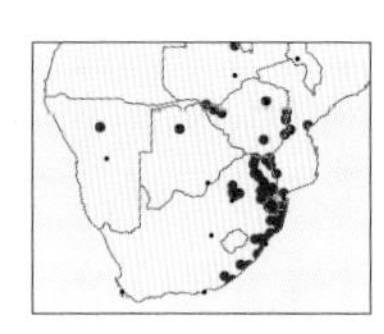

### 6 *Cerynea thermesialis* — Yellow Flag

Forewing 8–10mm. Wings rounded; body yellow, head and wings shaded with reddish-brown but some yellow always visible, black postmedian line jagged, usually with white spots, double black cell spot; hindwings same as forewings but with single cell spot. Six species in genus in region; very similar *C. ignealis* fawn, not yellow. **Biology:** Caterpillar host associations unknown. Adults throughout the year. **Habitat & distribution:** Forest and woodland; Eastern Cape, South Africa, north to tropical Africa.

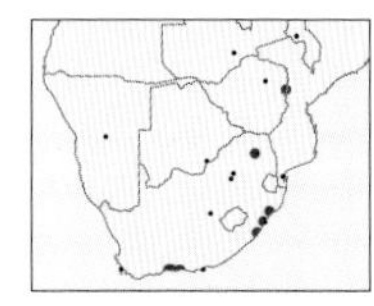

### 7 *Hypobleta viettei* — Long Flag

Forewing 8–10mm. Distinctive species with elongate falcate wings; ground colour fawn, sprinkled with reddish-brown scales, especially in median area, distinctive dark terminal patch in falcate curve; hindwings same as forewings but with black cell spot and dark spot on inner fold extending onto abdomen. Two species in genus in region. **Biology:** Caterpillar host associations unknown. Adults September–April. **Habitat & distribution:** Moist forest; Western Cape, South Africa, north to tropical Africa, nominate subspecies from Madagascar.

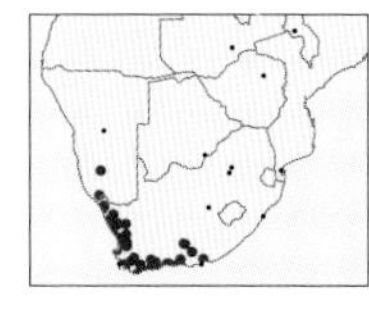

### 8 *Tegiapa virescens* — Green Tegiapa

Forewing 7–10mm. Distinctive, with black antemedial line jagged and postmedian line with branch around cell, ground colour white, variably dusted with green and black scales; hindwings white to grey. Eight species in genus in region. **Biology:** Caterpillar host associations unknown. Adults August–April. **Habitat & distribution:** Fynbos and succulent karoo; Cape provinces, South Africa, and southern Namibia.

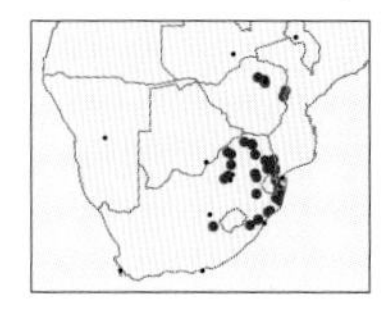

### 9 *Tegiapa goateri* — Goat Tegiapa

Forewing 6–8mm. Similar to *T. virescens* (above), but antemedial and postmedian lines incomplete, 3 distinct triangular black marks on costa and inner margin at line ends, wings white with sparse dusting of olive scales; hindwings grey with white postmedian line. **Biology:** Caterpillar host associations unknown. Adults September–May. **Habitat & distribution:** Grassland; Eastern Cape, South Africa, north to Zimbabwe.

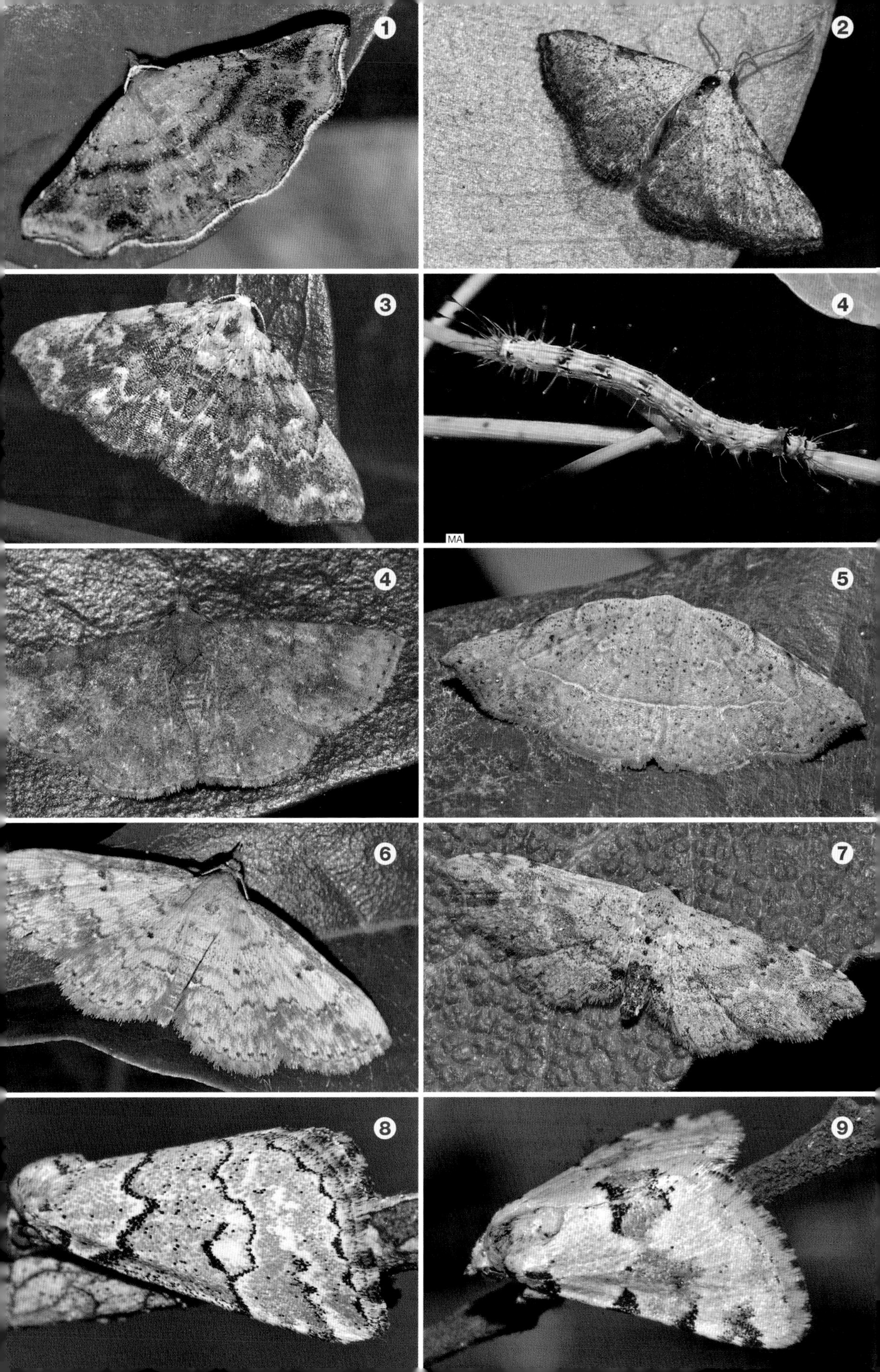
1
2
3
4
MA
4
5
6
7
8
9

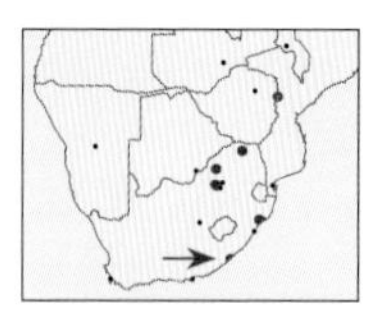

**1 *Tegiapa melanoleuca*** Melanic Tegiapa

Forewing 9–11mm. Larger and wings broader than other species in region, ground colour creamy white to pink, distinctive shape of postmedian line and black markings around margins; hindwings white to grey. **Biology:** Caterpillar host associations unknown. Adults throughout the year. **Habitat & distribution:** Forests; Eastern Cape, South Africa, north to Zambia.

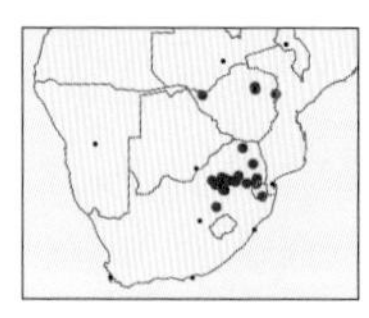

**2 *Tegiapa larentiodes*** Carpet Tegiapa

Forewing 7–8mm. Similar in appearance to Carpet Moth species (Geometridae: Larentiinae), wings held flat over body when at rest; ground colour orange to green, antemedial and postmedian lines parallel and jagged, forming medial area which is shaded with shiny black scales; hindwings plain cream. **Biology:** Caterpillar host associations unknown. Adults throughout the year. **Habitat & distribution:** Grassland and bushveld; Free State, South Africa, north to Zimbabwe.

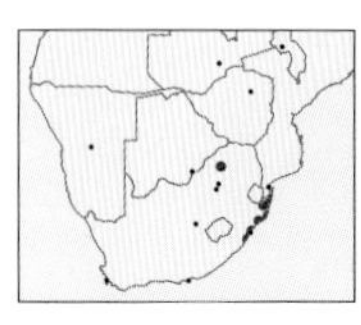

**3 *Marca proclinata*** Straight Lines

Forewing 7–10mm. Can be confused with *Mocis proverai* (p.356), but smaller and inner black band not along inner margin but along postmedian line; straw with 2 distinct black bars, one before postmedian and another beyond subterminal lines, small black cell spot; hindwings with black postmedian bar. One species in genus in region. **Biology:** Caterpillar host associations unknown. Adults throughout the year. **Habitat & distribution:** Forest and coastal bush; KwaZulu-Natal, South Africa, north to tropical Africa and Madagascar.

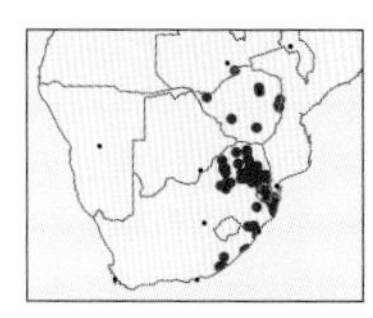

**4 *Syngatha latiflavaria*** Clear Falsegeo

Forewing 8–10mm. At rest, wings held flat and wide open; yellow, basal, median and costal area of wings covered in brick red scales, lower half of median area light pink in some individuals, remainder of wings bright yellow. Eight species in genus in region, often confused as belonging to the family Geometridae. **Biology:** Caterpillar host associations unknown. Adults September–April. **Habitat & distribution:** Variety of habitats; Eastern Cape, South Africa, north to Malawi and Angola.

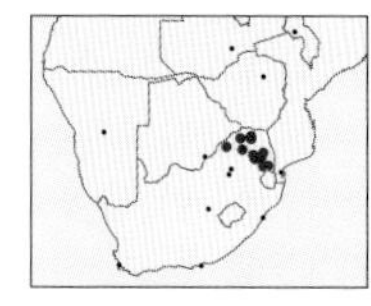

**5 *Syngatha pyrrhoxantha*** Muddled Falsegeo

Forewing 8–10mm. At rest, wings held flat and wide open; yellow, basal, median and costal area of wings with blood red scales, except for postmedian line, remainder of wings bright yellow, border between red and yellow muddled. **Biology:** Caterpillar host associations unknown. Adults November–March. **Habitat & distribution:** Bushveld; Mpumalanga and Limpopo, South Africa, and Ethiopia.

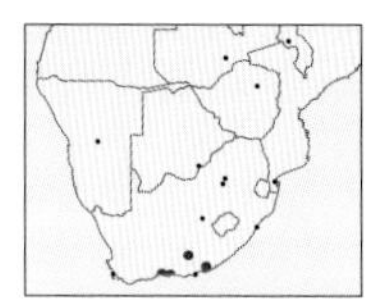

**6 *Syngatha argyropasta*** Ochre Falsegeo

Forewing 10mm. At rest, wings held flat and wide open; ground colour yellow, costal area of forewing ochre-brown, remainder of wings plum or mauve with variable light yellow margins, raised black scales around cell and on thorax. **Biology:** Caterpillar host associations unknown. Adults October–February. **Habitat & distribution:** Forest and kloofs; Western Cape and Eastern Cape, South Africa, and Malawi.

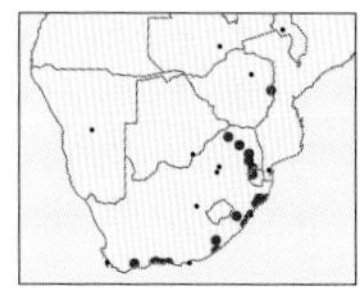

**7 *Paroruza subductata*** White-eye Falsegeo

Forewing 14–16mm. Large for subfamily; at rest, slightly falcate wings held flat and wide open; ground colour variable from bright orange to deep purplish-brown; forewings with black-lined white cell spot, black and greatly enlarged in some females, black-lined yellow postmedian line slightly curved, distinct on both wings. One species in genus, originally placed in Geometridae, often confused with *Pareclipsis* and *Idiodes* spp. (Geometridae). **Biology:** Caterpillar host associations unknown. Adults throughout the year. **Habitat & distribution:** Moist forests; Western Cape, South Africa, north to Zimbabwe.

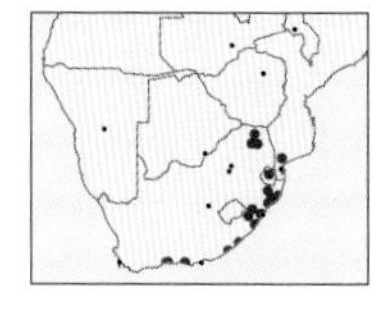

**8 *Eublemmistis chlorozonea*** Tent Falsegeo

Forewing 8–10mm. Characteristic resting posture with body at an angle to substrate and wings forming a tent; ground colour white with mustard yellow shading, no shading on antemedial and postmedian zigzag lines, 2 dark spots on forewing cell; hindwings similarly marked. Two species in genus in region; similar *E. aberfoylea* from eastern Zimbabwe is much smaller. **Biology:** Caterpillars feed on tree lichens, covering themselves with the lichens. Adults September–April. **Habitat & distribution:** Moist forests; Western Cape to Limpopo and Mozambique, also Tanzania, Uganda and Ethiopia.

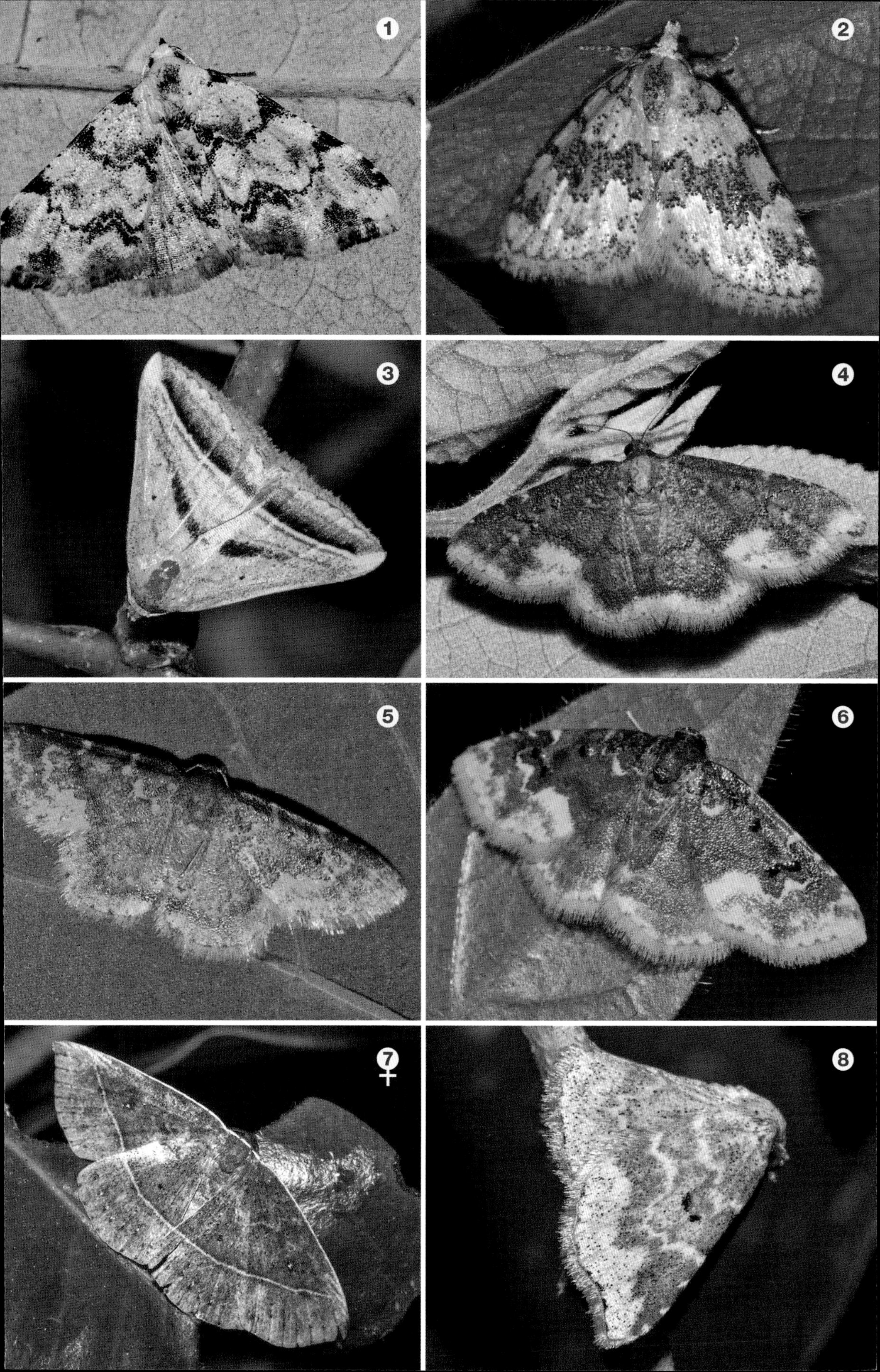
1
2
3
4
5
6
7♀
8

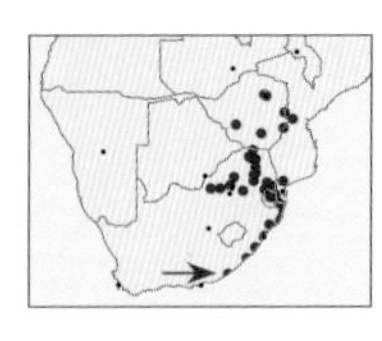

**1 *Oediblemma trogoptera*** Durban Tooth-wing

Forewing 10–11mm. Head, thorax and basal part of wings white; forewings with very broad red-brown median area, wing tip with grey crescent, forewings with deeply incised wing margin; hindwings with short tail, colouration varies from light to rich brown. One species in genus in region. **Biology:** Caterpillar host associations unknown. Adults October–April. **Habitat & distribution:** Forest and riverine bush; Eastern Cape to Malawi.

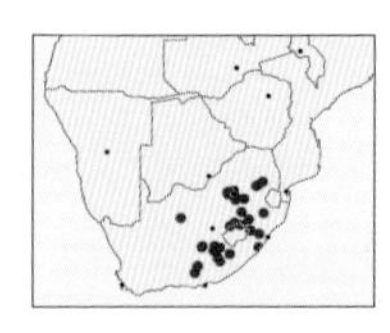

**2 *Phytometra sacraria*** Pastel Phytometra

Forewing 11–12mm. Elongate wings held tight against body; fawn to yellow, diagnostic postmedian line consists of a row of spots of variable intensity curved toward apex, terminal area rosy pink or of ground colour in some individuals, small black cell spot; hindwings plain straw. Twelve species in genus in region. **Biology:** Caterpillar host associations unknown. Adults August–May, peaking in summer months. **Habitat & distribution:** Grassland; Western Cape, South Africa, north to Mpumalanga, South Africa.

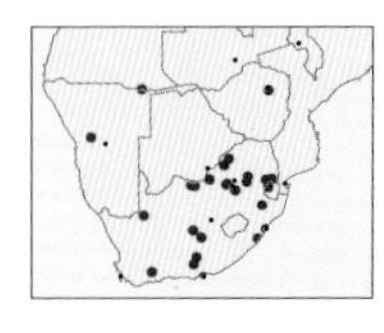

**3 *Phytometra fragilis*** Fragile Phytometra

Forewing 10–11mm. Head and frons dark brown, ground colour mustard yellow, with darker marginal band; forewings variably speckled with brown spots, sometimes forming an incomplete sinuous band before postmedian line; hindwings grey. Caterpillar an olive green, slender semi-looper. **Biology:** Caterpillars feed on *Polygala uncinata* (Polygalaceae). Bivoltine, adults September–May. **Habitat & distribution:** Variety of habitats; Western Cape north across the Afrotropical region.

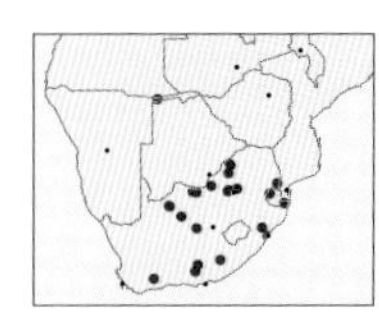

**4 *Phytometra subflavalis*** Black-end Phytometra

Forewing 10–12mm. Distinctive; head and frons dark brown, ground colour creamy yellow, with prominent dark marginal band; forewings sometimes finely speckled with tiny brown spots, often with no spots; hindwings cream with dark margin. **Biology:** Caterpillar host associations unknown. Adults December–May. **Habitat & distribution:** Variety of habitats; Eastern Cape, South Africa, north to tropical Africa, nominate subspecies from Madagascar.

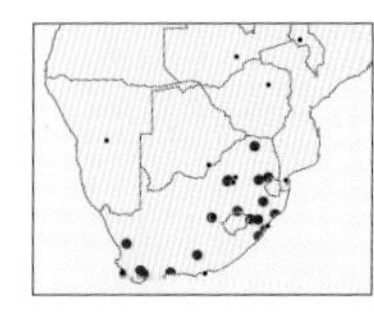

**5 *Phytometra helesusalis*** Curved-line Phytometra

Forewing 11–13mm. Head and frons of ground colour; ground colour variable, cream to coppery brown, diagnostic postmedian and subterminal lines curved and converging near apex, 2 brown spots around cell; hindwings plain cream. **Biology:** Caterpillar host associations unknown. Adults throughout the year. **Habitat & distribution:** Fynbos and grassland; Western Cape north to Limpopo, South Africa.

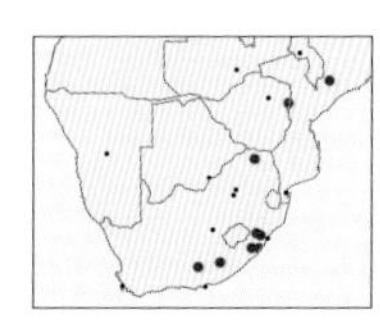

**6 *Phytometra heliriusalis*** Waveline Phytometra

Forewing 13–15mm. Larger than other species in genus; resting posture with wings flat over body; head and frons of ground colour; ground colour variable, reddish-brown to yellow, well-developed wavy postmedian line diagnostic, 2 brown spots around cell; hindwings grey. Can be confused with *Bertula* spp. (Herminiinae). **Biology:** Caterpillar host associations unknown. Adults throughout the year. **Habitat & distribution:** Forests and kloofs; Eastern Cape, South Africa, north to Zimbabwe and Mozambique.

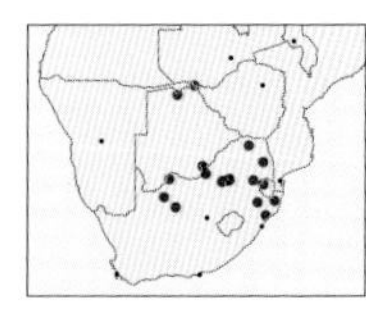

**7 *Radara subcupralis*** Turntail Radara

Forewing 12–13mm. Ground colour brown, head and thorax lighter brown; forewings variably mottled with shades of brown with postmedian line turned back toward costa and diagnostic wavy, often faint, subterminal line; hindwings uniform brown. Caterpillar thin, apple green with longitudinal white lateral line. Three species in genus in region; similar *R. vacillans*, a forest species, has subterminal line evenly curved, not wavy. **Biology:** Caterpillars feed on Stinging Nettles (*Tragia* spp.). Multivoltine, adults throughout the year. **Habitat & distribution:** Bushveld; KwaZulu-Natal, South Africa, north to tropical Africa.

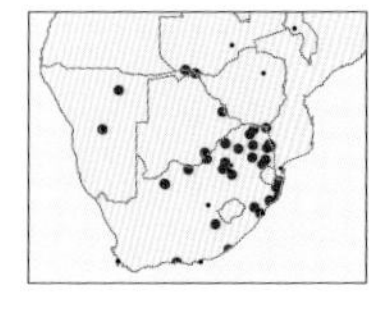

**8 *Rhesala moestalis*** Small Cloth

Forewing 6–8mm. Ground colour uniform light grey, wings covered in parallel fine brown lines; forewings with darker median band that forms a bar across wings at rest (band sometimes composed of variable broken dark spots), diagnostic postmedian line strongly curved outward before curving to costa. Caterpillar apple green, black setal bases with 4 longitudinal stripes and green head capsule. Three species in genus in region; similar *R. goleta* is much larger; *R. natalensis* has postmedian line straight before curving to costa. **Biology:** Caterpillars feed on Ana Tree (*Faidherbia albida*) and Hook Thorn (*Senegalia caffra*). **Habitat & distribution:** Forest, grassland and bushveld; Western Cape, South Africa, north to tropical Africa, also Madagascar and Saudi Arabia.

## Subfamily Calpinae Underwing Moths

Large, with relatively narrow wings and characteristic upturned palpi. The caterpillars are usually smooth, lacking hairs and are semi-loopers with 1 or 2 pairs of prolegs missing. The moths sometimes have brightly coloured hindwings, hence their name, adults often visit sweet sources such as rotting fruit, and can be attracted by sugaring tree trunks. There are 144 species in the region.

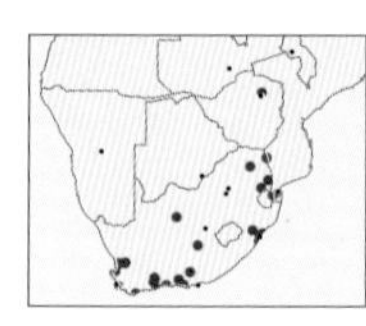

### 1 *Oraesia emarginata* Muddled Saddle Snout

Forewing 14–16mm. Unique resting posture with narrow wings deeply indented on inner margin, appearing like a saddle when at rest; forewings mottled in shades of brown and grey, dark area from middle inner margin to centre with white streak above; hindwings straw to grey. Caterpillar black with 2 dorsolateral lines of yellow, white and red marks. Four species in genus in region. **Biology:** Caterpillars feed on *Cissampelos mucronata* (Menispermaceae). Multivoltine, adults throughout the year. **Habitat & distribution:** Variety of habitats; Western Cape, South Africa, north to tropical Africa, and India.

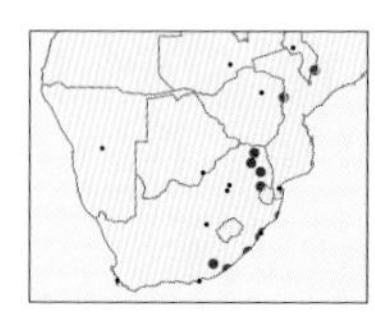

### 2 *Plusiodonta natalensis* Golden Saddle Snout

Forewing 16–18mm. Resting posture similar to *Oraesia emarginata* (above); upturned palpi and body brown; forewing inner margin in basal area and all of median area purplish-brown, rest of basal area a golden triangle, terminal area golden sheen with unique subterminal markings; hindwings dull grey. Five species in genus in region. **Biology:** Caterpillar host associations unknown. Adults throughout the year. **Habitat & distribution:** Moist forests; Eastern Cape, South Africa, north to tropical Africa.

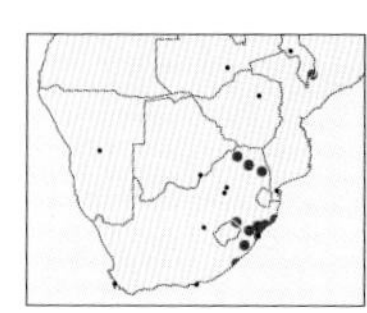

### 3 *Plusiodonta achalcea* Brown Saddle Snout

Forewing 16–17mm. Similar to *P. natalensis* (above), but with golden areas replaced by dark brown, basal triangular patch with white line and 2 white dots beyond postmedian line present in some individuals, colour can vary from light to dark brown; hindwings dull grey. Caterpillar a well-camouflaged mottled green, yellow and white semi-looper. **Biology:** Caterpillars feed on *Cissampelos torulosa* (Menispermaceae). Adults throughout the year. **Habitat & distribution:** Moist forests; Eastern Cape, South Africa, north to Tanzania.

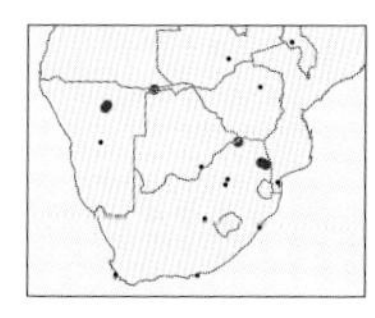

### 4 *Schalidomitra ambages* Two-tone Snout

Forewing 10–16mm. Abdomen yellow with black segmental bands; forewings white, lines black, coalesce at inner margin, 2 large black dots either side of cell, and row of smaller black marginal dots; hindwings cream with marginal black dots, pale postmedian line and cell spot. Caterpillar with alternating grey and black-and-white striped segments, body covered sparsely with long hairs. **Biology:** Caterpillars feed on Apple-leaf (*Philenoptera violacea*). **Habitat & distribution:** Lowland bushveld where apple-leaf trees grow; Mpumalanga, South Africa, north to Tanzania.

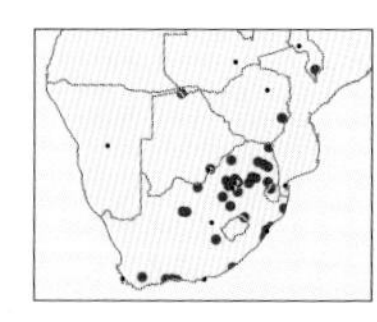

### 5 *Antiophlebia bracteata* Pearl Spangled

Forewing 16–25mm. Squat, grey or brown, with crest of raised scales on thorax; forewings grey-brown with lighter terminal area and with line of alternating large and small, silvery triangular markings; hindwings off-white. Antennae bipectinate. Caterpillar brown, twig mimic. One species in genus in region. **Biology:** Caterpillars feed on trees and shrubs in the family Ebenaceae. Bivoltine, adults August–May. **Habitat & distribution:** Variety of habitats; Western Cape, South Africa, north to tropical Africa.

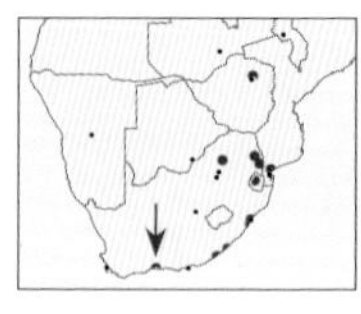

### 6 *Bareia incidens* Squinting Tabby

Forewing 15–19mm. Largely brown with a violet sheen in live individuals; forewings with darker apical region bearing a black and a white eyespot, short black bar halfway along forewings on costa; hindwings with a few wavy transverse lines. Caterpillar squat, green, covered in fine white lateral and black dorsal setae and with 3 cream longitudinal lines. **Biology:** Caterpillars feed on fig trees (*Ficus* spp.). Multivoltine, adults throughout the year. **Habitat & distribution:** Forest and bushveld where fig trees grow; Western Cape, South Africa, north to tropical Africa, also Madagascar.

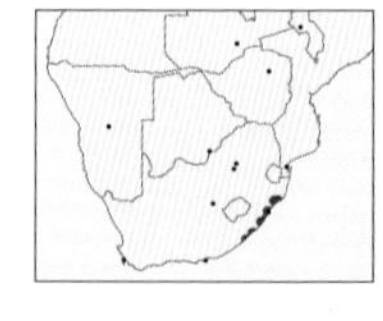

### 7 *Bonaberiana crassisquama* Purple Tabby

Forewing 15–16mm. Wings held apart exposing hindwings; greyish-brown with a violet sheen in live individuals; forewings with a diagnostic median complex of markings that vary in colour from light pink to deep purple, distinctive cream reinform before cell. Caterpillar a light green semi-looper. One species in genus in region. **Biology:** Caterpillars feed on *Monanthotaxis caffra* (Annonaceae). Adults throughout the year. **Habitat & distribution:** Moist lowland forest; Eastern Cape, South Africa, north to tropical Africa.

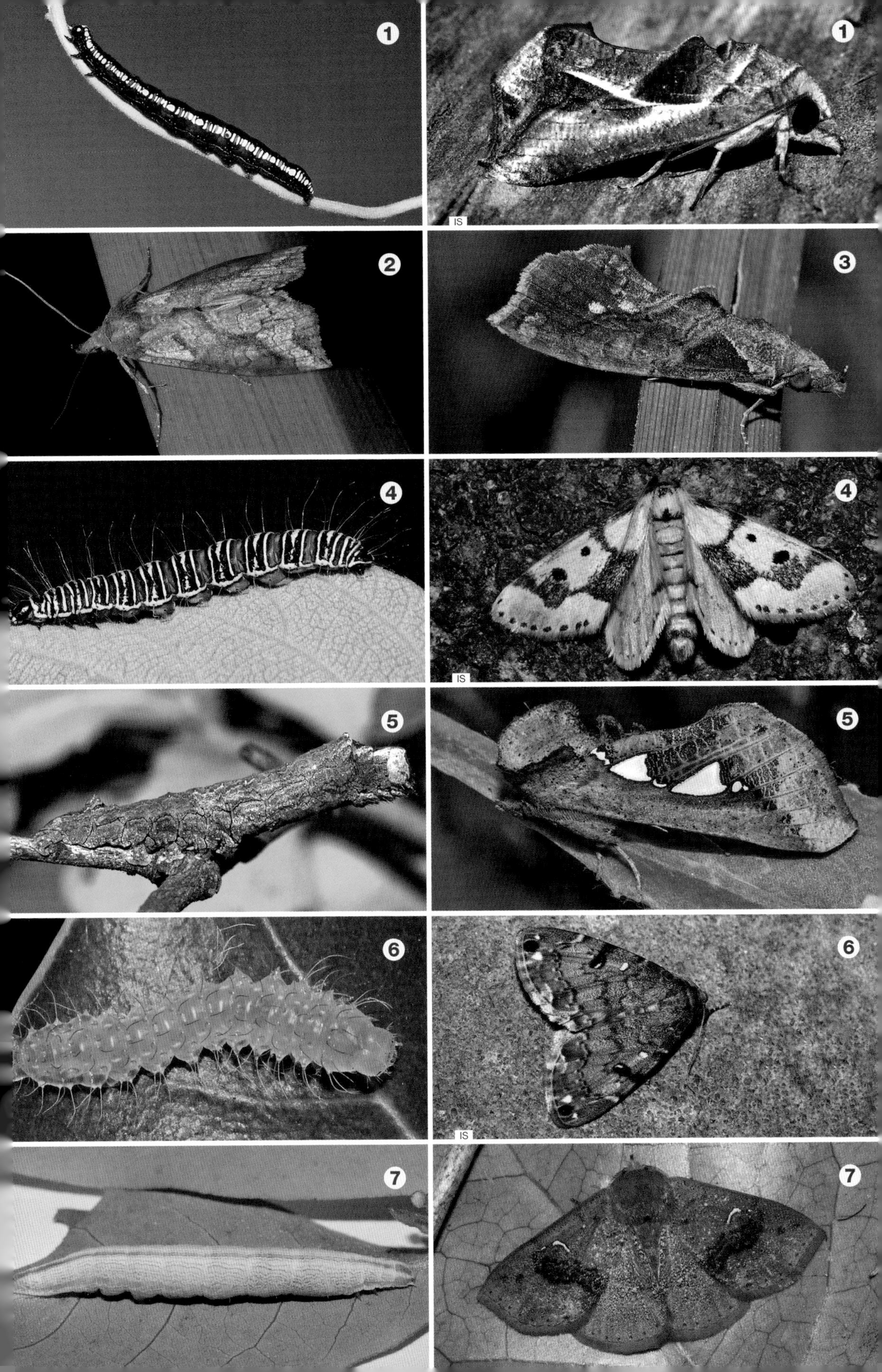
1
1
IS
2
3
4
4
IS
5
5
6
6
IS
7
7

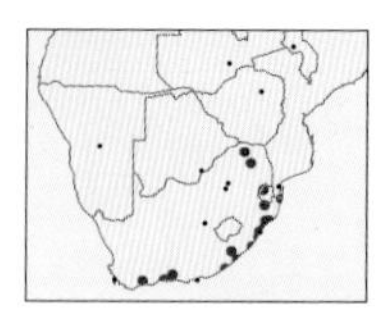

## 1 *Disticta atava* Dotted Ancestor

Forewing 12–16mm. Rests with body at angle to substrate with head downwards, prominent upturned snout, front legs held forward and often widely splayed at rest; fore- and hindwings similar, crossed by sinuous dark lines, postmedian separating the wings into tan basal and grey outer regions, forewings with 3 diagnostic black-ringed, white dots, the basal spot often indistinct. One species in genus in region. **Biology:** Caterpillar host associations unknown. Adults September–April. **Habitat & distribution:** Moist forests; Western Cape, South Africa, north to Mozambique.

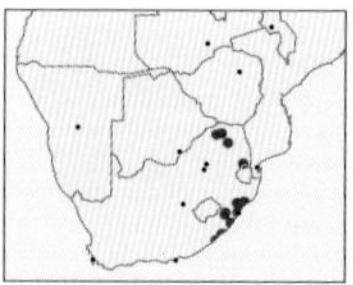

## 2 *Maxera zygia* Variable Ancestor

Forewing 14–17mm. Light grey to greenish-brown, lines and marks distinct; fore- and hindwings similar, postmedian line diagnostically curving inward on forewing costal region, in some forms consisting of a number of closely spaced dots. Caterpillar slender, apple or olive green with indistinct yellow longitudinal stripe. Five species in genus in region. **Biology:** Caterpillars feed on Coastal Climbing Thorn (*Senegalia kraussiana*). Adults throughout the year, active by day amongst the leaf litter in forest understorey. **Habitat & distribution:** Moist forests; Eastern Cape, South Africa, north to Mozambique and Zimbabwe.

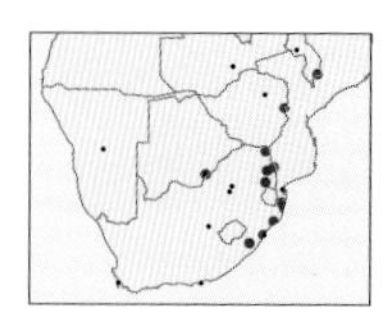

## 3 *Maxera marchalii* Grey Ancestor

Forewing 12–15mm. Variable, shades of grey, fading in dried specimens; diagnostic diagonal median band across fore- and hindwings, augmented in females by bands and rows of dots. Caterpillar pale dorsally, with black and white longitudinal stripes, white lateral line, dark ventrally. Similar *M. brachypecten* straw, not grey, and lacks prominent median band. **Biology:** Caterpillars feed on Apple-leaf (*Philenoptera violacea*). Adults throughout the year. **Habitat & distribution:** Lowland bushveld; KwaZulu-Natal, South Africa, north to tropical Africa and Madagascar, Indian Ocean islands.

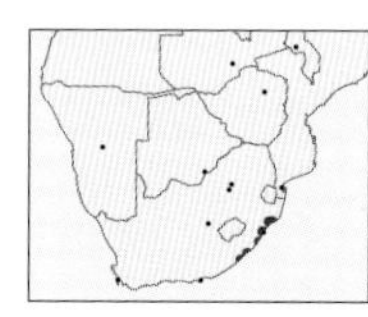

## 4 *Tatorinia fumipennis* Slender Snout

Forewing 17–18. Similar to *Maxera zygia* (above), and occupies same habitat on forest floor, but larger, with postmedian line diagnostically straight, ending before costa; exceptionally long, slender palpi; light to dark brown. One species in genus in region. **Biology:** Caterpillars feed on *Philenoptera sutherlandii* (Fabaceae). Adults throughout the year, active by day amongst the leaf litter in forest understorey. **Habitat & distribution:** Coastal bush and lowland forest; Eastern Cape, South Africa, north to Kenya.

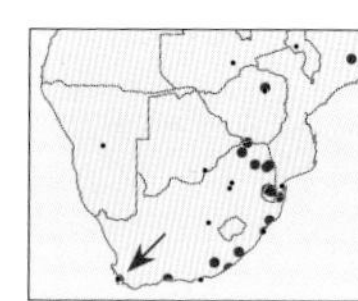

## 5 *Egnasia vicaria* Vicarious Snout

Forewing 11–15mm. Rests with wings behind its back and legs protruded; upper side has same markings as underside, but paler, colour varying from very pale to deep buff, wings with scalloped margins, hayline spots around cell and broad transverse band delineated by thin black lines, apical part of wings with darker band. Caterpillar light green with yellow sides and a yellow band on each segment. Two species in genus in region; similar *E. lioperas* has wings plain fawn with fine black dusting. **Biology:** Caterpillars feed on various Rubiaceae trees and shrubs. Multivoltine, adults throughout the year. **Habitat & distribution:** Variety of frost-free habitats where Rubiaceae trees grow; Western Cape, South Africa, north to tropical Africa, also Madagascar and most of sub-Saharan Africa.

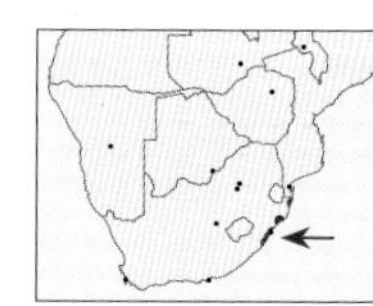

## 6 *Facidia vacillans* Cinnamon Snout

Forewing 20–23mm. Uniform cinnamon with bluish sheen fading to dark brown in dried specimens; forewings broad, rounded, lines punctuated with black and white spots, basal and antemedial line merge near costa, postmedian line incomplete ending before inner margin with 1 or 2 larger cream spots; hindwings with short, incomplete zigzag line. Caterpillar a grey twig mimic, resting with last part of abdomen bent and raised upward. One species in genus in region. **Biology:** Caterpillars feed on Dwaba Berry (*Monanthotaxis caffra*). Adults throughout the year. **Habitat & distribution:** Coastal bush and lowland forest; KwaZulu-Natal, South Africa, north to tropical Africa.

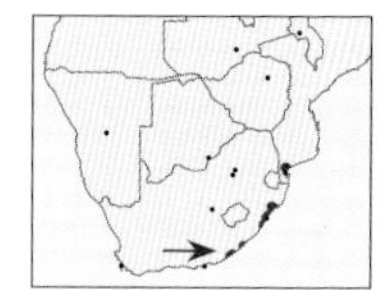

## 7 *Meliaba pelopsalis* Unreal Snout

Forewing 17–18mm. Adult rests at an angle, with head away from substrate, this together with raised setae on palpi and crest on thorax make it appear unlike any other moth; forewing colouration variable with dark diagonal brown line with silver arrow eyespot; hindwings uniform brown. Caterpillar thin, dull green with small yellow spots on head capsule and body. One species in genus in region. **Biology:** Caterpillars feed on Natal Guarri (*Euclea natalensis*). Adults remarkable leaf litter mimics. **Habitat & distribution:** Coastal bush and lowland forest; Eastern Cape northeastwards to Tanzania.

1
2
3
3
4
5
5
6
6
SB
7
7

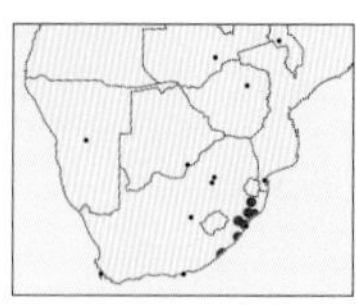

### 1 *Rhanidophora aurantiaca* — Chocolate Dice Moth

Forewing 23–25mm. Wings chocolate brown fading to pale yellow on edges in older specimens, forewing with 3 large oval white spots surrounded by indistinct darker scales together with a single subterminal row of distinct whitish dots diagnostic; hindwings orange. Caterpillar a semi-looper, striped in irregular black and white bars and with conspicuous black and white spatulate hairs. Similar *R. odontophora* is paler and has 2 rows of subterminal spots. Eight species in genus in region. **Biology:** Caterpillars feed on Black-eyed Susan (*Thunbergia alata*). Multivoltine, adults throughout the year. **Habitat & distribution:** Moist forests; Eastern Cape north to KwaZulu-Natal, South Africa, and East Africa.

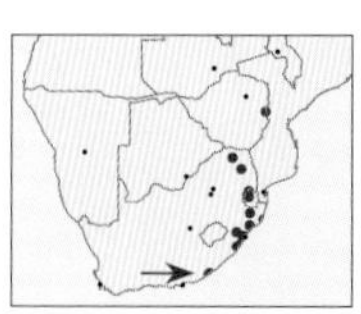

### 2 *Rhanidophora cinctigutta* — Gold Dice Moth

Forewing 17–20mm. Forewings cinnamon to gold, with 3 large white spots encircled with distinct black line, pair nearest wing apex not touching, indistinct row of white terminal spots in some individuals; hindwings uniform golden. Caterpillar a semi-looper with alternating black and white bands and long spatulate hairs. **Biology:** Caterpillars feed on various *Thunbergia* spp. Multivoltine, adults throughout the year. **Habitat & distribution:** Forests and gardens; Eastern Cape, South Africa, north to Ethiopia.

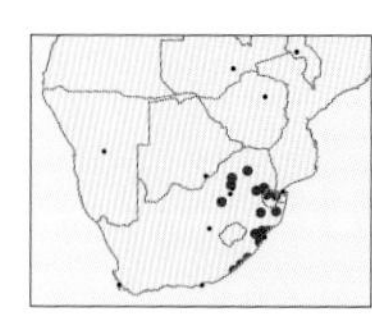

### 3 *Rhanidophora phedonia* — Silver Dice Moth

Forewing 14–19mm. Forewings silvery cream, similar to *R. cinctigutta* (above), but posterior pair of white spots touching or almost touching; hindwings uniform yellow. Caterpillar with alternating black and white-and-yellow wavy bands, thin white hairs arising from back. **Biology:** Caterpillars feed on various grassland *Thunbergia* spp. Bivoltine, adults October–March. **Habitat & distribution:** Grassland; Eastern Cape north to Limpopo, South Africa.

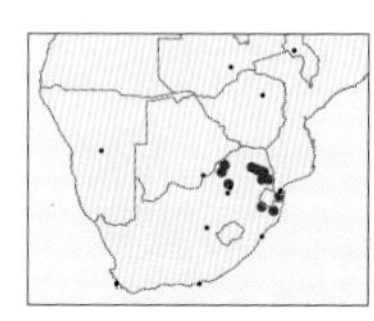

### 4 *Rhanidophora ridens* — Lead Dice Moth

Forewing 17–20mm. Forewings light lead-grey or flesh-coloured with 3 diagnostic yellow spots, the pair nearest the wing tip not touching; hindwings sulphur yellow with dark marginal band. Caterpillar a semi-looper with broad black and narrower white bands, and sparse, long, spatulate black hairs. Similar *R. flavigutta* has hindwings plain yellow. **Biology:** Caterpillars feed on *Thunbergia* spp. Bivoltine, adults October–April. **Habitat & distribution:** Bushveld; KwaZulu-Natal, South Africa, north to tropical Africa.

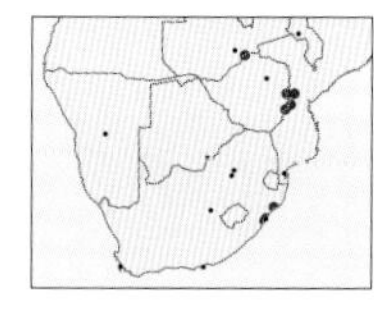

### 5 *Eudocima divitiosa* — Spotless Underwing

Forewing 39–46mm. Bulky; forewings with inner margin concave, displaying orange hindwing at rest, sinuous postmedian line ending on costa well before apex; forewing plain brown to mottled shades of black, silver and green markings; hindwings orange with black border, chequered white cilia and lacking any cell spot. Caterpillar, bizarre shape, creamy with 2 conspicuous eyespots. Four species in genus in region. **Biology:** Caterpillars polyphagous often on Menispermaceae. Adults throughout the year. **Habitat & distribution:** Tropical forests; KwaZulu-Natal, South Africa, north to tropical Africa.

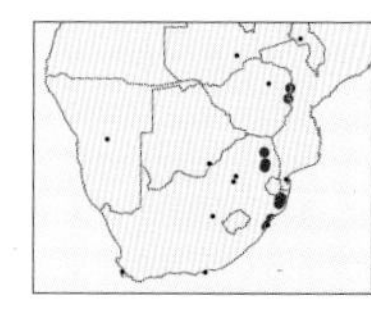

### 6 *Eudocima phalonia* — Comma Underwing

Forewing 38–44mm. Large, bulky; forewings with similar variation in colouration to *E. divitiosa* (above), but postmedian line ending on apex of wing; hindwings orange with black border, chequered white cilia and diagnostic comma-shaped cell spot, rarely enlarged spreading to inner margin. **Biology:** Caterpillars polyphagous often on Menispermaceae. Adults throughout the year. **Habitat & distribution:** Adults participate in the great autumn migrations, can then be found way out of habitat to west coast desert and in the middle of oceans; forest; KwaZulu-Natal, South Africa, north to tropical Africa.

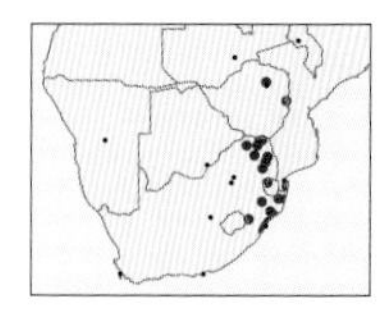

### 7 *Eudocima materna* — Dot Underwing

Forewing 35–45mm. Bulky head with violet-blue sheen; forewing white, finely striated with brown streaks, grizzled in grey, brown and white, often with silver or black panels defined by branching white streak; hindwings orange with black border, chequered white cilia and round black cell spot. Caterpillar charcoal with white speckling and a pair of orange and blue eyespots, holds up tail in bizarre posture. **Biology:** Caterpillars polyphagous often on Menispermaceae. Adults throughout the year. **Habitat & distribution:** Adults participate in the great autumn migrations, can then be found way out of habitat; forest; KwaZulu-Natal, South Africa, north to tropical Africa.

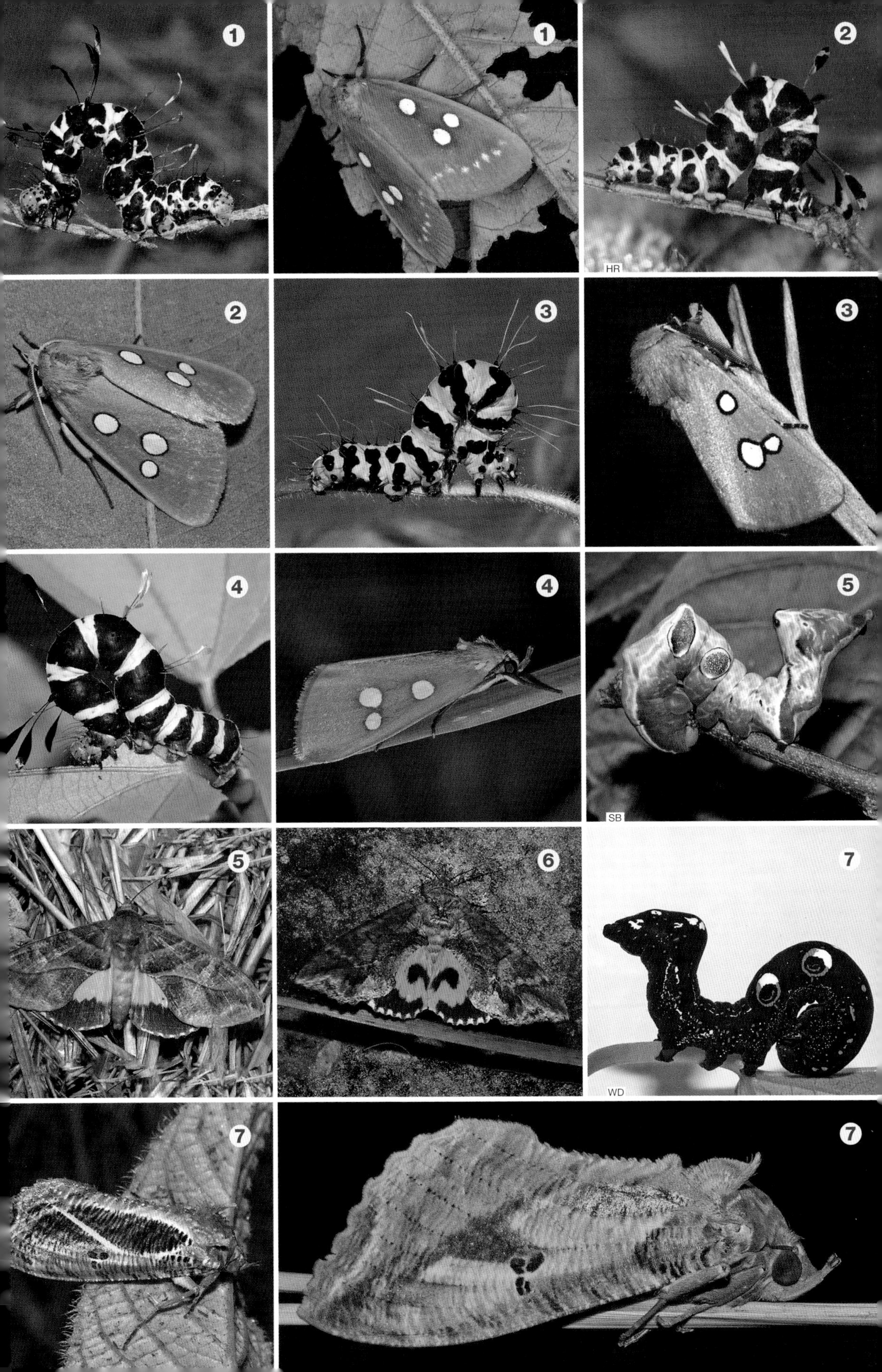

1
1
2
HR
2
3
3
4
4
5
SB
5
6
7
WD
7
7

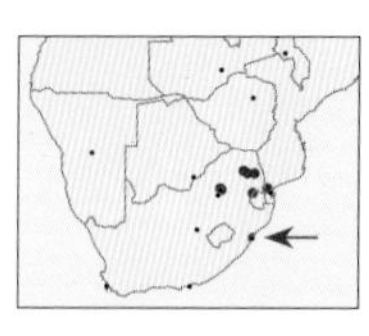

### 1 *Arsina vausema* — Black-heart Snout

Forewing 12mm. Brownish-grey; forewings with sinuous postmedian line, large dark heart-shaped marking near base of wings, which can vary from a comma shape to almost triangular, as well as 2 irregular black-ringed circles in cell; hindwings grey. Caterpillar a slender green semi-looper. One species in genus in region. **Biology:** Caterpillars feed on Quinine Tree (*Rauvolfia caffra*). Adults October–April. **Habitat & distribution:** Bushveld; KwaZulu-Natal, South Africa, north to Mozambique and in Kenya.

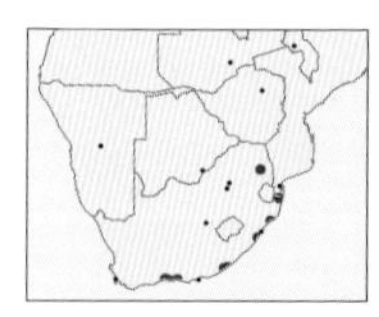

### 2 *Hondryches phalaeniformis* — Broken Snout

Forewing 12–14mm. Palps narrow and projecting upward; forewings with deeply scalloped margin, mustard yellow, densely covered in reddish-brown scales, antemedial and postmedian lines straight, then curved near costa lines and large black eyespot variably covered in grey scales; hindwings pale yellow. One species in genus in region, similar *Paralephana argyresthia* has postmedian line not curved before costa and no cell spot. **Biology:** Caterpillars feed on Cat-thorn (*Scutia myrtina*). Adults throughout the year. **Habitat & distribution:** Moist forests; Western Cape north to Limpopo, South Africa and Kenya.

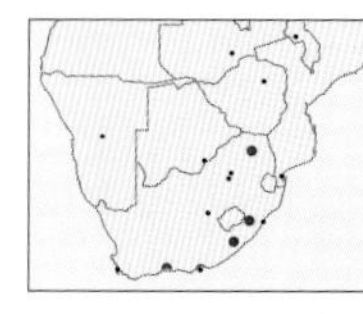

### 3 *Tolna limula* — Plain Forest Snout

Forewing 23–26mm. At rest, wings flat over body; head and body dark brown, wings crenulated, dark brown, jagged antemedial and postmedian lines projecting sharply outward before curving to costa forming a funnel band, median area with dark purple or silver scales and brown between on costa; hindwings dark brown. Caterpillar mottled brown, red and white. Eight species in genus in region. **Biology:** Caterpillars feed on Cape-hazel (*Trichocladus crinitus*). Adults throughout the year. **Habitat & distribution:** Cool moist forests; Western Cape north to Limpopo, South Africa, also reported from Kenya.

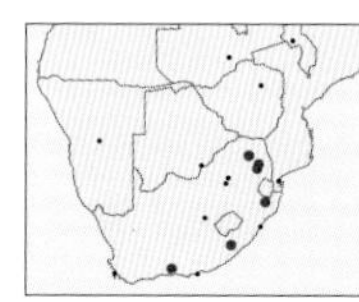

### 4 *Tolna variegata* — Coloured Forest Snout

Forewing 23–25mm. At rest, wings flat over body; wings crenulated, similar to *T. limula* (above), but with diagnostic prominent fawn and black collar over thorax and square patch midway on costa well defined with round eyespot, variable white lines and marks over wing; hindwings uniform charcoal. Similar *T. sypnoides*, from Zimbabwe northwards, lacks the square patch and has extensive white shading. **Biology:** Caterpillar host associations unknown. Adults throughout the year. **Habitat & distribution:** Cool moist forests; Western Cape north to Limpopo, South Africa.

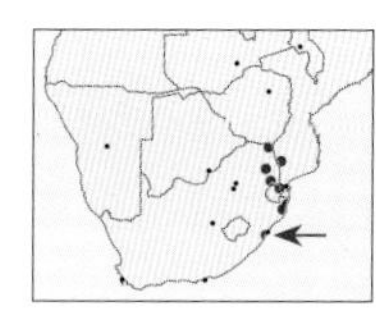

### 5 *Parafodina pentagonalis* — Pentagonal Snout

Forewing 14–15mm. Resembles Triangle species in Erebinae (below), but prominent upturned palpi define it as an underwing; deep chocolate brown with white frons and collar over thorax; at rest, yellow-orange antemedial and postmedian lines form a pentagonal shape that extends over thorax, filled with black, yellow marginal line; hindwings brown with curved cream line. Two species in genus in region. **Biology:** Caterpillar host associations unknown. Adults November–February. **Habitat & distribution:** Lowland forests and riverine vegetation; KwaZulu-Natal, South Africa, north to Kenya.

## Subfamily Erebinae Large Owls

Generally broad-winged nocturnal erebids with smaller upturned palpi than in Calpinae. Adults of many species often abundant in late summer, participating in the annual migrations. There are 298 species recorded from the region.

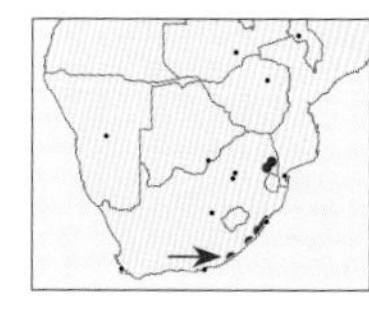

### 6 *Catephia amplificans* — White Underwing

Forewing 20–24mm. Grey to rich brown; forewings with large light and dark brown patches and indistinct transverse wavy lines; hindwings brown or grey with 2 large white crescents. Caterpillar mottled green with faint grey stripe running along body and small upright projections arising from end of body. Sixteen species in genus in region. **Biology:** Caterpillars feed on Lathberry (*Eugenia cordata*), Dune False-currant (*Allophylus natalensis*) and Water Berry (*Syzygium cordatum*). Adults throughout the year. **Habitat & distribution:** Forests; Eastern Cape north to Mpumalanga, South Africa.

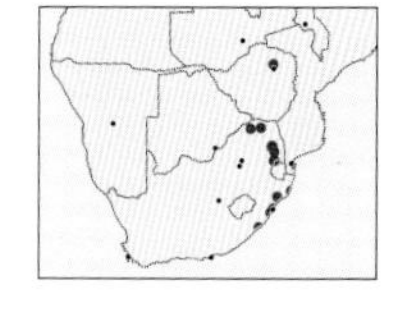

### 7 *Lacera alope* — Toothed Drab

Forewing 20–23mm. Body stout, resting with wings over body; fore- and hindwings irregularly scalloped to give a broken outline; wings brown with complex darker marbling and transverse black line enclosing cream patch on trailing edge of forewings; hindwings with cream and black transverse lines and darker margins. Caterpillar a pale green or brown semi-looper, with short red horns at end of abdomen. **Biology:** Caterpillars feed on Redwing (*Pterolobium stellatum*). **Habitat & distribution:** Forest and riverine bush; Eastern Cape, South Africa, north to tropical Africa.

1
1
IS
2
3
3
4
5
6
SB
6
IS
7
7

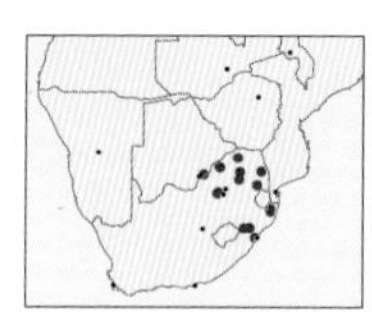

## 1 *Hypotacha retracta* — Hookthorn Drab

Forewing 12–13mm. Forewings grey, dusted with black scales, black antemedial and postmedian lines form a glass shape, sometimes with black spot, partially or completely obscured in some individuals. Caterpillar a perfect twig mimic of its host plant, includes the 'hook thorn' as a dorsal process. Eight species in genus in region. **Biology:** Caterpillars feed on Hook Thorn (*Senegalia caffra*). Adults August–May. **Habitat & distribution:** Bushveld and grassland; distribution matches that of host plant, KwaZulu-Natal, South Africa, north to Botswana, and Waterberg, Namibia.

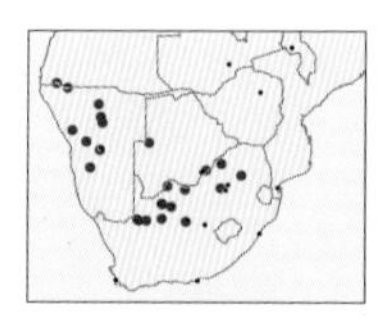

## 2 *Hypotacha isthmigera* — Hourglass Drab

Forewing 10–11mm. Similar to *H. retracta* (above), but smaller and greyer with less dark dusting and with distinct hourglass shape formed by antemedial and postmedian lines, which is obscured in some individuals. **Biology:** Caterpillar host associations unknown. Adults August–May. **Habitat & distribution:** Arid bushveld; Northern Cape, South Africa, north to most arid areas in Africa south of Sahara.

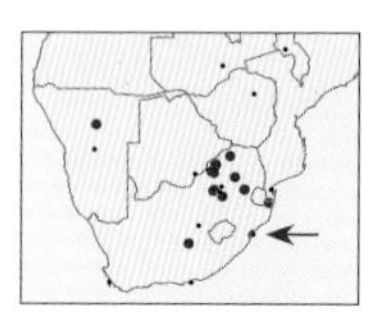

## 3 *Tachosa fumata* — Blackring Drab

Forewing 17–20mm. Similar in colouration and markings to *H. retracta* (above), but much larger; forewings mottled grey to brown, darker basally; 1 or 2 zigzag sinuous transverse bands with indistinct circle, sometimes obscured, between them. Caterpillar cryptic, pale green-grey-black with lichen-like projections arising from abdomen. Two species in genus in region. **Biology:** Caterpillars polyphagous on various trees. Adults throughout the year. **Habitat & distribution:** Forest, kloofs and bushveld; KwaZulu-Natal, South Africa, north to most of Africa.

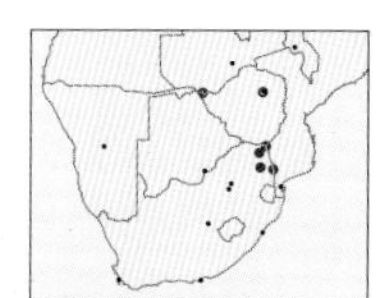

## 4 *Crypsotidia bibrachiata* — Bark Drab

Forewing 11–12mm. Forewings fawn, with variable black tip near apex leading to brown subterminal bar, antemedial line straight, postmedian line diagnostic, sharply turned back and around cell before costa; hindwings off-white with darker margins. Caterpillar elongate, uniform green. Three species in genus in region. **Biology:** Caterpillars feed on Ana Tree (*Faidherbia albida*). Adults October–April. **Habitat & distribution:** Bushveld; Limpopo, South Africa, north to Zimbabwe and Tanzania.

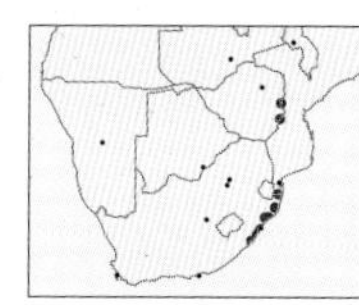

## 5 *Audea bipunctata* — Two-spotted Underwing

Forewing 22–25mm. Forewing light brown to almost black, lighter forms with 2 darker complete or incomplete bands; hindwings white, with dark marginal band. Grey caterpillar a bark mimic with fringes arising from side of body to reduce shadow. Eight species in genus in region. **Biology:** Caterpillars feed on Flat-crown (*Albizia adianthifolia*) and Peacock Flower (*Albizia gummifera*). Multivoltine, adults throughout the year. **Habitat & distribution:** Moist lowland forests; KwaZulu-Natal, South Africa, north to tropical Africa.

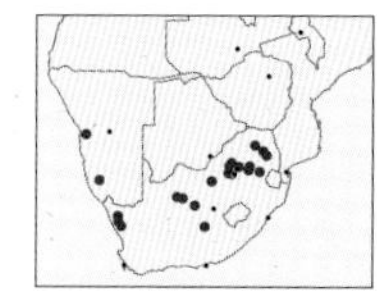

## 6 *Ctenusa varians* — Comma Drab

Forewing 19–29mm. Body and forewings deep cream to dark brown with thin, grey, curved subterminal band, postmedian line dotted when present; middle of forewings with large dark brown comma-shaped mark, variably extended to base in some individuals; hindwings white to cream with dotted postmedian. Caterpillar brown, with brown and dark grey stripes running along body. Three species in genus in region. **Biology:** Caterpillars feed at night on Sweet Thorn (*Vachellia karroo*), hide in bark crevices by day, take over 7 months to mature. Univoltine, adults July–November, primarily September. **Habitat & distribution:** Arid bushveld; Northern Cape, South Africa, north to tropical Africa.

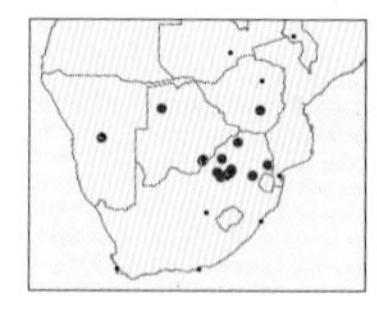

## 7 *Ctenusa pallida* — Dash Drab

Forewing 18–23mm. Similar to *C. varians* (above), but subterminal band straight, wings broader and forewing mark smaller and dash-shaped; variable, ground colour pink to fawn, some with lines well marked and others with lines almost absent. **Biology:** Caterpillar host associations unknown. Adults September–May. **Habitat & distribution:** Bushveld; Mpumalanga, South Africa, north to tropical Africa.

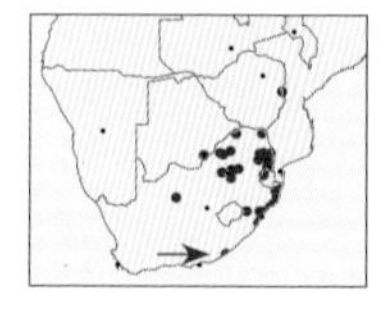

## 8 *Dysgonia angularis* — Angular Jigsaw

Forewing 14–18mm. Forewings grey to dark brown, with antemedial line evenly curved inward and postmedian line twice curved with sharp point, median area muddled between cream and black areas; hindwings grey or brown. Caterpillar elongate, grey or fawn, small black dorsal tubercle near end of body. Fourteen species in genus in region. **Biology:** Caterpillars feed on leafflower species (*Phyllanthus* spp.). Multivoltine, adults throughout the year. **Habitat & distribution:** Bushveld and forest; Eastern Cape, South Africa, north to tropical Africa and Indian Ocean islands, including Madagascar.

1
1
2
3
3
4
4
IS
5
SB
5
6
6
IS
IS
7
8
8

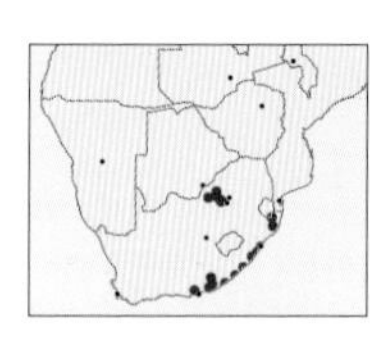

### 1 *Dysgonia properans* — Narrow Jigsaw

Forewing 14–18mm. Similar to other *Dysgonia* spp., but has antemedial line less evenly curved, median area clearly defined between dark and light areas; shape of postmedian line diagnostic, brown median area is narrow, ending in a point on inner margin; hindwings with white median band. **Biology:** Caterpillars feed on *Acalypha glabrata* (Euphorbiaceae). Adults September–March. **Habitat & distribution:** Forest and wooded kloofs; Eastern Cape, South Africa, north to North West and Angola and Tanzania.

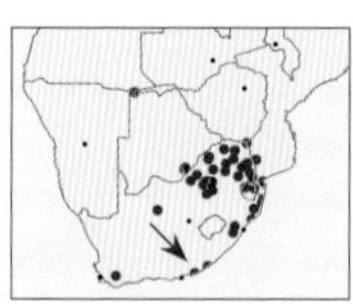

### 2 *Dysgonia torrida* — Broad Jigsaw

Forewing 16–22mm. Similar to other *Dysgonia* spp., but larger with dark areas darker brown, antemedial line evenly curved to just before costa where it is kinked, the median area clearly defined between dark and light areas; shape of the postmedian line diagnostic, brown median area is broad, ending in point on inner margin; hindwings with white median band. Caterpillar fawn, may have a pair of dorsal brown stripes running along body. **Biology:** Caterpillars polyphagous mainly on trees. Multivoltine, adults throughout the year. **Habitat & distribution:** Variety of habitats, including gardens; Western Cape, South Africa, north to tropical Africa, Arabia, Indian Ocean islands, Madagascar, southern Europe to Eurasia.

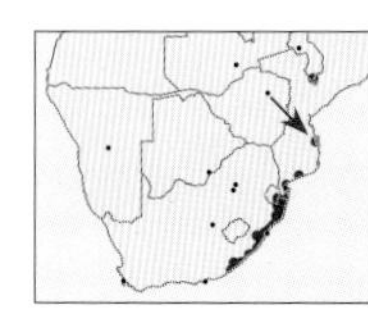

### 3 *Egybolis vaillantina* — Peach Moth

Forewing 21–28mm. Body and wings velvety blue-green, head and antennae yellow; forewings with basal orange bar outlined in black and 2 variable smaller orange spots halfway along wing; hindwings almost black with broad blue border. Caterpillar black to purplish, with white setae and variable yellow banding. **Biology:** Adults fly slowly by day, often around treetops. Caterpillars feed primarily on Dune Soapberry (*Deinbollia oblongifolia*), also other soapberries (*Sapindus* spp.). **Habitat & distribution:** Coastal bush and lowland forest; Eastern Cape, South Africa, north to Kenya.

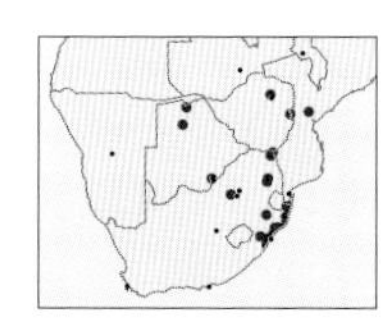

### 4 *Entomogramma pardus* — Divided Drab

Forewing 20–24mm. Mustard variably dusted with darker scales, yellow underside, diagnostic postmedian line sharply turned back to costa but extended yellow line complete the 'leaf midrib' when at rest, lighter veins and darker lines running back to costa completes the illusion. One species in genus in region. **Biology:** Caterpillar host associations unknown. Adults throughout the year. **Habitat & distribution:** Forests and riverine vegetation; KwaZulu-Natal, South Africa, north to tropical Africa.

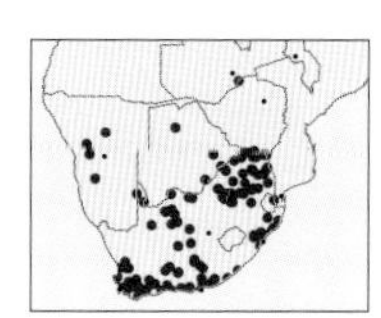

### 5 *Grammodes stolida* — Stolid Lines

Forewing 12–17mm. Members of genus have forewings brown with 2 bold cream bars, the outer one with tan edge and 3 black bars, subterminal reduced, median black bar tapered and diagnostically deeply indented; hindwings grey with diagonal white bar and distinct spot. Caterpillar grey with black and cream stripes running along body. Eight species in genus in region. **Biology:** Caterpillars polyphagous. Multivoltine, adults throughout the year. **Habitat & distribution:** Common in a variety of habitats, including forest, desert and transformed habitats; Western Cape north across Africa, Indian Ocean islands and Madagascar, southern Europe, Asia and Australia.

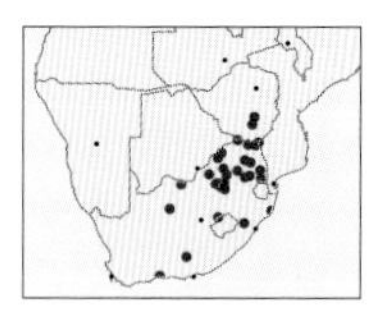

### 6 *Grammodes exclusiva* — Exclusive Lines

Forewing 12–14mm. Similar to *G. stolida* (above), but median black bar rectangular, not tapered, with slight kink on outer middle edge; hindwings lack white outer spot. **Biology:** Caterpillar host associations unknown. Adults throughout the year. **Habitat & distribution:** Variety of habitats; Western Cape north across Africa, and Madagascar.

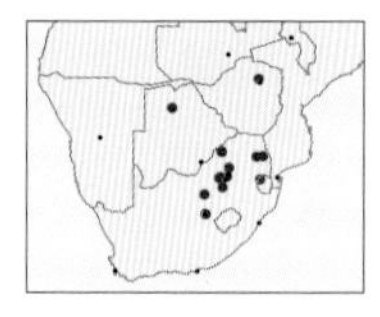

### 7 *Grammodes bifasciata* — Parallel Lines

Forewing 17–19mm. Much larger than *G. exclusiva* (above), and median black bar rectangular with sharp edges, no kinks, not tapered; hindwings with white bar but without white outer spot. Similar *G. congenita* has median black bar rounded, slightly oval and white lines browner. **Biology:** Caterpillars polyphagous. Adults throughout the year. **Habitat & distribution:** Variety of habitats; Free State north across Africa, Madagascar; described from Italy.

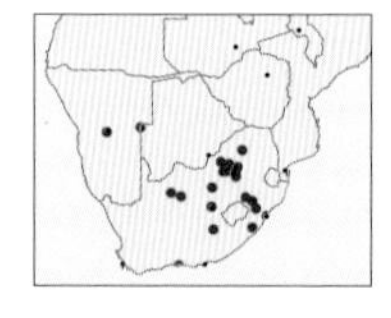

### 8 *Grammodes euclidioides* — Yellow Lines

Forewing 12–14mm. Similar to *G. stolida* (above), but median black bar gently tapered, only slightly kinked; hindwings with orange inner and broad black border with orange to yellow line. **Biology:** Caterpillar host associations unknown. Adults throughout the year. **Habitat & distribution:** Grassland and bushveld; Western Cape, South Africa, north to tropical Africa.

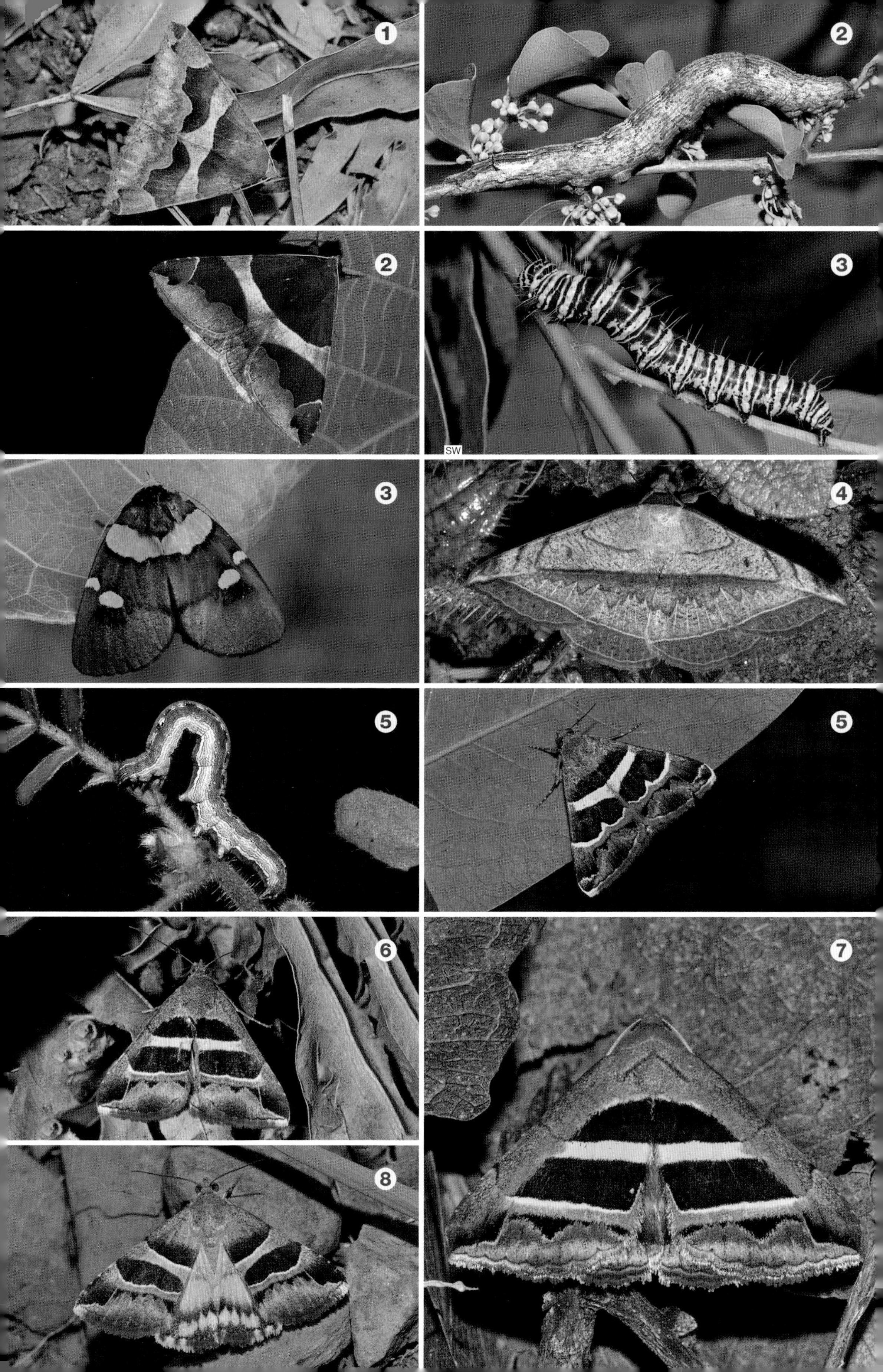
1
2
2
3
SW
3
4
5
5
6
7
8

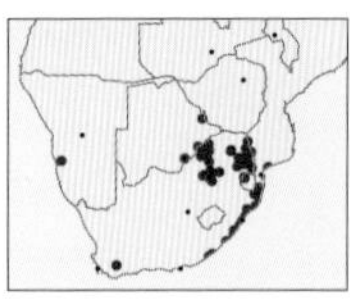

### 1 *Trigonodes exportata* — In-line Triangles

Forewing 15–20mm. Ground colour dusty grey; forewings dominated by 2 distinct black triangles bordered by cream on lower angle, diagnostically the top edge of the triangles are connected or if not connected in line with each other, dark brown shading in subterminal area; hindwings brown with faint line. One species in genus in region. **Biology:** Caterpillar host associations unknown. Adults throughout the year. **Habitat & distribution:** Variety of habitats; Western Cape, South Africa, north to Ethiopia.

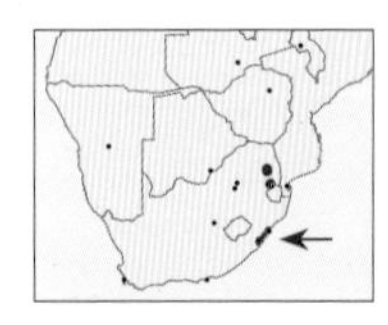

### 2 *Parachalciope mahura* — Out-of-line Triangles

Forewing 17–22mm. Superficially very similar to *Trigonodes exportata* (above), but the 2 triangles on forewings are not connected and upper edges are not in line with each other, usually a white line surrounds triangle, also larger and lighter in colour; hindwings uniform brown. Three species in genus in region. **Biology:** Caterpillar host associations unknown. Adults throughout the year. **Habitat & distribution:** Moist forests; KwaZulu-Natal, South Africa, north to tropical Africa.

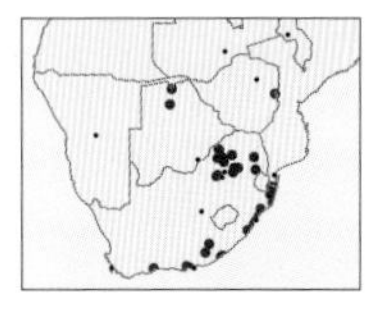

### 3 *Chalciope delta* — White Triangle

Forewing 13–15mm. Frons and cilia white; ground colour chocolate brown, lighter around margins of forewings, 3 diagnostic white lines forming a triangle, otherwise no markings; hindwings fawn shaded to brown. Three species in genus in region. **Biology:** Caterpillar host associations unknown. Adults throughout the year. **Habitat & distribution:** Warm grassland, coastal bush and bushveld; Western Cape, South Africa, north to East Africa and Madagascar.

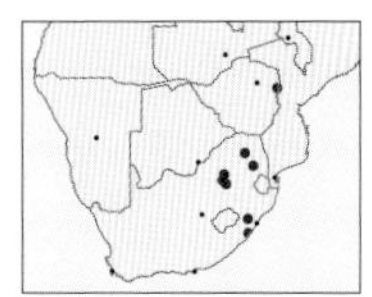

### 4 *Hypanua xylina* — Wood Wing

Forewing 27–28mm. Body light brown; forewings shaded light brown to dark brown, straight postmedian line diagnostic, antemedial twice kinked and subterminal crenulated, small cell spot and kidney-shaped spot beyond; hindwings charcoal. Caterpillar elongate, flesh-coloured with light marbling. Two species in genus in region. **Biology:** Caterpillars feed on River Bushwillow (*Combretum erythrophyllum*). Adults throughout the year. **Habitat & distribution:** Bushveld, and forest, where host plant grows near rivers; KwaZulu-Natal, South Africa, north to Kenya.

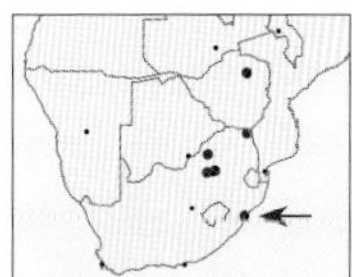

### 5 *Hypanua roseitincta* — Rosy Wood Wing

Forewing 23–26mm. Forewings uniform rosy pink, fading to brown in dried specimens, postmedian line curved inward before costa, small cell spot and thin kidney-shaped mark, diagnostic, distinct black and white arrow point on white dotted subterminal line; hindwings have tan shading to dark border. Caterpillar an elongate and slender light brown twig mimic with short dorsal process toward rear of body. **Biology:** Caterpillars feed on bushwillows (*Combretum* spp.). Bivoltine, adults October–April. **Habitat & distribution:** Forest and bushveld; KwaZulu-Natal, South Africa, north to tropical Africa.

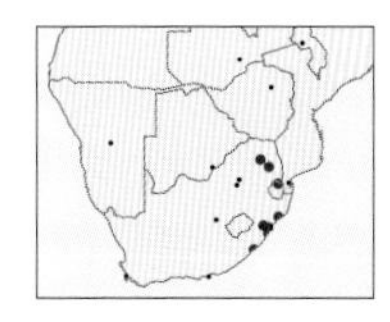

### 6 *Thyas rubricata* — Orange Wood Wing

Forewing 23–27mm. Head, thorax and rectangular forewings rufous; straight forewing lines slightly tapering out toward costa, elongate cell spot; hindwings orange to pink, brown on inner fold and incomplete black band, long androconial hairs on inner fold in males. Three species in genus in region. **Biology:** Caterpillars feed on *Combretum molle* (Combretaceae). Adults throughout the year. **Habitat & distribution:** Moist forests; Eastern Cape, South Africa, north to tropical Africa.

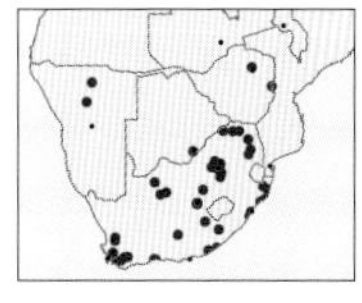

### 7 *Ophiusa tirhaca* — Green Wood Wing

Forewing 24–30mm. Head, thorax and forewings pale to richer lime green; terminal area light brown, postmedian line curved, spotted when present, 2 black markings on subterminal line; hindwings yellow-orange with or without dark brown spot or bar. Caterpillar a robust, grey to brown bark mimic, with faint, longitudinal lines. Fourteen species in genus in region. **Biology:** Caterpillars polyphagous on trees and herbs. Bivoltine, adults throughout the year. **Habitat & distribution:** Variety of habitats, including gardens; Western Cape, South Africa, north to tropical Africa.

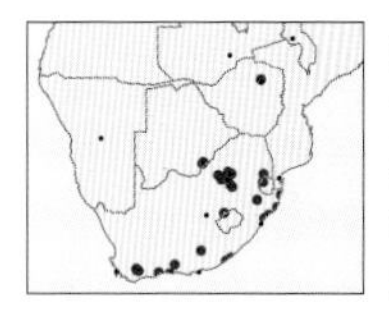

### 8 *Ophiusa dianaris* — Silver Wood Wing

Forewing 21–24mm. Similar to *O. tirhaca* (above), but forewings silvery grey with copper dusting, darker shading before subterminal line and lacks 2 black subterminal markings; hindwings pale yellow with broad black border. Caterpillar a pinkish-grey bark mimic, with extensive longitudinal lines. **Biology:** Caterpillars feed on Blue Guarri (*Euclea crispa*). Bivoltine, adults throughout the year. **Habitat & distribution:** Variety of habitats where *Euclea* trees grow; Western Cape, South Africa, north to tropical Africa.

1
2
3
4
4
IS
5
5
6
7
7
8
8

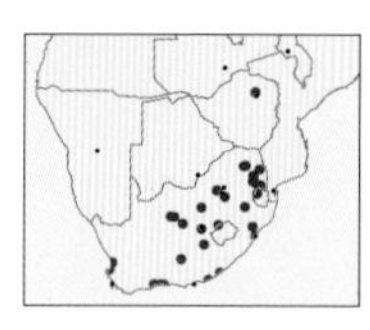

### 1 *Ophiusa selenaris* — Funnel Wood Wing

Forewing 17–20mm. Light brown, dusted with black scales, basal and terminal areas darker, antemedial line with sharp point and postmedian line curved to create funnel-shaped median area, orange to black, sometimes broken, subterminal line; hindwings fawn with indistinct black border. Caterpillar a pinkish-brown bark mimic, with longitudinal lines. **Biology:** Caterpillars feed on Ebenaceae, including *Euclea* and *Diospyros* spp. Bivoltine, adults throughout the year. **Habitat & distribution:** Variety of habitats; Western Cape, South Africa, north to Zimbabwe.

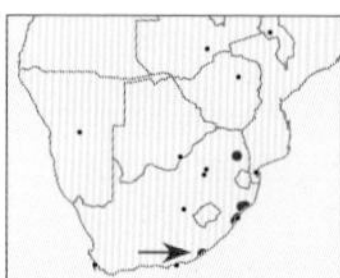

### 2 *Stenopis fumida* — Tail Wood Wing

Forewing 26–30mm. Light brown to dark brown, dusted with black scales, basal and terminal areas darker, diagnostic antemedial and postmedian lines irregularly crenulated, subterminal line dotted with orange and black or just black, sometimes some green, zigzag marginal line; hindwings brown with indistinct black border, white cilia, male has a short tail. Caterpillar slender greenish-brown. Four species in genus in region. Similar *S. hypoxantha* has postmedian line almost straight with 2 even curves before costa. **Biology:** Caterpillars feed on *Empogona lanceolata* (Rubiaceae). Bivoltine, adults throughout the year. **Habitat & distribution:** Lowland forest; Eastern Cape, South Africa, north to tropical Africa.

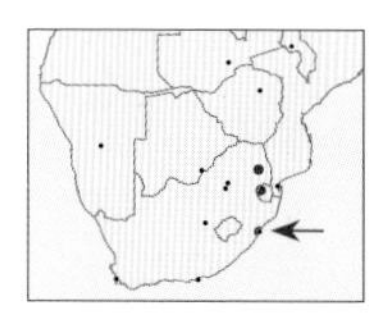

### 3 *Macaldenia palumbiodes* — Spider Eye

Forewing 15–16mm. Fore- and hindwings fawn-coloured, with diagonal darker postmedian line and darker wing tips; hindwings with short row of white dots and adjacent rufous patch, displayed when at rest and resemble spider eyes. Caterpillars elongate, green or greyish-green with dusting of black dots. One species in genus in region. **Biology:** Caterpillars feed on Horsewood (*Clausena anisata*). **Habitat & distribution:** Moist forest; KwaZulu-Natal, South Africa, north to tropical Africa.

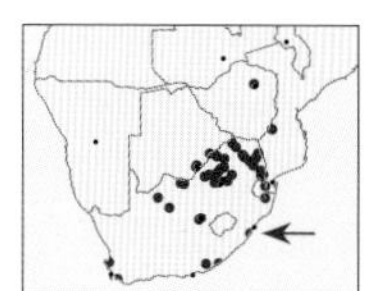

### 4 *Plecopterodes moderata* — Moderate Drab

Forewing 14–17mm. Very variable, or perhaps a complex of species, either grey with ochre band at tip of forewings and hindwings, sometimes with marginal dark band, or brown with darker bands at wing margins; variable diagonal bands on all wings and hindwings light orange at base, diagnostic postmedian line straight, then curved in toward costa. Caterpillar a slender, grey semi-looper with variable longitudinal lines and bands. Seven species in genus in region. **Biology:** Caterpillars feed on bushwillows (*Combretum* spp.). Adults throughout the year. **Habitat & distribution:** Grassland and bushveld where bushwillows grow; Western Cape, South Africa, north to tropical Africa.

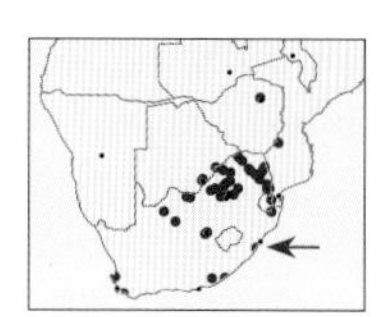

### 5 *Plecopterodes melliflua* — Whitespot Drab

Forewing 13–14mm. Differs from *P. moderata* (above), by being smaller and darker, with postmedian line speckled with white dots, antemedial and subterminal lines as well as black cell spot mostly also with white dots; hindwings vary from brown to dark orange. **Biology:** Caterpillar host associations unknown. Adults October–April. **Habitat & distribution:** Sandy bushveld; Northern Cape north to tropical Africa.

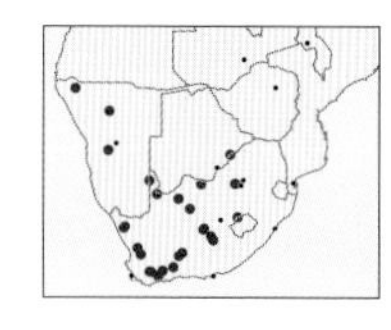

### 6 *Tytroca metaxantha* — Totem Drab

Forewing 13–15mm. Dark brown, diagnostic black, strongly curved antemedial and postmedian lines forming a characteristic rufus median area in males, obscured by a dense dusting of dark scales in females, terminal area almost white in some individuals with zigzag marginal line; hindwings orange with indistinct dark border. Two species in genus in region; similar *T. leucoptera* has antemedial line almost straight with slight kink near costa. **Biology:** Caterpillar host associations unknown. Adults August–May. **Habitat & distribution:** Karoo and arid bushveld; Western Cape, South Africa, north to Namibia.

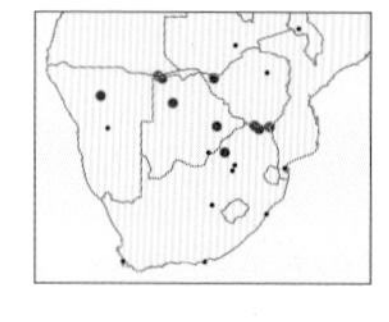

### 7 *Euphiusa harmonica* — False-totem Drab

Forewing 14–16mm. Similar to *Tytroca metaxantha* (above), but jagged lines and dark shading before postmedian line diagnostic, ground colour greenish-fawn with purplish sheen in live specimens; hindwings basally fawn shading to brown, small cell spot, straight antemedial line and partially white cilia. One species in genus in region. **Biology:** Caterpillar host associations unknown. Adults November–April. **Habitat & distribution:** Arid bushveld; Limpopo, South Africa to Namibia, also reported from Kenya and Oman.

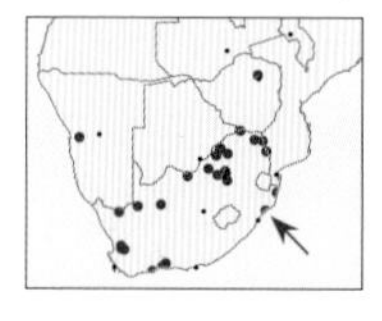

### 8 *Acantholipes circumdata* — Crooked Knob

Forewing 10–12mm. Very similar to *A. trimeni* (p.350), with similar variation but can be separated by the shape of the forewing median black mark encircling cell spot, which is curved and worm-like, not roughly triangular. Four species in genus in region. **Biology:** Caterpillars feed on Malvaceae. Multivoltine, adults throughout the year. **Habitat & distribution:** Variety of open habitats; Western Cape, South Africa, north to tropical Africa, also Arabian Peninsula.

1
1
2
2
3
3
IS
4
4
5
6
7
8

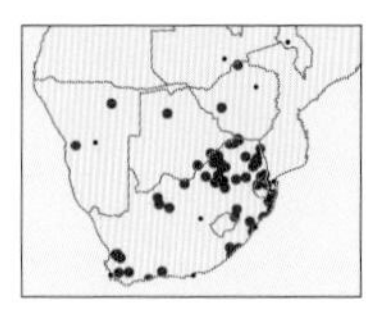

### 1 *Acantholipes trimeni* — Trimen's Knob

Forewing 10–12mm. Forewing with median black, roughly triangular mark encircling cell spot diagnostic; grey form has diagonal orange stripe on forewings bordered with black markings and hindwings with large orange crescent; brown forms with broad, tan diagonal band on all wings, bordered by dark brown areas. Caterpillar a slender, grey or fawn semi-looper, sometimes with black and white longitudinal bands. **Biology:** Caterpillars primarily feed on *Rhynchosia* spp., also other herbaceous Fabaceae. Multivoltine, adults throughout the year. **Habitat & distribution:** Variety of open habitats; Western Cape, South Africa, north to tropical Africa, also São Tomé and Príncipe.

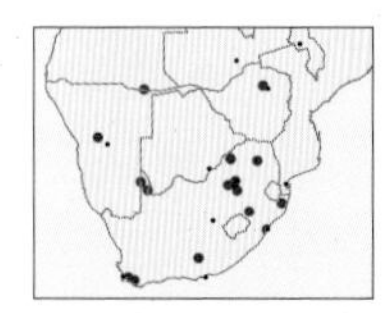

### 2 *Acantholipes namacensis* — Nama Knob

Forewing 9–11mm. Body and wings grey, dusted with black scales, terminal and median area brown in some individuals, lines wavy, subterminal line sometimes with white dots, orange dash in cell diagnostic, lacks median black mark prevalent in other species in genus. Can be confused with some *Plecoptera* spp. (see Anobinae pp. 318, 320). **Biology:** Caterpillars feed on *Rhynchosia monophylla* (Fabaceae). Adults September–May. **Habitat & distribution:** Variety of open habitats; Western Cape, South Africa, north to Zimbabwe and Angola, also reported from Ethiopia.

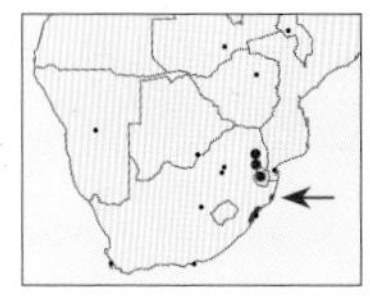

### 3 *Aedia virescens* — Green Eyes

Forewing 13–15mm. Forewings marbled grey or brown, with faint or distinct sinuous line across wings; hindwings brown or grey with distinct white spot near base. Caterpillar plump, uniform light tan, reddish or brown with dark segmental chevrons and 2 pairs of white spots. Three species in genus in region. **Biology:** Caterpillars feed on Tassel-berry (*Antidesma venosum*). Multivoltine, adults throughout the year. **Habitat & distribution:** Forests; KwaZulu-Natal, South Africa, north to Mozambique.

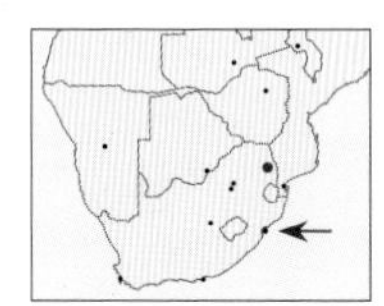

### 4 *Lyncestis mimica* — False Paint

Forewing 14–16mm. Serrated forewing margin diagnostic, very similar to some *Cucullia* spp. (see Noctuidae: Cuculliinae) ground colour grey with black streaks of varying length and thickness running along length of wing, 2 darker streaks near inner margin, most of forewing black in some individuals; hindwings white with broad black border. Caterpillar a plump leaf mimic, greenish-yellow with white longitudinal stripe and hump on top of end of body. **Biology:** Caterpillars feed on Boraginaceae. **Habitat & distribution:** Bushveld; KwaZulu-Natal, South Africa, north to Saudi Arabia.

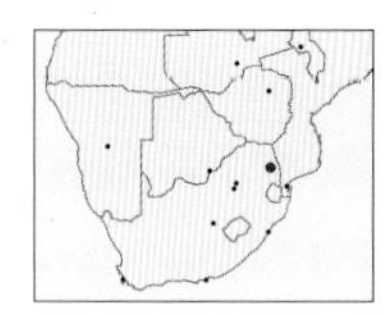

### 5 *Lyncestoides unilinea* — Martian

Forewing 17–21mm. Forewings light grey with prominent black line running length of wing in most specimens, only partially in some. Bizarre caterpillar tan and brown or grey, with remarkable coiled dorsal spire and 2 eyespots near thorax, and humped posterior. One species in genus in region. **Biology:** Caterpillars feed on Grey-leaved Saucer-berry (*Cordia sinensis*) and Sandpaper Saucer-berry (*C. monoica*). Adults December–January. **Habitat & distribution:** Woodlands, savanna and scrub; Limpopo, South Africa, and East Africa to Oman, described from India.

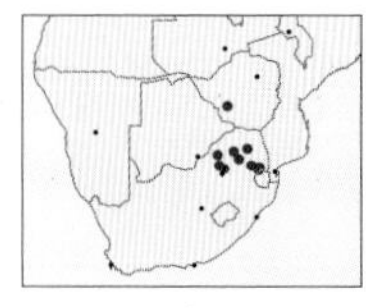

### 6 *Attatha superba* — Superb Custard

Forewing 15–18mm. Body and forewings pinkish-cream, distinctive basal, median and terminal black marks; shape of terminal mark at apex indented near costa and median mark with 1 rounded finger, diagnostic; rows of tiny black lines around lower margin; hindwings pinkish-orange with variable, incomplete black border. Three species in genus in region; similar *A. barlowi* has terminal mark sharply triangular not indented at costa and median mark with 2 sharp fingers. **Biology:** Caterpillar host associations unknown. Adults October–December. **Habitat & distribution:** Bushveld; Gauteng, South Africa, north to Zimbabwe.

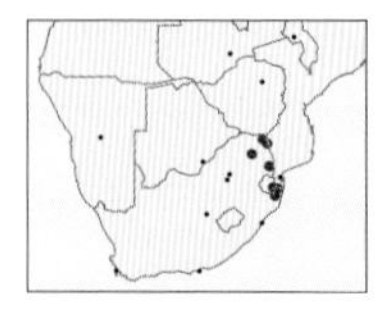

### 7 *Attatha attathoides* — Dappled Custard

Forewing 18–19mm. Similar to *A. superba* (above), but terminal mark curved inward from apex forming a funnel shape, median mark resembling a 2-finger point with sharper finger midway, 3 round black marginal spots, ground colour a soft creamy pink; hindwings a rich yellow. **Biology:** Caterpillar host associations unknown. Adults November–February. **Habitat & distribution:** Bushveld and woodland; KwaZulu-Natal, South Africa, north to tropical Africa, also Madagascar.

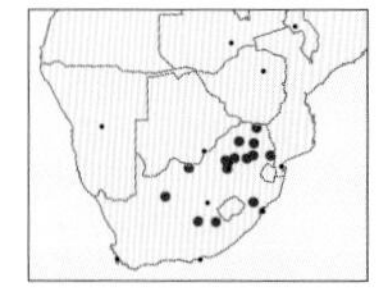

### 8 *Prionofrontia erygidia* — Grey Saw-wing

Forewing 12–14mm. Forewings brown, variably dusted with black scales, almost black in some individuals, lines indistinct or absent, conspicuous large black basal patch in females; hindwings fawn with indistinct darker border. Two species in genus in region. **Biology:** Caterpillar host associations unknown. Adults October–February. **Habitat & distribution:** Bushveld; Eastern Cape, South Africa, north to tropical Africa.

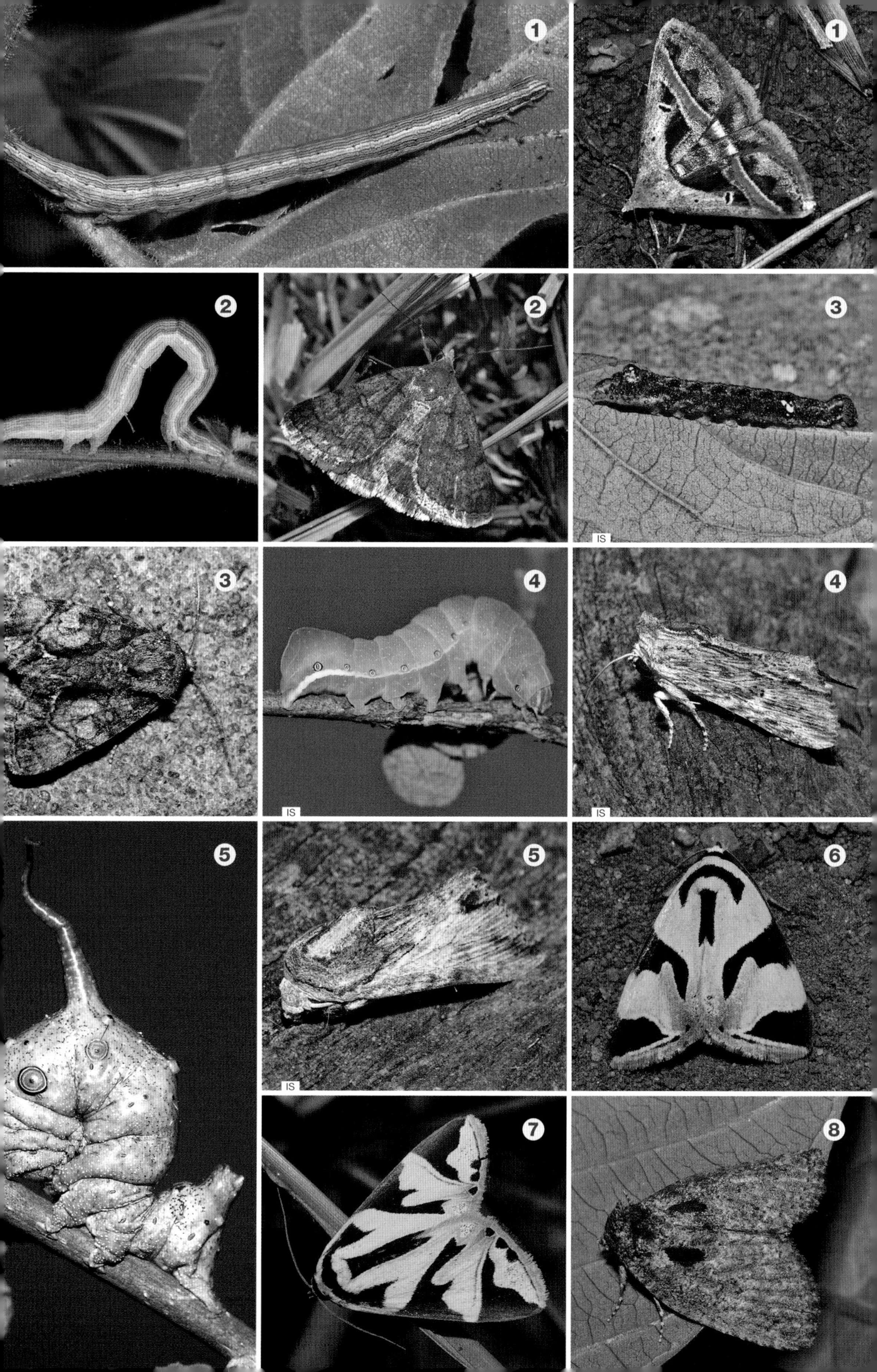

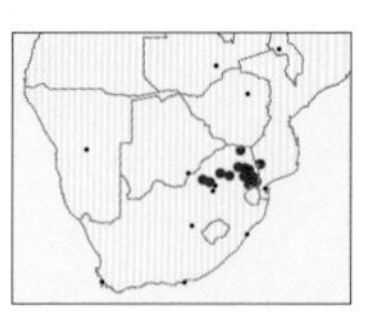

### 1 *Prionofrontia strigata* — Streaked Saw-wing

Forewing 12–13mm. Forewings grey or brown, antemedial line straight and postmedian line diagnostically kinked around cell with large cell spot, basal area darker, black streak from wing tip along half its length; hindwings pale with darker border. Caterpillar stocky, grey, with 2 short dorsal horns near rear of body. **Biology:** Caterpillar host associations unknown. Adults October–February. **Habitat & distribution:** Bushveld; Mpumalanga, South Africa, north to tropical Africa.

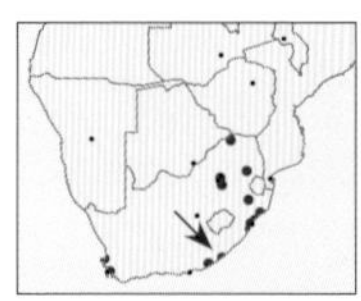

### 2 *Serrodes korana* — Footprint Saw-wing

Forewing 25–28mm. Large and robust, wing margins crenulated; fawn to dark brown, variably speckled with black, thin postmedian cream line, black costal spot with complex circle of markings in median area, diagnostic dark basal circle of black markings; hindwings tan with darker marginal border. Caterpillar robust, light brown, with 2 humps near head and dorsal projections at rear. Three species in genus in region. **Biology:** Caterpillars polyphagous on trees. **Habitat & distribution:** Moist habitats, including gardens; Western Cape, South Africa, north to tropical Africa, also Madagascar and Indian Ocean islands.

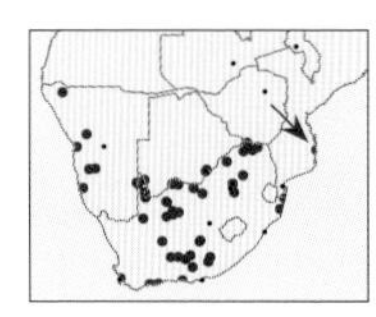

### 3 *Cerocala vermiculosa* — Vermiculous

Forewing 15–19mm. Light grey to cream, curved antemedial and postmedian line diagnostic, curving back to almost form a complete circle before veering straight to costa; basal, subterminal areas and median area around cell dark brown with sprinkling of silver scales; hindwings cream with broad, black band with white spot and 2 marginal white spots. Three species in genus in region; similar *C. contraria* is more uniform dark brown and postmedian line not as recurved. **Biology:** Caterpillar host associations unknown. Adults October–December. **Habitat & distribution:** Variety of arid and moist habitats; Western Cape, South Africa, north to Zambia, also Madagascar.

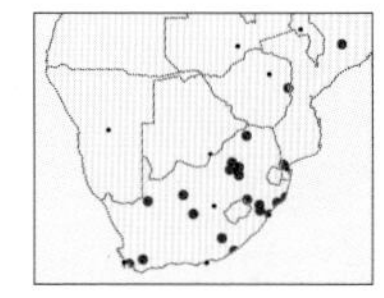

### 4 *Ulotrichopus variegata* — Thorn Underwing

Forewing 18–22mm. Adults bulky, forewings all-brown or all-grey with intricate markings or with brown bases and tips enclosing large pale central area in some specimens, conspicuous white-spotted, curved, subterminal line usually present; abdomen and hindwings orange, the latter with dark marginal band, with yellow tip. Caterpillar elongate, rufous or green with dorsal projections arising from posterior region. Seven species in genus in region. Similar *U. catocala* larger and has black lengthwise streaks on forewing. **Biology:** Caterpillars feed on thorn trees (*Vachellia* spp.). Multivoltine, adults throughout the year. **Habitat & distribution:** Various habitats where thorn trees grow; Western Cape, South Africa, north to Kenya.

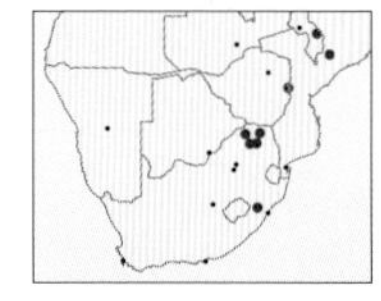

### 5 *Ulotrichopus pseudocatocala* — Lichen Underwing

Forewing 25–26mm. Similar to *U. variegata* (above), but larger and forewings more pointed; forewing variably mottled grey, black, white, brown, yellow and green; hindwings a deeper orange with black border extending to inner margin, without yellow tip. **Biology:** Caterpillar host associations unknown. Adults November–March. **Habitat & distribution:** Moist forests; KwaZulu-Natal north to tropical Africa.

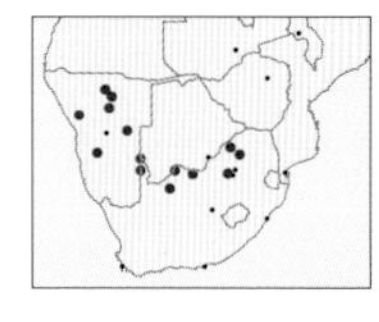

### 6 *Ulotrichopus tinctipennis* — Lichen Underwing

Forewing 16–20mm. Similar to *U. pseudocatocala* (above), but smaller and forewings without the white shading; hindwings cream to white, not orange, with a broader black border and with small white tip; also occupies a different habitat. **Biology:** Caterpillar host associations unknown. Adults November–March. **Habitat & distribution:** Dry bushveld; Northern Cape, South Africa, north to Namibia in region, also other dry bushveld areas in Africa north to the Sahara.

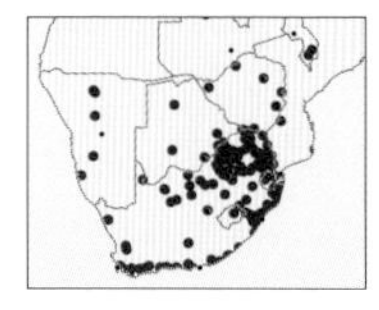

### 7 *Cyligramma latona* — Cream-striped Owl

Forewing 29–33mm. Prominent light to dark brown eyespot on forewings, cream postmedian line across all wings and shorter cream band across apical part of forewing; a series of diagnostic 'V'-shaped marks in terminal area of hindwings. Caterpillar a green or yellowish-green semi-looper with black longitudinal dots and white eyespot. Four species in genus in region; similar *C. magus* has a broken black wavy subterminal band, no 'V'-shaped marks. **Biology:** Caterpillars polyphagous, primarily on Fabaceae. Multivoltine, adults throughout the year, can swarm in late summer. **Habitat & distribution:** Usually in a variety of moist habitats, migrates to drier areas where it can establish itself in wet years; Western Cape, South Africa, north to tropical Africa and Madagascar.

1
1
IS
2
2
3
4
4
5
6
7
7

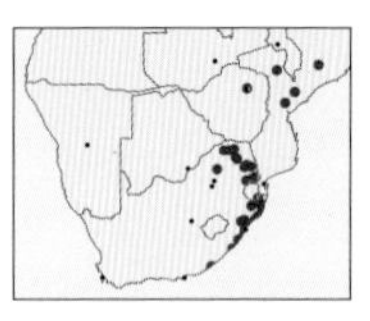

### 1 *Erebus walkeri* — Walker's Owl

Wingspan 55–64mm. Largest erebid in region; broad wings with scalloped margins, ground colour brown with several wavy lines across both wings, subterminal line a row of marks, 2 large eyespots with comma-shaped mark on forewing. Two species in genus in region; similar *E. attavistis*, from Zimbabwe northwards, is smaller, lighter and has lines thinner. **Biology:** Caterpillars reared on *Adenopodia spicata* (Fabaceae). Adults often congregate in caves or dark buildings during adverse weather. Adults throughout the year, can be abundant in late summer. **Habitat & distribution:** Lowland forest, woodland and coastal bush; Eastern Cape, South Africa, north to tropical Africa and Madagascar.

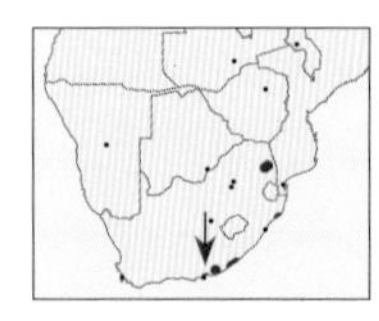

### 2 *Ericeia congressa* — Peppered Tabby

Forewing 20–22mm. Wings grey to tan, peppered with small spots, sinuous postmedian double band along outer edges, in darker specimens the band is subdivided into multiple bands; forewings slightly indented at tip. Caterpillar elongate, brown with darker longitudinal lines. Five very similar species described in genus in region. **Biology:** Caterpillars recorded from Monkey-thorn (*Senegalia galpinii*). Multivoltine, adults throughout the year. **Habitat & distribution:** Forests and wooded kloofs; Western Cape, South Africa, north to Kenya, also Madagascar and Indian Ocean islands.

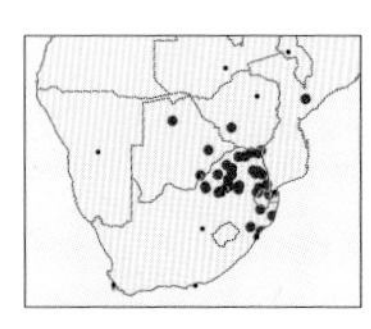

### 3 *Cometaster pyrula* — Faint Owl

Forewing 17–22mm. Male with pectinate antennae; cream with series of submarginal zigzag lines; forewings with comma-shaped black marking bordered with brown patches. Caterpillar a grey semi-looper with white longitudinal lines and small dorsal projections at rear of body. One species in genus in region. **Biology:** Caterpillars recorded from Scented-pod Acacia (*Vachellia nilotica*). Multivoltine, adults throughout the year. **Habitat & distribution:** Bushveld; KwaZulu-Natal, South Africa, north to tropical Africa.

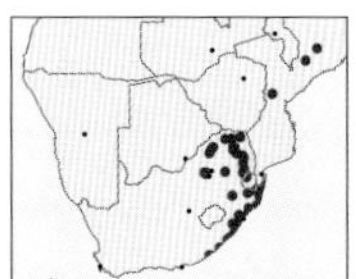

### 4 *Hypopyra capensis* — Red Tail

Forewing 28–35mm. Variable but straight, evenly curved, subterminal line over both wings, curved pointed forewings and pink to red abdomen and underside diagnostic; upperside of wings amber, peach, grey or dark brown, some forms with zigzag lines or darker patches on wings; hindwings similar to forewings but with orange-red inner and fore-margins; abdomen orange-red. Caterpillar elongate, light brown with darker blotches or grey. Four species in genus in region. **Biology:** Caterpillars feed on various Mimosoideae. Orange-red flash of body and underside of wings seen when disturbed and in flight, quickly settles on ground as cryptic leaf mimic. Multivoltine, adults throughout the year. **Habitat & distribution:** Forest and riverine vegetation; Eastern Cape,, South Africa, north to tropical Africa and Madagascar.

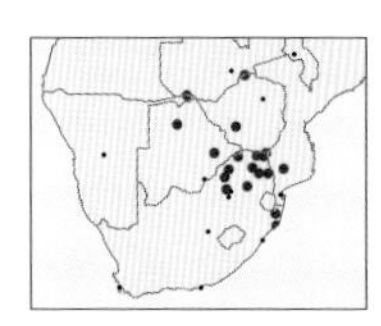

### 5 *Hypopyra carneotincta* — Exclamation

Forewing 20–24mm. Ground colour fawn, wings of males variably shaded in pastel pink colours, females finely dusted with dark scales; subterminal line straight, other lines curved, indistinct or absent, diagnostic black 'exclamation mark' on cell, connected in some specimens, black marks on costa. **Biology:** Caterpillar host associations unknown. Adults September–April. **Habitat & distribution:** Bushveld; KwaZulu-Natal, South Africa, north to Zimbabwe and Namibia.

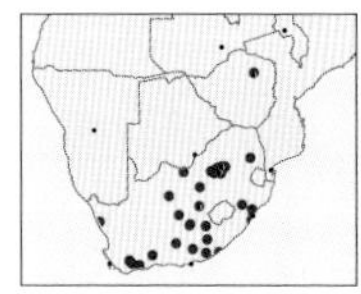

### 6 *Cuneisigna obstans* — Long Triangle

Forewing 15–17mm. Body and wings grey to cream; forewings with distinctive narrow black triangle on inner margin, and vertical undulating black bar separated from brown wing tip by thin white line; hindwings with grey shading to brown border. Caterpillar a tan semi-looper with median longitudinal dark line separating darker dorsal part of body. Three species in genus in region. **Biology:** Caterpillars feed on Sweet Thorn (*Vachellia karroo*). Bivoltine, adults throughout the year where conditions allow. **Habitat & distribution:** Various habitats where thorn trees grow; Western Cape, South Africa, north to Tanzania.

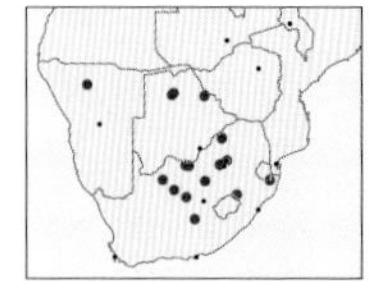

### 7 *Cuneisigna rivulata* — River Triangle

Forewing 13–15mm. Ground colour grey; body and forewing covered in diagnostic pattern of brown lines around a fawn and brown patch on inner margin extending to apex; hindwings cream with indistinct brown border. **Biology:** Caterpillar host associations unknown. Adults throughout the year. **Habitat & distribution:** Dry bushveld and grassland; Eastern Cape, South Africa, north to Kenya.

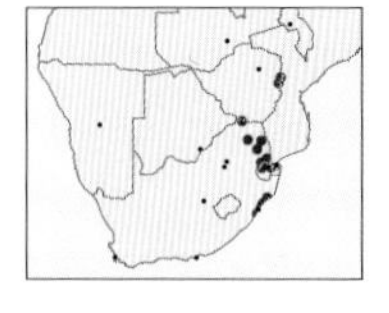

### 8 *Mocis conveniens* — Lined False-thorn

Forewing 17–19mm. Ground colour fawn, finely dusted with brown scales, antemedial and postmedian lines straight, ending on costa, 2 diagnostic large round marks on cell, lower one not coloured. **Biology:** Caterpillar host associations unknown. Adults throughout the year. **Habitat & distribution:** Forest and riverine vegetation; KwaZulu-Natal, South Africa, north to tropical Africa and Oman.

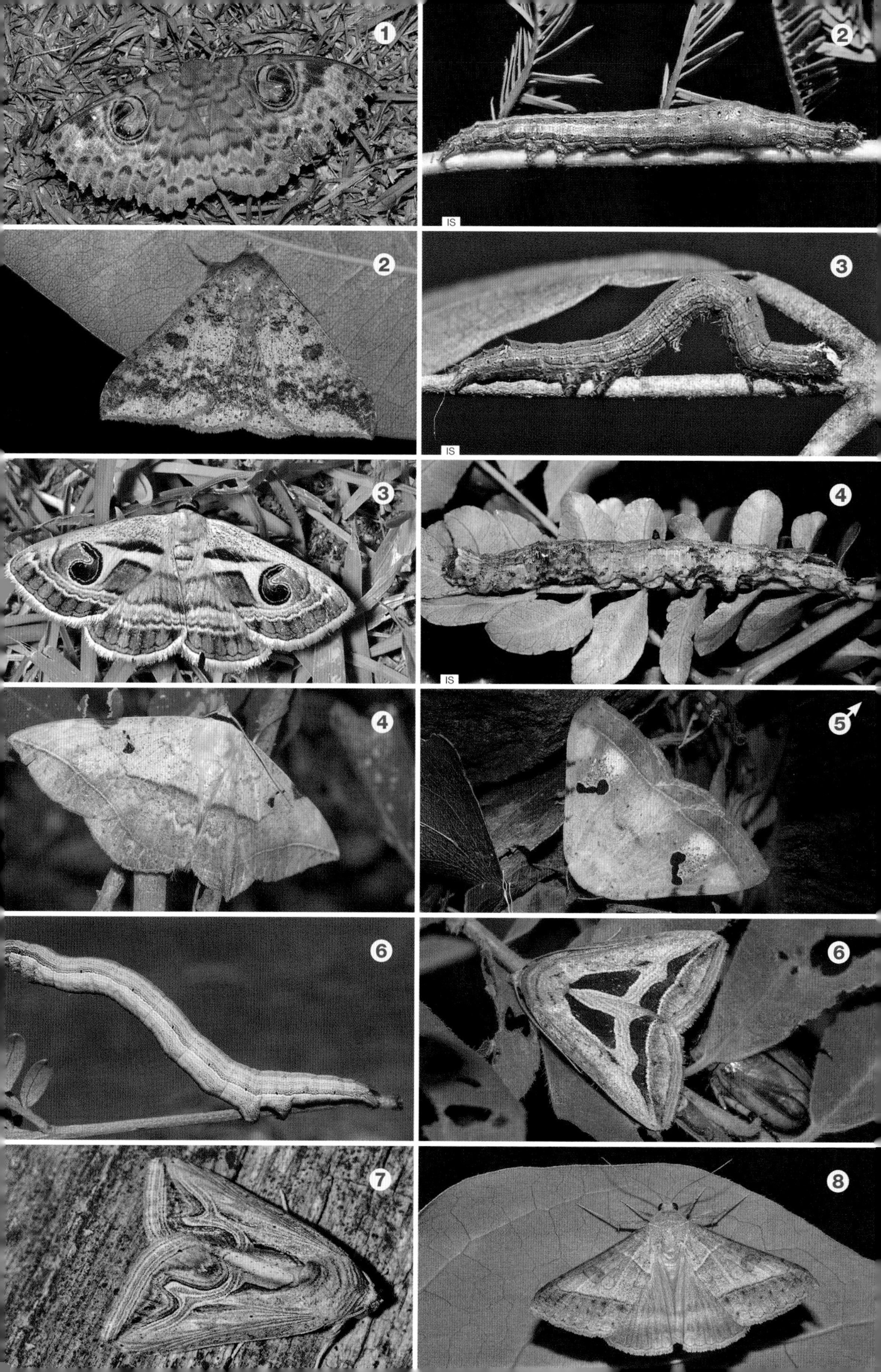
1
2
IS
2
3
IS
3
4
IS
4
5
6
6
7
8

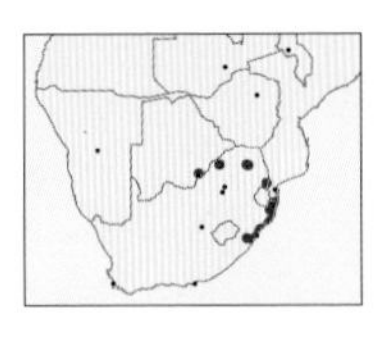

### 1 *Mocis proverai* — Geometric Lined False-thorn

Forewing 20–22mm. Can be confused with *Marca proclinata* (p.330), ground colour fawn, lines diagnostically straight and angled toward apex, postmedian black line and black line along inner margin; hindwings fawn with black spot on apex. Four species in genus in region. **Biology:** Caterpillar host associations unknown. Adults throughout the year. **Habitat & distribution:** Forest and riverine vegetation; KwaZulu-Natal, South Africa, north to tropical Africa and Oman.

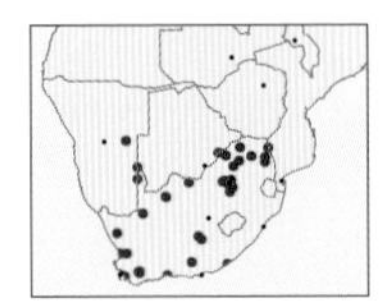

### 2 *Pandesma robusta* — Robust Tabby

Forewing 15–22mm. Stocky, wings fawn variably dusted with brown; fore- and hindwings with a series of straight scalloped transverse lines, hindwings dirty white at base darker toward edge. Caterpillar grey or brown (lighter along bottom of body and with a light tan broken band dorsally). Three species described in genus in region. **Biology:** Caterpillars feed at night on African thorn trees (*Vachellia* and *Senegalia* spp.), hide under bark by day. Multivoltine where conditions allow, adults throughout the year. **Habitat & distribution:** Variety of habitats where thorn trees grow; Western Cape, South Africa, north to Saudi Arabia.

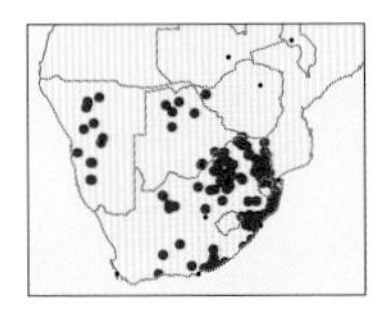

### 3 *Sphingomorpha chlorea* — Sundowner Moth

Forewing 26–31mm. Bulky, triangular outline at rest; body brown with broken longitudinal white dorsal line; forewings brown with diagnostic intricate pattern of marbling and streaking with iridescent scales; hindwings with light brown shading to dark marginal band. Caterpillar large and variable grey, brown or green, sometimes with broken white longitudinal stripes; in defensive posture body bent, exposing diagnostic red collar. Two species in genus in region. **Biology:** Caterpillars feed on African thorn trees (*Vachellia* spp.) and Wild Syringa (*Burkea africana*). Multivoltine, adults throughout the year, abundant in late summer, participating in annual migrations. **Habitat & distribution:** Variety of habitats where thorn trees grow; throughout Africa, Indian Ocean islands and India.

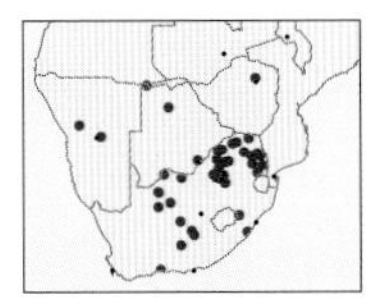

### 4 *Pericyma atrifusa* — Lined Shades

Forewing 13–16mm. Very variable in colouration, but pattern of lines diagnostic, especially shape of postmedian line, body and wings grey to brown to black; forewings with several transverse brown bars (sometimes faint), often extensive subterminal white markings, all wings with waved lines and scalloped at margin; hindwings with 1 major bold and several fainter bars. Caterpillar a pale green semi-looper with diagnostic wavy lateral line. Three species in genus in region. **Biology:** Caterpillars feed on hook thorn trees (*Senegalia* spp.). Bivoltine, adults September–April. **Habitat & distribution:** Variety of habitats where hook thorns grow; Western Cape, South Africa, north to tropical Africa and Saudi Arabia.

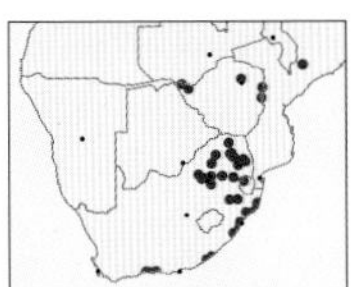

### 5 *Pericyma mendax* — Brown Shades

Forewing 14–20mm. Very variable in colouration, but pattern of lines diagnostic, especially shape of postmedian line, body and wings light tan to dark brown; forewings generally waved, with crenulate border, wing often marked by broad transverse bar, or broad horizontal dark bar on fore-margin with white panel below, median area sometimes pure white; hindwings barred. Caterpillar green, with diagnostic lateral line represented as broken zigzag and longitudinal white stripes. **Biology:** Caterpillars oligophagous on various Mimosoideae. Multivoltine, adults throughout the year. **Habitat & distribution:** Relatively moist habitats where Mimosoideae grow; Western Cape, South Africa, north to tropical Africa, Madagascar and Indian Ocean islands.

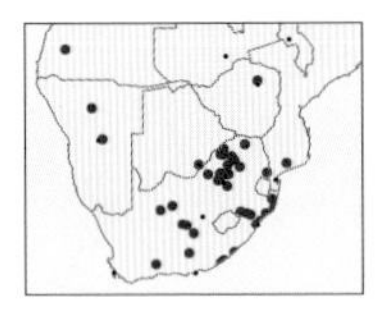

### 6 *Cortyta canescens* — Wavy Shades

Forewing 16–19mm. Stout, fore- and hindwings as well as abdomen grey, fawn or brown with multiple white and grey transverse, waved lines, giving grizzled appearance. Caterpillars off-white or light brown, thin and elongate, with peppering of small black spots and 2 small dorsal tubercles near end of body, mimicking thorns. Five species in genus in region. **Biology:** Caterpillars feed on thorn trees (*Vachellia* spp.), but also on Weeping Wattle (*Peltophorum africanum*). **Habitat & distribution:** Bushveld and riverine vegetation where thorn trees grow; Western Cape, South Africa, north to tropical Africa.

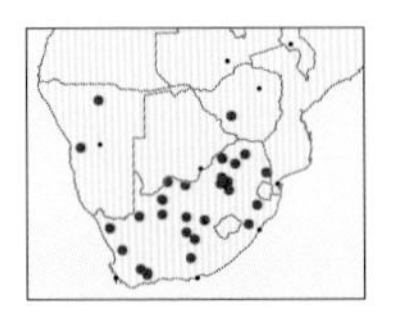

### 7 *Homaea cortytoides* — Sombre Shades

Forewing 14–17mm. Bulky, wings and body brown and rufous; fore- and hindwings with crenulate margins and a series of black transverse zigzag lines, rufous colour diagnostic. Caterpillar greyish or brown with dense black spots. One species in genus in region. **Biology:** Caterpillars feed on various species of African thorn trees (*Vachellia*). **Habitat & distribution:** Dry habitats where *Vachellia* trees grow; South Africa north to North Africa and Saudi Arabia.

1
2
2
3
3
3
4
4
5
5
6
6
7
7

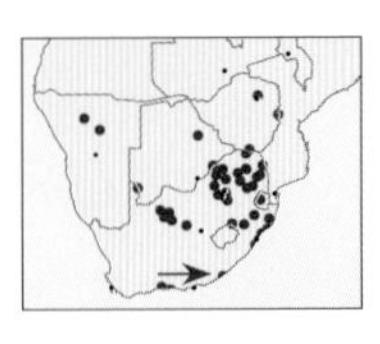

## 1 *Achaea catella* — Banded Achaea

Forewing 21–31mm. Thickset, body and wings grey to brown; forewings with broad transverse pale band bordered by sinuous black lines, shape of scalloped postmedian line together with black and white underside diagnostic; hindwings largely black with transverse white bar and 3 or 4 white patches along margin. Caterpillar variable, a grey to brown semi-looper, with a pair of short dorsal projections arising from the rear of the body, and a black band that is exposed when disturbed, 2 white dots on black head diagnostic. Twenty-seven species in genus in region. **Biology:** Caterpillars polyphagous on a wide range of woody plants. Multivoltine, adults throughout the year where conditions allow. **Habitat & distribution:** Variety of habitats, including transformed habitats; Western Cape, South Africa, north to the rest of Africa and associated islands.

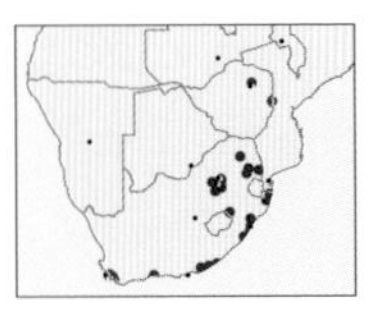

## 2 *Achaea echo* — Echo

Forewing 25–30mm. Bulky, body and wings brown or rufous; forewings with 2 dark brown transverse bands, shape of light terminal patches and antemedial line diagnostic; hindwings tan or brown. Caterpillar cryptic, a light to dark grey or brown semi-looper, with 2 short dorsal horns near rear of body. **Biology:** Caterpillars polyphagous. Multivoltine, adults throughout the year. **Habitat & distribution:** Variety of moist habitats, including gardens; Western Cape, South Africa, north to tropical Africa and Madagascar.

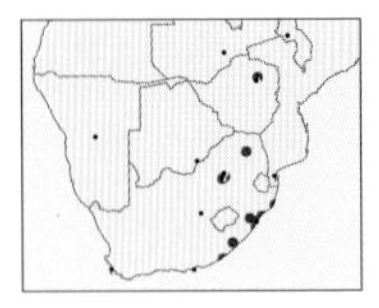

## 3 *Achaea finita* — Finite Achaea

Forewing 25–29mm. Bulky, body and wings light to dark brown; forewings darker toward cream margin, with faint to distinct wavy diagnostic antemedial and postmedian lines and often 2 small black dots on cell; hindwings grey or brown, darker at margin. Caterpillar uniform light or dark brown, with striped head capsule and cream mid-dorsal stripe. **Biology:** Caterpillars polyphagous. Multivoltine, adults throughout the year. **Habitat & distribution:** Variety of moist habitats, including gardens; Eastern Cape, South Africa, north to tropical Africa and Madagascar.

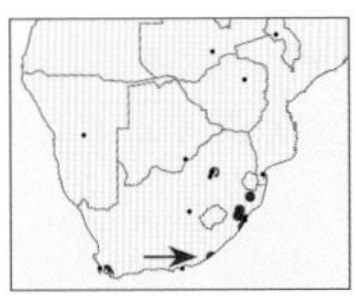

## 4 *Achaea indeterminata* — Indeterminate Achaea

Forewing 20–22mm. Bulky, body and wings tan to dark brown; all wings uniform brown with darker broad band toward margins, with 2 thin, sinuous, black and white transverse lines. Caterpillar a variable semi-looper, very pale green, or brown with numerous longitudinal lines and stripes, short pair of dorsal horns near rear of body. **Biology:** Caterpillar recorded from various native leguminous hosts (Fabaceae), and Red Beech (*Protorhus longifolia*). **Habitat & distribution:** Variety of moist habitats, including gardens; Western Cape, South Africa, north to tropical Africa.

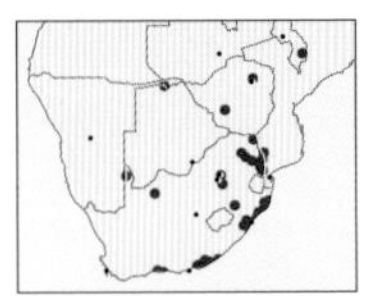

## 5 *Achaea lienardi* — Lienard's Achaea

Forewing 25–30mm. Highly variable in colour (grey through tan and ochre) and pattern, best determined by shape and curve of postmedian line in combination with the slightly curved antemedial line; forewings grey or brown with subapical black patch, and broad brown or whitish transverse band, base of wing occasionally with dark triangular triangle; hindwings with dark shading to black, with 3 white marginal patches. Caterpillar a semi-looper, with tan and reddish stripes, or grey to brown body; short dorsal projections at rear of body. **Biology:** Caterpillars polyphagous. Multivoltine, adults throughout the year, can swarm in good years in autumn. **Habitat & distribution:** Variety of moist habitats, including gardens; Western Cape, South Africa, north to tropical Africa; participates in annual autumn migrations when adults can be found far west of its normal range.

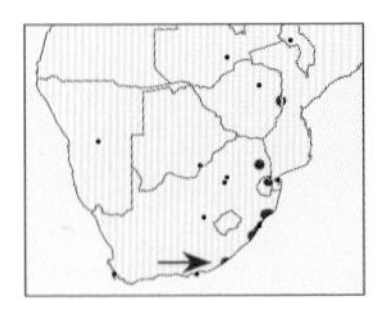

## 6 *Achaea praestans* — Painted Achaea

Forewing 26–28mm. Forewing similar to *A. lienardi* (above), but postmedian line continues to costa without curving back and antemedial line wavy; forewings in variable hues of pastel colours, median cream bar, with eyespot conspicuous when present; hindwings have diagnostic large yellow patch on inner curve, rest of wing rufous to brown. **Biology:** Caterpillar recorded on *Cassipourea gummiflua* (Rhizophoraceae). Adults throughout the year. **Habitat & distribution:** Moist forests; Eastern Cape, South Africa, north to East Africa and Madagascar.

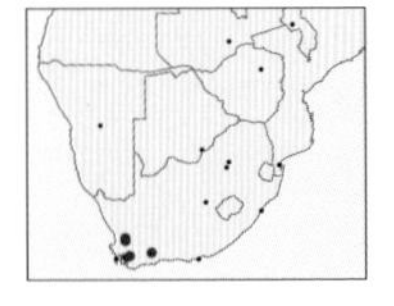

## 7 *Euonychodes albivenata* — White-veined Drab

Forewing 12–14mm. Distinctive species, wings rounded, may be confused with slug moths (see Limacodidae); ground colour coppery brown; forewing postmedian line black with white dots, veins and costal edge white; hindwings grey with similarly marked postmedian line. One species in genus in region. **Biology:** Caterpillar host associations unknown. Adults September–October. **Habitat & distribution:** Fynbos; Western Cape, South Africa.

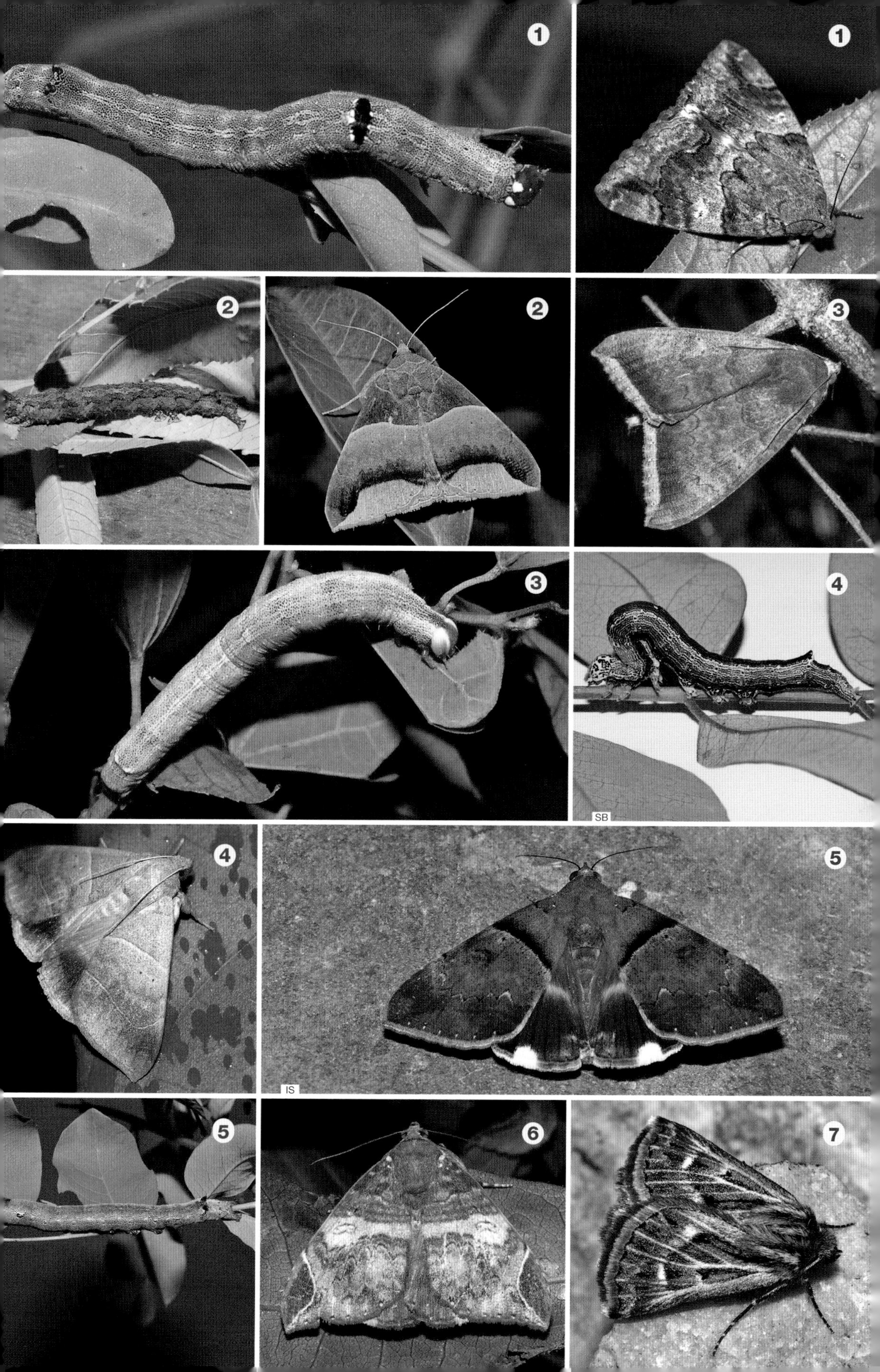
1
1
2
2
3
3
4
SB
4
5
IS
5
6
7

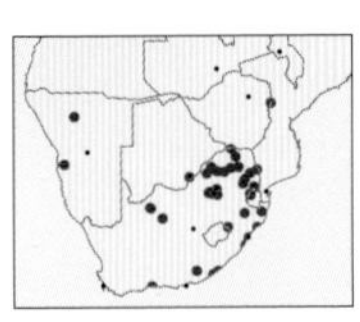

## 1 ***Anomis flava*** — Orange Cotton Moth

Forewing 13–15mm. Body and wings orange to gold and brown; forewings brown or gold, concave below apex, large basal gold triangle with honeycomb patterning, white cell spot, antemedial and postmedian lines meet on inner margin; hindwings cream or grey. Slender caterpillar, green with rows of white speckles. Eleven species in genus in region. **Biology:** Caterpillars feed on cotton plants (Malvaceae), including introduced weeds and crops. Multivoltine, adults throughout the year. **Habitat & distribution:** Variety of habitats where Malvaceae grow; spread globally following human habitation.

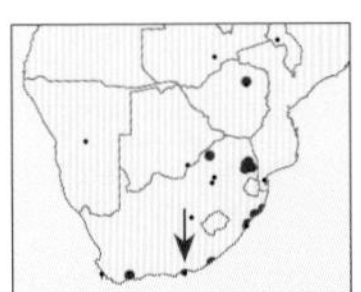

## 2 ***Anomis leona*** — Horned Cotton Moth

Forewing 16–18mm. Male has elongate, upturned palpi with long setae that extend over thorax; shape of postmedian line that is only slightly curved around cell diagnostic, forewings tan to dark brown, often darker toward apex, with several zigzag transverse lines and russet or yellow trailing edge; hindwings grey. Caterpillar a green and elongate semi-looper, with yellow lateral stripe. **Biology:** Caterpillars feed on various cotton plants (Malvaceae), including introduced weeds and crops. Multivoltine, adults throughout the year. **Habitat & distribution:** Variety of relatively moist habitats where Malvaceae grow; Western Cape, South Africa, north to Arabian Peninsula.

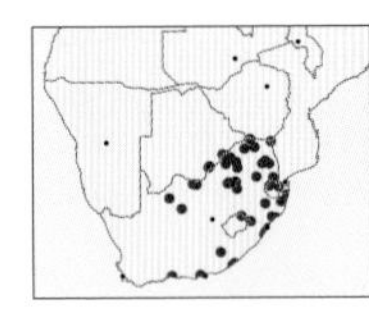

## 3 ***Anomis sabulifera*** — Brown Cotton Moth

Forewing 13–18mm. Similar to *A. leona* (above), but postmedian line that is curved widely around cell and kinked twice before costa is diagnostic, light tan to dark brown; forewings with broad, faint to dark transverse band, with light tan apical band; hindwings tan or dark brown. Caterpillar similar to that of *A. leona* but yellow patches at rear of body incompletely fused to form yellow band. **Biology:** Caterpillars feed on various cotton plants (Malvaceae), including introduced weeds and crops. Multivoltine, adults throughout the year. **Habitat & distribution:** Variety of habitats where Malvaceae grow; Western Cape, South Africa, north to tropical Africa, also oceanic islands and Arabian Peninsula.

# Subfamily Hypocalinae

This subfamily is in need of taxonomic revision; it is not known how many species occur in the region. Caterpillars form shelters by spinning leaves together, feeding inside, much like Crambidae and some Tortricidae.

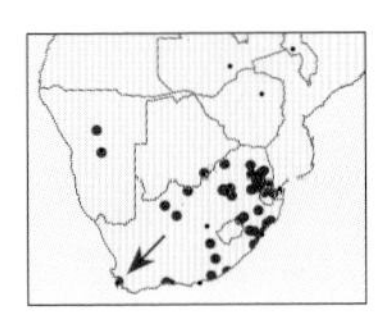

## 4 ***Hypocala plumicornis*** — Variable Yellow Underwing

Forewing 14–19mm. Bulky, variable, abdomen banded brown and yellow or grey; forewings light to dark grey, sometimes with central black-ringed crescent and reddish line along lower wing margin, extensive black shading in some specimens; hindwings black with variable degree of bright yellow markings. Caterpillar green to yellow with black head capsule while in shelter, turns black with yellow or orange spots in an aposematic prepupal stage before leaving shelter to pupate in soil. **Biology:** Caterpillars feed on *Diospyros* spp. and *Euclea natalensis* (Ebenaceae). Multivoltine, adults throughout the year, peaking in summer. **Habitat & distribution:** Variety of habitats where Ebenaceae grow; Western Cape north through sub-Saharan Africa and Madagascar, also Asia and Oceania.

# Subfamily Tinoliinae

It is not certain whether this is a natural (monophyletic) group within the Erebidae. There are 5 species in the region.

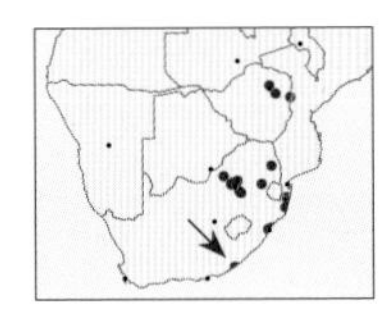

## 5 ***Calesia xanthognatha*** — Yellow-headed Widow

Forewing 20–22mm. Body and wings brown; fore- and hindwings with lighter brown margins, antemedial and postmedian lines faint, dark brown band, pair of small white spots on forewing, inner one may be obsolete, crenulated yellow subterminal line. Caterpillar a semi-looper, white with grey, finely striped dorsum, legs and membrane between segments crimson. Five species in genus in region. **Biology:** Caterpillars feed on bush violets (*Barleria* spp.). Adults October–May. **Habitat & distribution:** Bushveld; Eastern Cape, South Africa, north to Zimbabwe, Kenya and Saudi Arabia.

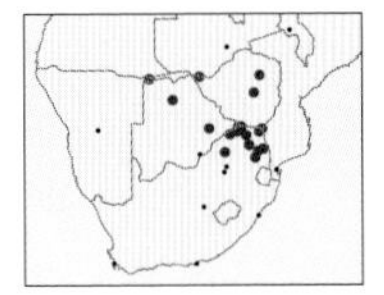

## 6 ***Calesia zambesita*** — White Chain

Forewing 16–18mm. Distinctive; orange head and legs, body and wings coppery grey with red abdomen and row of conspicuous white subterminal spots, as well as white dots and a dash lengthwise on forewings, merged in some individuals; hindwings of ground colour with row of faint white spots. **Biology:** Caterpillar host associations unknown. Adults November–April. **Habitat & distribution:** Bushveld; Mpumalanga, South Africa, north to East Africa.

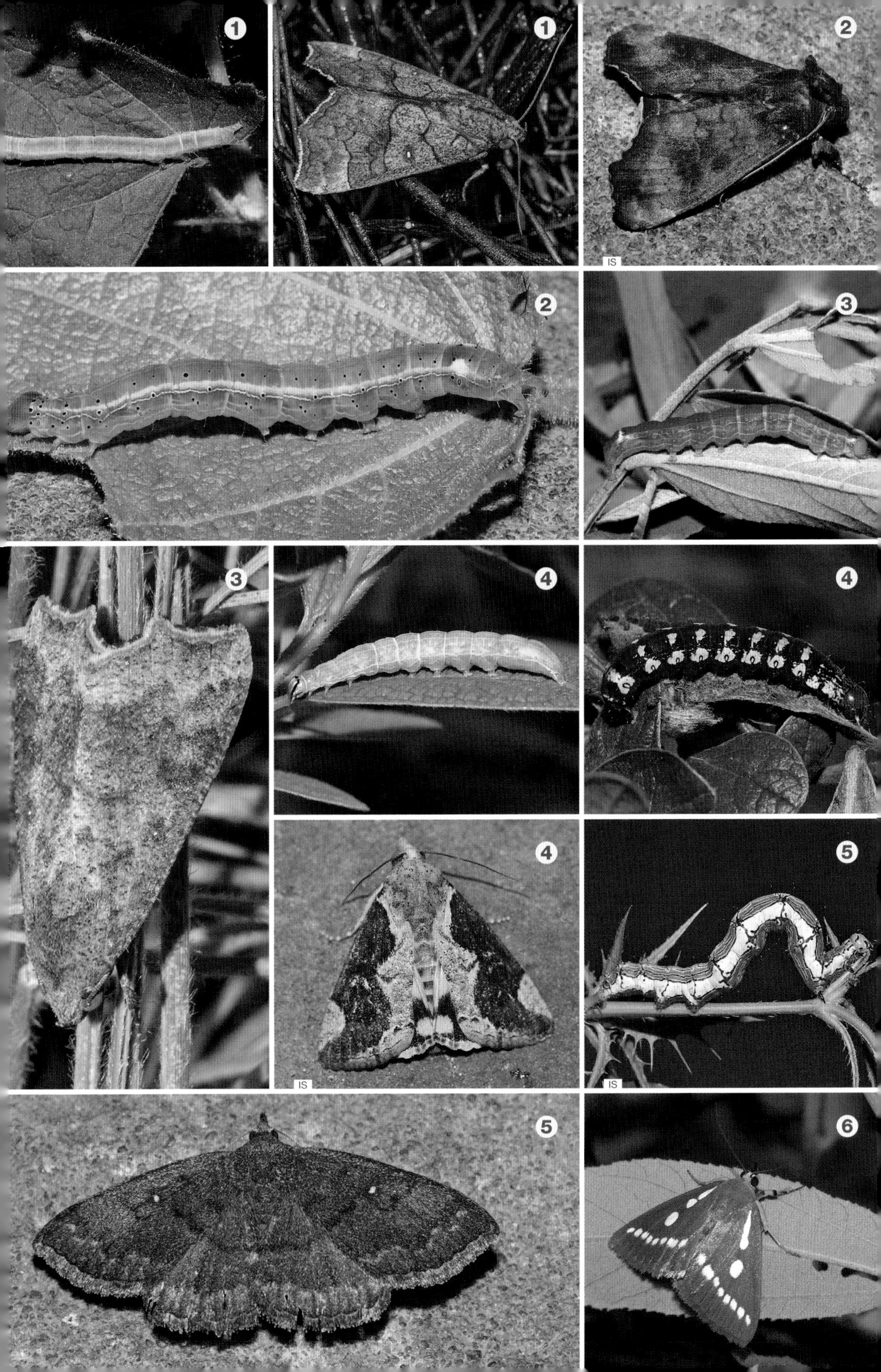
1
1
2
IS
2
3
3
4
4
4
IS
5
IS
5
6

## Subfamily Toxocampinae

The claspers of the male genitalia have an ancestral form, present in certain Noctuidae. At one stage they were included in the Erebinae. There are 9 species in the region.

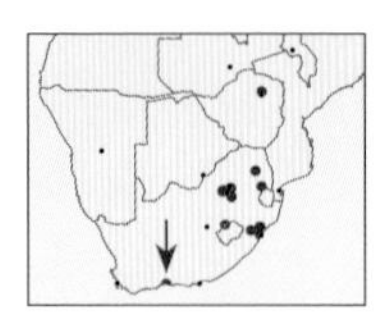

### 1 *Exophyla multistriata* — Stinkwood Drab

Forewing 17–23mm. Buff, tan or darker brown; forewings uniformly tan, with diagnostic single black spot and somewhat darker wing tips; hindwings uniform tan with faint brown border. Caterpillar a pale green semi-looper with intricate pattern of thin cream lines and dots. Four species in genus in region. **Biology:** Caterpillars feed on various white stinkwoods (*Celtis* spp.) and Pigeon Wood (*Trema orientalis*), all Cannabaceae. Multivoltine, adults throughout the year. **Habitat & distribution:** Forests and kloofs where host trees grow; Western Cape, South Africa, north to East Africa.

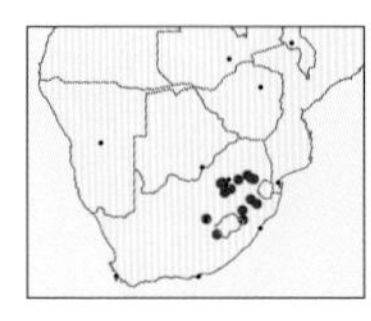

### 2 *Tathorhynchus plumbeus* — Grassland False-noctuid

Forewing 14–16mm. May be confused with Noctuidae due to elongate wings; sexually dimorphic; female grey, finely speckled with dark scales with brown shading toward apex, diagnostic comma-shaped cell spot; hindwings brown with darker border; male with forewing almost completely black, or completely black with black hindwings. Caterpillar brown with paler longitudinal red and cream stripes. Four species in genus in region, similar *T. homogyna* is more extensively covered in dark spots and has large round cell spot and male is like female. **Biology:** Caterpillars feed on *Tephrosia macropoda* (Fabaceae). Adults November–April. **Habitat & distribution:** Highveld grassland; Eastern Cape north to Mpumalanga, South Africa, also reported from East Africa.

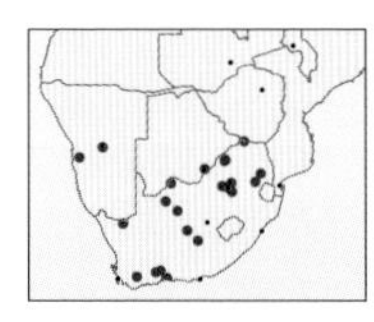

### 3 *Tathorhynchus exsiccata* — Bar False-noctuid

Forewing 12–14mm. Most similar to *Axylia annularis* (Noctuidae, p.440); tapered black mark on forewings with orange dot and double black dots on either side diagnostic, variable black streaks between veins and larger basal mark, sexes similar; hindwings fawn with broad brown border. **Biology:** Caterpillars recorded on Lucerne (*Medicago sativa*, Fabaceae). Adults November–April. **Habitat & distribution:** Dry habitats; Western Cape north across Africa to Middle East.

## Subfamily Pangraptinae Pangraptines

The caterpillar prolegs are only reduced on abdominal segments 3 and 4. Adults often resemble Geometridae, but have longer labial palps. Their closest relatives appear to be Herminiinae, Arctiinae and Aganainae. There are 6 species in the region.

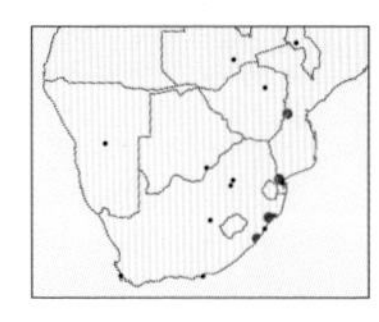

### 4 *Pangrapta melacleptra* — Melanic Tabby

Forewing 11–14mm. Male with well-developed pectinate antennae, broad rounded wings with upturned palpi, rests with wings behind back; wings crenulated and all lines crenulated orange, wings mottled with different shades of brown; hindwing median area with black shading, eyespot with grey scales on both wings. One species in genus in region. **Biology:** Caterpillar host associations unknown. Adults October–March. **Habitat & distribution:** Moist tropical forest; KwaZulu-Natal, South Africa, north to Mozambique.

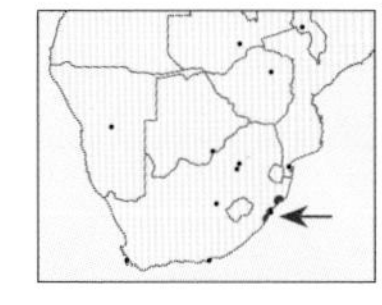

### 5 *Episparis xanthographa* — Dead Leaf Tabby

Forewing 18–19mm. Body and wings light tan to rufous-brown, dead leaf mimic, with scalloped wing margins; pale to rich yellow postmedian line running across fore- and hindwings; forewing with tiny white cell spot and faint or distinct black-ringed ocellus, surrounded by mauve halo; hindwings similar to forewings, all wings mauve at margins in darker form. Two species in genus in region. **Biology:** Caterpillar host associations unknown. Adults recorded March and April. **Habitat & distribution:** Humid forests; KwaZulu-Natal, South Africa, north to Kenya.

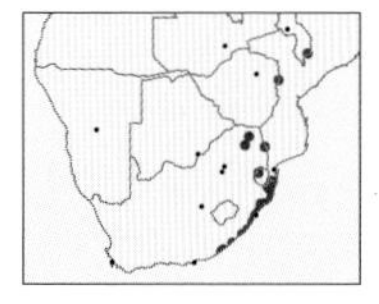

### 6 *Gracilodes caffra* — Orange False-head

Forewing 14–19mm. Grey forelegs held forward, tan, yellow or brown; fore- and hindwings with broad brown or russet apical band, and rest of wings yellow to orange; 3 diagonal lines cross fore- and hindwings, clear window in forewing cell in males; hindwings with conspicuous round, white eyespots on inner angle, creating a false head when adult at rest. Caterpillar translucent green. Two species in genus in region; similar *G. nysa* is smaller, brown and postmedian line is crenulated. **Biology:** Caterpillars feed on *Canthium ciliatum* (Rubiaceae). Multivoltine, adults throughout the year. **Habitat & distribution:** Humid forests and riverine bush; Eastern Cape, South Africa, north to tropical Africa.

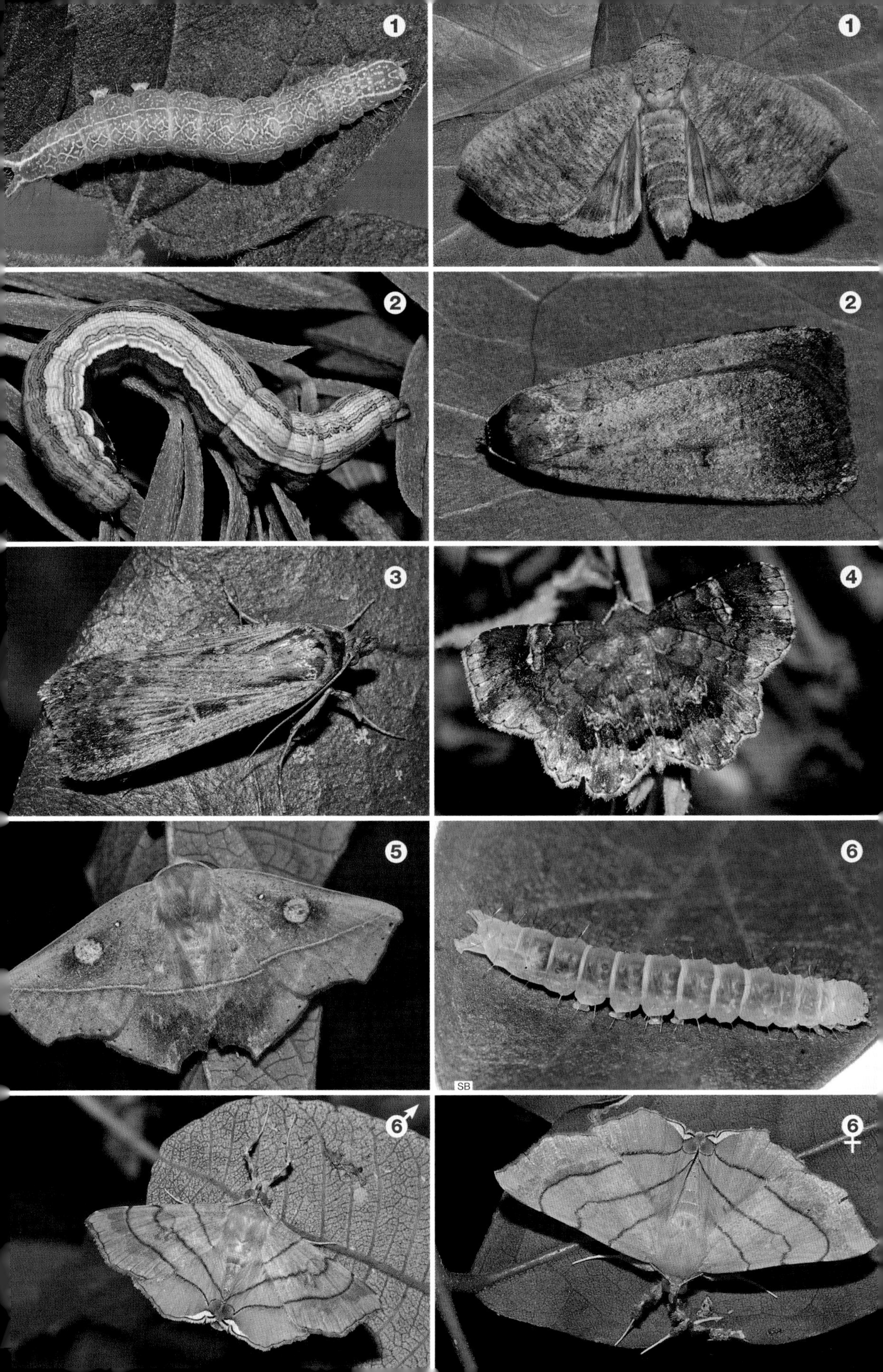
1
1
2
2
3
4
5
6
SB
6 ♂
6 ♀

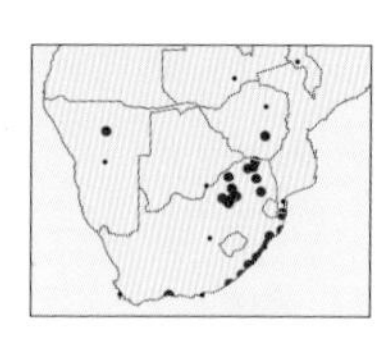

### 1 *Taviodes subjecta* Subjected Tabby

Forewing 19–21mm. Rests with wings held backward; broad, rounded wings with upturned palpi; upper and underside wing markings and colouration differ, upper side fawn to dark brown with all lines crenulated across both wings, underside fawn to deep orange with conspicuous straight postmedian line. One species in genus in region. **Biology:** Caterpillars feed on *Psydrax* spp. (Rubiaceae). Adults throughout the year. **Habitat & distribution:** Forest and bushveld; Western Cape, South Africa, north to tropical Africa.

## Subfamily Thiacidinae

This is a recently established subfamily with uncertain relationships. At one stage it was even placed within Notodontidae and Lasiocampidae. Species are univoltine, adults appear in late spring. There are 17 species in the region.

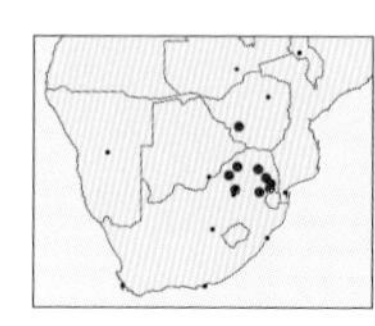

### 2 *Thiacidas nigrimacula* Mottled Thiacidas

Forewing 17–20mm. Body squat; forewings grey or brown, with white mottling, large circular white marking at wing apex, diagnostic broad sinuous bar across middle of wing; hindwings cream with dark waved transverse line and blotch. Caterpillar squat, swollen thorax with tubercles, head with forward projecting brown bristles, body with sparse bristles, cream ventrally and brown dorsally. Sixteen species in genus in region. **Biology:** Caterpillars feed on Wild Pear (*Dombeya rotundifolia*). Univoltine, adults October–November. **Habitat & distribution:** Bushveld; Mpumalanga, South Africa, north to Zimbabwe.

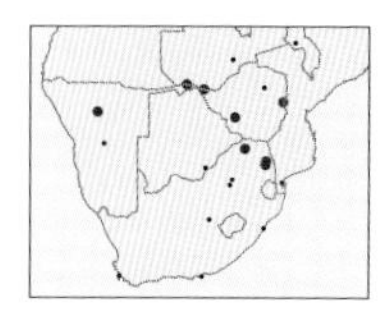

### 3 *Thiacidas roseotincta* Ocellated Thiacidas

Forewing 18–20mm. Thorax grey (occasionally brown); forewings pink-brown, with scalloped margins and 2 thin sinuous russet lines, postmedian line enclosing a grey ocellus at the trailing edge of the wing, grey at base; hindwings white, darkening at edge. Caterpillar grey, with fine black speckles and sparse long white hairs, and reddish bar on first few abdominal segments. **Biology:** Caterpillars feed on African myrrh (*Commiphora* spp.). Univoltine, adults October–November. **Habitat & distribution:** Bushveld; Limpopo, South Africa, north to East Africa and Arabian Peninsula.

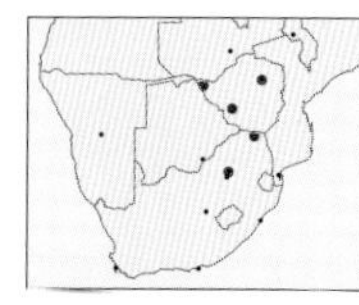

### 4 *Thiacidas cookei* Cooke's Thiacidas

Forewing 20–21mm. Ground colour fawn, body and forewings heavily dusted with black and orange scales, black antemedial and postmedian lines faint, crenulated; hindwings fawn with broad grey border and orange margin. **Biology:** Caterpillar host associations unknown. Univoltine, adults October–November. **Habitat & distribution:** Bushveld; Gauteng, South Africa, north to Zimbabwe.

## Subfamily Arctiinae Tiger Moths, Lichen Moths and Wasp Moths

Formerly in their own family, their closest relatives are Litter Moths (Herminiinae) and Aganainae. Tiger moths are usually nocturnal and often have ears that can detect bat echolocation calls – on detection of bat calls they either dive or produce their own ultrasound clicks, which may either confuse the bat sonar system or advertise their toxicity (by day their warning or aposematic colouration serves this function). These tympanal organs are located on the metathorax and are a feature of the subfamily. Other unifying features are the structure of the larval hairs, wing venation and the presence of anal glands in females. Caterpillars typically polyphagous specialists, able to sequester plant toxins, passing them on to the adults which often have aposematic colouration. They specialise in sequestering different chemicals from different plants and therefore caterpillars move around from plant to plant. The very hairy caterpillars ('woolly bears') can often be seen basking, to accelerate digestion. There are 394 species in the region.

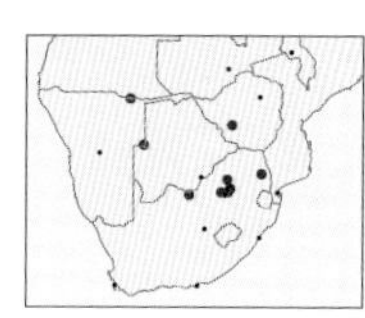

### 5 *Afrospilarctia flavidus* Orange-collared Yellow

Forewing 20–24mm. Forewings chalky yellow with paler hindwings; body yellow with distinctive orange collar on thorax, abdomen striped in black; legs and antennae black. Woolly bear caterpillar black, covered in dense white or black bristles. Four species in genus in region, many of these yellow species can only be reliably identified through dissection. **Biology:** Caterpillars polyphagous, individuals feeding on many plants. Adults October–February. **Habitat & distribution:** Bushveld; Northern Cape, South Africa, north to Namibia, Angola and Nigeria.

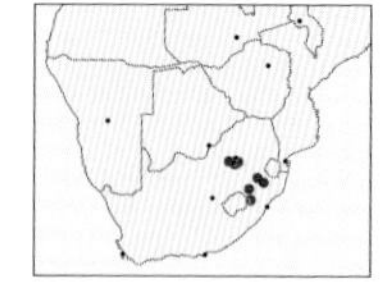

### 6 *Afrospilarctia dissimilis* Twinspot Yellow

Forewing 14–16mm. Body orange; forewings bright yellow, diagnostic black spot in cell (sometimes 2) and another black spot further up near costa; hindwings paler, females shaded in grey. **Biology:** Caterpillars polyphagous, mostly on herbaceous plants. Adults November–March. **Habitat & distribution:** Highveld grassland; KwaZulu-Natal, South Africa, north to Gauteng, South Africa.

1
1
2
2
3
3
IS
4
5
IS
5
6

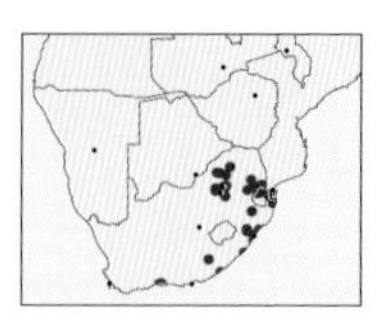

### 1 ***Saenura flava*** — Plain Yellow

Forewing 16–20mm. Wings and body uniform sulphur yellow; abdomen yellow with line of black spots dorsally and laterally. Caterpillar black with white dorsal stripe interrupted by orange bands on each segment, lateral cream stripes bear clusters of orange and black bristles. One species in genus in region. **Biology:** Caterpillars polyphagous, need to feed on many plants to obtain chemicals for defence, reproduction and nutrition. Multivoltine, adults throughout the year, peaking in summer. **Habitat & distribution:** Moist open habitats; Western Cape, South Africa, north to tropical Africa.

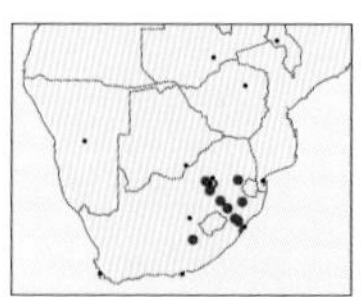

### 2 ***Logunovium scortillum*** — Border Yellow

Forewing 14–16mm. Body, base of forewings and hindwings white, rest of forewing yellow in male, white in female, with diagnostic black border on costa; hindwings without markings. One species in genus in region. **Biology:** Caterpillars feed on sedges (Cyperaceae). Adults October–March. **Habitat & distribution:** Near wetlands where sedges grow; Eastern Cape, South Africa, north to tropical Africa.

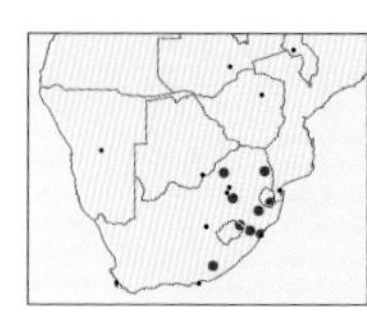

### 3 ***Popoudina leighi*** — One-spot Yellow

Forewing 16–17mm. Antennae black in male with bipectinate cream stalk, wings broad; wings and body cream to yellow, more intense toward head, wings generally unmarked, apart from small black mark in cell edge of forewing, wings of female buff with variable thin black streaks. Abdomen with row of black spots. Caterpillar black dorsally and cream ventrally, bearing stiff white and black bristles. Six species in genus in region. **Biology:** Caterpillars feed on grasses. Adults October–February. **Habitat & distribution:** Grassland; Eastern Cape to Limpopo, South Africa.

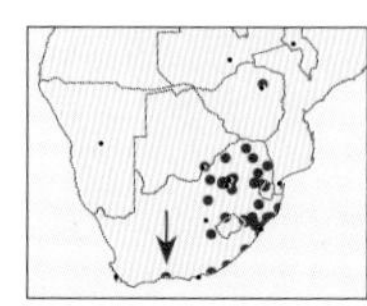

### 4 ***Popoudina linea*** — Lined Yellow

Forewing 16–20mm. Forewings and antennae pale yellow or tan, veins of forewing variably lined with black, always with stronger central streak forked terminally; antennae pectinate in males; body densely covered in tan hairs, thorax with longitudinal stripes and abdomen with 3 lines of black dots; hindwings a deeper yellow. Caterpillar brown, central and lateral pale yellow stripes. Similar *Acantharctia latifasciata* has black lines between veins, not on veins. **Biology:** Caterpillars feed on Guinea Grass (*Panicum maximum*). Adults August–April, peaking in summer. **Habitat & distribution:** Grassland and bushveld; Western Cape, South Africa, north to Zimbabwe, also reported from East Africa.

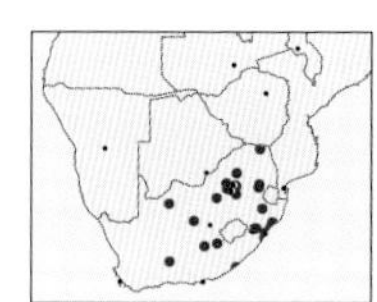

### 5 ***Leucaloa eugraphica*** — Graphic Yellow

Forewing 15–23mm. Adults much smaller from arid areas; similar to *Eyralpenus diplosticta* (below) with body and wings pale yellow to cream but with 2 zigzag, black (sometimes broken or almost absent) bands that each meet when wings closed; hindwings yellow with single black dot in cell. Caterpillar brown with central and lateral cream stripes; whorls of sparse, long white hairs on body, becoming stiffer and black and orange (on middle part of body); reddish head capsule. Three species in genus in region. **Biology:** Caterpillars polyphagous. Multivoltine, adults September–April, peaking in summer. **Habitat & distribution:** Variety of habitats; Western Cape to Limpopo, South Africa.

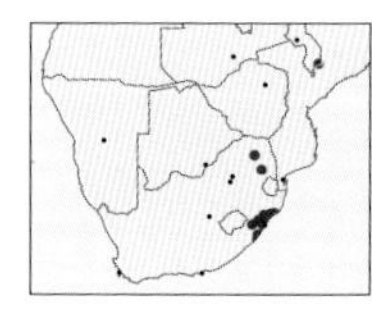

### 6 ***Eyralpenus diplosticta*** — Track Yellow

Forewing 15–18mm. Body and wings yellow; legs and antennae black; each forewing with 3 sinuate bands of streaks and dots , each band meeting when wings closed; hindwings yellow with single black cell spot; abdomen yellow with central line of black dots. Caterpillar has 2 final instar forms, one dark brown with clusters of strong, long bristles in dense whorls on each segment, second with central orange band of bristles. Four species in genus in region. **Biology:** Caterpillars polyphagous. Multivoltine, adults September–April. **Habitat & distribution:** Moist forests; KwaZulu-Natal north to tropical Africa.

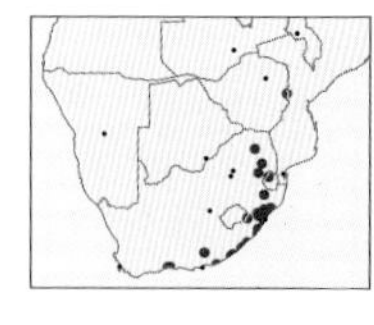

### 7 ***Eyralpenus testacea*** — Many-spot Yellow

Forewing 15–18mm. Similar to *E. diplosticta* (above), but paler with more spots on wings, including a subterminal row of spots, prominent larger black spot on costa near apex diagnostic, legs and antennae black; hindwings cream with single black cell spot. Similar *E. scioana* lacks prominent black spot on costa and forewings orange to fawn. **Biology:** Caterpillar host associations unknown. Multivoltine, adults August–May. **Habitat & distribution:** Moist forests; Western Cape, South Africa, north to tropical Africa.

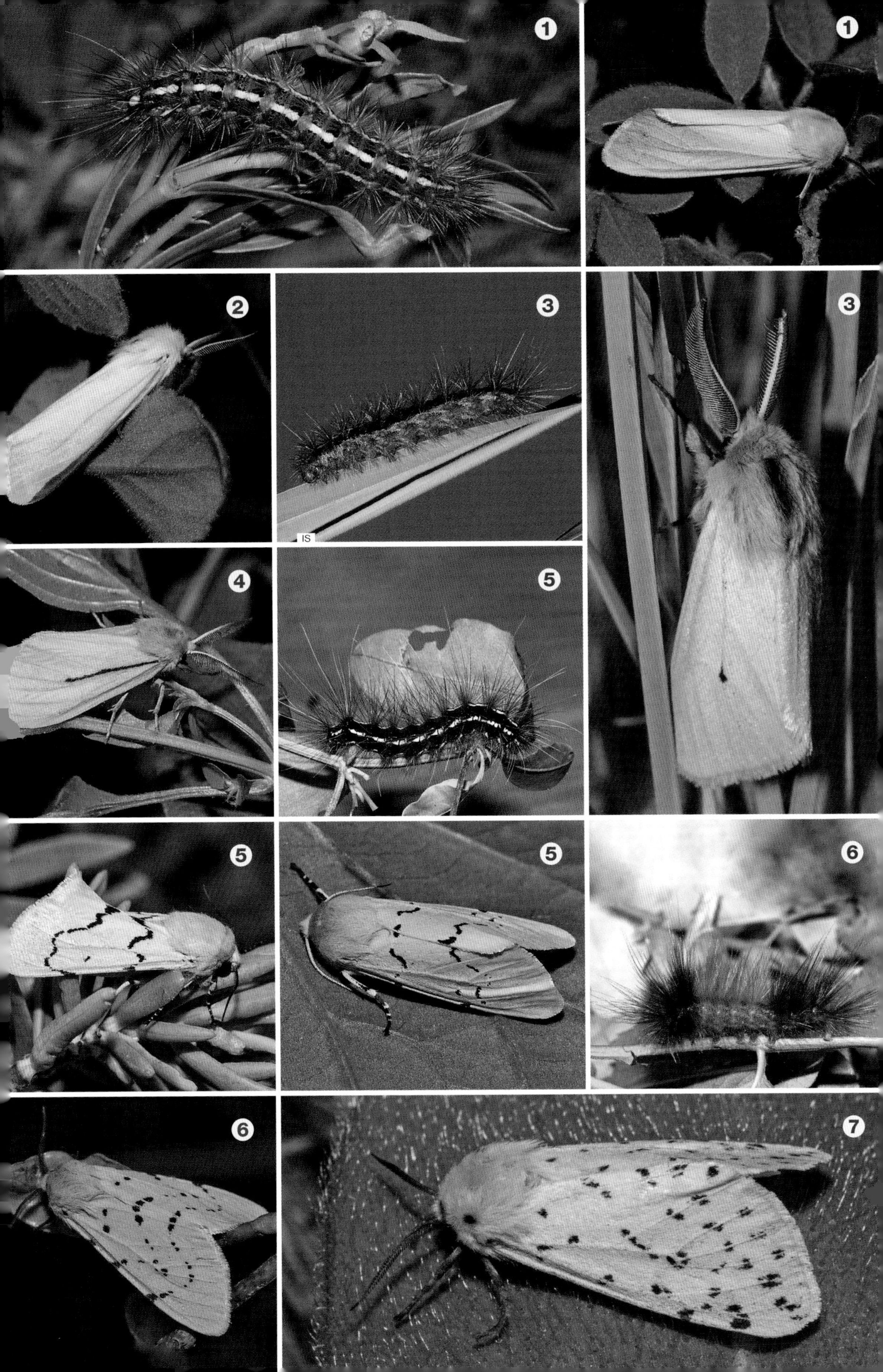
1
1
2
3
3
IS
4
5
5
5
6
6
7

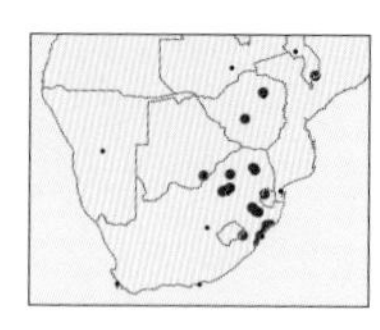

## 1 *Afromurzinia lutescens* — Translucent Ermine

Forewing 17–25mm. Body pale orange; forewings translucent grey with pale orange wing veins; hindwings translucent yellow; abdomen orange with black rings on each segment; thorax with grey patches and red collar. Caterpillar a dark woolly bear with mottled grey body, each segment with yellow markings and whorls of dense, stiff black hairs bearing longer white bristles toward end of body. **Biology:** Caterpillars polyphagous, needing to feed on many plants to obtain chemicals for defence, reproduction and nutrition. Multivoltine, adults throughout the year. **Habitat & distribution:** Variety of habitats; KwaZulu-Natal, South Africa, north to tropical Africa.

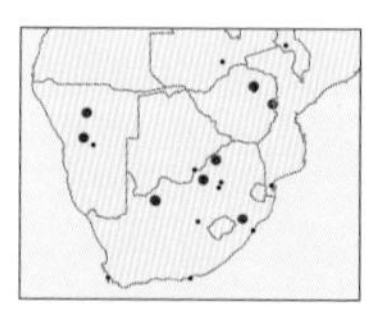

## 2 *Micralarctia punctulatum* — Dotted White

Forewing 20mm. Body and wings bright white; thorax pilose with transverse yellow bar, abdomen yellow with line of black spots, ending in a white tuft; forewings bright white variably scattered with small dark dots; hindwings with large cell spot and variable marginal dots. Two species in genus in region. **Biology:** Caterpillars recorded from Sweet Potato (*Ipomoea batatas*). Adults October–March. **Habitat & distribution:** Dry bushveld; KwaZulu-Natal, South Africa, north to tropical Africa.

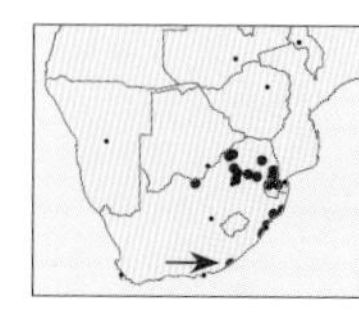

## 3 *Ustjuzhania lineata* — Lined White

Forewing 14–22mm. Specimens from arid areas smaller; body and wings uniform bright white, wing veins variably grey or black; abdomen yellow with line of black dots. Caterpillar black, with white dorsal and lateral stripes and covered in fairly long white and black hairs arising from orange tubercles. Two species in genus in region. **Biology:** Caterpillars polyphagous. **Habitat & distribution:** Bushveld; Eastern Cape, South Africa, north to tropical Africa.

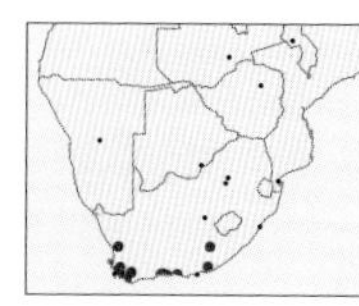

## 4 *Rhodogastria amasis* — Cape Tri-coloured Tiger

Forewing 25–30mm. White, 2 brown stripes on thorax; forewings with silvery grey shading between veins, black basal, antemedial and postmedian lines present, well to weakly marked; abdomen red to yellow above with black bands on each segment, white below; hindwings pale yellow with black cell spot and at least 1, usually several, black marginal spots. Caterpillar black with ginger bristles and black bristles. Two species in genus in region. **Biology:** Caterpillars obligatorily polyphagous, needing to feed on many plants to obtain chemicals for defence, reproduction and nutrition. Adults curl up in defensive posture when disturbed. Multivoltine, adults throughout the year. **Habitat & distribution:** Variety of habitats, including gardens; Cape provinces, South Africa.

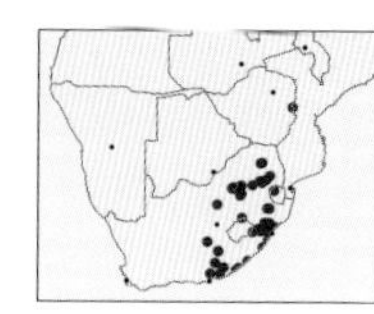

## 5 *Rhodogastria similis* — Tri-coloured Tiger

Forewing 26–35mm. Often confused with similar *R. amasis* (above), but forewings pure silvery white without grey shading, lines orange with only a few black spots or absent, abdomen fiery red above with black dorsal spots, not bands; hindwings orange-yellow without marginal black spots (very seldom one small spot). Caterpillar black with ginger bristles in first part of body and black bristles on rest of body, bristles arising from white tubercles. **Biology:** Caterpillars obligatory polyphagous specialists, needing to feed on many plants to obtain chemicals for defence, reproduction and nutrition. Multivoltine, adults throughout the year. **Habitat & distribution:** Variety of habitats, including gardens; Eastern Cape, South Africa, north to Zimbabwe; distribution overlaps with *R. amasis* in the Eastern Cape.

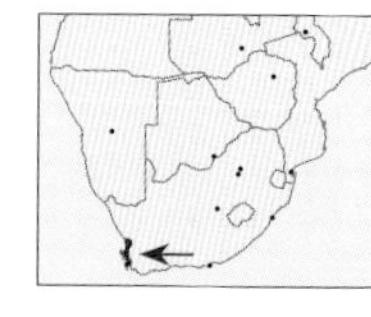

## 6 *Dasyarctia grisea* — Grey Tiger

Forewing 20–25mm. Body and forewings grey, thorax pilose, abdomen yellow with dorsal line of black spots, femora of legs and underside of forewings along costa red; forewings with 3 indistinct darker bars; hindwings white to grey, sometimes with marginal dots. One species in genus in region. **Biology:** Caterpillar host associations unknown. Adults July–September. **Habitat & distribution:** Strandveld vegetation, Western Cape and Northern Cape, South Africa.

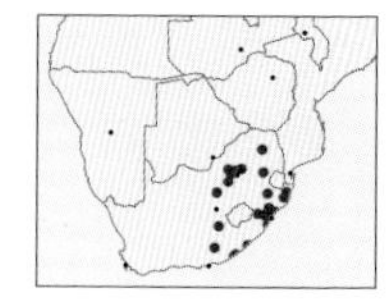

## 7 *Teracotona submacula* — Bark Tiger

Forewing 19–24mm. Variable; thorax greyish-brown with 2 brown lines, abdomen yellow, sometimes with 2 lines of black dots; ground colour cream, usually grey at base and then variably dusted in shades of brown, sometimes almost black, lines ill defined or absent, basal area to cell normally clear of dark scales, single black cell spot usually present; underside of each wing with black cell spot, inner flushed red; whole hindwing red in females. Caterpillar black with black bristles, apart from broad regions bearing ginger bristles. Four species in genus in region. **Biology:** Caterpillars polyphagous. Multivoltine, adults throughout the year. **Habitat & distribution:** Forests and riverine vegetation, including gardens; Eastern Cape north to Limpopo, South Africa, also reported from East Africa.

1
1
2
3
SB
3
4
4
5
5
6
7
7

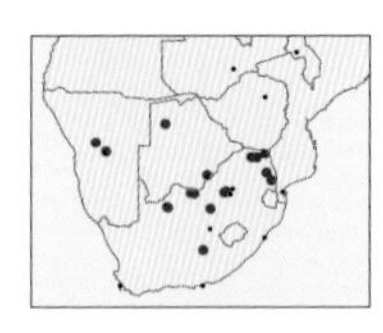

## 1 *Teracotona rhodophaea* — Lichen Tiger

Forewing 18–22mm. Similar to *T. submacula* (p.368), but ground colour white and dusted with black-grey scales, which are less extensive, usually leaving some white areas clear, lines usually clear; lines on thorax red; hindwings white, pink in females. Caterpillars similar. **Biology:** Caterpillars polyphagous. Bivoltine, adults October–March. **Habitat & distribution:** Bushveld and karoo habitats; Eastern Cape, South Africa, north to Namibia, also reported from other arid environments further north.

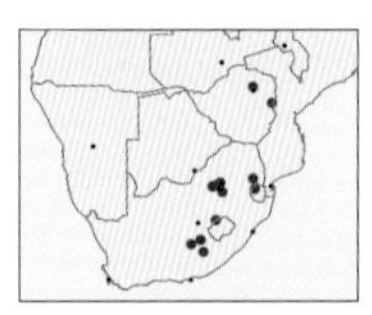

## 2 *Teracotona metaxantha* — Redspot Tiger

Forewing 21–24mm. Similar to other *Teracotona* spp. (above), but forewings dusted with only 1 distinctive shade of brown, prominent red cell spot diagnostic; hindwings yellow with black cell spot, pinkish in females. **Biology:** Caterpillars polyphagous. Bivoltine, adults October–February. **Habitat & distribution:** Grassland; Eastern Cape, South Africa, north to Zimbabwe.

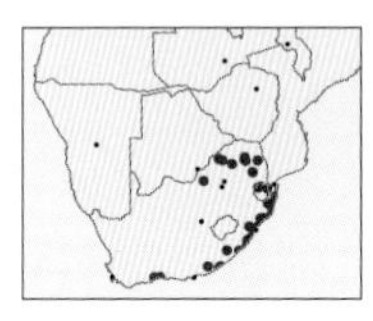

## 3 *Amerila bauri* — Baur's Frother

Forewing 27–28mm. Cream; thorax with 4 black dots in a row and another 4 dots arranged in a square, abdomen tan or pink; legs tan or red; forewings cream becoming browner toward apex with large, brown-edged translucent panels at apex; hindwings cream with fainter panels. Caterpillar uniform green. Ten species in genus in region; very similar *A. madagascariensis* has more extensive hyaline area of forewing, fewer dark brown markings and is restricted to moist forests. **Biology:** Caterpillars polyphagous; adult males pass on plant toxins to females during mating to coat eggs for protection; when disturbed, adults in genus exude foul-smelling sap with 2 'froth balls' either side of thorax. Multivoltine, adults throughout the year. **Habitat & distribution:** Only species in genus that is also found outside of moist forests in bushveld; Western Cape, South Africa, north to Zimbabwe.

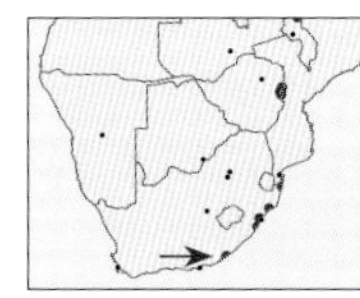

## 4 *Amerila bubo* — White Frother

Forewing 25–30mm. Wings and body immaculate white, with light brown veins; abdomen and legs yellow; thorax white with black dots. Caterpillar short, body yellow or brown with sparse, long white hairs arising from brown or black tubercles. **Biology:** Caterpillars polyphagous. Multivoltine, adults throughout the year. **Habitat & distribution:** Moist tropical forests; Eastern Cape, South Africa, north to tropical Africa.

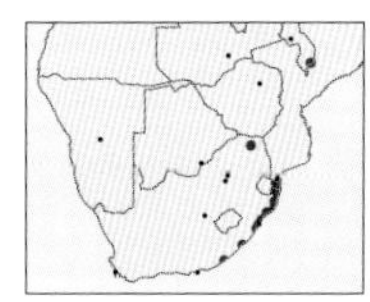

## 5 *Amerila lupia* — Small Red-tailed Frother

Forewing 18–25mm. Thorax and wings off-white with variable dark shading over wing veins; thorax with small black dots, abdomen orange or crimson. Caterpillar elongate, pale green with grey line running along length of back; body covered in sparse, long white hairs. **Biology:** Caterpillars polyphagous. Multivoltine, adults throughout the year. **Habitat & distribution:** Moist forests and lowland riverine vegetation; Eastern Cape, South Africa, north to Kenya.

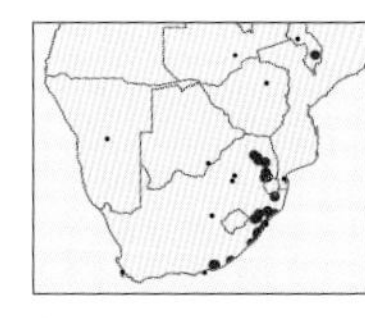

## 6 *Amphicallia bellatrix* — Beautiful Tiger

Forewing 28–34mm. Variable, lines present or absent, or with different shapes, especially north of region; body and wings deep orange, forewings with blue-black bands; hindwings usually with 2 bands and with large black blotches at tips. Caterpillar banded black and white. **Biology:** Caterpillars feed on rattlepods (*Crotalaria* spp.). Likely Müllerian mimic of *Callioratis* cycad moths (Geometridae, p.184). Multivoltine, adults throughout the year. **Habitat & distribution:** Forests and wooded kloofs; Eastern Cape, South Africa, north to Kenya.

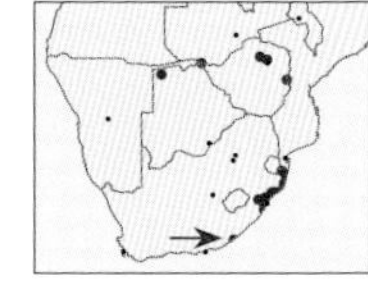

## 7 *Alytarchia amanda* — Cheetah

Forewing 18–26mm. Deep orange; forewings with white-ringed, black dots in male, not white ringed in female; hindwings orange with black oval markings and black border, sometimes incomplete, male with short tail. Two species in genus in region; similar *A. leonina* has silver lines connecting forewing black dots. **Biology:** Caterpillars recorded from various species of rattlepod (*Crotolaria* spp.). Multivoltine, adults throughout the year. **Habitat & distribution:** Occurs sporadically from South Africa northwards to Ghana and United Arab Emirates, also Madagascar and Indian Ocean islands.

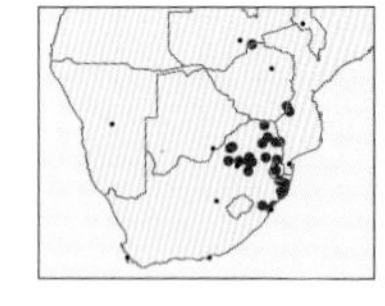

## 8 *Secusio strigata* — Orange Tiger

Forewing 16–19mm. Orange; thorax with black and cream lines; forewings cinnamon brown with veins white from base to median area, row of variable white oblique patches, merged in wet-season forms to almost absent in extreme dry-season forms; hindwings in females sometimes with pale brown border. Five species in genus in region. **Biology:** Caterpillar host associations unknown. Adults diurnal, but also attracted to light, throughout the year. **Habitat & distribution:** Grassland and bushveld; KwaZulu-Natal, South Africa, north to East Africa.

1
1
2
2
3
3
4
SB
4
5
5
MA
6
SW
6
7
8

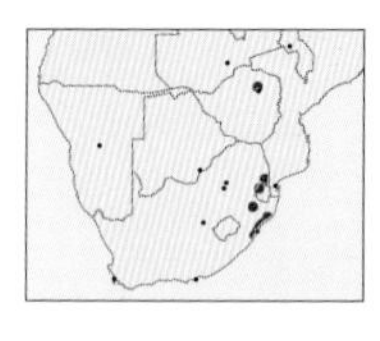

**1 *Creatonotos punctivitta*** Pink Ermine

Forewing 16–21mm. Thorax and wings pinkish-cream, abdomen deep pink with dorsal line of black spots; forewings pink to cream with thin, black stripe that ends in a few separated dots; hindwings white with grey dot on anal margin. One species in genus in region. **Biology:** Caterpillar host associations unknown. Adults throughout the year. **Habitat & distribution:** Moist grassland; KwaZulu-Natal, South Africa, north to tropical Africa.

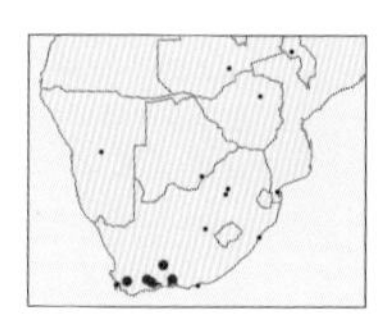

**2 *Cymaroa grisea*** Red-collar Tiger

Forewing 16–20mm. Ground colour orange, fading in dried specimens; forewings variably dusted in fine grey to brown scales, with darker zigzag lines; hindwings orange with cell spot and broken subterminal line; abdomen pale yellow with black spots. Woolly bear caterpillar black and stubby, covered in clusters of either black or white bristles. **Biology:** Caterpillars feed on exotic Ivy-leaved Toadflax (*Cymbalaria muralis*) and Bietou (*Osteospermum moniliferum*). Adults display a conspicuous red collar when disturbed, otherwise retracted. Bivoltine, adults August–October and March–April. **Habitat & distribution:** Dry fynbos and karoo habitats; Western Cape and Eastern Cape, South Africa.

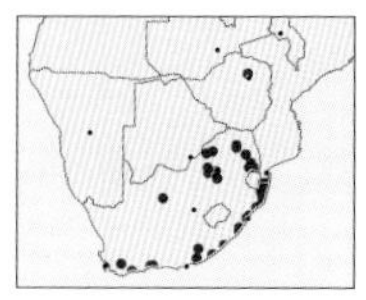

**3 *Diota rostrata*** Senecio Tiger

Forewing 14–16mm. Variable; wings off-white, grey or buff, with darker zigzag bands and sometimes with small black dots at base; thorax with black dots, abdomen yellow with a row of black dots. Caterpillar bicoloured: pale and dark green or grey and white, covered in clusters of white bristles, pair of long forward-projecting darker hairs near head. One species in genus in region. **Biology:** Caterpillars feed on *Senecio* spp. (Asteraceae). Multivoltine, adults throughout the year. **Habitat & distribution:** Variety of habitats where *Senecio* spp. grow; Western Cape, South Africa, north to tropical Africa.

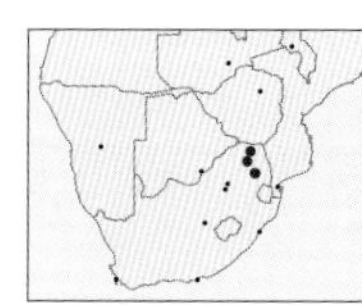

**4 *Galtara doriae*** Doria's Lace

Forewing 20–22mm. Body grey, thorax grey with numerous black and yellow dots, abdomen yellow with line of black dots; wings broadly rounded; forewings white with diagnostic pattern of brown patches and shading; hindwings hyaline with light brown border. Four species in genus in region. **Biology:** Caterpillars recorded feeding on *Senecio*. Adults September–January. **Habitat & distribution:** Moist forests; Mpumalanga, South Africa, north to tropical Africa.

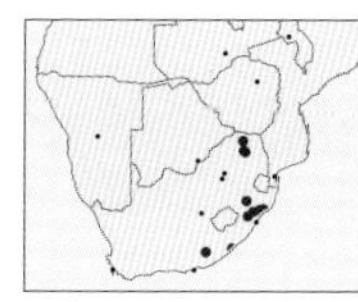

**5 *Galtara pulverata*** Pale Lace

Forewing 19–21mm. Body and wings grey to cream; thorax with 3 lines each of 3 black streaks, abdomen dull white; forewings dirty white with diagnostic light and brown patches (intensity varying across specimens), hindwings uniform white with a few faint marginal spots. Similar *G. purata*, from southern Cape forests, is densely dusted with brown scales, markings small and faint when present. **Biology:** Caterpillar host associations unknown. Adults September–April. **Habitat & distribution:** Moist forests; Eastern Cape north to Limpopo, South Africa.

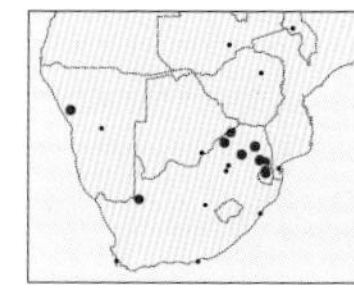

**6 *Galtara nepheloptera*** Grey Lace

Forewing 14–17mm. Superficially similar to *Digama ostentata* (Aganainae p.308), wings narrow for genus, body and wings grey; thorax with 3 lines of black spots, abdomen yellow with black dorsal spots; forewings grey with diagnostic brown patches and black dots, white areas along costa variable, specimens from arid areas paler; hindwings pale yellow to cream without markings. **Biology:** Caterpillar host associations unknown. Adults September–March. **Habitat & distribution:** Bushveld and desert; Northern Cape, South Africa, north to Tanzania.

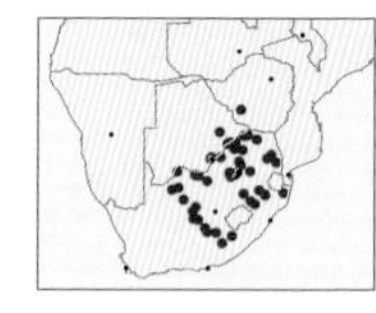

**7 *Paralacydes arborifera*** Branched Ermine

Forewing 13–20mm. White with variable brown, bifurcating bands running along forewings; hindwings white with black cell spot and occasionally with a few other black spots; thorax striped and yellow abdomen has 3 lines of black dots running along its length. Caterpillar grey with black, white and yellow stripes running along body; orange setal bases, white and black bristles on each segment, with some longer white hairs. Six species in genus in region. **Biology:** Caterpillars polyphagous on low-growing plants. A species adapted to summer rainfall, multivoltine, adults throughout the year, peaking in summer. **Habitat & distribution:** Grassland and bushveld; Eastern Cape, South Africa, north to tropical Africa.

1
2
AM
2
3
3
4
5
6
7
7

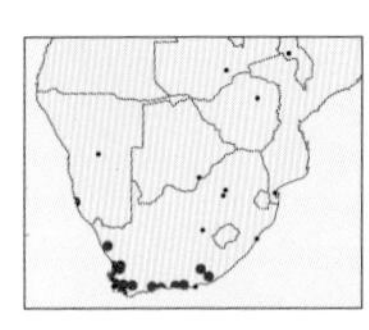

**1 *Paralacydes vocula*** Cape Branched Ermine

Forewing 13–15mm. Similar to *P. arborifera* (372), but forewings marked with broad, dark brown branching markings; hindwings with variable brown blotches, spots and lines; thorax brown with 2 white stripes. Caterpillar black with white stripe; whorls of bristles and hairs on each segment, forming short white stripe topped with longer black tufts. **Biology:** Caterpillars polyphagous on low-growing plants. Similar *P. bomfordi* has marks on forewings covering most of wing and hindwing uniform brown. A species adapted to winter rainfall, multivoltine, adults throughout the year, peaking in winter. **Habitat & distribution:** Fynbos, succulent karoo and desert; Western Cape, South Africa, north to Namibia.

**2 *Pseudophragmatobia parvula*** Cape Ruby Tiger

Forewing 13–14mm. Bulky body, brown abdomen striped with pink and orange circular bands; forewings reddish-brown with small black marking near mid-margin, hindwings pink with subterminal band and central dark spot. Caterpillar black, with whorls of short black bristles, and a thin, white and orange stripe central stripe. **Biology:** Caterpillars feed on Gifbol (*Drimia filifolia*), Jakkalsstert (*Anthospermum aethiopicum*) and Bitou (*Osteospermum moniliferum*). Adults June–January. **Habitat & distribution:** Fynbos; Western Cape, South Africa.

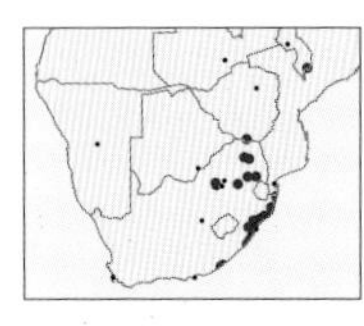

**3 *Podomachla apicalis*** White Bear

Forewing 22–26mm. Brown-black with white or cream-coloured diagonal band; hindwings white with broad black border; all wings sometimes with white tips at margin; thorax black with white dots. Caterpillar banded black and white or cream, red head and tufts of forward projecting setae. One species in genus in region. **Biology:** Caterpillar feed on various daisy species such as Fireweed (*Crassocephalum crepidioides*) and Canary Creeper (*Senecio deltoideus*). Multivoltine, adults throughout the year. **Habitat & distribution:** Moist forests and riverine vegetation; Eastern Cape, South Africa, north to tropical Africa.

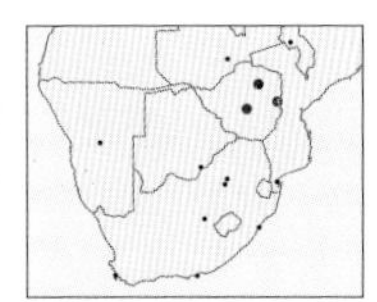

**4 *Seydelia ellioti*** Elliot's Tiger

Forewing 17–22mm. Forewings chocolate brown, criss-crossed with broad pink to yellow lines that form cross at wing apex; hindwings and abdomen largely pink or yellow; thorax plush and pilose, chocolate brown with medial pink or yellow stripe; abdomen pink with line of brown spots. One species in genus in region. **Biology:** Caterpillar host associations unknown. **Habitat & distribution:** Forest and woodland; Zimbabwe, South Africa, north to tropical Africa.

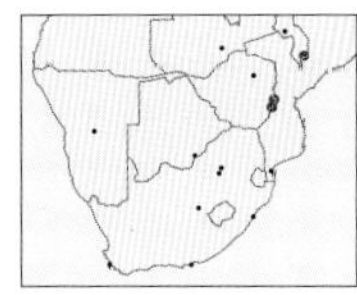

**5 *Karschiola holoclera*** Yellow-band Tiger

Forewing 27–30mm. Ground colour black, red and yellow collar, on thorax, with 2 black central dots that appear like false eyes; abdomen black with apical third red, white spots below; forewings pure black with yellow band across; hindwings black with large white basal spot, males with long cream androconial setae. One species in genus in region. **Biology:** Caterpillar host associations unknown. A diurnal species, courting male flutters around female displaying its androconia before mating. **Habitat & distribution:** Forest and woodland; Zimbabwe north to tropical Africa.

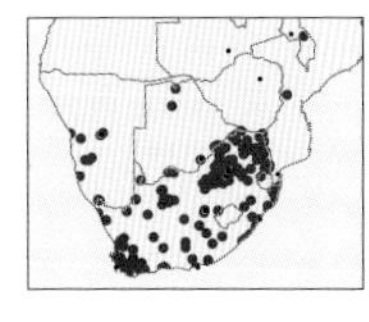

**6 *Utetheisa pulchella*** Crimson-speckled Footman

Forewing 15–21mm. Head yellow with black spots; abdomen white; forewings white, evenly marked with small black and pink-crimson panels. Caterpillar white with thin or broad irregular lateral stripes and segmental spots, incomplete orange bars on each segment, covered in long white hairs. **Biology:** Caterpillars obligatory polyphagous specialists, needing to feed on many plants to obtain chemicals for defence, reproduction and nutrition. Often abundant in disturbed fields, flying by day. **Habitat & distribution:** All habitats that support vegetation; most parts of Africa, Madagascar, the Middle East, as well as in Europe, Central Asia and western Indo-Malaysia.

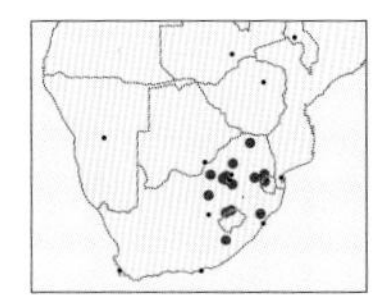

**7 *Ilemodes astriga*** Buttered False-footman

Forewing 15–20mm. Body and forewings shiny white in both sexes; forewings with small black cell spot, underside yellow; hindwings vivid yellow with small cell spot, male has short tail with androconial setae in anal fold; legs pale yellow, thorax with a few small black dots. Three species in genus in region. **Biology:** Caterpillar host associations unrecorded. **Habitat & distribution:** Forest and closed canopy bush; Eastern Cape, South Africa, north to tropical Africa.

1
1
2
2
AM
3
4
5
6
6
7

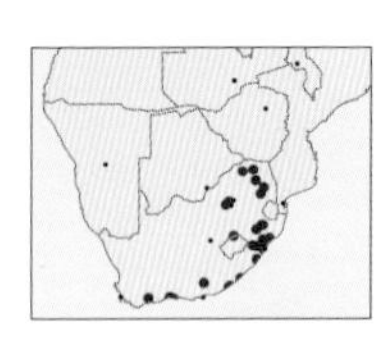

**1 *Ilemodes heterogyna*** Broad Buttered False-footman

Forewing 19–22mm. Thorax white with 4 black dots arranged in a square, abdomen yellow; sexually dimorphic; female forewings white, charcoal stripe running along length near inner margin; male forewings dusted with brown scales apart from white inner margin; hindwings bright yellow with black cell spot, males with short tail and androconia in anal fold. Caterpillar finely mottled yellow and black with large orange setal bases, setae of varying length. **Biology:** Caterpillars feed on algae and mosses, scraped off leaves. **Habitat & distribution:** Moist forests and kloofs; Western Cape, South Africa, north to tropical Africa.

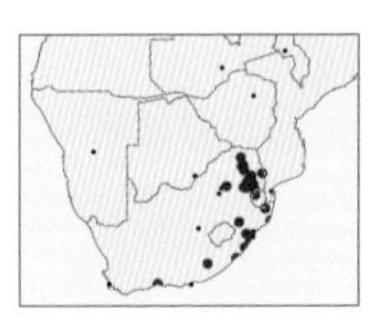

**2 *Cyana marshalli*** Marshall's Brown Lines

Forewing 10–13mm. Body and wings white; wings oval, forewings with terminal area shaded in grey and incomplete black lines of varying intensity; hindwings uniform white. Caterpillar fairly squat, grey-brown and covered in long and dense brown bristles. Eleven species in genus in region. **Biology:** Caterpillars feed on lichens. Adults throughout the year. **Habitat & distribution:** Moist forests and closed-canopy riverine vegetation; Western Cape north to Limpopo, South Africa.

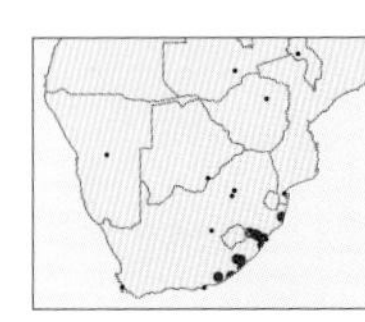

**3 *Cyana rhodostriata*** Red Brown Lines

Forewing 10–13mm. Very similar to *C. marshalli* (above), but can be distinguished by the lack of grey shading and by the red colour interspersed in the black lines; hindwings pure white. **Biology:** Caterpillar host associations unknown. Adults September–May. **Habitat & distribution:** Moist forests; Eastern Cape north to KwaZulu-Natal, South Africa.

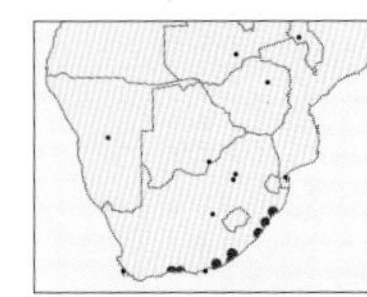

**4 *Cyana capensis*** Orange Lines

Forewing 10–12mm. Body and wings shining white, legs orange; forewings with 3 evenly spaced bands each comprising a thin orange and black line, 2 black spots in median area; hindwings uniform white. **Biology:** Caterpillar host associations unknown. Adults December–April. **Habitat & distribution:** Moist forests; Western Cape north to KwaZulu-Natal, South Africa.

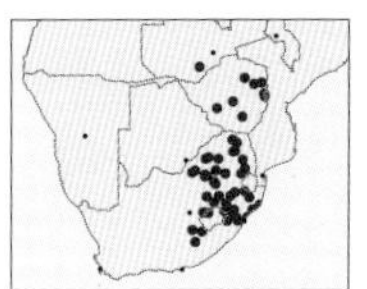

**5 *Cyana pretoriae*** Straight Red Lines

Forewing 13–17mm. Body and wings shining white, thorax with scarlet spots, legs striped yellow and white; forewings with 4 evenly spaced scarlet bands, straight postmedian band diagnostic, single and then paired black spots between middle bands, terminal band scarlet, faint or sometimes absent in female; hindwings uniform white. **Biology:** Caterpillars feed on lichens. Adults throughout the year. **Habitat & distribution:** Moist places in forest, grassland and bushveld; Eastern Cape, South Africa, north to Kenya.

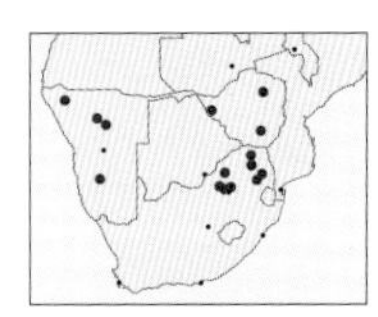

**6 *Cyana ignifera*** Kinked Red Lines

Forewing 12–15mm. Similar to *C. pretoriae* (above), but postmedian band diagnostically kinked toward black spot on forewing and terminal line comprising 3 brown spots; hindwings uniform white. **Biology:** Caterpillar host associations not recorded. Adults September–April. **Habitat & distribution:** Forest and bushveld, seems to be better adapted to arid conditions than other *Cyana* spp.; Gauteng, South Africa, north to Angola.

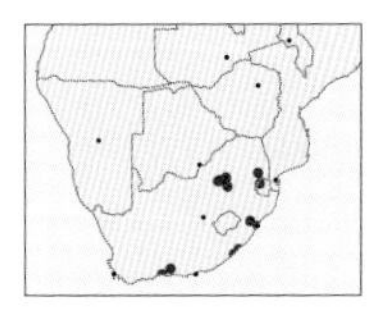

**7 *Cyana rejecta*** Lineless Yellow

Forewing 11–14mm. Unlike any other *Cyana* spp., with legs, antennae, head, body, forewings and hindwings a rich yellow, otherwise no markings apart from the 2 conspicous black forewing dots. **Biology:** Caterpillar host associations not recorded. Adults September–May. **Habitat & distribution:** Moist habitats with closed-canopy vegetation, including gardens; Western Cape north to tropical Africa and Madagascar.

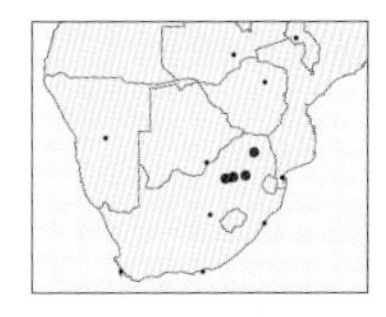

**8 *Carcinopodia argentata*** Silver Footman

Forewing 15–22mm. Body and wings cream-coloured, orange collar on thorax; forewings shining white or cream, antemedial and postmedian lines set back with distinctive additional curve toward apex, diagnostic antemedial line with distinct loop from inner margin and another near costa; hindwings yellow with grey basal shading in female. Three species in genus in region; similar *C. aurantiaca* has antemedial line without loops. **Biology:** Caterpillar host associations unknown. Diurnal species mostly active in afternoon, adults October–January. **Habitat & distribution:** Grassland; Gauteng, Mpumalanga and Limpopo, South Africa.

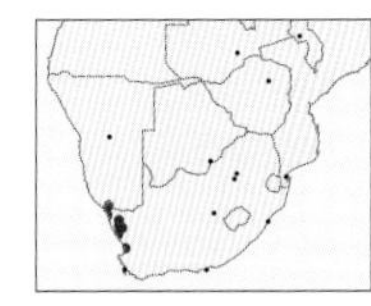

**9 *Eilema trichopteroides*** Cryptic Footman

Forewing 15–16mm. Cryptic; body and forewings dark brown densely covered in fawn scales, less dense in some areas leaving ill-defined dark blotches; hindwings uniform fawn. Three species in genus in region. **Biology:** Caterpillar host associations unknown. Adults August–October. **Habitat & distribution:** Arid winter-rainfall areas; Western Cape and Northern Cape, South Africa.

1
1♂
2
2
3
4
5
6
7
8
9

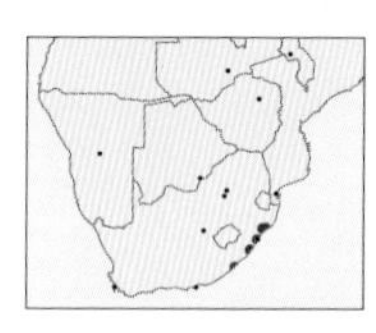

## 1 *Pseudocragia quadrinotata* — Sun Footman

Forewing 11–13mm. Orange; forewings with 2 black dots oblique with costa, costa black basally; hindwings canary yellow. Caterpillar light brown with black, dorsal longitudinal stripe, 3 dark brown bars on head, middle and rear of body, which is covered in dense, long straw-coloured hairs. One described species in genus in region. **Biology:** Caterpillars recorded from Lecanoromycetes lichens. Adults December–May. **Habitat & distribution:** Moist forests; Eastern Cape, South Africa, north to east Africa.

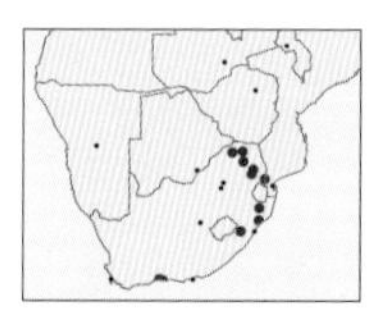

## 2 *Lamprosiella eborella* — Glossy Footman

Forewing 13–16mm. Head and frons orange, thorax grey, abdomen yellow; forewings glossy white, which appears silvery when viewed from an angle, variable slight orange edge on costa, otherwise no markings; hindwings canary yellow with no markings. Four described species in genus in region. **Biology:** Caterpillar host associations unknown. Adults throughout the year. **Habitat & distribution:** Cool, moist forests; Western Cape north to Limpopo, South Africa.

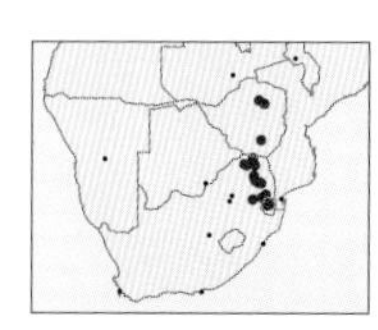

## 3 *Oedipygilema aposema* — Blacktip Footman

Forewing 13–17mm. Head, frons, thorax and forewings deep orange; forewings with a broad black terminal tip; hindwings colour split in two by straight line, upper wing orange and lower wing canary yellow, whole wing with broad black border. One species in genus in region. Superficially similar *Lepista pandula* has ground colour yellow on both wings. **Biology:** Caterpillar host associations unknown. Adults September–March. **Habitat & distribution:** Lowland forests and bush; Eswatini north to Zimbabwe and Mozambique.

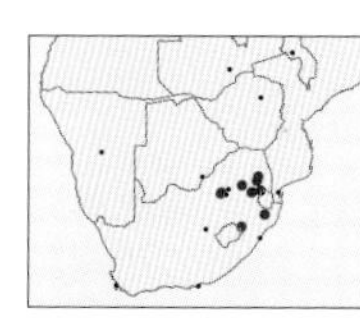

## 4 *Lomilema dilatans* — Fawn-grey Footman

Forewing 14–18mm. Head orange, thorax and forewings grey, fawn costal streak of forewing widens toward apex with fawn setae; hindwings creamy yellow without markings. Two species in genus in region. Several superficially similar species in different genera, but all have fawn costal streak parallel, not widening toward apex. **Biology:** Caterpillar host associations unknown. Adults August–April. **Habitat & distribution:** Moist forests and riverine vegetation; Eastern Cape, South Africa, north to Zimbabwe.

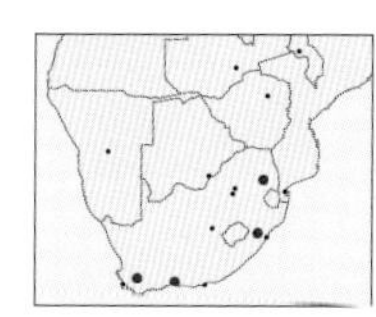

## 5 *Didymonyx infumata* — Tri-spot Footman

Forewing 13–14mm. Head orange, thorax and forewings charcoal grey; forewings with orange border and 3 distinctive black spots, 1 in cell and 2 beyond; hindwings orange with black cell spot, variably dusted with dark scales, area around anal angle clear of dusting; underside of both wings distinctly orange and black. One species in genus in region. **Biology:** Caterpillar host associations unknown. Adults actively fly by day, congregating in certain areas. Adults February–March. **Habitat & distribution:** Fynbos and fynbos remnant vegetation; Western Cape to Mpumalanga.

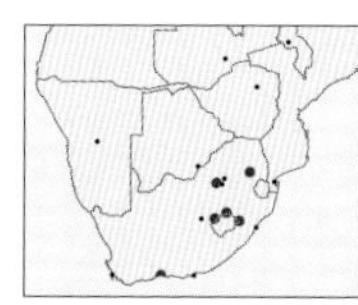

## 6 *Pusiola interstiteola* — Creamstreak Footman

Forewing 9–11mm. Species in this genus rather broad-winged; head orange, thorax dark brown, forewings grey with cream streaks along veins; hindwings fawn with central brown streak across from base. Three species in genus in region. **Biology:** Caterpillar host associations unknown. Adults February–March. **Habitat & distribution:** Montane fynbos and grassland; Western Cape to Mpumalanga, South Africa.

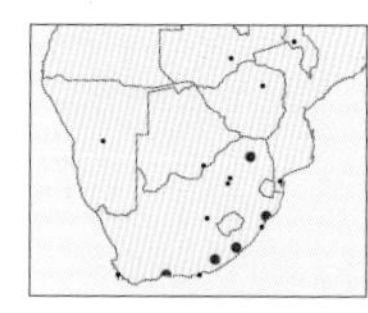

## 7 *Exilisia bipuncta* — Uncertain Footman

Forewing 10–11mm. Body and wings creamy white, variably dusted with dark brown scales, darker dusting in antemedial and postmedian line areas, 2 distinct dark black spots either side of cell; hindwings uniform white to grey. One species in genus in region, tribal placement uncertain. **Biology:** Caterpillar host associations unknown. **Habitat & distribution:** Moist forests; Western Cape to Limpopo, South Africa, also Madagascar.

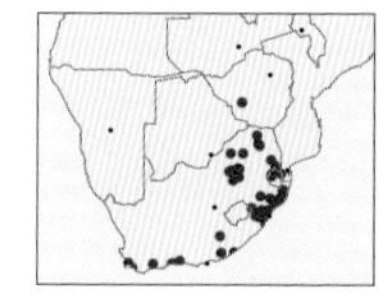

## 8 *Siccia caffra* — Speckled Grey Footman

Forewing 7–11mm. Creamy white; forewings covered in evenly spaced black dots. Caterpillar a cryptic, dull green lichen mimic with longitudinal white stripe and covered in short white bristles. Six species in genus in region; similar *S. melanospila* has evenly curved postmedian line, made up of black dots. **Biology:** Caterpillars recorded from Lecanoromycetes rock lichens. Multivoltine, adults throughout the year where conditions allow. **Habitat & distribution:** Variety of moist habitats, including gardens; Western Cape, South Africa, north to Mozambique and Zimbabwe.

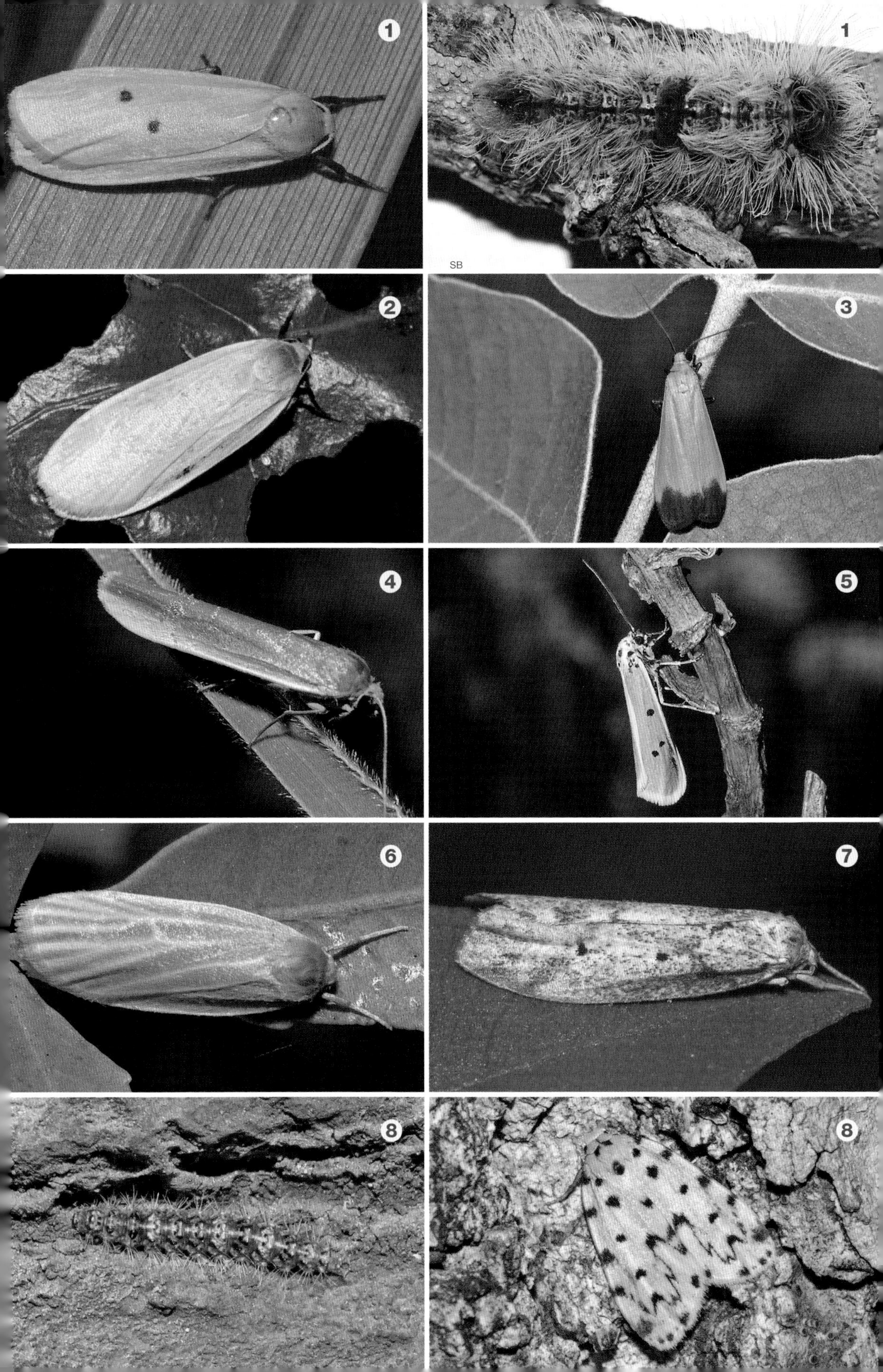
1
1
SB
2
3
4
5
6
7
8
8

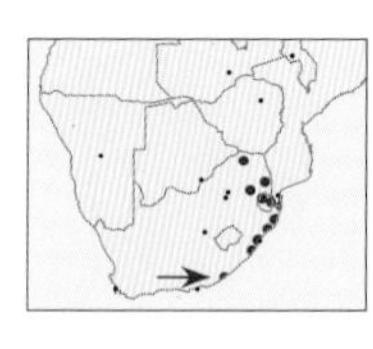

### 1 *Tumicla sagenaria* — Crossed Footman

Forewing 9–15mm. Body, ground colour orange to yellow; forewings criss-crossed by a number of broad purple-brown lines of varying intensity but with a distinctive pattern. Caterpillar with distinct alternating white and black longitudinal bands with extensive, long white hair fringes along sides and dorsum. **Biology:** Caterpillars probably feed on Lecanoromycetes lichens. **Habitat & distribution:** Lowland moist forests; Eastern Cape, South Africa, north to tropical Africa.

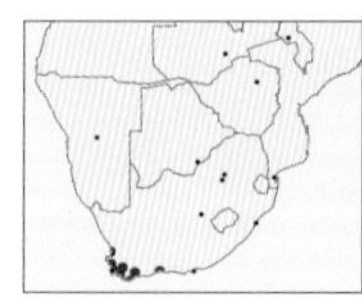

### 2 *Amata cerbera* — Heady Maiden

Forewing 14–17mm. Body satiny, iridescent blue-black, with 4 diagnostic scarlet abdominal bands, antennae with white tips; fore- and hindwings satiny blue-black with almost transparent white windows that vary in size. Caterpillar black, covered in short, dense ginger-brown and black bristles. Sixteen species in genus in region; similar, but allopatric from Limpopo, *A. schellhorni* has forewings narrower and basal hindwing window more square. **Biology:** Caterpillars polyphagous on herbaceous plants including grass, can withstand long periods with no food. Diurnal, adults September–May. **Habitat & distribution:** Open, often transformed habitats; Western Cape, South Africa.

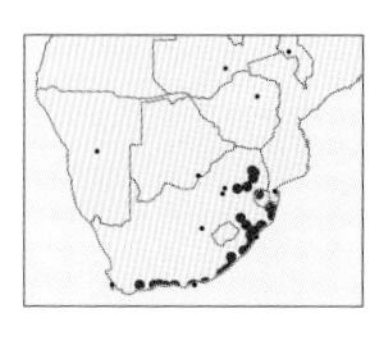

### 3 *Amata kuhlweinii* — Cool Maiden

Forewing 13–16mm. Similar to *A. cerbera* (above), but lacks the scarlet abdominal bands with only 1 spot at base of abdomen, which varies from orange to red. Caterpillar black, densely covered in black and russet bristles, head capsule orange. **Biology:** Caterpillars recorded from canary creepers (*Senecio tamoides* leaves, *S. macroglossus* flowers). Diurnal adults September–May. **Habitat & distribution:** Forest and kloofs; southern Cape north to Limpopo, South Africa.

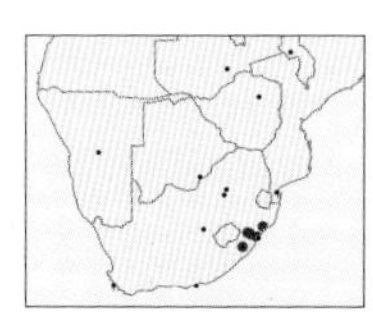

### 4 *Amata longipes* — Long-legged Maiden

Forewing 11–14mm. Body and wings black, with iridescent blue sheen; forewings with 5 opaque windows arranged as 2 triangles, and white wing apex; much smaller hindwings with 2 windows; abdomen with diagnostic single white bar near thorax, and white patches along sides. Similar *A. pseudosimplex* has abdominal bar yellow-orange. **Biology:** Caterpillar host associations unknown. Diurnal, adults October–April. **Habitat & distribution:** Moist forests; KwaZulu-Natal, South Africa.

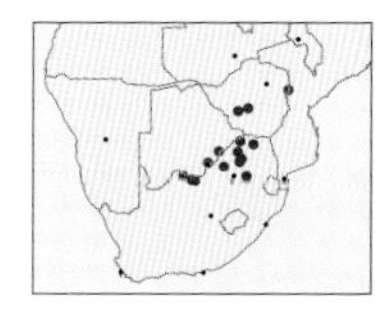

### 5 *Asinusca atricornis* — Orange-spot Maiden

Forewing 9–11mm. Body and wings black, with purplish sheen; forewings with 5 opaque windows characteristically arranged and no white wing apex; much smaller hindwings with 2 windows with 2 round spots, basal one orange and outer one hyaline; abdomen black with no markings. **Biology:** Caterpillar host associations unknown. Diurnal, adults November–April. **Habitat & distribution:** Bushveld; North West, South Africa, north to Zambia.

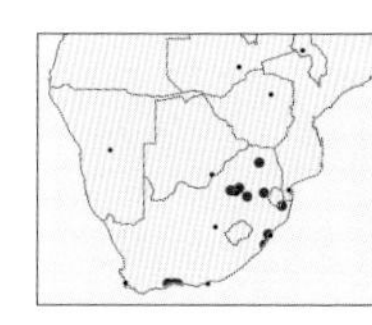

### 6 *Ceryx fulvescens* — Yellow Sleeved Maiden

Forewing 9–12mm. Thorax and abdomen dark brown, with a row of yellow triangles on abdominal segments and yellow tufts along body; long and triangular forewings brown with large hyaline windows dusted with yellow scales; hindwings similar but smaller. Caterpillar black, with thin white intersegmental bands, sparsely covered in long, white hairs. Six species in genus in region. **Biology:** Caterpillars feed on wet, decaying grass and other plant material. Diurnal, adults throughout the year. **Habitat & distribution:** Grassy wetlands, including gardens; Western Cape, South Africa, north to tropical Africa.

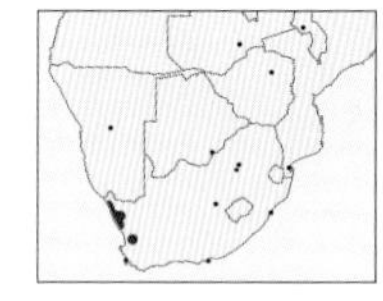

### 7 *Aizoa namaqua* — Wingless Maiden

Forewing 15–16mm. Males with squat, grey and white pilose body and charcoal grey wings with large white windows and iridescent sheen; females with minute vestigial wings and plump body; red and white abdominal bands in both sexes. Caterpillar squat, with orange bars along top of body and densely covered in short white bristles. **Biology:** Caterpillars polyphagous on succulent perennial plants. Univoltine, adults August–October. **Habitat & distribution:** Succulent karoo habitats; Western Cape and Northern Cape, South Africa.

### 8 *Euchromia amoena* — Pleasant Hornet

Forewing 21–23mm. Body black, with orange or red thorax; cream and red or yellow bands on abdomen; wings long, narrow and triangular, black with large yellow to orange panels. Caterpillar squat, pink-red with dense tufts of black hairs along sides and long tufts of black hairs at head and rear of body, top of body with short white tufts of hair. Two species in genus in region. **Biology:** Caterpillars polyphagous. Diurnal, adults throughout the year, males often seen congregating feeding on wilted plants. **Habitat & distribution:** Primarily in coastal forest and bush; Eastern Cape, South Africa, north to East Africa, also Madagascar.

1
1
2
SW
3
4
5
6 ♂
7
7 ♂
7 ♀
SB
8
8

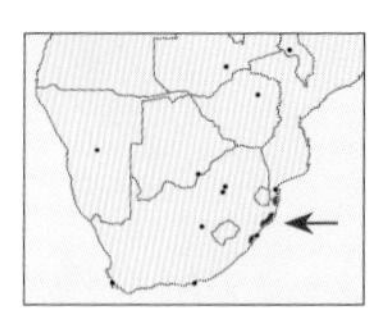

**1 *Euchromia folletii*** Splendorous Hornet

Forewing 22–25mm. Body black, thorax red, head metallic blue; abdomen with metallic blue bands, and 2 orange-red bands; wings long and triangular, black, with large yellow panels and a few metallic blue markings. Caterpillar squat, yellow, with sparse, short tufts of black hairs along the side of the body, and a pair of longer black hair tufts above head. **Biology:** Caterpillars polyphagous. Diurnal, adults throughout the year. **Habitat & distribution:** Primarily in coastal forest and bush; KwaZulu-Natal, South Africa, north to tropical Africa, also Madagascar.

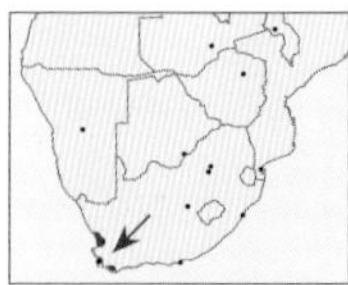

**2 *Eutomis minceus*** Trimen's Maiden

Forewing 17–19mm. Body and wings black with white markings; thorax pilose, black with white lateral patch, abdomen black with white segmental bands; fore- and hindwings iridescent blue-black with numerous opaque white panels. One species in genus in region. **Biology:** Caterpillar host associations unknown. **Habitat & distribution:** Strandveld habitats; Western Cape, South Africa.

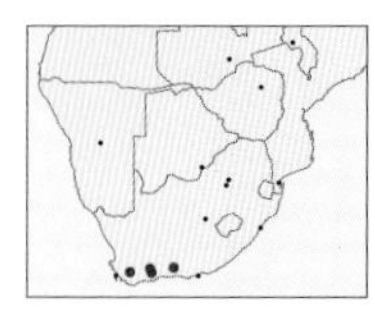

**3 *Cacoethes polidamon*** Golden Maiden

Forewing 17mm. Black with golden-yellow markings; thorax pilose, black with yellow dorsal patch in some specimens, abdomen black with golden-yellow segmental bands; fore- and hindwings iridescent green-black with diagnostic opaque yellow panels; hindwings with large yellow panel incorporating a diagnostic black spot, female probably apterous. Two species in genus in region; similar *C. fulvatrum* has fewer and smaller panels in forewing and lacks the black spot in hindwing panel. **Biology:** Caterpillar host associations unknown. Adults diurnal but also attracted to light, August–October. **Habitat & distribution:** Inland mountain fynbos habitats; Western Cape and Eastern Cape, South Africa.

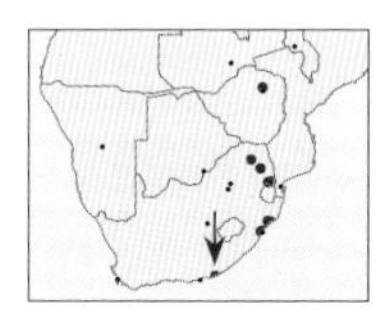

**4 *Pseudonaclia puella*** Girl Maiden

Forewing 8–14mm. Body brown, abdomen brown to yellow, with line of black markings; wings very long and narrow; forewings light brown to black with large white or golden-yellow panels; hindwings short, yellow or orange, with dark irregular border. Caterpillar brown, fairly smooth, with darker markings on each segment. **Biology:** Caterpillars feed on decaying plant material, females go deep into leaf litter to oviposit in suitably moist places. Diurnal, adults August–April, peaking November. **Habitat & distribution:** Moist forest and riverine vegetation; Eastern Cape, South Africa, north to tropical Africa.

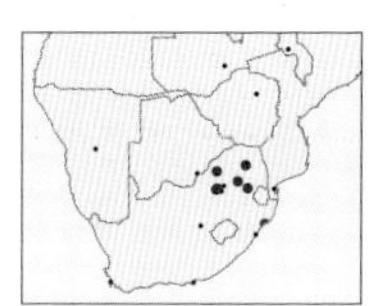

**5 *Metarctia jansei*** Sombre Matron

Forewing 13–15mm. Body squat, dark to pale brown, thorax densely pilose brown; forewings dark pinkish-brown fading toward apex, paler in some specimens, distinctive brown curved streak beyond cell on forewings; hindwings pale fawn with pinkish tinge. Caterpillar brown or black, covered in clusters of short, pale bristles. Thirteen species in genus in region. **Biology:** Caterpillars polyphagous, often seen walking on the ground in gardens. Adults throughout the year. **Habitat & distribution:** Variety of habitats; KwaZulu-Natal, South Africa, north to Kenya.

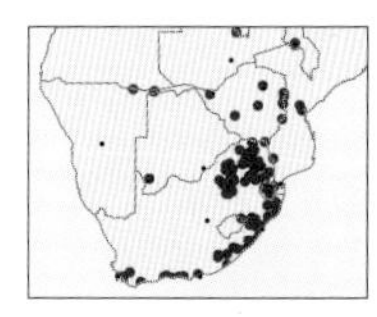

**6 *Metarctia rufescens*** Reddish Matron

Forewing 16–18mm. Similar to *M. jansei* (above), but larger and darker; forewings uniform dark reddish-brown with darker brown blotch (rather than streak) beyond cell, diagnostic yellow streak along subcostal vein, when present; hindwings uniform pinkish-brown. **Biology:** Caterpillars recorded from spiderworts (*Tradescantia*) and grass roots. Adults throughout the year. **Habitat & distribution:** Variety of relatively moist habitats, including gardens; Western Cape,, South Africa, north to tropical Africa.

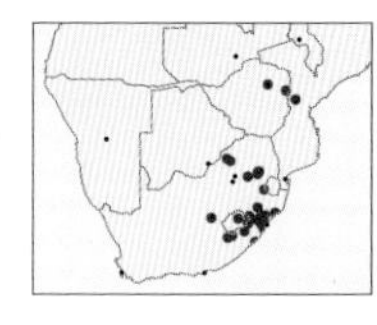

**7 *Metarctia lateritia*** Laterite Matron

Forewing 17–27mm. Body and wings intense rose-coloured; forewings diagnostic grey with rose-coloured veins, hindwings uniform pink; thorax and abdomen pilose, the latter banded in black; conspicuous pectinate black antennae. Similar *M. cinnamomea* has body and wings entirely rosy pink with no grey on forewing. **Biology:** Caterpillar host associations unknown. Adults often noticed on walls of buildings, October–April. **Habitat & distribution:** Variety of relatively moist habitats, including gardens; Eastern Cape, South Africa, north to tropical Africa.

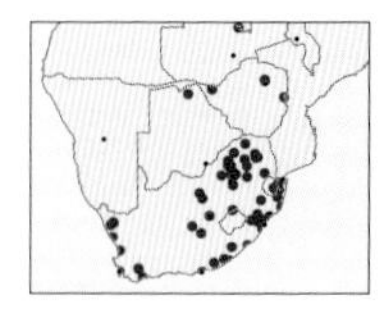

**8 *Apisa canescens*** Greyling

Forewing 13–24mm (small and large forms). Thorax pilose, grey-brown, abdomen white; forewings pearly grey-white, paler at base, dark grey along costa; hindwings reduced, white. Some forms are generally darker. Caterpillar cream-white with rows of tufts of very long, brown hairs. Two species in genus in region. **Biology:** Caterpillars polyphagous, sometimes occur in caves. Adults throughout the year. **Habitat & distribution:** Wide variety of habitats; Western Cape, South Africa, north to Arabia.

1
1
2
DM
3
4
DS
4
5
5
6
7
8

**1 *Automolis bicolora*** Twin-set Matron

Forewing 13–14mm. Thorax pilose, sulphur yellow with diagnostic black central area, abdomen banded yellow and grey; forewings charcoal grey with slightly hyaline centre, narrow yellow costal region diagnostic; hindwings hyaline with grey border. Four species in genus in region. **Biology:** Caterpillar host associations unknown. Adults January–March. **Habitat & distribution:** Moist grasslands; Eastern Cape, KwaZulu-Natal and Lesotho, also recorded from Zimbabwe.

**2 *Automolis meteus*** Orange-fronted Matron

Forewing 17–18mm. Sexually dimorphic; flightless female plump, short-winged, pale yellow-brown or golden; male with golden-yellow body, conspicuous pectinate antennae, abdomen banded yellow and black; forewings charcoal grey with diagnostic golden costal region and border; hindwings golden-yellow base with rest of wing charcoal. **Biology:** Caterpillars recorded from pigeon woods (*Trema* spp.), *Acalypha* and *Lasiosiphon*. Adults throughout the year. **Habitat & distribution:** Variety of habitats, mostly grassland; Western Cape north to Limpopo, South Africa, also reported from Somalia.

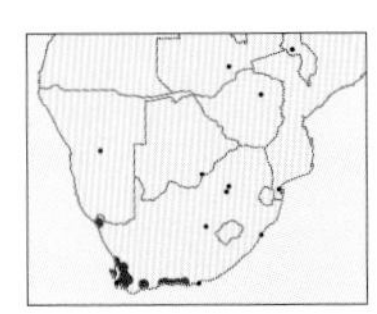

**3 *Automolis crassa*** Brown Matron

Forewing 15–17mm. Body and all wings brown; forewings with veins slightly blackened and costa black in most individuals; hindwings uniform brown. **Biology:** Caterpillar host associations unknown. **Habitat & distribution:** Variety of habitats; Western Cape and Northern Cape, South Africa.

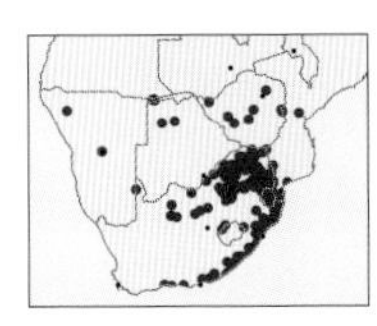

**4 *Thyretes caffra*** Bar Maiden

Forewings 11–15mm. Tan to black, thorax with 2 pairs of white patches but uniform grey between diagnostic, abdomen brown with a row of yellow markings, one on each segment; forewings tan or black, with broad line of mostly coalesced white or yellow windows; hindwings small, white or yellow with black mark on trailing edge. Caterpillar brown, with segmental black bands with long, white hairs arising from tubercles. Four species in genus in region. **Biology:** Caterpillars recorded on Sweet Thorn (*Vachellia karroo*). Adults throughout the year, peaking in summer. **Habitat & distribution:** Variety of habitats; Western Cape, South Africa, north to tropical Africa.

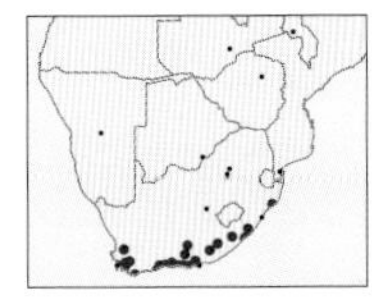

**5 *Thyretes hippotes*** Equine Maiden

Forewing 15–20mm. Similar to *T. caffra* (above), but more robust; thorax white and pilose, with 2 black lines; abdomen white, with black-circled yellow spot on each segment; forewings brown, with extensive white panels, basal panel does not extend to below next panel, hindwings reduced, with central white panels and brown margins. Caterpillar black, with tufts of white bristles and red head capsule. **Biology:** Caterpillars polyphagous. **Habitat & distribution:** Fynbos and open habitats; Western Cape north to coastal KwaZulu-Natal, South Africa, described from Angola.

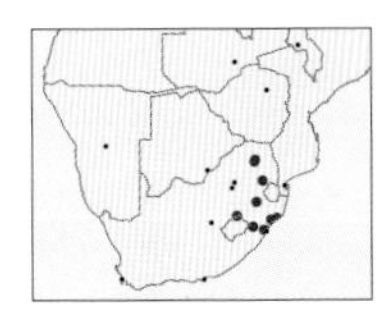

**6 *Thyretes montana*** Mountain Maiden

Forewing 16–21mm. Large pectinate antennae; bulky, thorax pilose, straw-coloured and marked with black lines, abdomen white, segments outlined by black areas and each segment bears a pale yellow spot; wings dark brown, with extensive creamy opaque panels, forewing basal panel extended to below next panel, diagnostic. **Biology:** Caterpillar host associations unknown. Adults November–January. **Habitat & distribution:** Moist grasslands; KwaZulu-Natal north to Limpopo, South Africa.

## Subfamily Lymantriinae Tussock Moths

Tussock moths have caterpillars heavily cloaked in secondary hairs. Many have a series of dorsal patches of bristles alternating with tufts of longer hairs ('pencils'). When these patches are absent, they are replaced by an eversible gland. These, irritating, hairs are often incorporated into the pupal silk. Newly hatched caterpillars remain covered by these hairs and feed gregariously. Other species protect the egg mass by secreting a froth that hardens over the eggs. Adult moths, many of which are diurnal, do not feed, and there is considerable sexual dimorphism, flightless females are common. Caterpillars are mostly polyphagous. There are 232 species in the region.

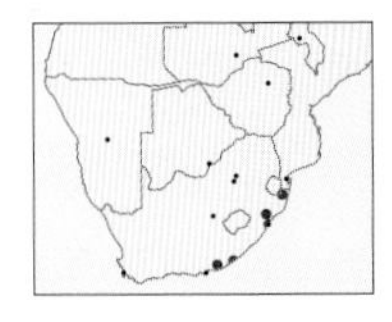

**7 *Aroa difficilis*** Ornate Vapourer

Forewing 15–18mm. Forewings dark grey, with broad, sinuate light grey median area (more conspicous in females) separating basal grey area from black terminal regions, wing bordered by black crenulated terminal line; hindwings black with 3 bold orange streaks. Caterpillar grey, becoming black, hairy, paired tufts of grey hair on dorsum, followed by pairs of bright blue short bristles, red head capsule. Four species in genus in region. **Biology:** Caterpillars reared on Bushveld Bluebush (*Diospyros lycioides*). Males diurnal, adults August–March. **Habitat & distribution:** Grassland; Eastern Cape and KwaZulu-Natal, South Africa.

1
2♂
3♂
4
5
5
6
7

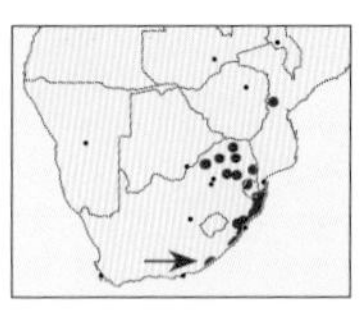

### 1 ***Aroa discalis*** — Banded Vapourer

Forewing 14–20mm. Male has yellow-brown forewings with broad oblique yellow or orange band, and black hindwings with broad orange streak; female with fore- and hindwings pale yellow, with diagnostic oblique, rust-coloured postmedian and subterminal lines across forewings and black cell spot. Caterpillar squat, heavily cloaked in long, black hairs, pale yellow on dorsal and ventral parts of body. **Biology:** Caterpillars feed on bristle grass (*Setaria*). Males diurnal, females nocturnal, adults August–May. **Habitat & distribution:** Forest and woodland; Eastern Cape, South Africa, north to tropical Africa.

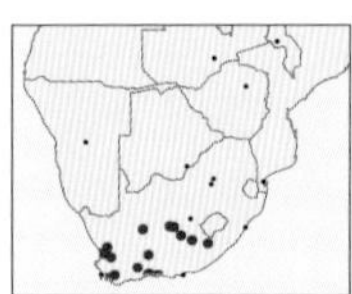

### 2 ***Bracharoa dregei*** — Brown Vapourer

Forewing 8–12mm. Forewings of male rusty to dark brown with yellow border, 2 black cell spots often merged; wings with broad charcoal border, no cell spot; female wingless and with reduced legs, body banded with pinkish hairs. Caterpillar squat, black bristles with 4 clumps of short, ginger bristles, sides of body bear extensive tufts of white bristles. Some forms have very long grey tufts of hair dorsally. Four described species in genus in region. **Biology:** Caterpillars polyphagous. Multivoltine, adults August–May. **Habitat & distribution:** Variety of open habitats; Cape provinces, South Africa.

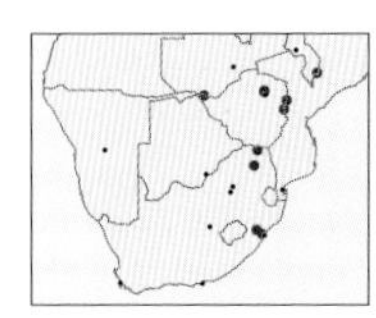

### 3 ***Bracharoa mixta*** — Mixed Vapourer

Forewing 9–13mm. Male with pale to dark brown forewings, transverse black and chalky white bar and variable white marking near wing apex; hindwings grey or rufous-brown, forelegs long and hairy; female wingless. Caterpillar black dorsally, with 4 clumps of pale yellow bristles, clumps of short orange bristles on grey sides of body; very long black hairs arising from sides of body and tufts of hair on head and rear of body. **Biology:** Caterpillars polyphagous on trees. Multivoltine, adults August–May. **Habitat & distribution:** Forests, woodland and cultivated areas; KwaZulu-Natal, South Africa, north to tropical Africa.

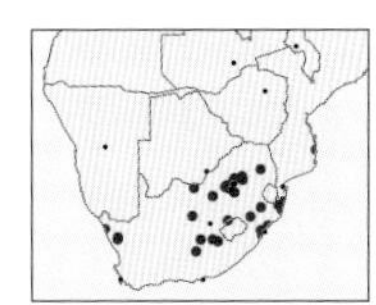

### 4 ***Bracharoa quadripunctata*** — Twin Vapourer

Forewing 10–12mm. Male sulphur-yellow, with 2, sometimes 3, small black spots on forewing and black antennae; hindwing colour as forewings with black cell spot and border; female apterous with vestigial legs, covered in cream, woolly hairs. Caterpillar black dorsally with 4 clumps of short bristles, (ginger or black), and another black tuft at end of body, sides of body with dense, long projecting clumps of white hairs. **Biology:** Caterpillars polyphagous, mainly on herbaceous plants. Multivoltine, adults throughout the year. **Habitat & distribution:** Grassland; Eastern Cape, South Africa, north to tropical Africa.

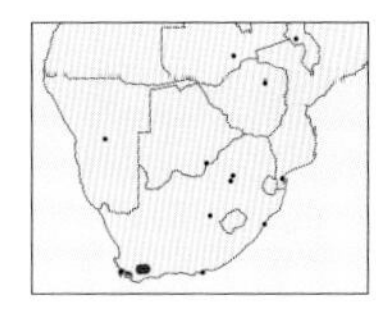

### 5 ***Bracharoa tricolor*** — Tiger Vapourer

Forewing 10–14mm. Male with highly pilose grey body, forewings brown to mustard yellow, with 2 sinuous, black transverse bands and row of black spots along border, hindwings black with large crescentic, yellow marking; female wingless and with vestigial legs, abdomen covered in bands of fine white hairs. Caterpillar black dorsally, with tufts of long white hairs; sides of body orange, bearing long black hairs; earlier instars darker. **Biology:** Caterpillars feed on various ice plants (Aizoaceae). **Habitat & distribution:** Succulent karoo habitats; Western Cape, South Africa.

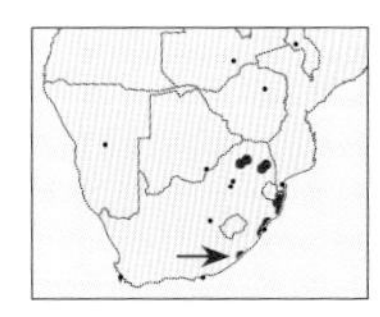

### 6 ***Cropera testacea*** — Coated Orange Peel

Forewing 13–21mm. Wings short and rounded, body and wings uniform orange, with diagnostic black cell spot on all wings; female paler than male and with tuft of hair at end of abdomen. Caterpillar pale orange, darker on back with paired tuft of black hairs; tufts of long, white hairs project from sides of body. Four species in genus in region. **Biology:** Caterpillars feed on grasses. Adults August–May. **Habitat & distribution:** Bushveld and coastal grassland; Eastern Cape, South Africa, north to tropical Africa.

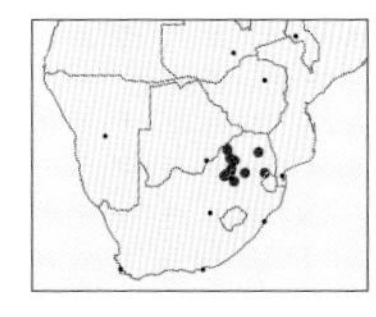

### 7 ***Cropera sericea*** — Old Orange Peel

Forewing 16–21mm. Similar to *C. testacea* (above), with wings short and rounded, but extremely setose body and wings of male and female entirely uniform cream coloured without markings. Can easily be confused with some females of *Euproctis subalba* (p.390), which are much smaller. **Biology:** Caterpillar host associations unknown. Adults August–March, peaking in February–March. **Habitat & distribution:** Grassland and bushveld; Gauteng north to Limpopo, South Africa.

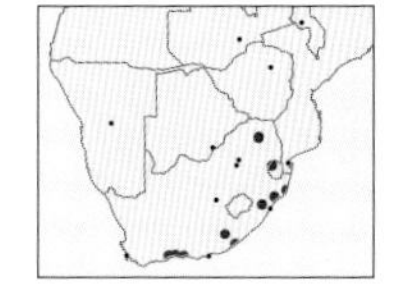

### 8 ***Dasychira octophora*** — Forest Puzzle

Forewing 13–18mm. Similar to *D. bryophilina* (p.388), but wings broader and occupies a different habitat; ground colour pure white, seldom slightly off-white but not rich cream, cell spot evenly rounded and of ground colour, lines not connected on inner margin; hindwings white with incomplete grey subterminal line. **Biology:** Caterpillar host associations unknown. Adults September–April. **Habitat & distribution:** Moist forests; Western Cape, South Africa, north to Zimbabwe.

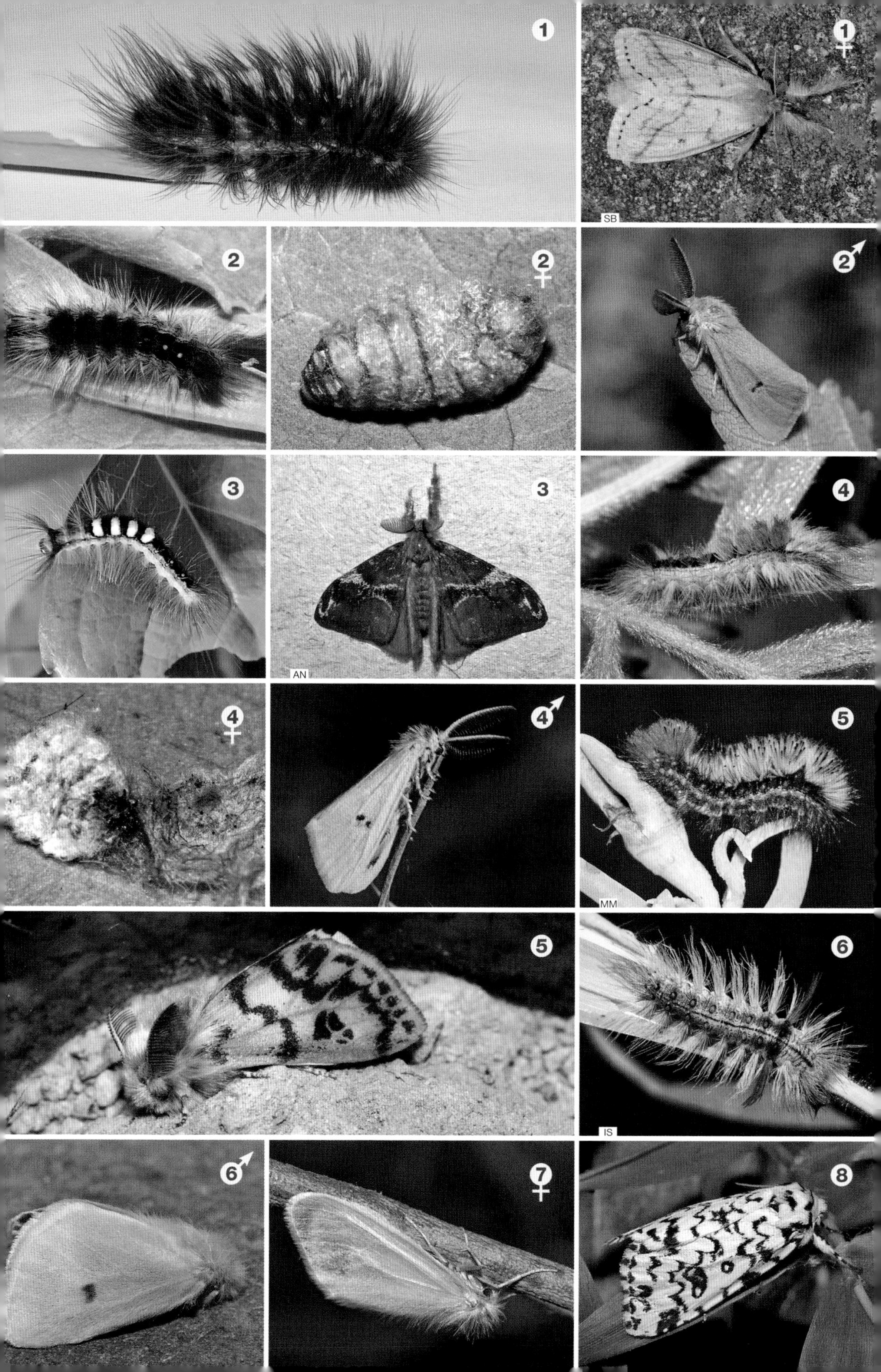
1
1♀
SB
2
2♀
2♂
3
3
AN
4
4♀
4♂
5
MM
5
6
IS
6♂
7♀
8

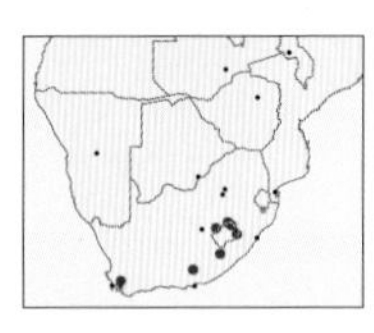

**1 *Dasychira bryophilina*** Fynbos Puzzle

Forewing 14–20mm. Individuals from high altitudes larger; rich cream, pilose thorax with some black setae, forewing antemedial and postmedian lines connected on inner margin, diagnostic uneven black ringed, different from ground colour, white spots on base, around cell and along postmedian line in median area, chequered cilia; hindwings white with variable grey blotches, sometimes completely grey. Three species in genus in region. **Biology:** Caterpillar host associations unknown. Adults November–March. **Habitat & distribution:** Fynbos and Afroalpine vegetation; Western Cape, South Africa, north to Lesotho.

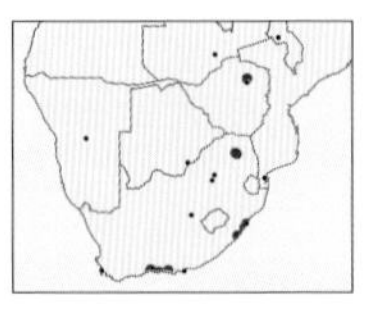

**2 *Dicranuropsis vilis*** Grey Puzzle

Forewing 16–22mm. Similar to *Dasychira* spp. (above), but ground colour charcoal grey, basal and terminal areas of forewings variably dirty white, large round black-ringed cell spot and kidney-shaped marking beyond diagnostic; hindwings charcoal grey. One species in genus in region. **Biology:** Caterpillar host associations unknown. Adults throughout the year. **Habitat & distribution:** Moist forests; Western Cape, South Africa, north to Zimbabwe.

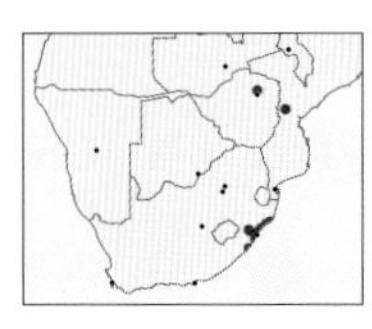

**3 *Eudasychira georgiana*** Georgiana's Tussock

Forewing 16–30mm. Thorax pilose, silvery white; abdomen largely orange with row of black dots dorsally, last few segments white, legs white and hairy; forewings snowy white, with incomplete thin zigzag lines, veins grey, diagnostic elongate black-bordered mark on costa; hindwings white. Caterpillar variable – ventral parts green or black, white or yellow lateral stripe, body covered in tufts of long white hairs, usually with 4 upright tufts of pale yellow hairs near head and sometimes with black tuft near rear. Four species in genus in region. Multivoltine, adults throughout the year. **Biology:** Caterpillars polyphagous on trees. **Habitat & distribution:** Moist forests; Eastern Cape, South Africa, north to tropical Africa.

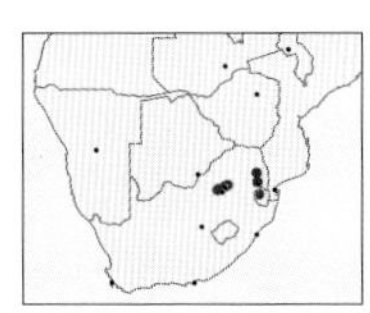

**4 *Eudasychira poliotis*** Hoary Tussock

Forewing 17–25mm. Stout, short brown antennae with white shaft, abdomen orange limbs covered in long white hairs; rest of body and forewings white, densely dusted with grey scales, diagnostic straight lines; hindwings pure white. Caterpillar flesh-coloured with reddish dorsum, 4 brown tufts of erect bristles, and long white hairs extending laterally from body. **Biology:** Caterpillars feed on various *Parinari* spp. (Chrysobalanaceae). Bivoltine, adults October–March. **Habitat & distribution:** Moist forest and rocky grassland where *Parinari* spp. grow; Eswatini north to tropical Africa.

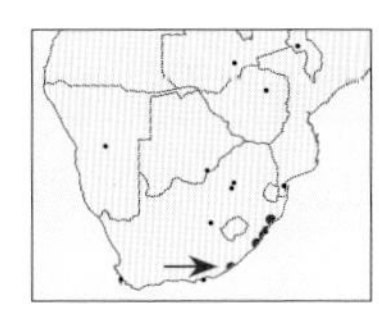

**5 *Eudasychira proleprota*** Earthy Tussock

Forewing 18–23mm. Sexually dimorphic; male with pointed wings, body and forewings brown with cream mottling, lines indistinct, hindwings brown; female forewings creamy brown with zigzag black lines, variable dusting of greenish scales, blue scales in cell, black 'V'-shaped marks and marginal dots; abdomen orange; hindwings orange with earthy brown border. **Biology:** Caterpillars polyphagous. Bivoltine, adults October–March. **Habitat & distribution:** Moist tropical forests; Eastern Cape to Zimbabwe, also Kenya.

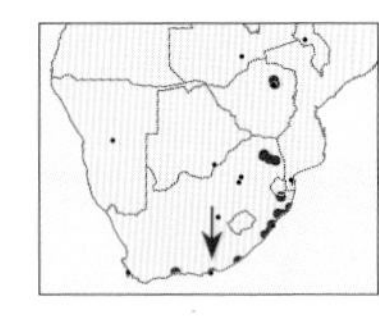

**6 *Euproctis aethiopica*** Peppered Euproctis

Forewing 10–12mm. Pale to deep yellow; forewings with paler diagonal bands broadly peppered with black scales in varying densities, less so in females; hindwings uniform cream to yellow. Caterpillar distinctive, black with short black bristles and white 'I'-shaped marking on back, with central red line; pair of longer hair tufts project forward from orange head region. Twenty-four species in genus in region. **Biology:** Caterpillars polyphagous on woody hosts. Multivoltine, adults throughout the year. **Habitat & distribution:** Humid habitats, including gardens; Western Cape, South Africa, north to tropical Africa.

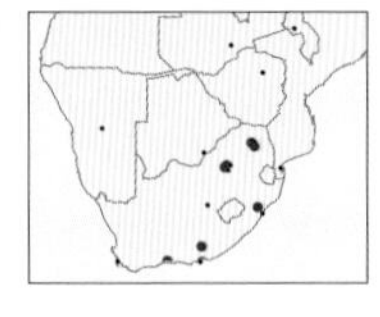

**7 *Euproctis hardenbergia*** Blackbar Euproctis

Forewing 12–15mm. Similar to *E. aethiopica* (above), but in male antemedial and postmedian bands much more densely covered in black scales, sometimes connected along inner margin, in female bands much reduced, ground colour a rich orange yellow; hindwings yellow, often dusted with dark scales in males. **Biology:** Caterpillars polyphagous on various low-growing plants. Adults throughout the year. **Habitat & distribution:** Forests; Western Cape north to Limpopo, South Africa.

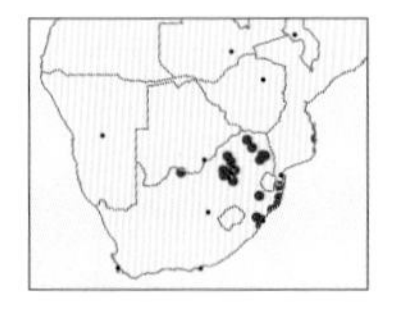

**8 *Euproctis bicolor*** Bi-coloured Euproctis

Forewing 13–18mm. Forewings yellow, with single red dot and incomplete, speckled transverse band; hindwings deep yellow; female paler, tuft of black hairs at end of abdomen. Caterpillar with line of dense, white bristles, interrupted by 2 black patches, sides of body covered in tufts of dense white bristles. **Biology:** Caterpillars polyphagous on plants, mosses and lichens. Bivoltine, adults throughout the year. **Habitat & distribution:** Grassland and bushveld; KwaZulu-Natal, South Africa, north to Zambia.

1
2
3
4
4
5 ♀
6
6 ♂
IS
IS
6 ♀
7 ♀
7 ♂
8
8

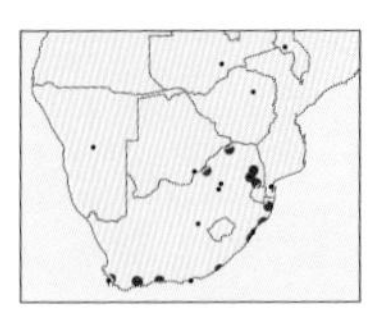

**1 *Euproctis nigripuncta*** Black-dotted Euproctis

Forewing 14–18mm. Yellow to mottled yellow and creamy white, legs hairy; forewings with broken, speckled and reddish diagonal bar, ending in black-ringed red cell spot, dotted black subterminal line variable but always some spots present; hindwings uniform cream. Caterpillar closely resembles that of *E. bicolor* (p.388), but tends to have more white setae. **Biology:** Caterpillars polyphagous. **Habitat & distribution:** Open habitats and forest; Western Cape north to Mpumalanga, South Africa.

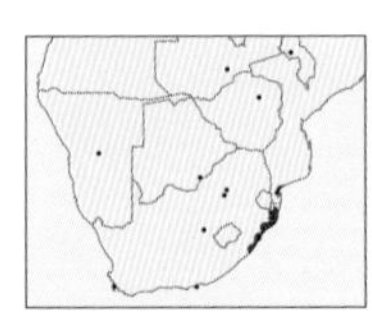

**2 *Euproctis punctifera*** Fine-dotted Euproctis

Forewing 15–18mm. Similar to *E. nigripuncta* (above), body deep orange, with a few faint diagonal lines, an incomplete, speckled bar, and a row of dots along edge of wing; in female wings pinkish-orange; hindwings and underside uniform bright orange. Caterpillar similar to that of *E. bicolor*, but has a few reddish dorsal bristle clusters toward rear. **Biology:** Caterpillars polyphagous. Adult males diurnal, flying in forest understorey, multivoltine, throughout the year. **Habitat & distribution:** Lowland forest and coastal bush; KwaZulu-Natal, South Africa, north to Mozambique.

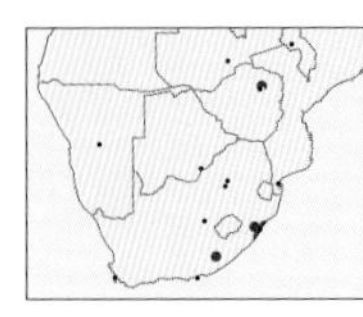

**3 *Euproctis rufopunctata*** Red Dotted Euproctis

Forewing 14–23mm. Wings and body cream-white, legs hairy; forewings with scattered blood red dots, one at fore-margin and others along edge of wing; hindwings white. Caterpillar reddish, with a pair of ginger bristle clusters near head and another at rear of body, long white hair clusters extend from sides of body, and orange hairs from head. **Biology:** Caterpillars polyphagous on trees. Prone to sporadic population explosions, which crash soon afterwards. Multivoltine, adults September–May. **Habitat & distribution:** Forests and gardens; Eastern Cape, South Africa, north to tropical Africa.

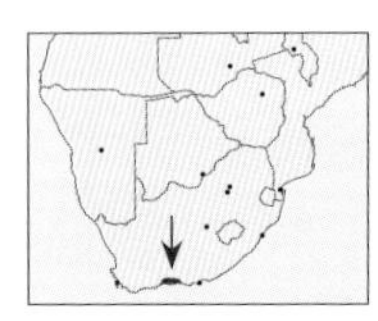

**4 *Euproctis haematodes*** Red-eye Euproctis

Forewing 15–20mm. Similar to *E. rufopunctata* (above), but with 2 large black-bordered red spots, one in cell and one subterminal, sometimes a third subterminal one, body and wings very hairy, cream or white; forewings cream and white; hindwings uniform white. Caterpillar densely covered in white setae with 2 black dorsal tufts and one on anal claspers. **Biology:** Caterpillars polyphagous on trees. Adults October–January. **Habitat & distribution:** Moist forests; southern Cape, South Africa.

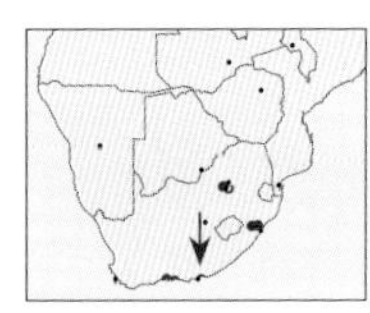

**5 *Euproctis subalba*** Ghost Euproctis

Forewing 11–14mm. Hairy, cream-white; forewings lacking markings, sometimes with a few small grey scattered spots; hindwings white. Caterpillar similar to that of *E. aethiopica* (p.388), but white dorsal stripe does not have side bars, and the thorax and head capsule are brown, not red and black. **Biology:** Caterpillars polyphagous on trees. Adults October–March. **Habitat & distribution:** Forests and riverine vegetation, including gardens; Western Cape north to Gauteng, South Africa.

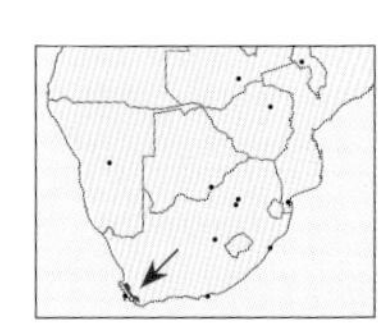

**6 *Euproctis arenacea*** Sandy Euproctis

Forewing 18–19mm. Body, wings and legs orange and hairy; forewings deep orange and marked with patches of dull purple speckles or raised scales, leading to rough appearance of wing; hindwings deep orange, sometimes with black speckling. Caterpillar squat, dull orange-brown, with segmental bands of short white hairs, and short cluster of black bristles near head. **Biology:** Caterpillars recorded from Christmas berries (*Passerina*). Adults April–September. **Habitat & distribution:** Winter-rainfall strandveld and fynbos; Western Cape, South Africa.

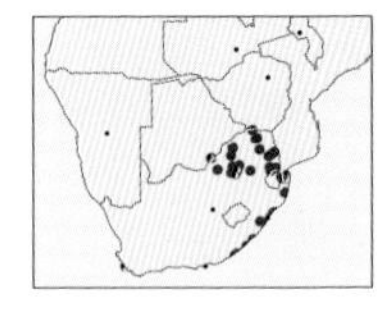

**7 *Euproctis fasciata*** Banded Euproctis

Forewing 11–19mm. Body, legs and wings hairy, pale to deep yellow; forewings with diagnostic antemedial and postmedian orange lines with dark purple speckling between them, apical parts of wing may have similar purple speckling; hindwings pale yellow. Caterpillar similar to that of *E. punctifera* (above), but larger with fewer red bristle clusters and denser white bristle stripe dorsally. **Biology:** Caterpillars polyphagous. Multivoltine, adults throughout the year. **Habitat & distribution:** Variety of habitats; Eastern Cape, South Africa, north to tropical Africa and Saudi Arabia.

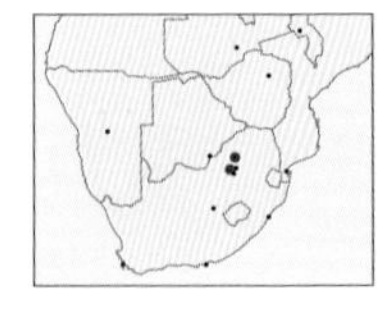

**8 *Pteredoa monosticta*** One Spot

Forewing 13–16mm. Body and forewings cream, forewings with single orange dot in cell; hindwings white without any markings. Caterpillar brown, with dense tufts of tan bristles, and 2 pairs of short, black bristle clumps after thorax, pair of short, black hair tufts project from head. **Biology:** Caterpillars feed on Wild Pear (*Dombeya rotundifolia*) and other Malvaceae. **Habitat & distribution:** Bushveld; North West, South Africa, north to tropical Africa.

1
1
2
JuB
2 ♂
2 ♀
3
SW
3 ♀
4
4
5
5
6
AM
6
7
7
8
8

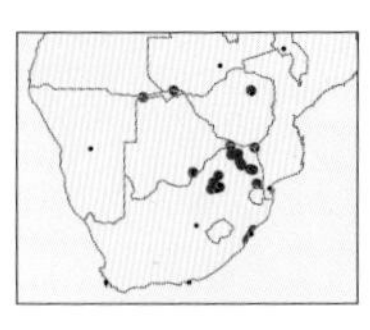

### 1 *Hemerophanes libyra* — Variegated Vapourer

Forewing 15–23mm. Male thorax brown with hair tuft; forewings variegated orange-brown and chalky white, with dark brown discal spot and inner margin of wing base; hindwings dark brown; female with cream forewings and orange hindwings. Caterpillar squat and pale cream, covered in dense tufts of ginger bristles, with longer bristles at rear, and a pair of blackish bristles arising from head, dorsum has a black-bordered cream stripe. Two species in genus in region. **Biology:** Caterpillars feed on various *Combretum* spp. Multivoltine, adults throughout the year where conditions allow. **Habitat & distribution:** Bushveld and forest where *Combretum* spp. grow; KwaZulu-Natal, South Africa, north to tropical Africa.

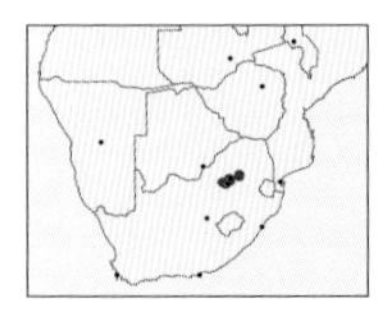

### 2 *Pseudobazisa perculta* — Small Bordered Orange

Forewing 12–14mm. Male orange; conspicuous black, pectinate antennae; fore- and hindwings similar – orange, veins in border region of wing apex black; female unknown. One species in genus in region. **Biology:** Caterpillar host associations unknown. Adult males diurnal, October–March. **Habitat & distribution:** Grassland; Gauteng to Mpumalanga, South Africa, also reported from DRC and Zambia.

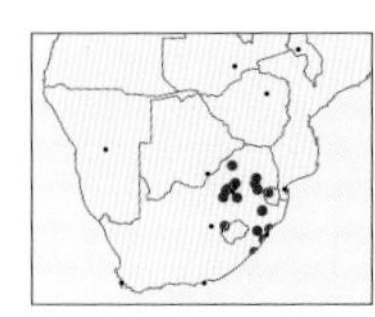

### 3 *Lacipa gemmata* — Pearl Lacipa

Forewing 13–15mm. White; thorax white with orange margins; forewings shiny white, with lines represented as orange dots, 2 or 3 closely spaced small black dots on postmedian line; hindwings pure white. Caterpillar pale yellow, segments after middle of body swollen, uniform yellow, outlined in black, medium-length white hairs extend from sides of body; a pair of thin, black antenna-like hair tufts extend from the head. Eleven species in genus in region. **Biology:** Caterpillars recorded from Small Ursinia (*Ursinia nana*). Bivoltine, adults October–June, peaking November and February. **Habitat & distribution:** Grassland; Eastern Cape north to Limpopo, South Africa, also reported from tropical Africa.

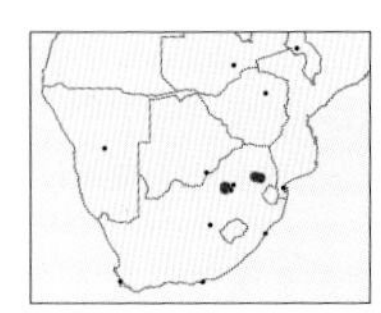

### 4 *Lacipa sexpunctata* — Six-spot Lacipa

Forewing 10–13mm. White; orange antemedial line sharply recurved before costa, postmedian line straight but curved slightly before costa, mostly 6 black spots around cell, line of well-defined marginal black dots; hindwings cream to yellow with no markings. Similar *L. gracilis*, from bushveld areas, has only 1 (sometimes 2) black spots around cell. Caterpillar cream or yellow and covered in medium-length bristles, and a pair of antenna-like clusters of long hairs arise from head; pair of large black markings on each abdominal segment, 2 red bristle clusters near end of body. **Biology:** Caterpillars polyphagous on herbaceous plants. **Habitat & distribution:** Highveld grassland; Gauteng and Mpumalanga, South Africa. Also reported from Democratic Republic of the Congo.

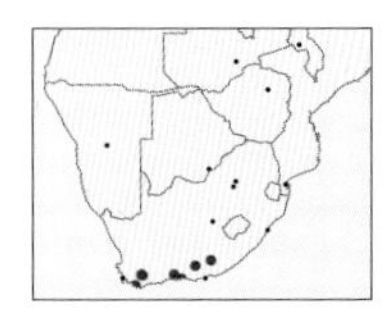

### 5 *Lacipa picta* — Embroidered Lacipa

Forewing 9–12mm. White; forewings with 3 lines diagnostically crenulated and bordered with black scales, distinct row of black marks in median area, line of black marginal dots. Similar *L. pulverea*, from highveld grasslands, has lines less crenulated, not bordered by black scales and black marks in median area broken. Caterpillar black with thin white-edged crimson line running length of body, whorls of black or white bristles, and pair of antenna-like clusters of long hairs arise from head. **Biology:** Caterpillars feed on flowers of low-growing plants. Adults October–April. **Habitat & distribution:** Succulent karoo and dry fynbos; Western Cape and Eastern Cape, South Africa.

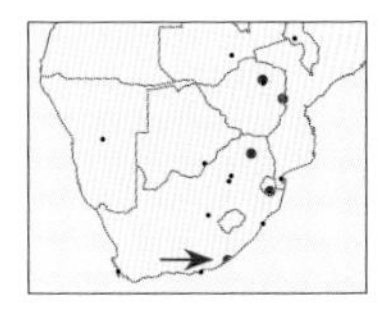

### 6 *Laelia curvivirgata* — Creeper Tussock

Forewing 14–18mm. Body, legs and wings hairy, light grey; forewings mottled brown and with cream transverse band, which has thin extensions to base and apex of wing; incomplete ocellus and large black mark at inner margin of this band; hindwings cream with irregular grey border. Caterpillar grey, covered in clusters of long, ginger bristles; pair of long, black antenna-like hairs projecting from head and rear of body, series of 4 black bars with central white spots on first abdominal segments. Sixty-four species in genus in region. **Biology:** Caterpillars feed on Traveller's Joy (*Clematis brachiata*). Adults throughout the year. **Habitat & distribution:** Moist forests; Eastern Cape, South Africa, north to tropical Africa.

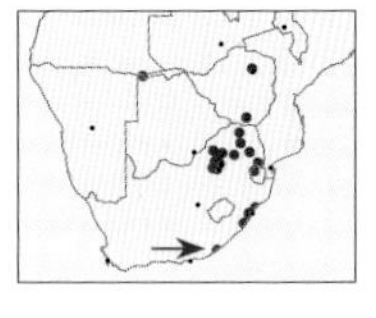

### 7 *Laelia extorta* — Extorted Tussock

Forewing 15–22mm. Forewings pale brown with darker brown or black mottling, concentrated at wing base; hindwings uniform cream, shape of postmedian line and clear median area with only cell spot diagnostic. Caterpillar grey or black, with very long, fine black or white hairs projecting from sides of body, 4 paired clusters of brush-like, black or brown bristles present on middle abdominal segments. **Biology:** Caterpillars feed on various *Ficus* spp. Bivoltine, adults September–April. **Habitat & distribution:** Forest and bushveld where fig trees grow; Eastern Cape, South Africa, north to tropical Africa.

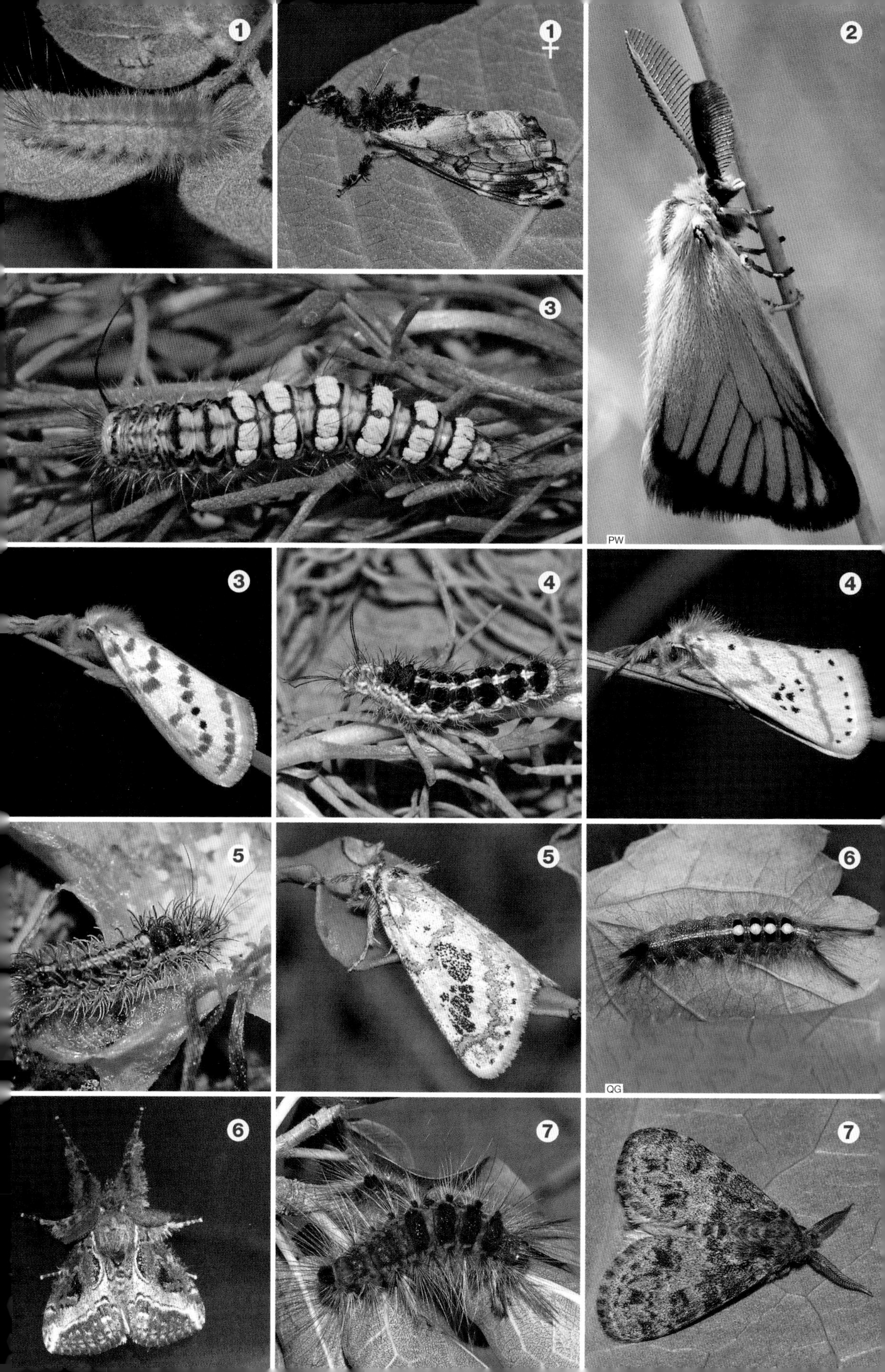
1
1 ♀
2
PW
3
3
4
4
5
5
6
QG
6
7
7

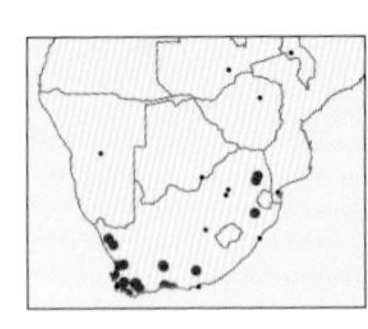

**1 *Laelia fusca*** Olive Tussock

Forewing 17–24mm. Wings and body grey, tan, brown or pale green; legs hairy; forewings marbled with a series of faint, transverse zigzag lines, line of black dots at edge of wing. Caterpillar green or orange and covered in clusters of dense, white bristles; white dorsal stripe running along length of body, 4 patches of orange bristles rising from anterior part of body, and long tufts of white or orange hair projecting from head and rear. **Biology:** Caterpillars feed on a variety of tree species. Adults throughout the year. **Habitat & distribution:** Fynbos and renosterveld habitats and fynbos remnants on mountains; primarily Cape provinces, also on various mountains up to Limpopo, South Africa.

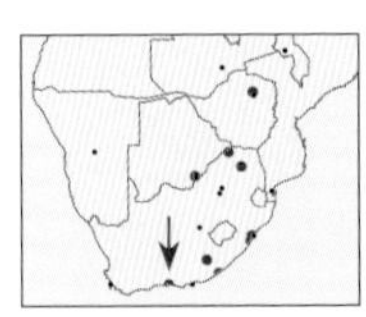

**2 *Laelia lunensis*** Lunate Tussock

Forewing 14–16mm. Body and wings grey to tan, legs hairy; forewings grizzled tan, shape of antemedial double line and postmedian lines diagnostic, distinctive curved black reniform and white patch in median area; hindwings white. Caterpillar cream or orange, with long, white bristles projecting from sides of body, 4 white pale yellow or green bristle tufts in middle of body. **Biology:** Caterpillars recorded from Dune False Currant (*Allophylus natalensis*) and Marula (*Sclerocarya birrea*). Adults throughout the year. **Habitat & distribution:** Forest and bushveld; Eastern Cape, South Africa, north to Zimbabwe.

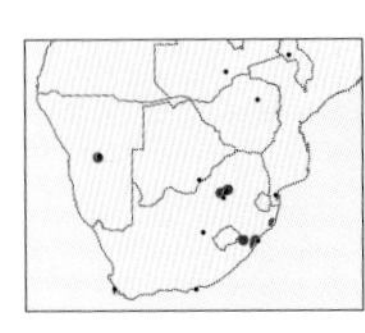

**3 *Laelia rocana*** P-Tussock

Forewing 15–18mm. Similar to *L. lunensis* (above), but has single antemedial line and postmedian line evenly curved from inner margin, large round ring around cell completing the letter 'P' on forewing; thorax, forewing basal area and along costa white, variably mixed with green and brown scales, some individuals all brown; hindwings white with broken subterminal band. **Biology:** Caterpillars polyphagous on bushveld and forest trees. Adults August–March. **Habitat & distribution:** Forest and bushveld; KwaZulu-Natal, South Africa, north to tropical Africa.

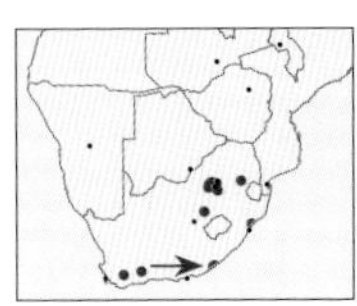

**4 *Laelia municipalis*** Municipal Tussock

Forewing 14–16mm. Body and wings cream; legs hairy; forewings variably dusted with grey scales, completely grey in some individuals, lines made up of diagnostic red dots, sometimes indistinct, with a black patch in cell, marginal dots; hindwings pure to dirty white. Caterpillar similar to that of *L. lunensis* (above), but body rust-red and the 4 pairs of pale yellow bristles enclosed by black border, narrow white stripe runs along top of body. **Biology:** Caterpillars feed primarily on Sweet Thorn (*Vachellia karroo*) but also Hook Thorn (*Senegalia caffra*). Adults October–April. **Habitat & distribution:** Habitats where thorn trees grow; Western Cape, South Africa, north to Tanzania.

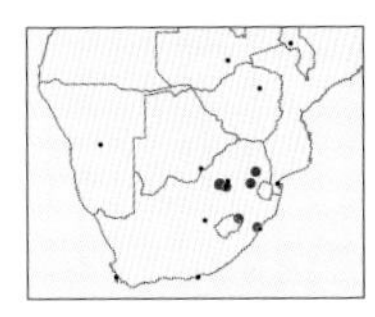

**5 *Laelia pyrosoma*** Fiery Tussock

Forewing 18–24mm. Thorax and wings pale buff or glossy white, abdomen white, banded orange-red and with orange tufts of hair; forewings uniform buff or shining white. Caterpillar covered in clusters of very dense and long, whitish bristles, with shorter patches of red bristles, each cluster emanating from fiery red base, and a red line running along the side of the body; pair of long, antenna-like hair tufts projecting from head and rear of body. **Biology:** Caterpillars feed on *Protea* spp. Adults throughout the year. **Habitat & distribution:** *Protea* grassland; KwaZulu-Natal, South Africa, north to Zimbabwe.

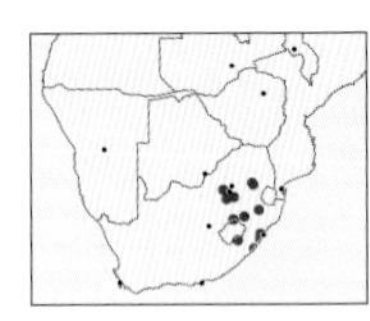

**6 *Laelia bifascia*** Bi-line Tussock

Forewing 13–15mm. Cream, thorax covered in dense setae; forewings dusted with black scales, 2 prominent black longitudinal lines, subterminal row of dots variable; hindwings white; female wings elongate and rounded. Very similar *L. angustipennis*, from the Cape provinces, paler with reddish-fawn ground colour. **Biology:** Caterpillars feed on grass. Adults August–March. **Habitat & distribution:** Grassland; Eastern Cape, South Africa, north to Kenya.

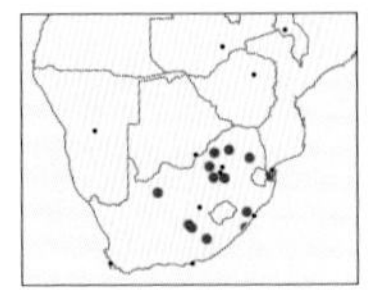

**7 *Laelia xyleutis*** Fawn Tussock

Forewing 16–19mm. Body and forewings fawn in male, cream in female, forewings finely dusted with brown scales and with prominent row of black subterminal spots; hindwings pale. Several similar species; *L. subrosea* has forewings rosy pink. **Biology:** Caterpillar host associations unknown. Adults throughout the year. **Habitat & distribution:** Grassland and bushveld; Eastern Cape north to Limpopo, South Africa, also reported from Arabian peninsula.

1
1
2
2
3
4
4
5
5
6
7

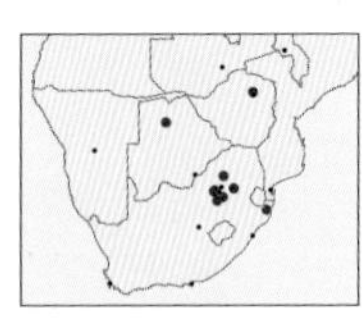

**1 *Lymantriades xanthosoma*** Yellow Tail

Forewing 8–12mm. Abdomen yellow, black segmental markings; forewings dark grey, with thin, black lines and a row of marginal dots; hindwings light grey. Caterpillar resembles that of *Lacipa gracilis*, with white dorsal stripe with a pair of short, red bristle patches, and short black hair tufts projecting from head. Can easily be confused with the superficially similar *Leptolepida rattus* (Notodontidae, p.298), but yellow abdomen and resting posture are diagnostic. One species in genus in region. **Biology:** Caterpillars feed on thorn trees (*Vachellia* spp.). Adults September–May. **Habitat & distribution:** Bushveld and grassland where thorn trees grow; Northern Cape, South Africa, north to tropical Africa.

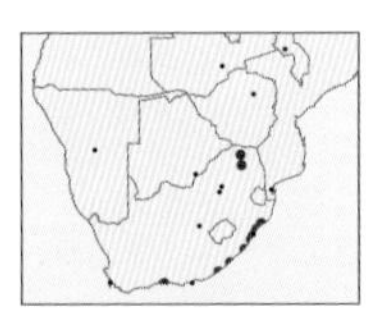

**2 *Creagra liturata*** Black-vein Satin

Forewing 17–24mm. Antennae, body and wings white, frons orange, legs black tipped; forewings with veins black in median and subterminal areas, otherwise no markings; hindwings white. One species in genus in region. **Biology:** Caterpillars recorded on *Hibiscus* (Malvaceae). Adults September–April. **Habitat & distribution:** Moist forests; Western Cape, South Africa, north to Kenya.

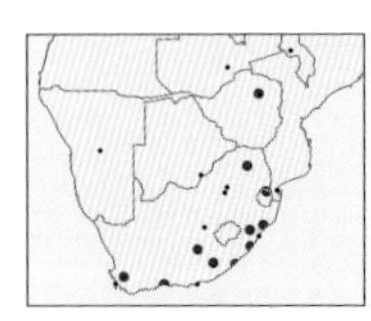

**3 *Marblepsis flabellaria*** Translucent Satin

Forewing 22–27mm. Two colour forms; one body orange, fore- and hindwings orange-yellow, fading in dried specimens, shining, translucent between veins, without markings; second form creamy translucent white; intermediates occur. Caterpillar covered in very long and dense yellow hairs, pupa retains hairs. **Biology:** Caterpillars polyphagous on trees. Adults August–May. **Habitat & distribution:** Forest, woodland and riverine vegetation; Western Cape, South Africa, north to Zimbabwe, also reported from further north.

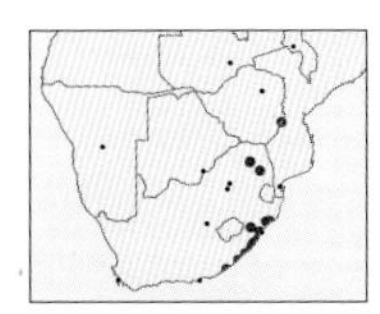

**4 *Marblepsis melanocraspis*** Black-base Satin

Forewing 13–20mm. Similar to *M. flabellaria* (above), but smaller with body and wings brilliant white and with thin black line along basal half of costa, otherwise no markings. **Biology:** Caterpillars feed on *Cassipourea gummiflua* (Rhizophoraceae). Males fly actively by day on forest edge, adults August–May. **Habitat & distribution:** Moist forests; Eastern Cape north to Zimbabwe, also reported from further north.

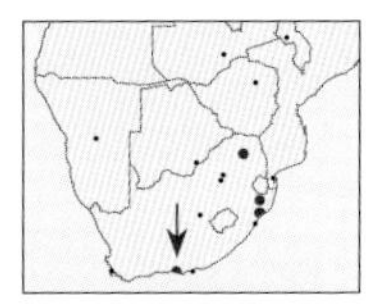

**5 *Olapa nigricosta*** Black-tip Satin

Forewing 19–22mm. Similar to *M. melanocraspis* (above), but larger, head and front part of thorax orange and a thick black line along the length of costa extending round the apex of the wing and distinct black tip on underside of forewings, otherwise no markings. Two species in genus in region. **Biology:** Caterpillar host associations unknown. Males diurnal, often seen flying with *M. melanocraspis*, October–February. **Habitat & distribution:** Moist forests; Eastern Cape north to Limpopo, South Africa, also reported from equatorial forest block.

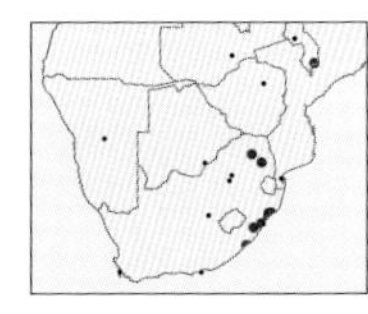

**6 *Ogoa simplex*** Giant Orange Tussock

Forewing 29–38mm. Butterfly-like appearance, yellow-orange, black pectinate antennae; fore- and hindwings similar, grey sinuous postmedian line on both wings, diaphanous with partially darkened veins, females have darkening of wing toward apex. Caterpillar black, elongate, with a pair of yellow or white stripes, very long clusters of hairs extending from sides of body, white, and arranged in elaborate tufts and fans near head and rear of body. **Biology:** Caterpillars polyphagous. **Habitat & distribution:** Moist forests and coastal bush; Eastern Cape, South Africa, north to Kenya.

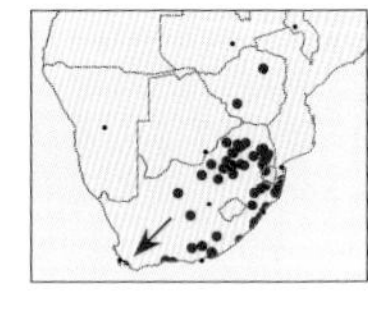

**7 *Morasa modesta*** Modest Gypsy

Forewing 14–22mm. Sexually dimorphic; male body and wings brown, thorax pilose, abdomen bicoloured pink and then yellow; forewings with a white antemedial bar with pointed expansion and central black cell dot, white curved postmedian; hindwings pale pink with dark border; female forewings brown with red speckling, white postmedian line and white area around cell. Caterpillar light brown with several broken black lines along sides, covered in long bristles, head capsule pink with 2 black bars and pair of long black hair tufts project from head. **Biology:** Caterpillars feed on various *Searsia* spp. Multivoltine where conditions allow, adults throughout the year. **Habitat & distribution:** Variety of habitats where *Searsia* spp. grow; Western Cape, South Africa, north to Ethiopia.

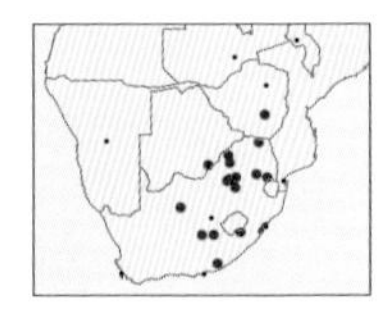

**8 *Polymona rufifemur*** Scarlet-tail Gypsy

Forewing 15–20mm. Often confused with females of *Morasa modesta* (above), but lacks white subterminal line and white marks around cell; abdomen scarlet with black dorsal dots and orange tip; hindwings pink with black border in female, white in male. Caterpillar brown with ginger bristles. One species in genus in region. **Biology:** Caterpillars feed on *Searsia*. Bivoltine, adults August–May. **Habitat & distribution:** Variety of habitats where *Searsia* spp. grow; Eastern Cape, South Africa, north to Ethiopia.

1
1♀
2
3
3
4
5
6
6
7
7♂
8
8

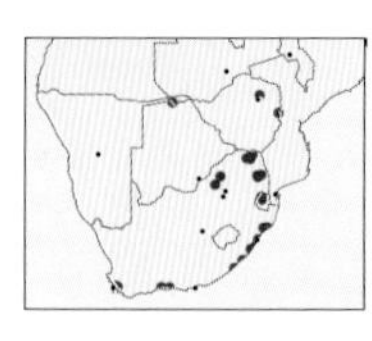

## 1 *Naroma varipes* — Fig Tussock

Forewing 12–22mm. Squat, pilose, normally white, sometimes cream; forewings satiny white, area around inner margin often grey, with row of small dots along central vein, basal area of wings and thorax yellow in some specimens. Caterpillar light tan, with long, white decumbent (lying flat) hairs extending from sides of body, grey stripe along sides and pair of long, black hair tufts extending from head. **Biology:** Caterpillars gregarious on figs (*Ficus* spp.). Multivoltine, adults throughout the year. **Habitat & distribution:** Frost-free forest and riverine vegetation; Western Cape, South Africa, north to tropical Africa.

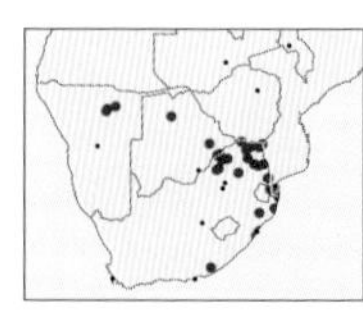

## 2 *Palasea albimacula* — White-barred Gypsy

Forewing 15–22mm. Thorax and abdomen yellow; forewings grey, with single, broad patch of contiguous white oval cells running across wings; hindwings pinkish-cream. Caterpillar black with 2 pairs of pale yellow lines, whorls of long bristles arising from orange tubercles extend from sides of body, long pair of black hair tufts arise from head. One species in genus in region. **Biology:** Caterpillars gregarious on *Commiphora* spp. Bivoltine, adults August–May. **Habitat & distribution:** Arid bushveld where corkwoods grow; Eastern Cape, South Africa, north to tropical Africa.

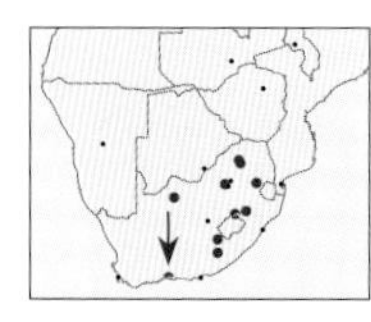

## 3 *Psalis africana* — Bicoloured Tussock

Forewing 18–20mm. Forewings divided along length into a tan bar along leading edge and grey bar along trailing edge; hindwings cream. Caterpillar yellow dorsally with black intersegmental lines, 4 clumps of short brown bristles near middle of body and similar group of 2 clumps near rear, short white bristle clumps extend from sides of body and a pair of long black hair tufts project from head. **Biology:** Caterpillars feed on various grasses. Adults October–April. **Habitat & distribution:** Grassland and transformed habitats with grass; Western Cape, South Africa, north to tropical Africa.

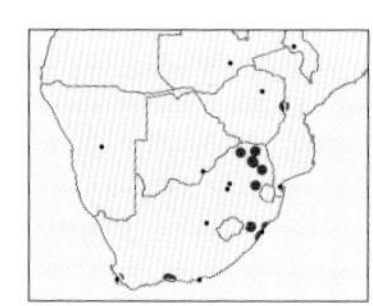

## 4 *Rhypopteryx rhodalipha* — Rosy Arches

Forewing 15–20mm. Ground colour of wings and body white, sometimes with rosy tinge; forewings with black zigzag lines, some median and terminal areas shaded black, especially in female, red to orange 'V'-shaped cell spot diagnostic, often obscured in females; hindwings white with incomplete grey border or no border. Ten species in genus in region. **Biology:** Caterpillar host associations unknown. Adults August–May. **Habitat & distribution:** Moist forests; Western Cape, South Africa, north to tropical Africa.

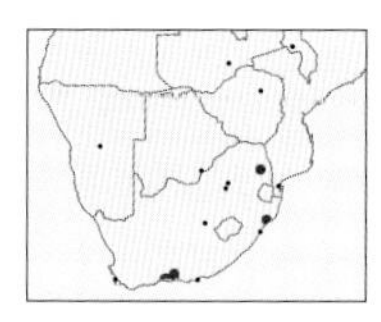

## 5 *Lymantria kettlewelli* — Cinnamon Arches

Forewing 15–20mm. Sexually dimorphic; male body and forewings cinnamon brown with black zigzag lines and black spots, some speckling of cream scales, cell mark of ground colour, hindwings unmarked, charcoal black; female lines as in male but ground colour white, hindwings white with incomplete grey border. **Biology:** Caterpillar host associations unknown. Adults September–May. **Habitat & distribution:** Cool, moist forests; Western Cape north to Mpumalanga, South Africa.

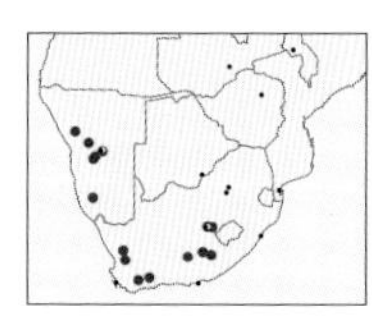

## 6 *Tearosoma aspersum* — White-edge Tussock

Forewing 13–15mm. Thorax grey with erect black and white hairs; forewings mouse brown with variable chalk white bar along costa, delineated by speckled black lines, transverse black bar across base of wing; abdomen orange with black segmental bands; hindwings yellow variably dusted with brown scales and antemedial and postmedian lines, sometimes broken. Caterpillar black dorsally, with short orange bristles and a pair of red bristle patches toward rear of body, body covered by whorls of long white bristles, black at head and rear of body. Two species in genus in region. **Biology:** Caterpillars feed on karoo honey-thorn bushes (*Lycium* spp.). Adults September–May. **Habitat & distribution:** Karoo and desert habitats where *Lycium* bushes grow; South Africa and Namibia.

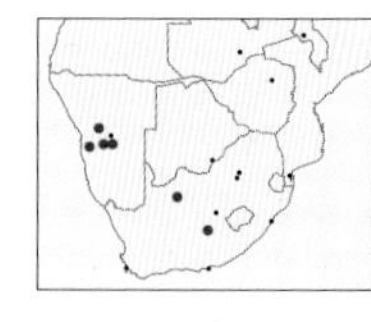

## 7 *Lymantica cidariensis* — Copper Tussock

Forewing 12–16mm. Ground colour coppery brown, frons and front of thorax copper, rest of thorax covered in black setae; forewings copper, variably dusted with dark scales and black undulating antemedial and postmedian lines; hindwings fawn with copper border. One species in genus in region. **Biology:** Caterpillar host associations unknown. Adults June–August and January. **Habitat & distribution:** Karoo, Kalahari and desert habitats; South Africa and Namibia.

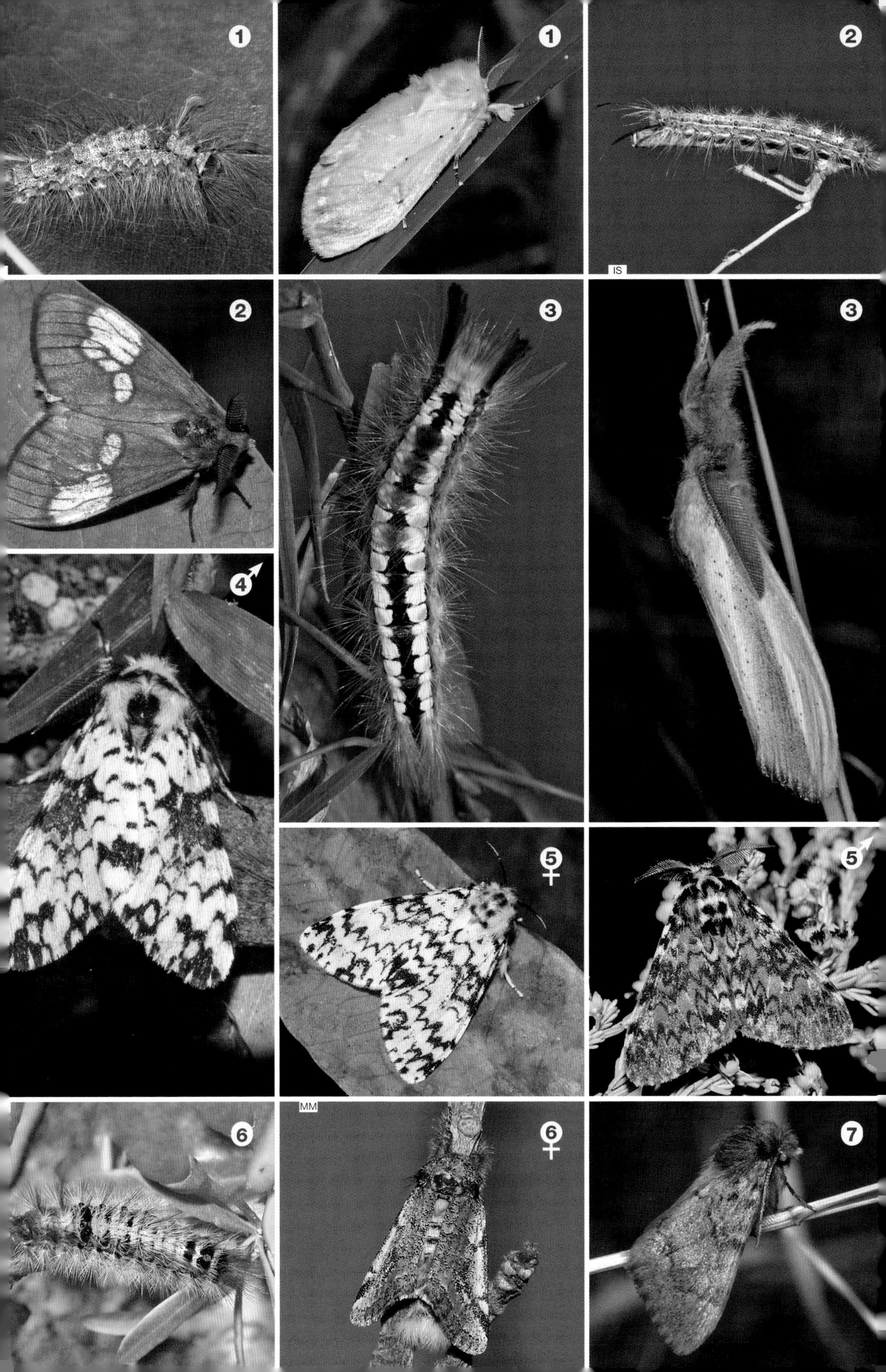
1
1
2
IS
2
3
3
4♂
5♀
5♂
MM
6
6♀
7

## FAMILY EUTELIIDAE Turntails

**Small to medium-sized moths with short, thick abdomen, often ending in a forked prong. Resting posture characteristic, wings held down and back and often folded, abdomen cocked upward and sometimes twisted to the side. Colour typically a complex pattern of browns, yellows and whites, mimicking dead leaves. Two subfamilies in region. Over 900 species globally, with 48 in the region.**

### Subfamily Euteliinae

Largely tropical, across savanna and semi-arid parts. Caterpillars short, slug-like, with 4 pairs of prolegs and mostly associated with Anacardiaceae. Pupation in a cocoon in soil or rotten wood. Adults often colourful with wings folded in pleats along their length.

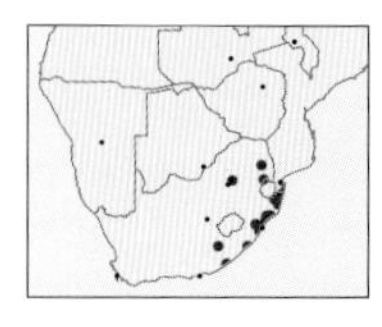

**1 *Caligatus angasii*** Venus Turntail

Forewing 20–27mm. Body light brown, abdomen broad, twisted laterally and with projecting flanges in male; forewings intricately patterned in pinkish-brown with clear white markings; hindwings white with broad pinkish-brown border. Caterpillar not described. One species in genus in region. **Biology:** Caterpillars polyphagous on various Anacardiaceae. Multivoltine, adults throughout the year. **Habitat & distribution:** Forest and bushveld; Eastern Cape, South Africa, north to tropical Africa.

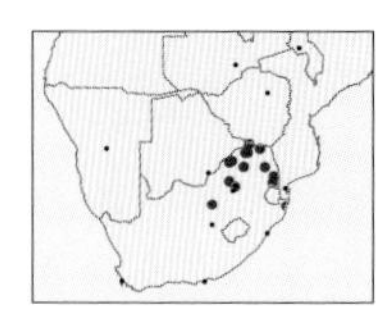

**2 *Colpocheilopteryx callichroma*** Colourful Turntail

Forewing 11–14mm. Forewings crossed by transverse yellow and brown bands, with conspicuous, white-bordered brown triangle near apex, wings held folded and obliquely back and away from broad, upturned abdomen; hindwings cream with dark brown transverse lines and broad brown margin. Two species in genus in region. **Biology:** Caterpillar host associations unknown. Adults October–March. **Habitat & distribution:** Bushveld; KwaZulu-Natal, South Africa, north to tropical Africa and Madagascar.

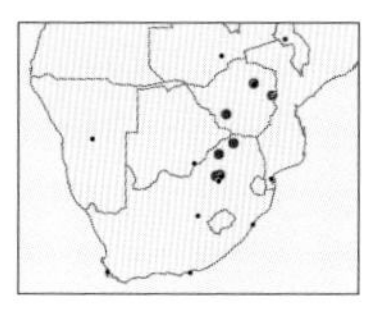

**3 *Colpocheilopteryx operatrix*** Worker Turntail

Forewing 13–16mm. Forewings marked in bands and wavy lines of yellow, purple and pinks; hindwings pale with broad pinkish-brown margin; upturned abdomen ends in a short forked setal tuft. **Biology:** Caterpillar host associations unknown. Adults September–March. **Habitat & distribution:** Bushveld; Gauteng, South Africa, north to Congo.

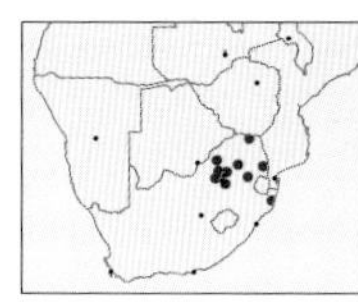

**4 *Marathyssa albidisca*** Whitespot Turntail

Forewing 10–14mm. Body dark grey to black with white dorsal spots with black cross on thorax, distinct white spot at end of abdomen, terminating in short prongs. Forewings dark grey with transverse wavy black lines and a single red cell spot; white oblique streak across wing which is folded in when at rest; hindwings white with black border. Caterpillar undescribed. Three similar species in genus in region. **Biology:** Caterpillars recorded to feed on Marula (*Sclerocarya birrea*). Bivoltine, adults September–February. **Habitat & distribution:** Bushveld; KwaZulu-Natal north to Limpopo, South Africa.

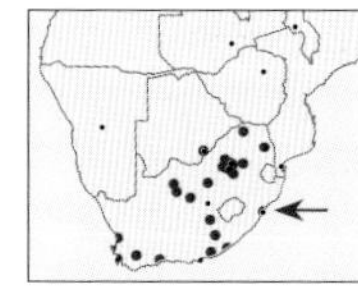

**5 *Eutelia adulatrix*** Flattened Turntail

Forewing 12–14mm. Body brown, forewings banded in cream and brown, a characteristic dark blue-grey patch on inner margin; very broad, upturned abdomen ending in 2 short tail-like projections. Caterpillar usually light green with white longitudinal stripes, pinkish in pre-pupal phase. There are 22 species in genus in region. **Biology:** Caterpillars feed primarily on various *Searsia* spp. but also on other Anacardiaceae and corkwoods (*Commiphora* spp.). Bivoltine, adults September–May. **Habitat & distribution:** Various habitats where *Searsia* spp. grow; Western Cape, South Africa, north to Oman and Saudi Arabia.

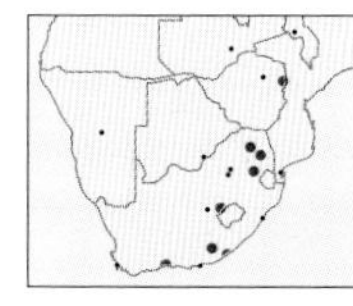

**6 *Eutelia bowkeri*** White-lined Turntail

Forewing 14–17mm. Similar wing pattern to *E. adulatrix* (above), but with more contrasting black and cream tones and no blue patch. Caterpillar pale green, with white longitudinal stripes. **Biology:** Caterpillars feed on various species of Anacardiaceae. Bivoltine, adults September–May, peaking November and February. **Habitat & distribution:** Moist cool forests; southern Cape, South Africa, north to tropical Africa.

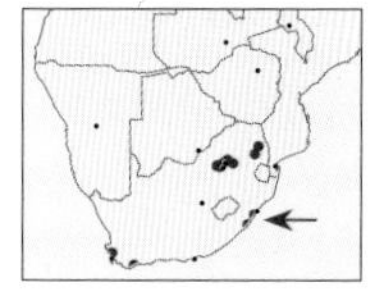

**7 *Eutelia catephioides*** Black Turntail

Forewing 11–16mm. Body and forewings dark grey with black lines and patches, hindwings white with broad dark border. Caterpillar green, yellow or red, strongly marked with bright transverse stripes, a yellow stripe and line of black dots laterally. **Biology:** Caterpillars feed on various species of Anacardiaceae. Multivoltine, adults throughout the year. **Habitat & distribution:** Various habitats where Anacardiaceae grow; Western Cape, South Africa, north to tropical Africa.

1
2
3
4
5
5
6
6
7
7

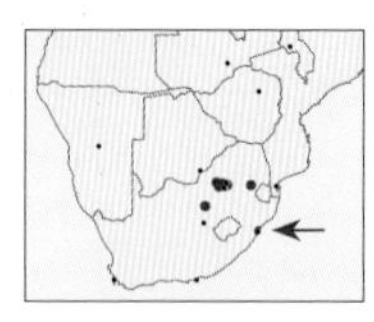

### 1 *Eutelia rivata* — Brown Turntail

Forewing 12–15mm. Body brown with darker brown markings; forewings brown with lines inconspicuous; hindwings pale brown, whitish toward base. Caterpillar green, speckled with yellow and yellow lateral line. **Biology:** Caterpillars feed on Taaibos (*Searsia pyroides*). Probably bivoltine, adults September–March. **Habitat & distribution:** Various habitats where host plants grow; KwaZulu-Natal to Limpopo, South Africa.

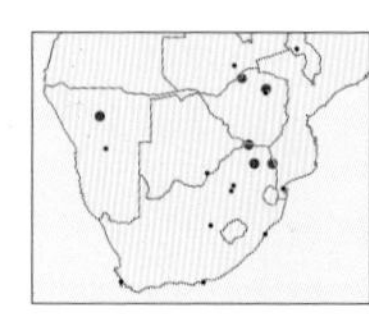

### 2 *Eutelia polychorda* — Varied Turntail

Forewing 11–14mm. Colour variable, ranging from pale yellow to red, wings crossed by dark-edged paler sinuous lines; hindwings white to cream with darker border. **Biology:** Caterpillars recorded on *Brachystegia*, Mopane (*Colophospermum mopane*) and *Baikaea* spp. Sometimes found in huge numbers. Adults November–April. **Habitat & Distribution:** Bushveld and forest; Limpopo north to tropical Africa.

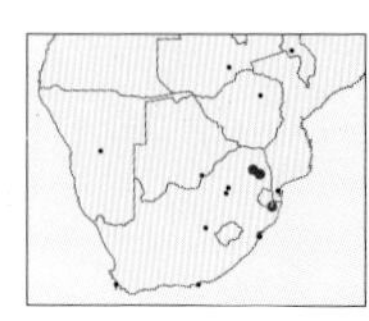

### 3 *Penicillaria ethiopica* — Mottled Turntail

Forewing 11–15mm. Light to dark brown with variably mottled paler bands at base and tips of forewings. Caterpillar pale with wavy longitudinal orange stripes and brown head capsule. One species in genus in region. **Biology:** Caterpillars feed on various trees with woolly leaves including Transvaal Milkplum (*Englerophytum magalismontanum*), Red Bushwillow (*Combretum apiculatum*) and Velvet Bushwillow (*C. molle*). Caterpillars conceal themselves with elaborate covering of plant hairs. Multivoltine, adults throughout the year. **Habitat & distribution:** Forests; KwaZulu-Natal, South Africa, north to Tanzania.

## Subfamily Stictopterinae

Occur in tropical forests. The caterpillars are smooth and slender, with full set of prolegs; the moths with reticulate patterns on forewings and translucent patch at base of hindwings. At rest, wings furled around body, mimicking twig.

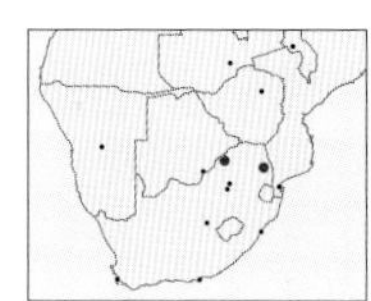

### 4 *Stenosticta grisea* — Grey Colon

Forewing 7–8mm. Grey, elongate forewings marked centrally with a pair of small adjacent black dots and with row of smaller dots along distal margin; hindwings pale basally, becoming dark toward distal margin, which is marked by central notch. Caterpillars stout, orange or black with yellow blotches, with 3 white bands and a row of black dots along sides. Four species in genus in region. **Biology:** Caterpillars feed on *Commiphora* spp. Bivoltine, adults September–November. **Habitat & distribution:** Dry bushveld where *Commiphora* trees grow; Limpopo, South Africa, north to Ethiopia and Somalia.

# FAMILY NOCTUIDAE Owlet Moths, Cutworms, Army Worms, Semi-loopers

**This used to be the largest lepidopteran family, but reduced in size when many of the subfamilies were moved to the Erebidae (p.306), which is now the largest. All members have quadrifid forewing venation (where the cubital vein in the forewing appears to have 4 branches). They are mostly dull coloured, but some have complex wing patterns. They have a trifid venation (3 medio-cubital veins) of the hindwing, which is of a uniform colour, or has a simple colour pattern. All members have a unique structure in the ear (located on the thorax) that prevents blood-feeding mites from entering the ear cavity. This ear (the tympanum) allows the moth to detect the echolocation calls of hunting bats, and to employ evasive flight manoeuvres. The majority of species are nocturnal and have thread-like antennae. The caterpillars are usually drab brown or green, cylindrical, smooth, plump with a distinctive hump toward the rear. They are mostly herbivorous, although some feed on other caterpillars. Certain species are able to store plant toxins in their bodies, and some (notably the subfamily Cuculliinae) have warning colouration. Many are polyphagous due to symbiosis, with gut bacteria that break down plant chemical defences. Some species have sporadic population explosions due to a reduction of natural parasitoids (an imbalance often caused by human activities or drought) and can then occur in enormous numbers, when they are termed 'armyworms', consuming large amounts of vegetation, including crops. There are approximately 12,000 species worldwide, and over 950 regional species.**

## Subfamily Oncocnemidinae

Night-flying moths, sometimes darkly coloured but often with whitish forewings patterned with black lines.

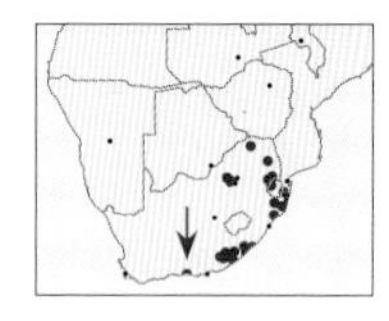

### 5 *Anedhella interrupta* — Short Rayed Gothic

Forewing 10–12mm. Forewings grey with diagnostic incomplete, parallel, dark brown-edged cream stripes, veins and apical border marked with thin brown lines, trailing edge with cinnamon-brown patches; hindwings grey. Four species in genus in region. **Biology:** Caterpillar host associations unknown. Adults September–April, primarily November. **Habitat & distribution:** Forest and riverine vegetation; Western Cape north to Limpopo, South Africa, also reported from Tanzania.

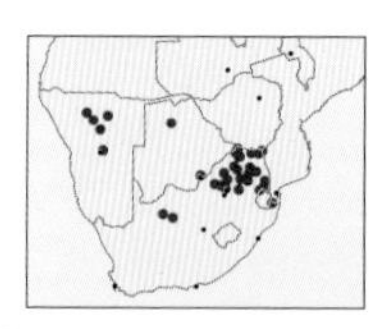

**1 *Anedhella rectiradiata*** Straight Rayed Gothic

Forewing 11–12mm. Forewings with diagnostic complete median brown-edged cream stripe, 2 small cream circles on either side of this stripe, remainder of wing grey with parallel brown streaks; hindwings brown with darker margins. **Biology:** Caterpillar host associations unknown. Adults October–February, primarily December. **Habitat & distribution:** Bushveld; Northern Cape, South Africa, north to Tanzania.

## Subfamily Bryophilinae

Small to medium-sized moths whose caterpillars feed on mosses and lichens. At rest, the broad and cryptically coloured wings are held roof-like over the body, camouflaged to resemble moss and lichen on tree bark.

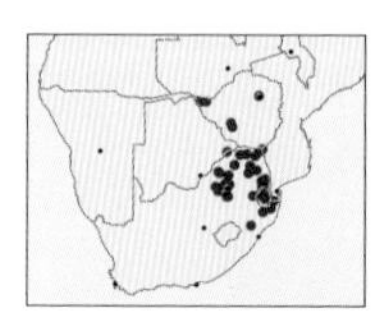

**2 *Cryphia leucomelaena*** Grey Lichen

Forewing 9–11mm. Forewings mottled white, brown and grey, irregular, black-edged brown basal and median bands containing 2 circles, bounded by variegated white bands, chequered cilia; hindwings speckled grey, darker margins. Three species in genus in region. **Biology:** Caterpillar host associations unknown. **Habitat & distribution:** Grassland and bushveld; KwaZulu-Natal, South Africa, north to Kenya.

## Subfamily Eriopinae

The caterpillars are green or brown and feed largely on ferns. The adults have leg segments fringed with hairs or scales, and the forewings usually bear 2 small dark spots.

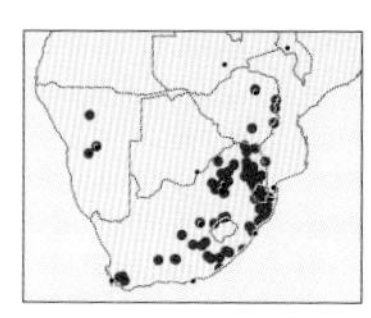

**3 *Callopistria latreillei*** Carpeted Tangle

Forewing 9–12mm. Forewings mottled grey and brown, circular cream spot near base, dentate black and white transverse line toward apical region of wing; hindwings grey or brown. Seven species in genus in region. **Biology:** Caterpillars reported to feed on various ferns. Adults throughout the year, peaking in summer. **Habitat & distribution:** Varied, including transformed habitats; Western Cape, South Africa, north to Europe, also Madagascar and Indian Ocean islands.

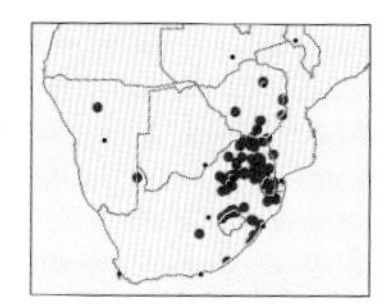

**4 *Callopistria yerburii*** Yerbur's Tangle

Forewing 12–13mm. Forewings with pointed apex, variegated brown and tan, with complex geometric panels, thin black and white basal line distinctly kinked before costa diagnostic; hindwings brown. Similar *C. natalensis* has basal line rounded, not kinked. **Biology:** Caterpillars reported to feed on maidenhair (*Adiantum*) and birds' nest (*Asplenium*) ferns. Adults throughout the year, peaking late summer. **Habitat & distribution:** Varied; Eastern Cape north to Arabian Peninsula, also Madagascar and Indian Ocean islands.

## Subfamily Acronictinae

The caterpillars lack secondary setae on the head capsule, and have a dorsal hump on the body. The forewings usually have a dark streak-like marking originating at the base of the tornus (lower end of the wing). The eyes are free of hairs, but the head and thorax are hairy.

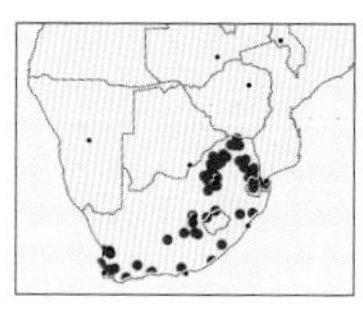

**5 *Acronicta transvalica*** Curved Dagger

Forewing 13–15mm. Forewings with indistinct antemedial and postmedian lines, diagnostic discontinuous curved black streak from wing base to inner curve of forewing, 2 black-ringed cream marks around cell. Five species in genus in region. **Biology:** Caterpillar host associations unknown. Adults throughout the year, peaking in early summer. **Habitat & distribution:** Variety of mostly wooded habitats; Western Cape, South Africa, north to Namibia, Zimbabwe and Mozambique.

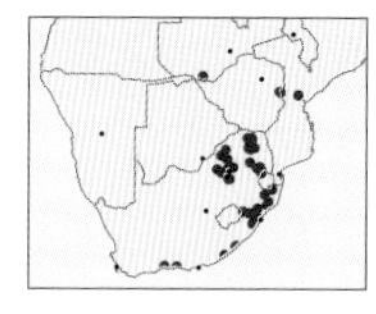

**6 *Acronicta verbenata*** Blue Dagger

Forewing 13–14mm. Ground colour diagnostic, blue-green with some pale brown shading in some individuals; white antemedial and postmedian lines dentate with black borders; hindwings white with grey border. Similar to *Nyodes* spp. (p.424) but wings shorter and blue-green colour diagnostic. **Biology:** Caterpillar host associations unknown. Adults throughout the year, peaking in summer. **Habitat & distribution:** Forests and riverine vegetation; Western Cape, South Africa, north to Zambia and Mozambique.

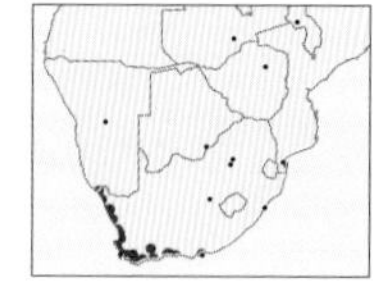

**7 *Araea indecora*** Mossy Dagger

Forewing 10–14mm. Wings elongate; individuals from arid areas much paler than those from moist areas; thorax dull green; forewings greyish-cream, with indistinct black markings and in some forms green cast, distinct basal spot, sometimes orange, and 2 eyespots oblique from each other diagnostic, apical border with row of small black dots; hindwings cream with darker border. One species in genus in region. **Biology:** Caterpillar host associations unknown. Adults throughout the year, peaking in spring. **Habitat & distribution:** Succulent karoo and fynbos; Cape provinces, South Africa.

1
2
3
4
5
6
7

## Subfamily Plusiinae

At rest, the wings held steeply roof-like over the body, forming a 'V' shape; large tuft of hairs on the thorax.

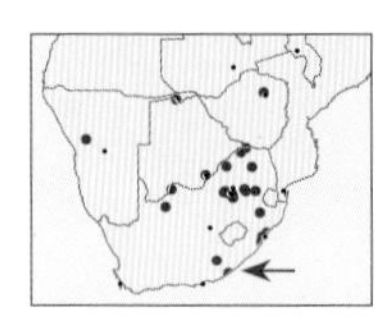

### 1 *Chrysodeixis acuta* — Silver U

Forewing 14–17mm. Upright tuft of thoracic hairs, males with diagnostic tufts of androconial hairs arising midway on side of abdomen; forewings shades of brown with golden or silver shading, pale silver marking near wing apex, a pair of contiguous white cell spots, postmedian line diagnostically kinked toward cell spot; hindwings grey or brown with pale fringe of cilia. Caterpillar pale green, with fine white lines along top of body, and a wider white stripe running along sides. Two species in genus in region. **Biology:** Caterpillars polyphagous, preferring herbaceous plants. Multivoltine when conditions allow, adults throughout the year. **Habitat & distribution:** Variety of habitats, often in transformed areas; Eastern Cape, South Africa, north to Saudi Arabia and westwards to Mauritania, also Madagascar.

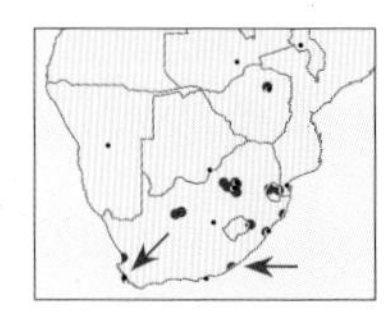

### 2 *Chrysodeixis chalcites* — Golden Silver U

Forewing 15–17mm. Similar to *C. acuta* (above) but male tufts of androconial hairs arise from end of abdomen and postmedian line is kinked away from cell spot; generally paler with more golden scaling and less silver. Caterpillar a pale green semi-looper, with a fine white lateral stripe, and pigmented spiracles. **Biology:** Caterpillars polyphagous, on herbaceous soft-tissue plants, often on *Solanum*. Multivoltine when conditions allow, adults throughout the year. **Habitat & distribution:** Variety of habitats, mostly transformed; Western Cape, South Africa, north to Europe, also Madagascar and Indian Ocean islands.

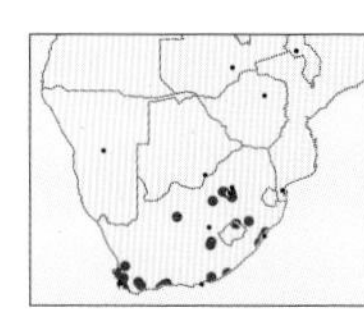

### 3 *Cornutiplusia circumflexa* — Circumflex

Forewing 16–18mm. Forewings long, pointed apex, variegated grey, cream and brown, pale transverse band near wing apex, pinkish 'U'-shaped marking at base of wing; hindwings brown or grey, darker at margin. Caterpillar pale green, with fine white lines running along length of body. One species in genus in region. **Biology:** Caterpillars polyphagous, mostly on herbaceous plants. Multivoltine when conditions allow, adults throughout the year. **Habitat & distribution:** Variety of habitats, mostly in transformed areas; Western Cape, South Africa, north to Europe, also Madagascar and Indian Ocean islands.

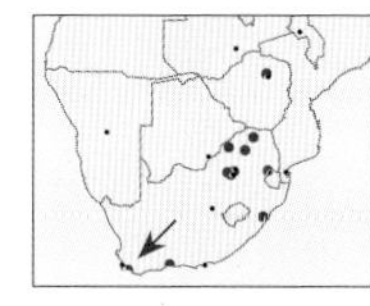

### 4 *Ctenoplusia furcifera* — Two-spot Plusia

Forewing 13–15mm. Tuft of erect thoracic hairs; forewing central dark grey patch in median area diagnostic, marbled grey or brown; hindwings dull brown, darker margin. Caterpillar plump, green or cream semi-looper, with distinctive black spots on each segment, head capsule with lateral black stripe. Nineteen species in genus in region. **Biology:** Caterpillars polyphagous, on herbaceous plants. Multivoltine when conditions allow, adults throughout the year. **Habitat & distribution:** Variety of habitats, mostly in transformed areas; Western Cape, South Africa, north to India, also Madagascar and Indian Ocean islands.

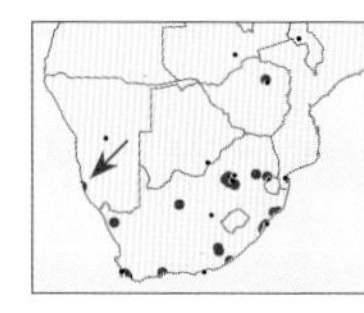

### 5 *Ctenoplusia limbirena* — Silver U-tail

Forewing 13–16mm. Erect tufts of hair on thorax; forewings mottled grey or brown, with diagnostic white 'U'-shaped marking touching (sometimes fused with) contiguous white spot. Caterpillar a plump, pale green semi-looper with pigmented spiracles. **Biology:** Caterpillars polyphagous, mostly on herbaceous plants. Multivoltine when conditions allow, adults throughout the year. **Habitat & distribution:** Variety of habitats, mostly in transformed areas; Western Cape, South Africa, north to Arabian Peninsula, also Madagascar and Indian Ocean islands.

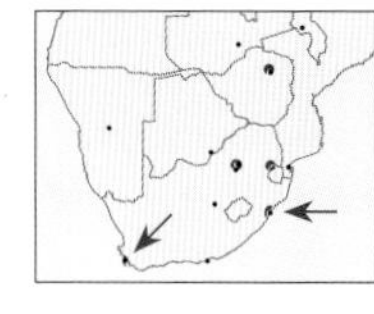

### 6 *Ctenoplusia molybdina* — Grey-banded Plusia

Forewing 15–16mm. Forewings mottled grey, with shape of dark brown patch with whitish band in middle of wing diagnostic; hindwings brown or grey with margin of white cilia. Caterpillar a plump, pale or olive green semi-looper, with a prominent white lateral stripe and faint, thin white lines running along the body. **Biology:** Caterpillars polyphagous, mostly on herbaceous plants. Multivoltine when conditions allow, adults throughout the year. **Habitat & distribution:** Variety of habitats, mostly in transformed areas; Western Cape, South Africa, north to Kenya, also Madagascar and Indian Ocean islands.

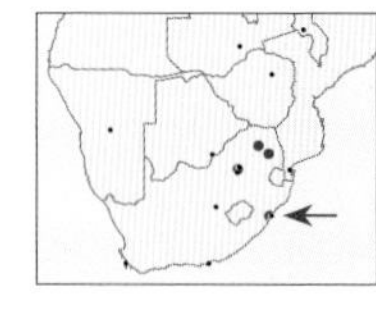

### 7 *Ctenoplusia phocea* — Bark Plusia

Forewing 15–19mm. Forewings grey, light or dark brown, with median black stripe, blotch or bar, and diagnostic dark brown subterminal band; hindwings grey or brown. Caterpillar plump, pale green with thin, cream lateral stripe. **Biology:** Caterpillars recorded from Sunbird Bush (*Metarungia longistrobus*). Adults November–April. **Habitat & distribution:** Moist forest and riverine vegetation; KwaZulu-Natal, South Africa, north to tropical Africa.

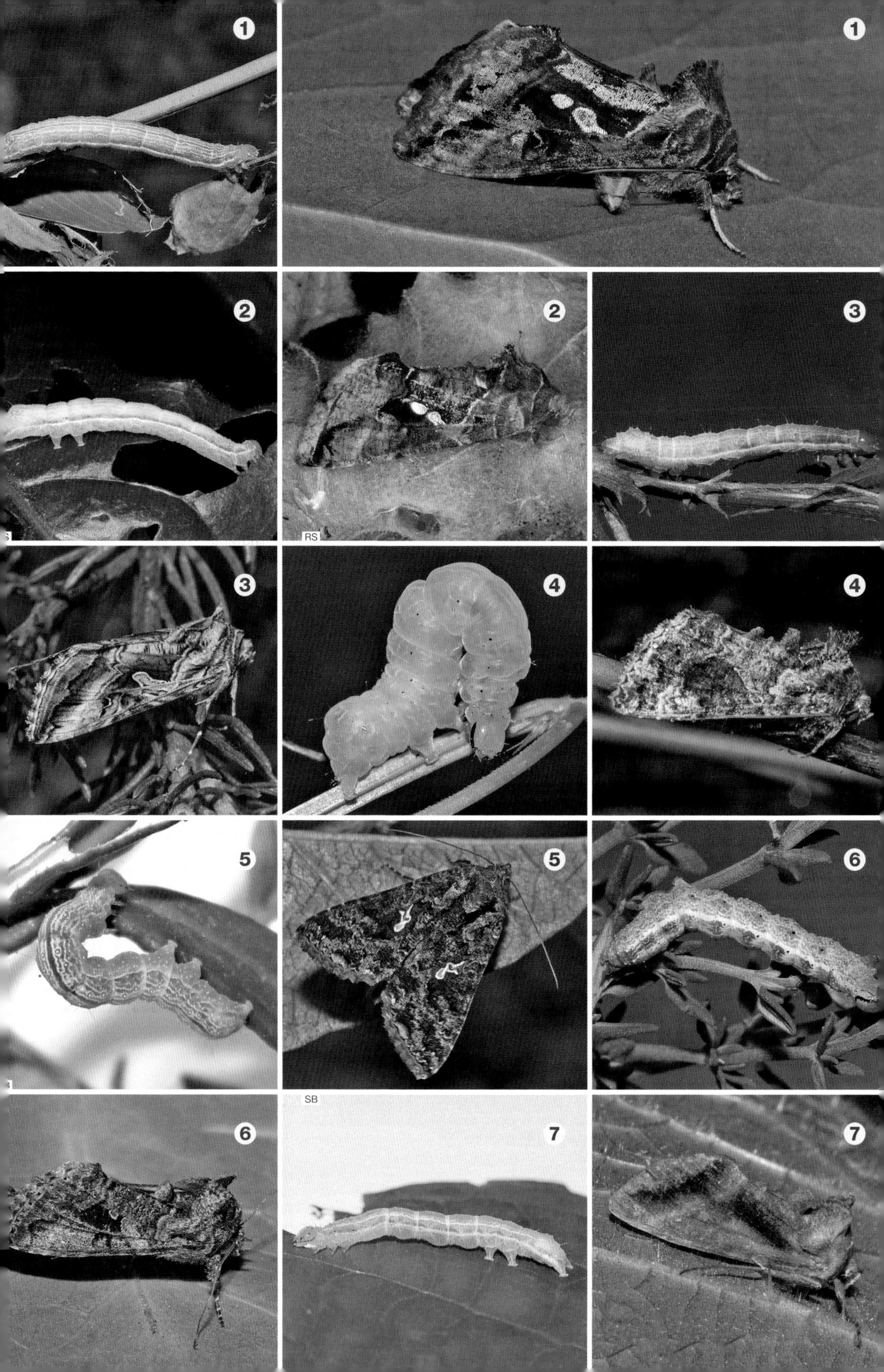

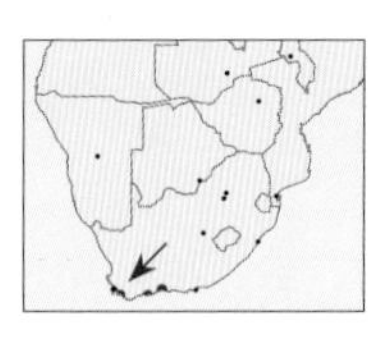

**1 *Ctenoplusia lavendula*** Fynbos Plusia

Forewing 15–17mm. Forewings various shades of grey, cilia chequered grey and white, tufts of thoracic hairs, basal, all lines characteristically developed, subterminal line forms diagnostic wavy band with black arrow mark; hindwings pale grey with veins darkened and brown border. Caterpillar orange with black and white bands. **Biology:** Caterpillar host associations unknown. Adults November–April. **Habitat & distribution:** Fynbos habitats; Western Cape, South Africa.

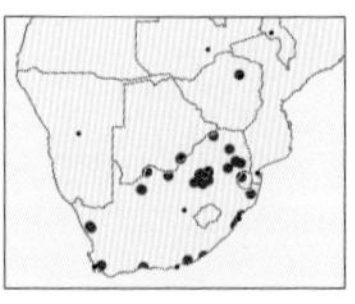

**2 *Vittaplusia vittata*** Streaked Plusia

Forewing 15–18mm. Forewings variable, grey to various shades of brown, with diagnostic short, black-lined cream streak; hindwings grey or brown with basal area straw. Caterpillar pale green, with lateral cream stripe. One species in genus in region. **Biology:** Caterpillars polyphagous, mostly on herbaceous plants. Multivoltine when conditions allow, adults throughout the year, common participant in late summer migrations. **Habitat & distribution:** Variety of habitats, mostly in transformed areas; Western Cape, South Africa, north to Arabian Peninsula.

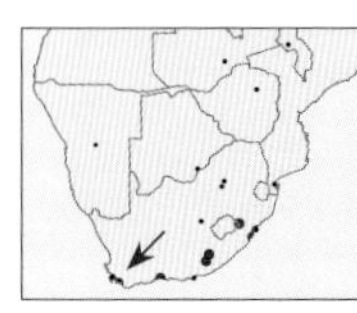

**3 *Thysanoplusia angulum*** Silver Angle

Forewing 16–18mm. Tufts of erect hairs on thorax; forewings brown with diagnostic short, sinuous white marking and a scribling of white lines on costa; hindwings dirty grey. Caterpillar pale green dorsally, with numerous fine lines running along body, sides and ventral part of body lime green, segments defined by pale yellow bands. Ten species in genus in region. **Biology:** Caterpillars recorded on Wild Rosemary (*Eriocephalus africanus*). Adults September–May. **Habitat & distribution:** Moist fynbos, forest and grassland; Western Cape north to KwaZulu-Natal, South Africa, also reported from Kenya.

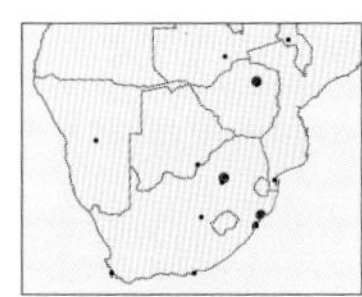

**4 *Thysanoplusia arachnoides*** Cobweb Plusia

Forewing 15mm. Ground colour mottled brown and grey, thorax with erect tufts of scales; forewings cream, grey and brown, veins marked in white, dark lunate mark at apex enclosed by white lines, with oblique, sinuous, white diagonal line in middle of wing; hindwings dirty grey or brown. Caterpillar yellow-green with numerous white spots and crinkles. **Biology:** Caterpillars feed on various Lamiaceae. Adults August–May. **Habitat & distribution:** Grassland; KwaZulu-Natal, South Africa, north to tropical Africa, also Madagascar.

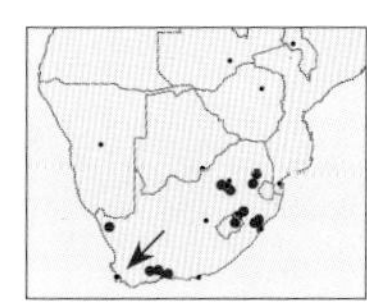

**5 *Thysanoplusia exquisita*** Exquisite Plusia

Forewing 13–17mm. Forewings brown with fine black and brown stippling, basal and subterminal lines red, grey and white diagnostic; hindwings orange with brown border. Caterpillar pale blue-green, with broad cream stripes and fine white lines on top of body. **Biology:** Caterpillars feed on various *Senecio* spp. (Asteraceae). Bivoltine, adults active by day but also attracted to light, throughout the year. **Habitat & distribution:** Variety of open habitats where *Senecio* spp. grow; Western Cape, South Africa, north to Arabian Peninsula, also Madagascar.

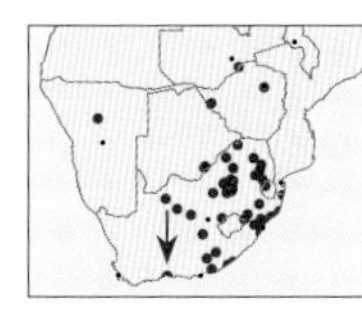

**6 *Thysanoplusia orichalcea*** Golden Plusia

Forewing 16–19mm. Forewings brown, with large metallic gold triangular panel; hindwings brown, with extensive dark marginal band. Caterpillar olive green dorsally with numerous fine white lines running length of body, lateral black and cream stripe and light green ventral parts. **Biology:** Caterpillars polyphagous on herbaceous plants. Multivoltine when conditions allow, adults throughout the year, common participant in late summer migrations. **Habitat & distribution:** Variety of habitats, mostly in transformed areas; Western Cape, South Africa, north to Arabian Peninsula, also Madagascar and Indian Ocean islands.

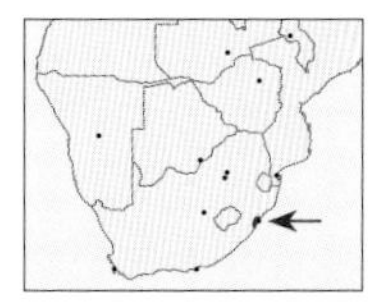

**7 *Trichoplusia roseofasciata*** Harlequin Plusia

Forewing 11mm. Similar to *T. sestertia* (below), but lacks the white teardrop marking on forewings; forewings pink and grey basally, followed by brown transverse band, complex pink, grey and brown band at apex. Caterpillar a green semi-looper with lemon-green dorsal parts and darker ventral parts. **Biology:** Caterpillars recorded from Misty Plume Bush (*Tetradenia riparia*). **Habitat & distribution:** Tropical forest; KwaZulu-Natal, South Africa, and further north to tropical Africa.

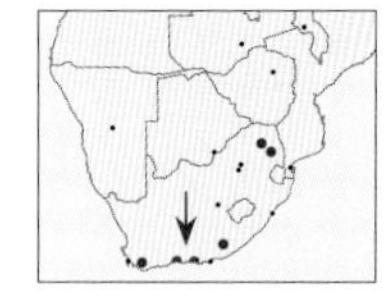

**8 *Trichoplusia sestertia*** Masked Plusia

Forewing 15mm. Thorax with erect tuft of scales, frons black; forewings marbled grey, pink and brown, with large, white mask-shaped square marking in middle with another diagnostic teardrop marking; hindwings brown, darker at margin. Caterpillar green with numerous white crenulated longitudinal lines. Eight species in genus in region. **Biology:** Caterpillars feed on *Plectranthus* spp. (Lamiaceae). Adults November–March. **Habitat & distribution:** Moist forests; Western Cape, South Africa, north to tropical Africa.

1
1
2
AM
2
3
MB
3
4
4
5
5
6
SB
6
7
SB
7
8
8

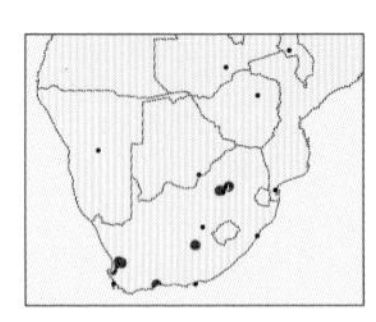

**1 *Trichoplusia ni*** Cabbage Looper

Forewing 17mm. Forewings mottled grey and brown, with a white dot and 'U'-shaped marking on central black patch; hindwings a diagnostic rusty brown, darkening toward margin. Caterpillar a green semi-looper, with several fine white lines running along back, and white lateral stripe. **Biology:** Caterpillars polyphagous, mostly on herbaceous soft-tissue plants. Multivoltine when conditions allow, adults throughout the year. **Habitat & distribution:** Variety of habitats, mostly in transformed areas; Western Cape, South Africa, north to Europe, also Madagascar and Indian Ocean islands.

## Subfamily Eustrotinae

The number and size of prolegs is reduced in mature caterpillars. Wings of adults are held partially roof-like over body. Many species have broad transverse bands running across the forewings.

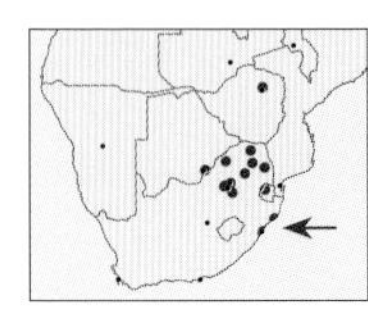

**2 *Aconzarba decissima*** Decisive White Bar

Forewing 8–11mm. Forewings marbled chocolate brown at base, separated from variegated brown apex by jagged, cream diagonal band, distinctive round cell spot; hindwings grey-brown, darkening at margin, with a fringe of white cilia. Two species in genus in region. Caterpillar green, densely packed with short white setae and a few longer ones. **Biology:** Caterpillars feed on Veld Justicia (*Justicia protracta*). Adults September–April. **Habitat & distribution:** Bushveld; KwaZulu-Natal, South Africa, north to Arabian Peninsula, also Madagascar and Comoros.

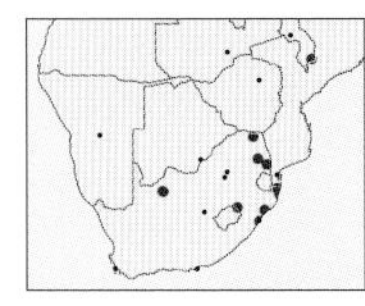

**3 *Ozarba plagifera*** Variable Ozarba

Forewing 8–10mm. Forewings variable, rich brown to pale brown with dentate transverse lines and a diagnostic large, black to brown subapical spot on costa, or grey basally, with brown central region and tan apical parts. Caterpillar pale green with pigmented spiracles and pale yellow segmental bands. **Biology:** Caterpillars feed on Creeping Foxglove (*Asystasia gangetica*). Adults September–May. **Habitat & distribution:** Forest and bushveld; KwaZulu-Natal, South Africa, north to Saudi Arabia.

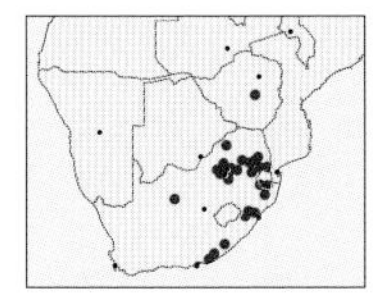

**4 *Ozarba semipurpurea*** Rusty Ozarba

Forewing 8–9mm. Forewings rusty brown basally, median area cream to grey, lines slightly crenulated and going around black cell spot; hindwings pale brown darkening toward margin. Similar *O. ferruginata* is larger, median area uniform cream and lines are smoother; *O. jansei* has median and terminal area smooth grey and distinctive kink in antemedial line. **Biology:** Caterpillar host associations unknown. Adults August–March. **Habitat & distribution:** Grassland and bushveld; Eastern Cape, South Africa, north to Ethiopia.

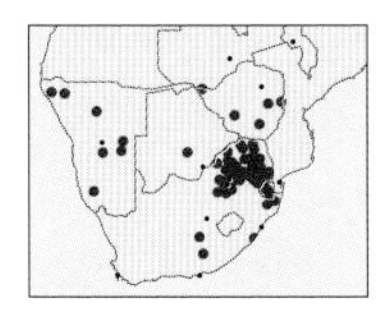

**5 *Ozarba accincta*** Pinkeye Ozarba

Forewing 8–10mm. Distinctive species, with diagnostically curved postmedian line, 2 dark median patches with pink eyespots on outer dark patch; hindwings orange with black border. Similar *O. corniculans* is larger, has 1 median dark patch on inner margin and has translucent brown hindwings. **Biology:** Caterpillar host associations unknown. Adults throughout the year, peaking October–March. **Habitat & distribution:** Bushveld and grassland; Eastern Cape, South Africa, north to Ethiopia.

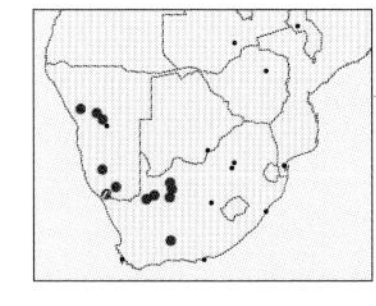

**6 *Ozarba gonatia*** Pool Ozarba

Forewing 9–11mm. Similar to *O. accincta* (above), but diagnostic antemedial black band complete from inner margin to costa with a distinctive widening in middle, median and subterminal areas cream to grey and subterminal line complete or broken; hindwings orange-brown. Similar *O. genuflexa* has antemedial band stopping before costa. **Biology:** Caterpillar host associations unknown. Adults nocturnal but also active by day, September–April. **Habitat & distribution:** Karoo and desert habitats; Western Cape, South Africa, north to Namibia.

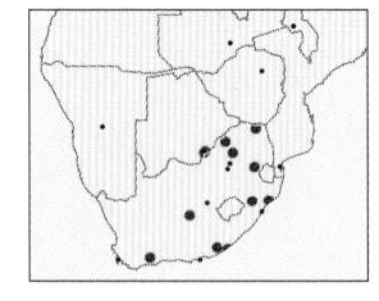

**7 *Ozarba bipartita*** Partitioned Ozarba

Forewing 8–10mm. Distinctive; head, thorax and basal half of forewings plain pinkish-cream, antemedial line a straight uneven black band, median and terminal areas charcoal grey, dusted with pale grey scales; hindwings charcoal grey. **Biology:** Caterpillar host associations unknown. Adults September–April. **Habitat & distribution:** Karoo, bushveld and grassland; Eastern Cape, South Africa, north to Zimbabwe.

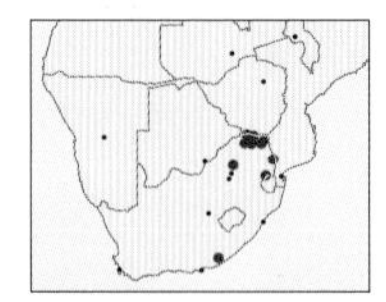

**8 *Ozarba orthozona*** Orange-underwing Ozarba

Forewing 9–11mm. Distinctive; head, thorax and forewings charcoal grey with markings obscure except for distinctive cream median band, postmedian gently curved; hindwings orange with brown border. **Biology:** Caterpillar host associations unknown. Adults September–April. **Habitat & distribution:** Bushveld; Eastern Cape north to Limpopo, South Africa.

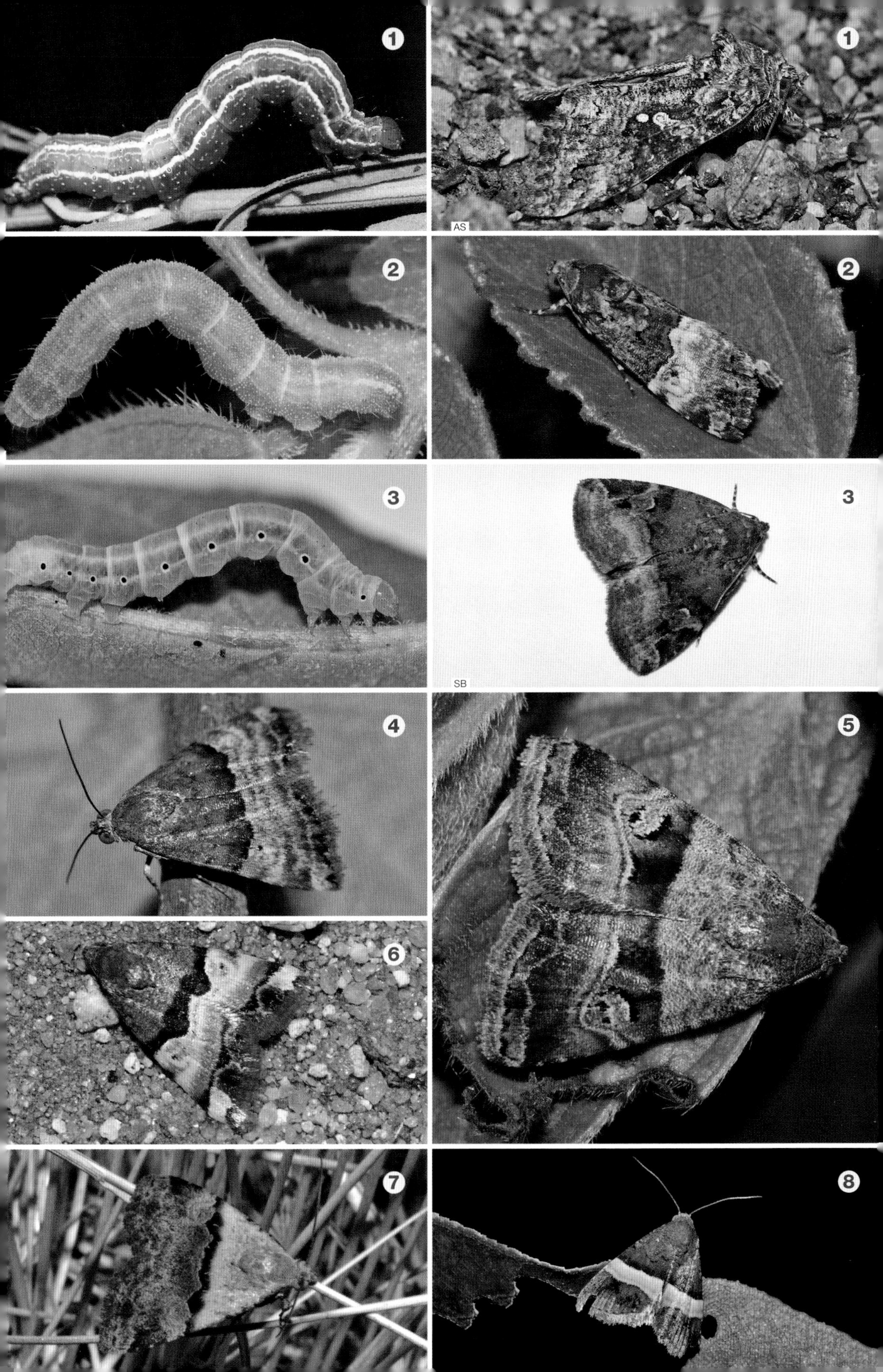

1
1
AS
2
2
3
3
SB
4
5
6
7
8

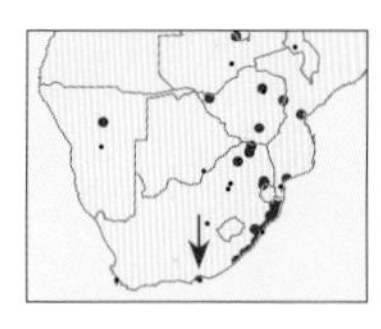

### 1 *Ozarba abscissa* — Black-underwing Ozarba

Forewing 9–12mm. Forewings dark brown with bold diagonal cream median band, postmedian straight; similar to *O. orthozona* (p.410), but hindwings charcoal grey to black. Caterpillar a grey semi-looper with fine black lines running along length of body, and segmental orange spots on side of body. **Biology:** Caterpillars feed on Creeping Foxglove (*Asystasia gangetica*). Adults throughout the year. **Habitat & distribution:** Forest and bushveld; Eastern Cape, South Africa, north to Kenya.

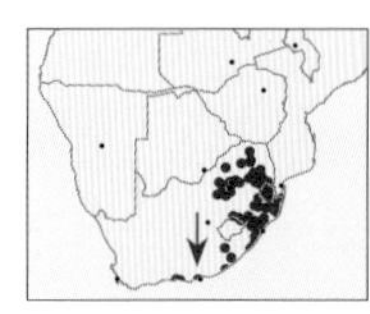

### 2 *Ozarba hypotaenia* — Cream-line Ozarba

Forewing 8–10mm. Distinctive; head, thorax and basal area of forewings rust brown, darker brown median area bordered by cream lines, terminal area grey to black, black and white eyespot; hindwings charcoal grey with white postmedian line and some white terminal markings. **Biology:** Caterpillar host associations unknown. Adults nocturnal but also active by day, throughout the year. **Habitat & distribution:** Moist grassland; Eastern Cape, South Africa, north to Kenya.

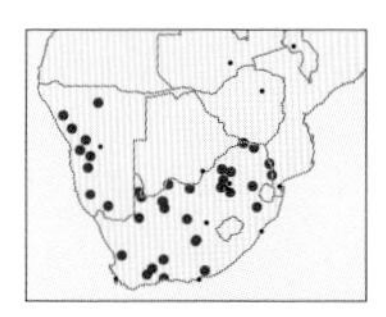

### 3 *Pseudozarba schencki* — Brown-band Ozarba

Forewing 8–9mm. similar to *Ozarba gonatia* (p.410), but diagnostic antemedial band broad and straight, distinctive thin postmedian line sharply curved around cell spot. Three species in genus in region. **Biology:** Caterpillar host associations unknown. Adults nocturnal but also active by day, September–April, can swarm in wet years. **Habitat & distribution:** Karoo, grassland, bushveld and desert habitats; Western Cape, South Africa, north to Angola.

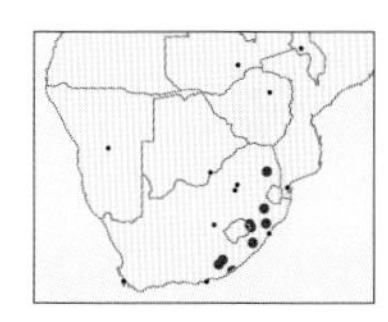

### 4 *Deltote olivula* — Olive-band Deltote

Forewing 11–12mm. Distinctive species; thorax and forewings white, olive median band broad at inner margin narrowing toward costa, orange cell spot, triangular mark on costa and olive mark on lower curve; hindwings cream, unmarked. Fourteen species in genus in region. **Biology:** Caterpillar host associations unknown. Adults October–December. **Habitat & distribution:** Moist forests; Eastern Cape north to Mpumalanga, South Africa.

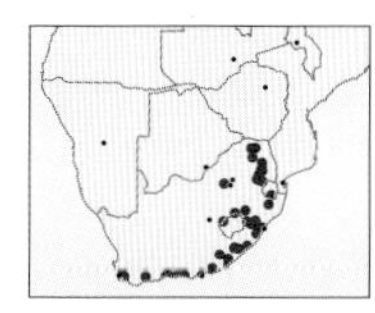

### 5 *Deltote varicolora* — Variable Deltote

Forewing 9–10mm. Very variable in colour but constant in markings; white markings in terminal and basal areas usually present, median area can vary from almost black, obscuring markings, to marble white or partially black and white, large eyespot with white dot diagnostic; hindwings white to grey. **Biology:** Caterpillar host associations unknown. Adults August–May, primarily in summer. **Habitat & distribution:** Moist forests and grassland; Western Cape, South Africa, north to Zimbabwe.

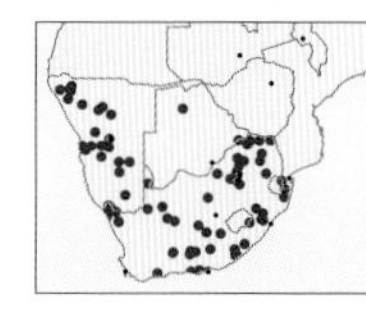

### 6 *Acontiola varia* — Carpet Orange

Forewing 9–11mm. Body and forewings mottled with grey and brown scales, antemedial and postmedian double lines diagnostic, variable black patches; hindwings orange with black border; underside of both wings bright orange. Several very similar species. Sixteen species in genus in region, many can only be separated through dissection. **Biology:** Caterpillar host associations unknown. Adults often seen by day flashing orange underside, October–December. **Habitat & distribution:** Grassland, karoo, desert and bushveld; Western Cape, South Africa, north to Angola.

## Subfamily Cuculliinae Sharks

Adults are generally cryptically coloured, and at rest the wings are held in a steep roof-like position. There is a large, forward projecting clump of thoracic hairs.

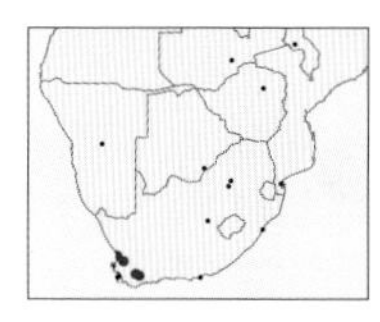

### 7 *Compsotata elegantissima* — Elegant Gothic

Forewing 14–18mm. Forewings tan with complex diagnostic pattern of geometric green markings, fades to black in dried specimens; hindwings cream, darkening toward margin with cell spot. Caterpillar stout, pale green with small black dots, red line running along the side of the body. One species in genus in region. **Biology:** Caterpillars feed on Fennel (*Foeniculum vulgare*). Adults January–June. **Habitat & distribution:** Fynbos and transformed habitats; Western Cape, South Africa, also reported from Kenya and DRC.

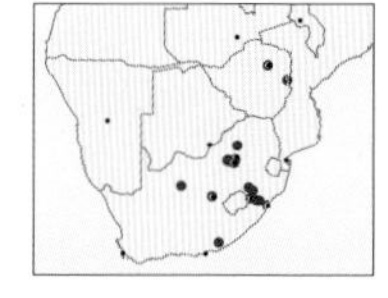

### 8 *Cucullia hutchinsoni* — Silver-streaked Shark

Forewing 16–18mm. Forewings with pointed apex, bold silver-white streak on broader golden-brown streak running across wing diagnostic, wing margins grey-brown; hindwings white. **Biology:** Caterpillar host associations unknown. **Habitat & distribution:** Grassland; Eastern Cape, South Africa, north to tropical Africa.

1
1
SB
2
3
4
5
6
7
7
8

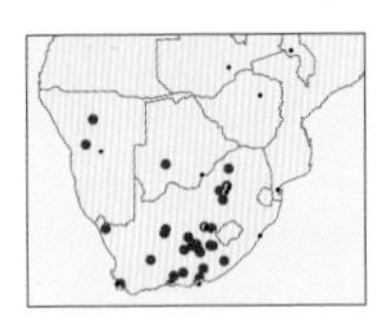

**1 *Cucullia brunnea*** Swift Shark

Forewing 15mm. Forewings narrowing toward apex, brown, mottled brown and grey, with broad, grey streak running through middle; hindwings cream with broad brown border. Caterpillar variegated white, brown, yellow and olive green, with short black hairs. Twenty-two species in genus in region. **Biology:** Caterpillars feed on Silver Stoebe (*Seriphium plumosum*). Adults throughout the year, peaking in summer. **Habitat & distribution:** Fynbos, karoo and grassland; Western Cape, South Africa, north to Namibia.

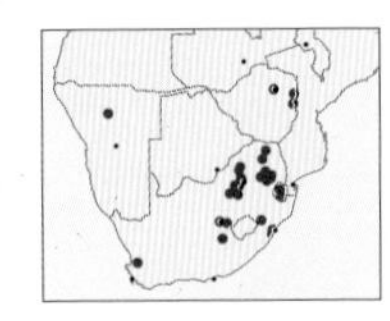

**2 *Cucullia chrysota*** Gold-streaked Shark

Forewing 15–17mm. Forewings have a golden sheen in live specimens, yellow fore-margin, brown streak, grey trailing edge, veins outlined black; hindwings cream with dark marginal band. Caterpillar olive green, with white line along length of body, separating lighter dorsum from darker green lower part. **Biology:** Caterpillars feed on Daisy Tea Bush (*Athrixia elata*). Adults throughout the year, peaking in summer. **Habitat & distribution:** Variety of habitats; Western Cape, South Africa, north to tropical Africa.

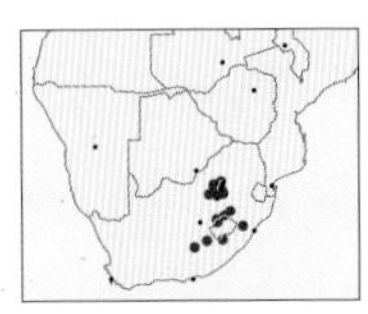

**3 *Cucullia pallidistria*** Brown-streaked Shark

Forewing 16–18mm. Similar to *C. chrysota* (above), but lacks the golden sheen and has a mustard brown streak edged by white extending almost the length of the wing, ground colour pale grey; hindwings white with broad brown border. **Biology:** Caterpillar host associations unknown. Adults September–April. **Habitat & distribution:** Grassland; Eastern Cape north to Gauteng, South Africa.

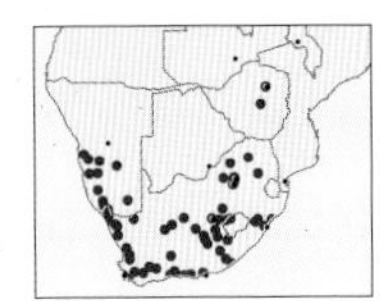

**4 *Cucullia terensis*** Grey Shark

Forewing 18–21mm. Forewings pointed at apex, grey, mottled in grey and black, veins black with wing appearing streaked, pair of black-circled spots may be present; hindwings white basally with large brown margin. Caterpillar grey with cream stripes along body, pairs of yellow markings along top of body, and black markings run along the side of body. **Biology:** Caterpillars recorded from Trailing African Daisy (*Osteospermum fruticosum*), Namaqualand Daisy (*Dimorphotheca sinuata*) and *D. spectabilis.* Adults throughout the year, peaking in summer. **Habitat & distribution:** Variety of habitats where daisies with fleshy leaves grow; Western Cape, South Africa, north to Namibia, Mozambique and Zimbabwe.

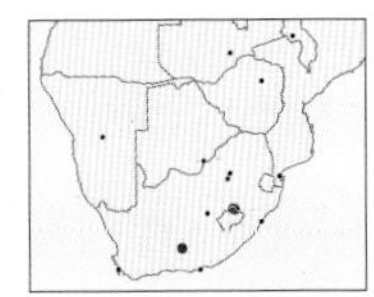

**5 *Diargyria argyrogramma*** Horse Diargyria

Forewing 12–14mm. Head, thorax and wings variegated white, grey and metallic gold or silver; forewings with sinuous gold or silver patches resembling a rocking horse when at rest, remainder of wing metallic grey, apical margin of wing with alternating patches of gold and white cilia; hindwings grey with white postmedian line. Four species in genus in region. **Biology:** Caterpillar host associations unknown. Adults in January. **Habitat & distribution:** Afroalpine vegetation at high altitudes along Great Escarpment; Lesotho and Eastern Cape, South Africa.

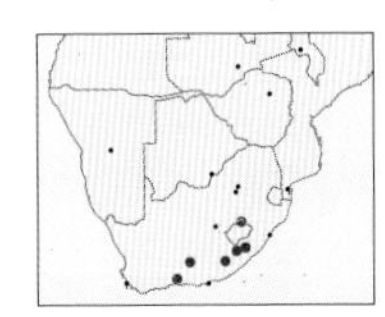

**6 *Diargyria argyrostolmus*** Dog Diargyria

Forewing 14–16mm. Head, thorax and wings variegated white; forewings white with diagnostic brown mark, resembling the outline of a dog, remainder of wing white or dusted with grey between white veins; hindwings white with variable dusting of grey scales, except on white postmedian line. **Biology:** Caterpillar host associations unknown. Adults January–March. **Habitat & distribution:** Mountain fynbos and grassland with fynbos elements; Western Cape to Lesotho.

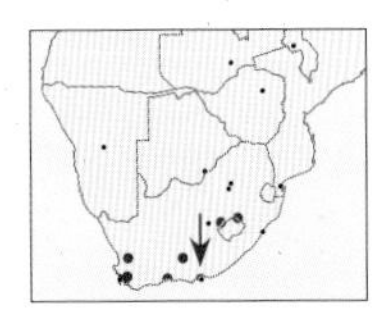

**7 *Diargyria argyhorion*** Rock Diargyria

Forewing 12–13mm. Head, thorax and wings variegated white and rusty brown; forewings white with diagnostic solid brown block, rest of wing variably dusted with brown scales; hindwings white with faint brown postmedian line. **Biology:** Caterpillar host associations unknown. Adults December–April. **Habitat & distribution:** Fynbos and grassland with fynbos elements; Western Cape to KwaZulu-Natal.

## Subfamily Agaristinae

Adults generally day-flying, often have clubbed antennae. Boldly marked, brightly coloured wings suggest they are distasteful to predators. May be confused with Tiger Moths (Arctiinae, p.364). Caterpillars are brightly coloured, often with a conspicuous hump on top of the body.

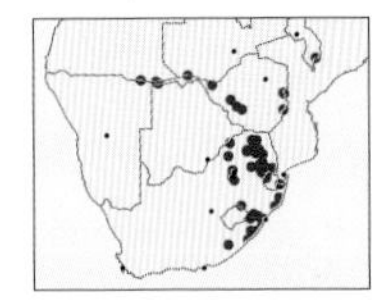

**8 *Aegocera fervida*** Fervid Tiger

Forewing 20–22mm. Forewings dark brown to black, dusted with white scales, broad cream bars; hindwings sulphur yellow or orange, with dark cell spot and brown marginal band. One species in genus in region. Adults September–May. **Biology:** Caterpillars recorded from wild grapes (*Rhoicissus*). **Habitat & distribution:** Forest and bushveld; Eastern Cape, South Africa, north to tropical Africa.

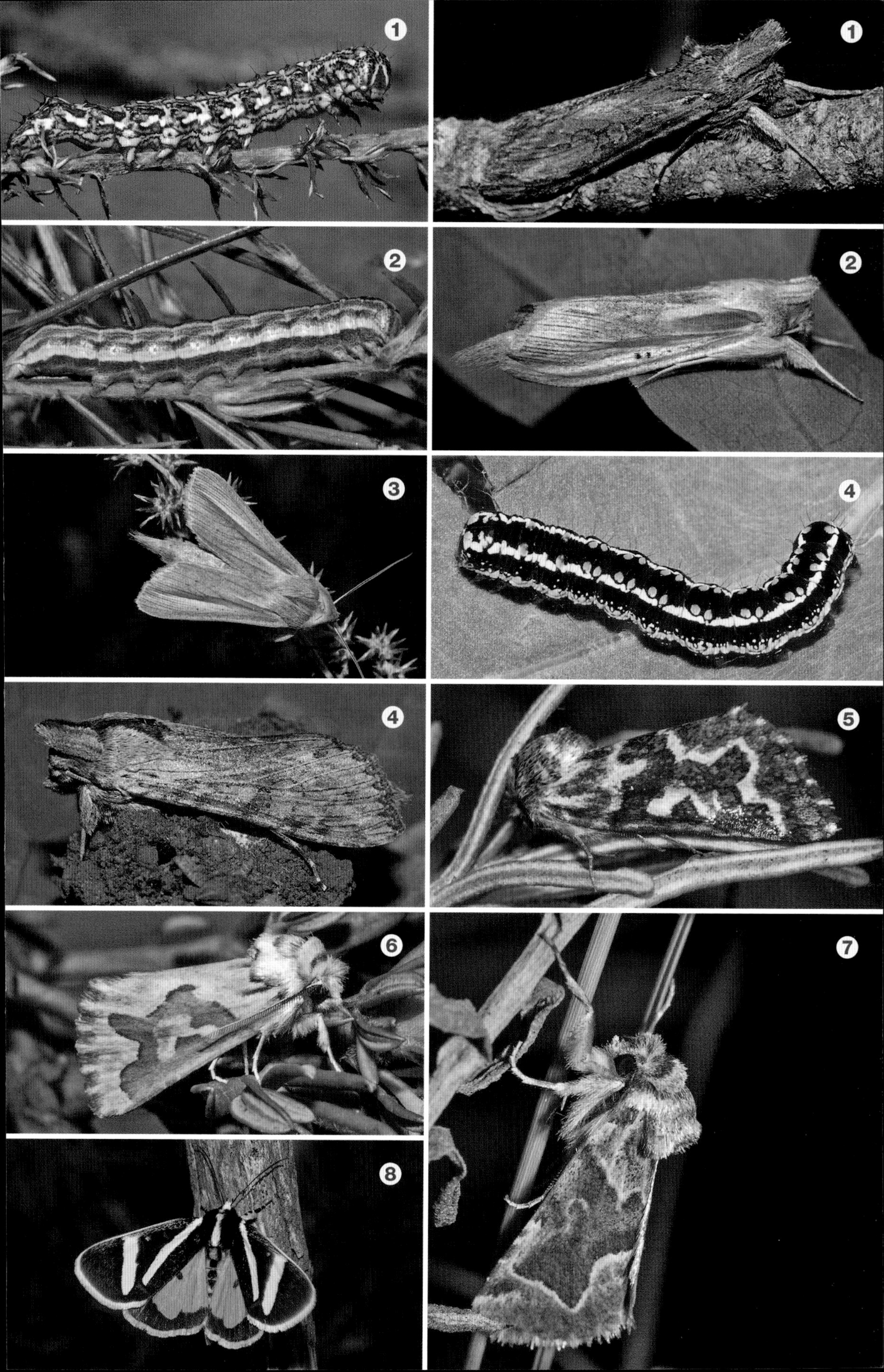
1
1
2
2
3
4
4
5
6
7
8

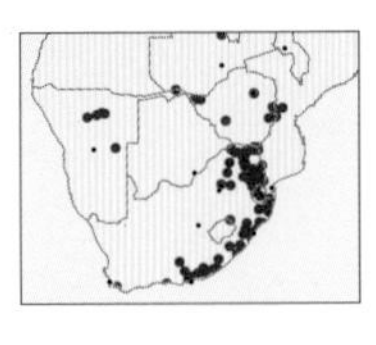

**1 *Agoma trimenii*** Trimen's False Tiger

Forewing 18–25mm. Variable, covered in thin white hairs; forewing brown or black, with 2 large, white or yellow markings; hindwings scarlet, orange or white, with black border. Caterpillar plump, cream with black-bordered, yellow bands, both head capsule and end of body orange. One species in genus in region. **Biology:** Caterpillars feed on various grape species (Vitaceae) . Adults October–April. **Habitat & distribution:** Bushveld and forest; Western Cape, South Africa, north to tropical Africa and Yemen.

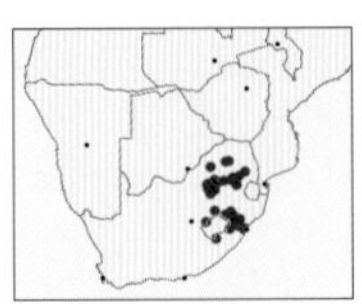

**2 *Brephos festiva*** Festive Red Tiger

Forewing 17–24mm. Ground colour of wings rust red, forewings with black-bordered cream markings, the postmedian one resembling the letter 'B', row of short oval terminal spots not touching median markings diagnostic; hindwings rust red with medium to large cream central marks, diagnostic lack of distinct terminal row of cream spots. Caterpillar pale yellow, head capsule and body orange, marbled black bands running along top and sides of body. Three species in genus in region. **Biology:** Caterpillars feed primarily on Tremble Tops (*Kohautia amatymbica*), but also on Wild Verbena (*Pentanisia angustifolia*), both Rubiaceae. Bivoltine, adults diurnal, September–February. **Habitat & distribution:** Grassland; Eastern Cape north to Limpopo, South Africa, also reported from Zimbabwe.

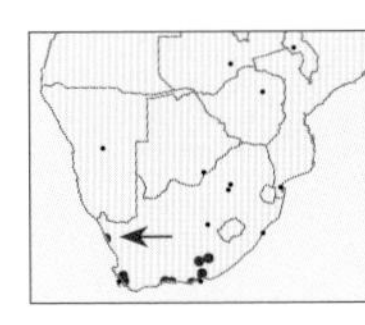

**3 *Brephos decora*** Decorated Red Tiger

Forewing 18–22mm. Similar to *B. festiva* (above), but forewings with white to cream markings (more or less extensive), row of elongate terminal spots normally touching other markings; hindwings have distinct row of terminal cream spots and only small central cream marks if present. **Biology:** Caterpillar host associations unknown. Adults diurnal, July–April. **Habitat & distribution:** Fynbos and grassy karoo habitats; Cape provinces, South Africa.

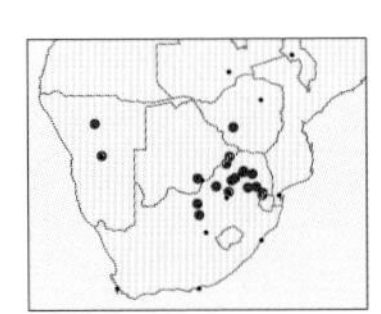

**4 *Paida pulchra*** Lemon Red Tiger

Forewing 18–22mm. Similar to *Brephos decora* (above), but forewings with unbroken lemon-cream band from base to postmedian, elongate row of terminal marks not touching median band, row of elongate marks on inner margin; hindwings of female with black cell spot and dark toothed subterminal band, hindwings of male with orange marginal dots, large orange central patch and long androconial hairs. One species in genus in region. **Biology:** Caterpillar host associations unknown. Adults diurnal, October–March. **Habitat & distribution:** Bushveld; North West, South Africa, north to Namibia and Zimbabwe.

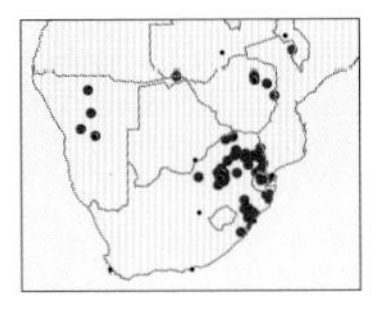

**5 *Heraclia superba*** Superb False Tiger

Forewing 28–36mm. Thorax black with white dots; forewings black, sometimes suffused grey, with variable but 2 basal, 2 median and 2 terminal pale yellow or orange patches; hindwings red or orange, with black border. Caterpillar pale white with thin black bands, orange patch at rear and covered in long white hairs. Six species in genus in region. **Biology:** Caterpillars feed on wild grapes (*Cissus*) and cultivated grapes (*Vitis*). Adults diurnal, September–April. **Habitat & distribution:** Forest and bushveld; Eastern Cape, South Africa, north to tropical Africa.

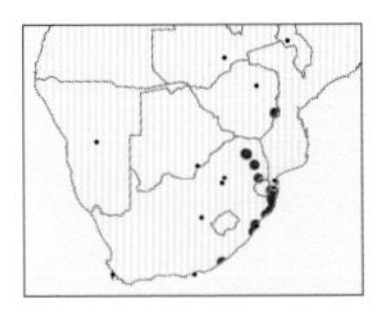

**6 *Heraclia africana*** African False Tiger

Forewing 29–35mm. Similar to *H. superba* (above), but forewings with only 1 basal, 2 median and 2 terminal yellow patches. **Biology:** Caterpillars feed on wild grapes (*Cissus*). Adults diurnal, September–April. **Habitat & distribution:** Forest and coastal bush; Eastern Cape, South Africa, north to tropical Africa.

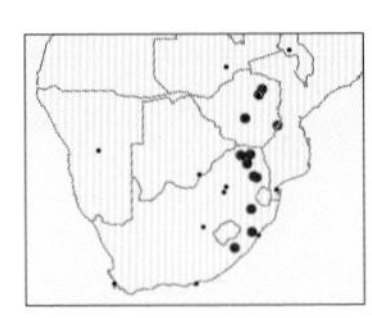

**7 *Heraclia butleri*** Butler's False Tiger

Forewing 21–30mm. Similar to *H. superba* (above), but forewings with 2 small basal patches, median patch elongate with small patch on inner margin in male, absent in female. **Biology:** Caterpillars feed on wild grapes (*Cissus*). Female diurnal, male attracted to light, October–March. **Habitat & distribution:** Forests; Eastern Cape north to tropical Africa.

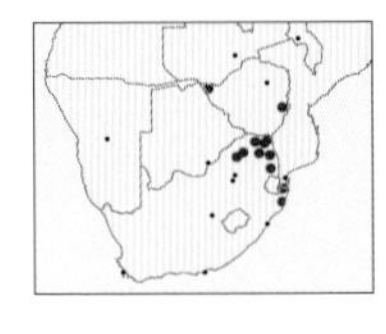

**8 *Lophonotidia melanoleuca*** Black Panther

Forewing 18–28mm. Bulky thorax and abdomen; basal part of forewings grey with black and tan transverse lines, white transverse band in middle of wing, black apex; hindwings white with broad, black marginal band. Caterpillar with warning colouration – banded black and orange, with tufts of plate-like white hairs. One species in genus in region. **Biology:** Caterpillars feed on Ivy-grape (*Cissus cornifolia*). Adults October–February. **Habitat & distribution:** Bushveld; KwaZulu-Natal north to Zimbabwe.

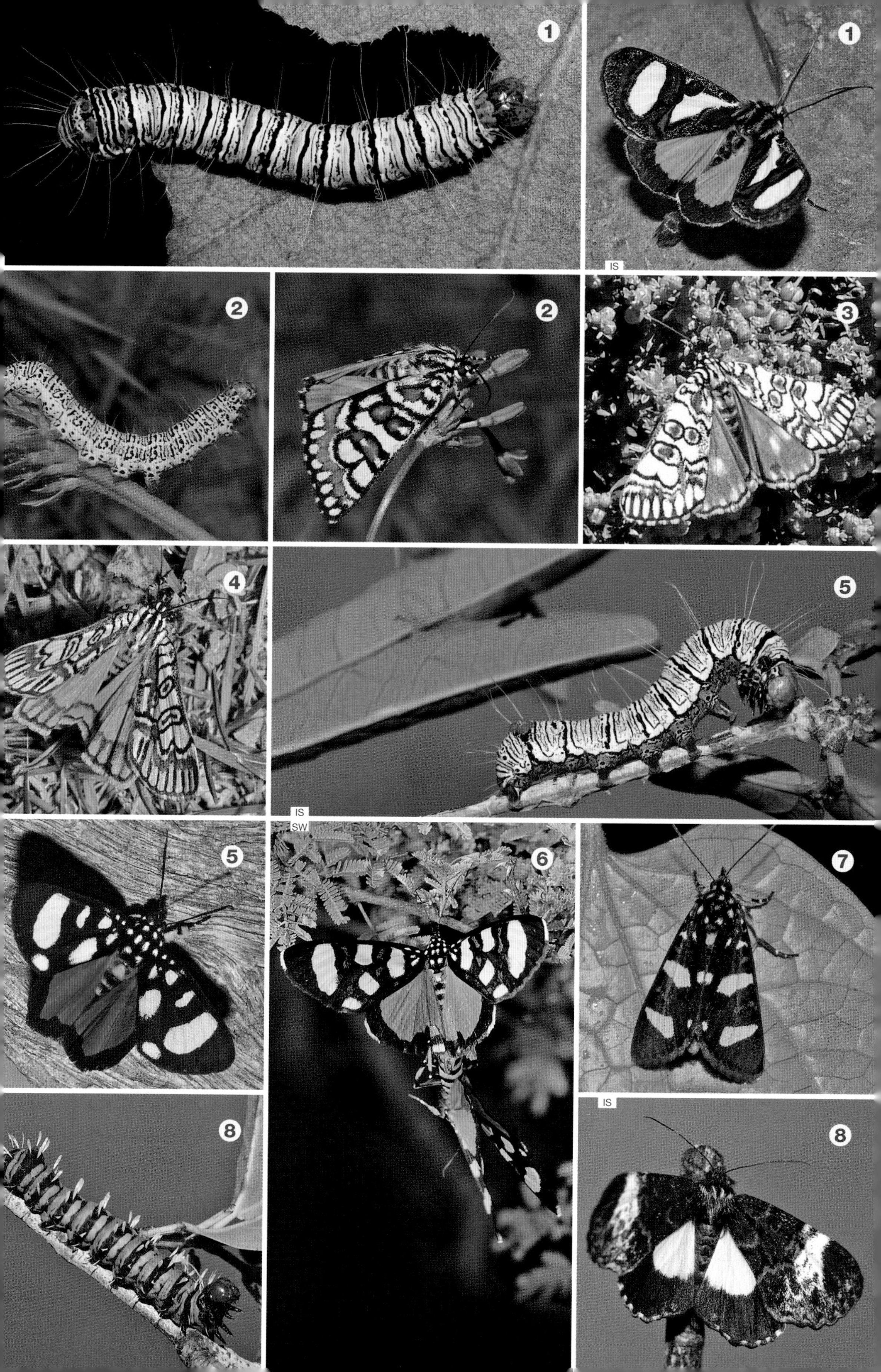
1
1
IS
2
2
3
4
5
IS
SW
5
6
7
IS
8
8

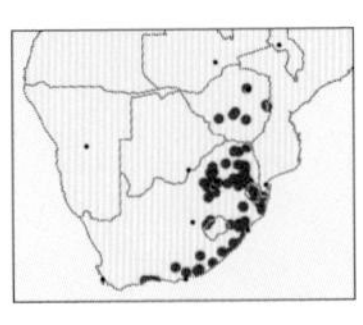

### 1 *Ovios capensis* — Cream Cake

Forewing 15–23mm. Distinctive; pilose thorax with a variety of colours; forewings speckled-grey on costa with 2 red oval markings, large median khaki-green marking running length of wing, surrounded by white, apical margin grey with red spots; hindwings yellow with row of large marginal black dots and black cell spot. One species in genus in region. **Biology:** Caterpillars feed on proteas. Adults throughout the year, primarily in summer. **Habitat & distribution:** Variety of habitats where proteas grow; Western Cape, South Africa, north to Kenya.

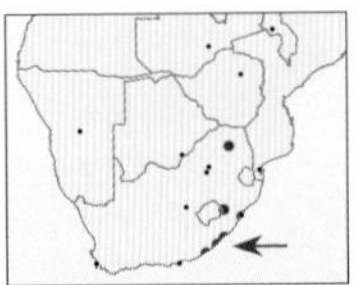

### 2 *Syfanoidea schencki* — Chocolate Cake

Forewing 16–20mm. Antennae clubbed; thorax and forewings chocolate brown, variably sprinkled with silver-white scales, veins white, 2 diagnostic large white marks, one from base and one on costa near apex; abdomen and hindwings orange with black cell spot and marginal band. One species in genus in region. **Biology:** Caterpillars feed on *Rhoicissus tridentata* (Vitaceae). Adults September–December. **Habitat & distribution:** Moist forests; Eastern Cape north to Limpopo, South Africa.

## Subfamily Acontiinae

Bird-dropping moths are small with narrow wings, mimicking the pearly uric acid droppings of birds.

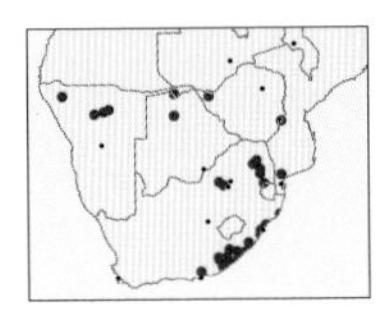

### 3 *Acontia antica* — Black-bar Droplet

Forewing 8–10mm. Forewings with large brown or black bar on inner margin stretching from wing base to apex, costal margin and leading edge cream, row of black marginal dashes; hindwings brown in females, cream with darker margin in males. Caterpillar a semi-looper, olive green dorsally and with yellow, marbled stripe running along length of body, with black bands at intervals. There are 62 regional species in the genus. **Biology:** Caterpillars recorded from Sun Hibiscus (*Hibiscus calyphyllus*). Mimics bird droppings, like most other *Acontia* spp., multivoltine, adults throughout the year. **Habitat & distribution:** Forest and bushveld; Eastern Cape, South Africa, north to Yemen, also Madagascar.

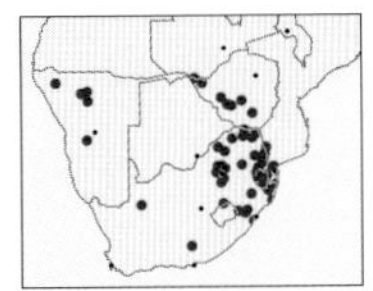

### 4 *Acontia porphyrea* — Waved Droplet

Forewing 12mm. Forewings pearly white, with faint grey bars at base, and mottled brown markings at apex and end of trailing edge; hindwings plain orange with no markings. Caterpillar ruby red, with orange dorsal markings on each segment. **Biology:** Caterpillars feed on Stinging Hibiscus (*Hibiscus engleri*). Adults October–May. **Habitat & distribution:** Grassland and bushveld; Eastern Cape north to Zimbabwe.

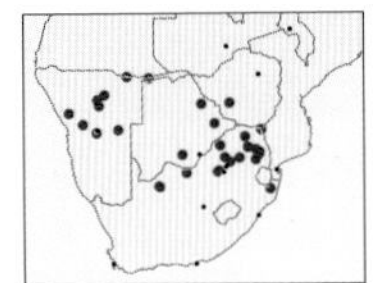

### 5 *Acontia conifrons* — Variegated Droplet

Forewing 8–10mm. Forewings pearly white along costa with variable extent of chocolate brown region on remainder of wing, darker in females, costa marked by faint, evenly spaced bars; hindwings white in male, brown in female. Caterpillar olive green, turning ruby red before pupating, orange spots on a white line that runs the length of the body. **Biology:** Caterpillars recorded from various mallows (Malvaceae), including exotic weeds. Bivoltine, adults November–April. **Habitat & distribution:** Bushveld; KwaZulu-Natal, South Africa, north to Namibia, also Tanzania.

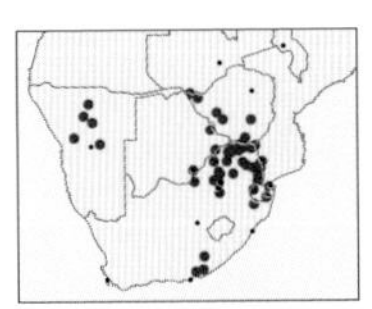

### 6 *Acontia discoidea* — Midnight Droplet

Forewing 14mm. Basal half of forewings pearly white and barred on leading edge, with 1 eyespot, a diagnostic 'U'-shaped black line, remainder of wing a jagged black terminal band; hindwings pale brown with dark border. Caterpillar a green semi-looper with faint orange blotches or diagonal white chevrons along body. **Biology:** Caterpillars recorded from native *Hibiscus*. Bivoltine, adults throughout the year, peaking November and March. **Habitat & distribution:** Bushveld; Eastern Cape north to Ethiopia.

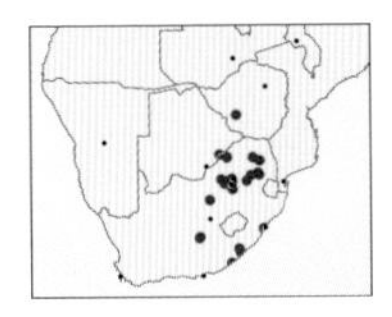

### 7 *Acontia tinctilis* — Morning Droplet

Forewing 11–12mm. Adult similar to *A. discoidea* (above), but forewings often with pink sheen, has 2 eyespots, the inner one a circle and the outer one an inverted comma sign; hindwings yellow with faint marginal band. Caterpillar similar to *A. tanzaniae* (p.420), but orange spots lateral, not dorsal. **Biology:** Caterpillars feed on various mallows. Bivoltine, adults September–April. **Habitat & distribution:** Grassland and bushveld; Eastern Cape, South Africa, north to Zimbabwe.

1
2
3
3
4
4
IS
5
5
6
6
7
7

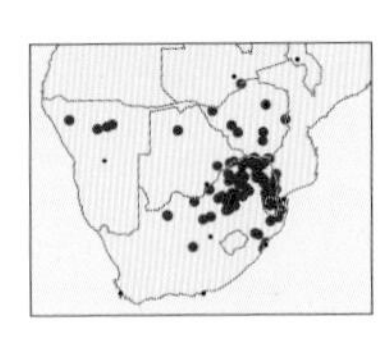

**1** ***Acontia gratiosa*** Short-winged Droplet

Forewing 7–10mm. Basal part of forewings creamy white, with pale olive bars along costa, remainder of wing variegated brown; hindwings orange with distinct black border. Caterpillar an olive green semi-looper with evenly spaced orange spots and numerous smaller yellow and black markings. **Biology:** Caterpillars feed on various honeycups (*Melhania* spp.). Multivoltine when conditions allow, adults throughout the year. **Habitat & distribution:** Suitable shady places where honeycups grow in bushveld and riverine vegetation; Northern Cape, South Africa, north to Angola and Zambia.

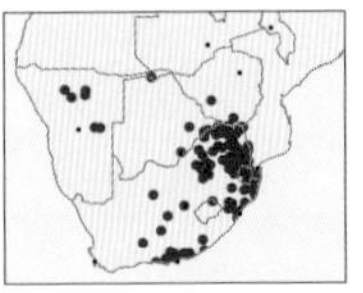

**2** ***Acontia guttifera*** Domino

Forewing 8–10mm. Sexually dimorphic; female with yellow, black-spotted thorax; forewings black with large yellow spots; hindwings yellow with dark border; male canary yellow with black marking on wing apex and inner margin, hindwings yellow with no or short dark border. Caterpillar with warning colouration, yellow with black segmental bands, orange head capsule, lateral spots and anal claspers, sharp black dorsolateral spines. **Biology:** Caterpillars feed on mallows (*Abutilon*). Multivoltine when conditions allow, adults September–May. **Habitat & distribution:** Variety of habitats where mallows grow; Western Cape, South Africa, north to Namibia, Zimbabwe and East Africa.

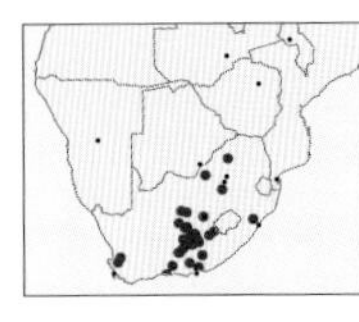

**3** ***Acontia mionides*** Karoo Droplet

Forewing 10–13mm. Narrow-winged; thorax pearly white with 3 faint brown bars; forewings white basally, with diagnostic large, white oblong patch on costa, remainder of wing brown with yellow-green shading, apex with 2 rows of small black dots; hindwings white with brown border in males, cream to brown in females. **Biology:** Caterpillar host associations unknown. **Habitat & distribution:** Karoo and grassland with karroid vegetation; Western Cape north to Free State, South Africa.

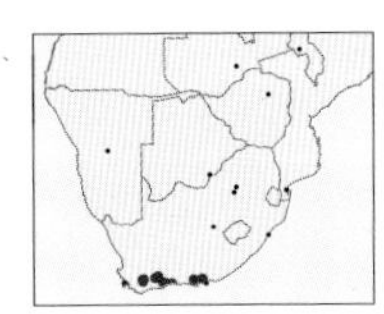

**4** ***Acontia accola*** Barred Rose

Forewing 10–12mm. Body and wings bright or pale pink or pale brown and pink; forewings variable in shades of pink and brown or deep or washed out pink, with diagonal, incomplete red or brown transverse bars, and subapical pink marking; hindwings grey-brown, with margin of pinkish cilia. **Biology:** Caterpillar host associations unknown. **Habitat & distribution:** Fynbos, renosterveld and succulent karoo; Western Cape and Eastern Cape, South Africa.

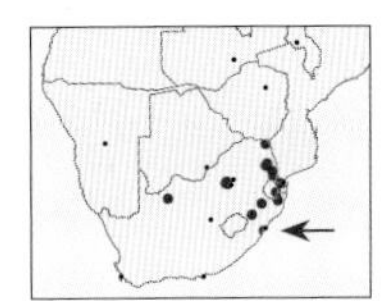

**5** ***Acontia tanzaniae*** Olive Droplet

Forewing 10–12mm. Forewings pearly white along costa with 2 orange crossbars, brown subterminal bar, diagnostic olive green full inner margin bar, 2 round eyespots; yellow hindwings diagnostic, greyish-yellow in some females, darkening at border. Similar *A. natalis* has white hindwings and bar on inner margin broken; *A. nephele* has white hindwings and lacks orange crossbars on costa. Caterpillar black with marbled yellow lateral band and orange dorsal spots, orange legs and head. **Biology:** Caterpillars recorded from various mallows, mostly on introduced fanpetal (*Sida*) weeds. Multivoltine where conditions allow, adults throughout the year. **Habitat & distribution:** Open habitats, common in transformed habitats; Eastern Cape, South Africa, north to Zimbabwe and East Africa

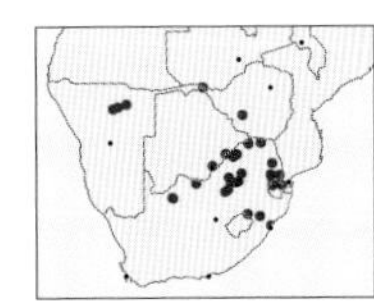

**6** ***Acontia transfigurata*** Mustard Droplet

Forewing 8–10mm. Variable; forewings cream, evenly mottled in olive green and black of varying intensity, margins bearing evenly spaced black bars and patches; hindwings white to grey, with darker margins. Caterpillar a grey or fawn-coloured semi-looper, with black patches along side of body, which bear small tubercles dorsally. **Biology:** Caterpillars recorded from various mallows, often on introduced Fanpetal (*Sida*) weeds. Multivoltine where conditions allow, adults throughout the year. **Habitat & distribution:** Bushveld, common in transformed habitats; KwaZulu-Natal, South Africa, north to tropical Africa, also Madagascar and Seychelles.

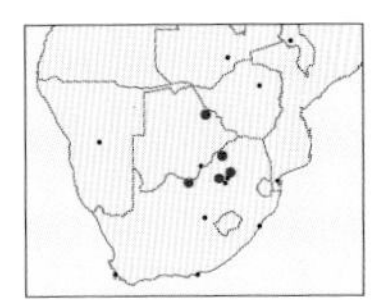

**7** ***Acontia bechuana*** Flame Droplet

Forewing 11–14mm. Unmistakable; colour from thorax to apex when at rest resembles a flame blending from pale yellow through orange to grey, black-dotted, curved postmedian line often obscured; hindwings cream in males, fawn in females. **Biology:** Caterpillar host associations unknown. Adults November–April. **Habitat & distribution:** Bushveld; North West to Botswana.

1
1
2
IS
2♂
2♀
3
4
5
5
6
6
7
7

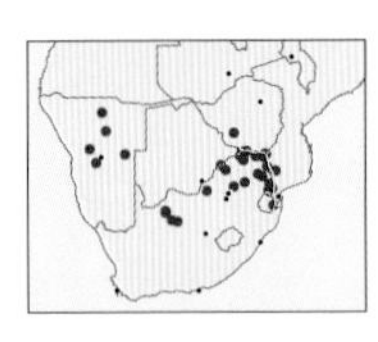

**1 *Acontia trimaculata*** Three-bar Droplet

Forewing 12–14mm. Forewings dark or olive brown, with grey cast toward apex, 3 white markings, diagnostic median marking curved toward apex, never connected to basal mark, and postmedian marking triangular with point slightly flattened in some individuals; hindwings white with black border in male, grey in female. **Biology:** Caterpillar host associations unknown. Adults October–April. **Habitat & distribution:** Bushveld; Northern Cape, South Africa, north to tropical Africa, also Saudi Arabia.

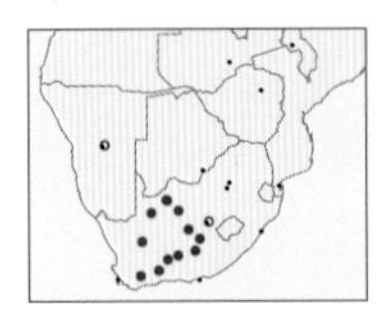

**2 *Acontia umbrigera*** Bold Droplet

Forewing 11–14mm. Similar to *A. trimaculata* (above), but brownish-grey and variably shaded with orange, fading in dried specimens, median mark curved to both basal and postmedian marks, often connected with basal mark, postmedian mark with curves, barely triangular. **Biology:** Caterpillar host associations unknown. Adults October–March. **Habitat & distribution:** Arid habitats; Western Cape, South Africa, north to Namibia.

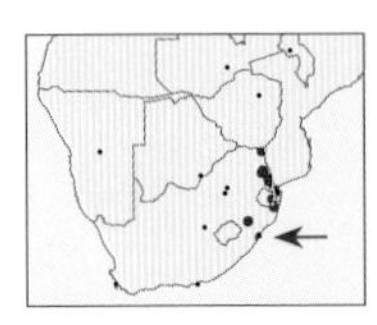

**3 *Acontia hampsoni*** Confusing Droplet

Forewing 14–15mm. Similar to *A. trimaculata* (above), but unmistakable with only 1 large white mark extending from white thorax reaching costa twice, row of grey marginal spots; hindwings white with broad brown border. Caterpillar swollen, with intricate scribblings of brown and white with black and orange dots. **Biology:** Caterpillars feed on Levant Cotton (*Gossypium herbaceum*, Malvaceae). Adults November–March. **Habitat & distribution:** Bushveld; KwaZulu-Natal, South Africa, north to tropical Africa.

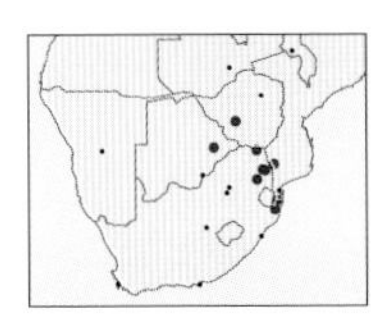

**4 *Thyatirina achatina*** Tunnel Droplet

Forewing 12–14mm. Unmistakble; thorax fawn and dark brown, forewings dominated by small basal and large anterior eyespots that generate a 3D tunnel effect; hindwings brown. Caterpillar black with uneven yellow and white crosslines, including on head, orange swollen setal bases. One species in genus in region. **Biology:** Caterpillars feed on Levant Cotton (*Gossypium herbaceum*, Malvaceae). Adults October–April. **Habitat & distribution:** Bushveld; KwaZulu-Natal, South Africa, north to Kenya.

## Subfamily Condicinae

The silk-spinning organs (spinnerets) of the mouthparts are generally reduced. Crochets of the prolegs are all of a uniform length.

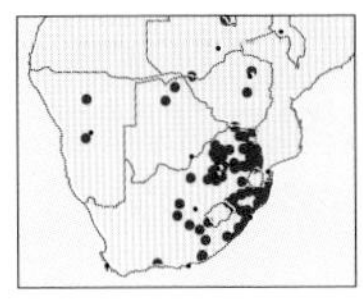

**5 *Condica capensis*** Footprint Owl

Forewing 15–17mm. Forewings light brown mottled with white dots or plain, diagnostic group of marks around cell resemble an animal footprint, dotted white postmedian line normally present; hindwings dirty white with grey border. Caterpillar green or black, with cream stripe along the side of the body. **Biology:** Caterpillars polyphagous. Multivoltine, adults throughout the year. **Habitat & distribution:** Varied, including transformed habitats; Western Cape, South Africa, north to Saudi Arabia, also Madagascar.

## Subfamily Hadeninae

The eyes of the adults are hairy, and the tibia lacks spines. Some have brightly coloured caterpillars that feed on poisonous bulb species.

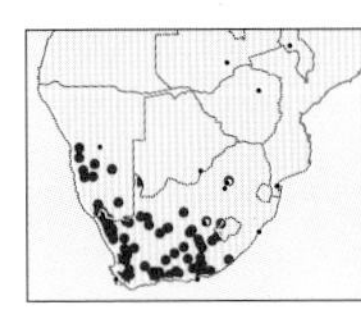

**6 *Cardepia definiens*** Vygie Brocade

Forewing 14–17mm. Forewings mottled grey and tan, with circular markings along fore-margin. Caterpillar pale apricot, with grey and apricot lines running length of body, separating lighter dorsum from brighter ventral region. One species in genus in region. **Biology:** Caterpillars feed on ice plants (Aizoaceae). Multivoltine where conditions allow, adults throughout the year. **Habitat & distribution:** Fynbos, karoo, desert and other habitats where ice plants grow; Western Cape, South Africa, north to Namibia.

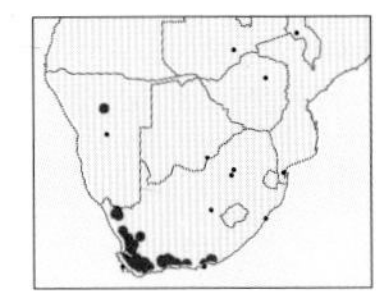

**7 *Diaphone eumela*** Cherry Spot

Forewing 16–20mm. Often confused with *D. mossambicensis* (p.424), these 2 species are best identified by caterpillars or distribution, forewings usually dark grey, sometimes lighter at costa, shape and position of black lines variable, in most individuals postmedian line is well away from central red crescent-shaped marking (cherry spot), but just touching in some individuals; hindwings white. Caterpillar pale yellow, with unevenly shaped or broken black bands incorporating dorsal and dorsoventral orange markings. Four species in genus in region. **Biology:** Caterpillars feed on various poisonous bulbs such as *Albuca*, *Drimia*, *Chlorophytum*, *Merwilla* and Chincherinchee (*Ornithogalum*). Bivoltine, adults throughout the year, peaking May–September. **Habitat & distribution:** Winter-rainfall areas, in fynbos and karoo; Cape provinces, South Africa and Namibia.

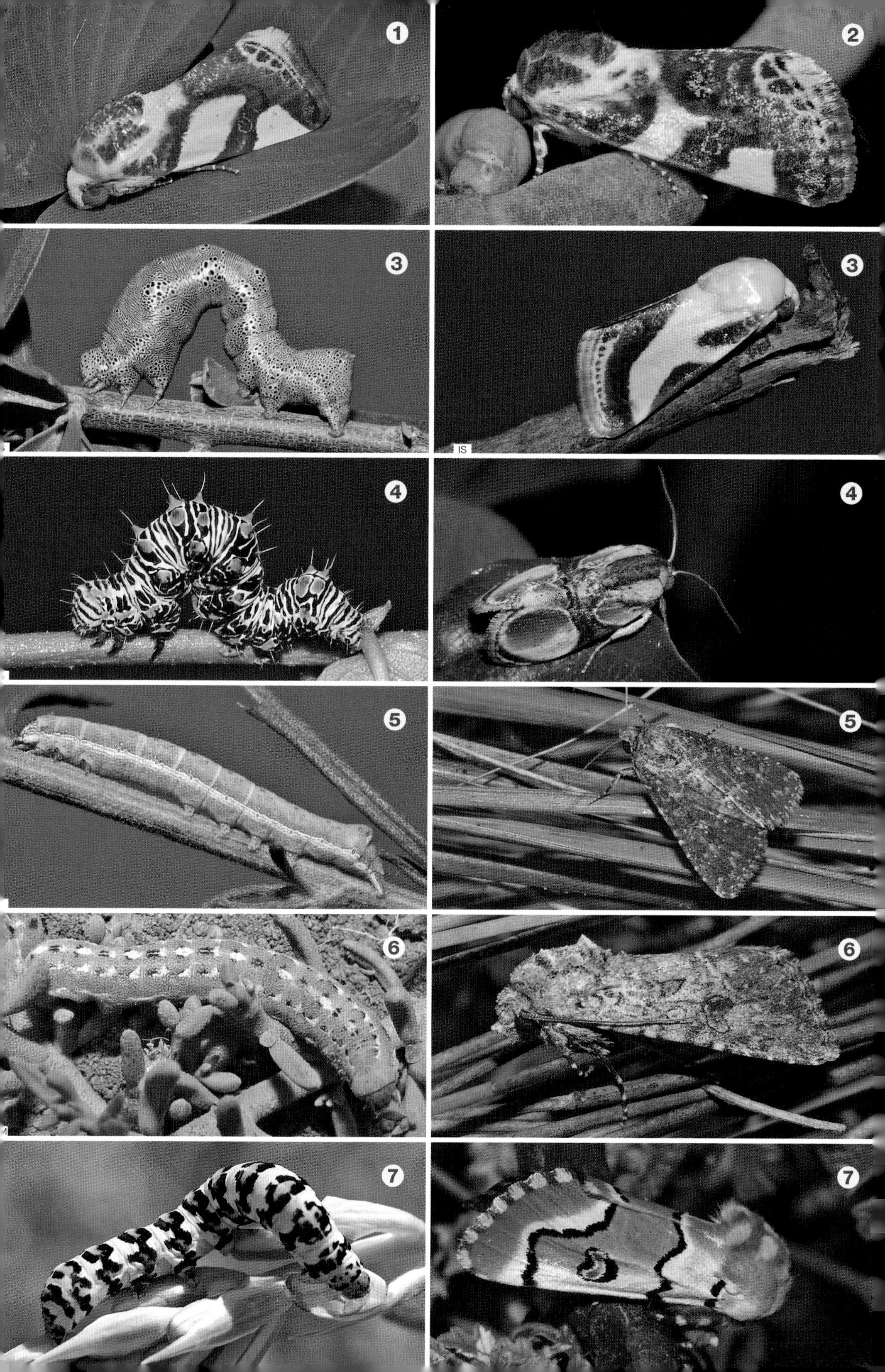
1
2
3
3
IS
4
4
5
5
6
6
7
7

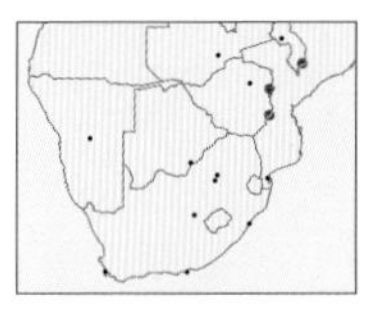

**1 *Diaphone lampra*** Peach Spot

Forewing 25–26mm. Differs from other *Diaphone* spp. by larger size, red tufts on thorax, evenly silver wings, yellow bands alongside black lines, uniform yellow cilia and yellow cell spot; hindwings charcoal grey. **Biology:** Caterpillar host associations unknown. Adults throughout the year. **Habitat & distribution:** Open higher altitudinal habitats; Zimbabwe north to Kenya.

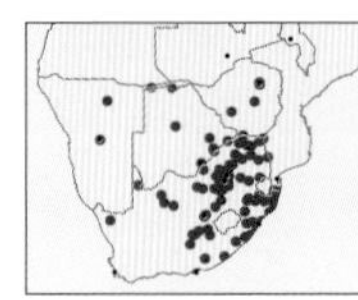

**2 *Diaphone mossambicensis*** Pale Cherry Spot

Forewing 15–17mm. Often confused with *D. eumela* (p.422), best identified by caterpillars or distribution; *D. mossambicensis* smaller and usually more white, postmedian line usually runs through 'cherry spot' or at least touches it; hindwings white. Caterpillar yellow or cream with distinct evenly shaped black bars on each segment, only dorsal orange markings on rear segments, no lateral orange marks. **Biology:** Caterpillars feed on various poisonous bulbs. Bivoltine, adults throughout the year, peaking October–February. **Habitat & distribution:** Summer-rainfall areas, in a variety of open habitats; Eastern Cape, South Africa, north to tropical Africa.

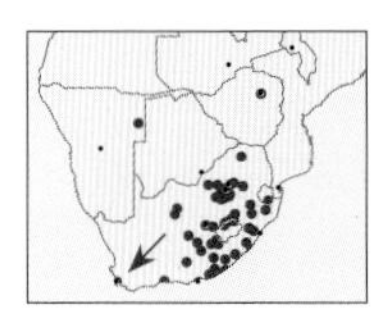

**3 *Odontestra bulgeri*** Bulger's Gothic

Forewing 15–16mm. Forewings with broad central transverse band bounded basally and apically by grey areas, oval and crescentic markings near fore-margin; hindwings white in male, grey-brown in female. Caterpillar lemon-green, with white-edged cream stripe running length of body. Five species in genus in region. **Biology:** Caterpillars recorded on besembos (*Thesium*) and alien Port Jackson Willow (*Acacia saligna*). Bivoltine, adults throughout the year, peaking October and March. **Habitat & distribution:** Variety of open habitats; Western Cape, South Africa, north to Zimbabwe, also East Africa.

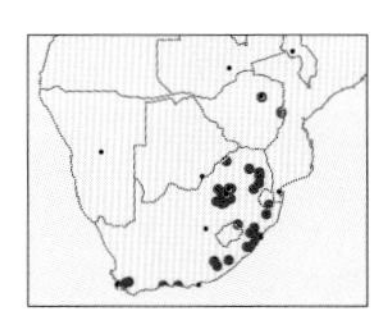

**4 *Odontestra speyeri*** Black Brocade

Forewing 15–18mm. Forewings dark grey or brown, fading to tan on inner margin, with short cream bar on black marking; hindwings white to cream. Caterpillar lime green, with thin cream line running along the length of the body, and small red marking on each abdominal spiracle in some individuals. **Biology:** Caterpillars polyphagous. Bivoltine, adults August–May, peaking October and January. **Habitat & distribution:** Open habitats; Western Cape, South Africa, north to tropical Africa, also Saudi Arabia.

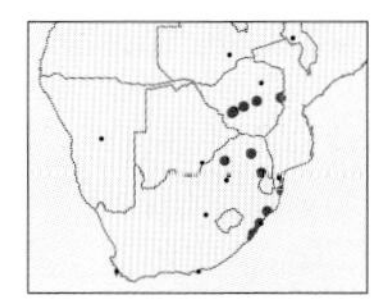

**5 *Odontestra vittigera*** Red Brocade

Forewing 13–18mm. Forewings with basal and apical grey bands, the outer band with dentate margins, broad median, brown band with dark markings and a large and a small cream marking, apical cream spot, trailing edge with cream bar; hindwings dirty grey. **Biology:** Caterpillar recorded on *Gladiolus*. Adults September–May. **Habitat & distribution:** Grassland; Eastern Cape, South Africa, north to tropical Africa.

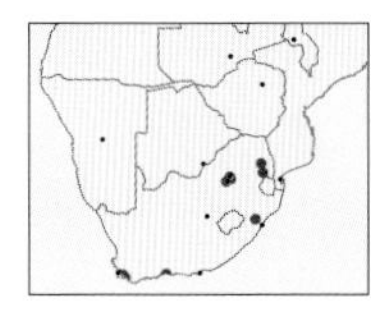

**6 *Neuranethes spodopterodes*** Agapanthus Borer

Forewing 12–15mm. Forewings mottled light to dark brown and grey, with cream band near wing base and tan bar near apex, black marking, and 2 cream oval markings near fore-margin; hindwings cream, with faint marginal band. Caterpillar white with rings of black spots. One species in genus in region. **Biology:** Feeds on various *Agapanthus* spp. and their hybrids, with a preference for the Drakensberg Agapanthus (*Agapanthus inapertus*). **Habitat & distribution:** Originally recorded from KwaZulu-Natal and Mpumalanga, it has now spread over most of South Africa via the nursery trade; only recorded from South Africa.

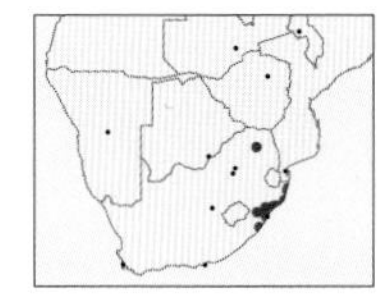

**7 *Nyodes acatharta*** Mottled Lichen

Forewing 15–17mm. Forewings mossy green mottled with grey, and crossed by a number of white, sinuous lines; hindwings cream with darker transverse band in male, uniform grey in female. Caterpillar black dorsally, with 2 lines of large, irregular yellow markings, ventral parts of body pale orange, covered in fine white hairs. **Biology:** Caterpillars feed on ferns, adult a bark mimic. **Habitat & distribution:** Moist forests; KwaZulu-Natal, South Africa, north to tropical Africa.

**8 *Nyodes lutescens*** Mossy Lichen

Forewing 18–20mm. Forewings mottled boldly in white and shades of green, incomplete sinuous lines running across wing; hindwings cream with faint transverse band. Caterpillar very similar to that of *N. acatharta* (above), circular yellow spots complete, covered in fine white hairs. **Biology:** Caterpillars feed on ferns. **Habitat & distribution:** Moist forests; Western Cape, South Africa, north to Zimbabwe, also Uganda.

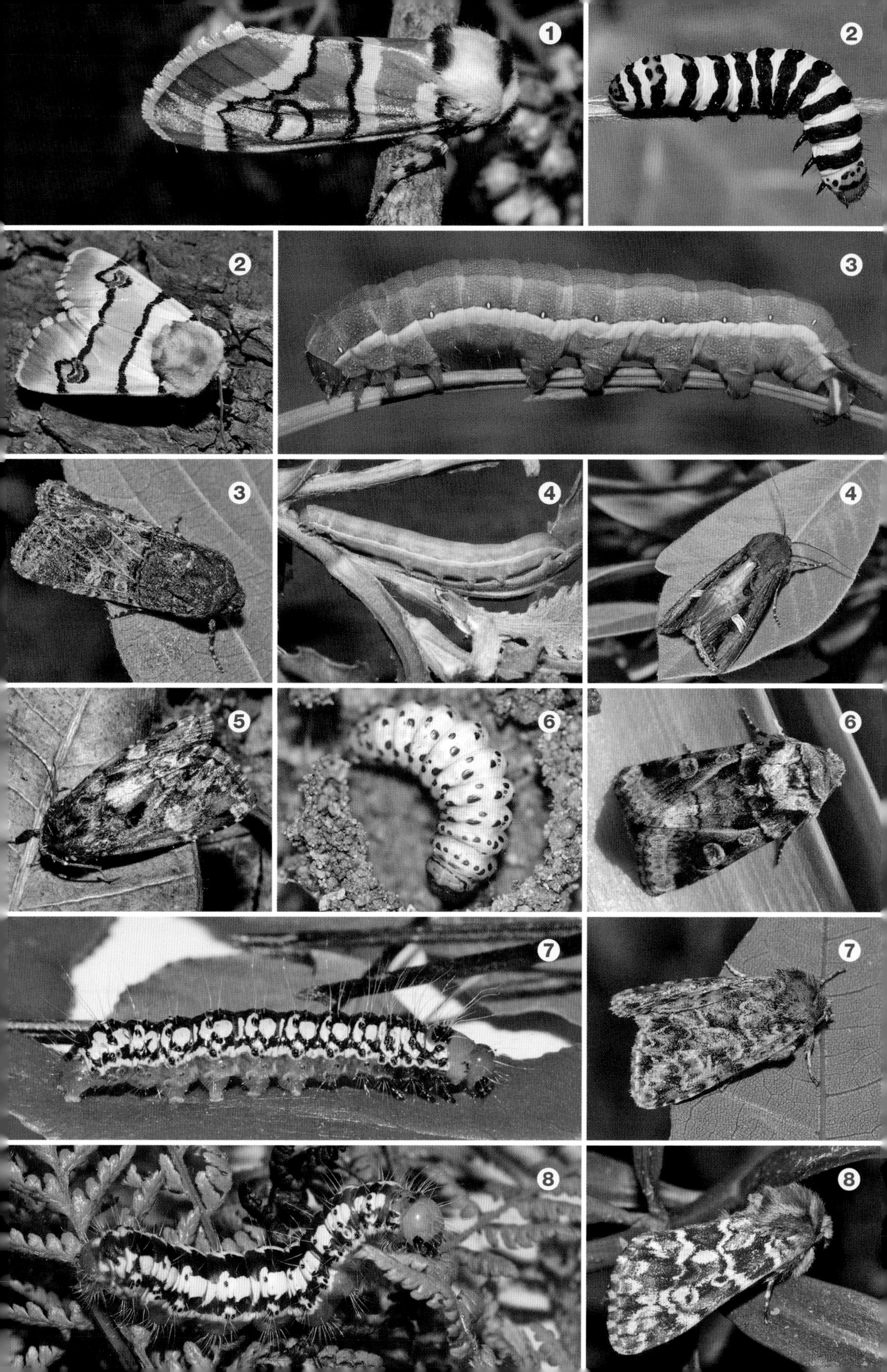
1
2
2
3
3
4
4
5
6
6
7
7
8
8

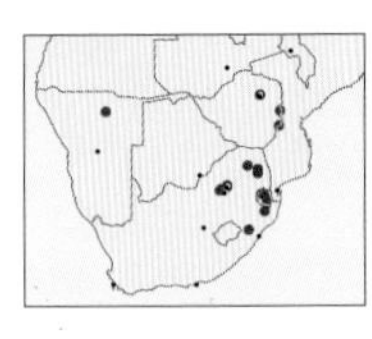

### 1 *Apospasta fuscirufa* — Sombre Brocade

Forewing 16–18mm. Forewings mottled grey and brown, with a cream marking near fore-margin; hindwings hyaline white with brown border. Caterpillar plump, tan or green with 2 faint, peach-coloured markings on each segment. **Biology:** Caterpillars polyphagous on soft new leaf and fruit, often in pods. Bivoltine, adults throughout the year, peaking September and April. **Habitat & distribution:** Forests and gardens; KwaZulu-Natal, South Africa, north to East Africa.

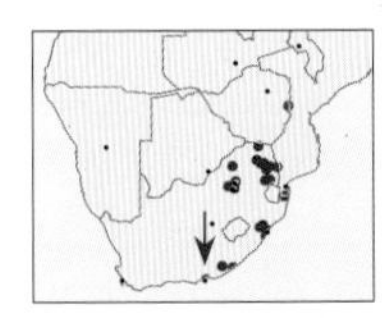

### 2 *Omphalestra mesoglauca* — Silvered Brocade

Forewing 14–16mm. Forewings variable, various shades of brown, but with diagnostic broad, silvery grey median band, containing 2 or 3 circular markings; hindwings tan, grey or brown. Caterpillar very pale fawn, dry grass mimic, with light cream lateral stripe bordered by row of very small black dots. Three species in genus in region. **Biology:** Caterpillars feed on grass (Poaceae) leaves and seeds. **Habitat & distribution:** Grassland and bushveld; Eastern Cape, South Africa, north to East Africa.

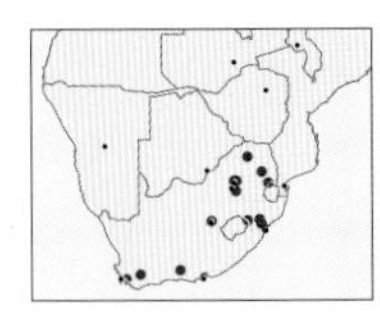

### 3 *Tycomarptes praetermissa* — Bar Brocade

Forewing 12mm. Tufts of scales on head and thorax and small white spot near abdomen; forewings brown with dark brown markings and pair of diagnostic white or white-ringed markings in middle of wing, terminal area not densely mottled; hindwings cream with darker margins. Caterpillar plump, tan or pinkish, with line of dark, square-shaped patches along back, and line of ringed dots running along side of body. Three species in genus in region. **Biology:** Caterpillars polyphagous on low-growing plants. Adults September–March. **Habitat & distribution:** Open habitats; Western Cape north to Limpopo, South Africa, and Kenya and Burundi.

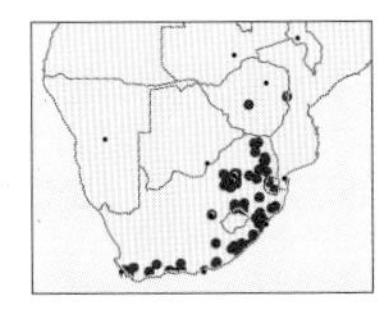

### 4 *Tycomarptes inferior* — Inferior Brocade

Forewing 12–13mm. Very similar to *T. praetermissa* (above), but more mottled, especially in terminal area, and pair of markings in median area smaller, not white or sometimes a cream lower half ring on outer mark; forewings fawn to dark brown, variable markings, indistinct dark transverse band in middle of wing; hindwings tan or brown, darkening toward margin. Caterpillar similar to that of *T. praetermissa*. **Biology:** Caterpillars polyphagous on low-growing plants. Adults throughout the year. **Habitat & distribution:** Moist habitats; Western Cape, South Africa, north to Ethiopia.

## Subfamily Bagisarinae

The caterpillars have toothed crochets (hooks) on the prolegs, and are semi-loopers whose movement resembles that of leeches. The larvae live on the underside of leaves (typically mallows, Malvaceae). The wings of adults have very fine, dust-like scales, and commonly have a large dark marking in the middle of the costal region.

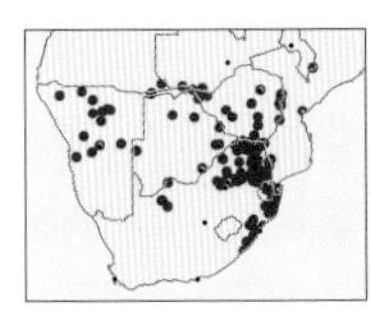

### 5 *Amyna punctum* — White Point

Forewing 14–18mm. Often confused with *A. axis* (below); forewings pointed, colouration variable, light to dark brown, mottled with poorly defined transverse bars, small incomplete white circle in middle of forewing, often with conspicuous white spot but absent in some individuals; hindwings uniformly dark brown in dark forms, pale grey in light-coloured forms. Caterpillar slender, charcoal grey or pale green with diagnostic black rings around setal bases. Four species in genus in region. **Biology:** Caterpillars feed on various *Croton* spp. Multivoltine, adults throughout the year. **Habitat & distribution:** Common in habitats where *Croton* trees grow; Eastern Cape, South Africa, north to tropical Africa.

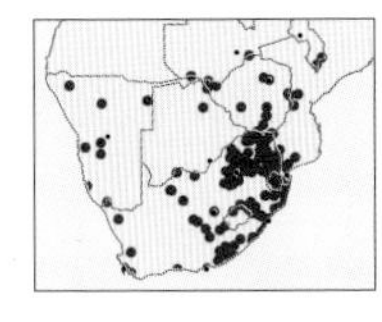

### 6 *Amyna axis* — Small White Point

Forewing 10–13mm. Similar to *A. punctum* (above), but smaller with wings more rounded, males have opaque patch on each wing, prominent white spot on forewings present or absent, not diagnostic for either species. Caterpillar green or brown with a darker dorsum and cream line running along the body, setal base not ringed. **Biology:** Caterpillars polyphagous. Multivoltine, adults throughout the year, abundant March–April. **Habitat & distribution:** Varied, including transformed habitats; Western Cape north across most of Africa, Madagascar and Indian Ocean islands, cooler parts of Asia and Australia.

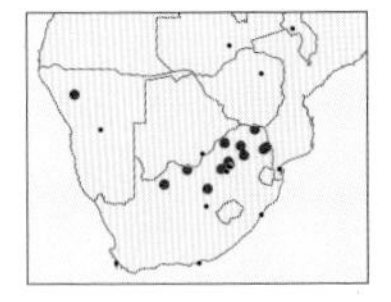

### 7 *Honeyia clearchus* — Two-spot Honeyia

Forewing 13–14mm. Dove grey to flesh coloured; forewings with dark-bordered cream lines, and 2 diagnostic black markings on costa; hindwings uniform cream or brown. Caterpillar blue-green with thin white stripe running along length of body. Four species in genus in region. **Biology:** Caterpillars recorded from White Raisin (*Grewia bicolor*). Adults November–March. **Habitat & distribution:** Bushveld; Northern Cape, South Africa, north to Saudi Arabia.

1
1
2
IS
2
3
MM
3
4
JuB
4
5
5
6
6
IS
7
7

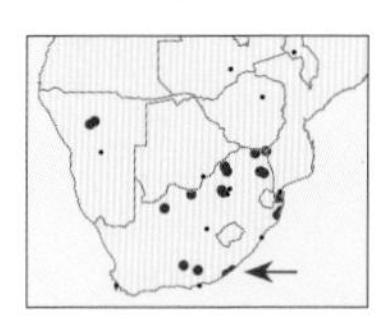

## 1 *Brevipecten cornuta* — Two-spot Crescent

Forewing 11–14mm. Cream to grey; forewings similar to that of *Honeyia clearchus* (p.426), but first bar on forewing brown and edged by a white crescent-shaped marking, and sharper transverse lines across wing; hindwings uniform cream or greyish. Caterpillar a pale green semi-looper, with row of large pinkish to whitish marks along back, bordered on each side by cream stripe. **Biology:** Caterpillars feed on various raisin bushes (*Grewia* spp.). Adults September–March. Five species in genus in region. **Habitat & distribution:** Bushveld where *Grewia* bushes grow; Eastern Cape, South Africa, north to Yemen.

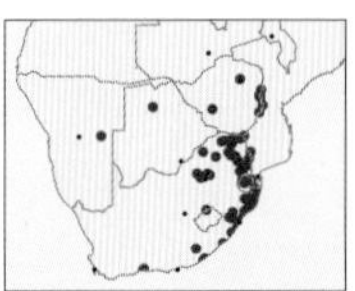

## 2 *Chasmina vestae* — Snow White

Forewing 14–18mm. Frons and foretibia orange and black; body and wings uniform pearly white; thorax pilose. Caterpillar apple green (sometimes with pink cast), with short hairs arising from small black dots, and 2 cream stripes running along top of body. Two species in genus in region. **Biology:** Caterpillars feed on Crossberry (*Grewia occidentalis*). Adults throughout the year, peaking November–March. **Habitat & distribution:** Forests and riverine vegetation where crossberries grow; Western Cape, South Africa, north to tropical Africa, also Madagascar.

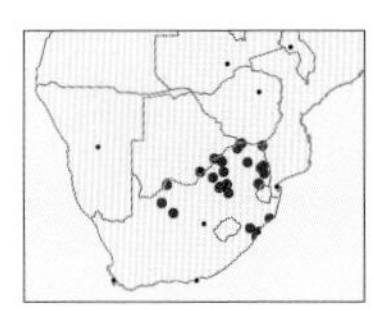

## 3 *Pardoxia graellsii* — Yellow Drab

Forewing 14–19mm. Ground colour creamy yellow; forewings with dark brown margin, variable brown streak from median area to margin and some dots and dusting of darker scales, plain forewing in some individuals; hindwings white with brown border. Caterpillar pale green with 2 lateral and 1 dorsal yellow stripe running along length of body, dorsal stripe bordered by black 'V'-shaped markings, body covered sparsely in white hairs. **Biology:** Caterpillars polyphagous on mallows (Malvaceae). Adults throughout the year, peaking December–March. **Habitat & distribution:** Varied, including transformed habitats, where mallows grow; KwaZulu-Natal, South Africa, north to Europe and Southern Asia.

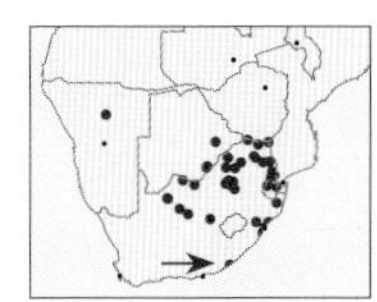

## 4 *Xanthodes albago* — Brown-tipped Drab

Forewing 10–14mm. Stocky body and shortish wings; forewings white to pale yellow, usually with darker margin before wing tip, diagnostic antemedial and postmedian lines, distinct brown cell spot in most specimens; hindwings cream with darker border. Caterpillar has 2 distinct forms: green with dorsolateral stripe; other with broad yellow lateral band broken by brown lines. Two species in genus in region. **Biology:** Caterpillars feed on various mallows (Malvaceae). Adults throughout the year, peaking November–March. **Habitat & distribution:** Varied, including transformed habitats; Eastern Cape, South Africa, north to Saudi Arabia, also Madagascar, Indian Ocean islands, Australasia, parts of Oriental and Palaearctic regions.

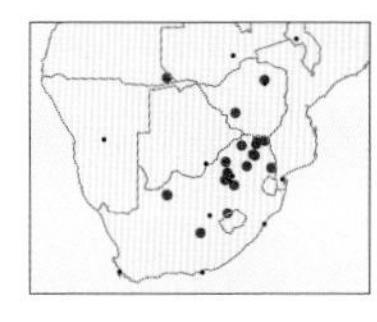

## 5 *Xanthodes brunnescens* — Brown Drab

Forewing 11–15mm. Forewings white, tan or brown, either with large, dark central streak running length of wing or in lighter forms only with small dark dusting of scales, often with white along costa, 2 short brown bars at trailing edge of wing; hindwings translucent brown with darker margin. **Biology:** Caterpillar host associations unknown. Adults September–March, peaking November–December. **Habitat & distribution:** Dry habitats; Eastern Cape, South Africa, north to Oman.

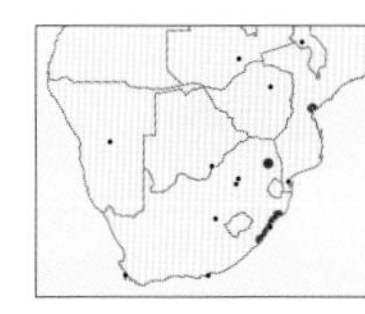

## 6 *Leocyma appollinis* — Yellow Jack

Forewing 15–18mm. Body orange, fades in dried specimens, wings sulphur yellow, forewings with 2 black dots in male, 1 in female; hindwings heavily setose on inner fold. Caterpillar mottled green and blue with black web-like lines and red spots on nodes. Two species in genus in region. **Biology:** Caterpillars feed on *Hibiscus*. Adults October–April. **Habitat & distribution:** Moist forests; KwaZulu-Natal, South Africa, north to Mozambique and Zimbabwe, also reported from Madagascar and Mauritania.

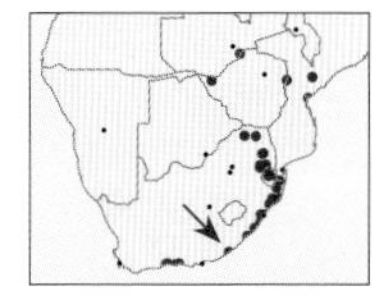

## 7 *Androlymnia torsivena* — Twisted Crepe

Forewing 11–14mm. Wings and body fawn; forewings short and broad with falcate apex, 2 diagnostic oblique brown markings on costa, and thin, brown transverse lines; hindwings uniform fawn. Caterpillar a thin, green semi-looper with broad cream stripe running length of body. One species in genus in region. **Biology:** Caterpillars feed on Crossberry (*Grewia occidentalis*). Adults throughout the year, peaking December–March. **Habitat & distribution:** Forests; Western Cape, South Africa, north to tropical Africa.

1
2
2
3
IS
3
4
4
5
6
6
7
7

## Subfamily Xyleninae

Large group of medium-sized, often drab moths, with squared-off forewings and grey to white hindwings. They are sometimes placed as a tribe within Hadeninae or Cuculliinae.

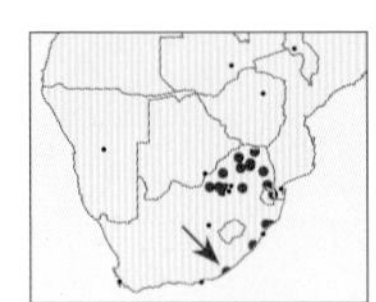

### 1 *Magusa viettei* — Swift Arches

Forewing 15–19mm. Similar to some *Cucullia* spp. (pp.412, 414) and *Neostichtis nigricostata* (p.436), with elongate wings and flat crest on thorax, but short white arrow dash in middle of forewing touching margin, diagnostic for genus; forewings very variable, cream, brown or grey or a mottling of all colours, often with a broad black stripe running along centre of wing; diagnostic hindwings white with brown marginal shading. Caterpillar cream with numerous variable black, green and white stripes running along length of body, a distinctive single yellow dorsal stripe, anterior and posterior ends with black spots. Two species in genus in region. Similar *M. versicolora* has the hindwings uniform dark brown. **Biology:** Caterpillars feed on various Rhamnaceae. Multivoltine, adults throughout the year. **Habitat & distribution:** Forest and riverine vegetation; Eastern Cape, South Africa, north to East Africa and Madagascar.

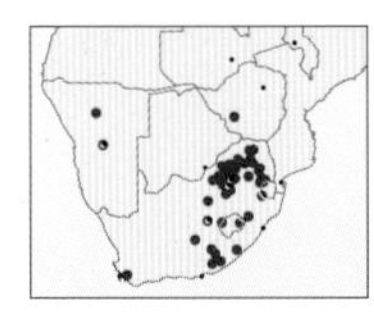

### 2 *Acrapex aenigma* — Enigmatic Acrapex

Forewing 11–12mm. Forewings pale to deep pinkish-brown, white central streak with darker line, veins outlined in cream; hindwings white. Eighteen species in genus in region, many can only be positively identified through dissection of male genitalia. **Biology:** Caterpillar host associations unknown. Adults October–April. **Habitat & distribution:** Primarily in grassland and bushveld; Western Cape, South Africa, north to tropical Africa.

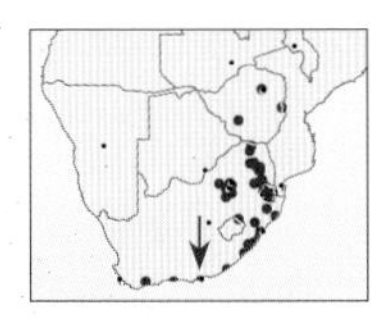

### 3 *Ariathisa abyssinia* — Black-U Dumpling

Forewing 15mm. Stocky, grey, pinkish or brown, with squared-off forewings; forewings mottled, with diagnostic central black marking surrounded by light grey area; hindwings hyaline white. Caterpillar stocky, brown, grey or pinkish, with a row of dark brown patches along body. **Biology:** Caterpillars polyphagous. Multivoltine, adults throughout the year. **Habitat & distribution:** Relatively moist, often disturbed, habitats, including cultivated areas; Western Cape, South Africa, north to tropical Africa and Saudi Arabia.

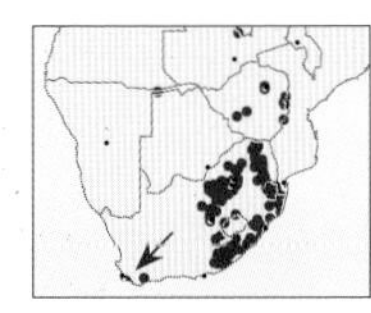

### 4 *Busseola fusca* — Maize Stalk Borer

Forewing 13–19mm. Forewing light or dark brown, with brown-edged cream stripe along apical region; hindwing cream in male, reddish in female. Caterpillar flesh-coloured with bands of small dark brown dots near head and at end of body. **Biology:** Caterpillars feed in leaf sheath, typically then boring into plant stem of larger grasses. Bivoltine, adults August–May, peaking November and February. **Habitat & distribution:** Grassland and cultivated areas; Western Cape, South Africa, north to most of sub-Saharan Africa.

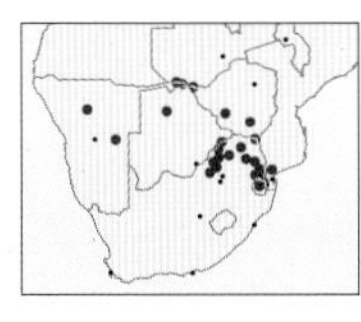

### 5 *Catamecia connectens* — Black-streak Silver

Forewing 11–13mm. Easily confused with *Hypotacha* spp. (Erebidae); forewings silver-grey with diagnostic broken row of black streaks and patches in centre of wing from base to margin; hindwings white with brown border. Caterpillar typical cylindrical noctuid, 2 forms: one brown, other green with dorsal and lateral streaks. One species in genus in region. **Biology:** Caterpillars feed on Leadwood (*Combretum imberbe*). Adults October–February. **Habitat & distribution:** Bushveld where leadwoods grow; Eswatini north to Zimbabwe and Namibia.

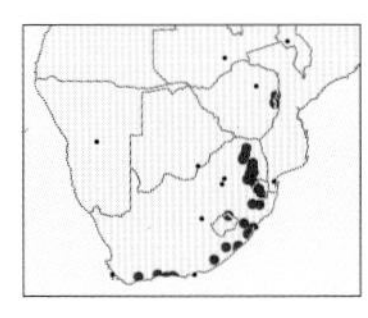

### 6 *Caradrina tenebrata* — Brown Crepe

Forewing 11–12mm. Stocky; forewings warm brown, with variably shaped group of white spots in cell, and 2 or 3 thin, transverse lines (not always present), faint grey marginal band; hindwings cream or brown with faint dark cell spot. Nine species in genus in region. **Biology:** Caterpillar host associations unknown. Adults throughout the year, peaking November and March. **Habitat & distribution:** Moist forests; Western Cape to Zimbabwe.

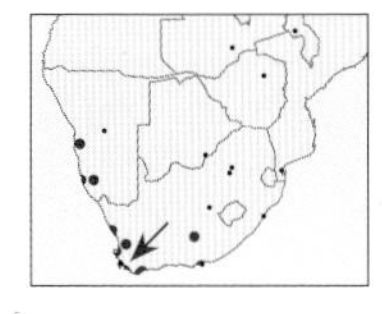

### 7 *Ectochela dicksoni* — Dickson's Ectochela

Forewing 19–20mm. Greyish-cream, dusted with grey, central black streak ending in faint black and orange marking around cell, indistinct, diagnostic evenly curved postmedian line, row of black marginal spots; hindwings white with marginal black spots in male, brown in female. Twenty-eight species in genus in region. **Biology:** Caterpillar host associations unknown. Adults throughout the year. **Habitat & distribution:** Dry fynbos, karoo and desert; Cape provinces, South Africa, north to Namibia.

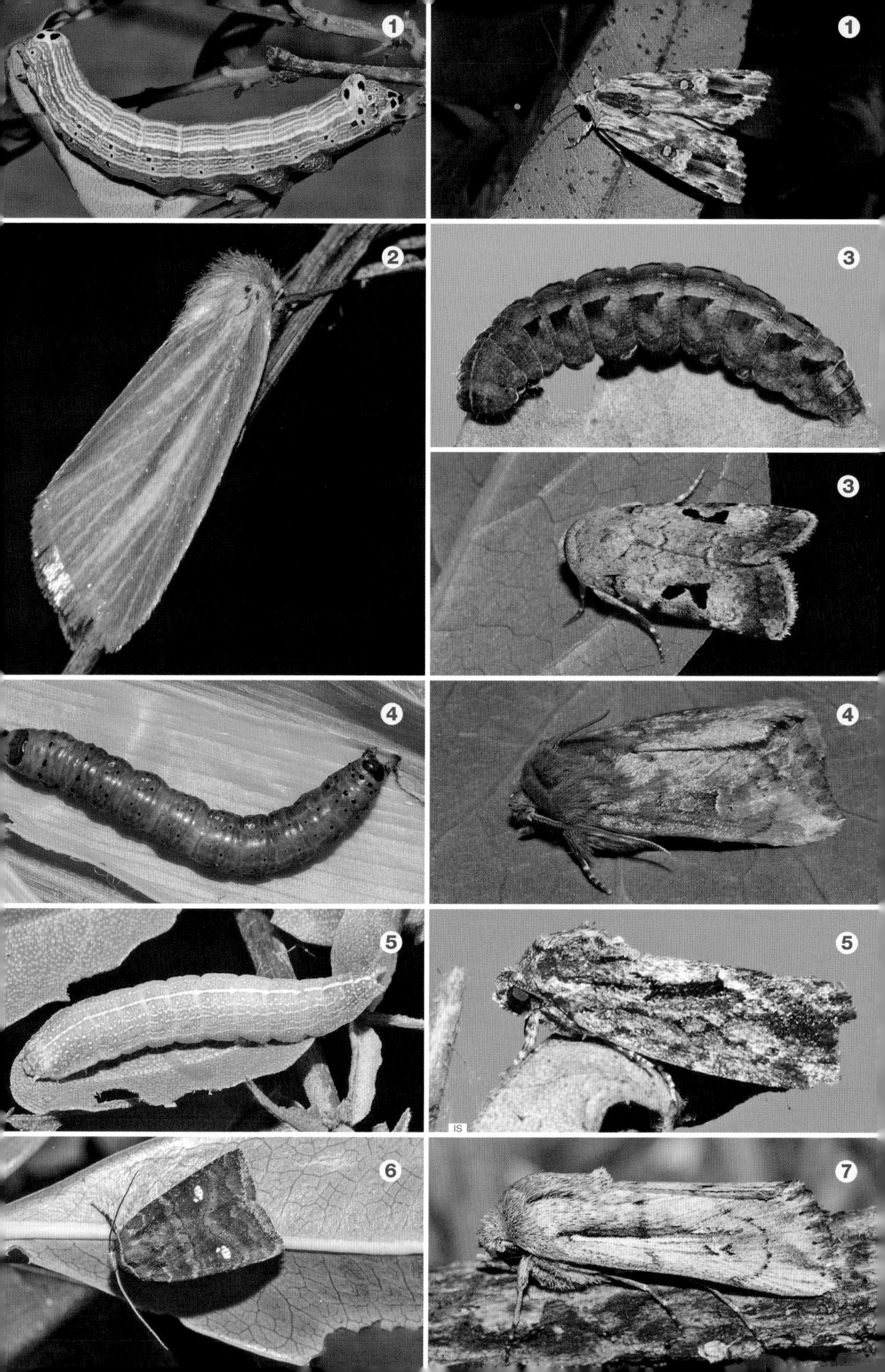
1
1
2
3
3
4
4
5
5
IS
6
7

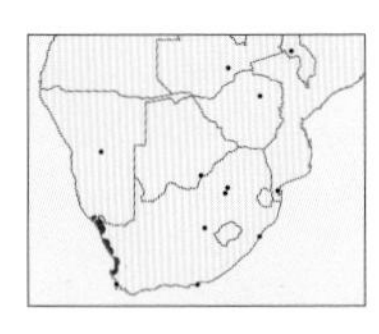

**1 *Ectochela pectinata*** West Coast Ectochela

Forewing 11–15mm. Similar to *E. dicksoni* (p.430), but smaller, more grey, lacks the central black streak and postmedian line not evenly curved; hindwings cream with broad brown border. **Biology:** Caterpillar host associations unknown. Adults August–October. **Habitat & distribution:** Coastal strandveld habitats; Western Cape and Northern Cape, South Africa.

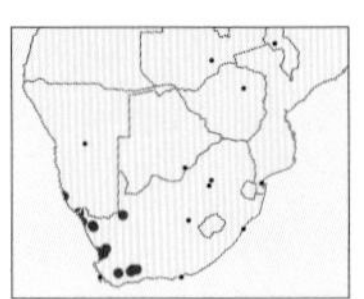

**2 *Ectochela nigrilineata*** Arctic Ectochela

Forewing 14–16mm. Forewings white, becoming grey toward apex, diagnostic veins yellow outlined in black, with black line running from wing base to margin, apical margin with alternating grey and white patches of cilia; hindwings white basally with broad brown border and white cilia. **Biology:** Caterpillar host associations unknown. Adults August–October. **Habitat & distribution:** Succulent karoo and desert; Western Cape, South Africa, north to Namibia.

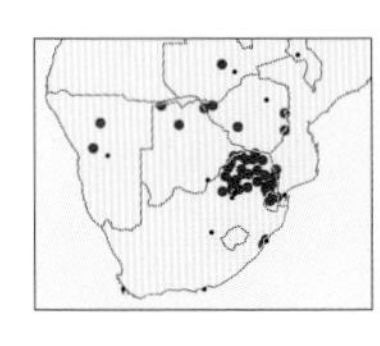

**3 *Cetola pulchra*** Cloaked Arches

Forewing 14–18mm. Wings with crenulated margins; forewings streaked brown with grey costa and inner margins; hindwings cream. Caterpillar dumpy, with swollen rear; dark red with orange markings, pronounced at rear. Three species in genus in region. **Biology:** Caterpillars feed on various Lamiaceae. Adults November–May. **Habitat & distribution:** Forest and bushveld; KwaZulu-Natal, South Africa, north to tropical Africa.

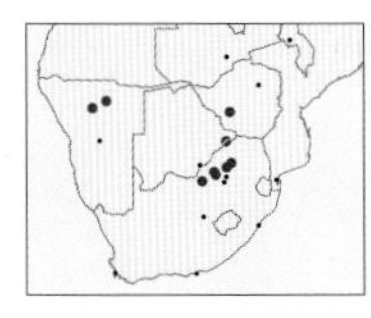

**4 *Cnodifrontia dissimilis*** Streaked Arches

Forewing 13–14mm. Forewings grey with dark brown streaks running along length of wing. Caterpillar cryptic leaf mimic of host plant, light tan with faint cream longitudinal stripes. One species in genus in region. **Biology:** Caterpillars feed on Camphor Bush (*Tarchonanthus camphoratus*). Adults throughout the year, peaking in summer. **Habitat & distribution:** Bushveld; North West, South Africa, north to Namibia.

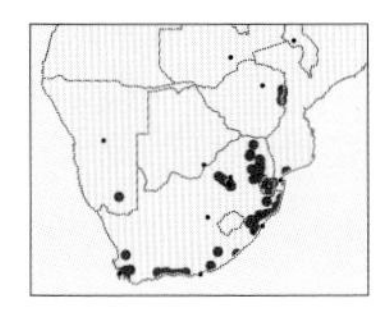

**5 *Conservula cinisigna*** Pink Angle Shades

Forewing 15–18mm. Forewings various shades of brown, with diagnostic central 'U'-shaped black marking and 'V'-shaped grey marking, numerous fine transverse lines run across length of wing; hindwings cream or grey, with darker border. Five species in genus in region. **Biology:** Caterpillar host associations unknown. Adults throughout the year, peaking October and March. **Habitat & distribution:** Fynbos, forest and riverine vegetation; Western Cape north to tropical Africa, also Madagascar and Indian Ocean islands.

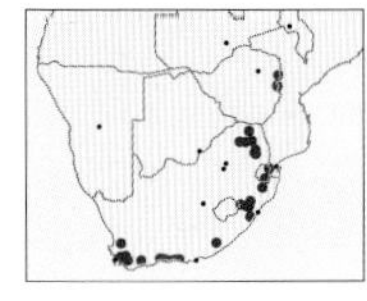

**6 *Conservula minor*** Clear Angle Shades

Forewing 14–17mm. Forewings copper brown, with 4 pink and brown transverse lines, broken into a diagnostic geometric pattern in middle of fore-margin of wing; hindwings cream, veins outlined in black. **Biology:** Caterpillar host associations unknown. Adults September–April, peaking October and March. **Habitat & distribution:** Moist fynbos and forest; Western Cape, South Africa, north to tropical Africa.

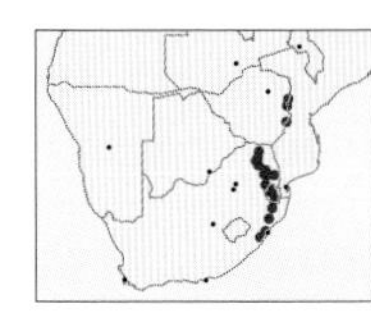

**7 *Conservula alambica*** Kidney Angle Shades

Forewing 15–18mm. Similar to *C. cinisigna* (above), but differs by the distinctive large black kidney-shaped mark on forewings; hindwings plain cream. **Biology:** Caterpillar host associations unknown. Adults throughout the year, peaking October and March. **Habitat & distribution:** Moist forest; KwaZulu-Natal, South Africa, north to tropical Africa.

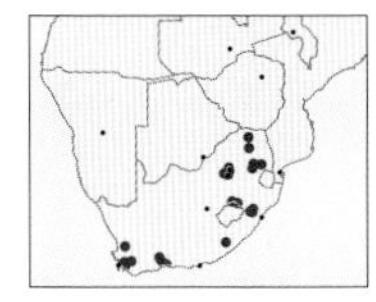

**8 *Euplexia augens*** Clouded Arches

Forewing 13–15mm. Forewings and thorax dark brown; diagnostic darker, straight postmedian line sharply stepped inward around cell spot to costa; forewings with wide, tan or white, terminal area, brown in some forms. Caterpillar plump, green or flesh coloured, with a diagnostic pair of small yellow spots at rear end of body. Three species in genus in region. **Biology:** Caterpillars polyphagous mostly on nightshades (*Solanum* spp). Multivoltine, adults throughout the year. **Habitat & distribution:** Variety of relatively moist habitats, often in gardens; Western Cape, South Africa, north to Mozambique.

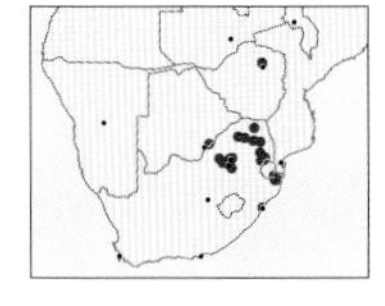

**9 *Cosmia natalensis*** Lined Speckles

Forewing 15–18mm. Distinctive; forewings short and broad, ground colour cream or fawn, densely speckled with reddish-brown, shape of lines diagnostic, postmedian kinked in female and evenly curved in male; hindwings hyaline white or cream. Two species in genus in region. **Biology:** Caterpillar host associations unknown. Adults September–May. **Habitat & distribution:** Forest and riverine vegetation; KwaZulu-Natal, South Africa, north to Zimbabwe and Uganda.

1
2
3
IS
3
4
4
5
6
7
8
8
9 ♀

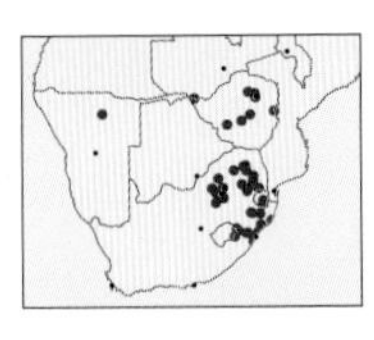

**1 *Ochrocalama xanthia*** Yellow Brown Shades

Forewing 13–14mm. Distinctive broad-winged species, forewing ground colour bone yellow with median area shaded with brown scales apart from 2 noctuid spots, cilia chequered; hindwings bone yellow in male, charcoal grey in female. One species in genus in region. **Biology:** Caterpillar host associations unknown. Adults October–April. **Habitat & distribution:** Grassland and bushveld; KwaZulu-Natal, South Africa, north to Zambia.

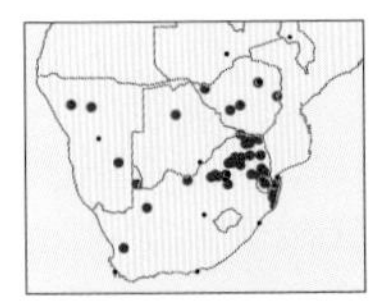

**2 *Cyclopera bucephalidia*** Small Pale Buff

Forewing 10–12mm. Broken twig mimic; forewings grey with apical tan spot circled in brown; hindwings cream. Caterpillar cryptic mimic of host plant, green with numerous white longitudinal stripes running along length of body. **Biology:** Caterpillars feed on Haakdoring (*Asparagus cooperi*). Bivoltine, adults throughout the year, peaking November and February. **Habitat & distribution:** Bushveld and grassland; Northern Cape, South Africa, north to Zimbabwe, Namibia and Botswana.

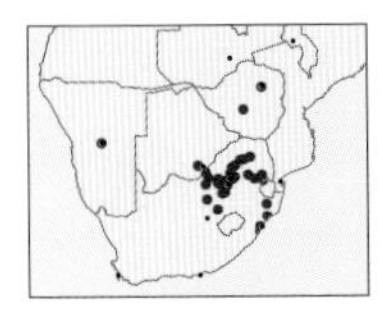

**3 *Selenistis annulella*** Cream Buff

Forewing 9–13mm. Similar to *Cyclopera bucephalidia* (above), but wing narrower toward apex, light tan with brown-edged white crescent, and darker wing margin. Caterpillar also similar, olive green with numerous longitudinal white stripes running along length of body. **Biology:** Host plant Bushveld Asparagus (*Asparagus laricinus*). Bivoltine, adults October–April. **Habitat & distribution:** Grassland and bushveld; KwaZulu-Natal, South Africa, north to Namibia, Botswana and Zimbabwe.

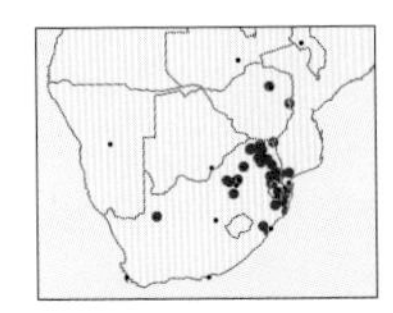

**4 *Diparopsis castanea*** Red Bollworm

Forewing 10–16mm. Dumpy, forewings ochre with 3 white transverse lines, the outer 2 enclosing a darker area and the inner bounding a reddish basal patch; hindwings pinkish. Caterpillar compact, flesh coloured with a row of red markings along back and sides, and orange head capsule. **Biology:** Caterpillars feed within developing fruits of various native mallows such as Wild Cotton (*Gossypium herbaceum*) and false hibiscus (*Cienfuegosia*). Adults October–April, peaking late summer. **Habitat & distribution:** Bushveld; KwaZulu-Natal, South Africa, north to tropical Africa.

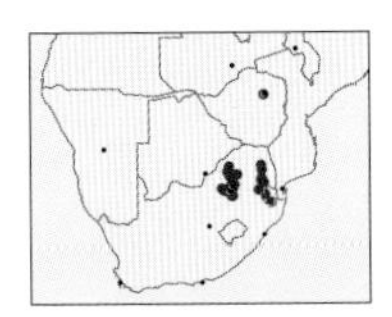

**5 *Ethiopica vinosa*** Claret

Forewing 11–12mm. Forewings uniform claret red-brown with weak postmedian line of cream dots and 2 pale spots surrounded by cream dots; hindwings white. Caterpillar olive green, with a broad, cream lateral line. Three species in genus in region. **Biology:** Caterpillars feed on Bloutee (*Oocephala staehelinoides*), mimicking host foliage. Bivoltine, adults October–April, peaking November and March. **Habitat & distribution:** Grassland; Eswatini north to Zimbabwe.

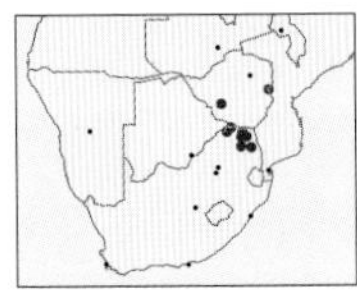

**6 *Ethioterpia lichenea*** Dirty Slender Lichen

Forewing 15–17mm. Forewings white, variably speckled with black, with dull green patches demarcated by black lines, creating diagnostic large white marks between; hindwings white with some grey dusting on margin, faint postmedian line in female. Three species in genus in region. **Biology:** Caterpillar host associations unknown. Adults November–December. **Habitat & distribution:** Lowland bushveld; Limpopo and Zimbabwe.

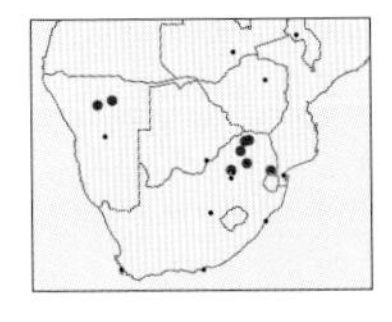

**7 *Ethioterpia marmorata*** Clean Slender Lichen

Forewing 13–14mm. Similar to *E. lichenea* (above), but smaller and lacks the black dusting on forewing, shape of white marks on forewings diagnostic; hindwings white with grey postmedian line and border. **Biology:** Caterpillar host associations unknown. Adults November–March. **Habitat & distribution:** Bushveld; Gauteng, South Africa, north to Namibia.

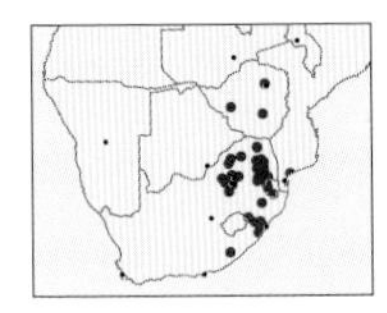

**8 *Lophotarsia ochroprocta*** Twig Grey

Forewing 11–16mm. Narrow-winged twig mimic, folds wings around body; forewings variably shaded in grey and dark to pale brown, straight basal line, postmedian curved when present, cell spot black ringed, black streaks in terminal area; hindwing white. Caterpillar bizarre, with small protruding head, which retracts into body, olive green, covered in short spines that exude sticky droplets, numerous white dots, orange and white lateral line, 10th segment narrowed to make anterior end with orange projections appear like a false head. Two species in genus in region. **Biology:** Caterpillar recorded from Blue Guarri (*Euclea crispa*). Adults October–March. **Habitat & distribution:** Forest and riverine vegetation; Eastern Cape, South Africa, north to tropical Africa.

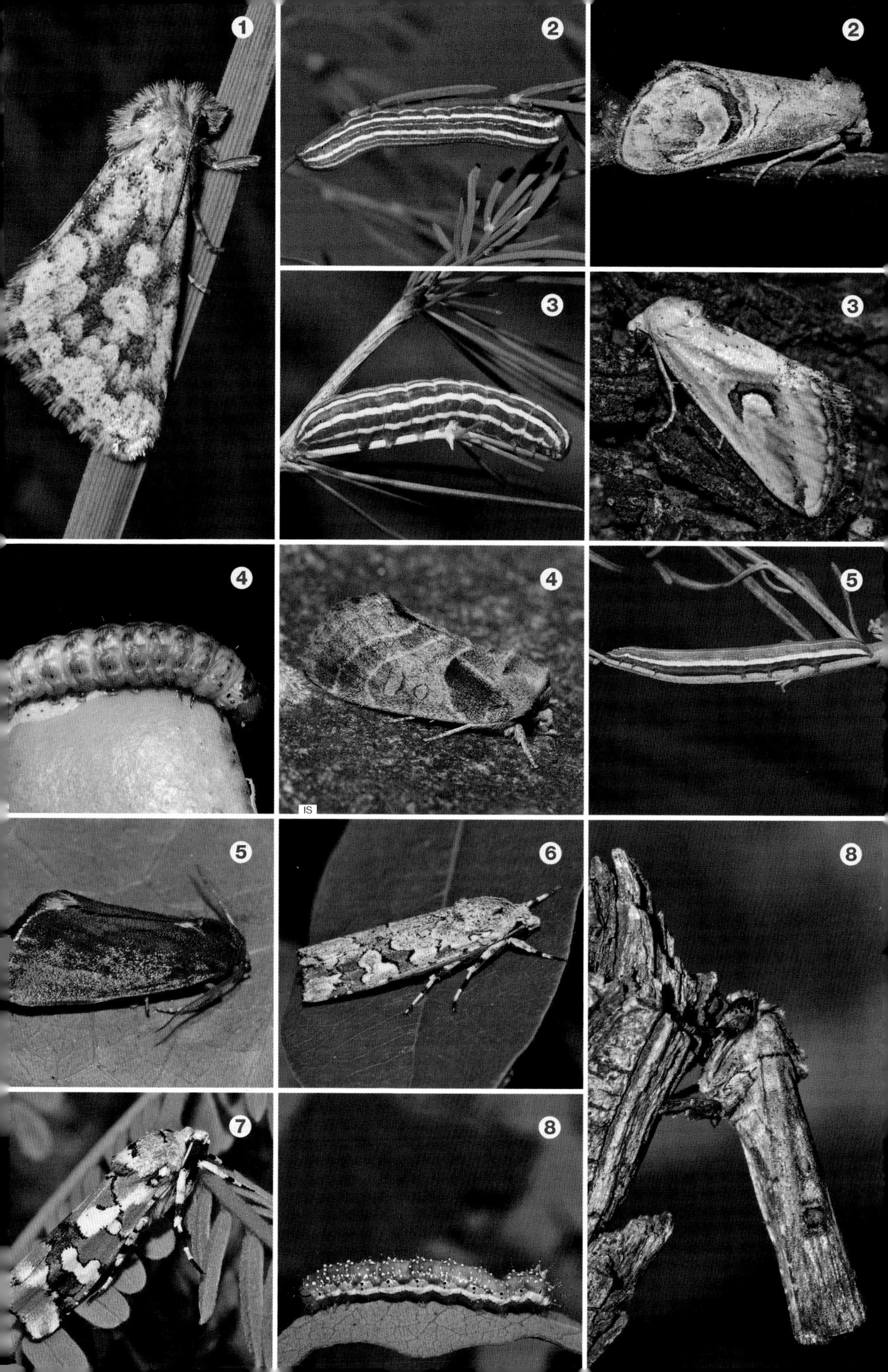

1
2
2
3
3
4
4
IS
5
5
6
8
7
8

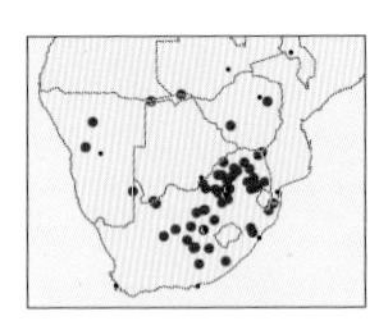

**1 *Euterpiodes pienaari*** Pienaar's Ribbons

Forewing 11–13mm. Forewings with elongate apex; brown, with large immaculate white panel edged by black and brown scalloped lines, and white apical margin; hindwings cream with submarginal and marginal dark lines. Nine species in genus in region. **Biology:** Caterpillar host associations unknown. Adults September–June, peaking in summer. **Habitat & distribution:** Grassland and bushveld; Eastern Cape, South Africa, north to Namibia, Botswana and Zimbabwe.

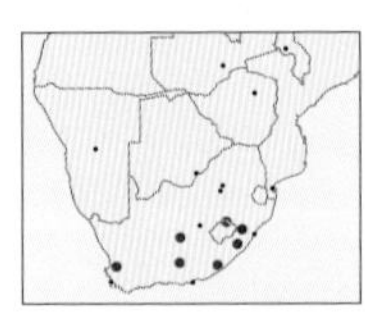

**2 *Stomafrontia albifasciata*** Long Ribbons

Forewing 16–17mm. Easily confused with some Crambinae (p.122) and *Argyrophora* (Geometridae pp.182, 184) species due to resting posture; forewings elongate, grey with central white streak that widens toward apex; hindwings uniform cream-white. Caterpillar typical noctuid with grey and orange longitudinal lines. One species in genus in region. **Biology:** Caterpillars feed on *Euphorbia*. Adults January–May. **Habitat & distribution:** Renosterveld and grassland with remnant fynbos vegetation; Western Cape east to Lesotho and KwaZulu-Natal.

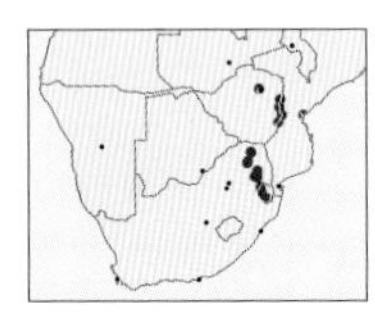

**3 *Neostichtis nigricostata*** Black Trim

Forewing 16–22mm. Often confused with some forms of *Magusa viettei* (p.430), but lacks the distinctive cream arrow dash on margin, the toothed, dark brown-edged costa is diagnostic; forewings cream or tan, and with woodgrain appearance; hindwings cream, dark border may be present. One species in genus in region. **Biology:** Caterpillar host associations unknown. Adults September–April. **Habitat & distribution:** Forests; Eswatini north to tropical Africa.

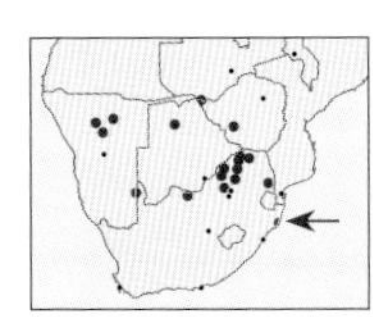

**4 *Paracroria griseocincta*** Broad Wing

Forewing 9–12mm. Head and thorax grey; forewings large and rounded, large tan and brown area rounded near apex and bounded by 2 brown lines, apical margin grey; hindwings white with grey subterminal and marginal lines. Two species in genus in region; similar *P. major* is darker and has hindwings grey, but can only be reliably separated by comparing male genitalia. **Biology:** Caterpillar host associations unknown. Apparent broken twig mimic, adults December–March. **Habitat & distribution:** Bushveld; Northern Cape, South Africa, north to tropical Africa.

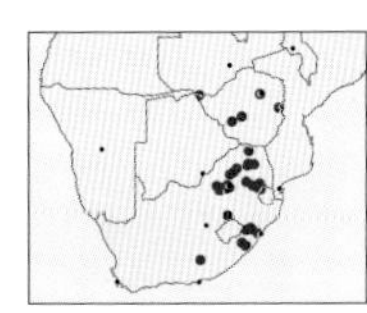

**5 *Phalerodes cauta*** Small Singe

Forewing 12–15mm. Forewings divided into tan and grey sections by longitudinal brown stripe running down middle of wing, variably coloured lines in terminal area; hindwings cream. Caterpillar cryptic mimic of host plant leaves, light grey with diagonal stripes. One species in genus in region. **Biology:** Caterpillars feed on Silver Vernonia (*Hilliardiella aristata*). Bivoltine, adults October–March. **Habitat & distribution:** Grassland where host plant grows; Eastern Cape, South Africa, north to Malawi.

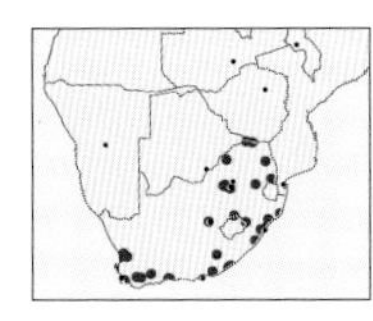

**6 *Spodoptera cilium*** Cape Lawn Moth

Forewing 10–15mm. Stocky; forewings uniform grey or tan, series of scalloped, transverse lines may be present, as well as small cream circle near fore-margin; hindwings translucent white. Caterpillar plump, variable; brown or pinkish with 2 dorsal rows of dark brown markings running along length of body, or pale green with broad light brown dorsal stripe. Seven species in genus in region. **Biology:** Caterpillars feed on the thatch of various grasses at soil level, often occurring in suburban lawns. Multivoltine, adults throughout the year. **Habitat & distribution:** Variety of relatively moist habitats; South Africa and most of Africa, Madagascar, Indian Ocean islands, parts of Europe and Asia.

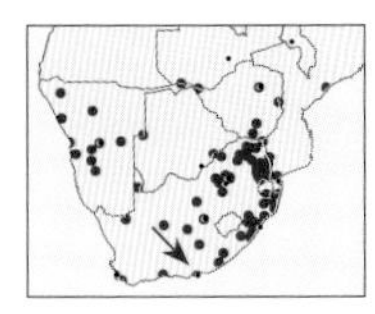

**7 *Spodoptera exempta*** African Army Worm

Forewing 12–16mm. Forewings grey or brown, with central zigzag, transverse band and elongate oval-shaped cream marking along fore-margin; hindwings cream or dirty white. Caterpillar in solitary phase pale green, brown or pinkish; in gregarious phase black, with 2 yellow dorsal stripes and broad yellow and black speckled bands running along sides of body. **Biology:** Caterpillars feed gregariously on various grasses, including crop grasses such as maize, sorghum and sugar cane. Outbreaks occur approximately every 10 years, and larvae can migrate to new areas once they have defoliated a feeding site. Multivoltine, adults throughout the year. **Habitat & distribution:** Variety of moist and arid habitats; South Africa, most of sub-Saharan Africa and Madagascar.

1
2
SM
2
3
4
5
5
6
6
IS
7
7

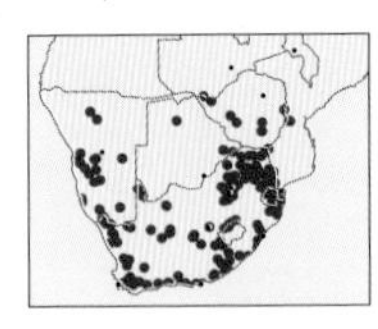

### 1 *Spodoptera exigua* — Lesser Army Worm

Forewing 10–15mm. Similar to *S. exempta* (p.436), but a little smaller; forewings grey to brown, females with oval and orange marking larger than in males. Caterpillar plump, green (darker dorsally) with cream or pink stripe running along the side of body. **Biology:** Caterpillars polyphagous on low-growing herbaceous plants. Multivoltine, adults throughout the year. **Habitat & distribution:** Ubiquitous, most common in transformed habitats; South Africa, most of Africa, Madagascar, Europe, North America and Oceania; thought to have originated from Southeast Asia.

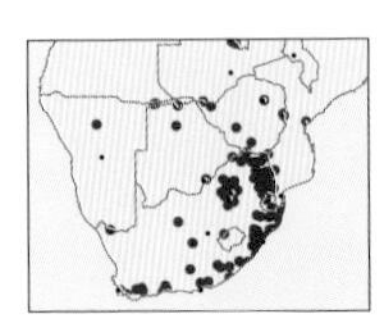

### 2 *Spodoptera littoralis* — Tomato Moth

Forewing 15–19mm. Forewings grey to brown, with complex network of light and dark markings, veins outlined in yellow, central diffuse brown band bordered by transverse grey patches in apical and basal regions. Caterpillar variable, light tan to grey, sometimes with a pair of dorsal black chevrons on each segment. **Biology:** Caterpillars polyphagous on herbs and trees. Multivoltine, adults throughout the year. **Habitat & distribution:** Variety of moist and arid habitats; South Africa and the rest of sub-Saharan Africa, also Madagascar and Indian Ocean islands.

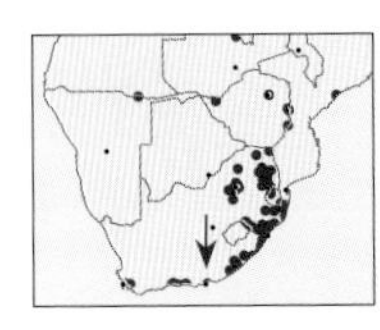

### 3 *Spodoptera triturata* — Lawn Moth

Forewing 15–18mm. Forewings tan, brown or grey, with faint zigzag, transverse lines, single dark marking next to cream spot in middle of wing, and another further toward the apex; hindwings translucent white. Caterpillar plump, grey, reddish-green, with a number of dull green longitudinal stripes, and often with 2 rows of dorsal, black chevrons on each segment. **Biology:** Caterpillars feed on roots of various grasses. Multivoltine, adults throughout the year. **Habitat & distribution:** Relatively moist habitats; South Africa and most of sub-Saharan Africa.

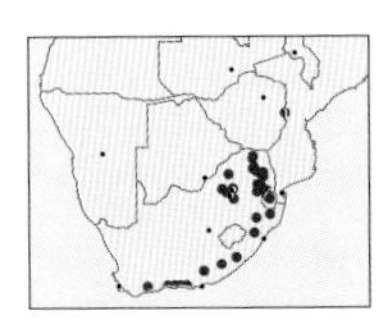

### 4 *Tracheplexia lucia* — Green Angle Shades

Forewing 16–19mm. Thorax and head mossy green; forewings dark brown, lighter along margins, green-edged antemedial and postmedian lines, green to yellow spotted subterminal line; hindwings charcoal brown. Three species in genus in region; similar *T. albimacula* has white cell spot. **Biology:** Caterpillar host associations unknown. Adults October–March. **Habitat & distribution:** Forest and riverine vegetation; Western Cape, South Africa, north to tropical Africa.

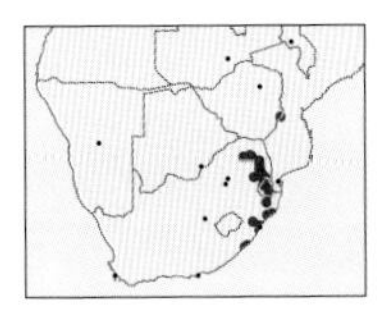

### 5 *Tracheplexia amaranta* — Rainbow Shades

Forewing 14–16mm. Thorax and head mossy green; forewings mottled with almost all the shades of the rainbow, fading in dried specimens, orange antemedial and postmedian lines touching at inner margin, 2 large spots, outer one distinctly yellow; hindwings brown with faint medial line. **Biology:** Caterpillar host associations unknown. Adults October–March. **Habitat & distribution:** Forests; Eastern Cape, South Africa, north to Zimbabwe.

## Subfamily Noctuinae

At rest, the wings are held flat and horizontally (not roof-like) over the stout body, which is covered densely with scales; the end of the forewing has a square, cut-off shape. The hind tibiae bear rows of spines. Includes the 'cutworms', caterpillars that feed at night off the lowest part of the plant stem, collapsing the plant, which they drag into their burrows to eat.

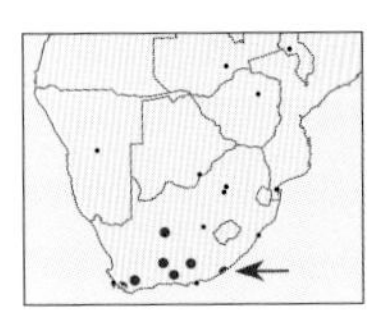

### 6 *Agrotis crassilinea* — Cream-bar Agrotis

Forewing 16–17mm. Forewings light brown, with median area a cream bar containing 2 large brown circular markings around cell, antemedial and postmedian double lines; hindwings white. Caterpillar plump, flesh coloured or pale green with a pair of black, dorsal stripes on each segment. Twenty species in genus in region. **Biology:** Caterpillars polyphagous on low-growing plants. Adults February–July. **Habitat & distribution:** Karoo habitats; Cape provinces, South Africa.

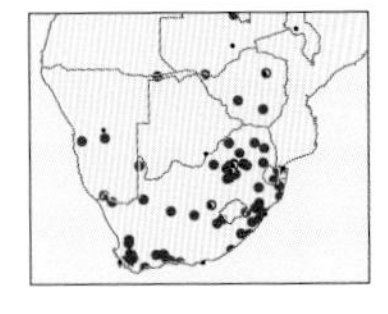

### 7 *Agrotis ipsilon* — Black Cutworm

Forewing 16–22mm. Forewings brown, variable degrees of black infusion on costa and middle of wing, 1 or 2 black dashes in apical region; hindwings translucent white with darker veins. Caterpillar appearing greasy, brown or ash grey with cream stripe along top of body. **Biology:** Caterpillars polyphagous on low-growing herbaceous plants. Multivoltine where conditions allow, adults throughout the year. **Habitat & distribution:** Most habitats, often disturbed; South Africa and patchily recorded across most of Africa, Madagascar, Indian Ocean islands and most other parts of the world.

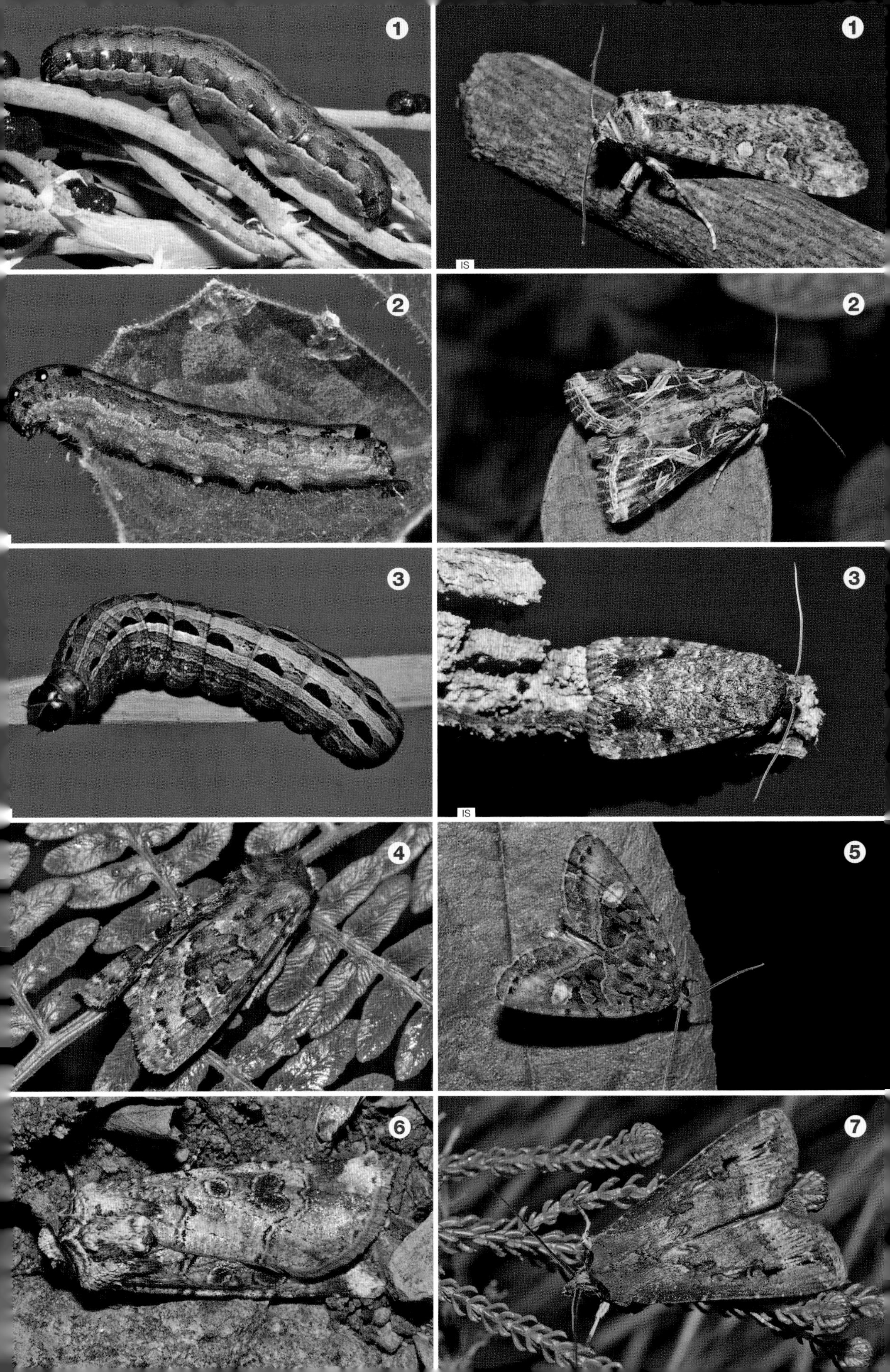
1
1
IS
2
2
3
3
IS
4
5
6
7

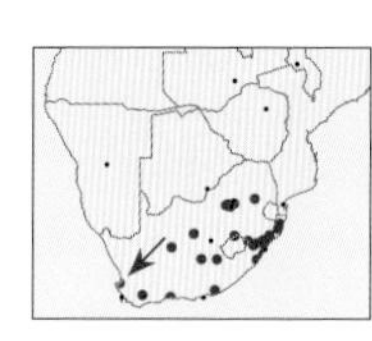

## 1 *Agrotis denticulosa* — Feathered Cutworm

Forewing 14–17mm. Forewings shades of brown, darker along costa, diagnostic inner teardrop marking around cell with large black tooth-like marking below. Caterpillar a thick-set, typical grey or brown cutworm. **Biology:** Caterpillars polyphagous on low-growing herbaceous plants. Like other *Agrotis* cutworms, the caterpillars hide by day in the soil, emerging at night to feed. Multivoltine where conditions allow, adults throughout the year. **Habitat & distribution:** Most habitats, often disturbed; South Africa eastwards to Saudi Arabia, and westwards to Nigeria, also Madagascar and Indian Ocean islands.

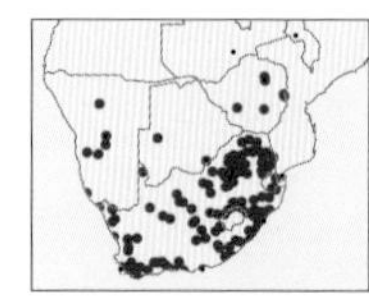

## 2 *Agrotis segetum* — Common Cutworm

Forewing 13–19mm. Forewings light to dark brown, with double, transverse lines near wing base, and 2 or 3 circular, crescentic and linear markings with dark outline; similar to *A. denticulosa* (above), but inner marking around cell is round not a teardrop. Caterpillar a plump grey or light tan cutworm with a line of small dark dots running along body. **Biology:** Caterpillars polyphagous on low-growing herbaceous plants. Multivoltine where conditions allow, adults throughout the year. **Habitat & distribution:** Most habitats, often disturbed habitats; South Africa and most of Africa, also Madagascar and Eurasia.

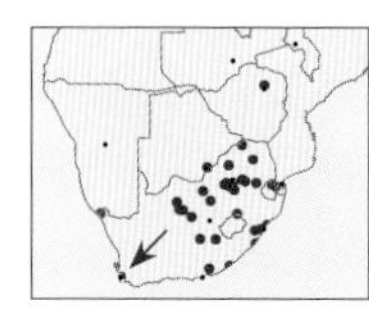

## 3 *Agrotis spinifera* — Spine Cutworm

Forewing 14–18mm. Distinctive; mottled in grey, brown and white, markings on forewings diagnostically spine-like; hindwings white. **Biology:** Caterpillars polyphagous on low-growing herbaceous plants. Multivoltine where conditions allow, adults throughout the year. **Habitat & distribution:** Most habitats, often disturbed habitats; South Africa and most of Africa, also Madagascar and Eurasia.

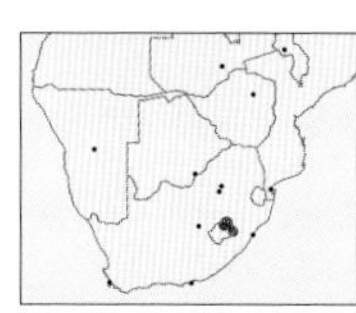

## 4 *Agrotis plumiger* — Alpine Cutworm

Forewing 12–14mm. One of several similar-looking species that inhabit very high altitudes in Lesotho and East African mountains; body and forewings variably mottled brown, white and black, curved postmedian line, white veins in terminal area and compact basal and median markings distinctive; hindwings plain grey. **Biology:** Caterpillar host associations unknown. Adults December–January. **Habitat & distribution:** Afroalpine vegetation over 2,000m, most common around 3,000m; Lesotho.

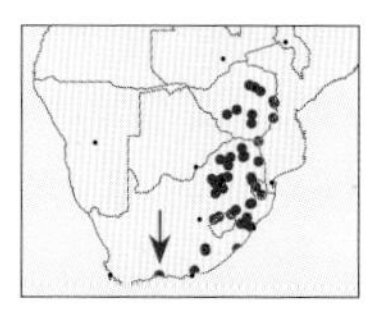

## 5 *Amazonides epipyria* — Grey-base Rustic

Forewing 13–15mm. Thorax pilose; forewings brown, grey base and darker along costa with evenly spaced white dots, 2 black-ringed circles either side of cell, wing veins brown toward wing apex; hindwings white, darker inner margin. Seven similar species in genus in region. **Biology:** Caterpillar host associations unknown. Adults October–April. **Habitat & distribution:** Forest and bushveld; Western Cape, South Africa, north to tropical Africa.

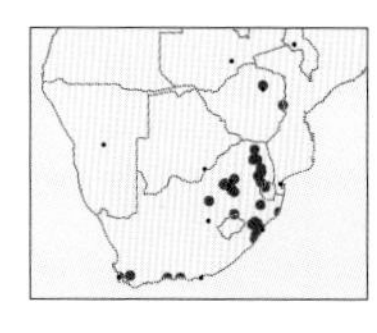

## 6 *Axylia annularis* — Ring Rustic

Forewing 1–12mm. Forewings tan, veins marked in black, lending streaked appearance to wing, diagnostic black streak with 2 small circular markings runs along middle of wing, indistinct, subapical row of small black dots may be present; hindwings white. Five species in genus in region. **Biology:** Caterpillar host associations unknown. Adults throughout the year. **Habitat & distribution:** Forest, riverine vegetation and gardens; Western Cape, South Africa, north to tropical Africa, also Madagascar.

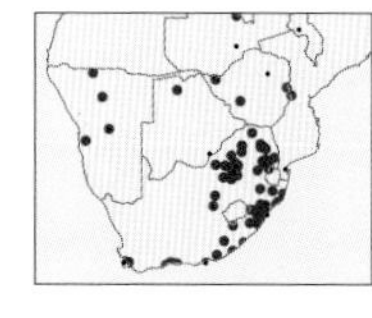

## 7 *Brithys crini* — Lily Borer

Forewing 13–18mm. Thorax silver-grey; forewings chocolate brown or grey, with lighter apical region and half-round cell spot; hindwings white to cream with dark border in female. Caterpillar yellow to brown with black segmental bands. One species in genus in region. **Biology:** Caterpillars feed on various poisonous lilies (Amaryllidaceae). Multivoltine, adults throughout the year where conditions allow. **Habitat & distribution:** Various habitats where Amaryllidaceae grow, often in gardens; Western Cape, South Africa, north to tropical Africa, also Madagascar, Indian Ocean islands, Mediterranean, parts of Asia and Australia.

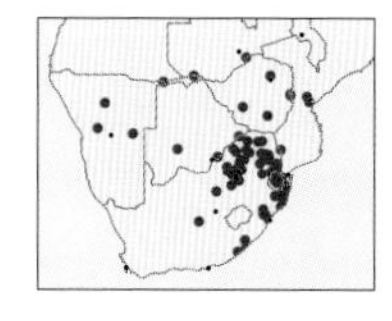

## 8 *Iambia inferalis* — Clouded Lichen

Forewing 12–14mm. Sexually dimorphic; forewings grey, brown or olive green, with diagonal brown bar at end of wing in male, apical region light cream, in female diagnostic curved postmedian line visible with large white median mark; hindwings dirty white in male, darkening at wing margin, grey-brown in female. Caterpillar plump, lime green speckled yellow, with yellow stripe running along length of body. Four species in genus in region. **Biology:** Caterpillars feed on Buffalo Thorn (*Ziziphus mucronata*). Bivoltine, adults September–April. **Habitat & distribution:** Bushveld; Eastern Cape, South Africa, north to Zambia.

1
1
2
2
3
4
5
6
6
7
7
8
8 ♀

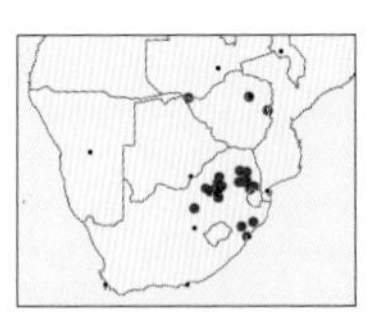

**1 *Iambia jansei*** Janse's Lichen

Forewing 14–15mm. Distinctive; wings short; marbled grey, brown, red and black, with diffuse, dark brown and grey diagonal bar before postmedian line, costa with small, evenly spaced brown bars; hindwings grey or tan. **Biology:** Caterpillar host associations unknown. Adults October–March. **Habitat & distribution:** Bushveld; KwaZulu-Natal, South Africa, north to Senegal.

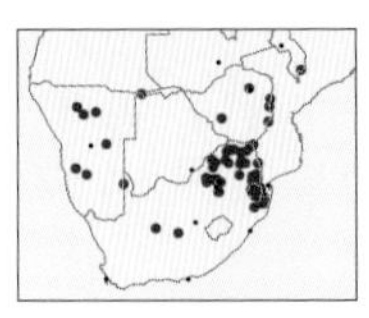

**2 *Iambiodes nyctostola*** Dark Lichen

Forewing 11–12mm. White frons; forewings mottled grey, brown and cream, with diagnostic lighter patch near apex and on costa; hindwings grey. Caterpillar plump, dull reddish-green. **Biology:** Caterpillar recorded from Large-fruited Bushwillow (*Combretum zeyheri*). Adults September-February. **Habitat & distribution:** Bushveld and riverine vegetation where bushwillows grow; Northern Cape, South Africa, north to Malawi.

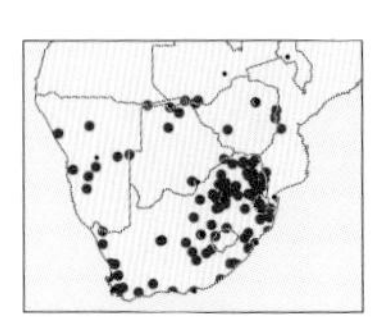

**3 *Leucania pseudoloreyi*** False Army Worm

Forewing 15–17mm. Forewing veins white, lending streaked appearance to wing, faint or dark brown median streak runs length of wing with distinct white spot, curved, postmedian row of dots sometimes visible; hindwings white or cream, with dark veins. Complex genus of about 15 species in region, in need of taxonomic revision. **Biology:** Caterpillars feed on grasses (Poaceae). Multivoltine, adults throughout the year, can be abundant in transformed habitats where natural balance is disrupted. **Habitat & distribution:** Ubiquitous, often in gardens and fields; Western Cape throughout Africa to Europe.

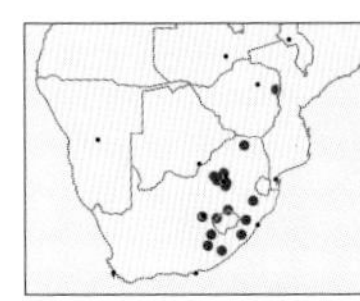

**4 *Meliana tenebra*** White-vein Meliana

Forewing 12–13mm. Distinctive; ground colour fawn, thorax with black streaks, forewings with black-dusted rusty streaks, veins white, row of black marginal dots; hindwings uniform grey. One species in genus in region. **Biology:** Caterpillar host associations unknown. Adults September–April. **Habitat & distribution:** Grassland; Eastern Cape, South Africa, north to Kenya.

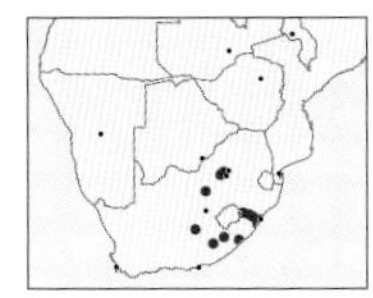

**5 *Mythimna melianoides*** White-lined Wainscott

Forewing 12–14mm. Thorax with brown streak that runs from base to margin and white mark running across it; forewings finely dusted with black scales, apical margin with row of small dark dots; hindwings white, with light brown streaks, row of marginal black dots may be present. A complex genus in need of revision, 30 species in region. **Biology:** Caterpillar host associations unknown. Adults throughout the year. **Habitat & distribution:** Grassland; Eastern Cape, South Africa, north to Ethiopia.

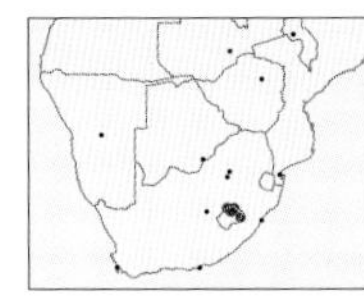

**6 *Mythimna corax*** Highland Wainscott

Forewing 13–16mm. Forewings grey-brown, veins white, costa cream, evenly spaced, black-edged white markings present, white dash on cell; hindwings grey with kidney-shaped cell spot. **Biology:** Caterpillar host associations unknown. Adults December–January. **Habitat & distribution:** Afroalpine vegetation above 2,000m; Lesotho and KwaZulu-Natal.

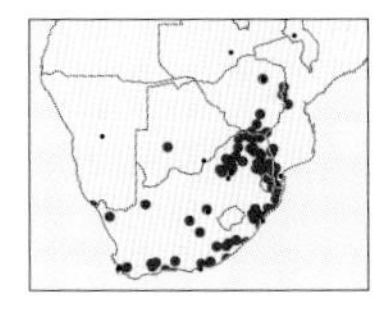

**7 *Mythimna natalensis*** Variable Wainscott

Forewing 13–17mm. Variable species; forewings variably mottled fawn to dark brown, dark area around cell always present to some extent, white below this in some, brown speckling and paler terminal area; hindwings hyaline white in centre with increasing shades of brown toward apex. Caterpillar cylindrical, brown to green. **Biology:** Caterpillars feed on grasses (Poaceae). Multivoltine where conditions allow, adults throughout the year. **Habitat & distribution:** Various habitats where grasses grow; Western Cape, South Africa, north to tropical Africa.

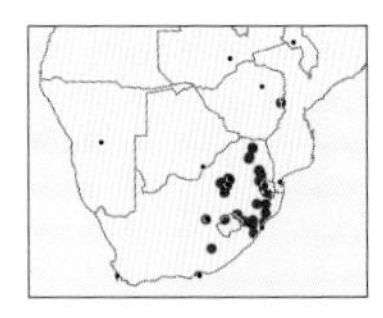

**8 *Leumicamia leucosoma*** Spiderweb

Forewing 12–15mm. Superficial resemblance to *Spodoptera littoralis* (p.438); forewings dark brown, broken up into geometric panels by network of white lines that form a cross in the apical region, and run along the wing margins; hindwings cream with darker veins. Two species in genus in region. **Biology:** Caterpillar host associations unknown. Adults throughout the year. **Habitat & distribution:** Grassland; Eastern Cape, South Africa, north to Tanzania, also reported from Madagascar.

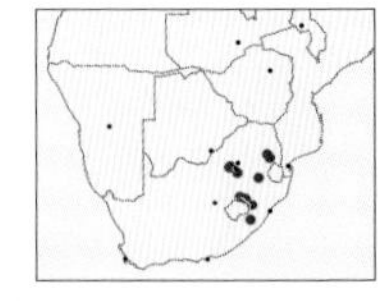

**9 *Lycophotia ecvinacea*** Frosted Cutworm

Forewing 17mm. Thorax has blue sheen in live specimens; forewing cinnamon to dark brown, with grey cast over most of the wing apart from apical region and circular patch in middle of forewing; hindwings white. Three species in genus in region. **Biology:** Caterpillar host associations unknown. Adults September–March. **Habitat & distribution:** Highveld grasslands; KwaZulu-Natal north to Mpumalanga, South Africa.

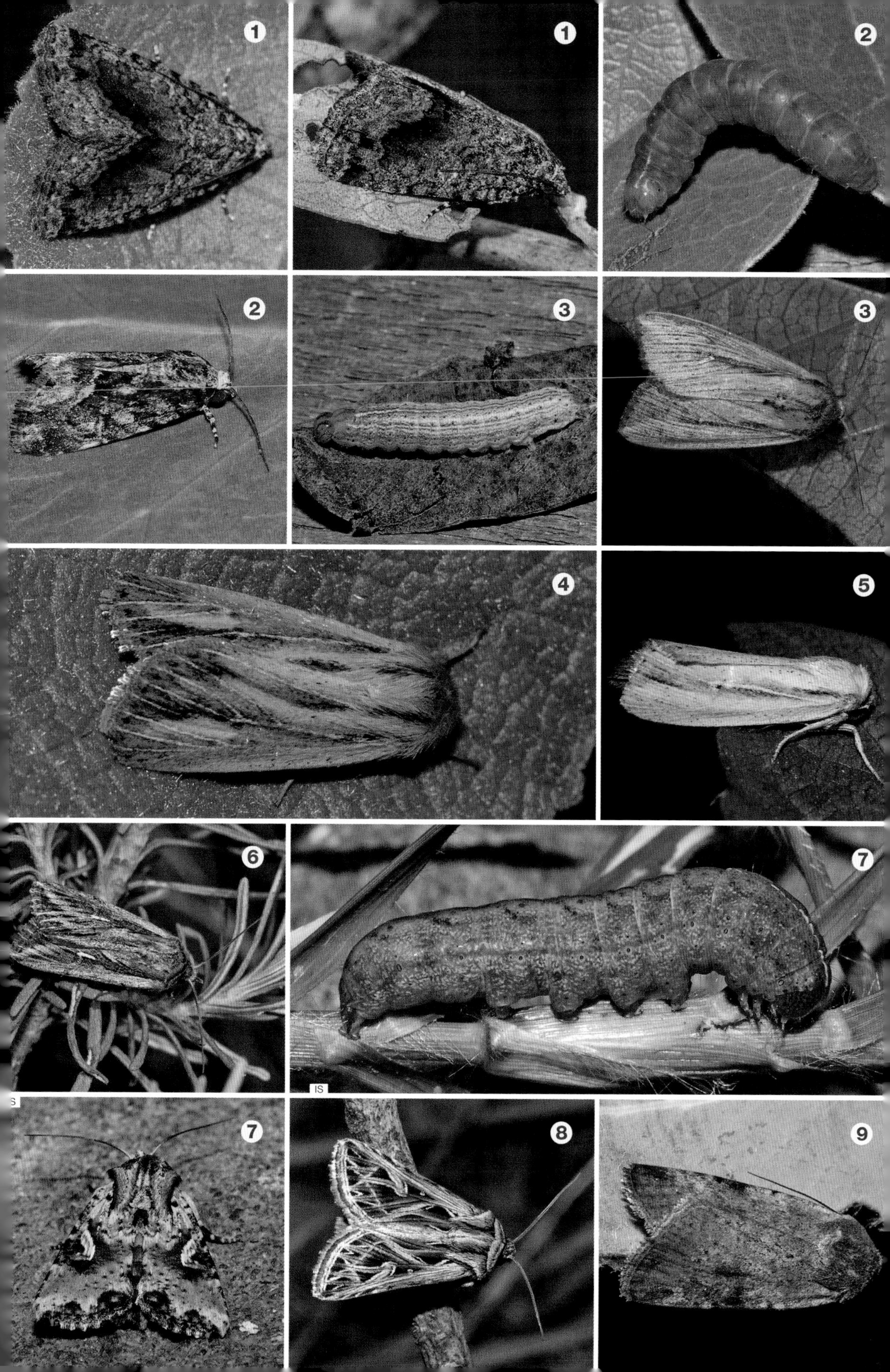
1
1
2
2
3
3
4
5
6
7
IS
7
8
9

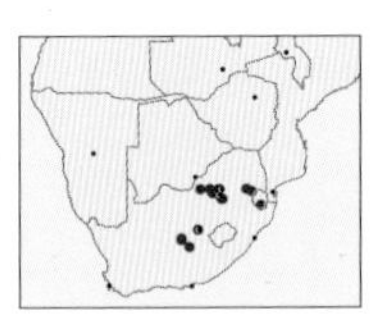

### 1 *Mimleucania perstriata* — Streaked Gothic

Forewing 14mm. Forewings tan and grey, veins black giving streaked appearance, with single black stripe running through middle of wing, 2 diagnostic sharp cream streaks in cell; hindwings white with grey border. Caterpillar whitish-green, with cream stripe running along length of body. One species in genus in region. **Biology:** Caterpillars feed on False Olive (*Buddleja saligna*). Bivoltine, adults October–April. **Habitat & distribution:** Bush and riverine vegetation where host plant grows; Eastern Cape north to Mpumalanga.

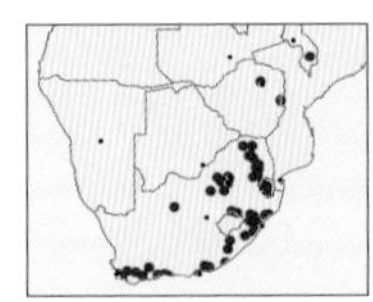

### 2 *Mentaxya albifrons* — White Quaker

Forewing 13–16mm. Frons white, forewings grey, some forms with broad, light brown median band; 3 wedge-shaped black markings spaced evenly along fore-margin, middle marking flanked by 2 oval spots, base of wing with diagnostic white cross; hindwings white. Caterpillar plump grey or tan cutworm. Nine species in genus in region. **Biology:** Caterpillars polyphagous on low-growing herbaceous plants. Multivoltine where conditions allow, adults throughout the year. **Habitat & distribution:** Relatively moist habitats, often disturbed habitats; South Africa and most of Africa, also Madagascar.

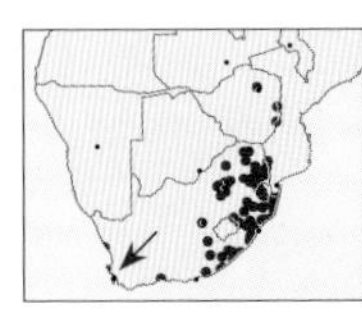

### 3 *Mentaxya ignicollis* — Lichen Quaker

Forewing 14–18mm. Thorax brownish-red with white to cream spot; forewings brown to greyish-green, with 3 evenly spaced wedge-shaped black marking along fore-margin, lacks white cross in base; hindwings white. Caterpillar plump, brown to green, with lateral line and small dark brown marking along side of body. **Biology:** Caterpillars polyphagous on low-growing herbaceous plants. Multivoltine where conditions allow, adults throughout the year. **Habitat & distribution:** Relatively moist habitats, often disturbed habitats; South Africa and most of Africa, also Madagascar.

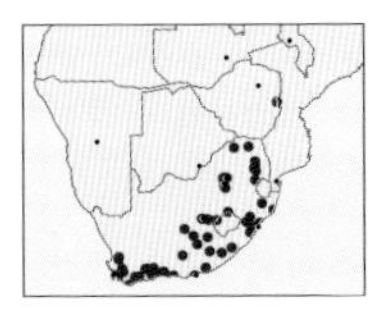

### 4 *Mentaxya muscosa* — Bark Quaker

Forewing 16–18mm. Forewings mottled brown or green; hindwings grey with faint grey band, and spot. Caterpillar variable; green, brown with green underside, or brown with a pair of dark, dorsal streaks on each segment. **Biology:** Caterpillars polyphagous on low-growing herbaceous plants. Multivoltine where conditions allow, adults throughout the year. **Habitat & distribution:** Relatively moist habitats, often disturbed habitats; South Africa, South Africa, north to tropical Africa.

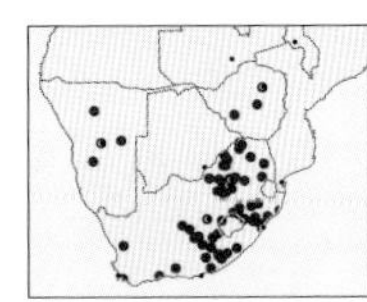

### 5 *Micragrotis intendens* — Tufted Rustic

Forewing 13–14mm. Thorax with small hair tufts; forewings tan to brown, lighter toward apex, outer spot large, black, often merged with inner spot, darker forms with oval dark brown and circular marking in middle of forewing; hindwings white. Caterpillar a tan cutworm. Nine species in genus in region. **Biology:** Caterpillar host associations unknown. Adults throughout the year. **Habitat & distribution:** Varied, including transformed habitats; Western Cape, South Africa, north to Uganda and Kenya.

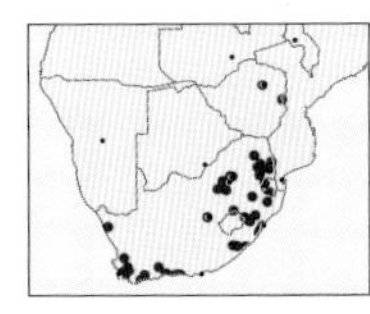

### 6 *Ochropleura leucogaster* — Roan

Forewing 15–17mm. Distinctive; thorax cream and dark brown; forewings with broad, cream fore-margin, followed by broad black stripe marked with grey oval and crescent-shaped markings, remainder dark brown with veins grey; hindwings white. Seven species in genus in region. **Biology:** Caterpillar host associations unknown. Adults throughout the year. **Habitat & distribution:** Varied, including transformed habitats; Western Cape, South Africa, north to Saudi Arabia, also Madagascar and Indian Ocean islands.

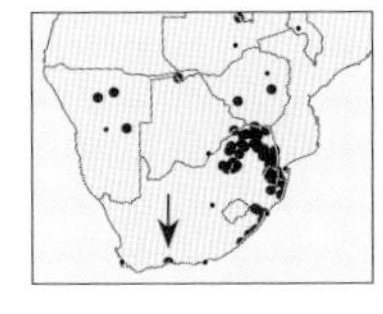

### 7 *Stenopterygia firmivena* — White Shades

Forewing 11–14mm. Forewings dark brown or dark grey, with variable, sometimes extensive, shading of white; hindwings dirty brown, margin kinked. Caterpillar honey-brown, evenly speckled in small white dots, and with lateral row of black-edged dots and another broken white row running along base of body. **Biology:** Caterpillars feed on *Ochna* spp. **Habitat & distribution:** Forests, hillsides, and bushveld where *Ochna* trees grow; Western Cape, South Africa, north to tropical Africa, also Yemen.

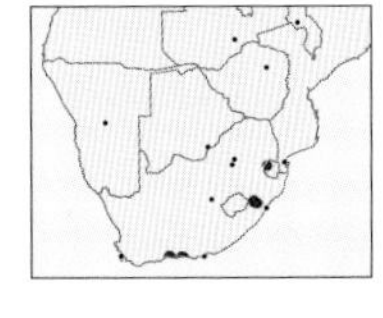

### 8 *Leucotrachea melanobasis* — Black Angle Line

Forewing 11–15mm. Distinctive; forewings brown, with chocolate brown basal band, bordered by sinuous white band with small black marking, 2 small chocolate brown subapical patches on fore- and trailing edges; hindwings chocolate brown. **Biology:** Caterpillar host associations unknown. Adults November–February. **Habitat & distribution:** Moist forests; Western Cape, South Africa, north to Eswatini.

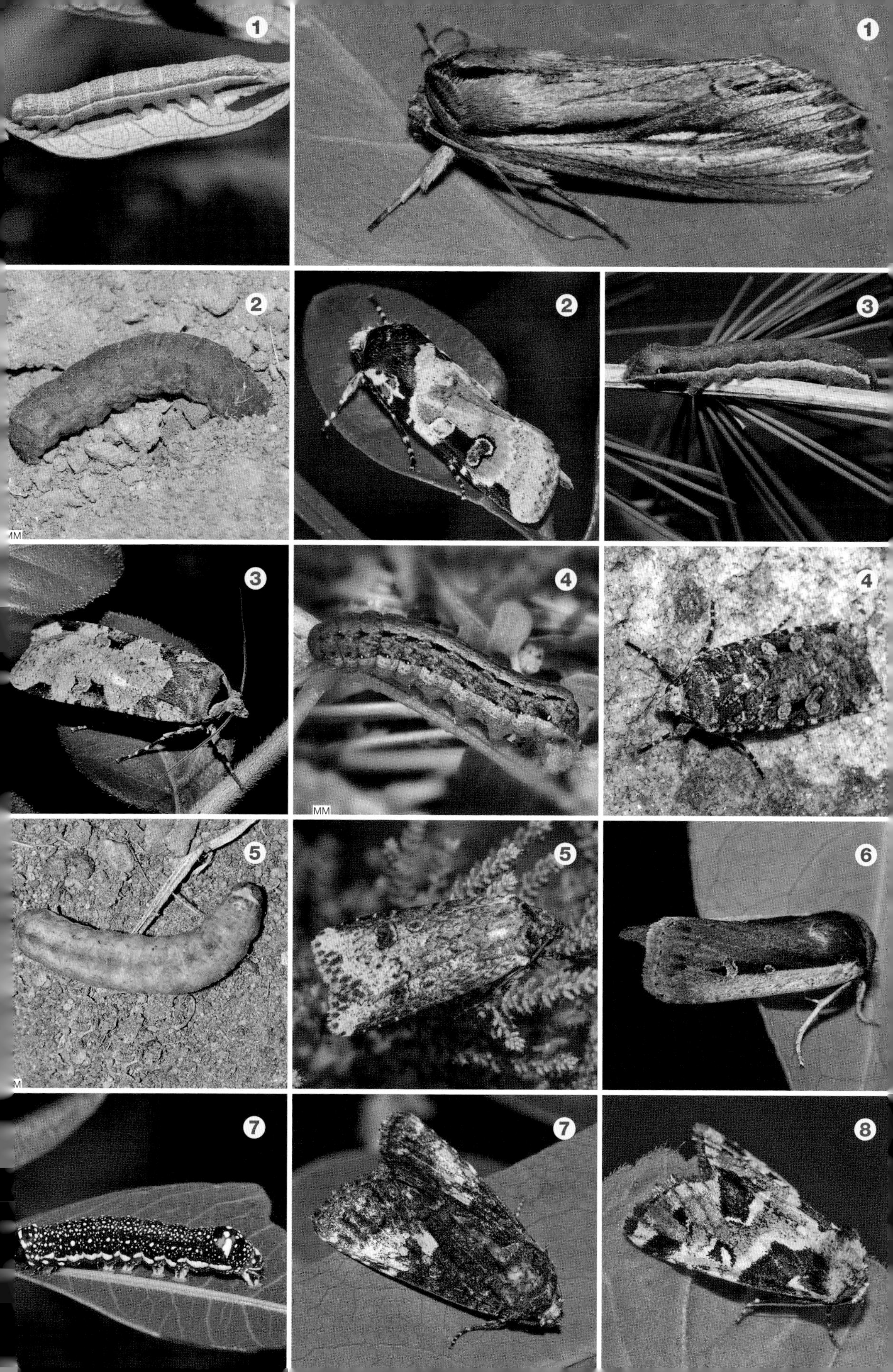
1
1
2
MM
2
3
3
4
MM
4
5
M
5
6
7
7
8

## Subfamily Heliothinae

The caterpillars have a spiny skin, and feed on leaves, and often fruits, seeds and flowers. Many feed on developing pods and seeds of crop plants. The adults frequently have orange or brown forewings and black-bordered hindwings.

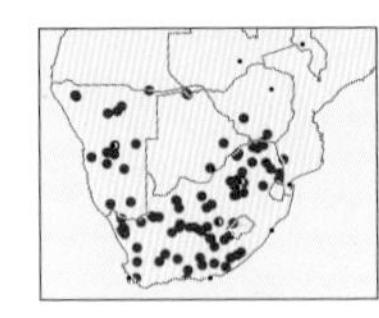

### 1 *Adisura aerugo* — Verdant Adisura

Forewing 10–11mm. Thorax and forewings uniform pale green to yellow in some individuals, forewings with thin orange border, incomplete white subterminal line diagnostic and brown cell spot; hindwings pale orange with dark brown border. Three species in genus in region. **Biology:** Caterpillar host associations unknown. Adults throughout the year, peaking September–May. **Habitat & distribution:** Variety of arid habitats; Western Cape, South Africa, north to Namibia and Zimbabwe.

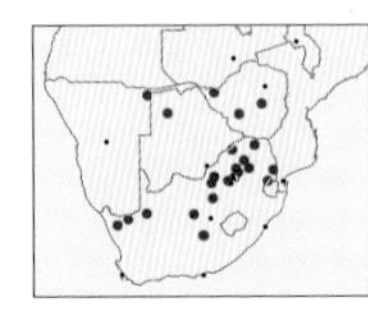

### 2 *Adisura straminea* — Pastel Adisura

Forewing 11–12mm. Forewings pointed at apex, variable, pinkish with black speckles and streaks, dotted postmedian line curving outward diagnostic; hindwings white with pinkish border. **Biology:** Caterpillar host associations unknown. Adults September–March. **Habitat & distribution:** Grassland and bushveld; Eastern Cape, South Africa, north to Namibia and Zimbabwe.

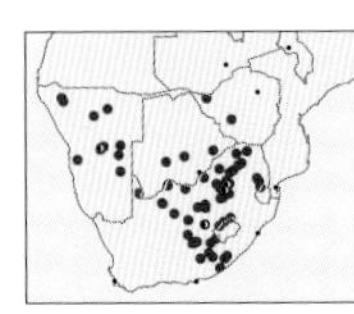

### 3 *Heliocheilus stigmatia* — Glazed Window

Forewing 11–14mm. Sexually dimorphic; forewings light brown, variably dusted with black scales, more so in females, 2 distinct eyespots either side of cell, partially obscured in males by 2 large, clear, androconial ribbed windows between veins; hindwings pale brown with darker border. Three species in genus in region. **Biology:** Caterpillar host associations unknown. Adults September–March. **Habitat & distribution:** Arid grassland and bushveld; Eastern Cape, South Africa, north to Namibia and Zimbabwe.

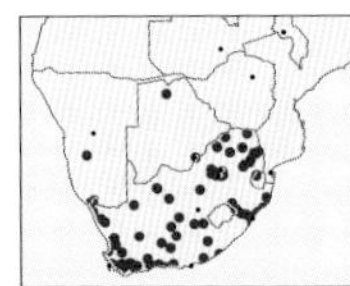

### 4 *Helicoverpa armigera* — African Bollworm

Forewing 14–19mm. Forewings fawn, tan, brown or pale green, with darker, transverse subterminal band, diagnostic row of black and white subterminal spots; hindwings dirty cream, with broad, darker marginal band. Caterpillar adapts colour to that of host plant, from green to red or a mixture of colours, with cream lateral stripe along length of body. Four species in genus in region; very similar *H. assulta* lacks the black and white subterminal spots. **Biology:** Caterpillars have exceptionally diverse host range, feeding on fruit, herbaceous weeds, soft seeds and other caterpillars. Multivoltine, adults throughout the year. **Habitat & distribution:** Most habitats, including transformed; throughout Africa, Madagascar, Indian Ocean islands, Europe, parts of Asia, Oceania and invasive in South America and Caribbean.

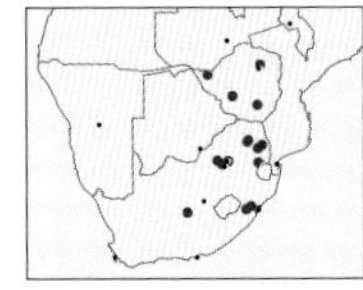

### 5 *Heliothis flavigera* — Yellow Rustic

Forewing 13–15mm. Thorax large, pink; forewings yellow with dark brown on costa, including circular grey cell spot, wing margin dark; hindwings pale yellow with darker marginal band. Caterpillar green with yellow stripes, darker forms flesh coloured with dorsal, segmental black bands and spots. Seven species in genus in region. **Biology:** Recorded from various herbs. Adults September–May. **Habitat & distribution:** Grassland and bushveld; Northern Cape, South Africa, north to tropical Africa and Madagascar.

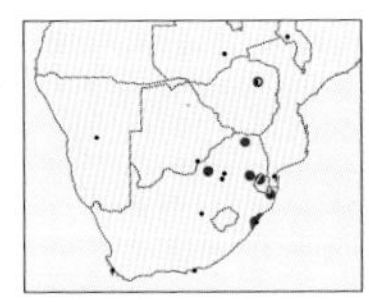

### 6 *Heliothis flavirufa* — Green Square Spot

Forewing 13–16mm. Similar to *H. flavigera* (above), but ground colour a greenish-yellow, has a dark marginal band and cell spot enlarged toward costa; hindwings white with thin brown marginal line. **Biology:** Caterpillars recorded on various Lamiaceae. Adults October–April. **Habitat & distribution:** Grassland, forest and bushveld; KwaZulu-Natal, South Africa, north to Zimbabwe.

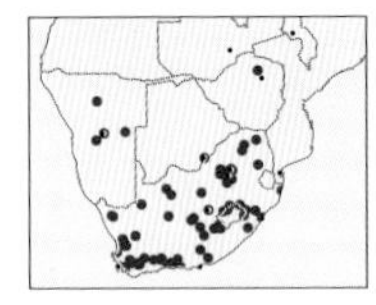

### 7 *Heliothis scutuligera* — Shield Rustic

Forewing 12–14mm. Body and forewings light brown, with cream marginal band and subterminal line of unconnected white dots and distinct cell spot diagnostic, longitudinal cream and reddish streaks along wing; hindwings creamy orange with black border. Caterpillar uniform pale or fresh green, with cream intersegmental bands. **Biology:** Caterpillars feed on various daisies (Asteraceae). Adults active during daytime but also at night, throughout the year. **Habitat & distribution:** Open habitats where Asteraceae grow; Western Cape, South Africa, north to Zimbabwe and Namibia.

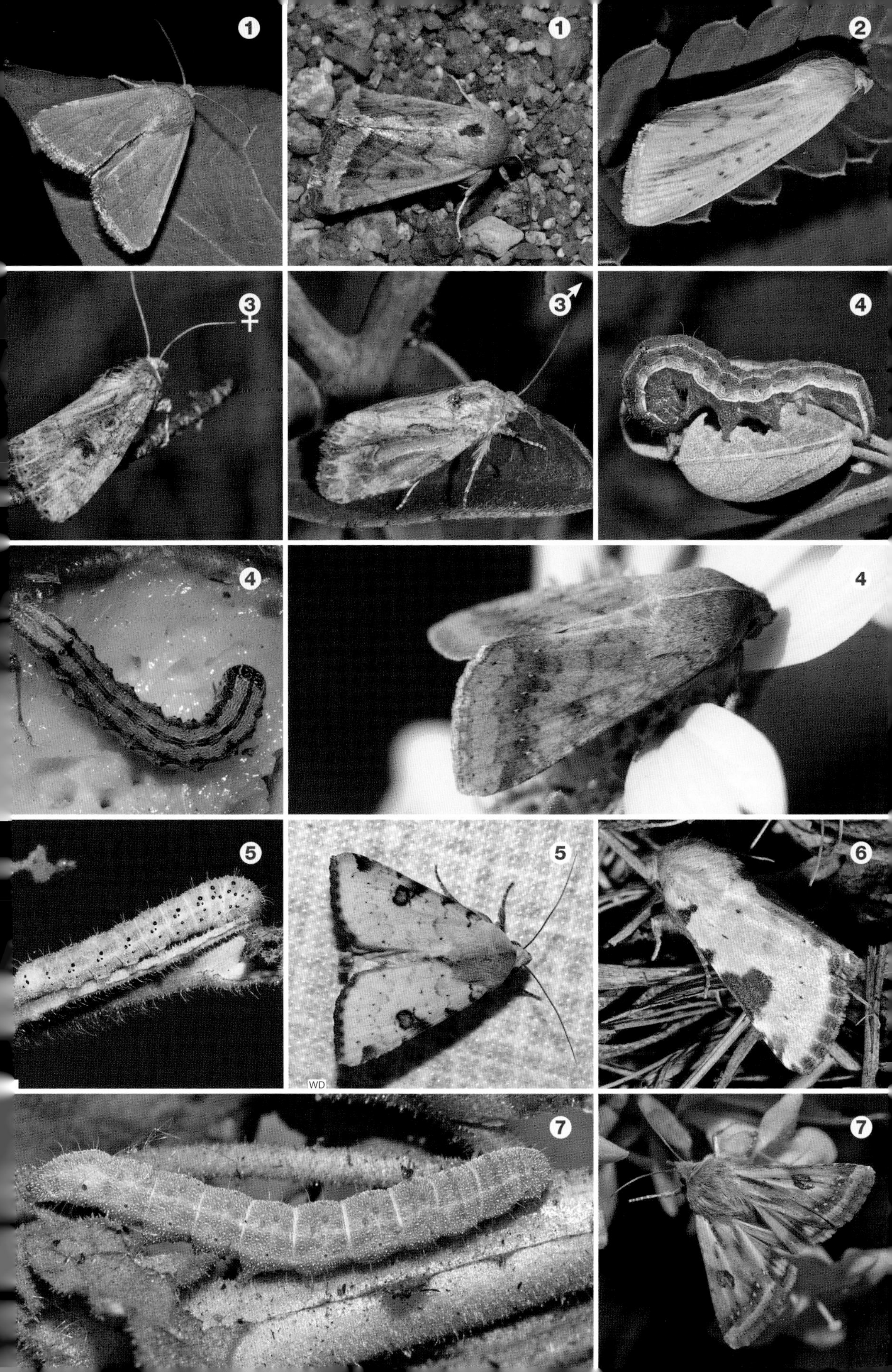
1
1
2
3
3
4
4
4
5
5
6
7
7
WD

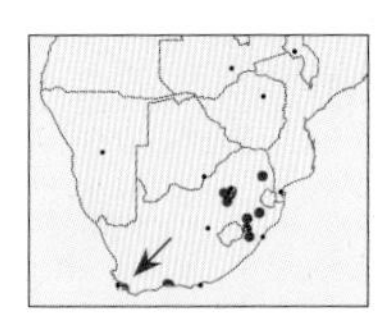

### 1 *Heliothis conifera* — Mask Rustic

Forewing 13–14mm. Similar to *H. scutuligera* (p.446), but cell spot extended to costa as a black patch, lacks the white subterminal dots and brown streaks across; darker basal patch on inner margin with 2 diagnostic curved black marks completes the 'face mask' effect when in resting posture; hindwings deep orange with black border and cell spot. **Biology:** Caterpillar host associations unknown. Adults October–April. **Habitat & distribution:** Fynbos and grassland; Western Cape north to Mpumalanga, South Africa.

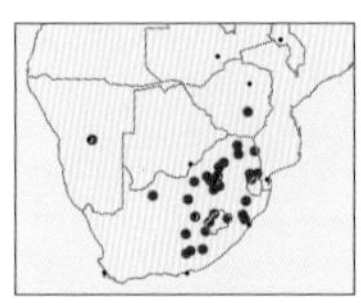

### 2 *Timora disticta* — Dotted Pink

Forewing 13–15mm. Thorax pink and pale yellow; forewings pink, pale yellow panel runs from base to apex of wing, 2 small white dots either side of cell, terminal area with white streaks between veins; hindwings white without markings. Ten species in genus in region; similar *T. albisticta* has a row of white dots comprising a curved postmedian line. **Biology:** Caterpillar host associations unknown. Adults November–April. **Habitat & distribution:** Grassland and bushveld; Eastern Cape, South Africa, north to Eritrea.

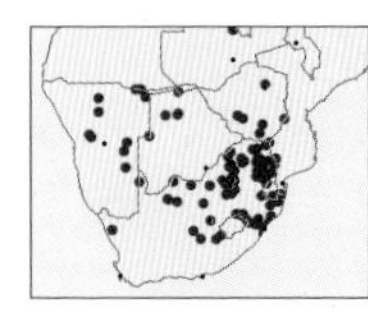

### 3 *Timora galatheae* — Golden Pink

Forewing 13–16mm. Similar to *T. disticta* (above), but lacks the 2 small white dots either side of cell, thorax and forewings carmine-pink; forewings with large triangular golden-yellow panel that runs from wing base to near apical margin of wing, margins pink; hindwings white. **Biology:** Caterpillar host associations unknown. Adults September–May. **Habitat & distribution:** Grassland and bushveld, Eastern Cape, South Africa, north to Oman.

## Unassigned Subfamily

The subfamily affiliation of the following species of Noctuidae is currently undefined.

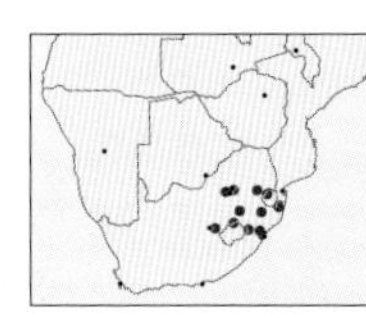

### 4 *Grammarctia bilinea* — Frowning Tigerlet

Forewing 12–15mm. Forewings cream-white, with 2 large black streaks along length, also smaller streaks in some individuals, and row of small black dots at wing apex; hindwings yellow, tinted brown near inner margin. Caterpillar grey, with orange abdominal rings, and dark grey stripe running along body, covered in short white hairs. One species in genus in region. **Biology:** Caterpillar feeds on Wild Garlic (*Tulbaghia acutiloba*). Adults September–March. **Habitat & distribution:** Grassland; KwaZulu-Natal north to Mpumalanga, South Africa, also reported from Zimbabwe and Mozambique.

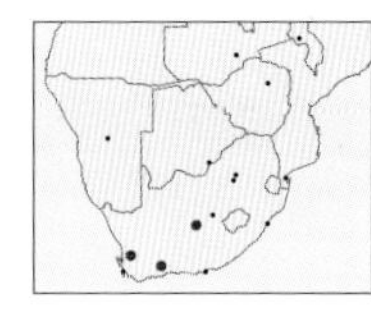

### 5 *Procanthia distantii* — Distant's Marvel

Forewing 14–16mm. Head and thorax pilose orange and grey; forewings pearly white with black-bordered orange basal, antemedial and postmedian lines, row of marginal black dots and variable; 'V'-shaped marking above cell, cilia orange; hindwings white with faint grey border and yellow cilia. Caterpillar blue-black with orange dorsolateral and yellow dorsal lines, setal bases blue. One species in genus. **Biology:** Caterpillar feeds on the moss *Gemmabryum dichotomum* (Bryaceae). Adults February–May. **Habitat & distribution:** Renosterveld and karoo habitats; Western Cape and Northern Cape, South Africa.

1
2
3
4
4
5
5
PW
MB

# LIST OF PHOTOGRAPHERS

Images used in this book are credited using the abbreviations below; uncredited images are those of the authors.

AA Adrian Armstrong
AC Andre Coetzer
AD Arthur Duke
AH Axel Hofmann
AHa Andrew Hankey
AM Andrew Morton
AN Anne Metcalf
AS Allison Sharp
AW Alberto Andrada/CC BY-SA 4.0, Wikimedia Commons
BH Brett Hurley
BS Ben Sale/CC BY-SA 4.0, Wikimedia Commons
BW Bart Wursten
CS CSIRO/CC BY-SA 4.0, Wikimedia Commons
DM Duncan MacFadyen
DS Dirk Stadie
GG George Gibbs
GH Gary Hoile
HR Hanna Roland
HV Hennie Vermaak
IS Ian Sharp
IT Ian Thomas
IU Ilia Ustyantsev/CC BY-SA 4.0, Wikimedia Commons
JB Jonathan Ball
JG Joe Grosel
JH Joseph Heymans
JJ John Joannou
JL Jeffrey W Lotz/CC BY-SA 4.0, Wikimedia Commons
JuB Julio Balona
KB Kate Braun
LP Lucia Phillips
MA Magda Botha
MB Magriet Brink
MG Martin Grimm
MM Marion Maclean
ND Neville Duke
PC Patrick Clement/CC BY-SA 4.0, Wikimedia Commons
PV Peter Vos
PW Peter Webb
QG Quartus Grobler
RO Rolf Oberprieler
RS Raimund Schutte
SB Suncana Bradley
SBa Stephen Ball
SJ Simon Joubert
SM Silvia Mecenero
SN Scott Nelson/CC BY-SA 4.0, Wikimedia Commons
SW Steve Woodhall
SY Syrio/CC BY-SA 4.0, Wikimedia Commons
TG Tony Gordon
TH Toby Hudson/CC BY-SA 4.0, Wikimedia Commons
WD Warren Dick
WM Wolfram Mey
WR Wolfachim Roland
XW xpda/CC BY-SA 4.0, Wikimedia Commons

*Zamarada deceptrix*

# FURTHER READING


Hacker, H.H. (Ed), 2019–2022. *Moths of Africa, Systematic and illustrated catalogue of the Heterocera (Lepidoptera) of Africa.* Vol 1–3: *Esperiana* verlag, Bad Staffelstein. (Vol 1, 816pp; Vol 2, 720pp.; Vol 3, 696pp.).

Hacker, H.H. (Ed), 2008–2016. *Esperiana Buchreihe zur Entomologie.* Vol 14–20, *Esperiana* verlag, Bad Staffelstein.

Hofmann, A.F. & Tremewan, W.G. 2017. *The Natural History of Burnet Moths. Part 1.* Museum Witt, Munich & Nature Research Center, Vilnius. 630pp.

Janse, A.J.T. 1930–1964. *The Moths of South Africa.* Vol 1–7: Transvaal Museum, Pretoria.

Kroon, D.M. 1999. *Lepidoptera of Southern Africa, Host-plants & Other Associations – A Catalogue.* The Lepidopterists' Society of Africa, 159 pp.

Krüger, M. 2020. *Checklist of the Lepidoptera of southern Africa, Metamorphosis* 31(2): 1–201.

Lampe, R.E.J. 2010. *Saturniidae of the World. Their Life Stages from the Eggs to the Adults.* Friedrich Pfeil, München, 368 pp.

Mecenero, S., Edge, D.A., Staude, H.S., Coetzer, B.H., Coetzer, A.J., Raimondo, D.C., Williams, M.C., Armstrong, A.J., Ball, J.B., Bode, J.D., Cockburn, K.N.A., Dobson, C.M., Dobson, J.C.H., Henning, G.A., Morton, A.S., Pringle, E.L., Rautenbach, F., Selb, H.E.T., Van der Colff, D. & Woodhall, S.E. 2020. *Outcomes of the Southern African Lepidoptera Conservation Assessment* (SALCA). *Metamorphosis* 31(4): 1–160.

Mitter, C., Davis, R.D. & Cummings, M. P. 2017. *Phylogeny and Evolution of Lepidoptera.* Annual Review of Entomology, 62: 265-283.

Murillo-Ramos, L., Brehm, G., Sihvonen, P., Hausmann, A., Holm, S., Ghanavi, H., Õunap, E., Truuverk, A., Staude, H., Friedrich, E., Tammaru, T., Wahlberg, N. 2019. *A comprehensive molecular phylogeny of Geometridae (Lepidoptera) with a focus on enigmatic small subfamilies. PeerJ* 7: e7386.

Nieukerken et al. 2011. *Order Lepidoptera.* In: Zhang, Z.-Q. (Ed.), *Animal biodiversity: An outline of higher-level classification and survey of taxonomic richness. Zootaxa* 3148: 212–221.

Oberprieler, R.G. 1995. *The Emperor Moths of Namibia.* Ekogilde, Hartbeespoort, 91 pp.

Picker, M., Griffiths, C. & Weaving, A., 2019. *Field Guide to Insects of South Africa.* Struik Nature, Cape Town. 527pp.

Pinhey, E.C.G. 1962. *Hawkmoths of central and southern Africa.* 139pp. Longmans Southern Africa (Pty) Ltd, Cape Town.

Pinhey, E.C.G, 1972. *Emperor moths of South and South central Africa*: 150pp. Struik (Pty) Ltd, Cape Town.

Pinhey, E.C.G. 1975. *Moths of Southern Africa.* 273pp. Tafelberg Publishers Ltd, Cape Town.

Prinsloo, G. L., & Uys, V. M. (Eds.) 2015. *Insects of Cultivated Plants and Natural Pastures in South Africa. Entomological Society of Southern Africa*, Pretoria, 786pp.

Pryzybylowicz, L. 2009. *Thyretini of Africa.* Entomonograph Vol 16. Apollo Books, Denmark. 70pp.

Schintlmeister, A., & Witt, T. 2015. *The Notodontidae of South Africa – including Swaziland and Lesotho (Lepidoptera: Notodontidae). Proceedings of The Museum Witt*, Volume 2. Museum Witt Munich & Nature Research Center Vilnius, 288 pp.

Scholtz, C., Scholtz, J. & De Klerk, H. 2021. *Pollinators Predators & Parasites – the ecological roles of insects in southern Africa.* Struik Nature, Cape Town. 448pp.

Sihvonen, P., Murillo-Ramos, L., Brehm, G., Staude, H. & Wahlberg, N. 2020. *Molecular phylogeny of Sterrhinae moths (Lepidoptera: Geometridae): towards a global classification. Systematic Entomology* 45: 606–634.

Staude, H.S. 1999. *An illustrated report of 510 geometrid moth taxa (Lepidoptera: Geometridae) recorded from 28 protected areas from the northern and eastern parts of South Africa. Metamorphosis Metamorphosis* 10 (3):97-148.

Staude, H.S., Mecenero, S., Oberprieler, R., Sharp, A., Sharp, I., Williams, M.C. & Maclean, M. 2016. *An illustrated report on the larvae and adults of 962 African Lepidoptera species. Results of the Caterpillar Rearing Group: a novel, collaborative method of rearing and recording lepidopteran life-histories. Metamorphosis* 27 Supplement: 330pp.

Staude, H.S., Maclean, M., Mecenero, S., Pretorius, R.J., Oberprieler, R.G., Van Noort, S., Sharp, A., Sharp, I., Balona, J., Bradley, S., Brink, M., Morton, A.S., Botha, M.J., Collins, S.C., Grobler, Q., Edge, D.A., Williams, M.C. & Sihvonen, P. 2020. *An overview of Lepidoptera-host-parasitoid associations for southern Africa, including an illustrated report on 2 370 African Lepidoptera-host and 119 parasitoid-Lepidoptera associations. Metamorphosis* 31(3): 380pp.

Woodhall, S., & Gray, L. 2015. *Gardening For Butterflies.* Penguin Random House South Africa, Cape Town, 187pp.

Zwier, J. H. H. 2022. *Aganainae of the World.* Museum Witt, Munich, 423pp.


**WEBSITES:**

Afromoths: www.afromoths.net

Barcode of Life: www.boldsystems.org

iNaturalist: www.inaturalist.org

Lepsoc Africa: www.lepsocafrica.org

Metamorphosis: www.metamorphosis.org.za

UK Moths: ukmoths.org.uk/thumbnails

Virtual Museum/ Lepimap: vmus.adu.org.za

# INDEX TO SCIENTIFIC NAMES

# INDEX TO COMMON NAMES